PHYSIOLOGY OF WOODY PLANTS

Third Edition

PHYSIOLOGY OF WOODY PLANTS

Third Edition

DR. STEPHEN G. PALLARDY

School of Natural Resources
University of Missouri
Columbia, Missouri

ELSEVIER

AMSTERDAM • BOSTON • HEIDELBERG • LONDON
NEW YORK • OXFORD • PARIS • SAN DIEGO
SAN FRANCISCO • SINGAPORE • SYDNEY • TOKYO
Academic Press is an imprint of Elsevier

Academic Press is an imprint of Elsevier
30 Corporate Drive, Suite 400, Burlington, MA 01803, USA
525 B Street, Suite 1900, San Diego, California 92101-4495, USA
84 Theobald's Road, London WC1X 8RR, UK

This book is printed on acid-free paper. ∞

Library of Congress Cataloging-in-Publication Data
Pallardy, Stephen G.
Physiology of woody plants / Stephen G. Pallardy.—3rd ed.
p. cm.
Rev. ed. of: Physiology of woody plants / Theodore T. Kozlowski, Stephen G. Pallardy. 2nd ed. c1997.
ISBN 978-0-12-088765-1
1. Woody plants—Physiology. 2. Trees—Physiology. I. Kozlowski, T. T. (Theodore Thomas), 1917– Physiology of woody plants. II. Title.
QK711.2.K72 2007
571.2—dc22
2007033499

British Library Cataloguing-in-Publication Data
A catalogue record for this book is available from the British Library.

ISBN: 978-0-12-088765-1

For information on all Academic Press publications
visit our Web site at www.books.elsevier.com

Printed in the United States of America
07 08 09 10 9 8 7 6 5 4 3 2 1

Theodore T. Kozlowski

Photo by Thad W. Sparks, Duke University, Durham, N.C.

Paul J. Kramer

This book is dedicated to Dr. Theodore T. Kozlowski and the late Dr. Paul J. Kramer (1904–1995), who pioneered the field of woody plant physiology.

Contents

CHAPTER

4

Reproductive Growth 87

CHAPTER

5

Photosynthesis 107

CHAPTER

6

Enzymes, Energetics and Respiration 169

CHAPTER

7

Carbohydrates 199

CHAPTER

8

Lipids, Terpenoids, and Related Substances 217

CHAPTER

9

Nitrogen Metabolism 233

CHAPTER

10

Mineral Nutrition 255

CHAPTER

11

Absorption of Water and Ascent of Sap 287

Preface

This book expands and updates major portions of the 1997 book on "Physiology of Woody Plants" (Second edition) by Theodore T. Kozlowski and Stephen G. Pallardy, published by Academic Press. Since that book was published there has been much new research that has filled important gaps in knowledge and altered some basic views on how woody plants grow. I therefore considered it important to bring up to date what is known about the physiology of woody plants.

This volume was written for use as a text by students and as a reference for researchers and practitioners who need to understand how woody plants grow. For all who use the book, it affords a comprehensive overview of woody plant physiology and a doorway to the literature for numerous specialized topics. The subject matter is process-focused, interdisciplinary in scope and should be useful to a broad range of scientists including agroforesters, agronomists, arborists, botanists, entomologists, foresters, horticulturists, plant molecular biologists, plant breeders, plant ecologists, plant geneticists, landscape architects, plant pathologists, plant physiologists, and soil scientists. It should also be of interest to practitioners who grow and manage woody plants for production of food and fiber.

The third edition of *Physiology of Woody Plants* retains the structure of the second. The first chapter emphasizes the importance of physiological processes through which heredity and environment interact to influence plant growth. The second chapter presents an overview of both form and structure of woody plants. Attention is given to crown form, stem form, and anatomy of leaves, stems, roots, and reproductive structures of angiosperms and gymnosperms. The third chapter describes patterns of vegetative growth of both temperate-zone and tropical woody plants. The fourth chapter characterizes the essentials of reproductive growth. Chapters five to thirteen describe the salient features of the important physiological processes involved in plant growth and development. Separate chapters are devoted to photosynthesis, respiration, carbohydrate relations, nitrogen relations, mineral relations, absorption of water, transpiration, and plant hormones and other signaling molecules.

No recommendations for use of specific management practices, experimental procedures and equipment, or use of materials are made in this text. Selection of appropriate management practices and experimental procedures will depend on the objectives of investigators and growers, plant species and genotype, availability of management resources, and local conditions known only to each grower. However, I hope that an understanding of how woody plants grow will help investigators and growers to choose research and management practices that will be appropriate for their situations.

A summary and a list of general references have been added to the end of each chapter. References cited in the text are listed in the bibliography. I have selected important references from a voluminous body of literature to make this book comprehensive and up to date. On controversial issues I attempted to present contrasting views and have based my interpretations on the weight and quality of available research data. As the appearance of exciting new reports must stand the scrutiny of the scientific community over time, I caution readers that today's favored explanations may need revision in the future. I hope that readers will also modify their views when additional research provides justification for doing so.

Many important botanical terms are defined in the text. For readers who are not familiar with some other terms, I recommend that they consult the "Academic Press Dictionary of Science and Technology" (1992), edited by C. Morris along with widely-available online dictionaries available on the Internet. I have used common names in the text for most well-known species of plants and Latin names for less common ones. Names of North American woody plants are based largely on E. L. Little (1979) "Check List of Native and Naturalized Trees of the United States," Agriculture Handbook No.41, U.S. Forest Service, Washington, D.C. Names of plants other than North American species are from various sources. Latin and common name indexes in the second edition have been removed in the third, as the abundant availability of Internet resources for cross-referencing have rendered those items largely superfluous.

I express my appreciation to many people who variously contributed to this volume. Much stimulation came from graduate students, colleagues, and collaborators in many countries with whom I have worked and exchanged information. I also express my appreciation to previous co-authors in earlier editions of this text, Drs. Paul J. Kramer and Ted Kozlowski, who pioneered the field of woody plant physiology and with whom I have been privileged to work.

Stephen G. Pallardy
Columbia, Missouri

CHAPTER

1

Introduction

Perennial woody plants are enormously important and beneficial to mankind. Trees are sources of essential products including lumber, pulp, food for humans and wildlife, fuel, medicines, waxes, oils, gums, resins, and tannins. As components of parks and forests, trees contribute immeasurably to our recreational needs. They ornament landscapes, provide screening of unsightly objects and scenes, ameliorate climate, reduce consumption of energy for heating and air conditioning of buildings, serve as sinks and long-term storage sites for greenhouse gases, and abate the harmful effects of pollution, flooding, and noise. They also protect land from erosion and wind, and provide habitats for wildlife. Shrubs bestow many of the same benefits (McKell et al., 1972). Unfortunately the growth of woody plants, and hence their potential benefits to society, very commonly is far below optimal levels. To achieve maximal benefits from communities of woody plants by efficient management, one needs to understand how their growth is influenced by heredity and environment as well as by cultural practices.

HEREDITARY AND ENVIRONMENTAL REGULATION OF GROWTH

The growth of woody plants is regulated by their heredity and environment operating through their physiological processes as shown in the following diagram.

Hereditary Potentialities
The fields of genetics and molecular biology
Selection and breeding programs, biotechnology
Potential rate of growth, size and longevity of trees
Type of xylem, canopy architecture, depth and extent of root systems

Environmental Factors
The fields of ecology, soil science, biometeorology, etc.
Radiation, temperature, mineral nutrients, water supply, carbon dioxide, competition, pests, silvicultural practices, etc.

↓

Physiological Processes and Conditions
The field of plant physiology
Photosynthesis, carbohydrate and nitrogen metabolism
Respiration, translocation
Plant water balance and effects of growth and metabolism
Growth regulators, etc.

↓

Quantity and Quality of Growth
The fields of arboriculture, forestry, and horticulture
Amount and quality of wood, fruit, or seeds produced
Vegetative versus reproductive growth
Root versus shoot growth

This scheme sometimes is called *Klebs's concept*, because the German plant physiologist Klebs (1913, 1914) was one of the first to point out that environmental factors can affect plant growth only by changing internal processes and conditions.

Woody plants show much genetic variation in such characteristics as size, crown and stem form, and longevity. Equally important are hereditary differences in capacity to tolerate or avoid environmental stresses; phenology and growth patterns; and yield of useful products such as wood, fruits, seeds, medicines, and extractives. Genetic variations account for differences in growth among clones, ecotypes, and provenances (seed sources) (Chapter 1, Kozlowski and Pallardy, 1997).

The environmental regime determines the extent to which the hereditary potential of plants is expressed. Hence, the same plant species grows differently on wet and dry sites, in light and in shade, and in polluted and clean air. Throughout their lives woody plants are subjected to multiple abiotic and biotic stresses of varying intensity and duration that, by influencing physiological processes, modify their growth. The important abiotic stresses include low light intensity, drought, flooding, temperature extremes, low soil fertility, salinity, wind, and fire. Among the major biotic stresses are attacks by insects, pathogens, and herbivores as well as plant competition and various activities of humans.

Both plant physiologists and ecologists routinely deal with stressed plants and/or ecosystems. However, the term *stress* has been variously interpreted. For example, it has been perceived to indicate both cause and effect, or stimulus and response. Hence, stress has been used as an independent variable external to the plant or ecosystem; that is, a *stimulus* that causes strain (Levitt, 1980a). In engineering and the physical sciences, stress generally is applied as force per unit area, and the result is strain. Some biologists consider strain to act as a dependent, internal variable; that is, a response caused by some factor (a stressor). This latter view recognizes an organism to be stressed when some aspect of its performance decreases below an expected value.

Odum (1985) perceived stress as a syndrome comprising both input and output (stimulus and response). The different perceptions of stress often are somewhat semantical because there is the implicit premise in all of them of a stimulus acting on a biological system and the subsequent reaction of the system (Rapport et al., 1985). In this book, and consistent with Grierson et al. (1982), stress is considered "any factor that results in less than optimum growth rates of plants," that is, "any factor that interrupts, restricts, or accelerates the normal processes of a plant or its parts."

Environmental stresses often set in motion a series of physiological dysfunctions in plants. For example, drought or cold soil may inhibit absorption of water and mineral nutrients. Decreased absorption of water is followed by stomatal closure, which leads to reduced production of photosynthate and growth hormones and their subsequent transport to meristematic sites. Hence, an environmental stress imposed on one part of a tree eventually alters growth in distant organs and tissues and eventually must inhibit growth of the crown, stem, and roots (Kozlowski, 1969, 1979). Death of trees following exposure to severe environmental stress, insect attack, or disease is invariably preceded by physiological dysfunctions (Kozlowski et al., 1991).

PHYSIOLOGICAL REGULATION OF GROWTH

To plant physiologists, trees are complex biochemical factories that grow from seeds and literally build themselves. Physiologists therefore are interested in the numerous plant processes that collectively produce *growth*. The importance of physiological processes in regulating growth is emphasized by the fact that a hectare of temperate-zone forest produces (before losses due to plant respiration are subtracted) about 20 metric tons of dry matter annually, and a hectare of tropical rain forest as much as 100 tons. This vast amount of biomass is produced from a relatively few simple raw materials: water, carbon dioxide, and a few kilograms of nitrogen and other mineral elements.

Trees carry on the same processes as other seed plants, but their larger size, slower maturation, and much longer life accentuate certain problems in comparison to those of smaller plants having a shorter life span. The most obvious difference between trees and herbaceous plants is the greater distance over which water, minerals, and foods must be translocated, and the larger percentage of nonphotosynthetic tissue in trees. Also, because of their longer life span, trees usually are exposed to greater variations and extremes of temperature and other climatic and soil conditions than are annual or biennial plants. Thus, just as trees are notable for their large size, they also are known for their special physiological problems.

Knowledge of plant physiology is essential for progress in genetics and tree breeding. As emphasized by Dickmann (1991), the processes that plant physiologists study and measure are those that applied geneticists need to change. Geneticists can increase growth of plants by providing genotypes with a more efficient combination of physiological processes for a particular environment. Plant breeders who do not understand the physiological functions of trees cannot expect to progress very far. This is because they recognize that trees receive inputs and produce outputs, but the actions of the genes that regulate the functions of trees remain obscure.

To some, the study of physiological processes such as photosynthesis or respiration may seem far removed from the practice of growing forest, fruit, and ornamental trees. However, their growth is the end result of the interactions of physiological processes that influence the availability of essential internal resources at meristematic sites. Hence, to appreciate why trees grow differently under various environmental regimes, one needs to understand how the environment affects these processes. Such important forestry problems as seed production, seed germination, canopy development, rate of wood production, maintenance of wood quality, control of seed and bud dormancy, flowering, and fruiting all involve regulation by rates and balances of physiological processes. The only way that cultural practices such as thinning of stands, irrigation, or application of fertilizers can increase growth is by improving the efficiency of essential physiological processes.

Some Important Physiological Processes and Conditions

Some of the more important physiological processes of woody plants and the chapters in which they are discussed are listed here:

Photosynthesis: Synthesis by green plants of carbohydrates from carbon dioxide and water, by which the chlorophyll-containing tissues provide the basic food materials for other processes (see Chapter 5).

Nucleic acid metabolism and gene expression: Regulation of which genes are expressed and the degree of expression of a particular gene influence nearly all biochemical and most physiological processes (which usually depend on primary gene products, proteins) (see Chapter 9 in Kozlowski and Pallardy, 1997; Weaver, 2005).

Nitrogen metabolism: Incorporation of inorganic nitrogen into organic compounds, making possible the synthesis of proteins and other molecules (see Chapter 9).

Lipid or fat metabolism: Synthesis of lipids and related compounds (see Chapter 8).

Respiration: Oxidation of food in living cells, releasing the energy used in assimilation, mineral absorption, and other energy-consuming processes involved in both maintenance and growth of plant tissues (see Chapter 6).

Assimilation: Conversion of foods into new protoplasm and cell walls (see Chapter 6).

Accumulation of food: Storage of food in seeds, buds, leaves, branches, stems, and roots (see Chapter 7; see also Chapter 2 in Kozlowski and Pallardy, 1997).

Accumulation of minerals: Concentration of minerals in cells and tissues by an active transport mechanism dependent on expenditure of metabolic energy (see Chapters 9 and 10).

Absorption: Intake of water and minerals from the soil, and oxygen and carbon dioxide from the air (see Chapters 5, 9, 10, 11, and 12).

Translocation: Movement of water, minerals, foods, and hormones from sources to utilization or storage sites (see Chapters 11 and 12; see also Chapters 3 and 5 in Kozlowski and Pallardy, 1997).

Transpiration: Loss of water in the form of vapor (see Chapter 12).

Growth: Irreversible increase in plant size involving cell division and expansion (see Chapter 3; see also Chapter 3 in Kozlowski and Pallardy, 1997).

Reproduction: Initiation and growth of flowers, fruits, cones, and seeds (see Chapter 4; see also Chapter 5 in Kozlowski and Pallardy, 1997).

Growth regulation: Complex interactions involving carbohydrates, hormones, water, and mineral nutrients (Chapters 3 and 13; see also Chapters 2 to 4 in Kozlowski and Pallardy, 1997).

Complexity of Physiological Processes

A physiological process such as photosynthesis, respiration, or transpiration actually is an aggregation of chemical and physical processes. To understand the mechanism of a physiological process, it is necessary to resolve it into its physical and chemical components. Plant physiologists depend more and more on the methods of molecular biologists and biochemists to accomplish this. Such methods have been very fruitful, as shown by progress made toward a better understanding of such complex processes as photosynthesis and respiration. Recent investigation at the molecular level has provided new insights into the manner in which regulation of gene activity controls physiological processes, although much of the progress has been made with herbaceous plants.

PROBLEMS OF FORESTERS, HORTICULTURISTS, AND ARBORISTS

Trees are grown for different reasons by foresters, horticulturists, and arborists, and the kinds of physiological problems that are of greatest importance to each vary accordingly. Foresters traditionally have been concerned with producing the maximum amount of wood per unit of land area and in the shortest time possible. They routinely deal with trees growing in plant communities and with factors affecting competition among the trees in a stand (Kozlowski, 1995). This focus has expanded in recent years to ecosystem-level concerns about forest decline phenomena, landscape-scale forest management, and responses of forest ecosystems to increasing atmospheric CO_2 levels. Many horticulturists are concerned chiefly with production of fruits; hence, they manage trees for flowering and harvesting of fruit as early as possible. Because of the high value of orchard trees, horticulturists, like arborists, often can afford to cope with problems of individual trees.

Arborists are most concerned with growing individual trees and shrubs of good form and appearance that must create aesthetically pleasing effects regardless of site and adverse environmental conditions. As a result, arborists typically address problems associated with improper planting, poor drainage, inadequate soil aeration, soil filling, or injury to roots resulting from construction, gas leaks, air pollution, and other environmental stresses. Although the primary objectives of arborists, foresters, and horticulturists are different, attaining each of them has a common requirement, namely a good understanding of tree physiology.

Physiology in Relation to Present and Future Problems

Traditional practices in forestry and horticulture already have produced some problems, and more will certainly emerge. It is well known throughout many developed and developing countries that the abundance and integrity of the earth's forest resources are in jeopardy. At the same time most people acknowledge legitimate social and economic claims of humans on forests. Hence, the impacts of people on forests need to be evaluated in the context of these concerns and needs, seeking a biologically sound and economically and socially acceptable reconciliation. Because of the complexity of the problems involved, this will be a humbling endeavor.

Several specific problems and needs that have physiological implications are well known. The CO_2 concentration of the atmosphere is increasing steadily, and may reach 460 to 560 ppm by the year 2050 (Watson et al., 2001). There is concern that such an increase could produce a significant rise in temperature, the so-called *greenhouse effect* (Baes et al., 1977; Gates, 1993; Watson et al., 2001). Mechanistic understanding of ecosystem responses, which has much to do with physiological processes, will be essential as scientists seek to predict and mitigate effects of climate change. We also need to know how other colimiting factors such as the supply of mineral nutrients interact with direct and indirect effects of increasing CO_2 concentrations in the atmosphere (Norby et al., 1986; Aber et al., 2001; Luo et al., 2004). Various species of woody plants may react differently to these stresses, thereby altering the structure, growth, and competitive interactions of forest ecosystems (Norby et al., 2001). Fuller understanding of the details of these interactions will be important in planning future plantations, especially where temperature and nutrient deficiency already limit growth. Air pollution also will continue to be a serious problem in some areas, and we will need to know more about the physiological basis of greater injury by pollution to some species and genotypes than to others.

There is much concern with rapidly accelerating losses of species diversity especially because a reduction in the genetic diversity of crops and wild species may lead to loss of ecosystem stability and function (Wilson, 1989; Solbrig, 1991). Diversity of species, the working components of ecosystems, is essential for maintaining the gaseous composition of the atmosphere; controlling regional climates and hydrological cycles; producing and maintaining soils; and assisting in waste disposal, nutrient cycling, and pest control (Solbrig et al., 1992; Solbrig, 1993). Biodiversity may be considered at several levels of biological hierarchy; for example, as the genetic diversity within local populations of species or between geographically distinct populations of a given species, and even between ecosystems.

Many species are likely to become extinct because of activities of people and, regrettably, there is little basis for quantifying the consequences of such losses for ecosystem functioning. We do not know what the critical levels of diversity are or the times over which diversity is important. We do know that biodiversity is traceable to variable physiological dysfunctions of species within stressed ecosystems. However, we have little understanding of the physiological attributes of most species in an ecosystem context (Schulze and Mooney, 1993).

It is well known that there are important physiological implications in plant competition and succession. Because of variations in competitive capacity some species exclude others from ecosystems. Such exclusion may involve attributes that deny light, water, and mineral nutrients to certain plants, influence the capacity of some plants to maintain vigor when denied resources by adjacent plants, and affects a plant's capacity to maximize fecundity when it is denied resources (Kozlowski, 1995; Picon-Cochard et al., 2006). Hence, the dynamics of competition involve differences in physiological functions and in proportional allocation of photosynthate to leaves, stems, and roots of the component species of ecosystems (Tilman, 1988; Norby et al., 2001).

Succession is a process by which disturbed plant communities regenerate to a previous condition if not exposed to additional disturbance. Replacement of species during succession involves interplay between plant competition and species tolerance to environmental stresses. Both seeds and seedlings of early and late successional species differ in physiological characteristics that account for their establishment and subsequent survival (or mortality) as competition intensifies (Bazzaz, 1979; Kozlowski et al., 1991). Much more information is needed about the physiological responses of plants that are eliminated from various ecosystems during natural succession, imposition of severe environmental stresses and species invasions.

There is an urgent need to integrate the physiological processes of plants to higher levels of biological organization. Models of tree stand- or landscape-level responses to environmental and biotic stresses will never be completely satisfactory until they can be explained in terms of the underlying physiological processes of individual plants. There have been relevant studies on specific processes (e.g., prediction of plant water status from models of hydraulic architecture) (Tyree, 1988), photosynthetic and carbon balance models (Reynolds et al., 1992), and models that integrate metabolism and morphology to predict growth of young trees (e.g., ECOPHYS, Rauscher et al., 1990; LIGNUM, Perttunen et al., 1998). However, much more remains to be done. Because of the complexity of this subject and its implications it is unlikely that the current generation of scientists will complete this task, but it must be undertaken.

Arborists and others involved in care of urban trees are interested in small, compact trees for small city lots and in the problem of plant aging because of the short life of some important fruit and ornamental trees. Unfortunately, very little is known about the physiology of aging of trees or why, for example, bristlecone pine trees may live up to 5,000 years, whereas peach trees and some other species of trees live for only a few decades, even in ostensibly favorable environments. We also know little about how exposure of young trees to various stresses can influence their subsequent long-term growth patterns, susceptibility to insect and disease attack, and longevity (Jenkins and Pallardy, 1995).

Horticulturists have made more progress than foresters in understanding some aspects of the physiology of trees, especially with respect to mineral nutrition. However, numerous problems remain for horticulturists, such as shortening the time required to bring fruit trees into bearing, eliminating biennial bearing in some varieties, and preventing excessive fruit drop. An old problem that is becoming more serious as new land becomes less available for new orchards is the difficulty of replanting old orchards, called the "replant" problem (Yadava and Doud, 1980; Singh et al., 1999; Utkhede, 2006). A similar problem is likely to become more important in forestry with increasing emphasis on short rotations (see Chapter 8, Kozlowski and Pallardy, 1997). The use of closely-spaced dwarf trees to reduce the costs of pruning, spraying, and harvesting of fruits very likely will be accompanied by new physiological problems.

The prospects for productive application of knowledge of tree physiology to solve practical problems appear to be increasingly favorable both because there is a growing appreciation of the importance of physiology in regulating growth and because of improvements in equipment and techniques. Significant progress has been made in understanding of xylem structure-function relationships, particularly with respect to how trees function as hydraulic systems and the structural features associated with breakage of water columns (cavitation) (Sperry and Tyree, 1988; Tyree and Ewers, 1991; Tyree et al., 1994; Sperry, 2003). There also has been significant progress in understanding of physiological mechanisms, including the molecular basis of photosynthetic photoinhibition and plant responses to excessive light levels (Demmig et al., 1987; Critchley, 1988; Ort, 2001), identification of patterns of root–shoot communication that may result in changes in plant growth and in stomatal function (Davies and Zhang, 1991; Dodd, 2005), and responses of plants to elevated CO_2 (Ceulemans et al., 1999; Long et al., 2004; Ainsworth and Long, 2005).

Recent technological developments include introduction of the tools of electron microscopy, molecular biology, tracers labeled with radioactive and stable isotopes, new approaches to exploiting variations in natural stable isotope composition, and substantial improvements in instrumentation. Precision instruments are now available to measure biological parameters in seconds, automatically programmed by computers. For example, the introduction of portable gas exchange-measuring equipment for studying photosynthesis and respiration has eliminated much of the need to extrapolate to the field data obtained in the laboratory (Pearcy et al., 1989; Lassoie and Hinckley, 1991). Widespread adoption of eddy-covariance analysis micrometeorological techniques employing fast-response intrared gas analyzers and three-dimensional sonic anemometers has extended the capacity for measurement of CO_2 and water vapor exchange to large footprints, allowing ecosystem-scale sampling (Baldocchi, 2003). Carefully designed sampling and analysis of stable isotopes of carbon, hydrogen, and oxygen has provided important insights into resource acquisition and use by plants, as well as partitioning of ecosystem respiration into autotrophic and heterotrophic components (Dawson et al., 2002; Trumbore, 2006).

Similarly, the tools afforded by progress in molecular biology have provided insights into regulation of plant structure at the level of the gene and its proximate downstream products, although much of this work has employed model plants like *Arabidopsis thaliana* and crop species. The first woody plant genome sequence (for *Populus trichocarpa*) just recently has been completed (Tuskan et al., 2006). The integration of molecular-level evidence into a coherent physiology-based model of plant growth and response to environmental factors is just beginning and is proving challenging (e.g., Sinclair and Purcell, 2005). Nevertheless, results of some of these studies and those available for woody plants have been incorporated, when relevant, in this edition. Ultimately, the advances at all levels of biological organization will surely lead us to a deeper understanding of how plants grow and result in better management practices.

In this book the essentials of structure and growth patterns of woody plants are reviewed first. The primary emphasis thereafter is on the physiological processes that regulate growth. I challenge you to help fill some of the gaps in our knowledge that are indicated in the following chapters.

SUMMARY

Trees and shrubs are enormously important as sources of products, stabilizers of ecosystems, ornamental objects, and ameliorators of climate and harmful effects of pollution, erosion, flooding, and wind. Many woody plants show much genetic variation in size, crown form, longevity, growth rate, cold hardiness, and tolerance to environmental stresses. The environment determines the degree to which the hereditary potential of plants is expressed. Woody plants are subjected to multiple abiotic and biotic stresses that affect growth by influencing physiological processes. Environmental stresses set in motion a series of physiological disturbances that ultimately adversely affect growth. Appropriate cultural practices increase growth by improving the efficiency of essential physiological processes.

Physiological processes are the critical intermediaries through which heredity and environment interact to regulate plant growth. The growth of plants requires absorption of water and mineral nutrients; synthesis of foods and hormones; conversion of foods into simpler compounds; production of respiratory energy; transport of foods, hormones, and mineral nutrients to meristematic sites; and conversion of foods and other substances into plant tissues.

A knowledge of physiology of woody plants is useful for coping with many practical problems. These include dealing with poor seed germination, low productivity, excess plant mortality, potential effects of increasing CO_2 concentration and global warming, environmental pollution, loss of biodiversity, plant competition and succession, and control of abscission of vegetative and reproductive structures.

Useful application of knowledge of the physiology of woody plants is favored by recent improvements in methods of measuring physiological responses. Research employing electron microscopy, molecular biology, isotopes, controlled-environment chambers, and new and improved instruments including powerful computers, is providing progressively deeper insights into the complexity and control of plant growth. These developments should lead to improved management practices in growing forest, fruit, and shade trees.

General References

Buchanan, B., Gruissem, W., and Jones, R. L., eds. (2000). *Biochemistry and Molecular Biology of Plants*. American Society of Plant Physiologists, Rockville, Maryland.

Carlquist, S. J. (2001). *Comparative Wood Anatomy: Systematic, Ecological, and Evolutionary Aspects of Dicotyledon Wood*. Springer, Berlin and New York.

Faust, M. (1989). *Physiology of Temperate Zone Fruit Trees*. Wiley, New York.

Fry, B. (2006). *Stable Isotope Ecology*. Springer, New York.

Gilmartin, P. M. and Bowler, C., eds. (2002). *Molecular Plant Biology: A Practical Approach*. 2 Vol. Revised edition. Oxford University Press, Oxford, New York.

Jackson, M. B. and Black, C. R., eds. (1993). *Interacting Stresses on Plants in a Changing Climate*. Springer-Verlag, New York and Berlin.

Jain, S. M. and Minocha, S. C., eds. (2000). *Molecular Biology of Woody Plants*. Kluwer Academic, Dordrecht, Netherlands.

Jones, H. G., Flowers, T. V., and Jones, M. B., eds. (1989). *Plants Under Stress*. Cambridge Univ. Press, Cambridge.

Katterman, F., ed. (1990). *Environmental Injury to Plants*. Academic Press, San Diego.

Kozlowski, T. T., Kramer, P. J., and Pallardy, S. G. (1991). *The Physiological Ecology of Woody Plants*. Academic Press, San Diego.

Landsberg, J. J. and Gower, S. T. (1994). *Applications of Physiological Ecology to Forest Management*. Academic Press, San Diego.

Lassoie, J. P. and Hinckley, T. M., eds. (1991). *Techniques and Approaches in Forest Tree Ecophysiology*. CRC Press, Boca Raton, Florida.

Lowman, M. D. and Nadkarni, N. M., eds. (1995). *Forest Canopies*. Academic Press, San Diego.

Mooney, H. A., Winner, W. E., and Pell, E. J., eds. (1991). *Response of Plants to Multiple Stresses*. Academic Press, San Diego.

Pearcy, R. W., Ehleringer, J., Mooney, H. A., and Rundel, P. W., eds. (1989). *Plant Physiological Ecology-Field Methods and Instrumentation*. Chapman & Hall, London.

Scarascia-Mugnozza, G. E., Valentini, R., Ceulemans, R., and Isebrands, J. G., eds. (1994). Ecophysiology and genetics of trees and forests in a changing environment. *Tree Physiol.* **14**, 659–1095.

Schulze, E.-D. and Mooney, H. A., eds. (1993). *Biodiversity and Ecosystem Function*. Springer-Verlag, Berlin and New York.

Smith, W. K. and Hinckley, T. M., eds. (1995). *Resource Physiology of Conifers*. Academic Press, San Diego.

Zobel, B. and van Buijtenen, J. P. (1989). *Wood Variation: Its Causes and Control*. Springer-Verlag, New York and Berlin.

CHAPTER

2

The Woody Plant Body

INTRODUCTION

The growth of woody plants is intimately linked with their form and structure. Knowledge of variations in form and structure is as essential to understanding the physiological processes that regulate plant growth as is a knowledge of chemistry. For example, crown characteristics have important implications for many physiological processes that influence the rate of plant growth and in such expressions of growth as increase in stem diameter and production of fruits, cones, and seeds. An appreciation of leaf structure is essential to understand how photosynthesis and transpiration are affected by environmental stresses and cultural practices. Information on stem structure is basic to an understanding of the ascent of sap, translocation of carbohydrates, and cambial growth; and a knowledge of root structure is important for an appreciation of the mechanisms of absorption of water and mineral nutrients. Hence, in this chapter an overview of the form and structure of woody plants will be presented as a prelude to a discussion of their growth characteristics and physiological processes.

Seed-bearing plants have been segregated into angiosperms and gymnosperms based on the manner in which ovules are borne (enclosed in an ovary in the former and naked in the latter). There is some molecular evidence that the gymnosperms are not a monophyletic group (i.e., they are not traceable to a common ancestor) and hence that the term *gymnosperm* has no real taxonomic meaning (Judd et al., 1999). However, other recent molecular studies have indicated that the gymnosperms are indeed monophyletic (Bowe et al., 2000; Chaw et al., 2000). Because of their many relevant morphological and physiological similarities, in this work I continue to use the term gymnosperm while recognizing the possibly artificial nature of this classification. Angiosperms have long been accepted as a monophyletic group (Judd et al., 1999).

CROWN FORM

Many people are interested in tree form, which refers to the size, shape, and composition (number of branches, twigs, etc.) of the crown. Landscape architects and arborists depend on tree form to convey a desired emotional appeal. Columnar trees are used as ornamentals for contrast and as architectural elements to define three-dimensional spaces; vase-shaped forms branch high so there is usable ground space below; pyramidal crowns provide strong contrast to trees with rounded crowns; irregular forms are used to provide interest and contrast to architectural masses; weeping forms direct attention to the ground area and add a softening effect to the hard lines of buildings.

The interest of foresters in tree form extends far beyond aesthetic considerations, because crown form greatly affects the amount and quality of wood produced and also influences the taper of tree stems. More wood is produced by trees with large crowns than by those with small ones, but branches on the lower stem reduce the quality of lumber by causing knots to form.

Tree fruit growers are concerned with the effects of tree form and size on pruning, spraying, exposure of fruits to the sun, and harvesting of fruits. Hence, they have shown much interest in developing high-yielding fruit trees with small, compact, and accessible crowns.

Variations in Crown Form

Most forest trees of the temperate zone can be classified as either excurrent or decurrent (deliquescent), depending on differences in the rates of elongation of buds and branches. In gymnosperms such as pines, spruces, and firs, the terminal leader elongates more each year than the lateral branches below it, producing a single central stem and the conical crown of the excurrent tree. In most angiosperm trees, such as oaks and maples, the lateral branches grow almost as fast as or faster than the terminal leader, resulting in the broad crown of the decurrent tree. The decurrent crown form of elms is traceable to loss of terminal buds (Chapter 3) and to branching and rebranching of lateral shoots, causing loss of identity of the main stem of the crown. Open-grown decurrent trees tend to develop shapes characteristic for genera or species (Fig. 2.1). The most common crown form is ovate to elongate, as in ash. Still other trees, elm for example, are vase-shaped. However, within a species, several modifications of crown form may be found (Fig. 2.2).

Because of the importance of crown form to growth and yield of harvested products, tree breeders have related productivity to "crown ideotypes" (types that are adapted to specific environments). For example, narrow-crowned ideotypes are considered best for densely spaced, short-rotation, intensively cultured poplar plantations, whereas trees with broad crowns are better for widely spaced plantation trees grown for sawlogs or nut production (Dickmann, 1985).

Tropical trees are well known for their wide variability of crown forms. The 23 different architectural models of Hallé et al. (1978) characterize variations in inherited crown characteristics. However, each tropical species may exhibit a range of crown forms because of its plasticity to environmental conditions. Plasticity of crowns of temperate-zone trees also is well documented (Chapter 5, Kozlowski and Pallardy, 1997).

The shapes of tree crowns differ among species occupying the different layers of tropical forests, with the tallest trees having the widest and flattest crowns (Fig. 2.3). In the second layer tree crowns are about as wide as they are high, and in the third layer the trees tend to have tapering and conical crowns. The shapes of crowns in the various layers of tropical forests also are influenced by angles of branching. In upper strata most major branches tend to be upwardly oriented, whereas in the third layer they are more horizontally oriented. The young plants of species that eventually occupy the upper levels of tropical forests and the shrub layers have diverse forms. Whereas many shrubs have a main stem and resemble dwarf trees, other shrubs (for example, members of the Rubiaceae) lack a main stem and branch profusely near ground level. Trees with narrow columnar crowns generally are associated with high latitudes and more xeric sites; broad or spherical crowns tend to occur in humid or moist environments (Landsberg, 1995).

Crown forms of tropical trees of the upper canopy change progressively during their development. When young they have the long, tapering crowns characteristic of trees of lower strata; when nearly adult their crowns assume a more rounded form; and when fully mature their crowns become flattened and wide (Richards, 1966; Whitmore, 1984).

Crown forms of tropical trees also are greatly modified by site. Species adapted to mesic sites tend to be tall with broad crowns, whereas species on xeric sites usually are short and small-leaved and have what is known as a xeromorphic form. Low soil fertility usually accentuates the sclerophyllous and xeromorphic characteristics associated with drought resistance, inducing thick cuticles and a decrease in leaf size. For more detailed descriptions of variation in structure of canopies of temperate and tropical forests see Parker (1995) and Hallé (1995).

FIGURE 2.1. Variations in the form of open-grown trees. (A) Eastern white pine; (B) Douglas-fir; (C) longleaf pine; (D) eastern hemlock; (E) balsam fir; (F) ponderosa pine; (G) white spruce; (H) white oak; (I) sweetgum; (J) shagbark hickory; (K) yellow-poplar; (L) sugar maple. Photos courtesy St. Regis Paper Co.

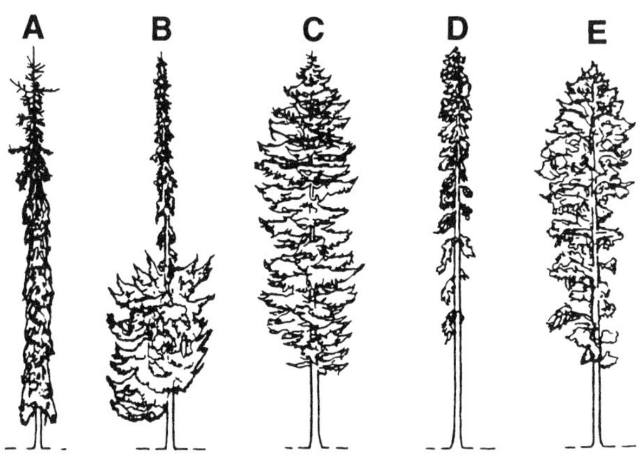

FIGURE 2.2. Variations in crown form of Norway spruce (A, B, C) and Scotch pine (D, E) in Finland. From Kärki and Tigerstedt (1985).

STEM FORM

Much interest has been shown in the taper of tree stems because of its effect on production of logs. Foresters prefer straight, nearly cylindrical stems, with little taper, and without many branches.

Tree stems taper from the base to the top in amounts that vary among species, tree age, stem height, and number of trees per unit of land area. Foresters quantify the amount of taper by a form quotient (the ratio of some upper stem diameter to stem diameter at breast height). The form quotient is expressed as a percentage and always is less than unity. Lower rates of stem taper and correspondingly greater stem volumes are indicated by higher form quotients. The form quotient is low for open-grown trees with long live crowns and high for trees in dense stands with short crowns.

FIGURE 2.3. Variations in crown form of trees occupying different layers of a tropical forest. From Beard (1946).

In dense stands the release of a tree from competition by removal of adjacent trees not only increases wood production by the retained tree, but also produces a more tapered stem by stimulating wood production most in the lower stem. When a plantation is established, the trees usually are planted close together so they will produce nearly cylindrical stems. Later the stand can be thinned to stimulate wood production in selected residual trees (Chapter 7, Kozlowski and Pallardy, 1997). Original wide spacing may produce trees with long crowns as well as stems with too much taper and many knots.

It often is assumed that tree stems are round in cross section. However, this is not always the case because cambial activity is not continuous in space or time. Hence, trees produce a sheath of wood (xylem) that varies in thickness at different stem and branch heights, and at a given stem height it often varies in thickness on different sides of a tree. Sometimes the cambium is dead or dormant on one side of a tree, leading to production of partial or discontinuous xylem rings that do not complete the stem circumference. Discontinuous rings and stem eccentricity occur in overmature, heavily defoliated, leaning, and suppressed trees, or those with one-sided crowns (Kozlowski, 1971b). In general gymnosperm stems tend to be more circular in cross section than angiosperm stems because the more uniform arrangement of branches around the stem of the former distributes carbohydrates and growth hormones more evenly. Production of reaction wood in tilted or leaning trees also is associated with eccentric cambial growth (Chapter 3, Kozlowski and Pallardy, 1997).

Some trees produce buttresses at the stem base, resulting in very eccentric stem cross sections. The stems of buttressed trees commonly taper downward from the level to which the buttresses ascend, and then taper upward from this height. Downward tapering of the lower stem does not occur in young trees but develops progressively during buttress formation.

Stem buttresses are produced by a few species of temperate zone trees and by many tropical trees. Examples from the temperate zone are tupelo gum and baldcypress. The degree of buttressing in baldcypress appears closely related to soil drainage, as buttressing is more exaggerated in soils that are inundated for long periods (Varnell, 1998). Whereas the buttresses of tupelo gum are narrow, basal stem swellings, the conspicuous buttresses of tropical trees vary from flattened plates to wide flutings (Fig. 2.4). The size of buttresses increases with tree age and, in some mature trees, buttresses may extend upward along the stem and outward from the base for nearly 10 m. Most buttressed tropical trees have three or four buttresses but they may have as many as 10. The formation of buttresses is an inherited trait and occurs commonly in tropical rain forest trees in the *Dipterocarpaceae*, legume families, and *Sterculiaceae*. Buttressing also is regulated by environmental regimes and is most prevalent at low altitudes and in areas of high rainfall (Kramer and Kozlowski, 1979).

VEGETATIVE ORGANS AND TISSUES

This section will briefly refer to the physiological role of leaves, stems, and roots, and then discuss their structures.

LEAVES

The leaves play a crucial role in growth and development of woody plants because they are the principal photosynthetic organs. Changes in photosynthetic activity by environmental changes or cultural practices eventually will influence growth of vegetative and reproductive tissues. Most loss of water from woody plants also occurs through the leaves.

FIGURE 2.4. Buttressing in *Pterygota horsefieldii* in Sarawak. Photo courtesy of P. Ashton.

Angiosperms

The typical foliage leaf of angiosperms is composed mainly of primary tissues. The blade (lamina), usually broad and flat and supported by a petiole, contains ground tissue (mesophyll) enclosed by an upper and lower epidermis (Fig. 2.5).

The outer surfaces of leaves are covered by a relatively waterproof layer, the cuticle, which is composed of wax and cutin and is anchored to the epidermal cells by a layer of pectin. The arrangement of the various constituents is shown in Figure 2.6. The thickness of the cuticle varies from 1 μm or less to approximately 15 μm. The cuticle generally is quite thin in shade-grown plants and much thicker in those exposed to bright sun. There also are genetic differences in the cuticles of different plant species and varieties. The structure and amounts of epicuticular waxes are discussed in Chapter 8.

Stomata

The mesophyll tissue has abundant intercellular spaces connected to the outer atmosphere by numerous microscopic openings (stomata) in the epidermis, consisting of two specialized guard cells and the pore between them (Fig. 2.7). Stomata play an essential role in the physiology of plants because they are the passages through which most water is lost as vapor from leaves, and through which most of the CO_2 diffuses into the leaf interior and is used in photosynthesis by mesophyll cells. In most angiosperm trees the stomata occur only on the lower surfaces of leaves, but in some

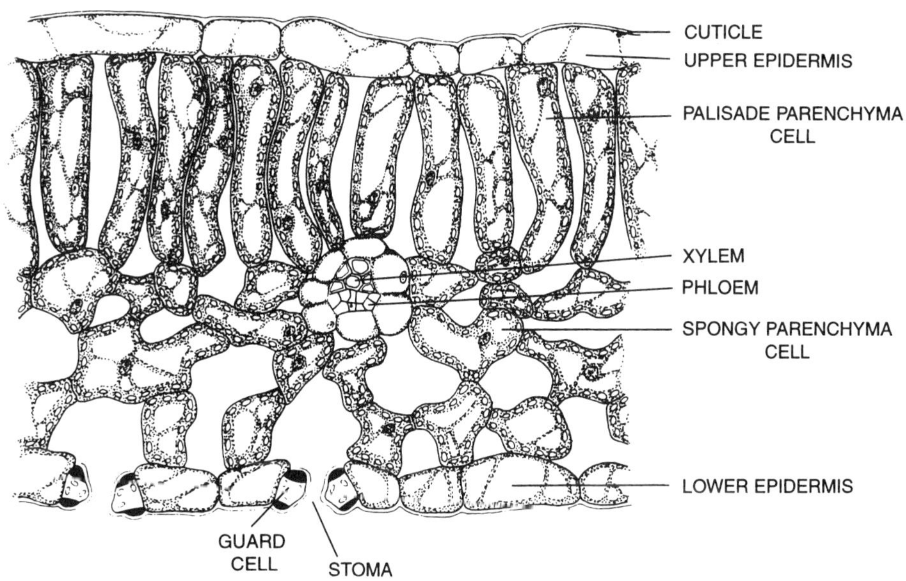

FIGURE 2.5. Transection of a portion of a leaf blade from a broadleaved tree. From Raven et al. (1992).

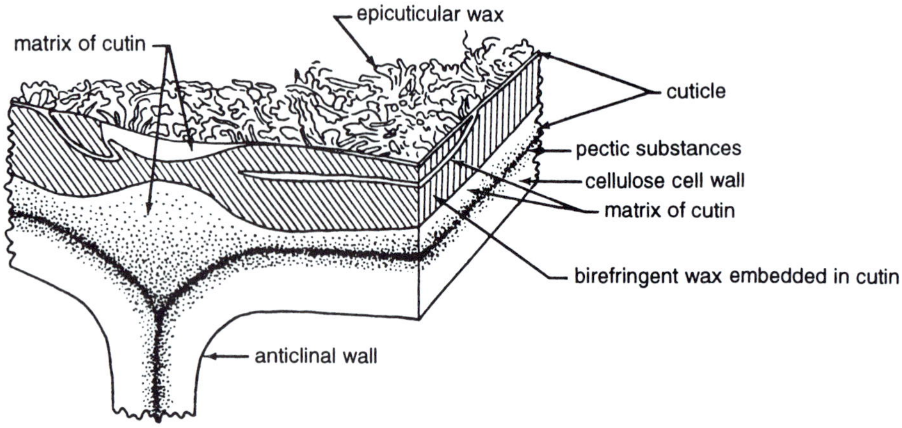

FIGURE 2.6. Diagram of outer cell wall of the upper epidermis of pear leaf showing details of cuticle and wax. From Norris and Bukovac (1968).

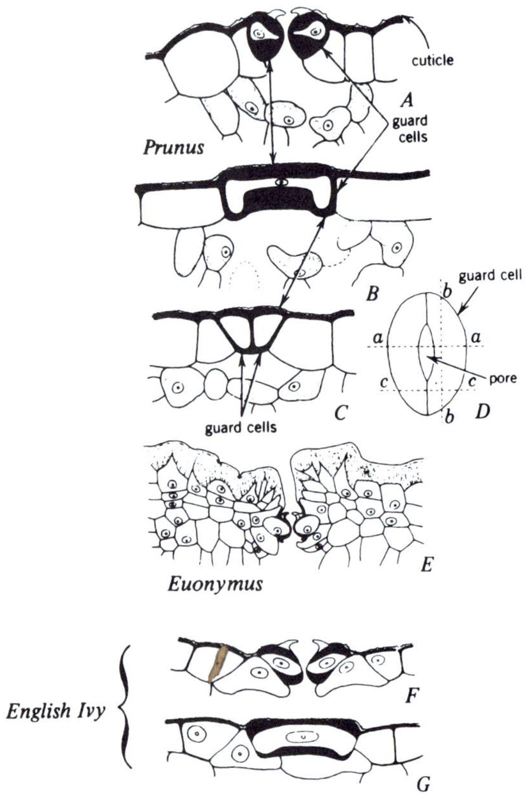

FIGURE 2.7. Stomata of woody angiosperms. A–C: Stomata and associated cells from peach leaf sectioned along planes indicated in D by the broken lines, aa, bb, and cc. E–G: Stomata of euonymus and English ivy cut along the plane aa. G: One guard cell of English ivy cut along the plane bb. From Esau (1965).

species, poplars and willows, for example, they occur on both leaf surfaces. However, when present on both leaf surfaces, the stomata usually are larger and more numerous on the lower surface (Table 2.1). In a developing leaf both mature and immature stomata often occur close together.

TABLE 2.1. Variations in Stomatal Distribution on Lower and Upper Leaf Surfaces of *Populus* Species[a]

	Stomatal density (no. cm^{-2})	
Clone	Lower surface	Upper surface
Populus maximowiczii x *P. nigra*	45,451 ± 1,003	4,216 ± 155
Populus maximowiczii	33,521 ± 868	7,730 ± 242
Populus trichocarpa	23,378 ± 581	5,013 ± 193
Populus deltoides	22,628 ± 408	18,693 ± 573
Populus nigra	20,450 ± 434	5,762 ± 313

[a]From Siwecki and Kozlowski (1973).

Of particular physiological importance are the wide variations in stomatal size and frequency that occur among species and genotypes. Stomatal size (guard cell length) varied among 38 species of trees from 17 to 50 μm and stomatal frequency from approximately 100 to 600 stomata/mm^2 of leaf surface (Table 2.2). Generally, a species with few stomata per unit of leaf surface tends to have large stomata. For example, white ash and white birch leaves had few but large stomata; sugar maple and silver maple had many small stomata. Oak species were an exception, having both large and numerous stomata. Both stomatal size and frequency often vary greatly among species within a genus, as in *Crataegus*, *Fraxinus*, *Quercus*, and *Populus* (Table 2.1).

Mesophyll

The mesophyll generally is differentiated into columnar palisade parenchyma cells and irregularly shaped spongy parenchyma cells. The palisade parenchyma tissue usually is located on the upper side of the leaf, and the spongy parenchyma on the lower

TABLE 2.2. Variations in Average Length and Frequency of Stomata of Woody Angiosperms[a]

Species	Stomatal length (μm)	Stomatal frequency (no. mm^{-2})
Acer negundo	21.6	233.9
Acer saccharinum	17.3	418.8
Acer saccharum	19.3	463.4
Betula nigra	39.4	281.3
Betula papyrifera	33.2	172.3
Catalpa bignonioides	23.2	328.6
Crataegus sp. *I*	22.3	399.1
Crataegus sp. *II*	37.4	221.4
Fraxinus americana	24.8	257.1
Fraxinus pennsylvanica	29.3	161.6
Ginkgo biloba	56.3	102.7
Gleditsia triacanthos	36.1	156.3
Hamamelis mollis	25.3	161.6
Juglans nigra	25.7	342.0
Malus sp.	23.8	219.5
Populus deltoides	30.4	163.4
Prunus serotina	30.5	306.3
Prunus virginiana	27.1	244.6
Quercus macrocarpa	24.0	575.9
Quercus palustris	30.9	530.4
Quercus rubra	26.7	532.1
Rhus typhina	19.4	633.9
Robinia pseudoacacia	17.6	282.1
Salix fragilis	25.5	215.2
Tilia americana	27.2	278.8
Ulmus americana	26.3	440.2
Vitis vinifera	29.7	120.5

[a]From Davies et al. (1973).

side. There may be only a single layer of palisade cells perpendicularly arranged below the upper epidermis or there may be as many as three layers. When more than one layer is present, the cells of the uppermost layer are longest, and those of the innermost layer may grade in size and shape to sometimes resemble the spongy parenchyma cells. When the difference between palisade and spongy parenchyma cells is very distinct, most of the chloroplasts are present in the palisade cells. Although the palisade cells may appear to be compactly arranged, most of the vertical walls of the palisade cells are exposed to intercellular spaces. Hence, the aggregate exposed surface of the palisade cells may exceed that of the spongy parenchyma cells by two to four times (Raven et al., 1992). Extensive exposure of mesophyll cell walls to internal air spaces promotes the rate of movement of CO_2 to chloroplasts, which are located adjacent to the plasmalemma (Chaper 5). A cell-to-cell liquid pathway for CO_2 would be much slower because the diffusion coefficient of this molecule in water is only 1/10,000 of that in air.

Carbohydrates, water, and minerals are supplied to and transported from the leaves through veins that thoroughly permeate the mesophyll tissues. The veins contain xylem on the upper side and phloem on the lower side. The small, minor veins that are more or less completely embedded in mesophyll tissue play the major role in collecting photosynthate from the mesophyll cells. The major veins are spatially less closely associated with mesophyll and increasingly embedded in nonphotosynthetic rib tissues. Hence, as veins increase in size their primary function changes from collecting photosynthate to transporting it from the leaves to various sinks (utilization sites).

Variations in Size and Structure of Leaves

The size and structure of leaves vary not only with species, genotype, and habitat but also with location on a tree, between juvenile and adult leaves, between early and late leaves, and between leaves of early-season shoots and those of late-season shoots.

The structure of leaves often varies with site. In Hawaii, leaf mass per unit area, leaf size, and the amount of leaf pubescence of *Meterosideros polymorpha* varied along gradients of elevation. Leaf mass per area increased, leaf size decreased, and the amount of pubescence increased from elevations of 70 to 2350 m. Pubescence accounted for up to 35% of leaf mass at high elevations (Geeske et al., 1994).

The structure of leaves also varies with their location on a tree. For example, the thickness of apple leaves typically increases from the base of a shoot toward the apex. Leaves near the shoot tip tend to have more elongated palisade cells and more compact palisade layers (hence comprising a higher proportion of the mesophyll tissue). The number of stomata per unit area also is higher in leaves located near the shoot apex than in leaves near the base (Faust, 1989).

Sun and Shade Leaves

There is considerable difference in structure between leaves grown in the sun and those produced in the shade. This applies to the shaded leaves in the crown interior compared to those on the periphery of the crown, as well as to leaves of entire plants grown in the shade or full sun. In general shade-grown leaves are larger, thinner, less deeply lobed (Fig. 2.8) and contain less palisade tissue and less conducting tissue than sun leaves. Leaves with deep lobes characteristic of the upper and outer crown positions are more efficient energy exchangers than shallowly lobed leaves (Chapter 12). Shade leaves also usually have fewer stomata per unit of leaf area, larger interveinal areas, and a lower ratio of internal to external surface.

Whereas leaves of red maple, American beech, and flowering dogwood usually had only one layer of palisade tissue regardless of the light intensity in which they were grown, shade-intolerant species such as yellow-poplar, black cherry, and sweetgum had two or three layers when grown in full sun but only one layer when grown in the shade (Jackson, 1967). Leaves of the shade-tolerant European beech had one palisade layer when developed in the shade and two layers when developed in full sun. Differentiation into sun- and

FIGURE 2.8. Sun leaves (left) and shade leaves (right) of black oak. From Talbert and Holch. (1957).

shade-leaf primordia was predetermined to some degree by early August of the year the primordia formed (Eschrich et al., 1989). As seedlings of the shade-intolerant black walnut were increasingly shaded, they were thinner, had fewer stomata per unit of leaf area, and had reduced development of palisade tissue (Table 2.3).

The plasticity of leaf structure in response to shading may vary considerably among closely related species. Of three species of oaks, black oak, the most drought-tolerant and light-demanding species, showed the greatest leaf anatomical plasticity in different light environments (Ashton and Berlyn, 1994). The most drought-intolerant species, northern red oak, showed least anatomical plasticity, and scarlet oak showed plasticity that was intermediate between that of black oak and northern red oak.

Increases in specific leaf area (amount of leaf area per gram of leaf dry mass) and decreases in its inverse, mass per unit leaf area, in response to shading have been shown for many species (e.g., LeRoux et al., 2001). The increases in specific leaf area often are accompanied by increased amounts of chlorophyll per unit of dry weight, but since shaded leaves are appreciably thinner than sun leaves, the amount of chlorophyll per unit of leaf area decreases (Lichtenthaler et al., 1981; Dean et al., 1982; Kozlowski et al., 1991).

Light intensity affects both the structure and activity of chloroplasts. The chloroplasts of shade plants contain many more thylakoids (Chapter 5) and have wider grana than chloroplasts of sun plants. In thylakoids of shade-grown plants there is a decrease in the chlorophyll a–chlorophyll b ratio and a low ratio of

TABLE 2.3. Photosynthetic and Anatomical Characteristics of Leaves of Black Walnut Seedlings Grown under Several Shading Regimes[a]

Shading treatment[b]	Light transmission (% of PAR)	Stomatal density (no. mm^{-2})	Palisade layer ratio[c]	Leaf thickness (μm)	Quantum efficiency (mol CO_2 fixed per mole PAR absorbed)
Control	100.0	290.0a	1.00a	141.3a	0.023ab
GL1	50.0 20.2	293.1a	1.44b	108.3b	0.023a
ND1	15.7	264.0b	1.25ab	104.9b	0.026ab
GL2	20.9 8.3	209.2c	1.85c	89.9c	0.030bc
ND2	3.3	187.0d	1.93c	88.2c	0.033c

[a]From Dean et al. (1982). PAR, Photosynthetically active radiation.

[b]The GL treatments consisted of green celluloid film with holes to simulate a canopy that had sunflecks. For GL1 and GL2, the upper value indicates the sum of 100% transmission through holes and transmission through remaining shaded area; the lower value indicates transmission through shade material only (GL1, one layer; GL2, two layers). ND1 and ND2 represent treatments with two levels of shading by neutral density shade cloth. Mean values within a column followed by the same letter are not significantly different ($p \leq 0.05$).

[c]Palisade layer ratio equals the cross-sectional area of palisade layer of control eaves divided by that of leaves of other treatments.

FIGURE 2.9. Variations in leaf form of juvenile, transitional, and adult leaves of English ivy. Photo courtesy of V. T. Stoutemyer.

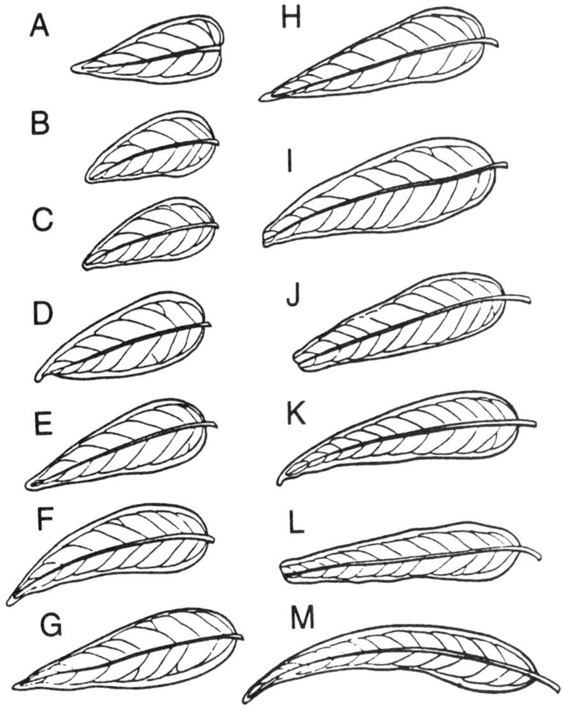

FIGURE 2.10. Series of leaves from a single tree of *Eucalyptus macarthuri*, showing the transition from juvenile (A) to adult (M) foliage. From Penfold and Willis (1961).

soluble protein to chlorophyll. Because leaves usually transmit only about 1 to 5% of the incident light, the structure of the chloroplasts on the shaded side of a leaf in sun may be similar to that of chloroplasts of plants on the forest floor (Anderson and Osmond, 1987).

Juvenile and Adult Leaves

Several studies show variations between juvenile and adult leaves in leaf form and structure of some woody plants. In English ivy, for example, the juvenile leaves are lobed and the adult leaves are not (Fig. 2.9). The shape of *Eucalyptus* leaves changes as the trees progress from juvenility to adulthood (Fig. 2.10). Tasmanian blue gum shows striking differences between its juvenile and adult leaves. The relatively thin juvenile leaves, which normally are borne horizontally, are sessile and cordate (E. D. Johnson, 1926). They have a pointed apex, are about twice as long as wide, and arranged in pairs at right angles to each other. The thick, spirally-borne adult leaves are sickle-shaped. Their petioles are twisted so they hang vertically. Adult leaves lack the heavy waxy coating found on juvenile leaves. In *Ilex aquifolium* trees the leaves vary from dentate in the juvenile zone to entire in the adult zone.

Trees that show a small and gradual change from the juvenile to adult condition are described as *homoblastic*; those with an abrupt transition are called *heteroblastic*. A feature of the New Zealand flora is the large number of heteroblastic species (some 200 species). The juvenile form of *Elaeocarpus hookeriana* has small, toothed, or lobed leaves, varying from obovate to linear; adult leaves are more regular, lanceolate to oblong and with crenate margins (Harrell et al., 1990). Gould (1993) described wide variations in the morphology and anatomy of leaves of seedling, juvenile, transitional, and adult phases of development of *Pseudopanax crassifolius*. Seedlings produced five leaf types, all small and thin, resembling the leaves of many shade plants. Juvenile leaves were long, linear (up to 1 m long), and sharply toothed. Adult leaves were much shorter and wider than juvenile leaves. Transitional leaves were morphologically intermediate between juvenile and adult leaves. The shapes of juvenile and adult leaves differed beginning early in their development (Clearwater and Gould, 1994).

Gymnosperms

Except for a few genera such as *Larix*, *Pseudolarix*, and *Metasequoia*, and some species of *Taxodium*, the leaves of gymnosperms are evergreen. Most gymnosperm leaves are linear or lanceolate and bifacially flattened but other shapes also occur. For example, the leaves of spruce and occasionally larch are tetragonal in cross section. Scalelike leaves are characteristic of *Sequoiadendron*, *Cupressus*, *Chamaecyparis*, *Thuja*, and *Calocedrus*. Broad, ovate, and flat leaves are found in *Araucaria*.

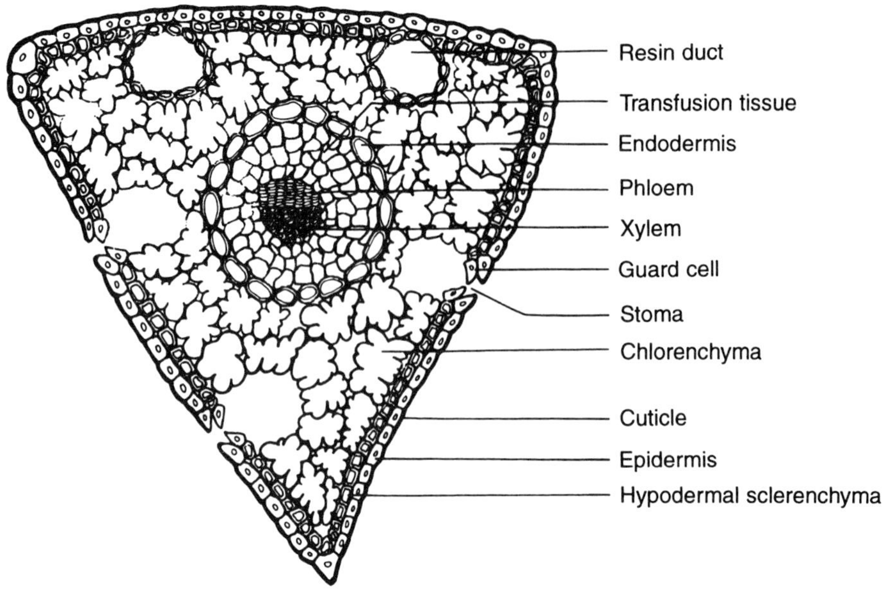

FIGURE 2.11. Transection of secondary needle of eastern white pine.

In leaves of *Abies, Pseudotsuga, Dacrydium, Sequoia, Taxus, Torreya, Ginkgo, Araucaria,* and *Podocarpus*, the mesophyll is differentiated into palisade cells and spongy parenchyma. The leaves of the last two genera have palisade parenchyma on both sides (Esau, 1965). In pines the mesophyll is not differentiated into palisade cells and spongy parenchyma (Fig. 2.11).

Pine needles, borne in fascicles, are hemispherical (two-needled species) or triangular (three- and more-needled species). Those of the single-needle pinyon pine are circular in cross section. Sometimes the number of needles per fascicle varies from the typical condition. This often is a response to unusual nutritional conditions, injury, or abnormal development.

In pines the deeply sunken stomata are arranged in rows (Fig. 2.12). Below the epidermis and surrounding the mesophyll is a thick-walled hypodermal layer. Parenchyma cells of the mesophyll are deeply infolded. The one or two vascular bundles per needle are surrounded by transfusion tissue consisting of dead tracheids and living parenchyma cells. The transfusion tissue functions in concentrating solutes from the transpiration stream and retrieving selected solutes that eventually are released to the phloem (Canny, 1993a). Two to several resin ducts also occur in pine needles. The cells of the endodermis, which surrounds the transfusion tissue, are rather thick walled. The epidermis of pine needles has a heavy cuticle. Considerable wax often is present in the stomatal antechambers of some gymnosperms (Gambles and Dengler, 1982) and not others (Franich et al., 1977).

In conifers the size and weight of needles vary with their position along a shoot. In Fraser fir, for example, needle length, weight, and thickness were maximal at the middle (50%) position (Fig. 2.13). Similarly in Sitka spruce the largest needles with most cells were located near the middle of the shoot (Chandler and Dale, 1990). In eastern hemlock needle size decreased and the number of needles per unit length of shoot increased with decrease in shoot length (Powell, 1992).

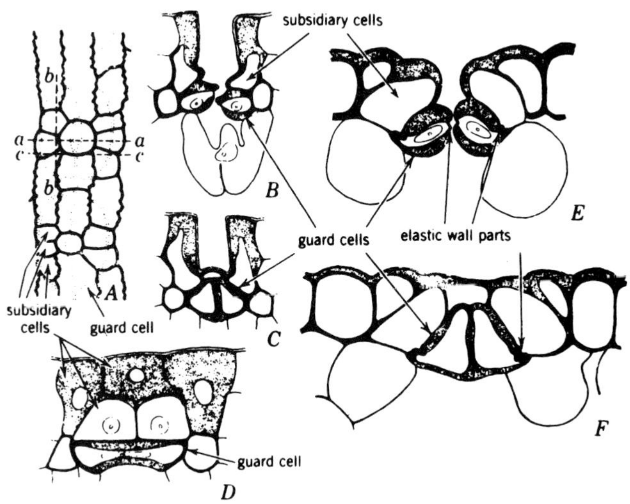

FIGURE 2.12. Stomata of gymnosperms. A: Surface view of epidermis with sunken stomata of *Pinus merkusii*. B–D: Stomata of pines; E, F: Stomata of *Sequoia*. The broken lines in A indicate the planes along which the sections were made in B–F: aa, B, E; bb, D; cc, C, F. From Esau (1965).

When grown in the shade gymnosperm needles usually show similar responses to those of angiosperm

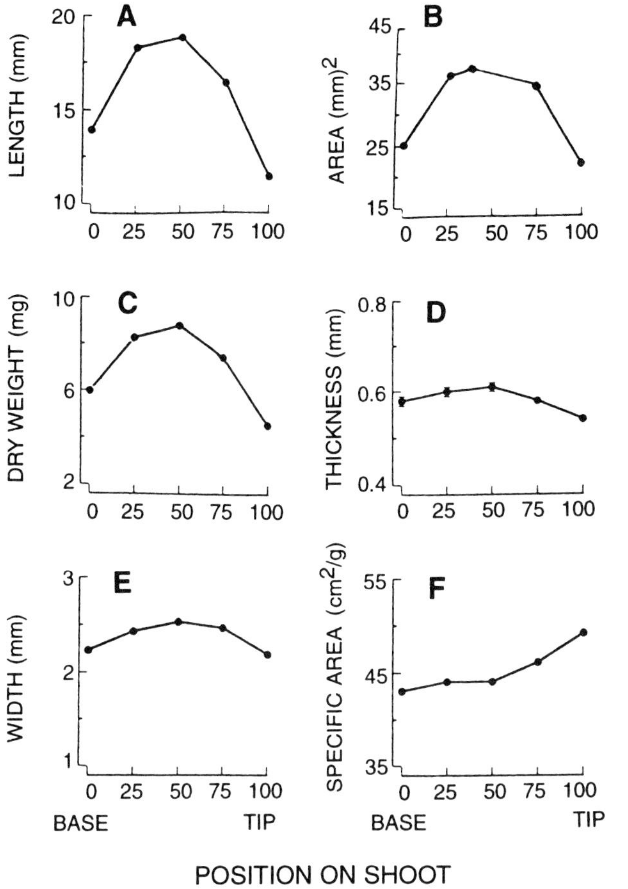

FIGURE 2.13. Variations in traits of needles of Fraser fir with position on shoots. A, length; B, surface area (one side); C, dry weight; D, thickness; E, width; and F, specific area. From Brewer et al. (1992).

leaves, being thinner, higher in chlorophyll content, and with reduced stomatal frequency when compared with sun-grown needles. Needles of western hemlock that developed in the shade were thinner, had a higher ratio of width to thickness, thinner palisade mesophyll, and a higher ratio of surface area to weight than those developed in full sun (Fig. 2.14).

STEMS

Stems of woody plants support the crown; store water, carbohydrates, and minerals; conduct water and minerals upward from the roots; and transport foods and hormones from points where they are synthesized to those where they are used in growth or stored for future use.

As may be seen in Figures 2.15 and 2.16, a mature tree stem typically consists of a tapering column of wood (xylem) composed of a series of layers or annual increments, added one above the other, like a series of overlapping cones, and enclosed in a covering of bark. At the apex of the stem and each of its branches is a terminal growing point where increase in length occurs. Between the bark and wood of the stem, branches, and major roots is the vascular cambium (hereafter called cambium), a thin, sheathing lateral meristem.

Sapwood and Heartwood

Young xylem or sapwood conducts sap (primarily water), strengthens the stem, and to some extent serves

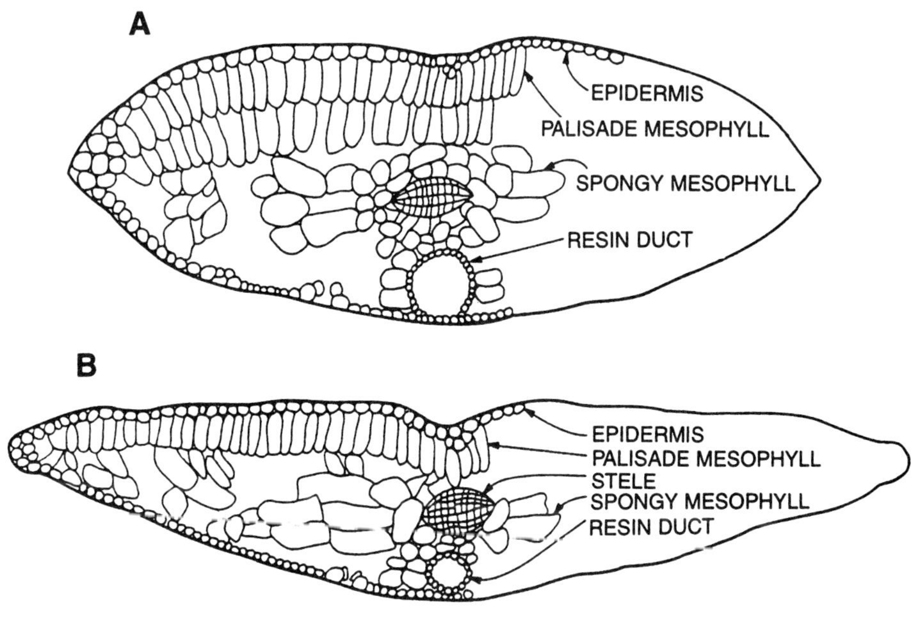

FIGURE 2.14. Variations in structure of western hemlock needles grown in sun (A) and in shade (B). Used with permission of the Society of American Foresters from Tucker, G. F., and Emmingham, W. H. (1977). Morphological changes in leaves of residual western hemlock after clear and shelterwood cutting. *For. Sci.* 23, 195–203.; permission conveyed through Copyright Clearance Center, Inc.

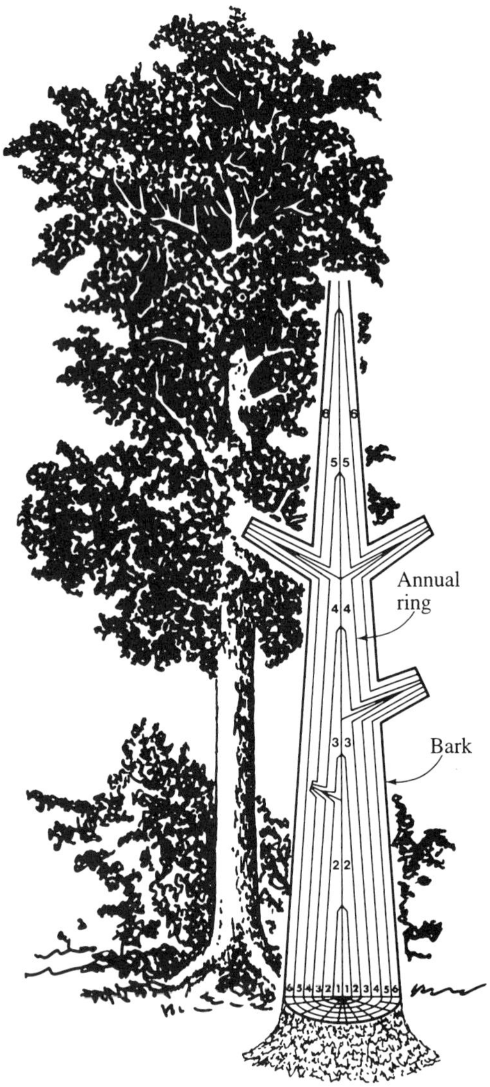

FIGURE 2.15. Diagrammatic median longitudinal section of a tree showing pattern of annual xylem increments in the stem and major branches.

FIGURE 2.16. Generalized structure of a tree stem showing orientation of major tissues including outer bark, cambium, sapwood, and heartwood. Photo courtesy of St. Regis Paper Co.

as a storage reservoir for food. The living parenchyma cells in the sapwood, which are very important because they store foods, consist of horizontally oriented ray cells and, in many species of woody plants, of vertically oriented axial parenchyma cells as well. On the average only about 10% of the sapwood cells are alive. As the xylem ages all the living cells die and the tissue often becomes darker, producing a central cylinder of dark-colored dead tissue, called heartwood, which continues to provide mechanical support but is no longer involved in physiological processes. This is demonstrated by the fact that old trees from which the heartwood has been destroyed by decay can survive for many years, supported by a thin shell of sapwood. The outline of the heartwood core is irregular and does not follow a specific annual ring either at different stem heights or in the cross section of the stem. Cross sections of stems of many species show a distinct transition or intermediate zone, usually less than 1 cm wide, surrounding the heartwood. In some species the transition zone, which typically is lighter in color than the heartwood, is not readily recognized or does not exist. Formation of heartwood is discussed in Chapter 3.

Xylem Increments and Annual Rings

The secondary vascular tissues consist of two interpenetrating systems, axial and radial. The axial components are oriented vertically; the radial components are oriented radially (rays) or horizontally.

In trees of the temperate zone the annual rings of wood (secondary xylem) stand out prominently in stem and branch cross sections. In gymnosperms the wood formed early in the season consists of large-diameter cells with relatively thin walls; hence the wood that forms early is less dense than the wood formed later. Because of the uniformity of composition of gymnosperm xylem, changes in cell wall thickness are correlated closely with changes in wood density. In angiosperms, however, the density of wood depends not only on cell diameter and wall thickness but also on the proportion of the various cell types present. This proportion is relatively constant within a species and even within many genera, although it varies within a season. However, the arrangement of cells and proportions of different cell types vary greatly among woods of different genera of angiosperms.

Because of the consistent periclinal (tangential) divisions of the cambial cells during the production of secondary xylem and secondary phloem (Chapter 3)

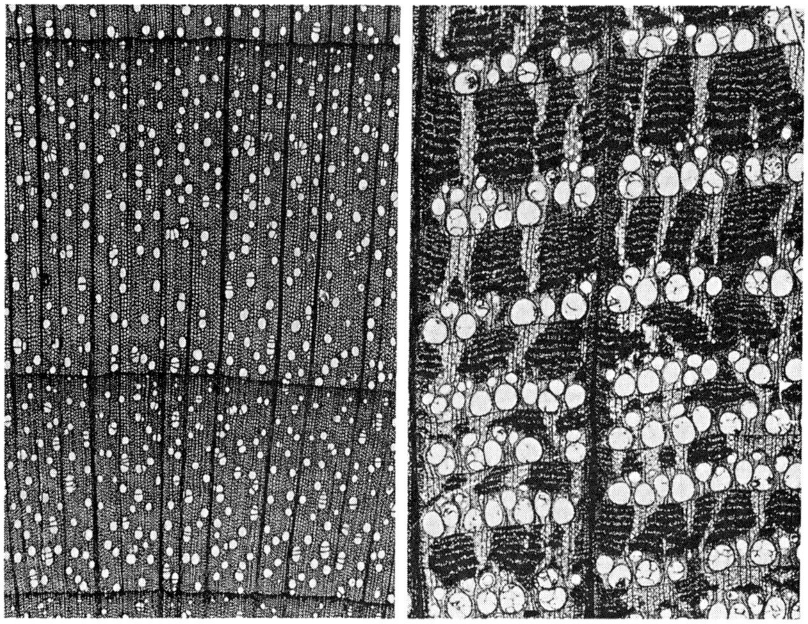

FIGURE 2.17. Stem transections showing variation in vessel diameters and distribution within annual growth increments of diffuse-porous species, silver maple (left), and a ring-porous species, white oak (right) (×50). Photo courtesy of U.S. Forest Service.

the young, undifferentiated xylem and phloem cells are regularly aligned in radial rows. In gymnosperms such a regular radial arrangement generally is maintained throughout differentiation of tracheids. In contrast, in angiosperms the early radial alignment of cambial derivatives in the xylem becomes obscured as some cells, such as vessel members, enlarge greatly and distort the position of rays and adjacent cells. Hence, it is not uncommon for a narrow ray to be bent around a large vessel. However, rays that are many cells wide generally are not distorted by the enlarging vessel elements. In some angiosperms, intrusive growth of fibers alters the radial arrangement of xylem cells.

As may be seen in Figure 2.17, angiosperms are classified as ring-porous or diffuse-porous. In ring-porous trees, such as oaks, chestnut, black locust, ashes, and elms, the diameters of xylem vessels formed early in the growing season are much larger than those formed later. In diffuse-porous trees, such as poplars, willows, basswood, maples, and birches, all the vessels are of relatively small diameter and those formed early in the growing season are of approximately the same diameter as those formed later.

Considerable variation exists in the nature of the outer boundaries of annual xylem growth increments. In areas of high rainfall and cold winter the boundaries between annual xylem increments as seen in cross sections of stems or branches are well defined in comparison to those in species growing in hot, arid regions. In the juvenile core of the stem of a normal tree, the transition from one year's xylem increment to another is gradual nearest the pith and becomes increasingly abrupt in the older wood. In old trees the lines of demarcation between xylem increments generally are very sharp. The width of annual rings often is materially reduced by drought and this fact has been used extensively to study climatic conditions in the past and even to date ancient structures (Fritts, 1976).

Earlywood and Latewood

The wood of low density usually (but not always) produced early in the season is called "earlywood." The part of the annual xylem increment that usually is produced late in the growing season and is of higher density than wood produced early in the season is called "latewood." There is much interest in earlywood–latewood relations of trees because they affect wood quality.

Earlywood and latewood have been used in the literature as synonyms for "springwood" and "summerwood," but the latter terms are really misnomers because either type of wood may be produced in more than one season in the same year. As early as 1937 Chalk suggested that the terms springwood and summerwood be abandoned, but they are still used despite their shortcomings. Glock et al. (1960) also objected to the terms earlywood and latewood because the latewood sometimes is found at the beginning of a growth layer or as fragments within an annual increment. They preferred to use the terms "lightwood" and "densewood," thereby emphasizing the structure of the tissues rather than the time when tissues form or their relative position within a growth layer or increment.

The boundary between the earlywood and latewood of the same ring can be very sharp or gradual. The boundary is sharp in hard pines, Douglas-fir, larch, and juniper. Ladefoged (1952) found an abrupt earlywood–latewood transition in ring-porous angiosperms and a gradual one in diffuse-porous species. Various arbitrary methods of clearly characterizing both earlywood and latewood cells have been advanced. One of the most popular standards for gymnosperms is that of Mork (1928), who considered a latewood tracheid to be one in which the width of the common wall between the two neighboring tracheids multiplied by two was equal to, or greater than, the width of the lumen. When the value was less than the width of the lumen the xylem was considered to be earlywood. All measurements were made in the radial direction. Mork's definition originally was applied to spruce xylem but has been adopted widely for general use with gymnosperm woods. This definition is not useful for angiosperm woods because there often are serious problems in distinguishing between earlywood and latewood.

Within an annual xylem increment the width of the earlywood band generally decreases and the width of the latewood band increases toward the base of the tree. In gymnosperms the earlywood tracheids are wider toward the stem base than near the top of the stem within the same xylem increment. The transition between the last earlywood tracheids and first-formed latewood tracheids of the annual increment also is sharper in the lower stem than in the upper stem.

Some tracheids fit the usual definition of latewood because of a decrease in their radial diameter, without appreciable change in wall thickness. Other tracheids, however, become latewood because of an increase in wall thickness without a change in diameter. Both dimensions show continuous change from the top of the stem toward the base until the latewood forms. In upper parts of stems "transition latewood" often forms, which cannot be conveniently classified as either true earlywood or latewood (Fig. 2.18).

Cambial growth of tropical trees is very diverse and appears to be strongly determined by heredity. In many species xylem may be added to the stem during most or all of the year. Hence, many tropical trees, especially those in continually warm and wet tropical climates, lack growth rings or have very indistinct ones. Examples are *Agathis macrophylla* in Melanesia, many tropical mangroves, and mango in India (Whitmore, 1966; Fahn et al., 1981; Dave and Rao, 1982). Other tropical species produce distinct growth rings, often more than one each year.

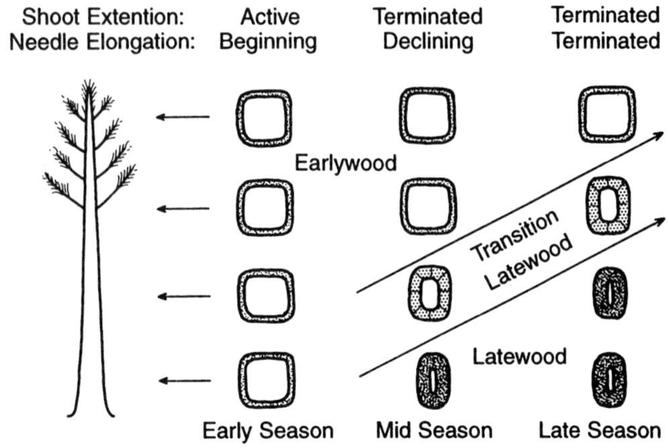

FIGURE 2.18. Seasonal variation in formation of earlywood, transition latewood, and latewood at different stem heights of a red pine tree. From Larson (1969).

The anatomical features that delineate growth rings in tropical woods vary greatly among species. In *Acacia catechu*, for example, growth rings are outlined by narrow bands of marginal parenchyma, and sometimes by thick-walled fibers in the outer latewood. The growth rings of *Bombax malabaricum* are identified by radially compressed fibers and parenchyma cells in the outer latewood. The xylem increments of *Shorea robusta* have many irregularly shaped parenchyma bands that sometimes are mistaken for annual rings.

Phloem Increments

The annual sheaths of mature phloem are much thinner than the increments of xylem because less phloem than xylem is produced annually. The total thickness of phloem in general also is limited because the old phloem tissues often are crushed, and eventually the external nonfunctional phloem tissues are shed.

In many woody plants the phloem is divided by various structural features into distinguishable growth increments. However, these are not as clearly defined as annual xylem increments. Often the structural differences of early and late phloem are rendered indistinguishable by collapse of sieve tubes and growth of parenchyma cells.

In some species the annual increments of phloem can be delineated because early phloem cells expand more than those of the late phloem. In pear, tangential bands of fiber sclereids and crystal-containing cells are characteristic boundaries of annual growth of phloem (Evert, 1963). Early and late phloem increments sometimes can be identified by features of phloem parenchyma cells. For example, phloem parenchyma cells produced early have little tannin and they collapse when the phloem eventually becomes nonfunctional. In contrast, the tannin-laden, late-phloem parenchyma cells become turgid. Hence, their appearance is useful in identifying the limits of annual increments. In some species the annual increments of phloem can be identified by the number of distinct zones of various types of cells.

It is especially difficult to identify the annual increments of secondary phloem in gymnosperms. Although differences occur in diameters of early and late sieve cells, these often are obscured by pressure from expanding parenchyma cells. In false-cypress (*Chamaecyparis*) and thuja (*Thuja*), the early formed fibers of an annual increment have thicker walls than the fibers formed later. The early phloem of the *Pinaceae* is made up almost wholly of sieve elements. As sieve elements collapse, they form a dark band that outlines the boundary of the annual increment. Using such criteria, Srivastava (1963) attempted to identify annual growth increments in the phloem of a variety of gymnosperms. The results were variable. Some species, including Jeffrey pine, blue spruce, Norway spruce, and European larch, had distinct growth increments. In a number of other species the boundaries of growth increments were not readily discernible, either because phloem parenchyma cells were scattered, or because a distinct line of crushed sieve cells could not be identified between successive bands of phloem parenchyma.

Both the proportion of conducting and nonconducting cells in the secondary phloem, as well as the cross-sectional area occupied by sieve elements in the conducting zone, vary widely among species, even in the same genus (Khan et al., 1992). The layer of phloem that has conducting sieve tubes is exceedingly narrow. For example, the layer of conducting phloem is only about 0.2 mm wide in white ash; 0.2 to 0.3 mm in oak, beech, maple, and birch; 0.4 to 0.7 mm in walnut and elm; and 0.8 to 1.0 mm in willow and poplar (Holdheide, 1951; Zimmermann, 1961). Because of distortions of tissues in the nonconducting phloem it is only in the narrow conducting zone that important characteristics of phloem tissues can be recognized. These include shapes of various phloem elements, presence of nacreous (thickened) walls, structure of sieve plates, and variations among parenchyma cells. After sieve elements cease functioning, several important changes may occur in the phloem including intensive sclerification, deposition of crystals, collapse of sieve elements, and dilation of phloem tissues resulting from enlargement and division of axial and ray parenchyma cells. The extent to which each of these changes occurs varies with species.

WOOD STRUCTURE OF GYMNOSPERMS

In most gymnosperm stems the longitudinal elements of the xylem consist mainly of tracheids and a few axial parenchyma and epithelial cells (Fig. 2.19). Axial parenchyma cells occur in the xylem of redwood and thuja but not in xylem of pine. The horizontally oriented elements, which are relatively few, include ray tracheids, ray parenchyma cells, and epithelial cells. Interspersed also are axially and horizontally oriented resin ducts, which are intercellular spaces of postcambial development rather than cellular elements. Resin ducts are a normal feature of pine, spruce, larch, and Douglas-fir. Horizontal resin ducts occur only in the wood rays and only in relatively few rays. In addition, traumatic resin ducts caused by wounding may occur together with normal resin ducts. Traumatic ducts also may be found in woods lacking normal resin ducts such as *Cedrus*, hemlock, and true firs.

Most resins are secreted into special ducts by the layer of parenchyma cells that surround them. The ducts are formed schizogenously (by separation of cells). The ducts are much-branched, so when one of

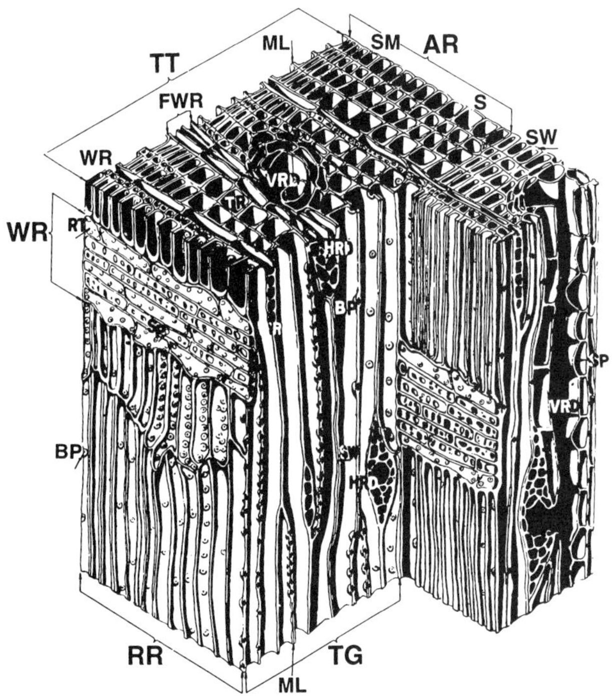

FIGURE 2.19. Anatomy of gymnosperm wood. TT: transection; RR: radial section; TG: tangential section; TR: tracheids; ML: middle lamella; S: earlywood; SM or SW: latewood; AR: annual ring; WR: wood ray; RT: ray tracheid; FWR: fusiform wood ray; SP: simple pit; BP: bordered pit; HRD: horizontal resin duct; VRD: vertical resin duct. Photo courtesy of U.S. Forest Service.

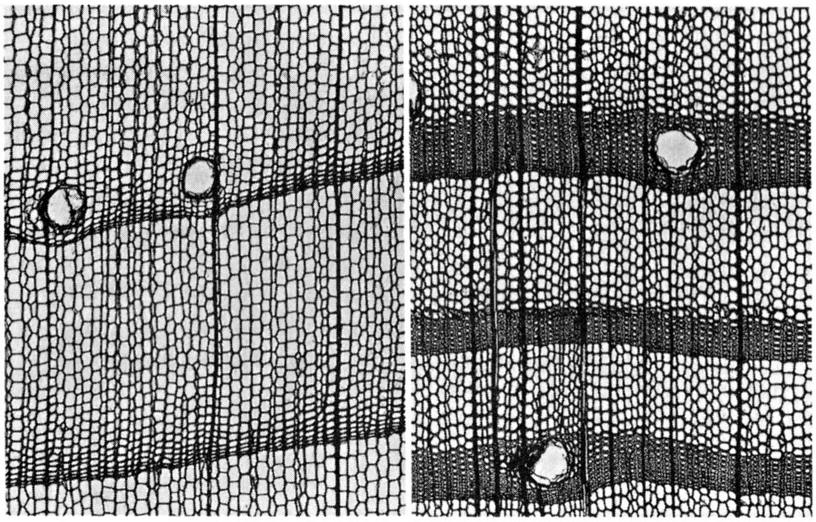

FIGURE 2.20. Variations in transition from earlywood to latewood in gymnosperms. Gradual transition of sugar pine (left) and abrupt transition in longleaf pine (right) (×27.5). Photo courtesy of U.S. Forest Service.

the branches is tapped or wounded, resin flows from the wounded area from long distances. Occasionally resins also are found in cell interiors and cell walls. They are not used as reserve foods and their role in metabolism of woody plants is uncertain. However, resins play an important role in defense against insects and fungi (Chapter 8).

Axial Elements

As much as 90% of the xylem of gymnosperms is made up of vertically oriented, overlapping tracheids, arranged in rather uniform radial rows. These four- to six-sided, thick-walled, tapering cells often are as much as 100 times longer than wide. They may vary in length from about 3 to 7 mm, but in most temperate zone gymnosperms they average 3 to 5 mm in length and 30 μm in diameter. Those formed early in the growing season are larger in cross section and have thinner walls than those formed later. The transition from large, earlywood tracheids to small, latewood tracheids may be gradual as in sugar pine, or it may be abrupt as in loblolly pine or longleaf pine (Fig. 2.20).

Walls of axial tracheids have various types of pits that facilitate transfer of liquids between adjacent cells (Figs. 2.21 and 2.22). Large bordered pits develop between adjacent axial tracheids, smaller bordered pits between axial tracheids and ray tracheids, and half-bordered pits between tracheids and ray parenchyma cells. Pits on tracheid walls occur predominantly on radial surfaces and tend to be concentrated near the ends of tracheids.

Bordered pit-pairs have a common membrane of primary walls and a middle lamella. In such a pit-pair the secondary wall of each adjacent cell arches over the

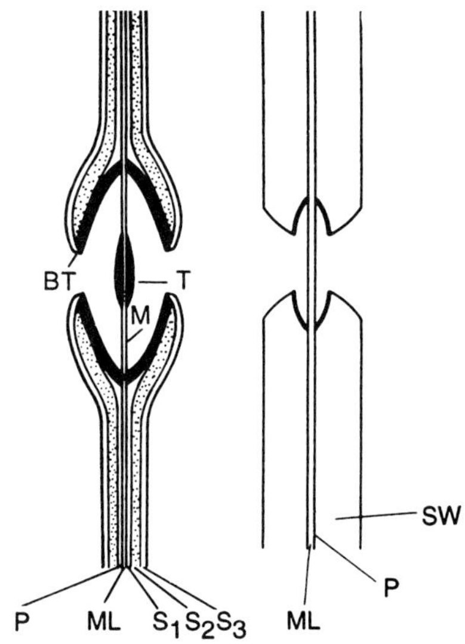

FIGURE 2.21. Pit of gymnosperm wood (left) and angiosperm wood (right). In the gymnosperm wood, ML: middle lamella; P: primary wall; S_1: outside layer of secondary wall; S_2: middle layer of secondary wall; S_3: inner layer of secondary wall; M: pit membrane (or margo); T: torus; and BT: initial border thickening. In the angiosperm wood, ML: middle lamella; P: primary wall; and SW: secondary wall. Reprinted with permission from *Botanical Review*, vol. 28, pp. 241–285, by A. B. Waldrop, Copyright 1962, The New York Botanical Garden.

pit cavity. In many gymnosperms the pit membrane consists of a disc-shaped or convex, lens-shaped thickening called the torus, surrounded by a thin margin, the margo (Fig. 2.21). The membranes of bordered pits are made up of cellulose strands that radiate, spokelike, from the torus to the margin of the pit cavity. Liquid moves more readily through the pores in the margo of the pit membrane when the torus is in a medial position. However, when a pit is aspirated (the

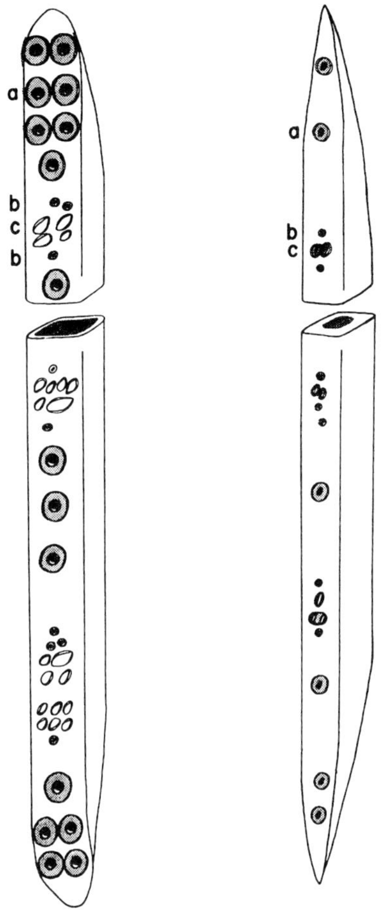

FIGURE 2.22. Earlywood and latewood tracheids of pine. a: Intertracheid bordered pits; b: Bordered pits to ray tracheids, c: Pinoid pits to ray parenchyma. Photo courtesy of U.S. Forest Service.

torus is pushed against the pit border) or when a pit membrane becomes encrusted with amorphous substances, the flow of liquid through the pit is restricted.

Pores in the membranes of bordered pits of gymnosperm xylem vary from less than 1 nm[1] to several nm in the same pit. The size of these pores is critical to maintaining function in water flow, as large pit pores are more likely to seed cavitation-inducing air bubbles into xylem elements under tension in both gymnosperms and angiosperms (Chapter 11). Pit diameter also varies greatly among species. The bordered pits of gymnosperms are more numerous and much larger in earlywood than in latewood of the same annual ring (Fig. 2.22). There appears to be much more resistance to water transport in latewood than in earlywood (Kozlowski et al., 1966) (see Chapter 11).

[1]1 nm or nanometer = 10^{-9} m, 10^{-6} mm, or 10^{-3} μm

When present in gymnosperms axial parenchyma occurs as long strands. Axial parenchyma is relatively abundant in redwood and baldcypress, sparse in larch and Douglas-fir, and absent in pines.

Horizontal Elements

Wood rays comprise the major horizontally oriented elements of gymnosperm wood. These ribbon-shaped aggregates of cells radiate in a stem cross section like wheel spokes. The rays play a very important physiological role in storage of carbohydrates and minerals and in radial translocation of water, minerals, and organic compounds.

Two types of rays occur in gymnosperms: (1) narrow rays, usually one cell wide (uniseriate), although some species have biseriate rays; and (2) wide fusiform rays when transverse resin ducts are present. In the majority of gymnosperms the narrow rays are only about 10 to 15 cells high but in some species, such as baldcypress, they may be up to 60 cells high.

Individual rays of gymnosperms are composed of ray parenchyma cells or of both ray parenchyma cells and ray tracheids, or solely of ray tracheids. Ray tracheids always occur in pine, spruce, larch, and Douglas-fir and are less commonly found in true firs, baldcypress, redwood, cedar, incense cedar, and junipers. When ray tracheids are present, they may occur in rows at ray margins or among layers of ray parenchyma cells.

Ray parenchyma cells have thin walls and living protoplasts when they are located in the portion of the ray that is in the sapwood. Ray tracheids have thick lignified walls. The ray tracheids of hard pines are described as dentate because of the tooth-like projections on their inner walls.

Fusiform rays, which may be found in pine, spruce, Douglas-fir, and larch, consist of marginal ray tracheids, ray parenchyma cells, and epithelial cells around a horizontally oriented resin canal. Fusiform rays are proportionally few in number and do not exceed 5% of the total number of rays present.

WOOD STRUCTURE OF ANGIOSPERMS

The axial system of angiosperm wood consists of tracheary elements (vessel elements, tracheids, fibers, and various kinds of parenchyma cells) (Fig. 2.23). The radial system consists of horizontal ray parenchyma cells. Although axial or horizontal resin ducts occur normally in various tropical angiosperms (Fahn, 1979), they are conspicuously absent in virtually all broad-leaved trees of the temperate zone.

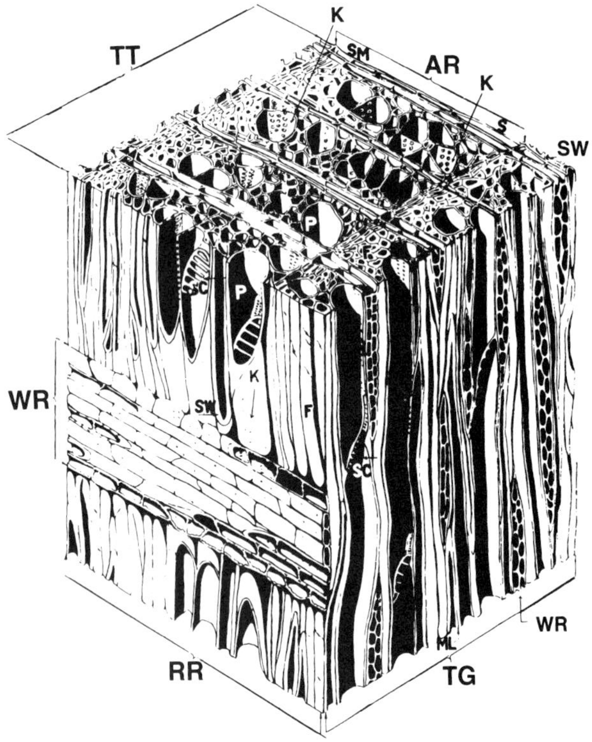

FIGURE 2.23. Anatomy of angiosperm wood. TT: transection; RR: radial section; TG: tangential section; P: vessel; SC: perforation plate at end of vessel; F: fibers; K: pit; WR: wood ray; AR: annual ring; S: earlywood; SM or SW: latewood; ML: middle lamella. Photo courtesy of U.S. Forest Service.

Axial Elements

There are more cell types in the wood of angiosperms than in that of gymnosperms. Most conspicuous are the vessel members, which are the chief water-conducting cells. Vessel members occur end-on-end, forming tubular conduits called vessels. The end walls of the vessel members partly or completely disintegrate.

In cross section most vessels are oval in shape but in woods of the more primitive species the vessels tend to be angular. Vessel arrangement is variable but fixed for species and therefore very useful in wood identification. For example, vessels may be solitary or arranged in various groups. In stem cross sections they occur in holly in chains and in elm and hackberry as groups in the latewood in concentric wavy bands.

The length of vessels, which varies greatly among species and in different parts of the same tree, is positively correlated with vessel diameter. However, the length of individual vessel members is negatively correlated with vessel diameter. The large-diameter earlywood vessels of ring-porous trees often are many meters long and some may be as long as the stem is high (Zimmermann and Jeje, 1981). Lianas have among the widest-diameter and longest vessels, up to 8 m long or more (Ewers, 1985). However, only a few of the vessels of woody plants belong to the longest length class. In shrubs and diffuse-porous species the longest vessels were about 1 m long, but most were much shorter, with the largest percentage less than 10 cm long (Zimmermann, 1983). Thus each woody plant has a range of vessel lengths and a much higher proportion of short than long vessels. In red maple both vessel diameter and length of the longest vessel increased from twigs to branches, down to the stem, and to the long roots (Zimmermann and Potter, 1982).

Tracheids are individual cells and they are smaller in diameter than vessels. Two types of tracheids occur in angiosperms, vascular tracheids, and vasicentric tracheids. Vascular tracheids are imperforate cells resembling small vessel members in form and position. Vasicentric tracheids are short, irregularly formed tracheids in the immediate proximity of vessels and do not form part of the definite axial rows.

The bulk of the xylem of angiosperms usually consists of fibers. Fibers somewhat resemble tracheids in that they are imperforate, but they have thicker walls, fewer pits, and smaller lumens. Xylem elements of angiosperms lack the orderly radial alignment characteristic of gymnosperm tracheids. The seemingly random distribution of elements in angiosperms results from extensive diameter growth of vessel members and intrusive growth of the fibers after they are laid down by the cambium. The expanding vessel members force other cells out of orderly alignment and cause narrow rays to bend around large vessels. This random arrangement also is partly caused by a lesser tendency for division of cambial initials opposite rapidly expanding vessels than of initials in a region where no large earlywood vessels form (Panshin et al., 1966).

In angiosperms, liquids move predominantly vertically through the vessels (Chapter 11). Lateral movement of liquids occurs through bordered and half-bordered pits between adjacent, overlapping vessels. The pits of angiosperms also may connect fibers to fibers, vessels to fibers, fibers to ray cells, and vessels to ray cells. The membrane of bordered pit-pairs of angiosperms is the compound middle lamella (primary wall plus middle lamella) of adjacent cells. This membrane is made up of randomly arranged microfibrils rather than centrally radiating ones as in gymnosperms (Coté, 1963).

The amount of axial parenchyma in wood of most angiosperms is considerably greater than in gymnosperms. In some tropical trees as much as half of the wood volume may consist of axial parenchyma. However, in most temperate zone trees axial parenchyma makes up less than 50% of the wood volume, and in some only a few percent. In poplar the amount of axial parenchyma is negligible. Patterns of arrangement of axial parenchyma vary greatly among species and are used in wood identification.

Horizontal Elements

Width, height, and spacing of rays vary much more in angiosperms than in gymnosperms. Usually rays are two to many cells wide and in oaks they may be up to 30 cells wide. Some species of angiosperms have rays of two size classes, with the smaller rays only one cell wide. A few genera of woody plants (alder, bluebeech, hazel) have "aggregate" rays consisting of groups of narrow, closely spaced rays with intervening tracheary elements. These aggregates often appear to be a single, very wide ray. Ray height also is extremely variable in angiosperms. The shortest rays are only a few microns high and the tallest ones may exceed 5 cm in height.

The rays of angiosperms are made up exclusively of parenchyma cells. Ray cells are variously classified. They may be radially oriented (procumbent) or vertically oriented (upright). Rays consisting entirely of procumbent or upright cells are classified as homocellular. Those containing both procumbent and upright cells are classified as heterocellular. The structure and size of pits in ray parenchyma cells, which may be simple to bordered, are used as diagnostic features in wood identification.

Differences in ray structure of angiosperms account for variations in the figure or grain of wood. For example, more ray tissue is exposed on quarter-sawn lumber (produced by cutting the wide faces in a radial-longitudinal plane of a log) than on flat-sawn lumber (produced by cutting the wide face tangentially to the annual rings) (Fig. 2.24). In quarter-sawn boards ray patterns vary from an inconspicuous figure in maple, with small rays, to the visually complex grain of oak, with conspicuously large rays.

BARK

The bark is structurally much more complex than the wood. Bark includes all tissues outside the cambium, including the inner living phloem and dead outer tissue (rhytidome). More specifically, in tissues that have gone into secondary thickening, bark tissues include primary and secondary phloem, cortex, and periderm. However, in stems not yet undergoing secondary thickening, only the primary phloem and cortex are included in the bark. The phloem plays an essential role in translocation of carbohydrates, and

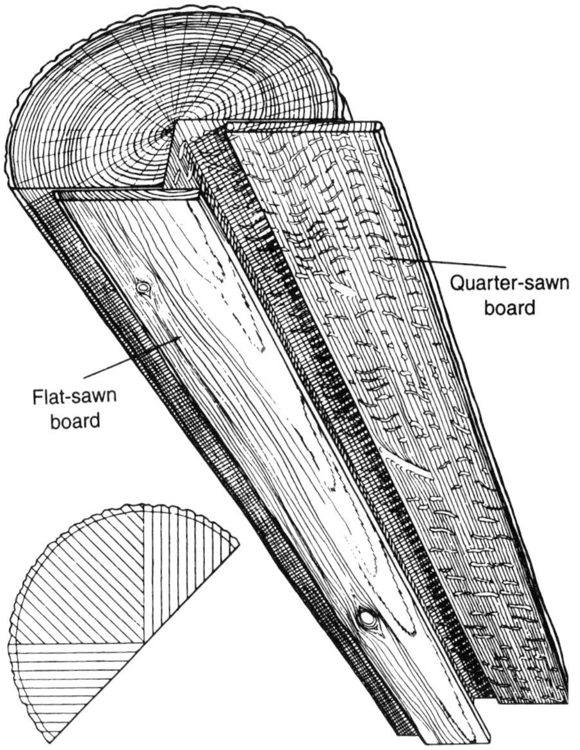

FIGURE 2.24. Variations in figure of quarter-sawn and flat-sawn lumber. For explanation see text. Photo courtesy of U.S. Forest Service.

the periderm reduces water loss and provides protection from mechanical injury.

In early times the phloem fibers of some trees, known as bast fibers, were used for cordage and matting. The name "basswood," often used for linden, refers to the fact that its bark was a good source of bast fiber. In the Pacific Islands the inner bark of *Broussonetia papyrifera* was used extensively to make tapa cloth, and linen, hemp, and jute are prepared from phloem fibers.

The cambium-produced secondary phloem is comprised of vertically and horizontally oriented systems of cells. In gymnosperms the secondary phloem is relatively simple, consisting only of vertically oriented sieve cells, parenchyma cells, and, often, fibers. The horizontally oriented, generally uniseriate rays contain only parenchyma cells or both parenchyma and albuminous cells. Vertical parenchyma often contains resins, crystals, and tannins. Distribution of Ca oxalate crystals is a useful diagnostic feature of barks of some broadleaved trees (Trockenbrodt, 1995).

The secondary phloem of angiosperms is much more complicated and variable structurally among species than the phloem of gymnosperms. Angiosperm phloem consists of vertically oriented sieve tube members, usually with companion cells, parenchyma cells, and fibers. The rays may be uniseriate, biseriate, or multiseriate. The axial and radial systems also may have sclereids, laticifers, secretory elements, idioblasts, and crystals.

ROOTS

The roots are important for anchorage, absorption of water and minerals from the soil, storage of reserve foods, and synthesis of certain growth hormones. Variations in the distribution and extent of roots are important because trees able to produce deeply penetrating and branching root systems absorb water and minerals from a larger soil volume than trees with more restricted roots (Kozlowski, 1971b).

Tree roots often spread laterally as far as or well beyond the width of the crown (Fig. 2.25). However, the extent of lateral spread of roots varies markedly with site and especially with soil type. For example, roots of fruit trees growing on sand extended laterally about three times as far as the crown; on loam about twice as far; and on clay about one and one-half times as far (Rogers and Booth, 1959, 1960).

Rooting depth varies greatly among species and often shows little relation to the size of the plant above ground. The effective rooting depth of tea bushes, for example, often is greater than that of the tall, overtopping trees. Most plants tend to develop a high concentration of roots in the surface soil, perhaps because it is well aerated, contains a higher concentration of minerals than deeper soil horizons and is sufficiently moist. There are discernible variations in root distributions across ecosystems (Table 2.4). As might be expected, roots are distributed more deeply in xeric ecosystems (desert, sclerophyll shrub communities, tropical grassland savannas), but also in temperate coniferous forests. Roots are shallowest in temperate grasslands and in cold-soil ecosystems (tundra and boreal forest). When compared across life forms, grasses exhibit the shallowest and shrubs the deepest root distribution patterns, with trees possessing intermediate depth of distribution (Fig. 2.26). Depth of rooting is discussed further in Chapter 11.

The root systems of woody plants consist of a framework of relatively large perennial roots and many small, short-lived branch roots. Forms of root systems often vary when the roots of different plants come in contact because roots may intermingle, graft, or avoid each other (Mahall et al., 1991). The large roots comprise most of the root biomass but account for little of total root length. In forest ecosystems, the fine roots (generally considered to be those roots <2 mm in diameter) account for most of the root length but little biomass at any given time. For example, in a 39-year-old Scotch pine stand the fine roots represented only 5% of the root weight but accounted for 90% of the root length (Roberts, 1976a). Ecosystems dominated by herbaceous species, especially grasses, tend to exhibit greater proportions of fine roots. Jackson et al. (1997)

FIGURE 2.25. Root system of 16-year-old Cox's Orange Pippin apple tree on Malling II rootstock. From Rogers and Head (1969), with permission of Horticulture Research International.

TABLE 2.4. Global Survey of Root Distribution Patterns from a Variety of Biomes. β values are derived from fitting the relationship between depth and relative root distribution. Larger values of β imply deeper rooting profiles. r^2 values indicate closeness of fit of the imposed function. Values of percent root biomass, root biomass, and root/shoot ratio are derived both from the authors' work and those from a broad sample of the relevant literature.[a]

Biome	β[b]	r^2	% Root biomass in upper 30 cm	Root biomass ($kg \cdot m^{-2}$)	Root/shoot ratio
Boreal forest	0.943	0.89	83	2.9	0.32
Crops	0.961	0.82	70	0.15	0.10
Desert	0.975	0.95	53	1.2, 0.4[c]	4.5, 0.7[c]
Sclerophyllous shrubs	0.964	0.89	67	4.8	1.2
Temperate coniferous forest	0.976	0.93	52	4.4	0.18
Temperate deciduous forest	0.966	0.97	65	4.2	0.23
Temperate grassland	0.943	0.88	83	1.4	3.7
Tropical deciduous forest	0.961	0.99	70	4.1	0.34
Tropical evergreen forest	0.962	0.89	69	4.9	0.19
Tropical grassland savanna	0.972	0.95	57	1.4	0.7
Tundra	0.914	0.91	93	1.2	6.6

[a]From *Oecologia*, A global analysis of root distributions for terrestrial biomes. Jackson, R. B., Canadell, J., Ehleringer, J. R., Mooney, H. A., Sala, O. E., and Schulze, E. D. (1996). **108**, 389–411, Table 1. © 1996 with kind permission from Springer Science and Business Media.

[b]$Y = 1 - \beta$ where Y is the cumulative root fraction (which varies from 0 to 1) from the soil surface to depth, d, and β is a fitted "extinction coefficient."

[c]For cold and warm deserts, respectively.

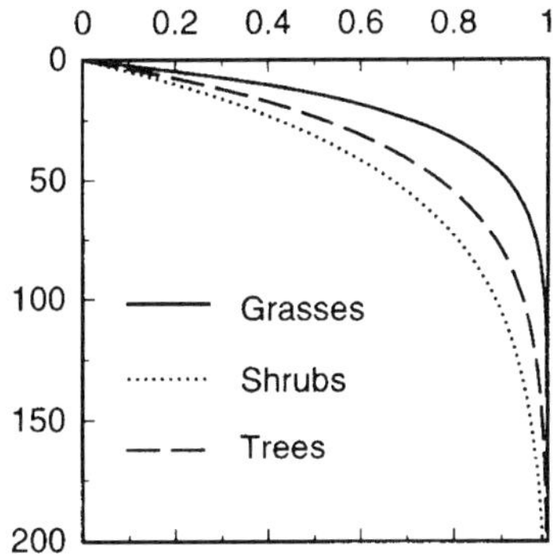

FIGURE 2.26. Composite fitted distribution of relative root depths for grass, tree, and shrub life forms across numerous biomes. From *Oecologia*, A global analysis of root distributions for terrestrial biomes. Jackson, R. B., Canadell, J., Ehleringer, J. R., Mooney, H. A., Sala, O. E., and Schulze, E. D. (1996). **108**, 389–411, Figure 5. © 1996 with kind permission from Springer Science and Business Media.

estimated that in savanna and temperate grassland ecosystems live fine roots constituted about 30 and 60% of total root biomass, respectively. In contrast, living fine root biomass in temperate and tropical forests ranged from 6.4 to 11.4% of total root biomass. Fine roots are subject to rapid turnover and may constitute a substantial sink for plant carbon, perhaps accounting for one-third of total allocation of photosynthate to roots (Nadelhoffer and Raich, 1992; Jackson et al., 1997).

Fine roots are responsible for most water and nutrient uptake and their number and longevity are quite sensitive to environmental conditions. Median root life span for fine roots of ponderosa pine seedlings was greatest when produced in the autumn as temperatures declined, and shorter at higher temperatures in the spring and summer (Fig. 2.27) (Tingey et al., 2000; Johnson et al., 2000). Fine root lifespan also increased as seasonal growth proceeded during the summer.

Figure 2.28 shows the cross sectional anatomy of jack pine roots at three stages of development. Young white roots possess an outer epidermis, cortex tissue beneath, and a vascular cylinder (stele) separated from cortex by an endodermis that is distinguished by Casparian bands and suberin lamellae that restrict the flow of water and ions within the apoplast (Chapters 8 and 11). The young roots of angiosperms are similar to those of gymnosperms in most respects, but many angiosperm species have a layer of cells adjacent to the epidermis, the exodermis, that possess Casparian bands and suberin lamellae (Perumalla et al., 1990; Peterson and Perumalla, 1990). The physiological role(s) of the exodermis is conjectural at this point; possibly, the tissue provides an additional selective barrier for water and ions moving into the root and a barrier to loss of materials (water and perhaps ABA and other solutes) when the water potential gradient between soil and root promotes

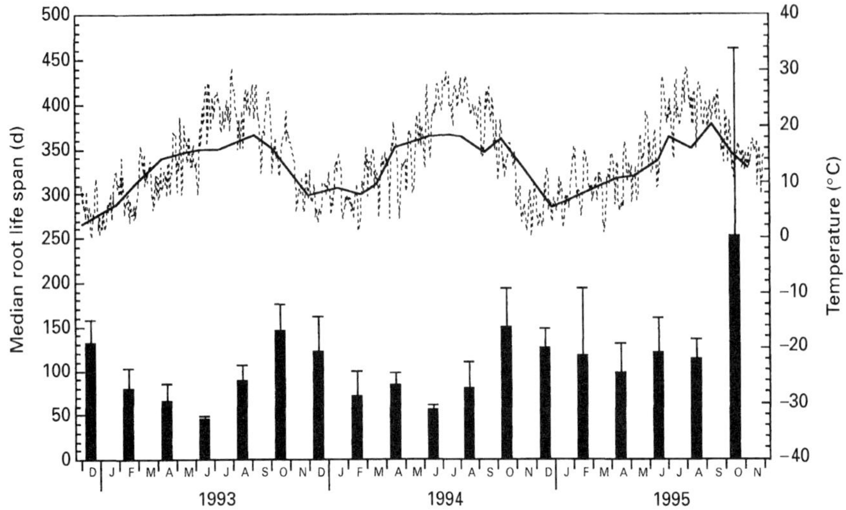

FIGURE 2.27. Median root life span of 18 *Pinus ponderosa* fine-root (<2 mm) cohorts produced over three years. Bar location indicates time of initial observation. Temperatures: soil (solid line), air (dashed line). From Tingey et al. (2000). Reproduced with permission of the New Phytologist Trust.

leakage to the rhizosphere (Hose et al., 2001). Decreased radial hydraulic conductivity was correlated with increased suberization in the exodermis, endodermis and adjacent tissues external to the endodermis in roots of *Agave deserti* exposed to drying soil (North and Nobel, 1991).

In pine roots of intermediate age, the cortex cells have died and likely are water-filled (Fig. 2.28). Older roots show secondary xylem and phloem tissues as well as an expanded pericycle and external cork cells. The epidermis, root hairs, and part of the cortex are destroyed by a cork cambium (Fig. 2.28; see also Chapter 3). Eventually the endodermis is lost because of cambial activity and the root consists of xylem, cambium, phloem, and a suberized layer in the outer surface of the phloem. These basic tissues are also seen in other gymnosperms and in angiosperms.

Adventitious Roots

Adventitious roots arise after injury to roots or from main stem, branch, or other tissues. Adventitious roots develop from preformed root primordia or from induced primordia by division of parenchyma cells, similar to the process of initiation of normal lateral roots. Prior to their emergence from the parent root, adventitious roots differentiate an apical meristem, root cap, and the beginning of a vascular cylinder.

Adventitious roots include those formed on stem and leaf cuttings and those produced by air layering. Adventitious aerial roots of many tropical trees are common. For example, *Ficus* produces free-hanging aerial roots that originate in the branches and undergo secondary thickening before they reach the soil. In many lianas roots also arise from aerial organs.

A number of species produce adventitious "stilt" roots that emerge from the main stem of the tree, bend downward, and enter the soil. Stilt roots often branch above ground and give rise to secondary and tertiary roots below ground (Fig. 2.29). The mangroves (*Rhizophora*) are good examples of trees that produce stilt roots but they also form in a number of trees found in fresh water swamps and rain forests. Genera that form stilt roots include *Pandanus*, *Clusia*, *Tovomita*, *Elaeocarpus*, *Xylopia*, *Dillenia*, *Eugenia*, and *Musanga*.

Root Tips

The root apex is covered by a thimble-like mass of living cells, the root cap (Fig. 2.30) that protects the root apical meristem and facilitates growth of roots through the soil. The root cap also regulates the responses of roots to gravity. Cells of the root cap form by continuous divisions of root cap initials located along the junction between the root cap and the apical meristem (Chapter 3). As a root elongates, peripheral cells of the root cap are shed. These discarded cells and the growing root tip are covered by a sheath of mucigel. Mucigel, a gelatinous material at the root surface, consists of natural and modified plant mucilages (Chapter 7), bacterial cells and their metabolic products, as well as colloidal, mineral, and organic matter. Mucigel is a product of the root-soil-microbial complex that lubricates the growing root and maintains contact between the root and soil, especially when roots shrink as they often do during the day (Rovira et al., 1979). Mucigel also may be important in promoting an active soil microflora (Sutton, 1980). As root cap cells are sloughed off, the root apical meristem forms new ones.

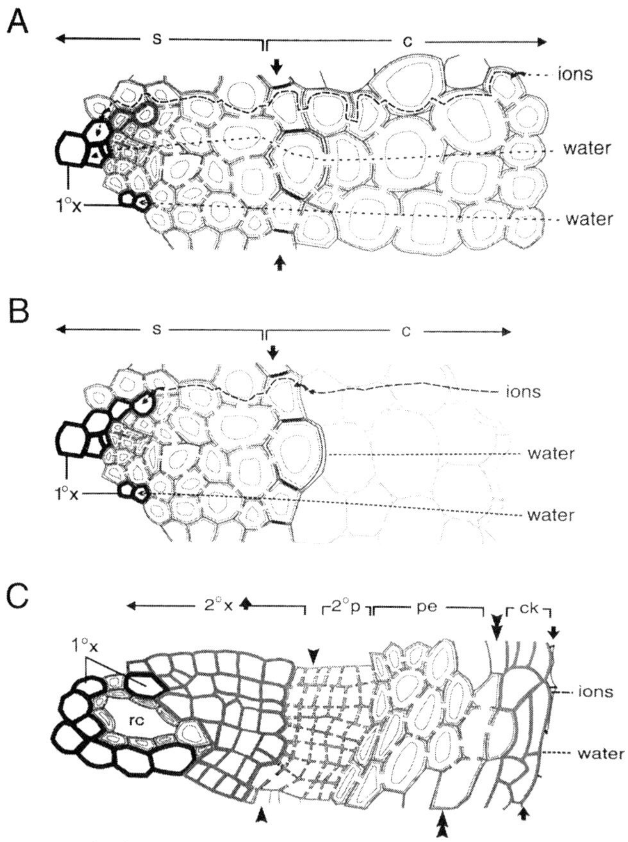

FIGURE 2.28. Cross sections of pine roots showing movement of ions and water from the soil across the cortex (c) and stele (s) and into the tracheids of the primary xylem (1° x). Plasmalemma and tonoplast membranes are indicated by dotted lines. In the endodermis (arrows) suberin lamellae are indicated by lines and Casparian bands by darkened primary cell walls. The center cell of the endodermis has a suberin lamella, flanking cells do not; thus the latter are passage cells. After secondary growth begins , a resin canal (rc) forms in the xylem, secondary xylem (2° x) and secondary phloem (2° p) are produced by the vascular cambium (single darts). The pericycle (pe) enlarges and cork (ck) cells form from the cork cambium (double darts). Endodermis remnants (arrows) remain external to cork cells. (A) White zone, water flow is through the transcellular path (Chapter 11), except at the endodermal cell with suberin lamellae where movement is symplastic. (B) Condensed tannin zone. Dead cortical cells are assumed to be waterlogged. (C) Cork zone. From *Plant and Soil*, Pine root structure and its potential significance for root function. Peterson, C. A., Enstone, D. E., and Taylor, J. H., **217**, 205–213, Figure 2. © 1999 with kind permission from Springer Science and Business Media.

Root Hairs

Many species of woody plants develop root hairs just above the zone of root elongation (Fig. 2.30). These tubular outgrowths are physiologically important because they increase absorption of water and mineral nutrients by increasing the root surface area (Cailloux, 1972; Itoh and Barber, 1983). They also increase adhesion between the soil and its surroundings (Hofer, 1991).

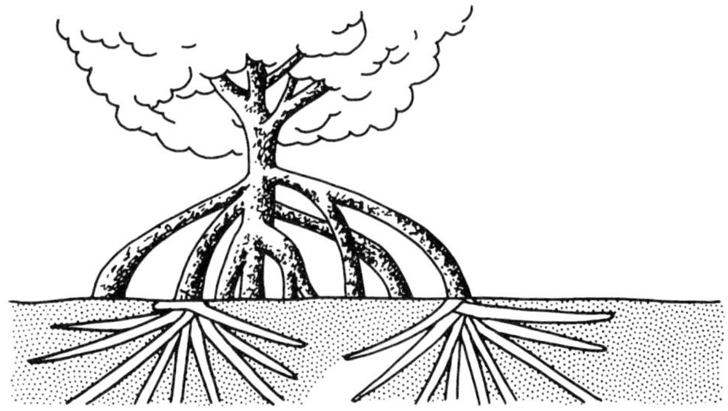

FIGURE 2.29. Stilt or prop roots of American mangrove (*Rhizophora*). From Scholander et al. (1955).

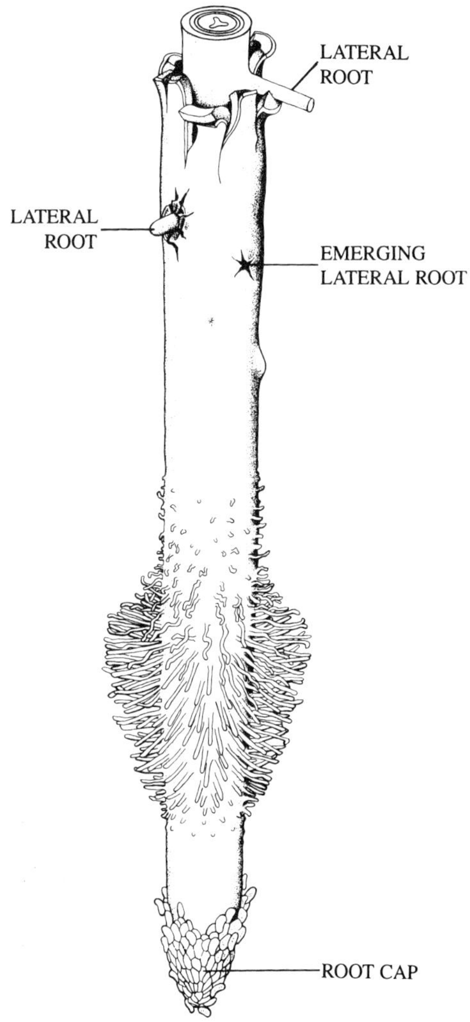

FIGURE 2.30. Portion of a root showing the spatial relation between the root cap and region of root hairs. The sites of emergence of lateral roots also are shown. New root hairs form just above the region of cell elongation as the older root hairs die. The root cap is covered by a sheath of mucigel, which lubricates the root as it grows through the soil. From Raven et al. (1992). Reprinted with permission.

In addition to their function in absorption, root hairs also may excrete liquids (Rougier, 1981). For example, the tips of root hairs of apple trees produced globules of liquid that increased in size for a few days and sometimes coalesced to form large drops that concealed the root hairs (Head, 1964; Rogers and Head, 1966).

Root hairs usually arise as protrusions from the external, lateral walls of epidermal cells, although in a few species they originate from cortical cells one or two layers beneath the epidermis. In conifers the root hairs of short roots arise from a surface layer of cells, whereas those of long roots arise from the second or third layer of cortical cells. The root hairs originate from the surface layer only where persistent root cap layers are absent (Bogar and Smith, 1965). Emergence of root hairs often follows inhibition of elongation of epidermal cells. When elongation of roots is arrested, as in compacted soil, long root hairs often are found close to the root tip.

Work with the *Arabidopsis* model plant system has shown that root hair development consists of two stages: (1) initiation of a bulge from the parent cell followed by (2) sustained tip growth (Schiefelbein, 2000). Wall loosening enzymes (and particularly xyloglucan endotransglycosylase, XET) are localized to areas where bulges will appear, and this enzyme and other wall growth factors, including expansins (Chapter 3), are found throughout the growing tip of root hairs (Vissenberg et al., 2001). Ethylene is necessary for root hair formation, as the inhibition of ethylene synthesis totally suppresses root hair formation in *Arabidopsis*. Mutants of *Arabidopsis* that are resistant to auxin action show reduced numbers of root hairs (Vissenberg et al., 2001).

The highly vacuolated and thin-celled root hairs vary in life span. Most of them live only a few hours, days, or weeks and are eliminated by changes of secondary thickening, including suberization and lignification. The zone of root hairs, usually 1 to 4 cm long, migrates because, as old root hairs die, new ones form regularly behind the growing point of an elongating root. Some trees, such as Valencia orange, may retain suberized or lignified root hairs for months or years (Hayward and Long, 1942). Such persistent root hairs appear to be relatively inefficient in absorption.

The number and size of root hairs are variable, depending on species and cultivar, type of root, and environmental factors, especially soil moisture content, soil texture, and soil salt concentration (Hofer, 1991). Root hairs generally vary in length between 80 and 1500 μm. The hairs on secondary roots or roots of high order usually are shorter than those on the main roots.

TABLE 2.5. Variation in Development of Roots and Root Hairs of Greenhouse-Grown, 7-Week-Old *Robinia pseudoacacia* and *Pinus taeda* Seedlings[a]

	Root length (cm)	Root surface area (cm^2)	Root hairs (no.)	Root hair surface area (cm^2)
Robinia pseudoacacia				
Primary	16.20	3.4466	1,166	3.6346
Secondary	115.62	15.7167	8,321	25.2172
Tertiary	30.60	3.1151	2,081	5.1759
Total	162.42	22.2784	11,568	34.0277
			520 root hairs cm^{-2}	
Pinus taeda				
Primary	6.45	2.7341	215	2.7973
Secondary	5.93	0.9683	371	2.0770
Total	12.38	2.7021	586	2.8743
			217 root hairs cm^{-2}	

[a]From Kozlowski and Scholtes (1948).

Roots of seven-week-old black locust seedlings grown in a greenhouse developed over 11,000 root hairs (520/cm^2), whereas those of loblolly pine of the same age had less than 600 root hairs (217/cm^2) (Table 2.5). At times many trees, such as avocado and pecan, lack root hairs. They also are absent on roots of some gymnosperms. The number of root hairs usually is higher in fast-growing than in slow-growing roots, with the latter group often free of root hairs. Root hairs often do not form in roots of plants growing in very dry or flooded soils. In general root hair formation is stimulated by environmental factors that decrease development of ectomycorrhizas (Marks and Kozlowski, 1973). However, root hairs are common on some endomycorrhizal roots (Lyford, 1975).

Suberized and Unsuberized Roots

Individual roots of trees may continue to grow for several weeks and produce lateral roots or they may stop growing after only a few weeks. The outer cortical tissues of roots remain unsuberized and white for a short time, which in apple may vary from one to four weeks during the summer and up to three months in the winter (Head, 1966). The cortical tissues then turn brown and degenerate. The remaining central cylinder may or may not undergo secondary thickening.

As root systems age an increasing proportion of the total root surface becomes suberized. Only the most recently formed roots are unsuberized and their total surface area is exceedingly small in comparison to the surface of the entire root system. Kramer and Bullock (1966) followed seasonal changes in proportions of growing and suberized roots in loblolly pine and yellow-poplar trees growing in North Carolina. The

TABLE 2.6. Seasonal Variation in Percentage of Surface Area in Unsuberized and Suberized Roots under a 34-Year-Old Loblolly Pine Stand in North Carolina[a]

Date	Growing tips (%)	Mycorrhizal (%)	Total unsuberized (%)	Total suberized (%)
March				
1	0.06	0.53	0.59	99.41
8	0.15	1.20	1.35	98.65
15	0.13	1.18	1.31	98.69
22	0.13	1.64	1.77	98.69
29	0.13	2.06	2.39	97.61
April				
8	0.43	2.27	2.70	97.30
13	0.39	5.09	5.48	94.52
22	0.53	2.77	3.30	96.70
30	0.34	5.30	5.64	94.36
May				
9	0.72	5.76	6.48	93.52
31	0.30	3.95	4.25	95.75
June				
7	0.25	6.06	5.31	93.69
17	0.48	3.05	3.53	96.47
24	0.38	3.00	3.38	96.62
July				
1	0.22	3.00	3.22	96.78
15	0.54	2.38	2.92	97.08
29	1.36	2.81	4.17	95.83
November				
11	0.61	2.84	3.55	95.83

[a]From Kramer and Bullock (1966).

surface area of growing, unsuberized roots under a loblolly pine stand usually amounted to less than 1% of the total root surface area (Table 2.6). It exceeded 1% at only one sampling time following a heavy rainfall in July. Most of the unsuberized root surface consisted of mycorrhizas, and the surface provided by growing tips plus mycorrhizal roots never exceeded 7% during the growing season. Thus from 93 to 99% of the root surface was suberized. This situation suggests that considerable absorption of water and minerals must occur through suberized roots (see Chapter 11).

Mycorrhizas

As are 90% of all terrestrial plants, the root systems of most woody plants are greatly modified by the presence of mycorrhizas (Fig. 2.31). These structures, formed by invasion of young roots by fungal hyphae, are symbiotic associations between nonpathogenic or weakly pathogenic fungi and living cells of roots. The relationship is one in which the tree supplies carbohydrates and other metabolites beneficial to the fungus; in turn the fungus benefits the tree by increasing the availability of mineral nutrients and water (see Chapters 10 and 11). Mycorrhizas also maintain soil structure, buffer against water stress, and protect plants from heavy metals (Lynch and Bragg, 1985; Pate, 1994). There also is some evidence that mycorrhizal fungi may protect the host tree from disease by utilizing excess carbohydrates, acting as a physical barrier, secreting fungistatic substances, and favoring protective organisms of the rhizosphere (Zak, 1964; Marx, 1969). Although many trees can be grown successfully without mycorrhizas under certain conditions such as very high soil fertility, they usually grow much better with mycorrhizas.

Mycorrhizas fall into three broad groups (Cairney, 2000):

- The ectotrophic forms, in which the fungus exists both inside and outside the root.
- The endotrophic forms, which exist entirely within the host cells. In endotrophic mycorrhizas the fungus always occurs in the cortical cells of the host roots and does not extend into the endodermis or stele.
- The ericoid type, which forms extensive coils of hyphae within epidermal cells and exhibits little penetration of soil adjacent to roots.

Evolutionarily, endotrophic mycorrhizas are the oldest type of association, appearing at the same time or soon after colonization of land by plants (ca. 475 million years ago). The early association of roots with fungal partners may reflect the poor absorbing capacity of the rudimentary root systems of early land plants. Early appearance of mycorrhizas also emphasizes the capacity of the fungus to make more available scarce nutrients, especially phosphorus (Cairney, 2000). Endotrophic mycorrhizas are the most common types in the plant kingdom but occur on only a few genera of woody plants such as *Liriodendron*, *Acer*, and *Liquidambar*.

Ectomycorrhizas appeared much more recently (200–250 million years ago) and are especially abundant in woody plants in the *Pinaceae*, *Fagaceae*, *Betulaceae*, *Salicaceae*, *Juglandaceae*, and a few other families. More than 5,000 species of fungi form ectotrophic associations on woody plants, with many species often found on a single tree. Whereas some fungi can form ectomycorrhizas with a wide variety of woody plants, others are family-, genus-, or even species-specific (Molina et al., 1992). Colonization of forest trees typically occurs by a succession of early and late fungi in mycorrhizal associations. As forest stands age and early-stage fungi are succeeded largely by late-stage fungi, diversity in fungal species increases greatly (Last et al., 1984; Dighton et al., 1986).

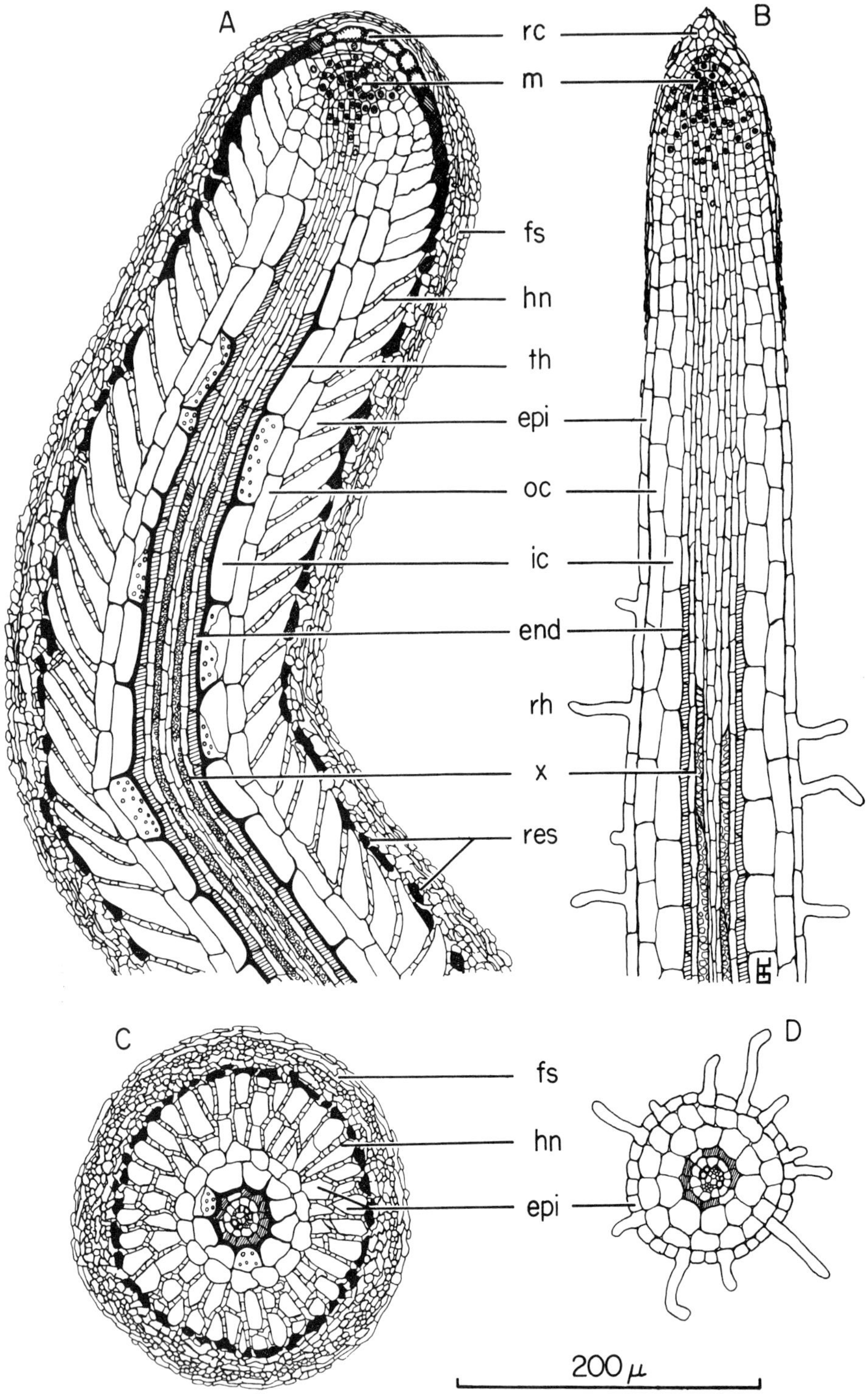

FIGURE 2.31. Anatomy of mycorrhizal and uninfected roots of *Eucalyptus*. A, B: Median longitudinal sections of mycorrhizal and uninfected root, respectively. C, D: Transverse sections through fully differentiated region of mycorrhizas and uninfected root, respectively. rc: rootcap; m: meristematic region; fs: fungal sheath or mantle; hn: Hartig net; th: thickened walls of inner cortex; epi: epidermis; oc: outer cortex; ic: inner cortex; end: endodermis (shaded to indicate extent of tannin impregnation); rh: root hair; x: lignified protoxylem; res: collapsed residues of cap cells. From Chilvers and Pryor (1965). The structure of Eucalypt mycorrhizas. Aust. J. Bot. **13**, 245–249.

In ectotrophic mycorrhizas the fungus produces a weft of hyphae on the root surface and the mycelia may form either thin, loosely woven tissue, tightly woven masses, or compacted pseudoparenchymatous structures. The fungus penetrates the cortex, forcing its way between the cortical cells without actually entering individual cells of the host. In root transections the cortical cells appear to be separated by a fungal net, called the Hartig net (Marks and Kozlowski, 1973).

Quantitative differences in structure of mycorrhizas and roots include: (1) lack of production of root hairs by mycorrhizas and some root hairs in uninfected fine roots; (2) limited rootcap tissue in mycorrhizas (rarely more than two cell layers between the apex and fungal sheath) and extensive rootcap tissue in uninfected roots; and (3) differentiation much closer to the apex in mycorrhizas than in uninfected roots. The morphology of mycorrhizas is similar to what might occur from slow growth as a result of unfavorable environmental conditions. Among qualitative differences ascribed to fungal infection of roots are thickening of the inner cortex and radial elongation of epidermal cells.

Ericoid mycorrhizas form in epidermal cells of delicate hair roots of plants from a number of families in the Order Ericales, including the Ericaceae. During the infection process, fungal hyphae dissolve portions of the cell wall and occupy an extensive fraction of the intracellular volume with the plant plasma membranes invaginating around each hyphal strand as it elongates within the cell (Smith and Read, 1997). Although the fungal component of ericoid mycorrhizas apparently does not form extensive hyphal networks in soil, the fungus (1) improves the acquisition of N and P by the plant, (2) makes more available (by secretion of lytic enzymes) the abundant organic sources of N and P characteristic of the humus-rich soil in which many erocioid plants grow, and (3) may indirectly protect the plant from toxic concentrations of metal ions (e.g., Zn) that accumulate in acid soils (Denny and Ridge, 1995).

REPRODUCTIVE STRUCTURES

Many botanists restrict the term "flower" to angiosperms. However, I will use this term more broadly, in line with horticulturists and foresters, and also refer to the cones or strobili of gymnosperms as flowers.

Angiosperms

Typical complete flowers of angiosperms bear four types of floral parts on their receptacles (Fig. 2.32). The lowermost of these are the sepals, which together make up the calyx. Above the sepals are the petals, collectively called the corolla. The sepals and petals together comprise the perianth. Inside the perianth are the pollen-producing stamens, collectively called the androecium, and the carpels which comprise the gynoecium. A flower may have one to several carpels. These usually consist of a lower, fertile part, the ovary, and an upper sterile part, the style. At the top of the style is the stigma on which the pollen grains land prior to fertilization of immature seeds or ovules. The ovule is important in reproduction because it is the site of megaspore formation and development of the female gametophyte called the embryo sac.

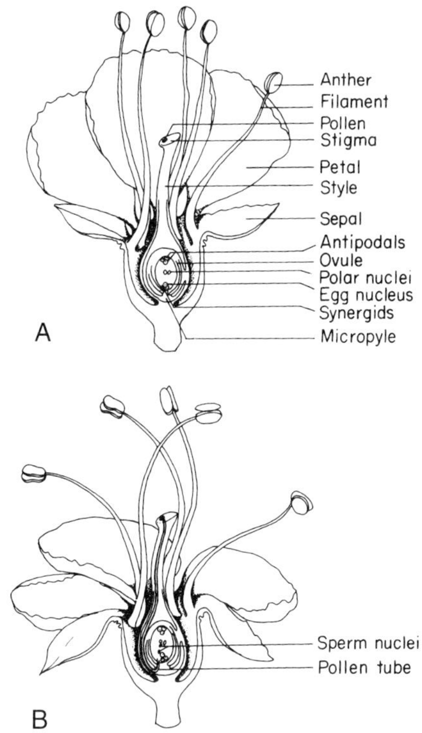

FIGURE 2.32. Typical flower before and after fertilization. A: Flower before fertilization; B: Flower shortly after union of sperm nucleus and egg nucleus. Some petals have fallen and stamens are withered. (From *The Ripening of Fruit*, J. B. Biale. Copyright © 1954 by Scientific American, Inc. All rights reserved.)

Woody plants show many examples of floral modifications and often lack parts of the complete flower. For example, flowers of poplars and black walnut lack a corolla and those of willows lack both calyx and corolla (Fig. 2.33). Another floral modification involves fusion of floral parts. In grape and rhododendron, for example, carpels are fused; in catalpa petals are fused; and in viburnum sepals are fused.

Whereas flowers of many fruit trees are showy, those of most forest trees are inconspicuous. Flowers of angiosperms are borne individually or, more commonly, in groups on various types of inflorescences.

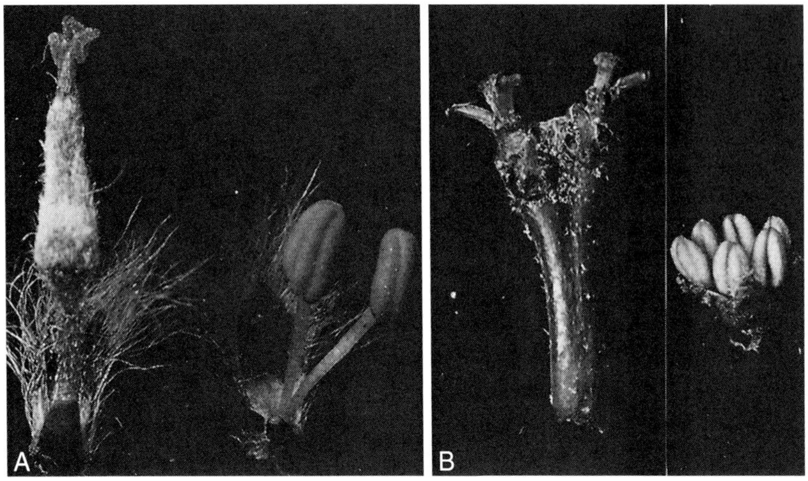

FIGURE 2.33. A: Pistillate and staminate flowers of willow; B: Pistillate and staminate flowers of red oak. Photo courtesy of W. M. Harlow.

Flowers of apple trees are produced in clusters of three to seven, usually five. As in apple, the pear flower bud opens into a terminal cluster of about five flowers. In olive the flowers are borne in paniculate inflorescences, each consisting of about 15 flowers. The inflorescences appear on shoots one or sometimes two years old. The flowers are either perfect, with functioning stamens and pistils, or staminate, with the pistil aborted.

In magnolia and yellow-poplar the flowers occur singly in leaf axils. In black cherry and striped maple they are borne in racemes; in poplars, birches, and alders in catkins; in buckeye in panicles; and in elder and viburnum in cymes.

Many woody angiosperms are monoecious, with staminate and pistillate flowers on the same plant as in birch and alder. Others, such as persimmon, poplar, and willow, are dioecious and bear staminate and pistillate flowers on separate plants. It should be obvious that a staminate tree will not produce seeds. If ornamental dioecious shrubs or trees such as holly are grown for their decorative fruits, care must be taken to plant some staminate trees along with the pistillate trees to insure pollination and production of fruits. Some tree genera, such as *Prunus*, *Magnolia*, and *Liriodendron*, have perfect flowers that present both stamens and pistils on a single flower, whereas others bear perfect flowers as well as both staminate and pistillate flowers (*Aesculus*, *Celtis*). Still another combination occurs in *Rhamnus* and *Fraxinus*, which have perfect flowers as well as either staminate or pistillate flowers. Some examples of flowers of woody angiosperms are shown in Fig. 2.34.

FIGURE 2.34. Flowers of woody angiosperms. A: Flowers of pear; B: Flowers of black locust; C: Staminate and pistillate flowers of black maple; D: Flowers of basswood; E: Pistillate flowers of eastern cottonwood; F: Staminate flowers of eastern cottonwood.

Gymnosperms

The gymnosperms produce seeds that are naked (i.e., they lack an enclosing, protective ovary wall). The calyx, corolla, stamens, and pistil are absent in gymnosperms. In most species the flowers consist of pollen-producing cones (staminate strobili) and seed-producing cones (ovulate strobili) (Figs. 2.35 and 2.36). In yews, however, a fleshy aril grows from the base of a single-stalked ovule to partially or completely enclose the seed.

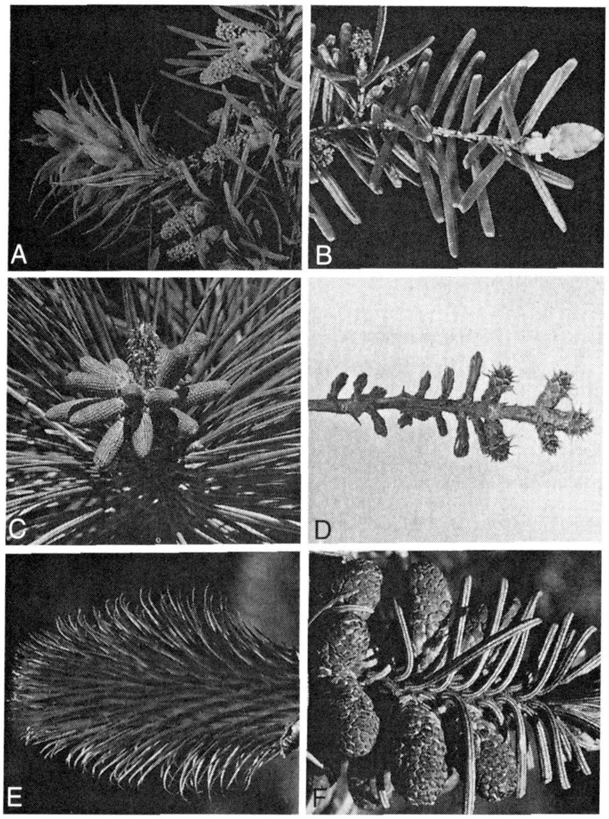

FIGURE 2.35. (A) pollen and seed cones of Douglas-fir; (B) pollen and seed cones of eastern hemlock; (C) pollen cones of slash pine shortly before shedding pollen; (D) seed cones of baldcypress; (E) receptive seed cone of noble fir; (F) pollen cones of noble fir, showing swollen pollen sacs about a day before shedding of pollen. Photos courtesy of the U.S. Forest Service.

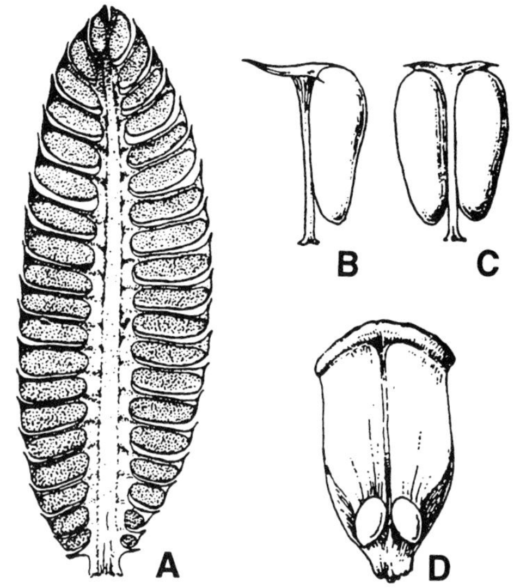

FIGURE 2.36. (A) Pollen cone of Austrian pine. (B, C) Side and abaxial views of microsporophyll; (D) ovuliferous scale with two ovules. From *Plant Morphology*, by Haupt (1953). Used with permission of McGraw-Hill Book Company.

Conifer cones may differentiate from a previously vegetative apex (hence they are located at shoot apices), or they may differentiate from newly formed axillary primordia that are undetermined until they differentiate into lateral vegetative shoots or cones (Owens and Hardev, 1990). Conifers are predominantly monoecious. In contrast, cycads and ginkgo are dioecious. Certain genera (e.g., *Juniperus*) vary from entirely monoecious to entirely dioecious, or mostly monoecious to mostly dioecious (Owens and Hardev, 1990).

SUMMARY

Crown form affects the appearance of trees; the amount and quality of wood, fruits, and seeds produced, and influences practices in managing stands of forest and fruit trees.

The leaf blade of angiosperms contains mesophyll or ground tissue enclosed by an upper and lower epidermis, each covered by a cuticle. The structure and amount of wax on the leaf surface vary with species, genotype, and environmental conditions. The stomatal pores that penetrate the leaf surface are physiologically important because atmospheric CO_2 diffuses through them into the leaf interior and is used in photosynthesis by mesophyll cells. Also, water is lost as vapor through the stomata.

The mesophyll tissue of leaves of broadleaved trees consists of columnar parenchyma cells and irregularly shaped spongy parenchyma cells. In a number of gymnosperms, but not in pines, the leaf mesophyll also is differentiated into palisade cells and spongy parenchyma. The mesophyll tissue is permeated by veins that transport water and minerals to, and photosynthates from, leaves.

Leaf shapes and structures are quite variable. Of special physiological importance are wide variations in stomatal size and frequency that occur among species and genotypes. Shade-grown leaves are larger and thinner, less deeply lobed, and contain less palisade and conducting tissue than sun-grown leaves. Leaf shapes change in many species as woody plants progress from juvenility to adulthood.

Stems of woody plants support the crown and transport carbohydrates, water, mineral nutrients, and growth hormones both upward and downward. The annual rings of wood in stems and branches of temperate zone trees are prominent because the cells of wood formed early in the season are larger in diameter, and have thinner walls than the wood cells formed later in the season. In many tropical trees increments of wood are added to the stem during most or all of the year. Such trees often lack annual growth rings or have indistinct rings.

The annual phloem increments are thinner than the wood (xylem) increments because less phloem than xylem is produced annually by division of cambial cells, old phloem tissues often are crushed, and external phloem tissues are shed.

Angiosperm woods are classified as ring-porous or diffuse-porous. In ring-porous trees (e.g., oaks) the diameters of xylem vessels formed early in the season are much larger than those of vessels formed late in the season. In diffuse-porous trees (e.g., maple) the vessels have small diameters and the diameters of those formed early vary little from those formed later.

The wood structure of gymnosperms is simpler than that of angiosperms. The axial elements of gymnosperm wood consist mostly of tracheids and a few parenchyma and epithelial cells. The walls of axial tracheids have pits that facilitate movement of water between adjacent cells. The few horizontally oriented elements in gymnosperm wood include ray tracheids, ray parenchyma cells, and epithelial cells. Both axially and horizontally oriented resin ducts are found in pine, spruce, larch, and Douglas-fir, but not in hemlock or the true firs.

Angiosperm wood consists of longitudinal components (vessel elements, tracheids, fibers, axial parenchyma, and epithelial cells). Horizontal elements include ray parenchyma and sometimes, epithelial cells. Axial or horizontal resin ducts occur in some tropical angiosperms but are virtually absent in temperate zone species.

The bark is much more complex than the wood. In tissues that have not undergone secondary thickening, the bark includes primary and secondary phloem, cortex, and periderm. The phloem consists of vertically oriented sieve tubes, parenchyma cells, and fibers.

The root system is composed of a framework of large perennial roots and many small, short-lived branch roots. The small (fine) roots are quite important for water and mineral absorption by plants and account for most of the root length but little root biomass. Death and replacement of fine roots occur simultaneously and both are sensitive to soil water and nutrient status as well as soil temperature.

The root tip is covered by a root cap that protects the root apical meristem. Absorption of soil water is facilitated by root hairs that develop just above the zone of root elongation. As the short-lived root hairs die in days to weeks, new ones form.

Specialized and modified roots include adventitious roots, aerial roots, knee roots or pneumatophores, and mycorrhizas (symbiotic associations between nonpathogenic or weakly pathogenic fungi and living root cells). The host plant supplies the fungus with carbohydrates and other metabolites, and the fungus increases absorption of soil water and mineral nutrients.

Woody plants show many floral modifications and often lack parts of complete flowers. Flowers of many fruit trees are showy; those of most forest trees are inconspicuous. Most woody angiosperms are monoecious (staminate and pistillate flowers on the same plant) but some are dioecious. The calyx, corolla, stamens, and pistil are absent in gymnosperms. In most gymnosperms the flowers consist of pollen-producing cones and seed-producing cones. Conifers are predominantly monoecious but cycads and ginkgo are dioecious.

General References

Carlquist, S. J. (2001). *Comparative Wood Anatomy: Systematic, Ecological, and Evolutionary Aspects of Dicotyledon Wood*. Springer, New York.

Dickison, W. C. (2000). *Integrative Plant Anatomy*. Academic Press, San Diego.

Esau, K. (1965). *Plant Anatomy*. Wiley, New York.

Fahn, A. (1979). *Secretory Tissues in Plants*. Academic Press, London.

Fahn, A. (1991). *Plant Anatomy, 4th ed*. Pergamon Press, Oxford.

Fengel, D. and Wegener, G. (1984). *Wood: Chemistry, Ultrastructure, Reactions*. Walter de Gruyter, Berlin and New York.

Hallé, F. (1995). Canopy architecture in tropical trees: A pictorial approach. In *Forest Canopies* M. D. Lowman and N. M. Nadkarni, eds., pp. 27–44. Academic Press, San Diego.

Hallé, F., Oldemann, R. A. A., and Tomlinson, P. B. (1978). *Tropical Trees and Forests: An Architectural Analysis*. Springer-Verlag, Berlin, Heidelberg, New York.

Higuchi, T. (1997). *Biochemistry and Molecular Biology of Wood*. Springer, New York.

Kozlowski, T. T. (1971). *Growth and Development of Trees*. Vols. 1 and 2. Academic Press, New York.

Kozlowski, T. T., Kramer, P. J., and Pallardy, S. G. (1991). *The Physiological Ecology of Woody Plants*. Academic Press, San Diego.

Juniper, B. E. and Jeffree, C. E. (1983). *Plant Surfaces*. Edward Arnold, London.

Larson, P. R. (1994). *The Vascular Cambium*. Springer-Verlag, New York and Berlin.

Marks, G. C. and Kozlowski, T. T. (1973). *Ectomycorrhizae*. Academic Press, New York.

Owens, J. N. and Harder, V. (1990). Sex expression in gymnosperms. *Crit Rev Plant Sci* **9**, 281–291.

Owens, J. N. (1991). Flowering and seed set. In *Physiology of Trees*, A. S. Raghavendra, ed., pp. 247–227. Wiley, New York.

Panshin, A. V., DeZeeuw, C., and Brown, H. P. (1966). *Textbook of Wood Technology*. McGraw-Hill, New York.

Parker, G. G. (1995). Structure and microclimate of forest canopies. In *Forest Canopies* M. D. Lowman and N. M. Nadkarni, eds., pp. 73–106. Academic Press, San Diego.

Romberger, J. A., Hejnowicz, Z, and Hill, J. F. (1993). *Plant Structure, Function and Development*. Springer-Verlag, Berlin.

Smith, S. E. and Read, D. J. (1997). *Mycorrhizal Symbiosis*. Academic Press, London, U. K.

Tsoumis, G. (1991). *Science and Technology of Wood—Structure, Properties, Utilization*. Van Nostrand, Reinhold, New York.

CHAPTER

3

Vegetative Growth

INTRODUCTION

This chapter will present an overview of the important features of the nature and periodicity of shoot growth, cambial growth, and root growth of trees of the temperate zone and tropics. Early increase in size or dry weight of plants, organs, or tissues is approximately linear. Eventually, however, various internal growth-controlling mechanisms induce departure from a linear relationship so that over a long period growth can best be characterized by a sigmoid curve. Seasonal and lifetime growth of shoots, roots, and reproductive structures generally conform to such a pattern (Evans, 1972).

Plants grow in height and diameter through the activity of meristematic tissues, which comprise a very small fraction of the plant body. The various parts of plants grow at different rates and often at different times of the year. For example, woody plants of the temperate zone fluctuate from a state of endogenously controlled, deep-seated winter dormancy to meristematic activity during the growing season. But even during the frost-free season, periods of growth alternate with periods of inactivity followed by recurrence of growth. Although many tropical trees grow more or less continuously throughout the year, albeit at varying rates, others do not, and form distinct xylem growth rings (Carlquist, 1988).

CELL AND TISSUE GROWTH

Plant growth depends on expansion of cells, so a discussion of basic processes is warranted. As expansive cell growth is intimately related to cell water relations, concepts discussed in Chapter 11 will be incorporated here. Cell growth begins when loosening processes cause walls to relax. This, in turn, causes turgor (and consequently Ψ) within cells to be reduced and induces a water potential gradient between the cell and its external environment (Cosgrove, 1993, 1997). Water movement into cells in response to this water potential gradient increases cell volume. For sustained growth, solutes must continuously be imported into growing cells so that the osmotic potential (Ψ_π) does not increase in the face of osmotic dilution by inflowing water. Solute import thus assures maintenance of the water potential gradient that will sustain inward water movement.

Volume growth of cells can be described by an equation that considers both cell turgor and cell wall properties:

$$\frac{dV}{dt}\frac{1}{V} = m(\Psi_p - Y) \qquad (3.1)$$

where $dV/dt\ 1/V$ = the rate of volume growth per unit of original volume, m = wall extensibility, Ψ_p = turgor potential (or pressure), and Y = a minimum turgor potential below which no growth occurs (the "yield threshold") (Lockhart, 1965; Boyer, 1985). Both m and Y depend on the physical properties of cell walls, but the biochemical basis of these parameters is not well understood. Although represented as constants, the values of m and Y are variable and reflect the dynamic metabolic activity in growing tissues (Passioura and Fry, 1992; Passioura, 1994). Pritchard (1994) suggested that m could be the result of enzyme cleavage of load-bearing bonds (or "tethers") within cell walls; Y might be associated with the tension necessary to "open" cleavage sites in the proper conformation for enzyme action.

Wall loosening may be related to the activity of specific wall enzymes that break the hemicellulosic tethers that connect cellulose microfibrils (Fig. 3.1). One such enzyme, xyloglucan endotransglycosylase (XET), is present in walls of growing cells and has been hypothesized as such a wall loosening enzyme because xyloglucan cleavage and chain transfer can be detected in growing cell walls (Smith and Fry, 1991; Tomos and Pritchard, 1994). However XET does not induce extension in isolated cell walls (Cosgrove, 1999). McQueen-Mason et al. (1992) reported that a crude protein fraction extracted from the growing region of cucumber hypocotyls and containing no XET activity could induce wall extension activity. This fraction did contain

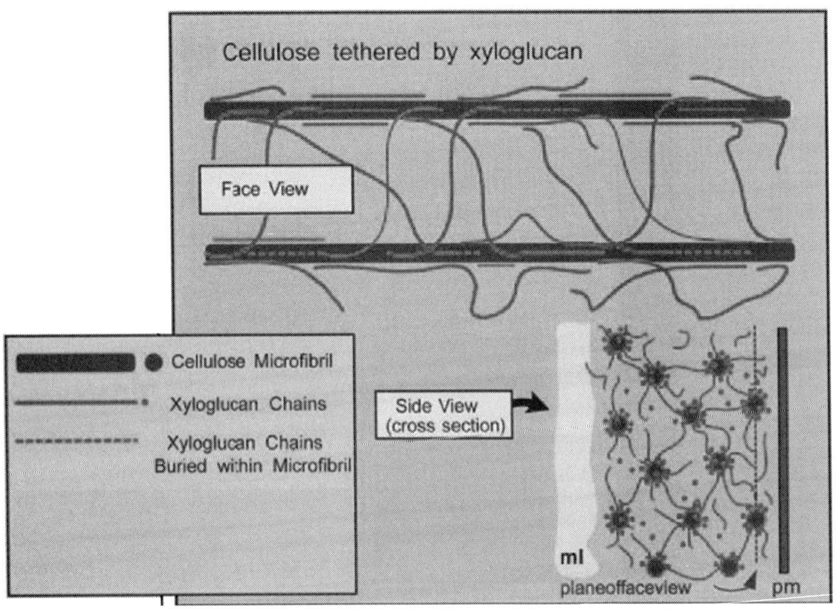

FIGURE 3.1. A model of cell wall structure showing cellulose microfibrils "tethered" by xyloglucan strands. ml, middle lamella; pm, plasma membrane. Used with permission of the American Society of Plant Biologists, from Cosgrove, D. (2001). Wall structure and wall loosening: A look backwards and forwards. Plant Physiol **125**, 131–134; permission conveyed through Copyright Clearance Center.

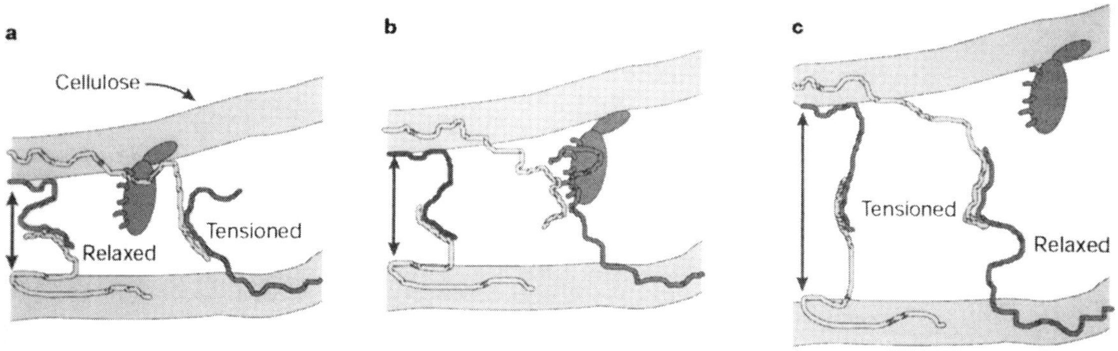

FIGURE 3.2. A model of the possible mechanism of action of expansin protein in wall-loosening. Cellulose microfibrils are tethered to one another by glycan tethers. Expansin is hypothesized to disrupt hydrogen bonding of glycan molecules to the microfibrils (a) or to each other (b). Cell turgor then causes a displacement of the tethers (c) and slippage in the point of adhesion between polymer molecules. Reprinted by permission from McMillan Publishers, Ltd.: *Nature* "Cosgrove, D. J. (2000). Loosening of plant cell walls by expansins. *Nature* **407**, 321–326." © 2000.

two active proteins, which the authors named *expansins*. These proteins, which have a molecular weight of about 26 kDa, induce extension in isolated cell walls under acid conditions, and application of expansins increases growth in living cells. While XET presumably promotes wall-loosening by cleavage of covalent bonds, expansins show no hydrolytic activity. Rather, they appear to promote slippage along load-bearing polysaccharide molecules, thereby functioning as "biochemical grease" to promote wall loosening (Fig. 3.2) (Cosgrove, 1998, 2000, 2001; McQueen-Mason and Cosgrove, 1995; Shieh and Cosgrove, 1998; Whitney et al., 2000).

Expansins are widely distributed among plants, including *Arabidopsis*, *Rumex*, cucumber, rice, and corn, and they also have been conclusively identified in loblolly pine (Hutchison et al., 1999), pear, and avocado (Rose et al., 2000). Expansins of grasses are found in vegetative tissues and pollen, with those in the latter apparently responsible for human allergic reactions to grass pollens (Cosgrove et al., 1997). Grass expansins generally do not promote wall loosening in other angiosperms, suggesting that the protein has been evolutionarily modified for specific cell wall structures. The expansins in grass pollen may promote softening of style tissues to facilitate the growth of the pollen tube. There also is some evidence that expansins are involved in fruit growth and softening (Rose et al., 2000). Interestingly, expansin activity appears to be a key determinant of organogenesis of leaves (Pien et al., 2001; Cleland, 2001) and perhaps roots (Hutchison et al., 1999).

Equation 3.1 assumes that growing cells are in direct contact with the water supply and hence that there is no resistance to water flow to the cells. In growing tissues this is clearly not the case, as expanding cells may be several cells removed from the xylem water supply. In this case, tissue growth can be described by an equation that is similar to Eq. 3.1 (Boyer, 1985):

$$\frac{dV}{dt}\frac{1}{V} = L(\Psi_o - \Psi_w) \qquad (3.2)$$

where dV/dt is the relative volume of water entering the tissues per unit time (≈ rate of volume growth), L is the hydraulic conductivity of tissues between growing cells and xylem water source, and Ψ_o and Ψ_w are the water potentials of the xylem water source and elongating tissues, respectively. When the growth rate is constant, these two equations can be set equal, and rearrangement provides a combined growth equation that characterizes the relationship between tissue growth and controlling factors:

$$\frac{dV}{dt}\frac{1}{V} = \frac{mL}{m+L}(\Psi_o - \Psi_\pi - Y) \qquad (3.3)$$

where $(\Psi_o - \Psi_\pi - Y)$ represents the "net osmotic force" for growth (i.e., the maximum possible turgor potential of the growing tissue reduced by the yield threshold turgor potential).

Thus, internal and external factors may influence growth of enlarging tissues by affecting some combination of:

1. ψ of the xylem water supply.
2. Osmotic potential (and hence turgor) of the growing tissue.
3. The yield threshold, Y.
4. Wall extensibility, m.
5. The hydraulic conductivity of the pathway through which water must flow to the growing tissues, L.

Sudden reduction in the Ψ of water supply, or other environmental changes such as dark-light transitions or root chilling (Pardossi et al., 1994), may reduce both

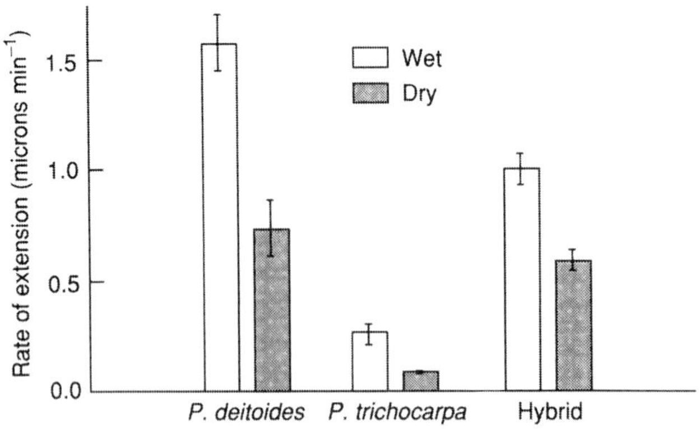

FIGURE 3.3. Effect of irrigation on cell wall extensibility of poplar. From Roden et al. (1990).

turgor and growth in the short term, but prolonged drought often leads to solute accumulation and osmotic adjustment in growing regions. As a result, turgor often is maintained, or at least does not drop below the yield threshold, in dehydrated growing tissues (Michelana and Boyer, 1982; Roden et al., 1990; Nonami and Boyer, 1990; Schultz and Matthews, 1993; Chazen and Neumann, 1994). There also is evidence that water stress influences wall extensibility, causing cell wall hardening. For example, growing leaves of three poplar clones did not lose turgor when irrigation water was withheld; however, cell wall extensibility was considerably reduced (Fig. 3.3), as was the rate of leaf extension (Roden et al., 1990). Similarly, Nonami and Boyer (1990a,b) observed substantial reductions in m in water stressed stem tissue of soybean seedlings; however, they also reported that hydraulic conductivity of the tissues was reduced.

There has been little study of the dynamic changes in parameters of Eq. 3.3 in water-stressed plants. Nonami and Boyer (1990a) asserted that the initial reduction in growth of soybean seedling stems was attributable to collapse of the water potential gradient between xylem and growing cells. Passioura and Boyer (2003) suggested that the declining water potential in the xylem actually reduces the propagation of tensile forces by compressed cells in the stem cortex to the epidermis (which is thought to be the limiting cell layer for extension growth). Cells in the cortex experience a reduction in turgor as xylem water potential declines. Responses subsequent to reductions in xylem water potential included relatively rapid reductions in m and (ψ_p – Y) and slower declines in L. All these parameters, except L, recovered at least partially during prolonged water stress. Hormonal growth regulators also play an important role in controlling growth during droughts. Thus, growth inhibition under water stress appears to be a complex phenomenon, subject to both direct biophysical and indirect metabolic control.

DORMANCY

Fully developed buds of woody plants of the temperate zone alternate from active growth during the warm season to cessation of growth during the cold season, the latter state often referred to as *dormancy*. After seasonal shoot growth ceases and buds form, the new buds first enter a reversible phase of inactivity (sometimes called quiescence or predormancy). Buds in this condition still have a capacity for growth, but the range of environmental conditions in which they can grow becomes progressively more narrow. The state of bud dormancy continues to deepen progressively until shoot apices cannot elongate even under the most favorable environmental conditions. Eventually, such deep dormancy is terminated, usually by exposure to cold, and transition to a quiescent state occurs followed by an active growth phase (Vegis, 1964).

Dormancy Concepts

Over the years more than 50 terms have been used to characterize the sequential phases of dormancy (Martin, 1991). Many of these terms are easily misunderstood and lack both precision and a physiological basis, thus leading to confusion. Recognizing this problem, Lang et al. (1985, 1987) simplified the terminology and defined dormancy as "any temporary suspension of growth of any structure containing a meristem." They also realized, as earlier emphasized by Romberger (1963), that there are three points of dormancy control: (1) the environment, (2) the shoot apex, and (3) the condition within the affected organ. Lang et al. (1987) introduced a nomenclature that differentiated clearly among these three points of control. Phasic dormancy phenomena were characterized in terms of (1) *ecodormancy* (regulated by environmental factors), (2) *paradormancy* (regulated by physiological factors outside the affected structure; e.g., apical dominance), and (3) *endodormancy* (regulated by physiological factors inside the affected structure) (Fig. 3.4).

The importance of this terminology lies in its emphasis on the condition or event that changes the state of dormancy and on where the condition or event is perceived. A further advantage of this nomenclature is that it should improve communication among researchers because it is based on physiological conditions, hence focusing attention on the causes of dormancy and its perception. Figure 3.5 summarizes the

DORMANCY			
	Ecodormancy	**Paradormancy**	**Endodormancy**
	Regulated by ***environmental*** factors	Regulated by ***physiological*** factors outside the affected structure	Regulated by ***physiological*** factors inside the affected structure
EXAMPLES	Temperature extremes Nutrient deficiency Water stress	Apical dominance Photoperiodic responses	Chilling responses Photoperiodic responses

FIGURE 3.4. Dormancy: A simple, descriptive terminology applied to regulatory factors and examples of plant dormancy. From Lang et al. (1987) *Hortscience* **22,** 371–377.

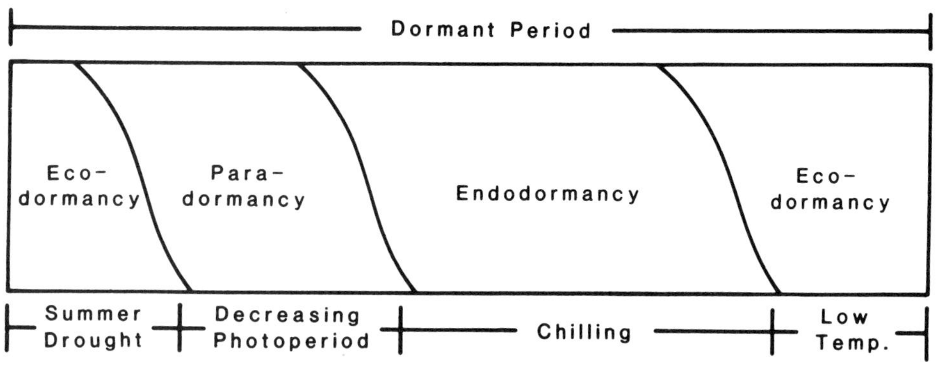

FIGURE 3.5. The relative contribution of the various types of dormancy during a hypothetical dormant period for an apical bud. From Lang et al. (1987) *Hortscience* **22,** 371–377.

sequential phases of dormancy as they are influenced by environmental factors. The stages of dormancy overlap to some degree and the beginning and end of each cannot be precisely fixed.

One of the essential attributes of dormancy is the suppression of cell division. Cellular and molecular level studies of bud dormancy have emphasized regulation of the cell cycle of mitosis (Fig. 3.6) as key to understanding control of cell division in dormancy induction, maintenance, and emergence. The cell cycle in eukaryotes has four phases:

- A synthetic or "S" phase during which DNA is duplicated and related proteins, including histones, are constructed
- Mitosis followed by cytokinesis, "M"
- Two interphase "gap" periods, G1 and G2, associated with cell growth

The G1/S and G2/M regions appear to be key regulatory checkpoints for the cell division process, with the cycle not proceeding without proper cell developmental progress. The regulation of cell cycle progression is complex with numerous signaling pathways and metabolic events. Especially important are several cyclin proteins that are characteristic of specific cell cycle stages and that bind to cyclin-dependent kinase

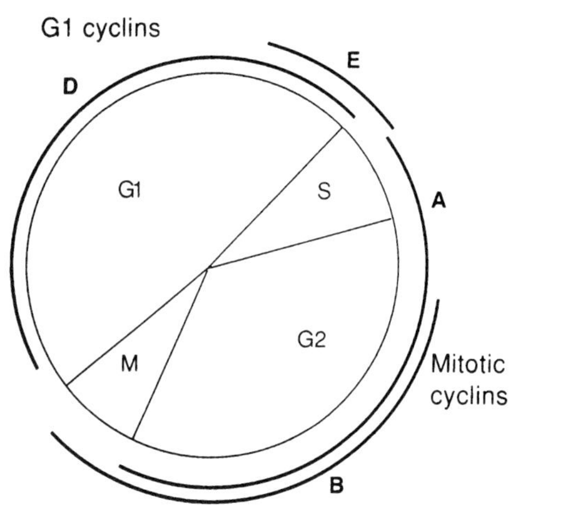

FIGURE 3.6. The cell cycle and relative abundance of various cyclins. Cyclins A and B are mitotic cyclins; D and E are G1 cyclins. Cells that are dividing generally spend more time in G1 and G2 than in S and M phases. This figure was published in *"Plant Growth and Development: Hormones and the Environment,"* Srivistava, L. M. Copyright Elsevier 2001.

enzymes and a variety of stage-specific substrates for kinase phosphorylation. These products along with other inhibitory proteins orchestrate the metabolic processes that propel the cell cycle. It appears that the cells of endodormant and paradormant buds largely are arrested in the G1 stage and possibly at the G2/M transition stage (Rohde et al., 2000; Anderson et al., 2001).

Hormonal Influences on Bud Dormancy

Induction, maintenance, and release from dormancy of vegetative buds are extremely complex and vary with the type of dormancy and buds. Plant hormones have long been thought to play key roles in dormancy and a large body of correlative evidence, with some contradictory reports, has accumulated in support of their participation (Rohde and Boerjan, 2001). Variation in tissue sensitivity to hormone concentration and multiple simultaneous hormone influences further complicate interpretation of such studies. Other important environmental and biochemical regulators, such as phytochrome-linked light effects and chemical signals such as sugars also appear to influence dormancy in vegetative buds (Anderson et al., 2001).

In many temperate species, particularly those with an indeterminate growth habit, the first detectable stage of dormancy is the cessation of shoot elongation induced by short photoperiods. This response appears to be phytochrome-mediated, as was indicated by manipulation of phytochrome gene expression in transgenic hybrid aspen trees (Olsen et al., 1997). In this study, hybrid aspen (*Populus tremula* × *P. tremuloides*) was transformed with DNA of the phytochrome A gene from oats. Preceding this gene in the transferred DNA was a promoter that directed overexpression of the gene. Overexpression prevented growth cessation at normal critical photoperiods and cold hardening and maintained GA_1 and IAA levels in stems that otherwise fell substantially under short photoperiods. Drought and exogenous ABA can also trigger growth cessation in many species (Barros and Neill, 1987) and elevated levels of ABA were associated with short days and further increased if accompanied by drought in downy birch (Rinne et al., 1994a,b).

For a long time it was believed that endodormancy of buds was largely controlled by a balance between growth-inhibiting hormones (e.g., ABA) and growth-promoting hormones (e.g., gibberellins, cytokinins). In sycamore maple most growth inhibitors were found in buds and leaves when the buds were not expanding; and the smallest amounts when the shoot apex was expanding (Phillips and Wareing, 1958). When the buds began to expand, there was a decrease in inhibitors, an increase in growth promoters, or both. In sycamore maple the release of bud dormancy was better correlated with an increase in gibberellins content of buds than with a decrease in growth inhibitors (Eagles and Wareing, 1964).

As noted earlier, it is unlikely that individual hormone levels alone control the induction or breaking of bud dormancy. It has been claimed that ABA is responsible for bud endodormancy and that breaking of dormancy by chilling causes a decrease in ABA. In some species the capacity for bud opening increases progressively downward on the plant axis (Powell, 1987). In broadleaved trees there often is less ABA in buds located in the lower crown than in those in the upper crown. Furthermore, the onset of dehydration of buds at the beginning of exposure to short days with a simultaneous increase in ABA suggests that ABA might be involved in induction of growth cessation and development of bud dormancy (Rinne et al., 1994a). However, the amount of ABA in buds declined about equally when plants were exposed to warm or cold temperatures (Mielke and Dennis, 1978; Barros and Neill, 1988). Furthermore, in apple trees a decrease in ABA began well before the buds were exposed to chilling (Fig. 3.7). The amount of endogenous ABA was similar in bay willow seedlings grown under short or long days (Johansen et al., 1986). Also no correlation was evident between the ABA content of buds and short-day-induced cessation of shoot growth of sycamore maple (Phillips et al., 1980). Martin (1991) concluded that ABA has no direct effect in controlling bud dormancy.

Gibberellins often have been implicated in dormancy release because their concentration in expanding buds often increases. In general GA-like activity increases prior to or concurrently with bud break and activity decreases as budset is approached. Determining the role of gibberellins in release of dormancy is complicated because most of the GAs in plants are inactive. Hence, the total amounts of GAs in buds during changes in dormancy release may be misleading.

Walser et al. (1981) concluded that GA applications could substitute for chilling in breaking dormancy. However, both Leike (1967) and Paiva and Robitaille (1978) observed that GA induced budbreak only after the buds had been chilled. It thus appears unlikely that GA can substitute for chilling in breaking endodormancy. As emphasized by Leike (1967), an increased concentration of GA in expanding buds may not be the cause, but rather the result of release of dormancy by chilling.

A role of cytokinins in dormancy release has been postulated because of increasing levels during bud growth and stimulation of bud growth by exogenous cytokinins. However, the weight of evidence does not support a primary role for cytokinins in release of buds from dormancy. Although cytokinins in buds increase during dormancy release (Domanski and Kozlowski,

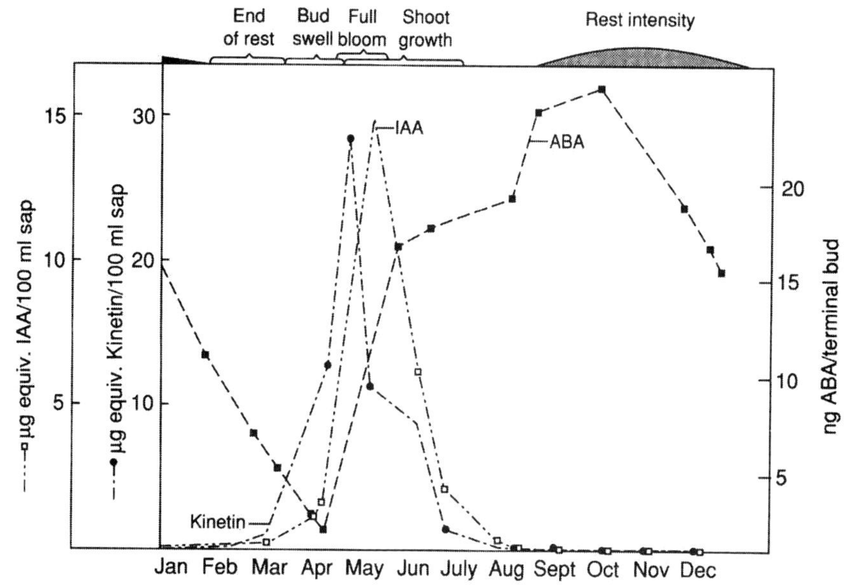

FIGURE 3.7. Changes in growth substances in apple trees during the annual growth cycle. From Seeley and Powell (1981).

1968), the increase is small (Wood, 1983; Young, 1989). Cytokinins apparently stimulate bud growth after release from dormancy but do not appear to directly control dormancy release. This conclusion is consistent with studies that show that applied cytokinins, at concentrations much higher than endogenous levels, induce growth of buds after they have been partially chilled (Shaltout and Unrath, 1983).

Ethylene does not appear to have an important regulatory role in release of dormancy of buds that have a chilling requirement. Stimulation of bud growth by exposure to ethylene was attributed to effects of ethylene after partial or full release from dormancy by chilling (Paiva and Robitaille, 1978; Wang et al., 1985). Ethylene probably was not involved in breaking dormancy of apple buds because ethylene production did not increase in buds in which dormancy was broken by removal of bud scales (Lin and Powell, 1981).

The release of bud dormancy is a sequential process, with each phase characterized by different metabolic and hormonal effects as shown for English walnut (Dathe et al., 1982). Current evidence indicates that control of bud dormancy is very complex and involves interactive effects of growth hormones and other compounds. Some of these interactions are still not fully understood and much more research will be needed before we fully understand the mechanisms involved. The attractive model of Saure (1988) suggests that release of bud dormancy may occur in the following sequential phases as it apparently does in seeds (Zarska-Maciejewska and Lewak, 1983):

Phase I. Removal of the primary cause of dormancy, largely involving the enzymes that are most active at chilling temperatures. These changes cause transition from endodormancy to ecodormancy.

Phase II. This catabolic phase is characterized by high metabolic activity, increasing GA activity, which promotes synthesis and/or activation of hydrolytic enzymes (e.g., amylases, proteases, ribonuclease), and development of a translocation system, causing mobilization of reserve carbohydrates essential for growth. During this phase, bud swelling indicates a transition from endodormancy to ecodormancy.

Phase III. This phase in characterized by additional increases in gibberellins and cytokinin activity, further mobilization of reserve carbohydrates, increasing auxin activity, promotion of mitosis, and complete release from dormancy. The end of this phase is characterized by bud break and rapid shoot growth.

SHOOT GROWTH

Growing shoots, which usually consist of a stem portion plus leaves, consist of nodes and internodes. Nodes are parts of stems or branches at which leaves are attached. Often the term node also is used to refer to the region of stem where long shoots or branch whorls are attached. Internodes are lengths of stem between two successive nodes.

Shoots elongate as the result of bud opening and expansion at the many growing points (apical meristems) distributed over the stem, branches, and twigs. During shoot growth the duration of expansion of internodes and of leaves often varies appreciably, depending at least in part on cell turgor, cell wall extensibility, and availability of structural and energy resources and hormonal growth regulators. The overall growth of a shoot involves division of cells of the apical meristem and their subsequent elongation, differentiation, and maturation. These phases are not sharply delimited and occur sequentially at varying distances from the tips of stems and branches. Almost all shoot extension is the result of internode elongation.

Meristematic activity in elongating shoots occurs at a short distance below the shoot apex. Shoot elongation is a highly organized process that involves continued cell divisions followed by subsequent expansion of cells, with cell division predominating. During growth and development of internodes, cell length increases only two to three times but cell numbers increase by 10 to 30 times. Shoot elongation of species with diverse growth patterns is highly correlated with the rate and duration of cell division. In several different species there was an initial three- to five-day period of rather uniform growth throughout entire internodes (Brown and Sommer, 1992). The center of growth then shifted progressively to the middle and later to the upper end so that an upward growth wave progressed from the base throughout the entire internode. Hence, the lowermost portion of an internode matured first. Very high rates of cell division in the pith of sweetgum internodes provided the driving force for internode growth and development (Brown et al., 1995a,b). In short shoots cell division in developing internodes is inhibited and little or no elongation occurs.

Bud Characteristics

A mature bud is an embryonic shoot, or part of a shoot, bearing at the tip the apical meristem from which it originated. Most lateral buds are initiated in leaf axils and arise in relatively superficial tissues. The initiation of buds involves cell division in cell layers in the leaf axil to form a bud protrusion as well as organization of the apical meristem.

Usually buds are classified as to location, contents, or activity (e.g., terminal, lateral, axillary, dormant, adventitious, vegetative, flower, or mixed buds). Each of these types may be further classified as active or dormant. Vegetative buds vary greatly in maturity. They may consist of little more than an apical meristem. More commonly, however, they contain a small mass of meristematic tissue, nodes, internodes, and small rudimentary leaves, with buds in their axils, all enclosed in bud scales. Flower buds contain embryonic

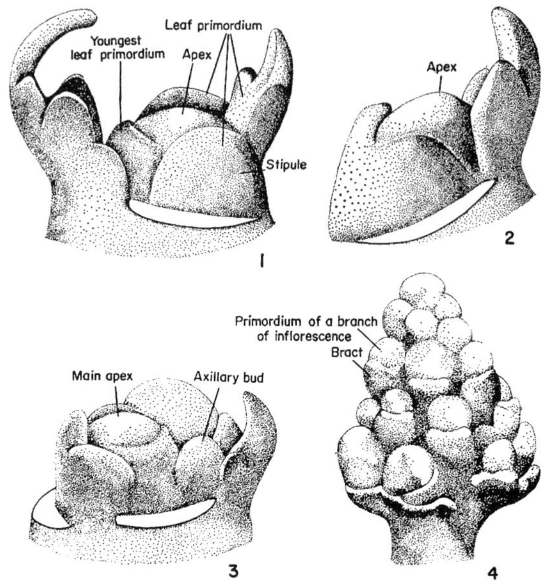

FIGURE 3.8. Vegetative shoot apex of grape. From Fahn (1967), with permission of Butterworth Heineman Ltd.

flowers and most also have some rudimentary leaves. Mixed buds contain both flowers and leaves.

During formation of vegetative buds, leaf primordia appear in upward succession. Hence, the largest and oldest leaf primordia are located at the bud base and the smaller rudimentary leaves occur toward the growing point (Fig. 3.8). The leaves form as a result of divisions in subsurface cells of apical meristems. As noted earlier, location of leaf initiation appears closely related to the presence of expansin protein in apical meristems (Pien et al., 2001). When leaves are first formed they occur close together and nodes and internodes are indistinguishable. Subsequently, as meristematic activity occurs between leaf insertions, internodes become recognizable as a result of intercalary growth (through activity of a meristem inserted between more or less differentiated tissue regions).

Dormant and Adventitious Buds

All the buds on a tree do not expand into shoots because some remain dormant, die, or produce flowers (Maillette, 1982a). Buds in the upper crown of European white birch trees had a much higher probability of expanding into shoots than did the buds on lower branches (Maillette, 1982b). Approximately 60% of the buds on each sessile oak shoot began to expand into shoots, but only about half of these formed branches (Harmer, 1992). The small buds on the lower branches of red pine trees did not expand into shoots (Kozlowski et al., 1973). In European white birch capacity for bud burst was high throughout the year in underground basal buds. In contrast, the terminal buds of short shoots remained dormant until October. Thereafter dormancy was gradually broken and capacity for bud opening was highest in March to April (Rinne et al., 1994a,b).

Some of the buds on a tree remain dormant, sometimes throughout the life of a woody plant. Dormant buds, originally developed in leaf axils, subsequently are connected to the pith by a bud trace. Branching of dormant buds occurs rather commonly. Buds that form irregularly on older portions of a plant and not at the stem tips or in the leaf axils are called adventitious buds. These form on parts of the root or stem that have no connection to apical meristems. They may originate from either deep or peripheral tissues. Unlike dormant buds, adventitious buds do not have a bud trace all the way to the pith.

Many new branches are produced following pruning of branches. Stump sprouts originate from root collars and the lower stem; epicormic branches of angiosperms and sprouts of gymnosperms following fire or injury arise from dormant rather than from adventitious buds. Root sprouts ("root suckers") arise from adventitious buds. Reproduction by root suckers is well known in trembling aspen and bigtooth aspen, but it also occurs in numerous other woody angiosperms. Cutting branches of trees so as to leave a stub stimulates release of dormant buds and formation of new branches from adventitious buds. The latter, especially, are only weakly attached to the stem for many years. The practice of topping, the systematic cutting of a tree crown back to large-stubbed branches, usually results in many such branches and predisposes trees to subsequent branch loss under wind and ice influence. Stubs often never heal over and provide an entry point for decay-inducing organisms. For this and other reasons, most professional arborists maintain that the practice of tree-topping should be avoided (Karlovich et al., 2000, Close et al., 2001).

Hormonal Influences on Shoot Growth

A number of investigators have shown that exogenous gibberellins increased height growth and internode elongation in a variety of woody angiosperms and gymnosperms. The effects of exogenous GA vary in detail and depend on the specific gibberellins applied; concentration, frequency, and method of application; and on the species and age of trees.

Most elongation of shoots results from cell expansion, with gibberellins playing a dominant role in the hormonal complex regulating shoot expansion. Evidence for this conclusion comes from an increase in shoot elongation following application of gibberellins, the counteracting effects of applied gibberellins on those of endogenous growth inhibitors or applied growth retardants, and correlation of shoot extension with endogenous gibberellins levels. Cell elongation following GA treatment requires recognition of GA by a receptor molecule, interaction of the activated receptor with the cell, and production of a wall-loosening factor or inhibition of a wall-stiffening factor (Jones, 1983). For example, there is evidence that exogenous application of GA increases the expression of genes coding for both expansin proteins and XET, both of which are thought to induce cell wall loosening (Chen et al., 2002; Lee and Kende, 2001, 2002). Work with herbaceous plants has shown that brassinosteroid compounds also promote stem growth, and are associated with enhanced levels of XET (Oh et al., 1998) (Chapter 13).

Applying GA_3 to the shoot apex of bay willow in the spring delayed growth cessation and shoot tip abortion. Both growth cessation and shoot tip abortion were induced by short days or application of growth retardants, and these effects were antagonized by GA_3 (Juntilla, 1976). Gibberellin-like activity in the apical part of bay willow shoots decreased prior to cessation of shoot growth, indicating that low GA activity in the shoot may be a prerequisite for cessation of shoot growth (Juntilla, 1982). The effects of specific gibberellins on shoot growth vary appreciably. For example, exogenous GA_{20} and GA_1 stimulated shoot elongation of bay willow under short-day conditions and could substitute for transfer of plants to long-day conditions, whereas GA_{53}, GA_{44}, and GA_{19} were inactive. Both GA_{20} and GA_1 were active even when applied at concentrations one-thousandth of the concentration of applied GA_{19} (Juntilla and Jensen, 1988).

Interactions between day length and gibberellins on shoot elongation have been well documented, with short day-induced growth cessation of many deciduous species prevented by GA_3.

Gibberellins may be synthesized in partially expanded leaves as well as in other organs. The inhibition of internode extension following removal of the young leaves of apple was partially or entirely eliminated by exogenous GA_3 (Powell, 1972). Removal of the young leaf just as it was unfolding not only inhibited elongation of the internode below it but also elongation of two internodes above it. Applied GA_3 completely replaced the leaf with regard to expansion of the internode below, but only partially substituted for the leaf in controlling growth of internodes above it. This suggested that the leaf supplied essential growth-controlling factors in addition to gibberellins.

In contrast to the stimulation of shoot growth in angiosperms by short-term treatments with GA, early experiments showed negligible or only slightly stimulatory effects of gibberellins on shoot growth of conifers. However, in these studies the lack of stimulation of shoot growth by applied gibberellins may have been due to insufficient amounts of GA, application of an inappropriate GA, poor absorption of GA, or overriding inhibitory factors (Ross et al., 1983). Several lines of evidence now show that endogenous gibberellins are very important in regulating shoot growth of conifers (Dunberg, 1976; Juntilla, 1991). Spray treatments of GA applied monthly for nine months increased height growth (observed for 18 months) of loblolly pine seedlings (Roberts et al., 1963). Both soil drenches (applied several times weekly) and foliar sprays (applied weekly for 3 months) of GA_3 increased height growth of balsam fir seedlings. The increase was associated with a change in the distribution of photosynthate rather than in the rate of photosynthesis (Little and Loach, 1975). Ross et al. (1983) cited studies that showed that exogenous gibberellins (including GA_3 and $GA_{4/7}$) stimulated shoot growth in 26 species of conifers.

If auxins play an important role in shoot elongation they should show a relation to the seasonal timing of growth. Ample evidence exists of such a relation for broadleaved trees. For example, the auxin content increased in the spring as apple shoots began to expand. Auxin levels declined during the growing season and the decrease was followed by slowing of shoot growth (Hatcher, 1959).

Auxin-induced cell wall acidification and loosening of cell walls, likely involving the action of expansins and other wall proteins (Cosgrove, 1997), dominates the rate of cell expansion with increase in cell size occurring as a result of a turgor-driven extension of cell walls. The site of auxin action may be on or within the cytoplasm. An auxin-binding protein, AXP1, has been associated with the plasma membrane, although it is much more abundant within the endoplasmic reticulum (Leyser, 2002). Auxin is also transported across the plasmalemma, and acts internally in the cytoplasm although modes of action are still largely unknown. Auxin influences at the molecular level are quite complex, involving multiple interconnected pathways of signaling. Recent studies have emphasized the regulation by auxin of protein degradation via the ubiquitination-26S proteasome pathway leading to gene activation, as well as stimulation of protein synthesis by transcription factors (Kepinski and Leyser, 2002; Tiwari et al., 2003).

Although applied auxins increased elongation of excised stem sections of some gymnosperms (Terry et al., 1982), auxins appear to be less important in intact trees. This is shown by negligible or inhibitory effects of applied auxins on shoot elongation (Heidmann, 1982; Ross et al., 1983). There are some exceptions, however, and Dunberg (1976) found that the auxin content of Norway spruce shoots was highly correlated with their rates of elongation. Some evidence for auxin involvement in regulation of stem elongation of gymnosperms comes from inhibition of height growth following application of triiodobenzoic acid (TIBA), an inhibitor of polar auxin transport (Little, 1970).

There has long been an important role assigned to auxin as a regulator of primary vascular system induction and development. Sachs (1981) proposed a "canalization hypothesis" that associated vascular bundle initiation with channeling of polar auxin flow from sites of synthesis to a narrow file of cells that promote further basipetal flow. Carrier proteins have been identified that are preferentially located at the basal end of auxin-transporting cells and may promote directed transport of IAA (Gälweiler et al., 1998). Application of auxin inhibitors to leaves results in exaggerated development of marginal veins in the region where auxin synthesis is likely intense (Mattson et al., 1999). Similarly, inhibitors tend to thicken vascular strands and tissues, suggesting that the polar transport and channeling of auxin into narrow cell files is disrupted. Mutants with defective auxin transport capacity show development and morphology similar to that induced by chemical inhibitors.

Abscisic acid (ABA) does not appear to have a major role in regulating shoot growth when the potential for growth is high. When apple seedlings were treated with ABA in January in a greenhouse (low growth rate), shoot growth stopped and terminal buds formed. However, a similar treatment in June (high growth rate) had little effect on shoot elongation (Powell, 1982).

Some investigators have implicated polyamines in regulation of shoot growth. For example, Rey et al. (1994) found correlations between high spermidine and spermine levels with rapid shoot growth and leaf expansion of hazel trees. They also reported that low spermidine and spermine levels, together with increasing amounts of putrescine, may be associated with induction of bud dormancy.

Leaf Growth

Trees bear several types of foliar appendages including cotyledons, foliage leaves, and cataphylls. Cotyledons or "seed leaves" are developed in the seed and contain or have access to stored materials (Chapter 2, Kozlowski and Pallardy, 1997). They generally differ in size and shape from the first foliage leaves. Cataphylls or "lower leaves," which usually are involved in storage, protection, or both, are represented by bud scales.

Angiosperms

Cell division predominates during the early stages of development of leaf primordia. Subsequently, a leaf achieves its final shape and size by both cell division and expansion, with the latter predominating. Leaves form only on shoot apices. The apex swells to form an undifferentiated leaf primordium. As noted earlier, location of leaf initiation appears closely related to the presence of expansin protein in apical meristems (Pien et al., 2001). Shortly thereafter cell division stops in the area of attachment and the leaf base is differentiated. The upper part of the primordium continues to divide and forms the blade. The petiole forms later from an intermediate meristematic zone. The various leaf parts, such as petiole, blade, sheath, and stipules are initiated soon after the primordium has formed.

Growth of a leaf is at first localized at the tip but this continues for a short time only and is followed by intercalary growth, which accounts for most of the increase in leaf length. The flattened form of the blade is initiated from meristems located along the two margins of the leaf axis.

When a leaf primordium achieves a critical length, the leaf blade begins to develop, together with increased mitotic activity from marginal initial cells on the flanks of the primordium (Dale, 1992). Although cell division declines with leaf age, it may continue until the blade reaches most of its final size. In the early stages of growth, the leaf blade is comprised of several layers of rather uniform cells. However, differentiation results in the formation of several layers of palisade cells and spongy parenchyma cells, initially as a result of breakdown of cellulose and protein materials in the walls adjacent to the future space (Jeffree et al., 1986). Later, intercellular spaces also form as the epidermal cells continue to expand for a longer time than the mesophyll cells do, pulling the mesophyll cells apart. Increase in height of palisade cells accounts for most of the increase in leaf thickness (Dengler et al., 1975). Formation of the vascular system also begins early during blade formation.

The epidermal tissue from which stomata originate is differentiated early in leaf development. Hence, most stomata are found on leaves in young buds, but stomata also may form late in leaf development. Stomatal density and stomatal index (number of stomatal guard cells as a percentage of the number of stomatal guard cells + epidermal cells) respond to environmental factors such as shading and CO_2 concentration of the air, and there is much evidence to indicate that this response is controlled by signals translocated from mature leaves (Allard et al., 1991; Lake et al., 2001). Stomatal density tends to increase as CO_2 concentration declines and irradiance increases. Brownlee (2001) suggested that stomatal density and patterning were determined by release of inhibitors of guard cell differentiation by the developing guard cells themselves because stomata rarely are found immediately adjacent to one another. The quantity or transport of inhibitor released from the guard cells might also respond to the external signal to determine stomatal density and index. The light environment of mature leaves also appears to have a large influence on whether axillary buds will produce sun or shade leaves regardless of the light environment during leaf development (Eschrich et al., 1989).

Gymnosperms

Three distinct types of foliar appendages form sequentially in gymnosperm seedlings. First are the cotyledons, which are present in the seed, then the primary needles of a young seedling, and finally the secondary needles, which form the permanent complement of leaves. Leaf growth in gymnosperms starts with foliar primordia located on the flanks of apical meristems. Both apical growth and intercalary rib meristem activity form the leaf axis, but apical growth is of short duration. The narrow leaf blade is initiated by marginal growth. In pines leaf growth lags shoot extension growth to some extent, giving the so-called candle appearance of new shoots. In conifers of other genera (e.g., *Picea* and *Abies*), leaf growth shortly precedes or accompanies shoot extension in a parallel fashion.

Seasonal Leaf Growth Characteristics

Several patterns of seasonal production of foliage leaves have been shown. Trees of some species achieve maximum leaf area early in the season and do not produce any more leaves during the year, whereas others add new leaves, either by continuous production and expansion of new leaf primordia, or by several intermittent "flushes" of growth involving recurrent formation and opening of buds during the growing season, followed by expansion of their contents.

The duration of expansion of individual leaves varies greatly among species. In angiosperms leaves generally expand in a matter of a few days to weeks (e.g., white birch and trembling aspen; Kozlowski and Clausen, 1966b). In apple trees all the spur leaves develop rapidly but the leaves on long shoots develop as the shoots elongate, which may require an additional 60 days. Expansion of leaves of gymnosperms is slower than that of angiosperms. Elongation of Scotch pine needles in England continued until early August

(Rutter, 1957). Leaves of evergreen angiosperms generally expand slowly. For example, citrus leaves expanded for 130 days (Scott et al., 1948). Even after leaves of evergreens and deciduous trees are fully expanded in area they often continue to thicken and increase in dry weight.

Leaf Area Index

Annual production of leaves by woody plants is enormous. Forests may produce up to 20 ha of leaves for each ha of land occupied (Dale, 1992). Because of its participation in biological processes, the leaf area of a tree or stand of trees has important implications in ecology of forests (Fownes and Harrington, 1990; Kozlowski et al., 1991). Hence much interest has been shown in the leaf area index (LAI, the ratio of projected leaf surface area of a plant or stand to ground surface area). The same units (m^2) are used for leaf area and ground area; hence LAI is a dimensionless measure of the amount of leaf cover. The LAI varies with tree species and genotype, plant size and age, spacing of trees, and factors that influence the number and size of leaves. Climate has a strong influence on both the maximum LAI a forest can develop and on its rate of development (Margolis et al., 1995).

As a forest grows, its LAI increases to a maximum and either stabilizes or decreases thereafter (Kozlowski et al., 1991). In forest stands net primary production (NPP: the sum of increases in biomass, litter production, and amount of biomass consumed by animals and microbial decomposers) is positively correlated with LAI up to some value. At higher LAI values NPP typically decreases, reflecting tree mortality and reduction in the rate of photosynthesis of lower, shaded leaves. The slow growth and long leaf life spans of conifer leaves are associated with their higher LAI values (up to 20) over those of deciduous forest stands (3 to 6) (Waring and Schlesinger, 1985), as well as later development of maximum LAI in conifer stands. In the temperate zone conifer stands typically achieve maximum LAI in 25 to 40 years as compared to five years in some deciduous forests (Bond, 1989; Woodward, 1995).

SHOOT TYPES AND GROWTH PATTERNS

Shoots generally are classified on the basis of location, development, or type of bud from which they are derived. With regard to location, shoots are classified as terminal leaders, laterals, or basal shoots. Coppice shoots arise from dormant buds near the base of a woody plant. As mentioned, root suckers arise from adventitious buds on roots. Some of the more important shoot types and their growth patterns will be discussed briefly.

Determinate and Indeterminate Shoots

In some woody plants such as pines, spruces, oaks, and hickories, shoot growth results from expansion of terminal buds on the main stem and its branches. After the terminal shoots elongate, there is a period of inactivity until new terminal buds form and expand. In such determinate (monopodial) species only one bud may expand on a shoot each year, or two or more may form sequentially and expand in the same year. In indeterminate (sympodial) trees the shoots do not expand from true terminal buds but arise from secondary axes. Sympodial growth often results when a reproductive structure occurs at the end of a branch or when a shoot tip aborts. In indeterminate species the subtending, pseudoterminal bud, which often is mistaken for a true terminal bud, is the same size as other lateral buds, but has a twig scar resulting from abortion of the shoot tip. Shoot tip abortion occurs commonly in such genera as *Betula*, *Carpinus*, *Catalpa*, *Corylus*, *Diospyros*, *Gleditsia*, *Platanus*, *Robinia*, *Salix*, *Tilia*, and *Ulmus*.

Epicormic Shoots

Dormant buds on the main stem or branches of trees often are stimulated by sudden exposure to light to produce epicormic shoots (also called water sprouts). There is much concern about epicormic shoots because they produce knots that greatly reduce the grade of lumber. Epicormic shoots occur much more commonly in angiosperms than in gymnosperms.

Inasmuch as species vary widely in the abundance of dormant buds produced, the tendency for epicormic sprouting often can be predicted. For example, oaks tend to produce many epicormic shoots and ashes produce few (Table 3.1). Within a species more

TABLE 3.1. Variations among Species in Production of Epicormic Shoots[a]

Number of epicormic shoots	Species
Very many	White oak, red oak
Many	Basswood, black cherry, chestnut oak
Few	Beech, hickory, yellow-poplar, red maple, sugar maple, sweet birch
Very few	White ash

[a]From Smith (1966) and Trimble and Seegrist (1973).

epicormic shoots are produced by young and small trees than by old and large trees. Suppressed trees tend to have more epicormic shoots than dominant trees of the same species (Bachelard, 1969). Formation of epicormic shoots also is influenced by the severity of thinning of forest stands. For example, white oak trees in heavily thinned stands produced an average of more than 35 epicormic shoots; in moderately thinned stands 21; and in unthinned stands 7 epicormic shoots, respectively (Ward, 1966).

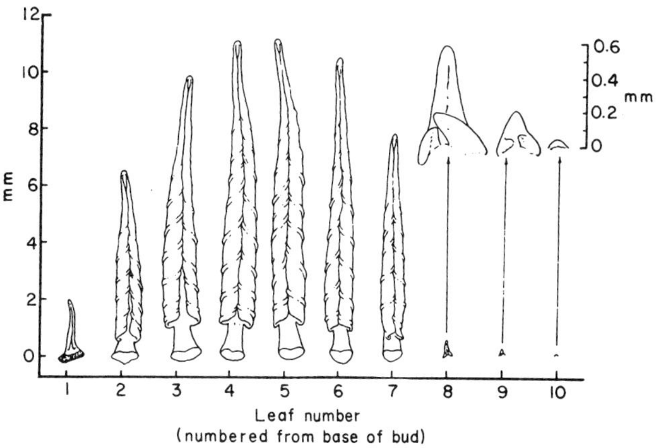

FIGURE 3.9. Contents of a winter bud of black cottonwood. The first left leaf aborted; the tree leaf primordia (8–10) are shown in an enlarged scale at upper right. From Critchfield (1960).

Preformed and Neoformed Shoots

Shoots may be formed by fixed (determinate) growth, free (indeterminate) growth, or a combination of both. Fixed growth involves the elongation of preformed stem units after a rest period. For example, winter buds of adult trees of many species of angiosperms and gymnosperms contain primordia of all the leaves that will expand during the following growing season. In such species shoot formation involves differentiation in the bud during the first year (n) and extension of the preformed parts into a shoot during the second year (n + 1) (Kozlowski, 1964). Examples of species exhibiting this pattern are mature trees of some northern pines (such as red pine and eastern white pine), as well as spruce, fir, beech, green ash, and some maples.

The numbers of preformed leaf primordia vary among genotypes and bud location on the tree (Remphrey and Davidson, 1994a), and with tree age. Provenance studies have shown genotypic variation in the development of buds of both gymnosperms and angiosperms. Variations in the number of leaf primordia produced often are linked to the latitude of plant origin (Cannell et al., 1976). The number of preformed leaves in buds of green ash varied among provenances but was subject to plastic modification, as shown by both site and year-to-year differences (Remphrey and Davidson, 1994b). All shoots of mature green ash trees were preformed, but in saplings considerable shoot neoformation occurred (Remphrey, 1989; Remphrey and Davidson, 1994a, 1994b).

In contrast to fixed (preformed) growth, free growth involves the elongation of a shoot by simultaneous initiation and elongation of new (neoformed) stem units (Pollard and Logan, 1974). In adult trees of certain species some of the shoots are fully preformed in the bud and other shoots are not. The preformed shoots produce early leaves only and generally expand into short shoots. Internodes of short shoots are only 1 to 2 mm long in ginkgo and form repeatedly year after year, giving rise to peg-like structures on twigs of this species. Preformed short shoots of striped maple are 1 to 2 cm long. The shoots that are not fully preformed in the winter bud are long shoots, also called heterophyllous shoots, which produce two sets of leaves: (1) early leaves, which are relatively well developed in the winter bud; and (2) late leaves, which expand from primordia present in the winter bud or, more commonly, from leaf primordia that continue to form and grow during the current year while the shoot is elongating (Fig. 3.9).

Such free growth resembles that shown by a number of herbaceous plants. The two sets of leaves produced are distinguishable since they may differ in morphology and anatomy. Early leaves generally are smaller than late leaves (Fig. 3.10). Examples of woody plants exhibiting free growth in some of their shoots are poplars, apple, white birch, yellow birch, sweetgum, Virginia creeper, several species of *Acer* and *Eucalyptus*, ginkgo, larch, eastern hemlock, and some tropical pines. Shoots of sugar maple with at least three long internodes were of two types: (1) shoots expanding three or four pairs of preformed leaves and then developing terminal buds, and (2) shoots expanding usually four pairs of preformed leaves and then several pairs of neoformed leaves before developing terminal buds (Powell et al., 1982).

Considerable caution is advised in rigidly classifying patterns of shoot development for taxonomic groups of woody plants that have a large number of species and occupy extensive ranges because they may exhibit diverse growth patterns (e.g., *Pinus* spp.) (Lanner, 1976).

Recurrently Flushing Shoots

Shoot growth of some temperate-zone pines such as loblolly and Monterey pines, most tropical pines, and

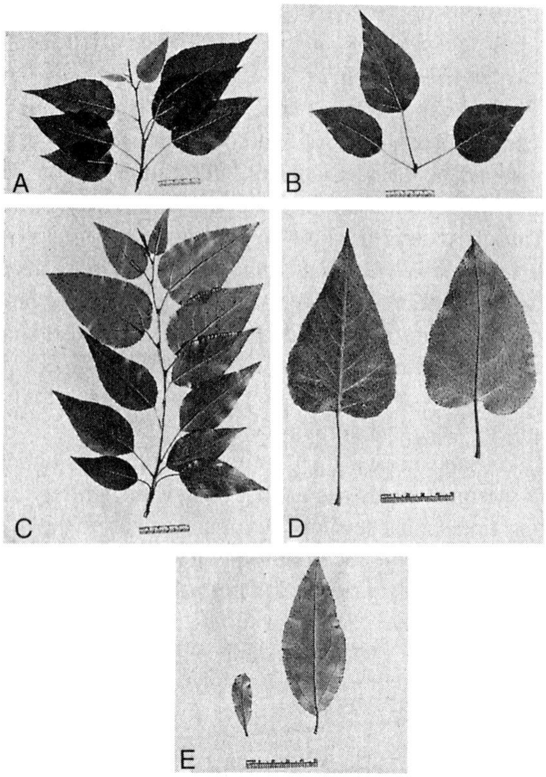

FIGURE 3.10. Leaf dimorphism in black cottonwood. A. Early stage (5 weeks after bud opening) in development of heterophyllous shoots. The six early leaves have almost completed expansion and the first late leaf is beginning rapid growth. B: A short shoot collected on same date as A. Three early leaves are mature and the terminal bud is beginning to form. C: Later stage in development of a heterophyllous shoot. The six early leaves are fully expanded and the first one or two late leaves nearly so. D: Early and late leaves. E: Leaves 1 and 7 of an adventitious shoot. From Critchfield (1960).

many tropical and subtropical angiosperms occurs in a series of waves or "flushes" during the growing season. Such growth involves elongation of more than one terminal bud per shoot each year. After the first bud with its fixed complement of leaves expands into a shoot, a second bud forms rapidly at the apex of the same shoot and this bud also expands shortly thereafter, thereby cumulatively extending the shoot. The second growth phase may be followed by additional waves of growth from even more buds formed and expanded sequentially at the tip of the same shoot. In southern pines of the United States, such as loblolly and longleaf pines, the first seasonal growth flush usually is the longest (Fig. 3.11).

The number of successive buds that form and expand on the same shoot during one growing season varies with individual trees, species and genotype, shoot location on the stem, and climatic conditions. The terminal leader and upper false-whorl shoots

FIGURE 3.11. One year's stem elongation in the recurrently flushing shortleaf pine. From Tepper (1963).

usually produce more buds and show more growth flushes than do lower branches. The terminal leader of an average adult loblolly pine tree does not elongate more than two or three times annually, but as many as seven successive elongations have been recorded in one summer for terminal leaders of some trees (Wakeley and Marrero, 1958).

Species that exhibit free growth of long shoots also may produce some short shoots by fixed growth, as in larch, ginkgo, and apple. Eastern hemlock trees produced both neoformed leaves and preformed leaves on shoots that developed from overwintered buds and reached a length of at least 6 cm (Powell, 1991). In some genera (e.g., *Pinus*), shoot growth of various species may vary from fixed to free. The pattern of shoot growth also may differ with tree age. For example, free growth may occur in seedlings of conifers that show only fixed growth as mature trees. Free growth has been reported in firs and spruces up to 10 years old (Pollard and Logan, 1976). The proportion of long shoots (free growth) to short shoots (fixed growth) changes with tree age, with the older trees generally producing more short shoots (Kozlowski and Clausen, 1966b). By the time trembling aspen trees were six years old, long shoots comprised only 13% the canopy and there were no long shoots at all in 52-year-old trees (Pollard, 1970).

Abnormal Late-Season Shoots

Some trees produce an abnormal late-season burst of shoot growth from opening of recently formed buds

FIGURE 3.12. Normal and abnormal late-season shoots of jack pine. A: Normal shoot and winter bud. B: Proleptic shoots; three lateral buds at the base of the terminal bud cluster lost dormancy and expanded after completion of normal seasonal growth. From Rudolph (1964). C: Typical Lammas shoot; all but the upper part of the Lammas shoot is bare of needles.

that are not expected to open until the following year. The two main types of late-season shoots are (1) Lammas shoots, which result from elongation of terminal buds (so-called because of their appearance about the time of Lammas Day (August 1) in the liturgical year of English Christians), and (2) proleptic shoots (also called summer shoots), occurring from expansion of lateral buds at the bases of terminal buds (Fig. 3.12). Lammas and proleptic shoots may be found alone or in combination. Such late-season shoots may be shorter or longer than the early shoots of the first growth flush. Abnormal late-season shoots have been reported for many genera, including *Quercus, Fagus, Carya, Alnus, Ulmus, Abies, Pinus*, and *Pseudotsuga*.

Lammas and proleptic shoots may be under strong genetic control (Rudolph, 1964) and often are stimulated to form by abundant late-season rainfall (Hallgren and Helms, 1988). Both Lammas and proleptic shoots may be injured during the winter because they do not always harden adequately to cold. Lammas and proleptic shoots also may cause poor stem form. Stem forking is caused by proleptic shoots if a Lammas shoot does not form, and also if a Lammas shoot forms and fails to survive the winter. If both Lammas and proleptic shoots form on a branch they often compete for apical dominance.

Less well known than Lammas or proleptic shoots are late-season sylleptic shoots, which form when axillary buds of elongating shoots develop into branches before the buds are fully formed. Sylleptic shoots may form earlier than Lammas or proleptic shoots (when the normal early shoots are still expanding). Hence, sylleptic shoots may not be recognized as such.

Apical Dominance

As mentioned in Chapter 2, the terminal leader of most gymnosperms elongates more than the lateral branches below it. This produces a more or less conical tree form, often described as having excurrent branching. A few gymnosperms lack strong apical dominance. For example, second-order branches of Norfolk Island pine apparently lack the inherent capacity to assume apical dominance, and removal of the apical shoot is not followed by formation of a new leader by one of the lateral branches as occurs in pine and spruce. Rooted cuttings of this species may grow horizontally for many years. Although many pines exhibit strong apical dominance for a long time, some species, such as Italian stone pine, often lose apical dominance rather early in life (Fig. 3.13).

FIGURE 3.13. Ten-year-old Italian stone pine tree showing lack of apical dominance. Photo courtesy of A. De Phillips.

The physiological mechanism of apical dominance, an example of paradormancy, is not well understood. It is clear that apical dominance has both genetic and developmental influences, as species vary in manifestation of the phenomenon and respond to release from dominance by apical buds differently depending on age (Srivistava, 2001). Polar basipetal transport of auxin in the stem and cytokinins transported in xylem sap likely play significant roles maintaining or overcoming bud quiescence, respectively, but exactly how is not known.

The occurrence of apical dominance is very important to foresters. When apical dominance is destroyed by invasion of the terminal leader of eastern white pine by the white pine weevil (*Pissodes strobi*), one of the lateral shoots of the first false-whorl eventually assumes dominance while other shoots in the same whorl are suppressed. However, because of competition among branches, considerable time elapses before one of the false-whorl shoots establishes dominance and others are suppressed. Meanwhile the tree is degraded as a potential source of logs because of the fork that develops in the stem as a result of the injury to the terminal leader.

Loss of apical dominance in branches often is fostered by Christmas tree growers (Chapter 7, Kozlowski and Pallardy, 1997). Many conifers have long internodes in the main stem and branches and these give the tree a spindly appearance. By "shearing" or cutting back current-year shoots or by debudding shoots, Christmas tree growers inhibit shoot elongation and stimulate expansion of dormant buds as well as formation and expansion of new buds. Thus new lateral shoots form along branches and produce densely-leaved, high-quality Christmas trees.

Maximum Height

Species vary greatly in the height attained by mature trees. Among the tallest trees are species of *Sequoia* and *Eucalyptus*. Mature height of a species usually is related more to its longevity than to annual rate of shoot growth, type of shoot produced, or seasonal duration of shoot elongation. Bigtooth and trembling aspens often grow rapidly but never achieve great height because they are short lived. By comparison, some relatively slow growing but long-lived trees such as white oak often are more than 30 m tall. Great height of some species, such as Douglas-fir and Sitka spruce, is attributable to fast growth of young trees and continuation of height growth for many years in these long-lived species. Koch et al. (2004) claimed that limits to height growth in *Sequoia sempervirens* (122–130 m) were set by water stress effects on leaf expansion and photosynthesis. Water stress was imposed by the sizable gravitational component of water potential and xylem path length frictional resistance to water flow.

SHOOT GROWTH IN THE TROPICS

Shoot growth of tropical woody plants is very diverse. In general it is intermittent, with shoots expanding in one to several growth flushes during the year. Examples are cacao, coffee, olive, citrus, rubber, tea, mango, and many species of forest trees. Rigid classification of growth patterns of tropical species is difficult because they vary widely in different regions. For example, species of *Thespasia* and *Duabanga* are considered evergrowing in Singapore but deciduous in India (Koriba, 1958).

In tropical climates characterized by wet and dry seasons flushing of shoots is seasonal. In Bahia, Brazil, for example, the major flush of shoot growth of cacao occurs in September and October, and two or three minor flushes occur between November and April. In citrus there usually are two major flushes of shoot growth and from one to three minor ones, depending on location and climatic conditions. Young trees of tropical species usually flush more often than old trees. For example, young trees of litchi and cacao exhibited more shoot growth flushes than old trees (Huxley and Van Eck, 1974).

Both internode elongation and leaf expansion of many tropical woody plants can be very rapid. Bamboo may grow up to 1 m/day. Examples of rapid height growth were given by Longman and Jenik (1974):

- *Terminalia superba*, 2.8 m/year
- *Musanga cecropioides*, 3.8 m/year
- *Ochroma lagopus*, 5.5 m/year

Such high rates of growth often are determined for open-grown trees or at the forest border. Within the forest community the rates tend to be much lower and they often decline rapidly with increasing age of trees. Tropical woody plants exhibit several patterns of leaf initiation in relation to leaf expansion. For example, in *Oreopanax* most leaf primordia are formed very shortly before leaves expand (Borchert, 1969). By comparison, leaf primordia of tea are produced more or less continuously but leaves expand during intermittent growth flushes. Still another pattern occurs in *Rhizophora*, with production of leaf primordia and leaf expansion well synchronized, but with the rate of leaf initiation varying according to season (Tomlinson and Gill, 1973).

Shoot growth of pines in the tropics may be similar to their growth in the temperate zone, or it may be quite different. Normal seasonal growth typically involves recurrent flushing of a succession of false-whorls of buds on the terminal leader and lateral branches. After a period of shoot elongation, growth ceases briefly and new terminal bud clusters form. Shortly thereafter the recently formed buds expand to further lengthen the terminal leader and lateral branches. Generally from two to four such growth flushes occur annually. In contrast, some pine trees develop abnormally and grow continuously as a result of failure to set a bud cluster, the central bud of which would increase height and the lateral buds of which

FIGURE 3.14. Five-year-old normally branched trees and branchless foxtail forms of Carib pine in Malaysia. From Kozlowski and Greathouse (1970).

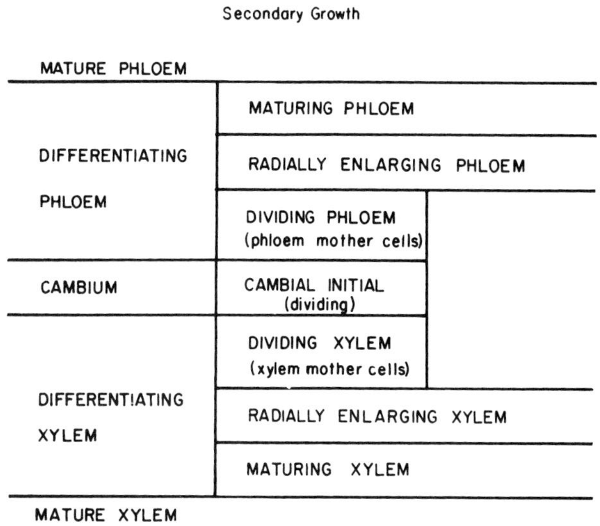

FIGURE 3.15. Terminology for describing cell types and tissues associated with cambial growth. From Wilson et al. (1966).

would elongate to form lateral branches (Lanner, 1964, 1966). Lloyd (1914) described such growth as *foxtailing* because the upper part of the abnormally elongating shoot had a conical or "foxtail" appearance (Fig. 3.14). This striking form of exaggerated apical dominance often produces trees with up to 6 m, and occasionally up to 13 m, of branchless stem. Foxtailing is a problem of variable degree wherever pines are grown in the tropics (Kozlowski and Greathouse, 1970).

CAMBIAL GROWTH

Increase in the diameter of tree stems occurs primarily from meristematic activity in the vascular cambium, a cylindrical lateral meristem located between the xylem and phloem of the stem, branches, and woody roots.

Over the years there has been spirited controversy about whether the term cambium should refer exclusively to a single layer of cambial initials or whether it should encompass both the cambial initials and their recent derivatives, the xylem mother cells and phloem mother cells. One problem is the difficulty of identifying the single (uniseriate) layer of initials. Although recognizing the existence of such a layer it often is useful to use the term "cambial zone" to refer to the entire zone of dividing cells (the uniseriate layer plus the xylem and phloem mother cells). The cambial zone in dormant trees may vary from 1 to 10 cells in the radial plane of the stem, but in growing trees the width is extremely variable. Bannan (1962) found the cambial zone to be 6 to 8 cells wide in slow-growing trees and 12 to 40 cells wide in fast-growing trees. A useful terminology for the various cell types and tissues involved in cambial activity is given in Fig. 3.15.

Cell Division in the Cambium

Two types of cell division occur in the cambium: additive and multiplicative. Additive division involves periclinal (tangential) division of fusiform cambial initials to produce xylem and phloem mother cells, which in turn divide to produce xylem and phloem cells. Multiplicative division involves anticlinal divisions of fusiform initials, which provide for circumferential expansion of the cambium.

Production of Xylem and Phloem

Following winter dormancy the cambium of temperate zone trees is reactivated to produce xylem inwardly and phloem outwardly. New annual increments of xylem and phloem are thus inserted between old layers of these tissues, causing the stem, branches, and major roots to increase in thickness.

Most investigators agree that cambial reactivation occurs in two stages, which involve change in appearance of the cambium (change in color, translucence, slight swelling) (Evert, 1960, 1963; Deshpande, 1967), followed by mitotic activity that produces cambial derivatives. As the second phase begins, the first few cell divisions may be scattered and discontinuous at different stem levels in large trees having buds on many lateral branches. Nevertheless, once seasonal cambial growth starts, the xylem growth wave is propagated downward beginning at the bases of buds (Wilcox, 1962; Tepper and Hollis, 1967).

Time of Growth Initiation and Amounts of Xylem and Phloem Produced

Undifferentiated overwintering xylem mother cells are rare except in very mild climates or under unusual circumstances (Larson, 1994). Photographs of stem transections taken during the dormant season typically show undifferentiated cambial zone mother cells abutting directly on mature xylem cells. In contrast, immature sieve elements or phloem parenchyma cells very commonly overwinter in partially differentiated states (Evert, 1960, 1963; Davis and Evert, 1968). These cells are the first to expand and mature the following spring (Larson, 1994).

Many studies suggest that cambial reactivation to produce phloem cells precedes xylem production. For example, in black locust phloem production began about a week before xylem production (Derr and Evert, 1967). In many diffuse-porous angiosperms and in gymnosperms phloem production occurs first. In trembling aspen, jack pine, red pine, and eastern white pine phloem production preceded xylem production by several weeks (Evert, 1963; Davis and Evert, 1965; Alfieri and Evert, 1968). In eastern larch, balsam fir, and black spruce much of the annual phloem increment was produced even before any xylem cells formed (Alfieri and Evert, 1973). For the first month and a half of cambial activity in pear trees, most of the cambial derivatives were produced on the phloem side (Fig. 3.16). By the middle of May, four to six rows of mature or partially differentiated sieve elements had formed. This amounted to approximately two-thirds of the total produced for the year (Evert, 1963). In horsechestnut the first cambial divisions to produce new phloem cells began five weeks before any xylem cells were cut off by the cambium (Barnett, 1992).

Patterns of cambial reactivation of tropical species are diverse (Fahn, 1990). In *Polyalthia longifolia*, phloem mother cells that went through the dormant period differentiated first. Later phloem cells formed by division of cambial initials. Subsequently phloem production stopped and xylem production began. Much later production of xylem stopped and phloem production resumed (Ghouse and Hashmi, 1978, 1979). *Avicennia resinifera* and *Bougainvillea* spp., which form successive cambia, produce alternating bands of xylem and phloem (Studholme and Philipson, 1966; Esau and Cheadle, 1969). In the evergreen species, *Mimusops elangi*, the first cambial derivatives formed on the xylem side (Ghouse and Hashmi, 1983).

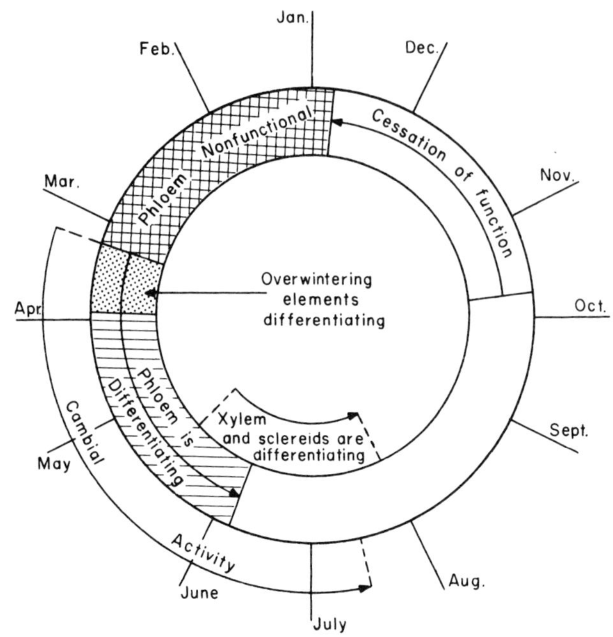

FIGURE 3.16. Seasonal changes in cambial activity of pear trees. From Evert (1960). Originally published by the University of California Press; reprinted by permission of the Regents of the University of California.

By the end of the growing season the number of xylem cells cut off by the cambium greatly exceeds the number of phloem cells produced. This is so even in species in which initiation of phloem production precedes initiation of xylem formation. In white fir the xylem and phloem cells were produced in a ratio of 14 to 1 (Wilson, 1963). In at least some species, xylem production is more sensitive than phloem production to environmental stress. Hence as conditions for growth become unfavorable, the xylem-phloem ratio often declines. In northern white cedar the xylem-phloem ratio fell from 15 to 1 to 2 to 1 as tree vigor decreased (Bannan, 1955). These relations apparently do not hold for certain subtropical species that lack recognizable annual growth rings in the phloem. In Murray red gum, for example, the ratio of xylem to phloem production changed little under different environmental conditions. A similar xylem-phloem ratio, about 4 to 1, was found for both fast-growing and slow-growing trees (Waisel et al., 1966).

Differentiation of Cambial Derivatives

After xylem and phloem cells are cut off by the cambial mother cells they differentiate in an ordered sequence of events that include cell enlargement, secondary wall formation, lignification, and loss of protoplasts (Fig. 3.17). These events do not occur stepwise, but rather as overlapping phases. For example, secondary wall formation often begins before growth

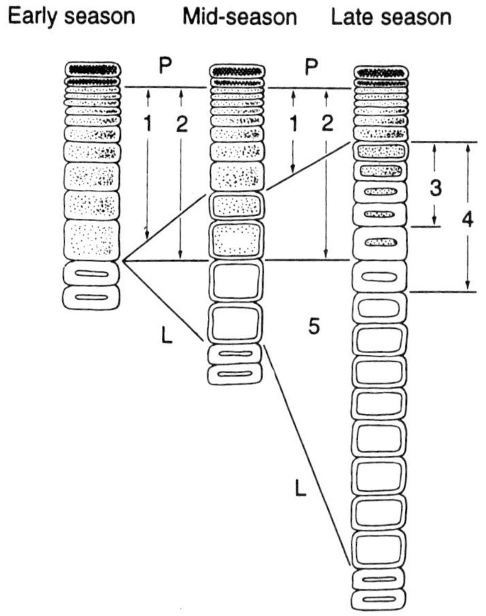

FIGURE 3.17. Variations in radial files of tracheids of red pine at different times during the growing season. 1: Primary wall zone; 2: Cytoplasm zone; 3: Flattened latewood cells; 4: More latewood cells; 5: Mature earlywood; P: Phloem; L: Latewood of the preceding year. From Whitmore and Zahner (1966).

of the primary wall ends. During cell differentiation most cambial derivatives are altered morphologically and chemically into specialized elements of various tissue systems.

Cambial derivatives that are cut off on the inner side of the cambium to produce xylem may differentiate into one of four types of elements: vessel members, fibers, tracheids, or parenchyma cells. Vessel members, tracheids, fiber tracheids, and libriform fibers develop secondary walls and the end walls of vessels become perforated, but the derivatives of ray initials change little during differentiation. However, the ray tracheids of gymnosperms are greatly altered as they develop secondary walls and lose their protoplasts. Changes in cell size also vary appreciably among different types of cambial derivatives.

The molecular and cellular events that direct xylem element growth and differentiation are complex. For example, Allona et al. (1998) indicated that hundreds of genes are expressed in immature xylem of young loblolly pine trees. Using expressed sequence tag (EST) data derived from this tissue and publicly available sequence database information, the authors identified cDNAs representing several classes of proteins that would be expected in growing regions, including cell wall proteins, lignin biosynthetic and carbohydrate metabolism enzymes and several proteins (e.g., protein kinases and transcription factors) that may regulate cell wall synthesis.

The cell cytoskeleton likely also plays a substantial role in xylem cell growth and differentiation. Chaffey et al. (1999, 2000) discussed the possible roles of microtubules (MT) and microfilaments in cell wall development of angiosperms. Microtubules in fusiform cambial initials are arranged randomly. As fibers begin to differentiate, MT become more organized and assume a dense, single-layered helical arrangement. This pattern persists as the cell passes from primary to secondary wall deposition and orientation of MT coincides with that of cellulose microfibrils leading to the widely-held idea that MTs somehow direct the deposition of cellulose, perhaps through positioning of cellulose synthase rosettes (Chapter 7). In vessel elements MT are associated with tertiary helical thickenings, pit borders, and edges of perforation plates. Microfilaments (MF) in fusiform intitals have a loose axial orientation that persists after helically-oriented wall deposition begins.

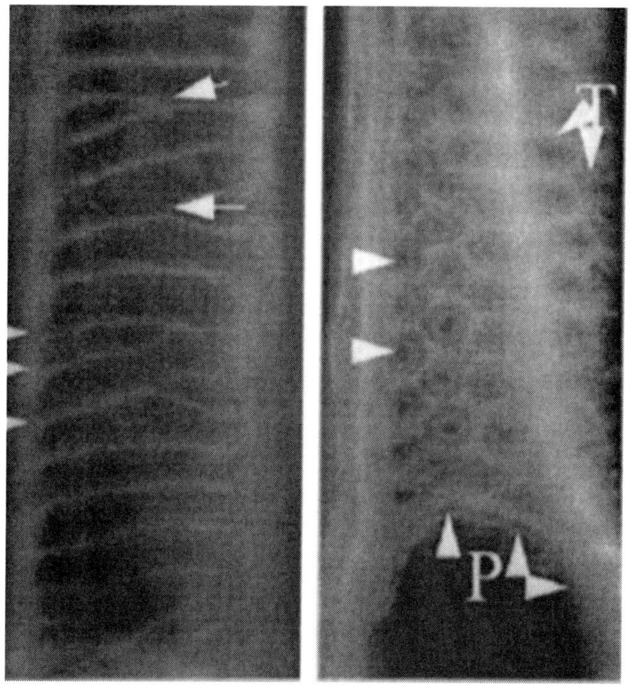

FIGURE 3.18. Epifluorescent images of the vessel elements in the late stage of secondary thickening, showing (left) narrow bands of microtubules (darts) and branching microtubules (arrows) and (right) the ring of microtubules associated with development of the perforation (P), rings of microtubules associated with bordered-pit development (darts, left), and strands of microtubules associated with tertiary thickening T. From *Planta*. A cytoskeletal basis for wood formation in angiosperm trees: The involvement of cortical microtubules. Chaffey, N., Barnett, J., and Barlow, P., **208**, 19–30, Figure 5. © 1999 with kind permission from Springer Science and Business Media.

As fibers undergo axial lengthening during this stage, MF may be involved in delivery of secretory vesicles that play a role in growth in the axial direction. In developing vessel elements, additional MF presences occur in rings that form early in the development of bordered pits, circular arrays at contact pit locations, linear arrays and cross-bridges that locate in the vicinity of tertiary helical ridges, circular arrangements that form at the edge of developing perforation plates, and meshwork structures in the plate opening proper (Fig. 3.18).

Increase in Cell Size

As noted earlier, growth begins with wall loosening, an event that causes turgor and cell Ψ to decline and creates an inward-directed Ψ gradient. As water moves in response to the gradient cells expand. The cambial derivatives that develop into vessel members expand rapidly both radially and tangentially but do not elongate appreciably. Tracheids and fibers undergo some radial expansion and elongate to varying degrees among different plant groups. In angiosperms the tracheids and fibers elongate greatly, whereas gymnosperm tracheids elongate very little. According to Bailey (1920), cambial derivatives of angiosperms elongated up to 500%; gymnosperm tracheids by only 20%. However, cambial initials are much longer in gymnosperms than in angiosperms. Hence, despite the relatively limited elongation of cambial derivatives of gymnosperms, they still are longer than those of angiosperms when both are fully expanded.

Growth in length of cambial derivatives is restricted to cell tips or at least to the apical zone. The increase in length involves intrusive growth and, when elongation of particular cells occurs, intercellular adjustments are necessary within a vertically static tissue. According to Wenham and Cusick (1975) an elongating cambial derivative secretes an enzyme that weakens the middle lamella between it and the cells adjacent to it. Where this occurs these cells, during their turgor-driven expansion, round off from each other at the corners. The tip of the elongating cell fills this space as it is created.

Hormonal Influences on Cambial Growth

Each of the major plant growth hormones is present in the cambial region and may be variously involved in the regulation of production, growth, and differentiation of cambial derivatives (Little and Savidge, 1987). As emphasized in Chapter 13 the effects of hormones are exerted by interactions among them as well as through second messengers that may or may not be plant hormones.

Auxins apparently exert a predominant role in regulating cambial activity (including mitosis of fusiform initials and differentiation of cambial derivatives) as shown by the following lines of evidence:

1. Both basipetal transport of auxin and cambial growth are reduced by disbudding of branches (Hejnowicz and Tomaszewski, 1969), defoliation (Kulman, 1971), or phloem blockage (Evert and Kozlowski, 1967; Evert et al., 1972).
2. Application of exogenous auxin to disbudded, defoliated, or phloem-severed plants increases production of xylem and phloem in gymnosperms (Larson, 1962; Little and Wareing, 1981; Little and Savidge, 1987), and angiosperms (Digby and Wareing, 1966; Doley and Leyton, 1968; Zakrzewski, 1983).
3. Production of cambial derivatives often is correlated with the amount of exogenous auxin up to an optimal level (Little and Bonga, 1974; Sheriff, 1983).
4. Exogenous auxin promotes enlargement of cambial derivatives and thickening of their walls (Larson, 1960, 1962b; Gordon, 1968; Porandowski et al., 1982; Sheriff, 1983).

As noted, basipetal, polar transport of auxin has been associated with cambial growth in both angiosperm and gymnosperm trees. A protein carrier molecule has been localized to the basipetal ends of cells conducting the auxin flux, and mutation of the gene coding for this protein causes localized proliferation of xylem near auxin sources similar to chemical inhibitors of transport (Gälweiler et al., 1998). The pattern of carrier protein distribution in developing shoots is self-reinforcing; that is, the initiation of auxin transport increases the localization of carriers to the transport pathway (Morris and Johnson, 1990). In addition to apical sources of auxin, leaves and the differentiating xylem tissues themselves may contribute to cambial auxin levels (Sundberg et al., 2000). Within the activated cambial region itself, a distinct radial profile of auxin concentrations develops on both xylem- and phloem-ward sides of the cambial zone that differs in form from that of dormant cambia, as was shown for *Pinus sylvestris* (Fig. 3.19). Similar profiles have also been observed in angiosperms (*Populus tremula* × *P. tremuloides*, Tuominen et al., 1997). Sundberg et al. (2000) suggested that the gradient of auxin concentration imposed across the radial cambial region served as a set of positional signals controlling cell growth.

Evidence for involvement of gibberellins in control of cambial growth is inconsistent. Exogenous GA_3 stimulated cambial growth and affected the anatomy of cambial derivatives in several broadleaved trees (Wareing et al., 1964; Savidge and Wareing, 1981). However exogenous gibberellins did not stimulate cambial activity in brittle willow (Robards et al., 1969),

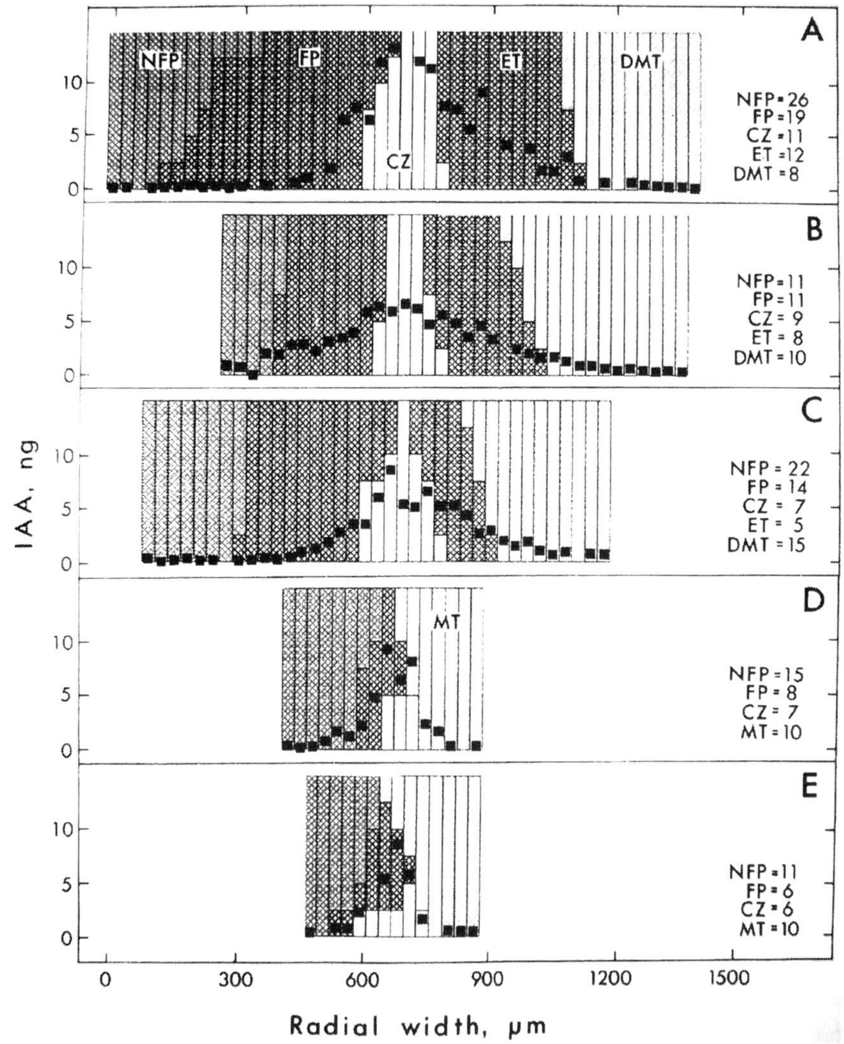

FIGURE 3.19. Radial distribution of IAA (black dots) in three *Pinus sylvestris* trees in late June (A–C) and in two different dormant trees in mid-January (D, E). Each section represents a 30 µm tangential section, and its composition of cell types from different developmental zones. MT, mature tracheids; NFP, nonfunctional phloem; FP, functional phloem; CZ, cambial zone; ET, expanding tracheids; DMT, differentiating tracheids. Number of radial file cells in each zone on right. From Uggla, C., Moritz, T., Sandberg, G., and Sundberg, B. (1996). Auxin as a positional signal in pattern formation in plants. *Proc. Natl. Acad. Sci. U.S.A.* **93**, 9282–9286. Copyright 1996 National Academy of Sciences, U. S. A.

apple (Pieniazek et al., 1970), or beech (Lachaud, 1983).

Gibberellins stimulated cambial growth in certain conifers (Little and Loach, 1975; Ross et al., 1983). Little and Savidge (1987) considered the role of gibberellins in regulation of cambial growth of conifers as unclear. Wang et al. (1992) reported that $GA_{4/7}$ increased tracheid production in Scotch pine seedlings, with the effect mediated through an increase in IAA in the cambial region.

A few studies showed that exogenous cytokinins, alone or together with IAA, accelerated production of xylem and phloem or ray tissues (Casperson, 1968; Philipson and Coutts, 1980a; Zakrzewski, 1983). Additionally, both auxin (NAA) and cytokinin (benzyladenine) were required for tracheary element differentiation in tissue cultures using the model plant *Zinnia* (Shinohara et al., 2000). In other studies exogenous cytokinins did not stimulate cambial growth (Zajaczkowski, 1973; Wodzicki and Zajaczkowski, 1974; Little and Bonga, 1974).

Although ABA appears to be involved in control of cambial growth there is some uncertainty about its precise role. Some evidence indicates that ABA inhibits cambial activity (Little and Eidt, 1968; Jenkins, 1974; Little, 1975). The formation of latewood has been attributed to increased ABA in the cambial zone (Jenkins and Shepherd, 1974; Wodzicki and Wodzicki, 1980). Late in the growing season ABA caused a decrease in tracheid enlargement in Monterey pine (Pharis et al., 1981). Nevertheless, actual involvement of ABA in the earlywood–latewood transition has been questioned (Little and Wareing, 1981; Savidge and Wareing, 1984).

Several studies suggest that ethylene plays a role in the hormonal complex that regulates the amounts of xylem and phloem cells produced and the anatomy of cambial derivatives. Involvement of ethylene in regulating cambial growth is emphasized by the following evidence:

1. Application of ethrel to stems accelerates xylem and phloem production in several species of trees, including Monterey pine (Barker, 1979), eastern white pine (Brown and Leopold, 1973; Telewski and Jaffe, 1986), Aleppo and Japanese red pines (Yamamoto and Kozlowski, 1987b,c), apple (Robitaille and Leopold, 1974), brittle willow (Phelps et al., 1980), American elm (Yamamoto et al., 1987), and Norway maple (Yamamoto and Kozlowski, 1987d).
2. Exogenous ethylene stimulates growth of bark tissues more than xylem tissues (Yamamoto and Kozlowski, 1987a,b; Yamamoto et al., 1987a,b,c).
3. Treatment of cuttings of Norway spruce seedlings with ethrel influences incorporation of carbohydrates into cell walls (Ingemarsson et al., 1991).

High concentrations of ethrel inhibited expansion of xylem cells while stimulating incorporation of cellulose in cell walls. Addition of small amounts of ethrel, which slightly stimulated ethylene emission, led to increases in the size of xylem cells, amounts of phloem tissue, and intercellular spaces in the cortex (Ingemarsson et al., 1991). The activity of several enzymes involved in lignin synthesis is stimulated in plants treated with ethylene (Roberts and Miller, 1983).

Eklund and Little (1995) reported that ethylene evolution was not specifically associated with IAA-induced tracheid production in balsam fir. Ethylene did not mimic the promotion effect of IAA on tracheid production. Ethylene could promote production of tracheids but only when its application was followed by unphysiologically high concentrations in the cambial region, which, in turn, induced accumulation of IAA. The weight of evidence suggests that ethylene has a synergistic effect on xylogenesis in the presence of auxin.

Cambial Sensitivity to Growth Hormones

The effects of hormone levels on cambial growth vary with seasonal changes in cambial sensitivity to the presence of hormones. Some evidence shows that the actual onset of cambial activity is not controlled by changes in IAA level alone. Uggla et al. (1996) noted that endogenous IAA levels in the dormant cambial zone of *Pinus sylvestris* trees were often as high as those in actively growing cambia (Fig. 3.19). Once the cambial cells regain capacity to respond to a stimulator, auxin levels and possibly radial profiles of auxin then appear to regulate cambial activity. The capacity of cambial cells to respond to IAA is restored before new cambial derivatives are cut off by the cambium (Little and Savidge, 1987; Lachaud, 1989). During the period of intense cambial activity, auxin concentrations control the intensity of mitosis, but not the duration of cambial growth (Little and Savidge, 1984; 1987; Lachaud, 1989).

Tracheid production was stimulated less by application of IAA to Scotch pine shoots late in the growing season than it was by earlier application (Wodzicki and Wodzicki, 1981). Little (1981) found similar rates of auxin transport in trees with resting or active cambia. Additionally, the auxin concentration at the beginning of seasonal cambial activity often is not very high and it may not decline markedly during the growing season. Such observations indicate that the sensitivity of cambial cells to IAA varies during the year and that an autumnal resting state reflects a lack of capacity of cambial cells to respond to IAA. The resting state presumably changes to a quiescent state in the winter and, after adequate chilling, cambial cells regain their capacity to respond to IAA. Creber and Chaloner (1984) emphasized that woody plants of the temperate zone develop an endogenous rhythm that is locked into the annual temperature and photoperiodic cycle. This rhythm is characterized by cellular changes that reflect variable responses to endogenous growth regulators that stimulate cambial activity during the part of the year that is most favorable for growth.

Responses of stem tissues to ABA also vary appreciably during the season. An ABA-mediated inhibition of cambial growth more likely reflects a change in cambial sensitivity to ABA rather than an increase in ABA (Lachaud, 1989).

Apical Control of Cambial Growth

Evidence for a dominant regulatory role of apically produced growth hormones on cambial growth comes from correlations between bud growth in the spring and initiation of xylem production below buds, basipetal migration of the cambial growth wave, arrested cambial activity in defoliated or disbudded trees or below phloem-blocking stem girdles, and initiation of cambial growth with exogenous hormones. Neither the roots nor stem tissues near the base of the stem can supply enough growth hormones to sustain normal cambial growth.

Initiation and Differentiation of Cambial Growth

After bud dormancy is broken by adequate chilling, apically produced growth-promoting hormones move down the branches and stem and provide the stimulus for initiation of seasonal cambial growth. Indoleacetic acid produced in developing leaves and expanding shoots is transported basipetally at a rate of about 1 cm hr^{-1} (Little and Savidge, 1987; Savidge, 1988).

The rate of production of cambial derivatives is influenced by the level of IAA, as shown in balsam fir (Little and Bonga, 1974), Sitka spruce (Little and Wareing, 1981), and Monterey pine (Sheriff, 1983). Regulation of division of cambial cells by apically produced hormones also is shown by a reduction in both auxin content and xylem increment below phloem blockages in branches and stems.

Assuming that differentiation of vascular elements is controlled to a considerable extent by basipetal auxin flow, Aloni and Zimmermann (1983) explained changes in vessel size and frequency along the stem axis by the following:

- Auxin concentration decreases progressively down the tree axis.
- Impeding auxin transport causes local increases in auxin.
- The distance from the source of auxin to differentiating vessels determines the amount of auxin they receive.
- The higher the amount of auxin the more rapid the rate of differentiation; hence, the duration of vessel differentiation increases from shoots to roots.
- Vessel diameters are determined by the rate of differentiation, with rapid differentiation resulting in narrow vessels and slow differentiation in wide vessels. The basipetal decrease in auxin is associated with a progressive increase in vessel diameters.
- The higher the auxin concentration, the higher the frequency of vessel elements. Hence the frequency of vessels decreases from shoots to roots.

When Aloni and Zimmermann (1983) blocked the downward flow of auxin in red maple stems, the vessels above the blockage (where auxin concentration was high) were narrow and vessel frequency was increased.

The formation and development of rays appear to be regulated by hormones, with ethylene prominently involved. Lev-Yadun and Aloni (1995) suggested that ethylene originating in the xylem flows outward, disturbing radial auxin flow, and regulates both the initiation of new rays and the expansion of existing rays. The formation of new rays following stem wounding (when large amounts of ethylene accumulate) suggests that ethylene is involved in the conversion of fusiform initials to ray initials. The importance of ethylene in regulating ray development also was emphasized by Yamamoto et al. (1987), who found that application of ethrel to stems of American elm seedlings was followed by formation of abnormally wide rays comprised of unusually large ray cells.

Formation of resin ducts appears to be regulated by growth hormones. Ethephon (ethrel), which increases production of ethylene, when applied to stems of Aleppo pine and Japanese red pine seedlings, stimulated production of resin ducts in the xylem (Yamamoto and Kozlowski, 1987b,c). It appeared that when internal ethylene reached a critical level it caused a change in cambial activity from production of mostly tracheids and a few resin ducts to a higher proportion of the latter. Resin ducts also may form in response to exogenous auxin. Resin ducts did not form until several weeks after auxins were applied to Aleppo pine stems (Fahn and Zamski, 1970), but formed shortly after ethephon was applied (Yamamoto and Kozlowski, 1987b). Hence, the effects of auxin on formation of resin ducts may involve auxin-induced ethylene formation (Fahn, 1988a). Oleoresin production in red pine stems was greatly stimulated by applied ethephon, indicating that ethylene may be a predominating hormonal factor in regulation of differentiation of traumatic resin ducts as well as oleoresin synthesis (Wolter and Zinkel, 1984).

Reaction Wood

A consequence of leaning of trees is a redistribution of the amount and nature of cambial growth on the upper and lower sides of the stem and of formation of abnormal "reaction" wood. Reaction wood, which is called "compression wood" in gymnosperms (Fig. 3.20) because it occurs on the lower (leeward) side, and "tension wood" in angiosperms because it occurs on the upper side of leaning stems, is well documented in trees on persistently windy sites (Savill, 1983). Reaction wood has several commercially undesirable characteristics such as a tendency toward shrinking, warping, weakness, and brittleness, which adversely affect its utilization.

Several investigators concluded that increased cambial growth on the leeward sides of tilted stems occurs because of a high auxin gradient that causes mobilization of structural and energy resources. Many investigators attributed formation of compression wood to high auxin levels (for review see Timell, 1986, Vol. 2). Induction of compression wood often follows application of IAA to buds as well as to stems. Compression wood that was induced by applied IAA could not be distinguished from naturally occurring compression wood by appearance or physical properties (Larson, 1969). Furthermore, application of inhibitors

FIGURE 3.20. Eccentric cambial growth and compression wood in ponderosa pine. Photo courtesy of U.S. Forest Service, Forest Products Laboratory.

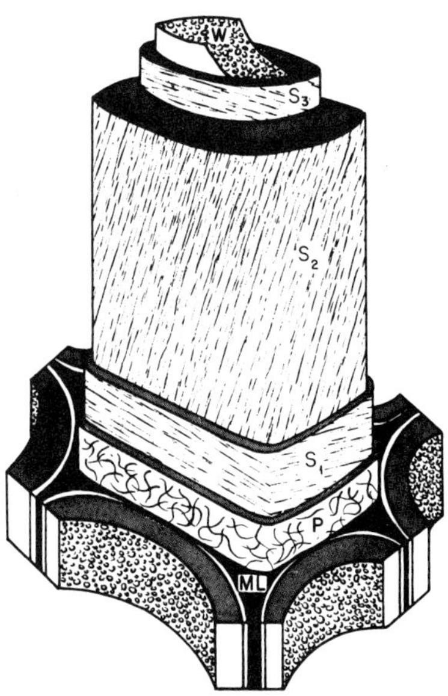

FIGURE 3.21. Diagram of cell wall organization of a mature tracheid. ML: middle lamella; P: primary wall; S_1, S_2, and S_3: secondary walls; W: warty layer. From Coté (1967), *Wood Ultrastructure*, University of Washington Press, Seattle, Washington.

of auxin transport to inclined stems stops formation of compression wood (Phelps et al., 1977; Yamaguchi et al., 1983).

Formation of tension wood is associated with auxin deficiency. This is shown by lower concentrations of auxin on the upper sides of tilted stems (Leach and Wareing, 1967), inhibition of tension wood formation by auxins applied to the upper sides of tilted stems, and induction of tension wood by applied auxin antagonists such as TIBA (2,3,5-triiodobenzoic acid) or DNP (2,4-dinitrophenol) (Morey, 1973). The few and small vessels in tension wood also are associated with auxin deficiency. Decreases in vessel frequency from the leaves to the roots are correlated with a gradient of decreasing auxin concentration (Zimmermann and Potter, 1982).

Cell Wall Thickening

The walls of most mature xylem cells consist of a thin primary wall and a thick secondary wall. The primary wall forms at cell division in the cambium and encloses the protoplast during surface growth of the cell. The secondary wall forms after completion of surface growth.

The primary wall is not lamellated and tends to have loosely packed microfibrils. The secondary wall is deposited in three layers (designated S_1, S_2, and S_3) on the primary wall (Fig. 3.21), and is made up of a fibrillar component, cellulose, as well as encrusting substances, primarily lignin, but also hemicellulose, pectin, and small amounts of proteins and lipids. The cellulose microfibrils are laid down by successive deposition of layer upon layer (apposition) in a pattern, as was noted earlier, that has a definite relationship with the cell cytoskeleton (Chaffey et al. 1999, 2000). In contrast, lignin and other encrusting substances are deposited within the cellulose framework by intussusception (Torrey et al., 1971). The amount of lignification varies among plant groups and species, cells, and different parts of the same cell. Mature gymnosperm tracheids and angiosperm vessels are heavily lignified but fiber tracheids and libriform fibers of angiosperms show little deposition of lignin.

Unlike most xylem cells that develop thick cell walls, most phloem cells remain soft-walled and eventually collapse or become greatly distorted. Phloem fibers, however, develop secondary walls.

Loss of Protoplasts

Final stages of maturation of xylem cells such as vessel elements and tracheids involve breakdown (autolysis) of protoplasm via programmed cell death. All the important organelles are present during secondary wall formation, but in the final stages of maturation the vacuolar membranes, cytoplasmic organelles, cytoplasm, and plasmalemma undergo autolysis, and the nucleus disintegrates, thus terminating the life of tracheid protoplasts. Lysis of protoplasts of differentiating xylem elements occurs rather rapidly. For example, in eastern hemlock tracheids it occurred in approximately four days (Skene, 1972); in Scotch pine tracheids in two to five days (Wodzicki and Brown, 1973).

Formation and Development of Rays

Some rays form near the pith from interfascicular parenchyma, connecting the pith with the cortex. Other rays originate from the cambium (Lev-Yadun and Aloni, 1995). The rays in the secondary xylem and phloem are produced by periclinal divisions of ray cell initials of the cambium. Ray development involves periodic changes in their number, height, and width as the tree grows.

Most new rays form in very young trees when peripheral expansion of the cambium is maximal. Thereafter the number of rays more or less stabilizes (Larson, 1994). In general ray height increases with tree age as a result of transverse divisions of ray cell initials, fusion of adjacent rays, or addition of segments from fusiform initials. When environmental stresses reduce the rate of cambial growth, the height of xylem rays may be reduced. The reduction occurs when rays are split by intrusion of fusiform initials into rays or as ray initials revert to fusiform initials. Some rays, especially small ones, simply disappear (Larson, 1994).

Ray widths in temperate-zone angiosperms often increase by anticlinal division of initial cells within rays or by merging of rays (Larson, 1994). The widths of multiseriate rays, especially, increase with increasing tree age in very young trees and stabilize thereafter. In tropical trees ray widths tend to increase progressively as trees age (Iqbal and Ghouse, 1985a). Ray widths in tropical trees also vary seasonally, increasing in the quiescent season and decreasing during the season of active growth when rays split (Iqbal and Ghouse 1985b; Larson, 1994).

Expansion of the Cambium

As a tree grows it becomes necessary for the cambial sheath to increase in area. This is accomplished by adding new cambial cells in two ways: (1) by increasing the length of the cambial sheath through addition of new cells from the procambium behind the root and stem tips, and (2) by increasing the circumference of the cambial sheath by anticlinal division of fusiform cambial cells, either by longitudinal division (in angiosperms with storied cambia) or by pseudotransverse division (i.e., division along an oblique anticlinal wall; found in angiosperms with nonstoried cambia and in gymnosperms). In addition a small percentage of fusiform cambial cells divide anticlinally to produce segments off the sides of the initials (Fig. 3.22). Some of the sister initials divide periclinally for a long time, whereas others fail by elimination, segmentation, or conversion to ray initials (Larson, 1994).

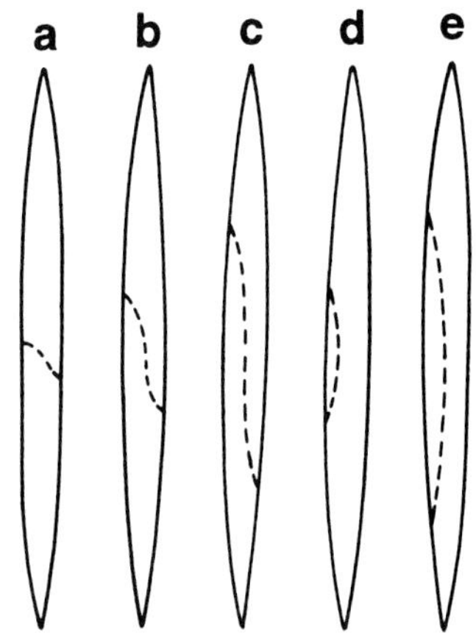

FIGURE 3.22. Tangential view of fusiform initial cells showing various types of anticlinal division involved in multiplication of cambial cells. (a–c) pseudotransverse division; (d, e) lateral division. From Bannan (1967).

The rate and distribution of pseudotransverse divisions of cambial initials change with age of trees. In rapidly-growing young trees pseudotransverse divisions occur frequently throughout the growing season and the rate of survival of new initials is very high. Under these conditions both the new fusiform initials and cambial derivatives are short. As trees age, however, pseudotransverse divisions occur less often and are confined to the latter part of the growing season, by which time most of the annual ring of wood has formed. Furthermore, the rate of survival of newly formed cambial initials declines in older trees. Since the surviving initials have more space into which to expand, longer fusiform initials and cambial derivatives gradually are produced until a maximum size characteristic for a species is reached.

Variations in Growth Increments

Cambial activity is not continuous in space or in time. It may be general over a tree at certain times and at other times, as during droughts, it may be localized. Hence, trees produce a sheath of xylem that varies in thickness at different stem and branch heights and, at a given stem height, it often varies in thickness on different sides of the tree.

Sometimes trees that form complete xylem increments in the upper stem do not form any annual xylem rings in the lower stem. Such "missing rings" are especially characteristic of very suppressed or old trees. Also, as branches undergo successive suppression by new branches above them, those in the lower stem often fail to produce xylem down to the point of juncture with the main stem.

Occasionally, more than one growth ring is produced in the same year (Fig. 3.23). Such "false" or "multiple" rings often occur when cambial activity stops during an environmental stress, such as drought, and resumes when the stress is alleviated. When this happens, alternations of earlywood and latewood are repeated. False rings also may result from injuries by insects, fungi, or fire.

The cambium may be dead or dormant on one side of a tree, leading to production of partial or discontinuous rings that do not complete the stem circumference. Discontinuous rings often are found in overmature, heavily defoliated, or suppressed trees, senescing branches, and stems of trees with one-sided crowns. In the last group, the ring discontinuities generally occur on the stem radius below the underdeveloped crown. Discontinuous rings are especially common in woody roots, which often are eccentric in cross section.

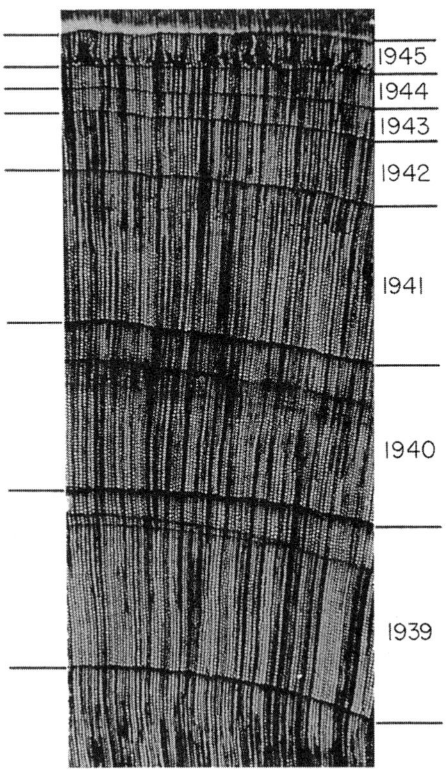

FIGURE 3.23. Multiple rings formed during 1939 and 1940 in a branch of Arizona cypress. From Glock et al. (1960).

Seasonal Duration of Cambial Growth

The time of year during which the cambium is active varies with climate, species, crown class, and different parts of stems and branches. In a given region seasonal cambial growth of evergreens as a group usually continues for a longer time than it does in deciduous trees (Winget and Kozlowski, 1965). The cambium of a suppressed tree may produce xylem for only a fraction of the time during which the cambium of an adjacent dominant tree of the same species remains active (Kozlowski and Peterson, 1962). In the same tree seasonal cambial growth below the shoot apex begins at about the same time that shoots begin to grow, but cambial growth often continues for a long time after shoot elongation ceases. Seasonal cambial growth continues for a longer time in the upper stem than in the lower stem. It should be remembered that cambial growth is very responsive to environmental stresses. For example, it often stops during a drought and resumes after a rain. The control of cambial growth is discussed in more detail previously and in Chapters 3 and 5 of Kozlowski and Pallardy (1997).

Anomalous Cambial Growth

Most information on cambial growth characteristics has been obtained from studies of temperate zone trees. Secondary growth of such species is considered to be "normal." In a number of species of tropical trees and lianas, cambial growth often deviates from the normal pattern. For example, Obaton (1960) reported anomalous cambial growth in 108 species of woody lianas in 21 families of plants in western Africa. Anomalous or atypical cambial growth may be found in some plants in which the cambium is in normal position. In other plants the cambium is atypically located. Often anomalous cambial growth is the result of unequal activity of various cambial segments, changes in amounts and position of xylem and phloem, or production and activity of successive concentric cambia. The various forms of anomalous cambial growth are difficult to classify into distinct groups because of their diversity and intergradation with normal forms of cambial growth. Anomalous cambial growth is discussed in detail by Carlquist (1988, pp. 256–277) and by Fahn (1990, pp. 397–407).

Sapwood and Heartwood Formation

Heartwood usually begins to form at a stem height of 1 to 3 m and it tapers from the height of initiation

TABLE 3.2. Variations among Crown Classes in Width of Sapwood at Different Stem Heights of Douglas-Fir[a]

Crown class	Average diameter at breast height (cm)	Mean width of sapwood (cm)		
		Top	Middle	Base
Dominant	29.0	3.81	3.12	4.39
Codominant	20.3	2.51	2.18	2.69
Intermediate	18.3	1.85	1.75	2.16

[a]From Wellwood (1955).

toward the crown and base of the tree. However, the degree of taper toward the crown varies in different trees. Heartwood also forms in the woody roots of many species but only in the region near the stem wood (Hillis, 1987). Once heartwood begins to form it increases in diameter throughout the life of the tree.

The amount and rate of sapwood and heartwood formation vary greatly with tree species, tree age, rate of growth, environmental conditions, and cultural practices. Heartwood formation usually begins in some species of *Eucalyptus* at about 5 years; in several species of pine at 15 to 20 years; in European ash at 60 to 70 years; and in beech at 80 to 100 years (Dadswell and Hillis, 1962). In a few species (e.g., *Alstonia scholaris*) heartwood may never form.

The width and volume of sapwood of a given species usually are greater in rapidly-grown than in slow-grown trees. For example, the sapwood bands were wider in dominant Douglas-fir trees than in suppressed trees (Table 3.2). Fast-grown eucalypts typically have wider than normal sapwood width (Hillis, 1987). There are exceptions, however. In both jack pine and tamarack the growth rate was not correlated with sapwood width, although the number of rings of sapwood was correlated with the sapwood growth rate (Yang et al., 1985).

Wound-Induced Discoloration

Many investigators have observed a central core of discoloration in tree stems following wounding or dying of branches. Wound-induced discoloration results from cellular changes and may or may not be associated with microorganisms. Such discoloration, which may be superimposed on normal heartwood, has been variously called heartwood, false heartwood, pathological heartwood, wound heartwood, and blackheart. Because of its similarity in color to heartwood, wound-induced discoloration has sometimes been considered to be an extension of normal heartwood into the sapwood. However, there are important differences between normal heartwood, formed from internal stimuli associated with aging, and the discoloration of sapwood induced by wounding.

Although normal heartwood continues to increase in diameter throughout the life of a tree, wound-induced discoloration of wood is limited to the diameter of the tree when it was wounded or the branch died. Normal heartwood has a similar color throughout the stem cross section and has a chemical composition that usually is constant in a given species. In injured and discolored wood the amount of extractives is higher than in the sapwood, the amounts of deposited substances that cause darkening are higher than in heartwood, and the extractable materials often differ quantitatively. Hart (1968) found that normal heartwood and discolored sapwood in the vicinity of wounds differed in color, water content, frequency of amorphous deposits, percent of material soluble in water or 1% NaOH, ash content, and pH. These differences emphasize that when cells die following injury that leads to discoloration of tissue, such tissue should not be considered an example of precocious development of normal heartwood.

Changes during Heartwood Formation

The most critical change during conversion of sapwood into heartwood is the programmed death of ray and axial parenchyma cells. Other important changes include a decrease in the metabolic rate and enzymatic activity, starch depletion, darkening and accumulation of extractives in the xylem, anatomical changes such as an increase in aspiration of pits in gymnosperms and formation of tyloses in angiosperms, and changes in moisture content.

Death of Parenchyma Cells

Patterns of mortality of parenchyma cells differ among species. Nobuchi et al. (1979) and Yang (1993) classified conifers into three broad types on the basis of differences in their patterns of death of parenchyma cells:

Type I: All parenchyma cells survived from the cambium inward to the transition zone. Examples are Japanese red pine, eastern white pine, and oriental arborvitae.

Type II: Some dead ray parenchyma cells were present in the middle or inner sapwood and the number of dead cells increased from the middle of the sapwood band toward the sapwood-heartwood boundary. Examples are Japanese cryptomeria, jack pine, and trembling aspen.

Type III: Some dead ray parenchyma cells were present in the outer sapwood and the number of dead cells increased toward the sapwood-heartwood boundary. Examples are balsam fir and black spruce.

Deposition of Extractives

During heartwood formation a wide variety of extractive substances, including tannins, dyestuffs, oils, gums, resins, and salts of organic acids accumulate in cell lumens and walls. Polyphenols, aromatic compounds with one or more hydroxyl groups, are among the most important heartwood extractives (Fig. 3.24). Deposition of extractives results in a dark-colored wood. However, the color of heartwood varies among species and with the types of compounds deposited. Heartwood is black in ebony, *Diospyros ebenum*, and *Dalbergia melanoxylon*; purple in *Peltogyne pubescens*; red in *Dalbergia variabilis* and *Caesalpinia* spp.; and yellow in *Chlorophora tinctoria* (Hillis, 1987) and *Cladrastis kentukea*. The color of the heartwood of true firs, hemlock, and poplar often does not differ from that of the sapwood. In the same genus some species form colored heartwood (e.g., *Betula alleghaniensis*) and others do not (e.g., *Betula verrucosa*).

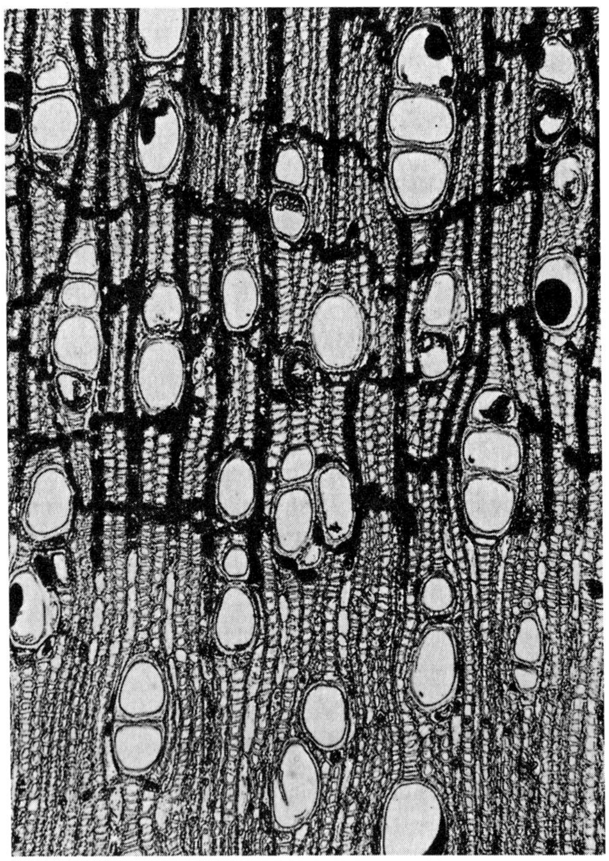

FIGURE 3.24. Cross section of stem of *Excoecaria parvifolia* showing formation of polyphenols at the sapwood-heartwood boundary. Photo courtesy of R. K. Bamber.

Two patterns of extractive deposition have been proposed. In the *Robinia*–Type (Type 1) extractives accumulate in the relatively narrow transition zone between sapwood and heartwood. In the *Juglans*-Type (Type 2) heartwood phenolic precursor compounds accumulate as the sapwood ages. These substances are subsequently converted in the transition zone (Burtin et al., 1998). In this second type, phenolic extractives are also synthesized from sugars in the transition zone (Beritognolo et al., 2002).

In *Robinia*-Type heartwood formation, extractives are synthesized from sucrose imported after conversion from starch in the outer sapwood (Magel et al., 1994). In the transition region sucrose synthase activity is elevated, suggesting that sucrose hydrolysis provides carbon skeletons for extractive synthesis. Phenyalanine ammonia lyase (PAL), the enzyme that constitutes the first committed step leading to flavonoid extractives, exhibits high activity in the transition zone, especially during the autumn peak of heartwood formation (Magel, 2000). A key downstream enzyme in the synthesis of flavonoid compounds (chalcone synthase, CHS) also has its highest autumnal activity in the same region. Molecular analysis has shown that the peak in CHS activity is associated with gene expression, whereas the elevated PAL activity seems more linked with post-translational modification or regulation of metabolite feedback inhibition of the enzyme (Magel and Huebner, 1997; Magel, 2000). *De novo* synthesis of phenolics in *Juglans*-Type heartwood formation appears to involve similar patterns of enzyme synthesis and regulation of CHS and PAL (Beritognolo et al., 2002).

Pit Aspiration

In gymnosperms, pits apparently aspirate (the torus is pushed against the pit border) where a tracheid wall is located between a tracheid containing water and another tracheid containing gas (Chapter 11). Aspiration of bordered pits in the innermost sapwood or outer part of the transition zone before a decrease in moisture content occurs is an important feature of heartwood formation. In Monterey pine the percentage of aspirated pits in the earlywood of the sapwood, middle of the transition zone, and the heartwood were 30, greater than 90, and 96%, respectively (Harris, 1954). In Japanese cryptomeria pit aspiration increased dramatically from 10 to 60% at the border of the sapwood and transition zone (Nobuchi and Harada, 1983). In addition to becoming aspirated near the sapwood-heartwood boundary, the bordered pit pairs often become encrusted with extractable materials, further decreasing permeability of the wood to fluids (Yamamoto, 1982).

Formation of Tyloses

Balloon-like structures called tyloses develop when turgor pressure causes part of the protoplast of a parenchyma cell to expand outward through a pit pair into the lumen of an adjoining cell. Tyloses are common in xylem vessels of many genera of angiosperms including *Populus, Rhus, Robinia, Morus, Sassafras, Catalpa, Juglans,* and *Quercus,* but they never occur in many other genera. Tyloses block water movement in vessels and limit water transport to that portion of the xylem where they are not prevalent.

Tyloses may be found in normal sapwood or in response to wounding, invasion by fungus pathogens, or virus infection (Beckman et al., 1953). Beckman (2000) proposed that pathogen-induced tyloses form under the influence of elevated auxin and perhaps ethylene levels so as to physically block fungal growth within xylem vessels. In many angiosperm genera the formation of tyloses is an important feature of the changeover of sapwood to heartwood. Hence, species that normally produce tyloses in the sapwood have more of them in the heartwood. During heartwood formation both tyloses and gums originate largely in the ray parenchyma cells rather than in axial parenchyma cells (Chattaway, 1949). The time of development of tyloses may vary in different climatic regions. In a cold temperate zone they began forming in August and stopped developing during the dormant and next growing season. In a warm temperate zone tyloses began forming in September and matured in December (Ishida et al., 1976; Fujita et al., 1978).

Moisture Content

In most gymnosperms the moisture content is higher in the sapwood than in the heartwood. Even within the sapwood there often is a steep moisture gradient, sometimes over one or two annual rings.

In angiosperms the moisture content across a stem cross section varies among species and with season (Chapter 12). In many species the moisture content of the heartwood differs little from that of the sapwood. However, in several genera (*Betula, Carya, Eucalyptus, Fraxinus, Juglans, Morus, Nyssa, Populus,* and *Quercus*) the heartwood contains more moisture than the sapwood, although this is not true of all species within these genera.

Variations in moisture content between the sapwood and heartwood are modified by seasonal influences on dehydration and rehydration of stem tissues. For example, in Japanese beech the moisture content was higher in the sapwood than in the heartwood during the winter but lower in the summer (Yazawa, 1960). By comparison, the moisture content of the heartwood of Manchurian ash, *Populus maximowiczii,* and harunire (*Ulmus davidiana*) was higher than that of the sapwood during both the winter and summer (Yazawa and Ishida, 1965; Yazawa et al., 1965).

Wounding and Wound Healing

Tree wounds are invasion routes for pathogenic organisms. Wounds result from broken branches, tops, or roots and by exposure of xylem as a result of mechanical wounds, animal wounds, fire wounds, and so on. The severity of the wound and vigor of the host influence the rate and effectiveness of plant response to wounding. Wounds that break the bark and only slightly injure the cambium generally heal rapidly.

The living portion of the sapwood shows a dynamic response to wounding and discolored wood containing various extractives forms around the area containing microorganisms. Such "protection wood" resists invasion by microorganisms. When chemical protective barriers are overcome by microorganisms, the host tree often responds by compartmentalizing the wounded tissues. Barriers to invasion include plugging of vessels in some species, formation of tyloses noted previously in others, and production of thick-walled xylem and ray cells by the cambium. These changes create a barrier wall that separates the wounded tissues from those formed after wounding. Invading microorganisms then spread along the path of least resistance, vertically through the compartmentalized tissues. If a tree is wounded again, another barrier wall forms and surrounds the inner compartments.

The physiology and biochemistry of the reactions to wounding and infection are complex and poorly understood. As mentioned, there are extensive changes in protein metabolism, increases in the number of mitochondria and in respiration, increases in enzymes, especially polyphenoloxidases and peroxidases, and in ethylene production in both the injured tissue and in adjacent uninjured tissue. There also often are changes in concentrations of hormonal growth regulators (Dekhuijzen, 1976). Some aspects of these responses are discussed by Uritani (1976).

Healing involves closure of a wound as well as walling off of infected and invaded tissues associated with the wound. Many large wounds on old trees never close; yet they heal from the inside. Healing of deep stem wounds involves sequential production of callus tissue and formation of a new vascular cambium by conversion of callus cells to cambial cells. A phellogen (cork cambium) also is regenerated during the wound healing process. Abundant callus formation usually is associated with healing of longitudinal frost cracks in tree stems. Such wounds may recurrently open and close in response to sudden temperature decreases and increases (Chapter 5, Kozlowski and Pallardy, 1997). During the rehealing phase, vertically oriented protrusions of abundant callus tissue, the "frost ribs," often develop along the edges of the wound.

Although the origin of callus may vary considerably among species, in most woody plants the vascular rays make the major and sometimes the only contribution to callus formation. Occasionally other components of the cambial zone contribute to production of callus tissue. Thus, wound callus may be produced by parenchyma of xylem rays and phloem rays, undifferentiated xylem cells, and cortical tissues.

The amount of callus formed during healing of stem wounds may vary with the size of the wound. Callus formation in shallow wounds sometimes is restricted or absent. The amount and rate of callus production following wounding also differ among species of plants. For example, callus was produced earlier and much more abundantly by injured *Populus* and *Acer* stems than by those of *Pyrus* (Soe, 1959). Formation of a new vascular cambium was independent of the amount or rate of callus production. Formation of a phellogen preceded regeneration of the vascular cambium. The new phellogen became active as soon as the callus pad was well developed.

Initiation of new cambium in wounded trees often has been associated with the original cambium at the edges of a wound as in *Hibiscus* (Sharples and Gunnery, 1933) and *Populus* (Soe, 1959). In some species, however, regeneration of a new cambium does not depend on the position or presence of an existing cambium at the sides of the wound. For example, in wounded Oriental trema stems, a new vascular cambium was differentiated in the middle of the callus (Noel, 1968).

The rate of closure of wounds is positively correlated with the rate of cambial growth. Wounds heal most rapidly in vigorous trees. Since cambial growth of trees in the north temperate zone occurs primarily during May, June, and July, wounds made prior to May heal rapidly; those made after July heal slowly. Wound shape has little effect on the rate of healing.

A number of wound dressings have been used on tree wounds over the years, including asphalt-type materials, shellac, house paints, and petrolatum. Their usefulness has been widely debated. Neely (1970) concluded that wound dressings had no significant effect on increasing the rate of wound healing. In fact, a petrolatum dressing reduced the rate of healing. Shigo and Wilson (1977) found no significant effect of several wound dressings on the rate of wound closure, vertical extensions of discolored and decayed wood, or presence of decay fungi. In addition, the dressings did not prevent infection by decay fungi. Applications of shellac or a resin emulsion with or without added fungicide to wounds of Norway maple, honeylocust, eastern white pine, and eastern hemlock did not systematically reduce discoloration or decay or hasten wound closing (Hudler and Jensen-Tracy, 2000). However, all these treatments reduced the length of the discoloration column of honeylocust trees wounded in June. Shigo and Wilson (1977) acknowledged that wound dressings have a strong psychological appeal but their usefulness in accelerating wound healing remains questionable.

In another approach, McDougall and Blanchette (1996) found that prompt wrapping of fresh wounds with polyethylene plastic sheeting for a week or two reduced dieback and promoted would closure in *Populus tremuloides* and *Acer rubrum* trees, but not in *Betula papyrifera*. Wrapping, even for periods as long as two years, did not increase colonization of trembling aspen wounds by pathogens. The authors suggested that plastic wraps might prevent desiccation of surface tissues, thus accelerating callus growth, or might otherwise act by elevating wound-induced ethylene in adjacent tissues.

ROOT GROWTH

The seed contains a radicle or root meristem in the embryo from which the first tap root develops. The first root branches and elongates to produce a ramified root system, or it may die back. Whereas lateral shoots on stems originate from peripheral tissues, lateral roots arise from the deep-seated outer layer of the stele known as the pericycle (Fig. 3.25). During initiation of lateral roots, several pericyclic cells become meristematic and divide periclinally to produce cells that then divide, both periclinally and anticlinally, to form a protruding lateral primordium that grows out through the endodermis, cortex, and epidermis. Before a lateral root breaks through the surface tissues of the main root, it develops a well-defined apical meristem and rootcap. Both digestion of surrounding tissue and mechanical pressure appear to be involved in the outgrowth of lateral roots through the cortex.

Studies of hormonal influences on root growth have emphasized the importance of auxin, ethylene, and cytokinins in root growth and morphology. Buildup of auxin stimulates cell division in the cell columns of root tips (Casson and Lindsey, 2003). A role for auxin

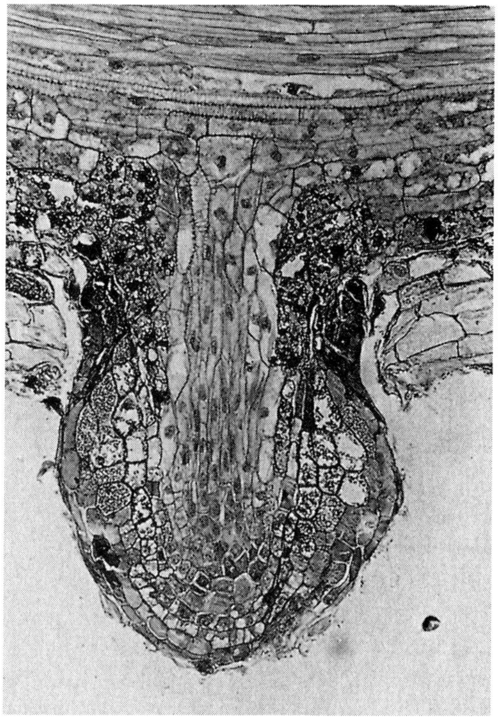

FIGURE 3.25. Late stage of formation of lateral root of red pine. Photo courtesy of H. E. Wilcox.

in promotion of *de novo* root formation is widely acknowledged and applied in production of rooted cuttings of woody plants (Kozlowski and Pallardy, 1997; Casson and Lindsey, 2003). Auxin routed through the root tip is also thought to be important in lateral root formation. Ethylene suppresses root growth and the fact that auxin stimulates the synthesis of ethylene provides the possibility of a negative feedback loop to control root growth. Ethylene also promotes the development of root hairs (Cao et al., 1999). Cytokinins at low concentrations tend to inhibit root elongation, possibly through stimulation of ethylene production (Cary et al., 1995).

The extent of branching and rebranching of both woody and nonwoody long roots is truly remarkable. Lyford (1975) calculated that a mature northern red oak tree formed a minimum of 500 million root tips. Rapid proliferation of roots also has been shown in very young woody plants. Three types of lateral root branches may form on woody long roots. The new branch may be a long root that eventually undergoes secondary thickening and becomes a part of the permanent woody root system. The second and most common type of branch roots are short roots. The third type develops when a short root lateral is converted to a long root. Branches of long roots usually are replacement roots following injury to a long root tip (Wilson, 1975). Injury also commonly occurs to nonwoody lateral roots and is followed by formation of replacement roots and forking.

Root Elongation

Tips of roots may be pointed in long roots and rounded in short roots. A longitudinal section of the end of a young root typically has four cell regions of different character (Fig. 2.30). At the tip is the protective cellular mass comprising the root cap. Behind it is the growing point, a meristematic region of small, thin-walled, cubical cells with dense cytoplasm. Mitotic figures often can be seen in this growing point, which usually is about 1 mm long. As the number of cells increases, some are added to the root cap and others to the region of elongation located above the meristematic zone. In this region the cells produced in the growing point rapidly increase in size, primarily in a longitudinal direction. Above the region of elongation is a zone of differentiation and maturation. Eventually the newly formed cells at the base of the region of elongation lose their capacity for further expansion and become differentiated into the epidermis, cortex, and stele.

Considerable variation may be found among species and different roots in the delineation of root zones. The root cap, for example, does not occur in mycorrhizal roots of pines (Chapter 2). The zone of differentiation often is difficult to measure because various types of cells are differentiated at different distances from the root tip. Furthermore, the distance from the apex at which cells differentiate is a function of the rate of root growth. Wilcox (1954) found that various elements of roots of noble fir matured closer to the apical initials in slow-growing than in fast-growing roots, and this generally is true.

The long horizontal roots of red maple radiated outward as much as 25 m from the base of a tree in a remarkably straight line. When deflected laterally by a barrier they curved back toward the original direction after passing the obstruction (Lyford and Wilson, 1964; Wilson, 1967). According to Head (1965), spiral growth is common in roots, but spiraling sometimes is confused with twisting. Wilson (1964) reported that a maple root was twisted more than four times in a distance of 22 m. Stone and Stone (1975b) reported some twisting in roots of red pine, but no spiraling. Stone and Kalisz (1991) summarized much data on vertical and horizontal extent of root systems.

The girdling of stems by roots is a common occurrence in transplanted trees. Johnson and Hauer (2000) reviewed the occurrence, effects, and treatment of stem girdling roots (SGRs) in ornamental trees. Nursery stock plants allowed to grow too long in containers

A B

FIGURE 3.26. (A) Stem girdling roots of a linden tree the base of which had been buried; (B) fallen trunk and base of a silver maple tree that fell under high wind. The base of the stem shows the compression point caused by a stem-girdling root. Photos courtesy of Ben and Gary Johnson.

frequently develop encircling roots near the container surface. If not corrected, or if planting of properly cultured plants involves curling of roots to fit a hole that is too small, stem girdling can result. Planting of nursery stock with the root-shoot junction below soil level, as commonly occurs, promotes upward growth of roots and makes stem girdling more likely. Such trees can have inherent structural weaknesses, as the girdling root compresses the stem, and are prone to snap when placed under loads, especially during windstorms (Fig. 3.26). Although difficult to attribute solely to SGRs, some affected trees also exhibit leaf dwarfing and scorch and branch dieback. In addition, there are possible physiological effects, including phloem blockage and decreased hydraulic conductivity of the stem (Hudler and Beale, 1981). Nursery production methods now often employ containers with vertical ribs and chemical coatings (e.g., copper hydroxide) that train the roots to grow vertically, or chemically prune them when they reach the edge of the container, respectively.

When seasonal root elongation ceases, roots often turn brown in a process called *metacutization*. This involves lignification and suberization of cell walls of the cortex and dormant root cap. Many roots retain a white root tip even though a metacutization layer is present. Presence or absence of a white root tip depends on how many layers of dead cells are cut off outside the metacutization layer.

Rate of Root Growth

The rate of root elongation of woody plants varies among species, genotypes, tree age, season, site, and environmental conditions. Roots may elongate from a fraction of a millimeter to well over 25 mm a day during the period of most active growth. Long roots of apple at East Malling grew 4 to 6 cm/week; those of cherry 7 to 8 cm/week (Head, 1973). According to Hoffmann (1966), a few roots of black locust and a species of poplar had exceptionally high growth rates of about 5 cm/day. Head (1965) and Lyr and Hoffmann (1967) found root elongation to be consistently greater during the night than during the day.

Seasonal Variations

The annual growth of roots involves two components: (1) elongation of existing roots and (2) initiation of new laterals and their subsequent elongation.

In the temperate zone root elongation usually begins earlier in the spring and continues later in the autumn than shoot elongation in the same tree (Fig. 3.27). There are exceptions, however. For example, initiation of root growth of green ash, Turkish hazelnut, and tree lilac in New York State followed the beginning of shoot growth. The late initiation of root growth was attributed to cold soil conditions (Harris et al., 1995). The time interval between the cessation of shoot elongation and of root elongation varies greatly among species. Root elongation may continue for many weeks in species whose shoots are fully preformed in the winter bud and expand rapidly. However, in species exhibiting both preformed and neoformed shoot growth (heterophyllous species) and recurrently flushing species, with shoots expanding for many weeks, root elongation may continue for only a slightly longer time than does shoot extension. In southern pines in the United

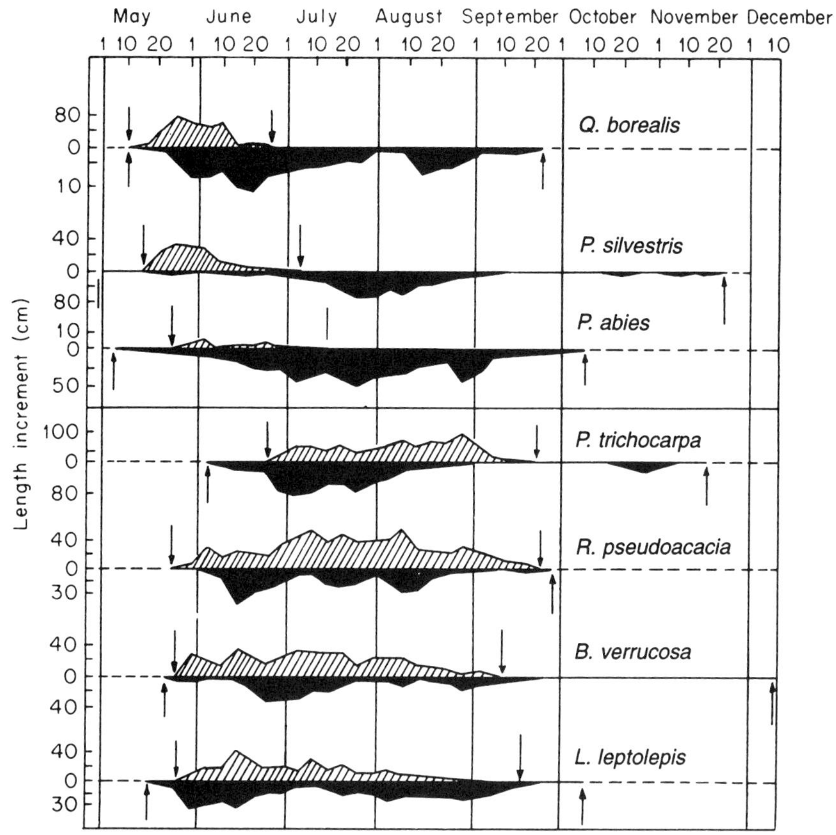

FIGURE 3.27. Variations in seasonal shoot and root growth characteristics for eight species of forest trees. Shading indicates shoot growth and solid black represents root growth. Seasonal initiation and cessation of growth are indicated by arrows. From Lyr and Hoffmann (1967).

States, root elongation typically occurs during every month in the year (Kramer, 1969, pp. 127–129). However, in the winter it is limited by low temperature; in the summer by dry soil.

Cambial Growth in Roots

Primary growth of some roots is followed by secondary growth involving formation of secondary vascular tissues by the cambium and of periderm by a phellogen (cork cambium). Secondary thickening may start during the first or second year. Stages in formation of the cambium and secondary growth of a woody root are shown in Fig. 3.28. At first some parenchyma and pericycle cells become meristematic and form a wavy cambial band on the inner edges of the phloem strands and outside the xylem. Eventually the cambium produces xylem in a complete cylinder. Shortly after the cambium forms, some of the pericycle cells divide to form the phellogen (cork cambium), which cuts off phelloderm tissue to the inside and cork to the outside. After cork formation begins, the cortex with its endodermis is shed and the tissue arrangement thereafter is similar to that in the stem.

Each year the cambium of roots of temperate-zone trees and shrubs produces xylem first in parts of the

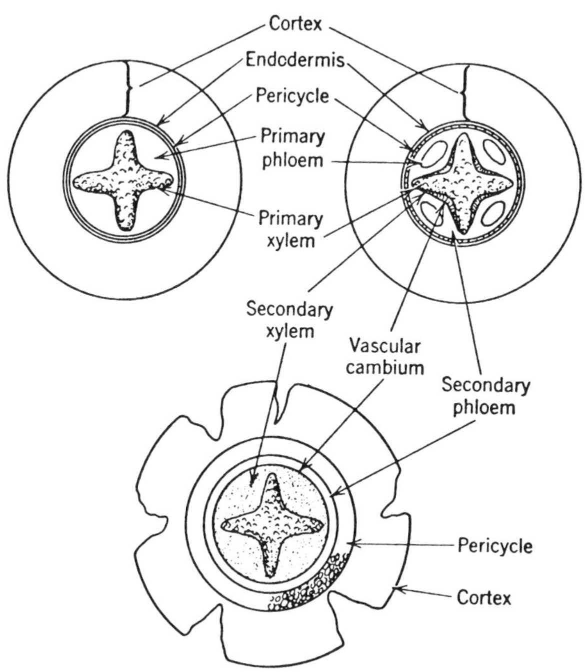

FIGURE 3.28. Secondary growth of a woody root, showing development of vascular cambium and production of secondary xylem and phloem. Enlargement by addition of secondary tissues crushes primary phloem and endodermis and splits off the cortex. After Esau (1965).

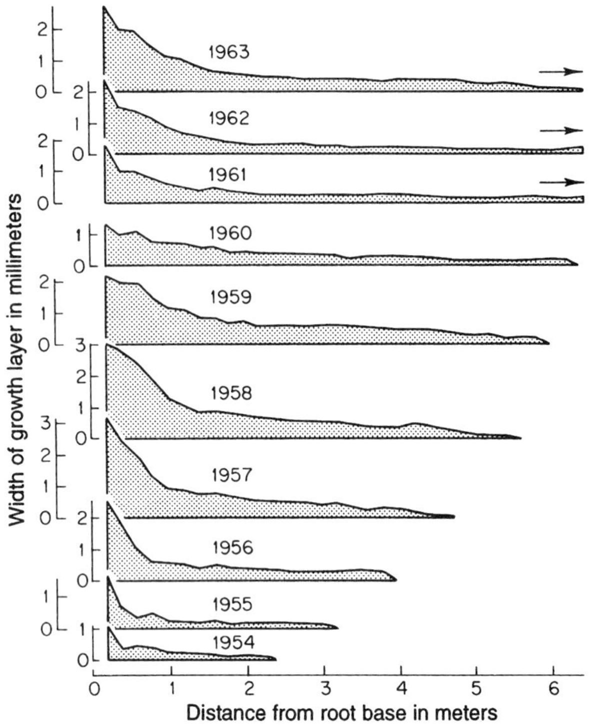

FIGURE 3.29. Annual xylem production along a main lateral root of red pine. From Fayle (1968).

FIGURE 3.30. Abundant knee roots in a stand of baldcypress in South Carolina. Photo courtesy of U.S. Forest Service.

perennial roots located near the soil surface and later in those in deeper soil layers. The downward migration of the cambial growth wave often is slower than in the stem. In orange trees, cambial activity occurred in the stem and branches in April and spread to the main root within two weeks. Subsequently the spread of cambial growth into the root system was slow and xylem production did not begin in lateral roots until late July, and in some small roots not until late September (Cameron and Schroeder, 1945).

Cambial growth is much more irregular in woody roots than in stems. It varies markedly along the length of the root and around its circumference. Maximum xylem production in roots typically occurs near the soil line. Hence, annual xylem increments taper rapidly below the soil line and gradually beyond to the root tip (Fig. 3.29). However, there may be departures from this pattern. For example, Head (1968) found that thickening of apple roots was irregular along the length of the root. Sometimes appreciable thickening began first in more distal parts of the roots and in some years there was no cambial growth at all. Young roots generally are circular in transection, but as they age xylem deposition around a root becomes more uneven. Hence old perennial roots tend to be very eccentric in cross section. False and double xylem rings abound in roots. The horizontal roots of many tropical species show much greater xylem production along the upper side than the lower one, leading to formation of buttresses (Chapter 2). Roots of baldcypress develop vertical knees (Fig. 3.30) because of very rapid cambial activity on the upper surface of roots (Whitford, 1956).

There is great variability in xylem production in different roots of the same tree. Usually there is greater growth eccentricity in the lateral horizontal roots than in vertical or oblique roots in the central portion of a root system.

SHEDDING OF PLANT PARTS

Woody plants recurrently shed buds, shoot tips, branches, prickles, cotyledons, leaves, stipules, bark, roots, and reproductive structures. Loss of certain plant parts has enormous implications in growth and development of the shedding plants, neighboring plants, and the environment. Natural shedding of plant tissues and organs alters plant form; provides the litter that becomes a major component of soil organic matter; accounts for drought tolerance of many plants in arid environments; removes injured, diseased, or senescent plant parts; and reduces competition for water and mineral nutrients within individual plants by removing the less vigorous leaves, branches, and fruits. However, shedding of plant parts also may be harmful by slowing plant growth, inhibiting seed germination by physical and chemical effects of litter (Chapter 2, Kozlowski and Pallardy, 1997), and causing loss of nutrients from ecosystems (Kozlowski, 1973).

Leaves

Leaves of woody plants are shed periodically by abscission, by mechanical factors, or a combination of both. True abscission is characterized by physiological changes that lead to formation of a discrete abscission layer at which separation of the leaf occurs (Fig. 3.31). Abscission of simple leaves occurs at or near the base of the petiole. In compound leaves separate abscission zones form at the bases of individual leaflets as well as at the base of the petiole of the whole leaf.

Physiological abscission is preceded by leaf senescence, which typically is induced by interactions of internal and external factors (Fig. 3.32). Initial perturbation may be caused by small changes in a few

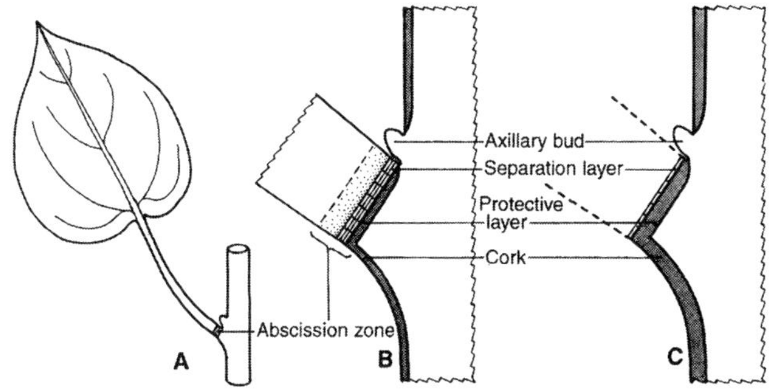

FIGURE 3.31. Abscission zone of a leaf. A: Leaf with the abscission zone located at the base of the petiole; B: Layers of the abscission zone shortly before leaf abscission; C: Layers of the abscission zone after leaf abscission has occurred. From Addicott (1970); with permission of McGraw-Hill.

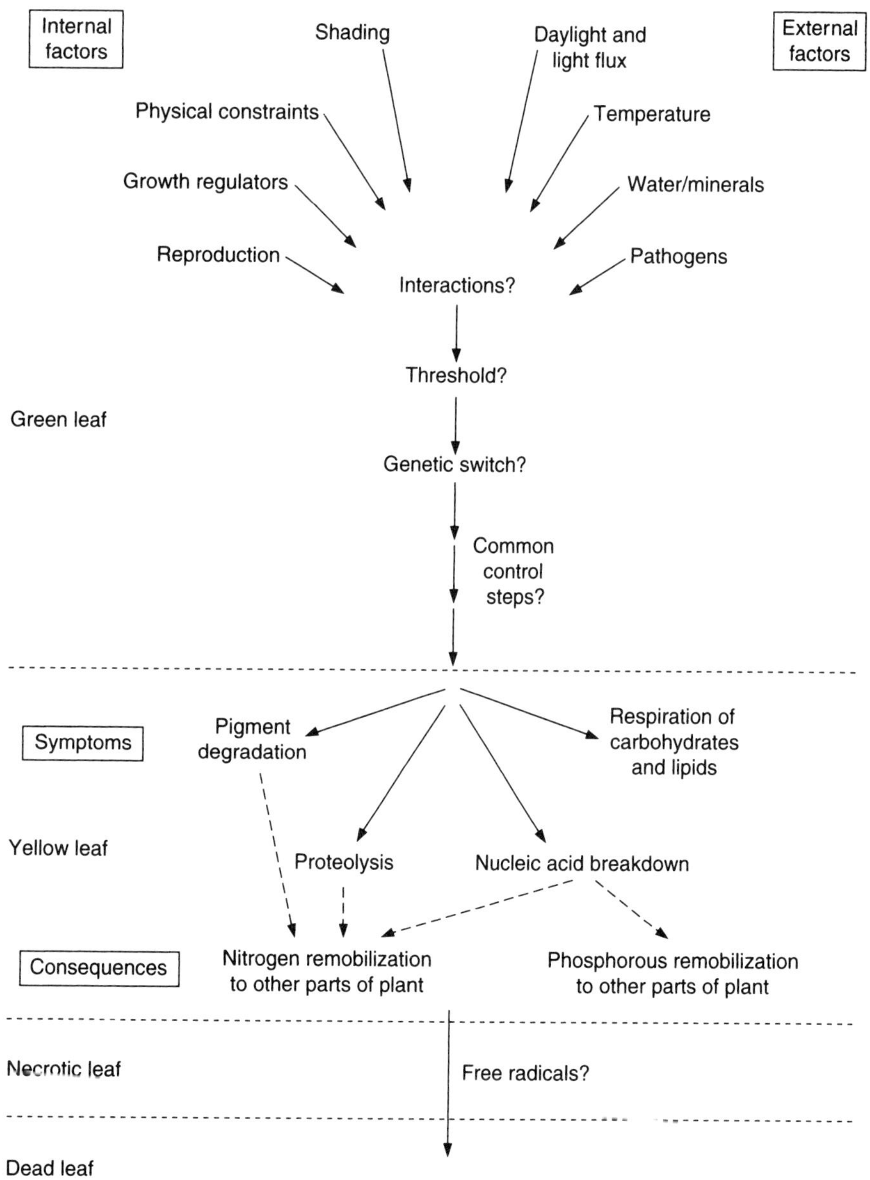

FIGURE 3.32. Factors involved in induction of leaf senescence. From Smart (1994).

controlling steps, followed by a cascade of secondary effects. Some investigators classified leaf senescence into three distinct types:

- Monocarpic senescence, which occurs as a consequence of reproduction
- Sequential senescence, which results from competition for resources between the older, lower leaves and upper, younger leaves
- Autumnal senescence, which may result from decreasing day length and temperature

The physiological control of abscission is discussed in more detail in Chapter 3 of Kozlowski and Pallardy (1997).

The leaves of most deciduous trees and shrubs of the temperate zone form an abscission layer during the season in which they expand and are shed in the summer or autumn of the same year. Exceptions are some marcescent species of oaks, beech, American hornbeam, and hophornbeam, which retain their leaves through the winter and shed them in the following spring. Although the leaf blades and most of the petiole of marcescent species die in the autumn, the cells in the abscission zone do not. In the following spring the processes of abscission are initiated and proceed similarly to those in leaves that were shed in the previous autumn.

Abscission involves both separation and protection. Separation of leaves occurs in one of two ways: (1) the middle lamella between two layers of cells in the abscission zone dissolves but the primary wall does not, or (2) both the middle lamella and primary wall between the two cells dissolve. Protection of the scar may involve suberization and lignification in addition to development of protective layers. After a leaf is shed, cells on the stem side of the abscission layer continue dividing to produce a corky protective layer (Addicott, 1982).

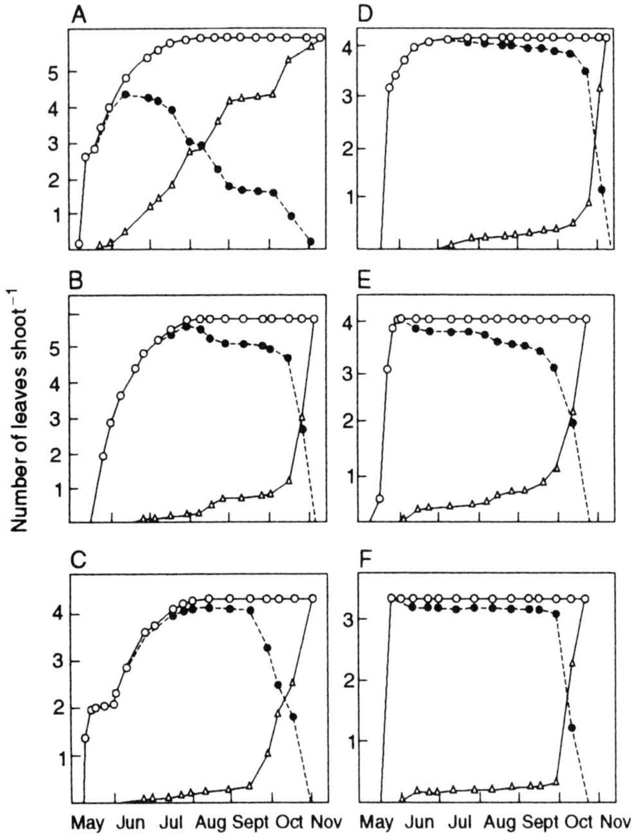

FIGURE 3.33. Seasonal changes in leaf number of six species of Betulaceae. (A) *Alnus hirsuta* (1979), (B) *Alnus pendula* (1979), (C) *Betula platyphylla* var. *japonica* (1978), (D) *Corylus sieboldiana* (1980), (E) *Ostrya japonica* (1979), (F) *Carpinus cordata* (1979). ○, Emergence curve, indicating cumulative mean number of emerged leaves per shoot; ●, survivorship curve indicating mean number of leaves actually persisting on the shoot; △, leaf fall curve indicating cumulative number of leaves abscised from the shoot. From Kikuzawa (1982).

Leaf Shedding of Temperate-zone Trees

In deciduous trees of the temperate zone the actual time and speed of annual leaf shedding vary appreciably among species and genotypes. However, individual trees may shed their leaves at any time during the growing season in response to injury or environmental stresses, especially drought.

Kikuzawa (1982) found wide variations in leaf survival among deciduous species of forest trees in Japan. The period of leaf retention was long for brittle willow, alders, birch, and Japanese maple (more than 190 days), and short for Manchurian ash (140 days). Species with leaves emerging more or less simultaneously also tended to shed most of their leaves at about the same time. Species with leaves emerging over a long time tended to retain some of their leaves later into the autumn. Examples of variations in patterns of leaf emergence and shedding of broad-leaved trees in Japan are shown in Fig. 3.33.

The leaves of evergreen conifers have a lifespan that varies from two to many years in different species. Leaves of hemlock are retained for three to six years, and those of spruces and true firs for seven to 10 years (Harlow et al., 1979). Leaves of *Araucaria* may remain alive for more than 30 years. Leaf retention varies greatly in different species of pines (Table 3.3): two years in longleaf pine; up to 12 years in Swiss stone pine (Nebel and Matile, 1992); and up to 45 years in bristlecone pine (Ewers and Schmid, 1981). The leaf life span of conifers varies with site and may be twice as long on some sites as on others. Longevity of leaves of evergreen conifers typically is shorter on fertile than

TABLE 3.3. Variations in Needle Retention of Species of *Pinus* Native to the United States and Canada[a]

Species	Needle retention (yrs)	Species	Needle retention (yrs)
Pinus aristata	10–20	*P. monophylla*	5–15
P. attenuata	4–5	*P. monticola*	2–4+
P. banksiana	2–3	*P. muricata*	2–3
P. cembroides	3–4	*P. palustris*	2
P. clausa	3–4	*P. ponderosa*	2–9
P. contorta	2–9	*P. pungens*	2–3
P. coulteri	3–4	*P. radiata*	3
P. echinata	2–5	*P. resinosa*	3–5
P. edulis	3–9	*P. rigida*	2–3
P. elliottii	2	*P. sabiniana*	3–4
P. flexilis	5–6	*P. serotina*	3–4
P. glabra	2–3	*P. strobus*	2–3
P. jeffreyi	5–9	*P. taeda*	2–5
P. lambertiana	2–3	*P. torreyana*	3–4
P. longaeva	10–45	*P. virginiana*	3–4

[a]Modified from Ewers, F. W., and Schmid, R. (1981). Longevity of needles fascicles of *Pinus longaeva* (bristlecone pine) and other North American pines. *Oecologia* **51**, 107–115, Table 1.

on infertile sites, in sunny than in shaded environments, and on cool than on cold sites (Reich et al., 1995).

Leaf Shedding of Tropical Trees

Leaf shedding tends to occur more or less continuously in tropical forests. However, there are small to very large seasonal peaks of leaf shedding and these vary with distribution of rainfall.

Reich (1995) presented an excellent review of the patterns of leaf shedding in South American tropical forests. In rain forests, with a majority of evergreen species, a small amount of leaf shedding occurs in each month of the year, with negligible seasonal variation. By comparison, cool temperate rain forests in Australia shed about half their leaves in the autumn. Both warm and subtropical rain forests show less significant leaf shedding in early spring (Lowman, 1986).

In South American tropical climates with distinct dry and rainy seasons, increasing severity of drought is associated with a progressively higher proportion of deciduous species. For example, in Jalisco, Mexico (ppt. 700 mm/yr), 96% of the species are deciduous, whereas in tropical rain forests (ppt. > 3500 mm/yr), only about 5% of the species are deciduous (Reich, 1995). The strongly deciduous species shed all or most of their leaves during the height of the dry season. Leaf shedding of some tropical trees is strongly influenced by heredity and is not influenced much by small climatic changes. The leaf lifespans of Amazonian tree species vary from 1.5 months to more than five years (Reich et al., 1991a). The longevity of leaves of Australian rain forest trees differs from about six months in *Dendroenide excelsa*, to one year in *Toona australis*, and to 12 or more years in *Doryophora sassafras* (Lowman, 1992).

An appreciation of the complexity and diversity of patterns of leaf shedding of tropical trees may be gained from the following classification of Longman and Jenik (1987):

1. Periodic Growth: Deciduous. Leaf shedding occurs well before bud break; leaf life span approximately 4 to 11 months. Trees may remain leafless for a few weeks to several months. Examples are *Cordia alliodora* in Costa Rica and *Terminalia ivorensis* in West Africa.
2. Periodic Growth: Leaf Exchanging. Leaf shedding is associated with bud break. The leaf lifespan is about 12 (or 6) months. Examples are *Terminalia catappa* and *Dillenia indica*.
3. Periodic Growth: Evergreen. Leaf shedding is complete well after bud break. The leaf lifespan is 7 to 15 months or more. Examples include *Clusia rosea*, *Mangifera indica*, and some pines.
4. Continuous Growth: Evergreen. Both formation and shedding of leaves occur continuously. The

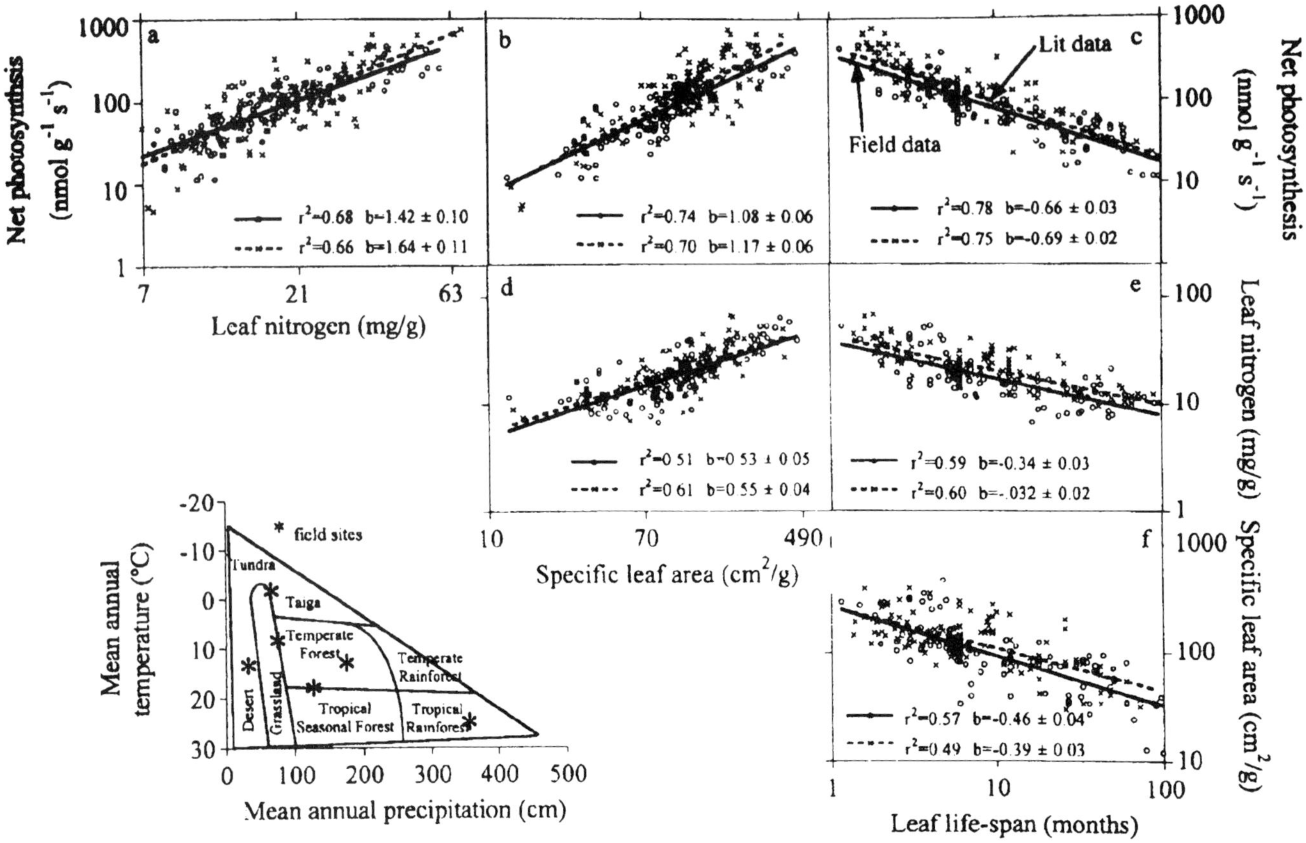

FIGURE 3.34. Log-log relationships (a–f) among mass-based photosynthetic capacity, specific leaf area (SLA), and leaf nitrogen concentration of young mature leaves and their expected life-span. Lines represent regressions (log y = a + b*log X) for species in two sets (-o-, author field data from 111 species of six biomes; -x-, global literature-derived data set). Coefficients of determination (r^2) and slopes (b values ± 1 standard error). Inset, global biome distribution in relation to annual temperature and precipitation and location (*) of field sites. From Reich. P. B., Walters, M. B., and Ellsworth, D. S. (1997). From tropics to tundra: Global convergence in plant functioning. *Proc. Natl. Acad. Sci. U.S.A* **94**, 13730–13734. Copyright 1997 National Academy of Sciences, U.S.A.

leaf lifespan varies from about 3 to 15 months. Examples include some palms, conifers, seedlings of many broad-leaved trees, and older trees of *Dillenia suffruticosa* and *Trema guineensis*.

Leaf Longevity and Apparent Constraints on Leaf Structure and Function

There is much interest in leaf longevity because of its importance to plant growth and plant responses to such environmental factors as light, water supply, nutrient supply, temperature, pollution, and herbivory. In general, stand level production of both evergreen conifers and broadleaved trees, expressed on a leaf area basis, decreases with increasing leaf life span (Reich et al., 1995).

Reich et al. (1997, 1999) have argued that common ecological and biophysical constraints result in convergent evolution in leaf structure and function. The authors analyzed both their own and a literature-based dataset of leaf life span, leaf N, photosynthesis and specific leaf area for numerous species from diverse biomes (Fig. 3.34). Consistent and statistically significant relationships among leaf life span, SLA, and mass-based leaf N and photosynthesis suggested that leaf structure and function can evolve only within certain limits without being maladaptive. Specifically, species with short life spans tend to have high mass-based leaf N and photosynthesis and high SLA (i.e., thin leaves). Those with long life spans tend to have opposite patterns of these attributes. These relationships were robust across life forms and biomes. The validity of these conclusions is supported by the lack of outlying species in their analysis. The authors suggest that there are no species with leaves that are thin and short-lived as well as having low photosynthetic rates because such leaves would likely not return to the plant the carbon and nutrient investment required to build them.

Similarly, thick leaves might not transmit sufficient light or permit sufficient CO_2 diffusion rates to justify large investment in N-rich photosynthetic machinery, and leaves with high N contents and long potential life might be so attractive to herbivores that their potential life spans would rarely be realized.

Branches

Shedding of branches sometimes up to 2.5 cm in diameter is characteristic of a number of species of trees and shrubs of the temperate zone and tropics. Such shedding influences crown form and reduces the number and size of knots in lumber, when dead branches are not shed, loose knots form that degrade lumber. The shedding of twigs and branches may occur through abscission by a process similar to that in leaf abscission, or natural pruning by death of branches and without formation of an abscission zone.

Cladoptosis

Shedding of branches by abscission, a process called *cladoptosis*, occurs through well-defined abscission zones and is preceded by a weakening of tissues and by periderm formation. Cladoptosis appears to be a response to adverse environmental conditions and aging effects that induce loss of branch vigor. Abscission of twigs does not occur in juvenile white oak trees but is common in mature trees (Millington and Chaney, 1973).

Cladoptosis is well documented in both temperate-zone and tropical trees. Certain conifers typically shed leafy branches. In a few species (e.g., baldcypress) all leafy branches are shed annually. In others only some of the branchlets of evergreen trees are periodically shed. *Agathis macrophylla* first abscises leaves and later the branches on which the leaves were borne (Addicott, 1991). Shedding of twigs in the temperate zone is characteristic of poplars, willows, maples, walnut, ashes, and oaks. In the tropics a strong tendency for shedding of branches has been shown by members of the following genera: *Antiaris, Albizzia, Canangia, Castilloa, Casuarina, Persea, Sonneratia,* and *Xylopia.* Branch abscission also occurs in several temperate-zone and tropical lianas, including *Ampelopsis cordata, Parthenocissus quinquefolia, Vitis* spp., and climbing species of *Piper*.

Natural Pruning

Death and slow shedding of lower branches without formation of an abscission layer occur in many forest trees growing in dense stands. The degree of such self-pruning varies appreciably among species. Whereas white oak and longleaf pine are good natural pruners, willow oak and white pine are not (Table 3.4; Fig. 3.35).

Natural pruning is preceded by sequential physiological senescence and death of branches low on the

TABLE 3.4. Variations in Natural Pruning of Lateral Branches of Various Species of Trees Growing in Dense Stands

Good natural pruners		Poor natural pruners	
Angiosperms	**Gymnosperms**	**Angiosperms**	**Gymnosperms**
Betula lenta	*Abies procera*	*Celtis laevigata*	*Abies balsamea*
Fagus grandifolia	*Larix laricina*	*Juglans nigra*	*Juniperus occidentalis*
Fraxinus americana	*Picea mariana*	*Quercus nigra*	*Juniperus virginiana*
Populus balsamifera	*Picea rubens*	*Quercus phellos*	*Larix occidentalis*
Populus tremuloides	*Pinus elliottii*		*Calocedrus decurrens*
Populus trichocarpa	*Pinus palustris*		*Picea engelmannii*
Prunus serotina	*Pinus resinosa*		*Pinus contorta*
Quercus alba	*Taxodium distichum*		*Pinus coulteri*
Quercus rubra			*Pinus monticola*
Salix nigra			*Pinus radiata*
			Pinus strobus
			Pinus virginiana
			Sequoiadendron giganteum
			Tsuga canadensis

FIGURE 3.35. Variations in natural pruning of forest trees. Good natural pruning of longleaf pine (top) and poor natural pruning of Virginia pine (bottom). After Fenton and Bond (1964), and U.S. Forest Service. From Millington and Chaney (1973), by permission of Academic Press.

stem. The dead branches are attacked by saprophytic fungi. Eventually they are shed because of their own weight or by the action of wind, rain, or whipping of adjacent trees. Natural pruning occurs faster in warm, humid regions than in cool dry ones because of greater activity of fungi in the former. Natural pruning also is favored by high stand density.

Bark

Bark tissues are routinely shed in a variety of patterns from both living and dead trees. In living trees the bark loosens naturally when cambial cells are dividing. However, the wood-to-bark bond begins to decrease appreciably several weeks before cell division occurs in the cambial cells and mother cells (at a time when the cambial cells are stretched). Once cell division begins the bond probably fails in the xylem mother cells. Early in the growing season the xylem-bark bond decreases progressively from the crown downward, and earlier in dominant than in suppressed trees. The reduction in the strength of the wood-bark bond is facilitated by fungi and bacteria that decompose the cambium (Kubler, 1990).

Much interest has been shown in seasonal variations in the xylem-bark bond because of the need for mechanical and chemical debarking of logs. Adhesion between xylem and bark varies with tree species, season, and conditions under which logs are stored. Lutz (1978) classified hardwood species of the southern United States on their resistance to bark peeling. Several species of hickory were rated as very difficult to debark; elm, white ash, and several oaks were intermediate; and sweetgum and southern pines were easily debarked. Berlyn (1964, 1965) noted that the strength of the wood–bark bond of conifers was only a third to half of that in broadleaved trees. Several pretreatments have been used to ease mechanical removal of bark, including steaming of logs, storing logs at 70 to 80°F in air at 100% relative humidity, immersing logs in hot water (190 to 200°F) for two hours, and applying pressure to the bark surface (Koch, 1985, Vol. II, pp. 1647–1648).

Patterns of Bark Shedding

Barks of different species vary from smooth to rough. The appearance of bark depends on the radial position of the first periderm, patterns of formation of subsequent periderms, and composition and arrangement of phloem cells (Borger, 1973).

In smooth-barked species the periderm remains in a superficial position for a long time. In trembling aspen, American beech, and lemon trees, the original periderm persists for the life of the tree. The outer bark of species with smooth bark consists of phellem cells (e.g., trembling aspen, American beech) or of phellem plus old phelloderm cells (e.g., lemon). Shedding of tissue from smooth-barked trees occurs slowly and is correlated with the rate of formation of phellem cells. In trembling aspen the outer phellem cells are shed individually or in small groups. In silver birch sheets of phellem several cell layers thick are shed as tangential pressures induce tearing in the outer phellem (Scott, 1950).

In species that develop scaly barks the first periderm may persist in a superficial position for a very long time. However, small additional periderms eventually form in patches below the first periderm, and they continue to form throughout the life of the tree. Hence a thick outer bark is produced that consists of alternating layers of periderms and tissues cut off by them in the form of flakes, as in Scotch pine, or in sheets, as in sycamore maple. In shagbark hickory periderms arise in long vertical strips; hence bark tissues are cut off in vertical, plate-like strips, reflecting the orientation of periderms, and a banded interlocking arrangement of numerous fibers in the phloem.

In species with furrowed barks the phloem contains abundant sclerenchyma tissue, especially fibers. Usually arcs of periderm arise in the outer phloem and the cells outside of the periderms die but do not separate because of the interlocking system of fibers. This results in very deep, furrowed, loose, and fibrous bark, as in redwood. The furrowed barks of willow, oak, elm, ash, and walnut owe their individuality of pattern to various proportions of cork cells and sclerified tissues. The soft bark of American elm contains large amounts of cork cells and little sclerified tissue, whereas the hard barks of oak and ash contain many sclerified cells and relatively little cork. The "stringy barks" of several species of *Eucalyptus* owe their patterns of shedding to a cohesive but loose fibrous rhytidome. The groups of eucalypts known as "ironbarks" contain large amounts of a hard, resinous substance, called *kino*, which prevents rupture of the rhytidome (Borger, 1973).

In species with "ring barks" (e.g., grape, clematis, honeysuckle, and species of Cupressaceae) the first periderm is initiated in very deep tissues. The second and subsequent periderms are concentric cylinders. This arrangement results in shedding of hollow cylinders of loose bark. Separation occurs through thin-walled cork cells but the rhytidome remains attached, giving the outer bark a characteristic shaggy appearance. For more details on bark formation and shedding see Borger (1973).

Roots

Root systems of woody plants consist of long-lived, large perennial roots and many short-lived small roots (Chapter 2). The growth of fine roots varies greatly with plant species, genotype, site, tree age, attacks by insects and fungi, and environmental conditions. Death and replacement of fine roots occur simultaneously. Annual estimates of turnover of fine roots range from 30 to 90% (Fogel, 1983). On poor sites fine roots may turn over two to five times annually (Cannell, 1989).

In healthy trees many of the small roots die shortly after they form. In apple trees, for example, small lateral roots lived only about a week (Childers and White, 1942). According to Kolesnikov (1966), the tips of main roots of seedlings of orchard trees die by the time the seedlings are two months old. Species appear to vary, however, in longevity of their small roots. In Norway spruce most absorbing rootlets usually lived for three to four years, only about 10% dying during the first year and 20% living for more than four years (Orlov, 1960). Head (1966) noted that black currant roots lived for more than a year. However, many of the

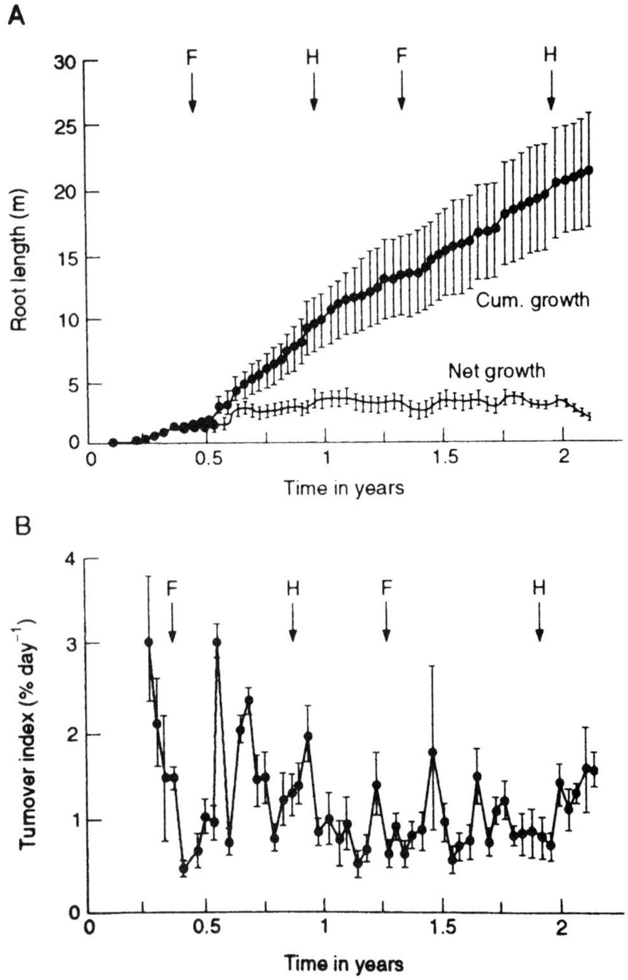

FIGURE 3.36. (A) Comparison between cumulative root growth and net length of roots per vine in kiwi. Cumulative death of roots per vine is the difference between these two curves. (B) Turnover index (average of relative growth and relative death rates per vine) as a function of time. F, 50% flowering; H, harvest. From Reid et al. (1993).

smaller so-called feeder roots of fruit and forest trees live less than a year.

A very high rate of turnover of fine roots was shown in a rhizotron for kiwifruit over a two-year period (Fig. 3.36). After an initial phase of rapid colonization of the repacked loam soil, the total length of roots changed little. However, the apparent stability of total root length obscured the rapid turnover of fine roots. Approximately 51% of these roots survived no more than 28 days; 69% died within 56 days, and only 8% survived for more than 252 days. In each of the two years of this study the cumulative length of roots grown was 2.75 times the maximum net length of roots that were visible. The turnover index (average of relative growth and relative death rate per vine) varied sharply over intervals less than four weeks long, but averaged 1.2% per day over the two years of the study (Fig. 3.36).

In the temperate zone the greatest mortality of small roots occurs during the cold months. In English walnut more than 90% of the absorbing roots were lost during the winter (Bode, 1959). According to Voronkov (1956) the dry weight of active roots of tea plants was about 12% lower in February than in the previous December. By early April, however, growth of new roots had more than made up for winter losses.

MEASUREMENT AND ANALYSIS OF GROWTH

Measurement of tree growth can be made by various methods, depending on the objectives of the investigator. Most important to foresters is measurement of the annual increment of wood produced by a stand of trees. This is calculated from measurements of bole diameter and height and is modified somewhat by the amount of bole taper. It usually is expressed as volume per unit of land area, but for some purposes is better expressed as weight per unit of land area. Standard texts on forest mensuration provide details of the methods of growth measurement and analysis (Husch et al., 2003). Additionally, Kozlowski (1971a,b) addressed dimensional growth of trees in detail, including measurement techniques relevant to physiologists.

Much interest has been shown in the "harvest index" of crop plants (i.e., that fraction of the final aboveground plant dry weight that constitutes the primary marketable product). This index is useful for annual crops, but less so for trees because much aboveground production is lost in shed organs before trees are harvested. A more useful concept in forestry is the "harvest increment," the increase in weight of the harvested part (i.e., wood) divided by the total aboveground increment, estimated for a period of one or a few years (Cannell, 1985).

Analysis of Growth

According to Ledig (1976) a better understanding of what constitutes optimum allocation or partitioning of growth among the various organs of a plant is one of the most important tasks of plant physiology. This requires a much more intensive analysis of plant growth than is provided by measuring growth in diameter and height. Information concerning the partitioning of growth among roots, the stem, branches, and leaves is necessary to an understanding of how various environmental and cultural practices affect growth.

Two approaches to analysis of growth and its components have emerged. The earlier approach, termed "classical growth analysis," employed relatively infrequent, large destructive harvests of sample plants over the course of an experiment from which dry weights and areas of various plant tissues were calculated. More recently, relatively frequent harvests of fewer plants have sometimes been employed in an approach that has been termed "functional" or "dynamic" growth analysis (Hunt, 1990). The former term did not arise because of its superiority with regard to plant mechanisms; rather from the use of mathematical functions to fit growth trends. Classical growth analysis has remained popular in research on woody plants because of its suitability to annual growth cycles of perennials, computational simplicity, and ease of interpretation. Functional growth analysis has some operational and theoretical advantages. However, it is computationally complex and depends greatly on success in fitting data with empirical functions. If properly applied, both classical and functional growth analyses reveal the underlying biology of growth. Radford (1967), Evans (1972), Hunt (1978, 1990), and Causton and Venus (1981) provide detailed comparisons of classical and functional approaches to growth analysis. Poorter and Garnier (1996) provided recommendations as to which method was better-suited to particular experimental objectives.

Hunt (1990) identified several classes of growth measurements and derived parameters common to both approaches:

I. Absolute Growth Rates: Simple rates of change involving one growth attribute and time (e.g., rate of total plant dry weight growth per day).
II. Relative Growth Rates: Rates of change involving one growth attribute and time, but placed on basis relative to initial size.
III. Simple Ratios: Ratios of two simple plant attributes that themselves may be similar (dry weights) or different (area and dry weight).
IV. Compounded Growth Rates: Rates of change comprising more than a single attribute. Primary examples are the rates of dry weight increase of an entire plant for a given leaf or land area.
V. Integral Durations: Cumulative areas under curves of plant attributes vs. time (e.g., leaf area duration).

Relative Growth Rate

It was established long ago that plant growth follows the compound interest law, the amount of growth made in a unit of time depending on the amount of biomass or size of plant at the beginning of the period. This fact led to development of the concept of relative growth rate (RGR) or measurement of increase in dry weight per unit of time per unit of growing material, often in grams per gram dry weight per week. This permits comparison of the effects of various factors in the environment on the rate of growth. The equation for calculating mean relative growth rate is:

$$\frac{\ln W_2 - \ln W_1}{t_2 - t_1} \tag{3.4}$$

W_1 and W_2 are the dry weights at the beginning and end of the sampling period, t_1 and t_2 are the dates of sampling. Hoffmann and Poorter (2002) have demonstrated that the replicate values that contribute to the numerator should be obtained by taking the mean of the ln-transformed plant weights rather than the logarithm of the mean plant weight. Only the former method provides an unbiased estimate if the variance of plant weight changes with time.

Allometric Formula and the Allometric Coefficient

Based on theory of allometric growth developed by Huxley (1932), the allocation of dry matter between two components of the plant body, C_1 and C_2, can be described by an exponential function:

$$C_1 = \alpha(C_2)^{\beta} \tag{3.5}$$

Integrated over time the allometric coefficient, β, describes the ratio between mean RGR of C_1 and C_2. This function allows changes in allocation patterns between plant organs that arise as a result of genetic differences, environmental factors, and time to be monitored by growth analysis. In practice paired measurements of components (e.g., shoot and root dry weight) are plotted as logarithmic transformations:

$$\log(\text{shoot weight}) = \alpha + \beta \log(\text{root weight}), \tag{3.6}$$

and coefficients are computed from linear regression of the relationship. The value of β may or may not be constant as the plant grows and plots of log transformed parameters may be broken into segments and each fitted for estimation of this parameter, if necessary (Causton and Venus, 1981). In this example, conditions that result in a comparatively higher value of α indicate inherently greater investments in roots whereas those that result in changes in β signify differences in the pattern of biomass allocation between shoots and roots that occur in response to treatments. Changes in β with time suggest altered sink strength between roots and shoot as plants develop. This approach has been used widely in studies of genetic

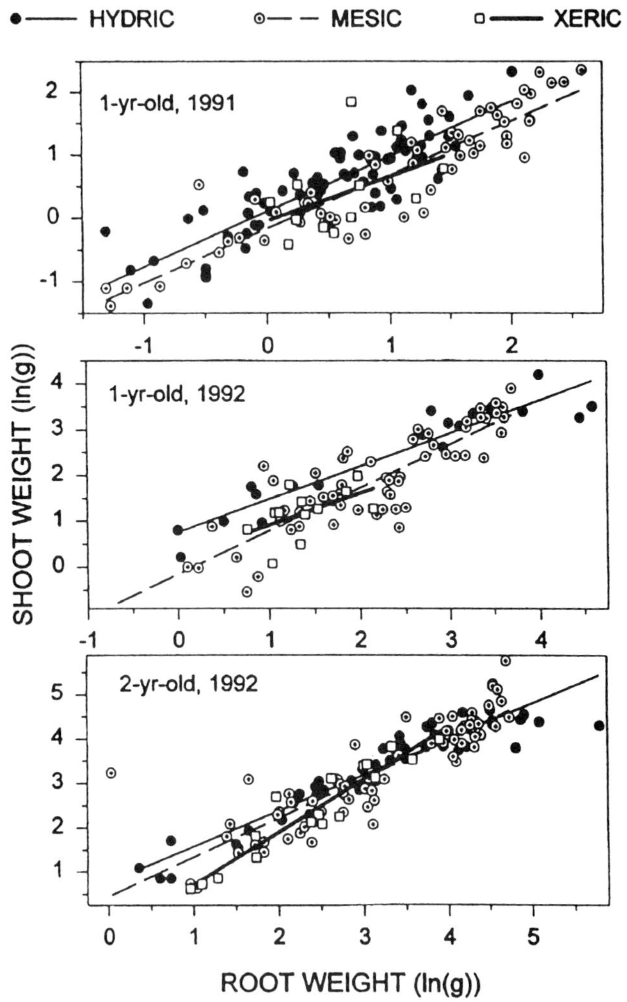

FIGURE 3.37. Allometric relationships between root and shoot weight for seedlings of 14 oak species sampled at different ages and in different years. Different symbols and lines represent oaks species with the indicated soil moisture preferences. From *Trees*, Seedling growth strategies and seed size effects in fourteen oak species native to different soil moisture habitats. Long, T. J., and Jones, R. H., **11**, 1–8, Figure 1. © 1996 with kind permission from Springer Science and Business Media.

and environmental influences on biomass allocation in woody plants, especially in seedlings. For example, Long and Jones (1996) demonstrated that small, young seedlings of *Quercus* from hydric habitats had inherently lower relative allocation to root growth than mesic and xeric *Quercus* species, but that this difference largely disappeared as seedlings grew larger (Fig. 3.37). Cromer and Jarvis (1990) employed allometric analysis to demonstrate that allocation of biomass to roots relative to leaves was reduced by enhanced N status in *Eucalyptus grandis* seedlings and that this difference was maintained for the duration of their experiment (i.e., β was constant with time/size).

Net Assimilation Rate and Other Growth Parameters

The components of growth often are used to identify key factors that explain differences in crop growth and productivity. Dry matter production basically depends on the allocation of carbohydrates to photosynthetic surface and the rate of carbon fixation (photosynthesis) per unit of leaf surface. An integrated estimate of the latter process less respiratory losses is expressed over the period t_1 to t_2 as net assimilation rate (NAR):

$$\mathrm{NAR}\,(\mathrm{gm^{-2}d^{-1}}) = \frac{W_2 - W_1}{t_2 - t_1} \times \frac{\ln L_{A2} - \ln L_{A1}}{L_{A2} - L_{A1}} \quad (3.7)$$

RGR and NAR have the following relationship:

$$\mathrm{RGR}(\mathrm{gg^{-1}d^{-1}}) = \mathrm{NAR}\,(\mathrm{gm^{-2}d^{-1}}) \times \mathrm{LAR}(\mathrm{m^2g^{-1}}) \quad (3.8)$$

In this expression LAR (leaf area ratio) is a Type III parameter that specifies the ratio of leaf area to total dry mass. The separation of RGR, which is a measure of inherent growth efficiency, into these components can identify whether a plant gains its growth superiority from greater photosynthetic performance (high NAR), greater relative allocation of photosynthate to leaf area (high LAR), or a combination of the two. Leaf Area Ratio can be further partitioned into two more Type III parameters to provide the following:

$$\mathrm{RGR} = \mathrm{NAR}\,(\mathrm{gm^{-2}d^{-1}}) \times \mathrm{SLA}(\mathrm{m^2g^{-1}}) \times \mathrm{LWF}(\mathrm{gg^{-1}}) \quad (3.9)$$

where SLA = specific leaf area and LWF = leaf weight fraction (leaf mass as a fraction of total plant mass). Application of this equation allows further separation of growth attributable to differences in leaf morphology from those attributable to greater total allocation of plant biomass to leaves. A spreadsheet program is available to compute these parameters from dry mass and leaf area data (Hunt et al., 2002).

Longevity of leaves also is important and retention of a large leaf surface able to carry on photosynthesis until late in the autumn is likely to increase dry matter production (Nelson et al., 1982; Nelson and Isebrands, 1983). From the standpoint of a canopy, this Type V measurement of Leaf Area Duration (LAD) would assume the following form:

$$\mathrm{LAD} = \int_{t1}^{t2} L_A\, dt \quad (3.10)$$

Physiologists therefore are interested in learning whether various factors affect growth chiefly by affecting the rate of photosynthesis per unit of leaf area, the leaf area itself, or the distribution of photosynthate among roots, stem, branches, and leaves. Farmer (1980)

TABLE 3.5. **Growth Characteristics of Six Deciduous Species during Their First Growing Season under Nursery Conditions**[a,b]

Species	Sampling period	Relative growth rate ($g\ g^{-1}\ week^{-1}$)	Leaf relative growth rate ($cm^2\ cm^{-2}\ week^{-1}$)	Net assimilation rate ($g\ m^{-2}\ week^{-1}$)	Leaf area partition coefficient ($cm^2\ week^{-1} \div g\ week^{-1}$)
Black cherry	1	0.585a	0.51a	27ab	170a
	2	0.509	0.39	32	105
	3	0.340	0.29	30	92
	4	0.302	0.25	34	69
	5	0.102	−0.04	17	−28
Yellow-poplar	3	0.664a	0.63a	46a	131a
	4	0.359	0.31	31	96
	5	0.161	0.07	19	35
Northern red oak	2	0.352b	0.24b	27b	78a
	3	0.200	0.16	21	72
	4	0.160	0.10	21	41
	5	0.097	0.03	19	13
White oak	2	0.164b	0.06b	22ab	25a
	3	0.174	0.16	25	64
	4	0.109	0.03	20	14
	5	0.138	0.08	35	22
Chestnut oak	2	0.275b	0.21b	25b	81a
	3	0.180	0.15	19	77
	4	0.140	0.08	18	35
	5	0.180	0.17	29	59
Bear oak	2	0.218b	0.11b	25a	39a
	3	0.250	0.20	37	48
	4	0.149	0.12	27	48
	5	0.150	0.10	33	29

[a]From Farmer (1980).
[b]Species with different letter suffixes are significantly different at the 0.05 level of probability.

employed classical growth analysis to compare first-year growth of seedlings of six deciduous species grown in the nursery (Table 3.5). The greater growth observed for yellow-poplar and black cherry seedlings as compared to several oak species was attributable primarily to preferential investment of photosynthate in leaf area by the former species. Net assimilation rates were not closely related to growth rate. Brix (1983) studied the effects of thinning and nitrogen fertilization on annual dry matter production per unit of leaf area and leaf mass in pole-sized Douglas-fir trees. Both photosynthetic efficiency of foliage and foliage biomass increased with treatments (Figs. 3.38, 3.39), with greater relative response to fertilization. Increases in photosynthetic efficiency of foliage tended to peak within a few years of treatment, whereas the response of foliage biomass was delayed and prolonged. Centritto et al. (1999) employed classical growth analysis to study the effects of elevated CO_2 on growth of small sitka spruce and cherry trees. Elevation of CO_2 to 700 ppm increased total dry weight substantially in both species. Much of the growth advantage under high CO_2 accrued early and was associated with

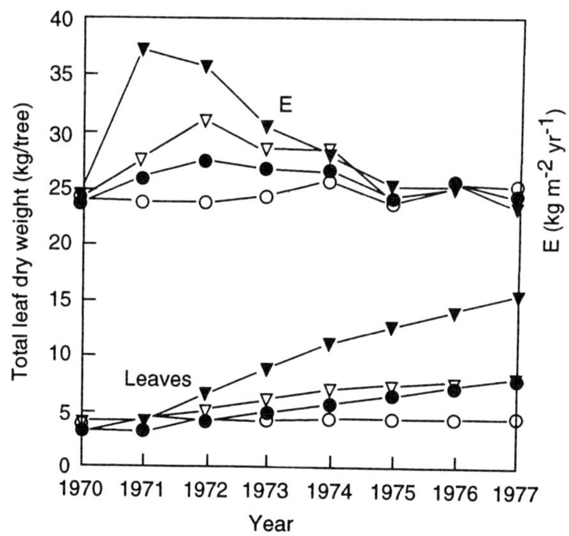

FIGURE 3.38. Response of leaf biomass and efficiency of stemwood production (E) of Douglas-fir saplings to thinning and fertilization in 1971. (○) no treatment; (●) stand thinned to one-third of original basal area; (▽) stand fertilized with 448 kg N/ha; (▼) stand thinned and fertilized. After Brix (1983).

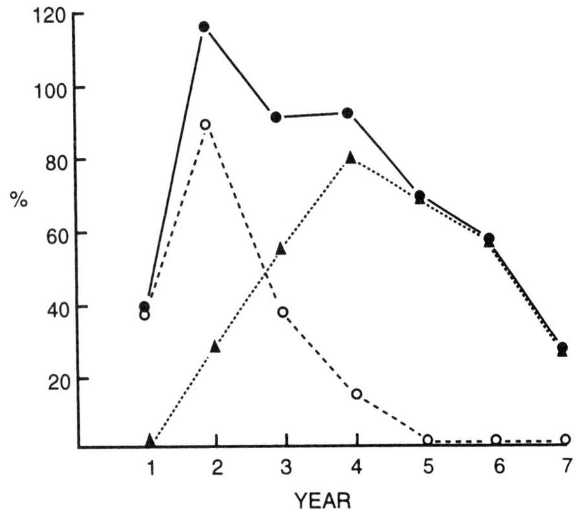

FIGURE 3.39. Stimulation of stemwood growth (percent above control) by thinning and fertilizer application to Douglas-fir trees. Total stem response (●) is partitioned into contributions (○) of photosynthetic efficiency (E) and (▲) foliage biomass in the years following treatment. From Brix (1983).

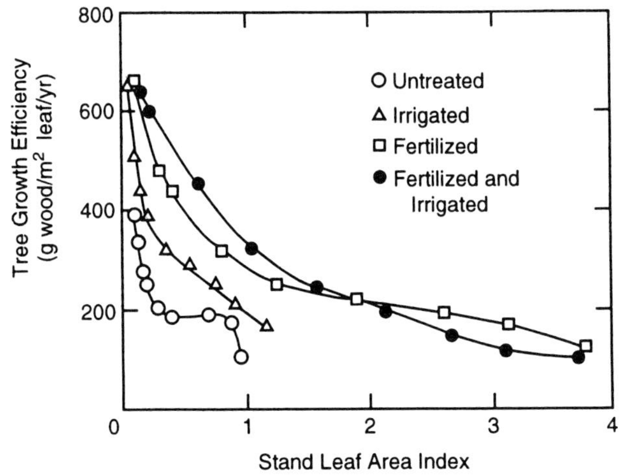

FIGURE 3.40. Tree growth efficiency response of Scotch pine saplings to ten years of fertilization and irrigation treatments. At low leaf area indexes, irrigated and fertilized trees show substantially greater growth efficiency. As stands approached maximum leaf area after 10 years growth efficiencies became similar, but leaf area index varied threefold. From Waring and Schlesinger (1985); originally from Waring (1985). (Reprinted from *For. Ecol. Manage.* **12**. Waring, R. H., Imbalanced forest ecosystems: Assessments and consequences, 93–112, © 1985 with kind permission of Elsevier Science-NL, Sara Burgerhartsraat 25, 1055 KV Amsterdam, The Netherlands.)

increased NAR. On the other hand, LAR and SLA exhibited a compensatory response over time, with plants in ambient CO_2 having greater allocation to leaf biomass and altered leaf morphology in a response that tended to offset the effects of persistently higher NAR in plants exposed to high CO_2.

As mentioned, there is considerable evidence that biomass production of forests is positively correlated with leaf biomass or leaf area index up to some optimum value. Beyond this there may be a decrease in net assimilation rate because of increasing shading of lower leaves (Waring and Schlesinger, 1985). For growth of forest stands it may be useful to consider a growth efficiency index that is similar in concept to the NAR but includes only stem wood growth (g wood m^{-2} leaf area yr^{-1}) (Waring et al. 1981). Waring and Schlesinger (1985) argued in favor of the value of this index because of the obvious economic importance of stem growth and its greater sensitivity to environmental stresses. The latter arises from a relative low priority for stem growth compared to new shoot and root growth. Competition and fertilizer studies have shown the effectiveness of the efficiency of stem wood growth in identifying environmental limitations on wood production (Fig. 3.40).

Limitations of Traditional Growth Analysis for Woody Plants

Growth analysis, which originally was developed for studies of annual herbaceous plants (Briggs et al., 1920 in Brix 1983), has limitations when applied to woody perennials. The accumulation of physiologically inert material in the plant body causes substantial reductions in RGR and LAR, and an ontogenetic decrease in the proportion of photosynthetic to respiratory tissue may reduce NAR. These responses have been termed "ontogenetic drift" (Evans, 1972) and, along with sheer logistical problems of measurements of large trees, have lessened the utility of growth analysis in forest stands. To some extent, this problem can be overcome in younger stands if treatment comparisons in growth parameters are made between trees of similar size rather than those of similar age. For example, when Britt et al. (1991) studied the influence of weed competition on loblolly pine seedling growth, they observed that RGR was higher during the first two years when weed competition was reduced, but by years 4 to 5 the trend was reversed. They concluded this was not a true treatment effect but rather the confounding ontogenetic effect of a much larger mean size of low-competition pine seedlings by years 4 to 5. When comparisons were made for plants of similar size rather than age, RGR was consistently higher when competition was reduced. Similar precautions are needed when employing the allometric formula to detect potential changes in allocation patterns (Ledig and Perry, 1965).

The complexity of growth of woody perennials discussed in this chapter also complicates application of classical growth analysis formulae, as the latter usually are based on simplistic underlying assumptions concerning the form of growth patterns between harvests (Radford, 1967; Evans, 1972).

Several alternate methods of growth analysis for large woody plants have been proposed, and some of these appear to have good utility. A general approach has been to remove the effect of the inert plant body by basing analysis on trends in annual production rates rather than total biomass. The growth efficiency (Waring et al., 1981) and photosynthetic efficiency (Brix, 1983) indexes described earlier are examples. Brand et al. (1987) proposed a general application of efficiency concepts to assessment of growth of any plant component. They recommended calculation of an index, Relative Production Rate (R_i), of any plant component as relative growth of annual increments according to the following general formula:

$$\overline{R_i} = \frac{\ln(y_i/y_{i-1})}{t_2 - t_1} \tag{3.11}$$

where $\overline{R_i}$ = mean R_i for some interval (usually one year) and y_i and y_{i-1} represent annual increments of current and previous intervals. As with most traditional growth analysis parameters, R_i can be resolved into yield components so that growth can be related to growth efficiency and allometry.

SUMMARY

Woody plants of the temperate zone alternate from active growth during the warm season to cessation of growth (dormancy) during the cold season. Control of and release from dormancy of buds has been variously correlated with plant hormone levels, but no convincing model of hormone control of dormancy has emerged. Plants grow through the activity of meristematic tissues, which constitute only a small fraction of the plant body. Shoot elongation generally arises from expansion of buds and shoot thickening through activity of the cambium. Shoots may be formed by fixed (determinate) growth, free (indeterminate) growth, or a combination of both. The pattern and periodicity of shoot growth vary greatly with species and genotype and with environmental influences. Particularly strong differences in shoot growth patterns often are observed between temperate and tropical species. Gibberellins and auxins have been implicated in control of shoot growth.

Increase in tree diameter arises from activity of the vascular cambium, a tissue that produces xylem inwardly and phloem externally. Cambial growth is not continuous in space or time. Temperate-zone trees exhibit seasonal activity of the vascular cambium, with reactivation in the spring and significant growing season influences, particularly those associated with drought. Auxins appear to play a predominant role in regulating cambial activity. Cambial growth of some tropical trees may continue for most or all of the year.

The amount and rate of sapwood and heartwood formation vary greatly with tree species, tree age, rate of growth, environmental conditions, and cultural practices. Important changes during conversion of sapwood into heartwood include death of ray and axial parenchyma cells; decrease in metabolic rates and enzyme activity; starch depletion; darkening of xylem and deposition of extractives; increase in pit aspiration; formation of tyloses; and changes in moisture content.

Root growth originates in the radicle of the seed from which the first tap root develops. Whereas lateral shoot growth originates in peripheral tissues, lateral roots arise from the vascular cylinder within an existing root. Root growth and rooting of cuttings is stimulated by auxins and inhibited by low concentrations of cytokinins and ethylene. Vigorous branching and root elongation growth are characteristic of most plants, but species vary widely in root system form and activity in response to environment. Turnover of roots, particularly fine roots, is substantial and consumes a significant portion of a plant's annual net primary productivity. Perennial roots possess a cambium, but diameter growth is more irregular than in stems both among and within roots.

Woody plants recurrently shed buds, shoot tips, branches, prickles, cotyledons, leaves, stipules, bark, roots, and reproductive structures. Natural shedding of plant parts alters plant form; produces litter that becomes soil organic matter; contributes to drought tolerance; removes infected or diseased plant parts; and reduces competition within plants for water and mineral nutrients. Shedding of leaves may be harmful to plants by slowing plant growth, inhibiting seed germination through effects of litter, and causing loss of nutrients from ecosystems. Leaves of shoots may persist from a few weeks to years, depending on species and environmental factors such as light, drought, nutrient supply, and herbivory. Leaf shedding may occur more or less continually, as in many tropical forests, or be quite closely timed to seasons as in temperate deciduous forests or seasonally dry tropical forests.

Physiologists have developed methods of growth analysis that permit dissection of basic growth patterns into parameters offering insight into the mechanisms of growth. Most useful have been techniques that estimate allometric coefficients and partition growth into components of growth efficiency per unit of leaf area and allocation of dry matter into new leaves. Conventional growth analysis has distinct limitations in application to perennial species because of a progressively increasing influence of a growing plant body on various parameters. A variety of growth analysis parameters not subject to ontogenetic influences on growth of woody plants has emerged, most of which employ measurement and analysis of growth increments rather than total cumulative growth.

General References

Addicott, F. T. (1982). *Abscission*. University of California Press, Berkeley.

Addicott, F. T. (1991). Abscission: Shedding of plant parts. In *Physiology of Trees*, A. S. Raghavendra, ed., pp. 273–300. Wiley, New York.

Barnett, J. R., ed. (1981). *Xylem Cell Development*. Castle House, Tunbridge Wells, England.

Bolwell, G. P. (1988). Synthesis of cell wall components: Aspects of control. *Phytochemistry* **27**, 1235–1253.

Cannell, M. G. R. (1989). Physiological basis of wood production: A review. *Scand J For Res* **4**, 459–490.

Carlquist, S. (1988). *Comparative Wood Anatomy: Systematic, Evolutionary, and Ecological Aspects of Dicotyledonous Wood*. Springer-Verlag, New York and Berlin.

Causton, D. and Venus, J. (1980). *The Biometry of Plant Growth*. Edward Arnold, London.

Dale, E. (1982). *The Growth of Leaves*. Arnold, London.

Dale, E. (1992). How do leaves grow? *Bioscience* **42**, 423–432.

Davies, P. J., ed. (1995). *Plant Hormones: Physiology, Biochemistry, and Molecular Biology*. Kluwer Academic, Dordrecht and Boston.

Esau, K. (1965). *Plant Anatomy*. Wiley, New York.

Evans, G. C. (1972). *The Quantitative Analysis of Plant Growth*. University of California Press, Berkeley.

Fahn, A. (1990). *Plant Anatomy*, 4th edition. Pergamon, Oxford.

Hunt, R. (1978). *Plant Growth Analysis*. (Studies in Biology Number 96). Edward Arnold, London.

Hunt, R. (1990). *Basic Growth Analysis: Plant Growth Analysis for Beginners*. Unwin Hyman, London and Boston.

Iqbal, M., ed. (1990). *The Vascular Cambium*. Research Studies Press, Taunton, England.

Kozlowski, T. T. (1971). *Growth and Development of Trees*, Vols. 1 and 2. Academic Press, New York.

Kozlowski, T. T. and Pallardy, S. G. (1997). *Growth Control in Woody Plants*. Academic Press, San Diego.

Larson, P. R. (1994). *The Vascular Cambium: Development and Structure*. Springer-Verlag, New York and Berlin.

Longman, K. A. and Jenik, J. (1987). *Tropical Forest and Its Environment*, 2nd Edition. Longman, New York.

Martin, G. C. (1991). Bud dormancy in deciduous fruit trees. In *Plant Physiology: A Treatise*. Volume 10. F. C. Steward, ed., pp. 183–225. Academic Press, San Diego.

Srivastava, L. M. (2002). *Plant Growth and Development: Hormones and Environment*. Academic Press, San Diego.

Steeves, T. A. and Sussex, I. M. (1989). *Patterns in Plant Development*. Cambridge Univ. Press, Cambridge and New York.

CHAPTER

4

Reproductive Growth

INTRODUCTION

Production of fruits and cones depends on successful completion of several sequential stages of reproductive growth, including:

- Initiation of floral primordia
- Flowering
- Pollination (transfer of pollen)
- Fertilization (fusion of male and female gametes)
- Growth and differentiation of the embryo
- Growth of fruits and cones to maturity
- Ripening of fruits, cones (strobili), and seeds

After seeds germinate, young woody plants remain for several years in a juvenile condition during which they normally do not flower. They then undergo complex physiological changes as they progress from the juvenile to a mature stage and finally to a senescent condition. In addition to lacking capacity for flowering during the juvenile stage, woody plants also may differ from adult plants in leaf shape and structure, ease of rooting of cuttings, leaf retention, stem anatomy, thorniness, and production of anthocyanin pigments. The length of the juvenile, nonflowering stage varies greatly among species. Some conifers remain in the juvenile stage for less than a year; others may retain juvenility for as much as 45 years or even throughout their life. For example, the ornamental retinosporas are juvenile forms of *Chamaecyparis* and *Thuja*, but because their appearance differs so much from the adult form these plants were erroneously classified in the genus *Retinospora*.

Once plants achieve the adult stage, their capacity for flowering usually is retained thereafter. However, it is important to distinguish between the capacity for flowering (achievement of the adult stage) and annual initiation of flower primordia. Juvenility has both desirable and undesirable aspects. A long juvenile phase is desirable when trees are grown for wood because no energy is directed to reproduction, and hence wood production is increased. By comparison, juvenility (lack of reproductive capacity) obviously is undesirable for fruit growers. Adult woody plants may possess a capacity for flowering but may not flower every year. Both environmental and internal conditions can control initiation of flower primordia after the adult stage has been achieved (see Chapters 4 and 6, Kozlowski and Pallardy, 1997).

Reciprocal Relations between Vegetative and Reproductive Growth

The size of a fruit or seed crop typically is correlated negatively with the amount of vegetative growth. This is dramatically shown in some biennially bearing fruit trees, with vegetative growth suppressed during the fruiting year and sometimes in the subsequent year or even years. In bearing peach trees, vegetative growth was suppressed over a three-year period (Proebsting, 1958). The size of an apple crop and cambial growth showed a negative linear correlation (Webster and Brown, 1980). A similar correlation was shown between crop yield and shoot growth of citrus trees (Sanz et al., 1987). Root growth commonly is reduced more than shoot growth by fruiting. When the fruit crop is unusually large, root growth may be reduced so much that the tree eventually dies (Smith, 1976). In forest trees, vegetative growth is greatly reduced during good seed years (Kozlowski, 1971b).

Biennial bearing in some species is associated with variations in production of flower buds; in other species, with abscission of flower buds. In apple, prune, and pecan trees the flower buds form in alternate years, presumably because heavy fruiting inhibits flower bud induction. In biennially bearing pistachio, however, flower buds form each year regardless of fruit load. Alternate bearing results from abscission of flower buds during the year that a heavy crop of fruit is produced (Crane et al., 1976). Irregular bearing of avocado results from excessive shedding of young fruits, resulting in a light crop, which is followed by a heavy crop in the next year.

The relationship between vegetative and reproductive growth often is linear over a considerable range, but not necessarily so when fruit crops are abnormally high or low. As shown in Figure 4.1, the growth-fruit yield relationship is more likely to be represented by the curve A-A′ rather than B-B′ (Proebsting, 1958). In some biennially-bearing species, the relation between reproductive growth and vegetative growth is positive. For example, pistachio trees produced their most extensive shoot growth in years in which they also had the heaviest crops of nuts. This correlation was associated with storage of more starch in nonbearing than in bearing branches (Crane and Al-Shalon, 1977).

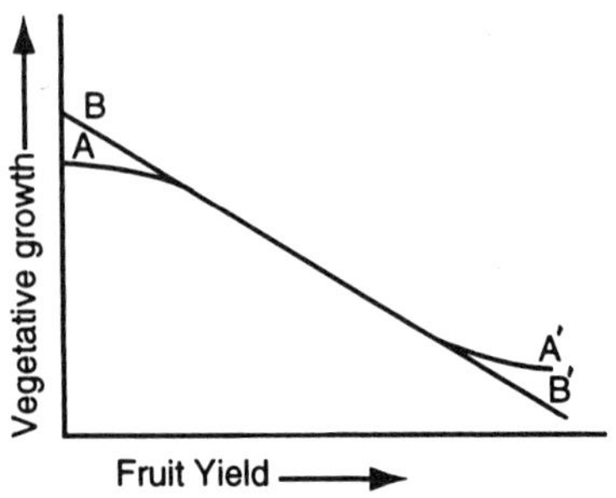

FIGURE 4.1. Relations between vegetative growth and reproductive growth under extremes of cropping. For explanation, see text. From Proebsting (1958).

SEXUAL REPRODUCTION IN ANGIOSPERMS

Several distinct stages of flowering in angiosperms have been described (Metzger, 1988), including (1) flower initiation (production of flower or inflorescence primordia), (2) evocation (processes occurring in the apex, before, and necessary for, formation of flower primordia), (3) flower formation (appearance of floral structures), and (4) flower development (processes occurring between flower formation and anthesis).

Flowering Periodicity

Floral initiation and development of woody plants of the temperate zone vary within and between years. Most deciduous broadleaved trees initiate flower buds in the summer or autumn preceding the spring in which the flowers open. The buds remain dormant over winter and anthesis occurs between nine and 12 months after initiation. Sweet orange is an exception, with flowers initiated as new shoots emerge during the late winter and early spring and anthesis occurs during the same spring (Mullins et al., 1989).

Floral induction in some species occurs throughout much of the summer. In pome fruits, flower bud initiation extends from the period of early fruit development almost to maturity, depending on the time of shoot growth cessation. Hence, while the current apple crop is growing, the following year's crop is beginning and developing. In stone fruits flower buds are initiated after fruit development in the same year is completed (Tukey, 1989).

The actual time of floral initiation is modified by weather, site conditions, and management practices. Hence, the time of floral initiation varies somewhat from year to year and also in different parts of the natural range of a species. The time of floral initiation also differs with shoot location on the tree. Flower buds of apple are initiated later in terminal buds of shoots than of spurs, and later in terminal buds of spurs on two-year-old shoots than those on three-year-old shoots (Zeller, 1955).

Flowering periodicity in the tropics is very diverse and differs greatly among species, within species, and within site (Borchert, 1983; Bawa, 1983). A major distinction often has been made between extended flowering and mass flowering of tropical trees. In extended flowering, the trees produce few flowers each day but continue to bloom for weeks to months. An example is *Muntingia calabura*, which flowers more or less continuously throughout the year (Frankie et al., 1974). In mass flowering, the trees produce many flowers each day but bloom for only short periods. For example, at irregular intervals of two to 10 years several species of the family *Dipterocarpaceae*, which dominate tropical rain forests, flower more or less simultaneously (Ashton et al., 1988). Dipterocarps typically flower for a few weeks to a few months after years without reproductive activity. Following a mass flowering event, enormous numbers of fruits and seeds ripen and are shed.

Some advantages have been attributed to extended flowering as well as to mass flowering (Bawa, 1983). Extended flowering (1) involves less risk of reproductive failure because of unfavorable weather or lack of pollinators, (2) adjusts the rate of flower production to match the resources available for fruit production, and (3) in outcrossing populations, increases fertilization of many mates and ensures pollen donors from many ecotypes. Mass flowering is advantageous to species consisting of relatively few plants because production of small numbers of flowers may not attract pollinators. Furthermore, mass flowering should protect plants from flower herbivores if the plants do not have other defense mechanisms.

Seasonal flowering patterns often vary among life forms in the same ecosystem. For example, trees in a temperate rain forest in Chile flowered later than shrubs. The flowering period also was longer in vines and hemiparasites than in trees and shrubs (Smith-Ramirez and Arnesto, 1994). Phenological differences in flowering also were shown among life forms in a seasonal dry forest in Costa Rica (Opler et al., 1980).

In nonseasonal tropical climates, flowering may be seen throughout the year, with some species blooming almost continuously. Flowering maxima occur at certain times of the year, and are much greater when there is a distinct dry season. In most seasonal tropical forests, flowering occurs chiefly in the dry season and often extends into the wet season. Variations in flowering periodicity of plants in a wet forest and a dry forest in Costa Rica are shown in Figure 4.2. Most wet forest plants exhibited several flowering episodes each year at three- to five-month intervals. Flowering of plants of the dry forest showed pronounced seasonal peaks, with most species flowering synchronously once or twice a year.

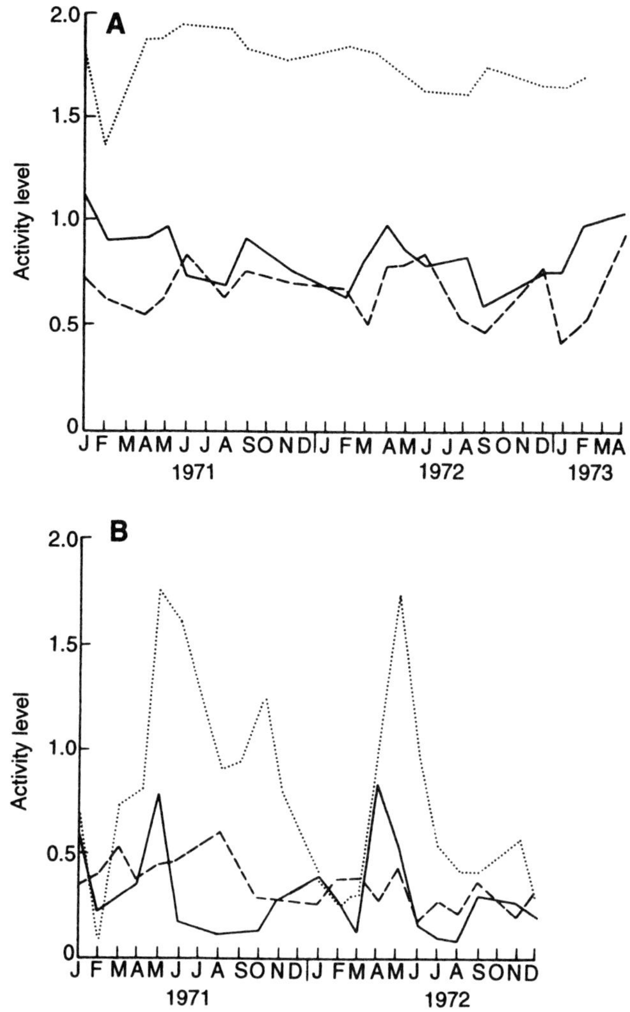

FIGURE 4.2. Variations in phenological patterns of treelets and shrubs in a young secondary forest at a wet site (A) and a mature forest at a dry site (B). Dotted line, leaf production; solid line, flower production; dashed line, mature fruit production. From Opler et al. (1980).

Flowering periodicity of the same species often varies with site. For example, in the lowland tropical rain forest *Hamelia patens* flowers continually; in the tropical dry forest it flowers annually but not in the dry season. *Ceiba pentandra* and *Andira inermis* show greater flowering frequency in the dry tropics than in the wet tropics (Newstrom et al., 1993).

Flowering periodicity often varies among species that occupy different layers of tropical forests. In British Guiana forests, plants of the upper story had two major flowering seasons, whereas flowering of shrubs was similar throughout the year (Richards, 1966). In a dry forest of Costa Rica most shrubs bloomed before and just after the flowering periods of trees (Frankie et al., 1974). Flowering periodicity also varied among species occupying different canopy layers of a semi-deciduous forest in Panama (Croat, 1969). Flowering phenology also varies between male and female plants of the same species. Male plants usually start flowering earlier and produce many more flowers than female plants do (Bawa, 1983).

Some idea of the extreme diversity of flowering in the tropics may be gained from the work of Newstrom et al. (1994), who classified flowering periodicity of woody plants at the La Selva Biological Station in Costa Rica into the following major classes:

Continual flowering: Plants flower more or less continually with sporadic brief breaks.
Subannual flowering: Plants flower in more than one cycle per year.
Annual flowering: Plants have only one major flowering cycle per year.
Supra-annual flowering: Plants have one flowering cycle over more than one year.

Newstrom et al. (1993) used other criteria such as duration of flowering, flowering amplitude, and date to flowering to further subdivide the major classes.

The longevity of individual flowers usually is quite short but differs with species, habitat, and environmental conditions. Tropical forest trees have predominantly one-day flowers; in other regions trees have longer-lived flowers. In temperate forests, the species that flowered in the spring, early summer, and late summer had flowers that lasted, on average, 6.8 days, 5.7 days, and 2.5 days, respectively (Primack, 1985). Flower longevity increased in the following order: dipterocarp forest, tropical dry forest, tropical rain forest, late-summer flowering temperate forest, early-summer flowering temperate forest, and spring-flowering temperate forest. Flower longevity is increased by cool day and night temperatures.

Pollination

The longevity of flowers is regulated by environmental conditions and more dramatically by pollination. Typically, a wave of increased potential for ethylene production passes through floral tissues after pollination and accelerates senescence and shedding of flowers (Stead, 1992). Many flowering plants produce far fewer mature fruits than flowers. Furthermore, fruits often contain fewer seeds than the number of ovules that were available for fertilization. This condition often is associated with an adequate pollen supply. In self-incompatible species pollinators may deposit so much of their own pollen on their stigmas that they become clogged and cannot be fertilized later. In addition pollen from the wrong species decreases fertilization. In self-compatible species, the ovules that are fertilized by self-pollen often produce seed that is not viable. Furthermore, many fruits and seeds abort even when compatible pollen is deposited on the stigma (Charlesworth, 1989).

Enormous amounts of pollen are produced by trees, particularly by wind-pollinated species. According to Faegri and Iversen (1975), a single anther of wind-pollinated birch may produce 10,000 pollen grains, whereas an anther of insect-pollinated maple produces only 1,000 grains. A wide range of vectors (e.g., wind, water, insects, birds, and mammals) is involved in pollen transfer. Pollen grains of birches, poplars, oaks, ashes, elms, hickories, sycamores, and conifers are dispersed by wind; those of basswood, maples, willows, and most fruit trees are distributed by insects, chiefly bees. A larger proportion of the pollen is wasted when dispersed randomly by wind than when distributed more systematically by insects moving from flower to flower. Wind pollination increases with both latitude and elevation; it is common in temperate, deciduous, and boreal forests but extremely uncommon in tropical rain forests, where insect pollination is more common. The tendency for wind pollination is high on remote islands, and higher in early than in late successional ecosystems (Whitehead, 1984).

The effectiveness of wind pollination depends on the number of pollen grains reaching receptive plants. Most airborne pollen comes to rest rather close to the tree that produced it. For example, Wright (1952, 1953) showed that pollen of some forest trees traveled far less than 100 meters.

Pollen is dispersed by both air turbulence and horizontal movement. Some air movement is necessary for shaking of anthers to release pollen. Dispersal of pollen depends on interactions between the terminal velocity of pollen grains (controlled largely by their density and size) and wind velocity. Small, light grains are dispersed more efficiently than large, heavy grains, which settle much faster. Pollen dispersal generally increases as the wind velocity increases and settling velocity decreases. The pollen dispersion distances vary considerably among species. For example the pollen of ash trees has a more rapid rate of fall and shorter dispersion distance than the pollens of poplar, elm, or walnut.

Fruit Set

The term "fruit set" has been used in the literature to refer to both initial and final fruit set. Initial fruit set occurs shortly after anthesis and is associated with the beginning of swelling of the ovary. Final fruit set refers to the number of fruits on a tree when the fruits and seeds are mature (Sedgley and Griffin, 1989). When a flower has been successfully pollinated, growth of the ovary is stimulated and floral parts such as stamens and petals usually wilt and abscise. Such changes, which characterize transformation of a flower into a young fruit, comprise initial fruit set.

On pollination a wave of biochemical activity precedes the pollen tubes along the length of the pistil. Typical changes include an increase in polysomes, variations in RNA and protein synthesis, as well as in amounts of sugar and starch and in respiration rate (Herrero, 1992). Except in apomictic and parthenocarpic species, flowers must be pollinated in order to set fruit. Marked differences in fruit setting among species and genotypes have been attributed to variations in pollination, fertilization, ovule longevity, flower structure, temperature, light intensity, competition among reproductive structures for resources (e.g., carbohydrates, hormones, and mineral nutrients), and competition for resources between reproductive and vegetative tissues (Dennis, 1979; Dennis et al., 1983; see Chapter 4 of Kozlowski and Pallardy, 1997). Young fruit trees often produce as many flowers (on a crown volume basis) as older trees but they produce smaller crops, partly because of lower fruit set. The age of wood on which flowers are borne also can influence fruit set. In apple trees the capacity of flowers to set fruit was lower on young than on older wood (Robbie and Atkinson, 1994).

Fertilization

A requirement for pollination is receptivity of the stigma when viable pollen reaches it. Such synchronization occurs within and among flowers on the same plant. In other cases, however, pollen shedding and stigma receptivity are not synchronized. For example, in individual plants of certain monoecious species (staminate and pistillate flowers on the same plant), pollen is released before the time of stigma receptivity; in other plants, pollen is released after stigma receptivity has ended. For example, sugar maple, which appears to have bisexual flowers but which is usually functionally monoecious, pollen is released before the time of female flower receptivity on a given tree (Gabriel, 1968). As other sugar maple trees in a population will be slightly different in phenological stage and have receptive female flowers at this time, outcrossing is favored. The pattern of temporal difference in pollen release and pistillate receptivity is termed dichogamy, with protoandry designating earlier pollen release and protogyny earlier pistillate receptivity. Similar differences may be found in some dioecious species among plants with unisexual flowers. Such differences in timing prevent self pollination. This is a valuable characteristic because progeny produced by cross pollination generally are more vigorous than those resulting from self pollination.

The essentials of the ovule inside of the ovary of angiosperms are the outer integuments, the micropylar opening opposite the stalk end, and the embryo sac, which occupies most of the ovule. Before fertilization the embryo sac generally contains eight nuclei. The three located at the micropylar end consist of the egg nucleus and two synergids. Two polar nuclei are located in the central part of the embryo sac, and the three nuclei at the end opposite the micropyle are the antipodals (see Fig. 2.32).

When viable pollen grains reach a receptive stigma they imbibe water and germinate, producing pollen tubes. After the pollen tube penetrates the stigmatic surface, it extends by tip growth and, after internal resources are exhausted, depends on free sugars, amino acids, glyco-lipids, lipids, polysaccharides, and proteins secreted into the extracellular matrix of the pistil tissues by transmitting tract cells (Swanson et al., 2004). Pollen tubes grow at various rates through the style and into the ovary, which contains the ovules. Guidance of the tube to the ovule is complex and not completely understood, but tissue gradients of amino butyric acid are involved in guiding the final stages of ovule approach. Although many pollen tubes may reach the ovary, only one enters the ovule. When the pollen tube reaches a synergid, growth ceases and the tube tip ruptures releasing sperm cells for double fertilization. The tube nucleus and generative nucleus of the pollen grain enter the pollen tube when it is formed. The generative nucleus undergoes an early mitotic division to produce two sperm. One of these fuses with the egg and the other with the two polar nuclei to form an endosperm nucleus. After fertilization is completed and a zygote formed, the remaining nuclei of the embryo sac, consisting of two synergids and three antipodals, usually degenerate.

Because pollen tubes grow very rapidly (sometimes several mm/hr) the time span between pollination and fertilization usually is rather short. According to Maheshwari (1950), in most angiosperms only 24 to 48 hours elapse between pollination and fertilization. However, this interval varies widely among species, and may be 12 to 14 hours in coffee (Mendes, 1941), three to four months in European hazel (Thompson, 1979), and 12 to 14 months in certain oaks (Bagda, 1948, 1952).

Postfertilization Development

After fertilization there is rapid growth of the endosperm and translocation of food into the enlarging ovule. Normal growth and differentiation of the embryo depend on the endosperm, which when mature is rich in carbohydrates, fats, proteins, and growth hormones. As the embryo increases in size, it draws on the contents of endosperm cells and in some species may consume nearly the whole endosperm.

Several lines of evidence emphasize the importance of the endosperm as a nurse tissue for embryo growth. At the time of fertilization, the embryo sac lacks an appreciable amount of food. As the endosperm grows, however, it accumulates enough food to supply the developing embryo. The zygote usually does not divide until after considerable endosperm growth takes place. Furthermore, the embryo develops normally only when the endosperm is organized. Should endosperm abortion occur, growth of the embryo is subsequently inhibited (Maheshwari and Rangaswamy, 1965).

Polyembryony

Multiple embryos sometimes develop in seeds. Polyembryony has been classified as either false or true. False polyembryony involves fusion of two or more nucelli or development of two or more embryo sacs within the same nucellus. In true polyembryony the additional embryos arise in the embryo sac either by cleavage of the zygote or from the synergids and antipodal cells. Adventive embryos, which also are examples of true polyembryony, arise from tissues outside the embryo sac (e.g., cells of the nucellus or integuments). Ultimately they enter the embryo sac where they grow to maturity. Adventive embryos differ from sexual embryos in having a lateral position, irregular shape, and lack of suspensor. They also are genetically different, the adventive embryos containing only genes from the maternal plant. In several species polyembryony has been linked to genetic causes and appears to result from hybridization (Maheshwari and Sachar, 1963).

Polyembryony has been of particular practical importance in propagating certain species of trees. For example, adventive embryos, which inherit characters of the maternal parent, have been used to provide genetically uniform seedlings of mango and citrus. In citrus they have been used as orchard stock on which grafts from other types have been made. In addition, citrus clones that have deteriorated after repeated vegetative reproduction have been restored to original seedling vigor with nucellar embryos (Maheshwari and Sachar, 1963). Such vegetative invigoration, or *neophysis*, may be traceable to hormonal influences of the embryo sac. Hofmeyer and Oberholzer (1948) raised better citrus seedlings from adventive embryos than from cuttings. The difference was attributed to infection of cuttings with virus diseases that were absent in nucellar embryos.

Apomixis

Although seeds usually arise from sexual reproduction, in a few species of woody plants they arise asexually. Plants in which seeds form without fertilization of the ovule (e.g., from diploid cells in the ovule) are called apomicts. Many species of hawthorns and blackberries in the eastern United States are apomictic derivatives. Most cultivated citrus varieties are facultatively apomictic by means of adventive embryony (Grant, 1981). Apomixis also is presumed to be present but not demonstrated in forest trees. Apomictic seeds often are indistinguishable from seeds of the same species resulting from sexual reproduction. Apomixis is of particular interest to geneticists because all seedlings produced by this process have the same genetic constitution.

Parthenocarpy

Ordinarily the development of mature fruits requires fertilization of the egg. In a few species, however, fruits are set and mature without seed development and without fertilization of an egg. Such fruits, called parthenocarpic fruits, are well known in some figs, pears, apples, peach, cherry, plum, and citrus. Parthenocarpy also occurs in several genera of forest trees including maple, elm, ash, birch, and yellow-poplar.

Some types of parthenocarpy require pollination and others do not. For example, fruit development in citrus and banana may occur without pollination. In other species of fruit trees, such as cherry and peach, seedlessness may occur because the embryo aborts before the fruit matures. In some species pollination stimulates fruit development, but fruits mature without the pollen tube reaching the ovule.

Growth of Fruits

The time required for fruit growth varies widely among species and genotypes. The period from anthesis to fruit ripening varies from about three weeks in strawberry to 60 weeks in Valencia orange. However, in fruits of many species this interval is about 15 weeks. Sparks (1991) found that the time of fruit growth of 46 pecan cultivars ranged from 19 to 33 weeks. Two distinct ecotypes were identified, those adapted to a growing season of 23 to 26 weeks and those adapted to about 30 weeks or more. Early maturity of nuts was associated with small nut size.

Various types of fruits grow at different rates and reach different sizes at maturity. For example, olives and currants grow very slowly (about 0.01–0.02 cm^3/day) (Bollard, 1970). It should be remembered, however, that growth rates of fruits vary greatly among seasons, environmental conditions, cultural practices, and even different fruits in the same crop.

Growth of fruits involves various degrees of cell division and cell expansion. During anthesis there is little cell division; after a fruit is set, however, it becomes an active carbohydrate sink, and many of its tissues become meristematic. In some fruits (e.g., currants, blackberries) cell division (except in the embryo and endosperm) is completed by the time of pollination, in others (e.g., apple, citrus) cell division occurs for a short time after pollination, and in still others cell division occurs for a long time after pollination. In avocado, cell division continues throughout the life of the fruit. In most species, however, increase in cell size makes the greatest contribution to total fruit expansion. In grape the increase in cell number accounts for a doubling of fruit size whereas increase in cell volume accounts for a 300-fold size increase (Coombe, 1976). The physiological regulation of fruit growth is discussed in Chapter 4 of Kozlowski and Pallardy, 1997.

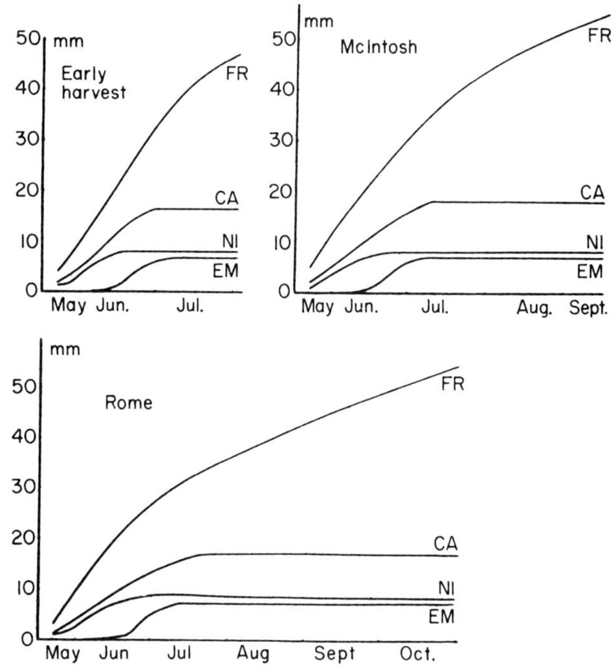

FIGURE 4.3. Increase in length of the embryo (EM), nucellus and integuments (NI), carpel (CA), and whole fruit (FR) of Early Harvest, McIntosh, and Rome apples from full bloom to fruit ripening. From Tukey and Young (1942). *Botanical Gazette* © University of Chicago Press.

Growth Curves

Growth curves of fruits are of two general types. The first is a simple sigmoid type in which there is initially an exponential increase in size followed by slowing down of growth in a sigmoid fashion (Fig. 4.3). This type of curve is characteristic of apple, pear, orange, date, banana, avocado, strawberry, mango, and lemon. The precise shape of the growth curve often differs somewhat with variety. The growth curve for development of Early Harvest apple fruits resembles a straight line, whereas that for each successively later ripening variety flattens as the season progresses. The rate of fruit growth also is a varietal characteristic, with early-ripening varieties growing faster than late-ripening ones (Tukey and Young, 1942).

The second type of growth curve, characteristic of stone fruits (e.g., cherry, peach, apricot, plum, olive, and coffee), as well as some nonstone fruits (e.g., grape and currant), is a double sigmoid type that depicts growth occurring in three stages (Fig. 4.4). In stone fruits during Stage I the ovary, nucellus, and integuments of the seed grow rapidly, but the embryo and endosperm grow little. During Stage II the embryo and endosperm grow rapidly, but the ovary does not increase much in size. Sclerification of the pit also begins and, by the end of Stage II, the embryo achieves full size and the amount of endosperm material increases greatly. Finally, in Stage III a new surge of ovary growth begins and continues to fruit ripening. The duration of the three growth stages is quite variable. Stage II may last only a few days in early-ripening varieties and about one month in late-ripening ones.

Neither of the two major types of growth curves is restricted to a particular morphological type of fruit. For example, growth of certain berries, pomes, simple fruits, and accessory fruits can be characterized by one or the other type of growth curve.

Fruit Ripening

Defining fruit ripening to satisfy everyone is very difficult. Consumers of fruits are interested in such aspects of ripening as taste, color, texture, aroma, and nutritional values of fruits. Growers and shippers of fruits are more interested in ripening as it relates to keeping quality and response to low temperature in storage. Here I will focus attention on ripening as defined by Watada et al. (1984), namely, "the composite of the processes that occur from the latter stages of growth and development through the early stages of senescence and that results in characteristic aesthetic

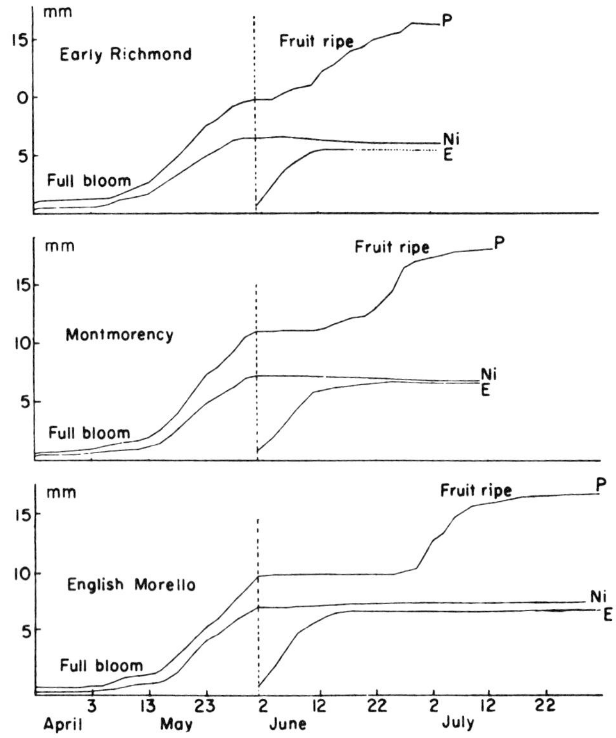

FIGURE 4.4. Growth curves of the pericarp (P), nucellus and integuments (Ni), and embryo (E) of three varieties of cherry ripening at different seasons. The times of rapid increase of embryo growth and related inhibition of growth of the nucellus and integuments and pericarp were similar for all varieties. Periods of rapid embryo increase also were similar, but rates of pericarp development were not. From Tukey (1935).

and/or food quality, as evidenced by changes in composition, color, texture, or other sensory attributes."

The ripening of fruits is characterized by a number of changes in color, texture, flavor, aroma, metabolism, and gene expression (Table 4.1; Fig. 4.5) that occur concurrently (Rhodes, 1980; Tucker and Grierson, 1987; Giovannoni, 2001; see Chapter 4 of Kozlowski and Pallardy, 1997).

The time of ripening varies with the developmental stage of fruits. In fruits with a single sigmoid pattern of growth, ripening usually occurs during the final phase of slow growth. In fruits with a double sigmoid growth curve ripening begins during Stage III; in grape and coffee (Fig. 4.6) near the beginning and in stone fruits toward the end of Stage III. Climacteric fruits (Chapter 6) normally enter the ripe edible stage at or shortly after the climacteric peak. Many fruits change color during ripening. The color of many ripe fruits such as apples, cherries, grapes, and blackberries is due to production of anthocyanidins and their glycosides (the anthocyanins) (Tucker and Grierson, 1987).

Much attention has been given to anthocyanin synthesis in developing apples. Young apples (<2 mm in diameter) have a background green color with a temporary peak of red color that disappears early but reappears during fruit ripening. In Orange Pippin apples anthocyanin pigments tripled during the month of ripening, chlorophyll concentration decreased four-fold, and carotenoids increased four-fold (Knee, 1972).

TABLE 4.1. Ripening Phenomena in Fruits[a,b]

Degradative	Synthetic
Destruction of chloroplasts	Maintenance of mitochondrial structure
Breakdown of chlorophyll	Formation of carotenoids and anthocyanins
Starch hydrolysis	Interconversion of sugars
Destruction of acids	Increased TCA cycle activity
Oxidation of substrates	Increased ATP generation
Inactivation of phenolic compounds	Synthesis of flavor volatiles
Solubilization of pectins	Increased amino acid incorporation
Activation of hydrolytic enzymes	Increased transcription and translation
Initiation of membrane leakage	Preservation of selective membranes
Ethylene-induced cell wall softening	Formation of ethylene pathways

[a]From Biale and Young (1981).
[b]TCA, tricarboxylic acid; ATP, adenosine triphosphate.

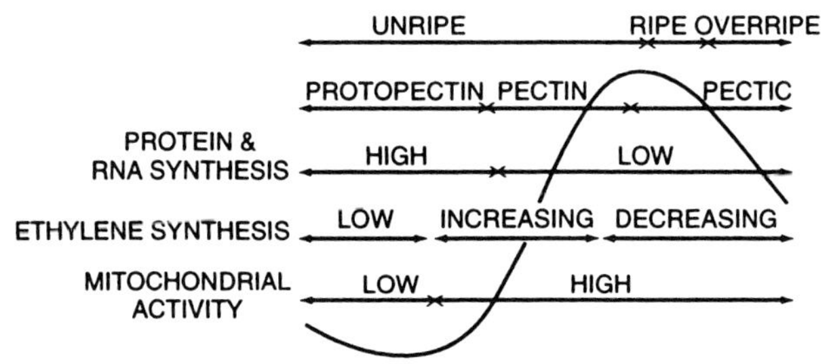

FIGURE 4.5. Changes in important ripening events as related to the climacteric pattern of fruits. From Biale and Young (1981).

Flavor

The development of flavor in fruits is influenced by both physical changes during development and by chemical constituents. Transformation of insoluble to soluble pectins occurs in cell walls of ripening fruits. Rupture of some of the cells follows and juices are released into spaces between cells thereby enhancing the taste of fruits.

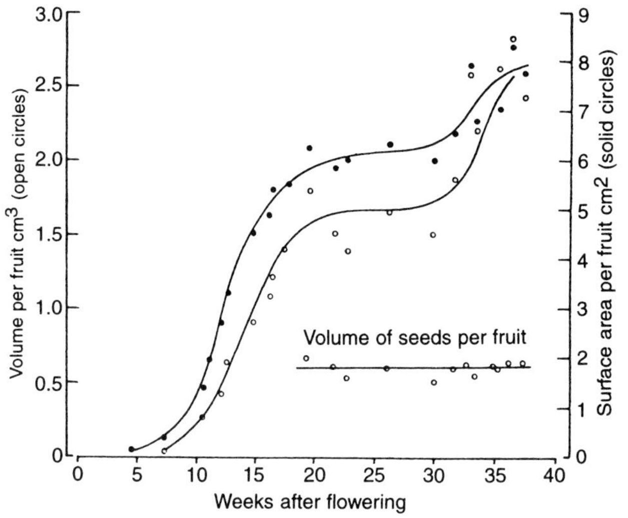

FIGURE 4.6. Increase in volume and surface area of coffee fruits and volume of the seeds. Ripening of fruits begins at the beginning of the second phase of expansion. From Cannell (1971).

Flavor is determined by a balance of sugars, acids, and astringent compounds. An increase in sweetness and a decrease in acidity usually accompany ripening. Most fruits contain starch that is converted to sugars as ripening proceeds. Sucrose, glucose, and fructose are the major sugars present in varying proportions in ripe fruits. In apples, all three sugars are present; in peaches and apricots sucrose predominates; and in cherry the bulk of the sugars consists of glucose and fructose. The sugar content of edible fruits varies from near 80% on a dry weight basis in grapes (Coombe, 1976) to between 9 and 20% in bananas, oranges, and apples.

The concentration of organic acids in fruits is relatively high, often amounting to 1 to 2% of fresh weight in apples and up to 4% in black currants. Organic acids probably are translocated into fruits from the leaves as inferred from decreases in acid contents of fruits following reduction of leaf area and higher acid contents in the center than in the periphery of grapes and apples (Nitsch, 1953). Malic and citric acids are the most common acids in fruits (malic in apples, apricots, bananas, cherries, peaches, pears, and plums; citric in citrus fruits, raspberries, and strawberries). Many other fruits contain mixtures of malic and citric acids. In grapes tartaric acid occurs in amounts as high as those of malic acid. Relatively small amounts of other organic acids also are found in fruits.

Flavor of fruits also is associated with aroma, which arises from greatly increased production of many volatile compounds during ripening, including organic acids, alcohols, esters, carbonyl compounds, hydrocarbons, and terpenoids. Total volatiles in fruits typically range from 1 to 20 ppm but in bananas may be as high as 300 ppm. Very small amounts of more than 200 different volatile compounds have been found in bananas. At least two major groups of precursors of volatile compounds have been identified. These include long-chain amino acids, leucine, isoleucine, and valine as well as linoleic and linolenic acids (Rhodes, 1980).

Softening of Fruits

The softening of fruits is related to changes in pectic compounds and hydrolysis of starch or fats. During fruit development, insoluble protopectin is laid down in primary cell walls and accumulates to high concentrations in some fruits, including apples, pears, and citrus fruits. Changes from insoluble pectins (protopectin and calcium pectate) to soluble compounds such as pectin, pectic acid, and pectinic acid are closely related to ripening. Between the green and ripe stages, soluble pectins in peaches may increase by a factor of two, in apples by 12, and in pears by 20. In overripe fruits, pectic substances tend to disappear and fruits soften greatly. In ripening citrus fruits, similar but smaller changes in pectic compounds occur. In the final stages of fruit senescence, protoplasts may disintegrate and cell walls collapse. The actual breakdown of fruits appears to be linked to disorganization and loss of control over enzymatic processes rather than to depletion of foods.

Softening of fruits occurs largely by dismantling of primary cell walls. Softening of fleshy tissues of fruits often results from changes in the integrity and cross-linking between cell wall polymers, especially those of the middle lamella, which are involved in cell to cell adhesion (Bartley and Knee, 1982). These polysaccharides often become depolymerized and more soluble during ripening (Redgwell et al., 1992; Hisawa et al., 2003). In ripening Spanish pears, dissolution of polysaccharides from the middle lamella and primary cell walls, together with other factors, determined the degree of fruit softening (Martin-Cabrejas et al., 1994).

In many climacteric fruits (Chapter 6), softening is correlated with ethylene production and synthesis of a variety of enzymes and proteins that could promote this process. For example, gene expression and protein levels of polygalacturonase, an enzyme that is responsible for depolymerization and solubilization of pectic compounds, increase dramatically during ripening in climacteric fruits (e.g., pear, peach, Fonseca et al., 2005; Trainotti et al., 2006). However, reduction in PG level by antisense RNA suppression of expression of the gene coding for PG does not prevent fruit softening

(Smith et al., 1988). Hence, it is likely that coordinated action by several enzymes, possibly including PG, pectin methylesterase, β-glucanases, β-galactosidases, xyloglucan endotransglycosylase, β-xylosidases, and expansins (noncatalytic proteins) that promote breaking and reforming of hydrogen bonds between cellulose microfibrils and hemicellulose tethers act to effect textural changes and softening during ripening (Giovanni, 2001; Hiwasa et al., 2003; Fonseca et al., 2005).

SEXUAL REPRODUCTION IN GYMNOSPERMS

The calyx, corolla, stamens, and pistil are absent in gymnosperms. The flowers consist of pollen cones and seed cones (staminate and ovulate strobili), which in most species are produced on the same tree (Fig. 2.35).

Pollen cones often are bright yellow, purple, or red when mature and consist of an axis bearing a series of scales, each of which bears two pollen sacs on the undersurface (Chapter 2). The sometimes colorful seed cones, which are larger and more persistent than the pollen cones, consist of an axis bearing ovulate scales, each of which is borne in the axis of a bract. In the Pinaceae two ovules appear as protuberances on the upper side of a scale. At the end of the ovule near the cone axis is an opening, the micropyle, through which pollen grains may enter.

Cone Initiation and Development

The cones of conifers may differentiate from previously vegetative apices (hence that are borne terminally on shoots). Alternatively, they may differentiate from newly formed axillary primordia that subsequently differentiate into either lateral vegetative shoots or cones. The time of apical determination, which varies among species, is near the end of rapid elongation of lateral shoots (Owens and Hardev, 1990).

The time of cone initiation varies among species and within a species in different environments, but cones normally are initiated in the season before pollination occurs. The precise time varies from early spring for Douglas-fir to early summer for western red cedar to autumn for certain pines. Cone development of all species requires many months. The order of morphological changes is similar in different environmental regimes, but the time of occurrence of developmental events varies.

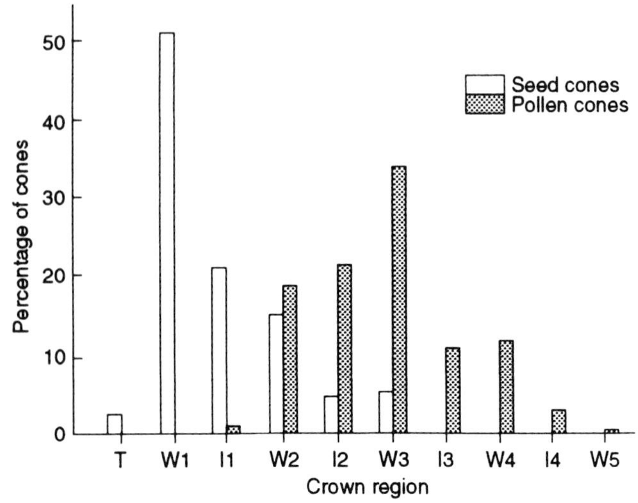

FIGURE 4.7. Vertical distribution of seed cones and pollen cones in crowns of eight- and nine-year-old white spruce trees. W1 indicates the uppermost nodal whorl, and I1 indicates the uppermost internode. From Marquard and Hanover (1984).

Pollen cones and seed cones are differentiated at different times and rates, with pollen cones usually forming first and developing faster. Seed cones generally are confined to the more vigorous shoots in upper branch whorls and internodes of the crown; pollen cones to the less vigorous shoots in the mid-crown region. However, seed cones and pollen cones are borne together in a transition zone of the crown (Fig. 4.7).

The distribution of cones in the crown varies with tree age. In six- to 18-year-old black spruce trees the seed cones were restricted to one-year-old shoots in the upper 25% of the crown. With increasing tree age the zone of cone distribution expanded to include older branches and small shoots on those branches. In 18-year-old trees approximately 25, 55, and 20% of the cones were borne on one-, two-, and three-year-old branches, respectively. Hence, the cones were concentrated across the tops of trees (Caron and Powell, 1992).

Three basic reproductive cycles have been described for temperate-zone conifers (Owens and Molder, 1984a,b,c; Owens and Smith, 1965). The most common one is a two-growing season cycle (Fig. 4.8) characteristic of spruce, larch, Douglas-fir, and hemlock. Pollination occurs in the spring or early summer of the second year and fertilization takes place a few weeks later. After fertilization, development of the embryo and seeds is continuous and rapid, with seeds released as early as late summer of the year of pollination.

A longer cycle characterizes reproduction of most pines, *Araucaria*, *Sciadopitys*, and *Sequoiadendron* (Fig. 4.9). Pollination takes place in the spring or early

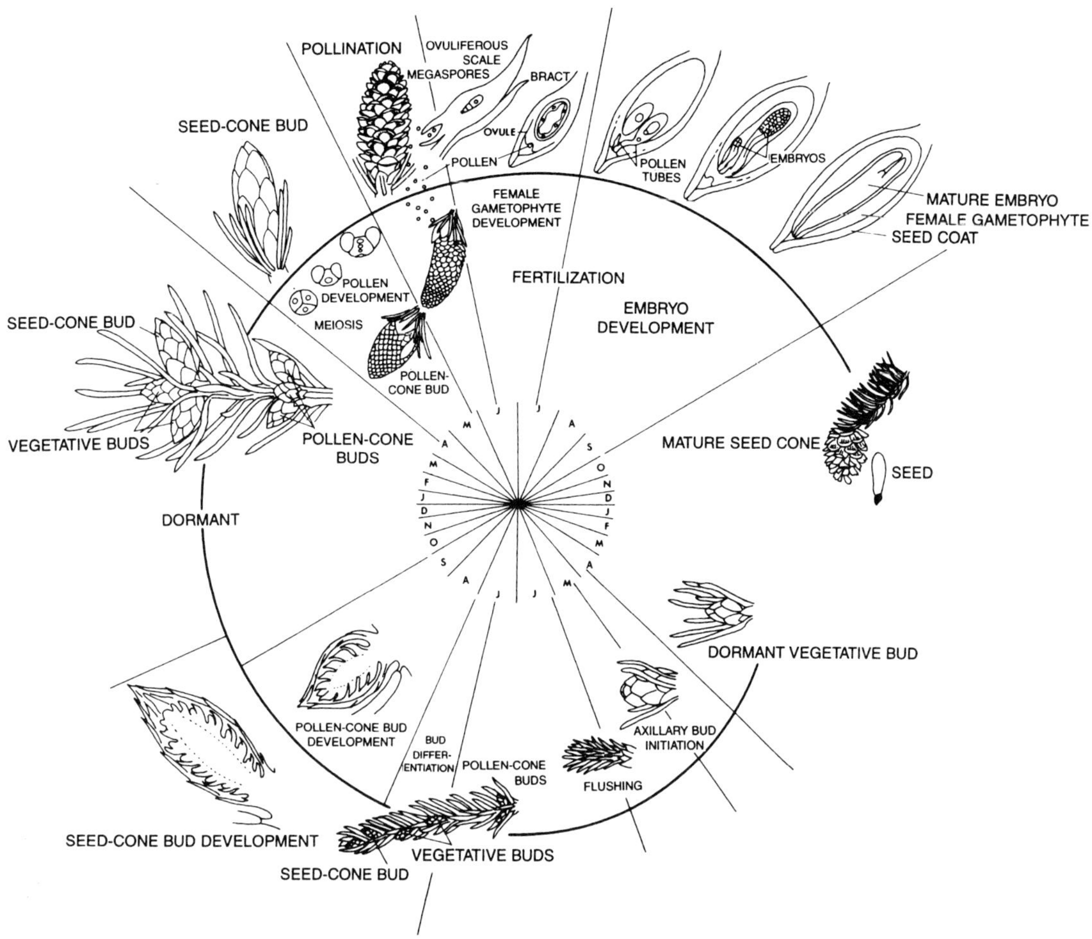

FIGURE 4.8. The reproductive cycle of white spruce. From Owens and Molder (1984c).

summer of the second growing season. Development of the pollen tube and ovule begins but ceases usually in midsummer. In the following spring development resumes and fertilization takes place. Both embryos and seeds are mature by autumn and seeds generally are shed in the year in which they mature. Exceptions are species such as lodgepole and jack pines with serotinous cones that may remain closed for many years. This minimum cycle requires approximately 27 months.

Another type of reproductive cycle is representative of a few conifers in the Cupressaceae such as *Chamaecyparis* and *Juniperus*. Pollination occurs in the spring or early summer of the second growing season and fertilization follows within a few weeks. Development of the embryo and seeds begins but is arrested in late summer or autumn. Both cones and seeds overwinter in a dormant state and their development is resumed in the spring of the third growing season.

There are a few variations of the three basic reproductive cycles. Deodar cedar has a cycle that is intermediate between the normal two- and three-growing season cycles of pines. Flowers are initiated in the summer, pollination occurs in autumn of the same year, fertilization takes place after winter dormancy, and seeds mature late in the second growing season (Roy Chowdhury, 1961).

The reproductive cycles of common juniper, and Italian stone pine, Chihuahua pine, and Torrey pine are combinations of the two- and three-growing season cycles. In common juniper, flowers are initiated before winter dormancy and pollination occurs in the following spring. Growth of the pollen tube and ovule development are arrested and they overwinter. Fertilization occurs in the third year. Immature embryos overwinter and complete their development in the fourth growing season (Owens and Blake, 1985).

Pollination and Fertilization

The amount of pollen produced and time of pollen shedding vary among species and years, and within a

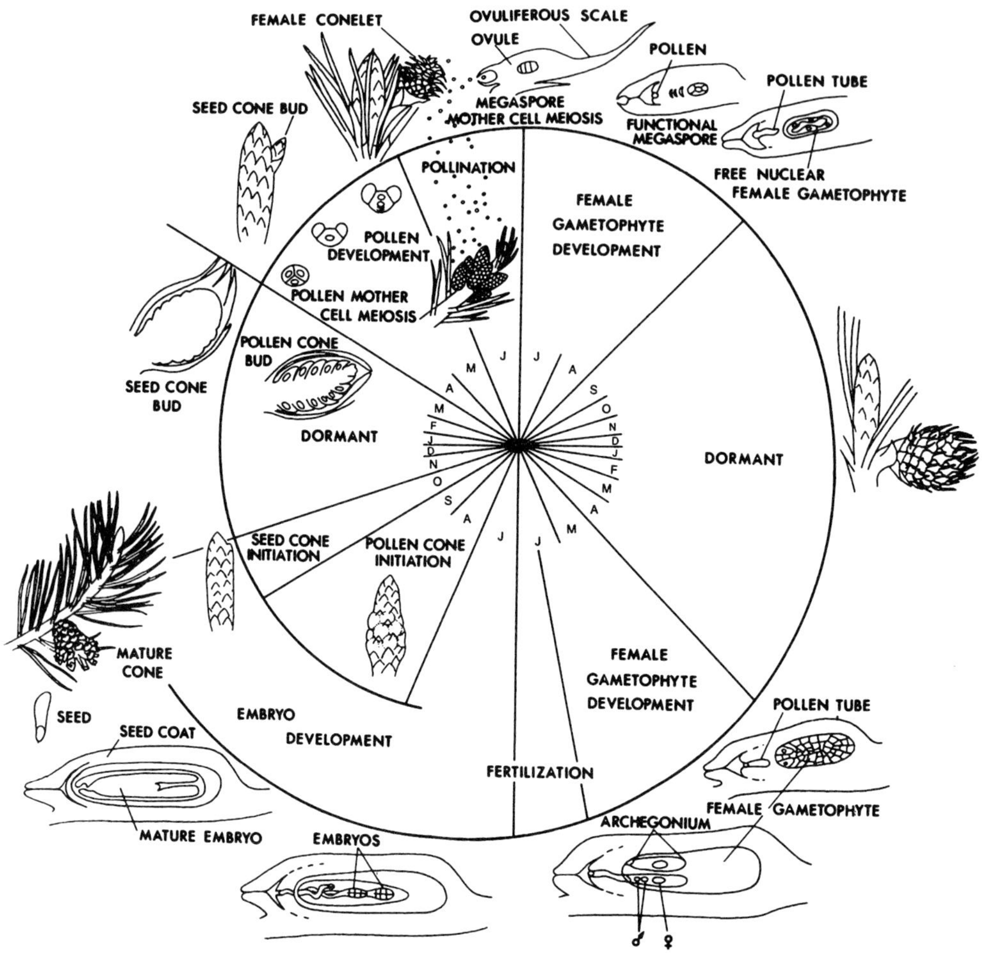

FIGURE 4.9. The reproductive cycle of lodgepole pine. From Owens and Molder (1984b).

species, they differ among stands, individual trees, and parts of trees. In Finland, shedding of Scotch pine pollen occurred during a five- to 10-day period. An individual stand shed its pollen in two or three days less than this; an individual tree in even less time. Shedding of pollen occurred for a longer time in Scotch pine than in Norway spruce or European silver birch (Sarvas, 1955a,b, 1962). During a year of a large crop, most pollen was shed during the middle of the flowering period. However, during years of light crops, a peak period of pollen production could not be identified.

The period of pollen receptivity of female flowers usually does not exceed a few days to a week. An exception is Douglas-fir, with an unusually long receptive period of about 20 days. Appreciable variation sometimes occurs in receptivity of flowers in different locations in the same tree. Since duration of the receptive period is influenced by prevailing environmental factors, especially temperature, humidity, and wind, it may be expected to vary from year to year.

Cross pollination occurs commonly in gymnosperms because the female flowers are concentrated in upper branches and male flowers in lower ones. The wind-transported pollen grains drift between the cone scales and contact ovules. Most gymnosperms exude a sugary "pollination drop" at the micropyle (*Abies*, *Cedrus*, *Larix*, *Pseudotsuga*, and *Tsuga* are exceptions). This fluid fills the micropylar canal during the receptive period. The pollination drop is secreted in the morning and disappears during the day. Exudation occurs for a few days or until the ovule is pollinated (Doyle, 1945; McWilliam, 1958). Pollen grains become incorporated in the fluid and, as the drop is withdrawn, pollen is sucked into the micropyle to contact the nucellus.

The pollen grains germinate to form a number of pollen tubes that grow downward into the nucellus.

As each pollen tube elongates, its generative cell divides to form a stalk cell and a body cell. The latter subsequently divides to form two sperm cells. The larger of these fuses with the egg nucleus within an archegonium and fertilization is completed. The other sperm cell disintegrates.

The time span between pollination and fertilization is extremely variable among gymnosperms. Pollen grains stay dormant on the nucellus for a few days in spruce, about three weeks in Douglas-fir, and nine months in *Cedrus*. In pines, fertilization occurs approximately 13 months after pollination (Konar and Oberoi, 1969).

Polyembryony

The presence of more than one embryo is a common feature of the Pinaceae. Two types of polyembryony occur in gymnosperms: (1) simple polyembryony, in which additional embryos result from fertilization of several archegonia in a gametophyte; and (2) cleavage polyembryony, in which several embryos arise from the splitting of embryonal cells of a single zygote. In cleavage polyembryony, it sometimes can be determined at a very early stage of development which of the embryos will be successful (determinate cleavage polyembryony). At other times, it is difficult to predict early which embryo will be successful (indeterminate cleavage polyembryony). All genera of Pinaceae exhibit simple polyembryony, and cleavage polyembryony occurs in *Pinus*, *Cedrus*, *Tsuga*, *Taxodium*, *Cryptomeria*, *Cunninghamia*, *Sequoia*, and *Podocarpus*.

Even though multiple fertilization takes place in the female gametophyte of pines, most mature seeds have only a single embryo. Occasionally, however, pines have more than one embryo per seed. The occurrence of multiple embryos appears to be higher in pines with large seeds than in those with small ones. Berlyn (1962) counted four embryos per mature seed in approximately one-third of the seeds of sugar pine and Swiss stone pine examined and in a few seeds of eastern white pine. In another study, Berlyn (1967) found eight embryos in one sugar pine seed.

Parthenocarpy

Development of unpollinated cones with fully formed but usually empty seeds (parthenocarpy) occurs in a number of genera of gymnosperms. Parthenocarpy is common in *Abies*, *Juniperus*, *Larix*, *Picea*, *Taxus*, and *Thuja* and also has been reported in *Chamaecyparis*, *Cryptomeria*, *Pseudotsuga*, and *Tsuga* (Orr-Ewing, 1957), but it rarely occurs in *Pinus*. For example, only 0.4% of Scotch pine cones had completely aborted ovules (Sarvas, 1962) and only one out of 76 developing cones of red pine had no developing ovules (Dickmann and Kozlowski, 1971).

Duration and Timing of Cone Development

Some idea of the similarities and differences in cone development of different gymnosperm genera in the temperate zone may be gained by comparing the timing of the significant events in reproductive growth of red pine and Douglas-fir.

In central Wisconsin, the cone primordia of red pine differentiated in one growing season, but the conelets are not externally visible until late May or early June of the following year. Pollination occurs in early June. The cones begin to enlarge after pollination and the scales close. By late July the cones have lengthened to 10 to 12 mm and little additional increase in length occurs during the rest of the first year (Lyons, 1956). Meanwhile, the pollen grains have germinated but the pollen tubes stop growing and are quiescent during late summer and winter. Megaspores form approximately three weeks after emergence of the seed cone. Successive cell and nuclear divisions of one of the megaspores result in an enlarged megagametophyte. A period of winter dormancy follows.

The cones resume growing early in the spring of the second season and attain their final length by early July. Growth of the megagametophyte also resumes in early spring, but the pollen tube does not grow until around mid-June of the second year when fertilization occurs. Embryo development then follows rapidly and seeds ripen by early September in Wisconsin (Dickmann and Kozlowski, 1969a).

Important events in growth of Douglas-fir cones throughout their 17-month developmental cycle at Corvallis, Oregon were studied by Owens and Smith (1964, 1965). Initiation of lateral vegetative, pollen cone, and seed cone primordia occurred during the second week of April. The reproductive tissues were formed about one month before vegetative buds opened and at the same time as the current season's seed cones opened. Cataphylls were initiated from early April to mid-July and bract initiation was continued from mid-July to early October. Scales were initiated early in September and continued until the cone became dormant early in November. Growth was resumed early the following March, and the cone buds burst approximately a month later. The cones achieved maximum size early in July. Maturation occurred in

TABLE 4.2. **Timing of Events in Development of Buds and Seed Cones (*Ovulate strobili*) of Douglas-fir near Corvallis, Oregon**[a]

Event	Date	Elapsed time from bud initiation (months)
All buds		
Lateral bud primordia initiated	Early April	Ovulate buds burst
Zonation becomes apparent	Mid-May	1.5
Cataphyll initiation complete; apical enlargement occurs; leaf, bract, or microsporophyll initiation begins	Mid-July	3.5
Ovulate strobili		
Beginning of scale initiation	Early September	5
All bracts initiated	Early October	6
All scales initiated and ovulate buds become dormant	Early November	7
Ovulate buds resume growth	Early March	11
Ovulate buds burst and pollination occurs	Early April	12
Fertilization	Early June	14
Elongation of strobilus complete	Early July	15
Maturation complete, strobilus opens, seeds released	Early September	17

[a]From Owens and Smith (1965).

FIGURE 4.10. Cones of slash pine in three stages of development (from left to right: 2 years, 1 year, and 1 month). Photo courtesy of U.S. Forest Service.

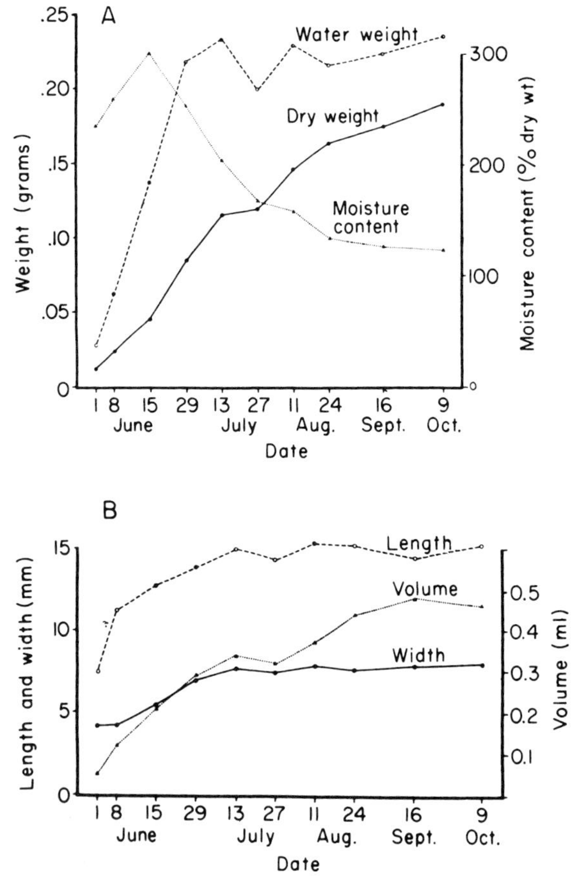

FIGURE 4.11. (A) Seasonal changes in dry weight, weight of water (per conelet), and percent moisture of first-year conelets of red pine. (B) Seasonal changes in length, width, and volume of first-year conelets. From Dickmann and Kozlowski (1969b).

July and August and generally was completed in September. The timing and duration of significant events are summarized in Table 4.2.

Increase in Size and Dry Weight of Cones and Seeds

In pines, the greatest increase in size and dry weight of cones occurs during the second year of their development (Fig. 4.10). For example, when growth of first-year conelets of red pine in Wisconsin ended they were only about one-fortieth the weight, one-thirtieth the volume, and one-third the length of mature cones at the end of their second year of development (Dickmann and Kozlowski, 1969a).

Seasonal patterns of increase in dry weight and size of first- and second-year cones of red pine in Wisconsin are shown in Figures 4.11 and 4.12. The dry weight of

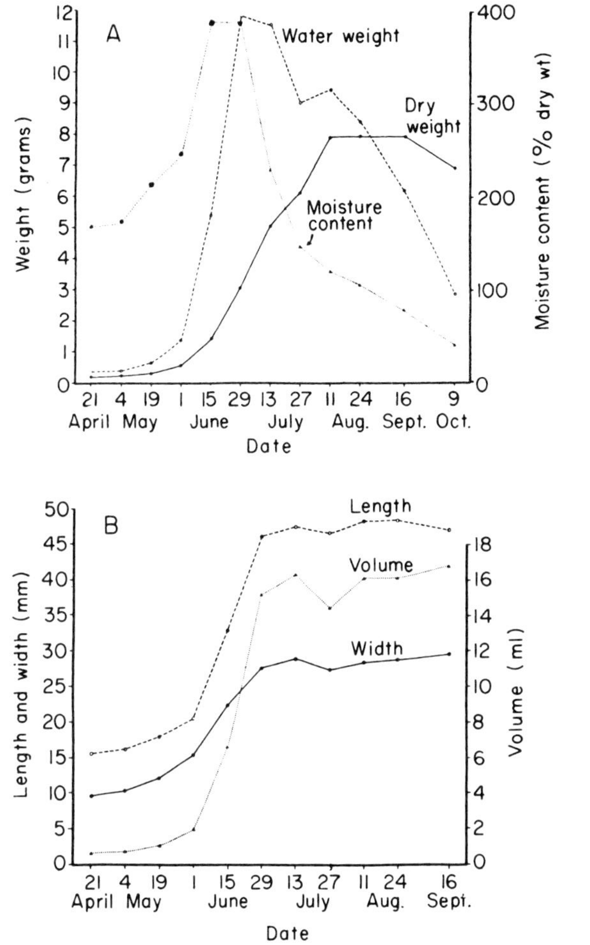

FIGURE 4.12. (A) Seasonal changes in dry weight, weight of water (per strobilus), and percent moisture of second-year cones of red pine. (B) Seasonal changes in length, width, and volume of second-year cones. From Dickmann and Kozlowski (1969b).

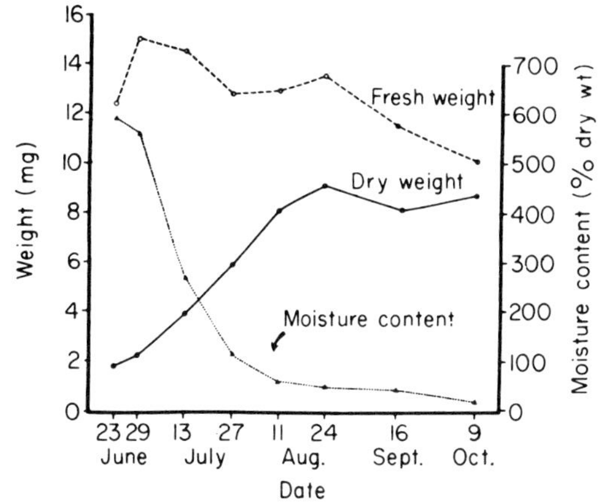

FIGURE 4.13. Changes in fresh weight, dry weight, and percentage moisture of developing red pine seeds from just after fertilization to maturity. From Dickmann and Kozlowski (1969b).

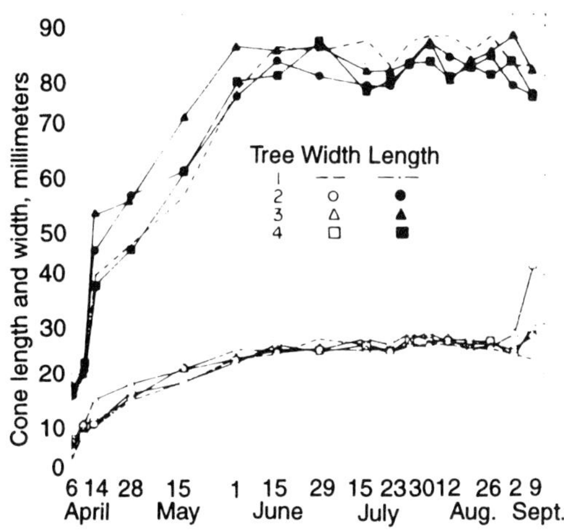

FIGURE 4.14. Changes in length and width of seed cones of four Douglas-fir trees near Corvallis, Oregon. From Ching and Ching (1962).

first-year cones increased at a steady rate until late September, by which time the average cone weighed slightly less than 0.2 g. Patterns of dry weight increment were similar during each of three successive growing seasons. The length and width of first-year cones increased until mid-July and changed little thereafter. At the end of the first season, an average cone was 8 mm in diameter and 14 to 15 mm long.

During the second year of development, red pine cones resumed growing in mid-April. Their dry weights increased slowly during May, rapidly during early June, and continued at a high rate until early August, when maximum dry weight was recorded. The dry weight of the average second-year cone increased from less than 0.2 g in April to nearly 8 g in August. Most increase in size of second-year cones occurred in June. The second-year cones reached maximum size about one month before maximum dry weight increase was recorded.

Seasonal changes in dry and fresh weights and moisture contents of fertilized ovules or seeds of red pine are shown in Figure 4.13. Dry weight of seeds increased rapidly from June until late August when it ceased. Moisture content (as percentage of dry weight) decreased sharply from a late June average of nearly 600% to 50% in late August. Subsequently, the moisture content of seeds decreased gradually to 17% by early October.

Changes in diameter and length of seed cones of Douglas-fir (which require one growing season for maturation) are shown in Figure 4.14. The length of cones increased rapidly to a maximum by June 1. Diameters of cones increased from 6 mm (excluding bracts) on April 24 to 24 mm on June 1, when maximum diameter was approached. Dry weight changes of

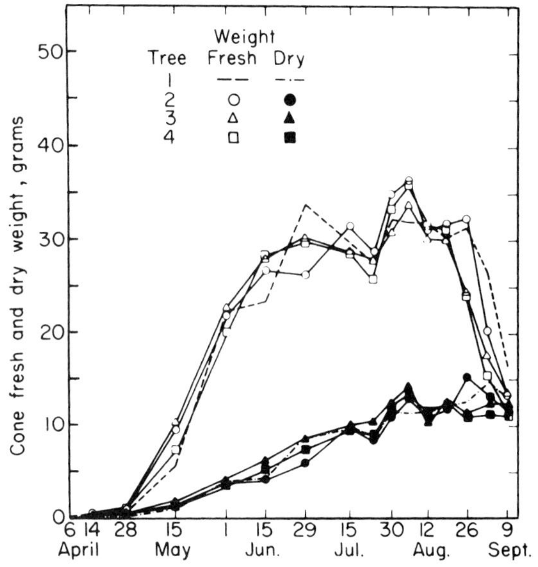

FIGURE 4.15. Changes in fresh and dry weights of seed cones of four Douglas-fir trees near Corvallis, Oregon. From Ching and Ching (1962).

cones followed a typical sigmoid pattern (Fig. 4.15) but the scales stopped gaining dry weight in July, whereas the dry weights of seeds increased until September. From early June to August the dry weight of cones increased, moisture content rapidly decreased, and dry weight of fertilized seeds increased as the seed enlarged.

MATURATION OF SEEDS

The problem of harvesting seeds when they show maximum germination capacity, either immediately or after a period of storage, is important to plant propagators because of the high costs of seed collection. A widely held view is that seeds are physiologically mature when they have achieved their maximum dry weight. At that time resources are no longer flowing into seeds from the mother plant. When they are physiologically mature the seeds of most species can be dehydrated to a low moisture content without losing viability (Harrington, 1972). Physiological maturity of seeds usually is considered necessary for attainment of full germination capacity. However, there are exceptions as when germinability is lowered by adverse environmental conditions such as drought, excess rainfall, flooding, or frost.

A distinction often is made between physiological maturity of seeds and morphological maturity. Seeds that have fully developed embryos are considered morphologically mature but may not be physiologically mature.

As a practical matter seed collectors generally judge maturity of fruits and seeds by a variety of visible indicators. Fruit maturation commonly is associated with development of orange color in persimmon, red in barberry, red in olive, and straw color or brown in tulip poplar. Color changes of cones and fruits are reasonably good indicators of seed ripeness in some gymnosperms but are most reliable when used by experienced collectors. The critical color changes in cones include transition to a brown color in jack pine, Douglas-fir, and sand pine; yellow-green with brown scale tips in red pine; golden brown with dark brown markings in Alaska-cedar; and purple in balsam fir (Schopmeyer, 1974). Color changes of cones were not considered useful in determining seed maturity of sugar pine, ponderosa pine, Scotch pine, hoop pine, white spruce, or western larch (Edwards, 1980).

The degree of embryo development sometimes has been used to determine when cones should be collected, provided the seeds are not immediately extracted from the cones and the cones are stored under conditions that favor continued ripening. Ching and Ching (1962) recommended that Douglas-fir cones should not be collected until the embryos are at least 90% elongated. Oliver (1974) found that 94% of the embryos of white fir seeds had reached maximum size when the cones were ripe enough to be harvested. Mercier and Langlois (1992) reported that white spruce seeds reached maturity in Quebec one or two weeks before the cones opened. Seeds collected at that time had high germination capacity and maintained viability even after 15 months of storage. The seeds were considered mature when (1) the embryo reached 90% of the embryo cavity length, (2) the moisture content of the cones was 63%, (3) cone specific gravity was 0.99 g cm^{-3}, or (4) when approximately 80% of the cones floated in a solution of methanol.

In some conifers, the germination capacity of seeds is very high well before the cones are mature as determined by color, weight, volume, specific gravity, or moisture content. For example, seeds ripening on Monterey pine trees had high germination capacity by the end of July in New Zealand, at which time the seeds could not easily be extracted from the cones (Rimbawonto et al., 1988a). In another study Monterey pine seeds harvested as early as April showed high germination capacity. The limiting factor for harvesting of seeds was not their immaturity but rather the difficulty of extracting them from the cones, which usually required six to nine weeks of drying (Rimbawonto et al., 1988b).

As a potential index of seed maturity, some investigators studied the amounts of chemical constituents of seeds that are relatively stable during the early part of the ripening season but change appreciably as seed maturity is approached. The amounts of reducing sugars in Douglas-fir seeds, which decreased as cones ripened, were one such index. In noble fir and red gum seeds, the amounts of crude fats were a better index of seed maturity (Rediske and Nicholson, 1965; Bonner, 1970). Despite these examples, determinations of biochemical changes during seed ripening have not been widely accepted as reliable tests for seed maturity (Edwards, 1980).

Maturity of pine seeds has been correlated with conductivity of seed leachate (conduction of electric current through deionized water after imbibition of seeds) and leaking of inorganic phosphorus and carbohydrates. Conductivity of leachate and leaking of inorganic phosphorus and carbohydrates of Scotch pine seeds decreased during seed ripening until near the end of physiological seed development (Sahlen and Gjelsvik, 1993). The conductivity of leachate and the amount of leached carbohydrates were four to eight and 10 to 15 times higher, respectively, in nongerminable than in germinable seeds.

ABSCISSION OF REPRODUCTIVE STRUCTURES

The final stage of flower senescence commonly is abscission. Shedding may involve the entire inflorescence, florets, or parts of flowers. When entire flowers are shed an abscission layer normally is formed as it does with leaves. However, cell division generally does not precede the shedding of petals. Before petals are shed a clear abscission layer does not form; shedding is caused by softening of the middle lamella associated with increasing activity of hydrolytic enzymes (Esau, 1965).

The site of abscission of fruits varies among species and between immature and mature fruits of the same species. Immature apple fruits abscise at the juncture of the pedicel and spur; those of sweet cherry between the pedicel and peduncle; those of pistachio at the juncture of the pedicel and lateral branches of the fruit cluster. Mature plum, sweet cherry, sour cherry, avocado, orange, and mango fruits abscise at the base of the fruit. After fruits of many tropical species form their stalks, they separate from them by a second abscission layer. Shedding of fruits together with their stalks also is common in forest trees including willow, poplars, basswood, black locust, and elm.

Cell wall changes that lead to separation of fruits (Baird and Webster, 1979) include the following:

- Hydrolysis or dissolution of the middle lamella
- Breakdown of all or part of the cellulose cell wall
- Mechanical breaking of dead elements

These changes may occur concurrently.

Abscission of immature fruits usually is preceded by cell division and development of a well-defined abscission zone. In contrast, abscission of mature fruits is not preceded by cell division and the abscission zone is indistinct. Abscission is characterized by loss of cell wall constituents and collapse of existing cells.

Abscission and Crop Yield

Heavy flowering is not always followed by a large crop of fruits or seeds. This discrepancy may be associated with deficient pollination or other factors such as adverse weather, attacks by insects and other pests, and deficiencies of carbohydrates, mineral nutrients, and hormones. In some species, the fruits from self-pollinated flowers are more likely to be shed prematurely than are fruits from cross-pollinated flowers.

Abscission of flower buds, flowers, and immature fruits commonly accounts for low crop yields. Some species produce mature fruits from as many as half the flowers they bear, others from only a small fraction (Table 4.3). Kapok trees may produce more than 1,000 flowers for each fruit that grows to maturity (Stephenson, 1981).

Very high rates of flower abscission are well documented. For example, 99% of the flower crop of olive trees may be lost by abscission (Weis et al., 1991). The size of the olive crop is determined by rapid abscission

TABLE 4.3. Variations in Initiation and Maturation of Fruits

Species	Female flowers that initiate fruits (%)	Female flowers that mature fruits (%)	Initiated fruits that mature (%)	Reference
Mesquite	0.3–3.5	0–0.2	0–11	Solbrig and Cantino (1975)
Orange	34.9–61.5	0.2–1.0	0.6–1.6	Erickson and Brannaman (1960)
Cacao	1.7	0.2	12	Proebsting (1934)

TABLE 4.4. Variations in Initiation and Maturation of Fruits of Washington Navel Orange at Riverside, California[a]

Diameter of ovary (mm)	Buds	Flowers	Fruits: With pedicel	Fruits: Without pedicel	Total
1 or less	59,635	674	1,002	12	61,323
2	22,790	14,939	13,953	344	52,026
3	10,804	15,295	25,043	541	51,683
4	3,114	2,321	13,469	589	19,493
5		6	3,853	460	4,319
6			2,399	634	3,033
7			1,250	643	1,893
8			586	450	1,036
9			417	663	1,080
10			179	462	641
11			77	357	434
12			37	275	312
13			20	194	214
14			7	137	144
15			7	118	125
16			3	79	82
17			2	71	73
18			2	55	57
19			2	40	42
20				32	32
21 or more				232	232
Total	96,343	33,235	62,308	6,388	198,274
Average no. of mature fruits per tree					419
Total flower buds per tree					198,693
Percent	48.5	16.7	31.4	3.2	Crop = 0.2

[a]From Erickson and Brannaman (1960).

of flowers and fruits during five to seven weeks after full bloom (Rappaport and Rallo, 1991). Up to 90% of the flowers of apple trees drop soon after petal fall. In California almost half the flower buds of orange trees dropped before they opened. Abscission of open flowers amounted to 16.7% of all the flower buds initiated. Small fruits also were shed throughout and after the flowering period. Only 0.2% of the flower buds eventually produced fruits (Table 4.4).

Abscission of flowers and immature fruits is a serious problem, especially with apples, pears, oranges, and grapefruits. There usually are three major periods of shedding of reproductive structures of fruit trees (Sedgley and Griffin, 1989):

1. Shedding of unfertilized flowers within two weeks of anthesis ("early drop").
2. Shedding of fertilized young fruits within two months of anthesis ("June drop" in the northern hemisphere; "December drop" in the southern hemisphere).
3. Preharvest shedding of full-sized but immature fruits.

Premature abscission of fruits of wind-pollinated species of forest trees is well known. For example, most of the potential acorn crop of white oak often abscises prematurely (Table 4.5). Williamson (1966) found that from early May to mid-July (the period of pollination, ovule development, and fertilization) almost 90% of white oak acorns were shed prematurely. In a seed orchard of Monterey pine, about half of all the seed cones aborted during the first year after pollination. Conelet drop started at the time of recep-

TABLE 4.5. Percentage of Abscission of White Oak Acorns at Various Development Stages[a]

	1962		1963	
Development stage	Starting date	Percent abscission	Starting date	Percent abscission
Pollination	April 30	55.6	April 24	28.4
Ovule development	May 17	10.6	May 20	37.9
Fertilization	June 4	15.8	June 6	18.1
Embryo development	July 6	16.7	July 3	10.5
Maturation	September 19	1.3	September 20	5.1
		100.00		100.00

[a]From Williamson (1966).

tivity, reached a peak four to six weeks later, and then decreased in intensity (Sweet and Thulin, 1969).

Many factors are involved in inducing abortion of fruits and seeds. Shedding of fruits varies with the time of fruit initiation, availability of metabolites and hormones, pollination intensity, number of developing seeds, and injury to fruits and seeds. Abortion of individual seeds varies with the time of seed initiation, position of the fruit, time of fruit initiation, position of the fruit in the inflorescence, and the pollen source (Lee, 1988). The sequential waves of abscission in fruit trees may result from different mechanisms. For example, the first wave of fruit drop of apple was attributed to poor pollination and fertilization; subsequent waves to competition between fruits and vegetative tissues for resources (Luckwill, 1953; Westwood, 1978).

Berüter and Droz (1991) showed that during the pre-June drop period of apples, a variety of treatments (shading, girdling proximal to the abscission zone, removal of seeds from attached fruits) induced abscission of fruits. When the same treatments were applied after June drop none of the fruits abscised. The data indicated that during the pre-June drop period, abscission was induced by blocking nutrient supply to the abscission zone of fruits. Subsequently the inhibition of abscission was related to the amount of stored carbohydrates in the young fruits.

SUMMARY

Production of fruits and cones requires successful completion of each of several sequential phases of reproductive growth including initiation of floral primordia, flowering, pollination, and fertilization as well as growth and ripening of fruits and cones. Reproductive growth usually is negatively correlated with the amount of vegetative growth. Biennial bearing occurs commonly and in some species is associated with variations in production of flower buds. In other species biennial bearing results from abscission of flower buds during the year that a heavy crop of fruit is produced.

Flowering periodicity varies within and between years. Most deciduous broadleaved trees of the temperate zone initiate flower buds in the summer or autumn before the spring in which the flowers open. The actual time of floral induction is modified by weather, site conditions, and management practices. Flowering periodicity in the tropics is diverse and varies among species and site. Extended flowering plants produce few flowers per day but continue to flower for months and even throughout the year. Mass flowering plants produce many flowers per day but bloom for only a short time. Flowering patterns often vary among life forms in the same ecosystem and between male and female plants of the same species.

Large differences in fruit set among species and genotypes result from variations in pollination, fertilization, ovule longevity, flower structure, temperature, light intensity, competition among fruits for resources, and competition for resources between reproductive and vegetative tissues. Development of mature fruits usually requires fertilization of the egg, but parthenocarpic fruits may be set and mature without fertilization.

Growth curves of fruits are of two general forms: (1) a curve with an initial exponential increase in size followed by slowing of growth in a sigmoid fashion (e.g., apple, pear, orange, date, banana, avocado, strawberry, mango, and lemon) and (2) a double sigmoid curve (e.g., stone fruits and other fruits such as grapes and currants).

Fruit ripening generally is associated with loss of chlorophyll, unmasking of other pigments, development of odor and flavor, softening, production of ethylene, and hydrolysis of insoluble pectins. Flavor is determined by a balance between sugars, acids, and astringent compounds. Flavor also is associated with aroma, the result of greatly increased production of volatile compounds (e.g., organic acids, alcohols, esters, carbonyl compounds, hydrocarbons, and terpenoids). Softening of fruits likely occurs by hydrolysis of cell walls by multiple enzymes and through the activity of noncatalytic proteins (expansins) that break and reform hydrogen bonds linking hemicelluloses and cellulose microfibrils.

In gymnosperms, the time of cone initiation varies among species and genotypes and in different environments. Normally cones form in the season before pollination occurs. Pollen cones not only form before seed cones, but they differentiate faster. Seed cones usually are borne on vigorous shoots in the upper crown; pollen cones on less vigorous shoots in the mid crown. Seeds of gymnosperms require one to three growing seasons to develop, depending on species.

The amount of pollen produced and time of pollen shedding vary among species and years. Pollen receptivity usually lasts for only a few days to a week. The time span between pollination and fertilization varies greatly among species.

A distinction often is made between morphological maturity and physiological maturity of seeds. When seeds have fully developed embryos and endosperm they usually are considered morphologically mature but may not be physiologically mature. Seed maturity has been judged by several indicators including color, degree of embryo development, amounts of chemical constituents, and conductivity of seed leachates.

Abscission of flower buds, flowers, and immature fruits often accounts for low crop yields. Reproductive structures typically are shed in three waves: (1) within two weeks of anthesis (early drop of unfertilized flowers), (2) within two months of anthesis (June drop of young fruits in the northern hemisphere; December drop in the southern hemisphere), and (3) preharvest shedding of full-sized but immature fruits. Cell wall changes leading to abscission of fruits include hydrolysis or dissolution of the middle lamella, breakdown of cellulose cell walls, and mechanical breaking of dead elements. Abscission is influenced by the time of fruit initiation, proximity of fruits to resources, pollination intensity, number of developing seeds, pollen sources, number of pollen donors, and injury to fruits and seeds. The sequential waves of abscission of fruits may result from different mechanisms.

General References

Addicott, F. T. (1982). *Abscission*. University of California Press, Berkeley.

Asker, S. E. and Jerling, L. (1992). *Apomixis in Plants*. CRC Press, Boca Raton, Florida.

Brady, C. J. (1987). Fruit ripening. *Annu Rev Plant Physiol* **38**, 155–170.

Giovannoni, J. (2001). The molecular biology of fruit maturation and ripening. *Annu Rev Plant Physiol Plant Mol Biol* **52,** 725–749.

Greyson, R. I. (1994). *The Development of Flowers*. Oxford Univ. Press, New York.

Halevy, A. H., ed. (1985). *Handbook of Flowering*. Vols. 1–5. CRC Press, Boca Raton, Florida.

Kozlowski, T. T., ed. (1973). *Shedding of Plant Parts*. Academic Press, New York.

Kozlowski, T. T. and Pallardy, S. G. (1996). *Growth Control in Woody Plants*. Academic Press, San Diego.

Lord, E. M. and Russell, S. D. (2002). The mechanisms of pollination and fertilization in plants. *Annu Rev Cell Dev Biol* **18,** 81–105.

Monselise, S. P., ed. (1986). *CRC Handbook of Fruit Set and Development*. CRC Press, Boca Raton, Florida.

Owens, J. N. (1991). Flowering and seed set. In *Physiology of Trees*, A.S. Raghavendra, ed., pp. 247–327. Wiley, New York.

Owens, J. N. (1991). Measuring growth and development of reproductive structures. In *Techniques and Approaches in Forest Tree Ecophysiology*, J. P. Lassoie and T. M. Hinckley, eds., pp. 423–452. CRC Press, Boca Raton, Florida.

Owens, J. N. and Hardev, V. (1990). Sex expression in gymnosperms. *Crit Rev Plant Sci* **9**, 281–298.

Sedgley, M. (1990). Flowering of deciduous perennial fruit crops. *Hort Rev* **12**, 223–264.

Sedgley, M. and Griffin, A. R. (1989). *Sexual Reproduction of Tree Crops*. Academic Press, London and New York.

Seymour, G. B., Taylor, J. E., and Tucker, G. A., eds. (1993). *Biochemistry of Fruit Ripening*. Chapman and Hall, London and New York.

Tucker, G. A. and Grierson, D. (1987). Fruit ripening. In *The Biochemistry of Plants*, Vol. 12, D. D. Davies, ed., pp. 265–318. Academic Press, New York.

CHAPTER

5

Photosynthesis

INTRODUCTION

Photosynthesis is the process by which light energy is captured by green plants and used to synthesize reduced carbon compounds from carbon dioxide and water. The importance of photosynthesis can scarcely be overemphasized because nearly all the energy entering the biotic portion of the biosphere is derived from photosynthesis. Field et al. (1998) estimated that land plants produce about 125×10^9 metric tons of dry matter per year (based on a 45% C content), of which about three-fourths was attributed to forests and savannas. Global average net primary productivity of forests exceeds that of the entire world by a factor of four, and that of agricultural land by a factor of more than two (Whittaker and Woodwell, 1971).

A great diversity of natural products of use to humans has its origin in the yield of photosynthesis produced by woody plants. The relationship of this yield with building materials made of wood is obvious. Additionally, much of the developing world depends on wood as a primary source of energy, while industrialized and developing nations intensively exploit ancient buried plant materials that have metamorphosed into deposits of coal and oil. Edible parts of woody plants ranging from fruits and seeds to ground bark and leaves used in seasoning contribute substantially to human nutrition and culinary pleasure. No matter how they are used, the energy and dry matter in plant products are or were made available for storage by the process of photosynthesis. Managed plant communities therefore often are directed toward improving the amount of photosynthesis per unit of land area and the efficiency with which the products of photosynthesis are converted into plant material.

Although most photosynthesis occurs in foliage leaves, some takes place in other green tissues, including cotyledons (see Chapter 2, Kozlowski and Pallardy, 1997), buds, stems, aerial roots, flowers, and fruits. In the vast majority of species the photosynthetic contributions of tissues other than cotyledons and foliage leaves are relatively unimportant. For example, photosynthesis by green fruits of locust, lemon, orange, avocado, grape, and plum generally is too low to contribute appreciably to their own growth. However, the photosynthetic contributions of reproductive structures vary greatly among species and sometimes are significant (Kozlowski, 1992). Fruit photosynthesis was estimated to contribute 15% of the total carbon requirement for development of rabbiteye blueberry fruits (Birkhold et al., 1992). Fruit photosynthesis supplied about half the carbon required during the first 10 days after bloom and 85% during the five days after petal fall. Pavel and De Jong (1993) estimated that photosynthesis of late-maturing cv. Cal Red peaches provided 3 to 9% of the weekly fruit carbohydrates early in the season and 15% in midseason. Photosynthesis of mature fruits contributed 3 to 5% of the total carbohydrate requirement of the fruits. The reproductive structures of Norway maple contributed almost two-thirds of the C required for seed production; those of red maple and sugar maple provided about one-third; but those of American elm, shagbark hickory, American sycamore, and bur oak less than 10% (Bazzaz et al., 1979). Photosynthesis of coffee fruits may account for up to one-third of their own dry weight (Cannell, 1975).

At certain stages of their development the green cones of conifers fix some CO_2. In most cases, however, photosynthesis of cones is not adequate to balance their high respiratory evolution of CO_2, and most of the carbohydrates needed for growth of cones are obtained from other sources (Dickmann and Kozlowski, 1970; Kozlowski, 1992). Refixation of CO_2 by Hinoki cypress cones compensated for only slightly more than half of the daily CO_2 loss by dark respiration (Ogawa et al., 1988).

Some photosynthesis occurs in the twigs and stems of certain plants, but usually is low, amounting to approximately 5% of the total in eastern cottonwood (Schaedle, 1975). In seedlings of some deciduous species the leafless twigs are photosynthetically active during the winter. In North Carolina the dry weight of leafless sweetgum seedlings increased appreciably during the winter, indicating a contribution of twig photosynthesis to their growth (Perry, 1971). In temperate and boreal zone trees maximum refixation rates of stem-respired CO_2 often exceed 50% (Pfanz et al., 2002) and likely contribute significantly to the carbon balance of forest stands, especially those that are young (Ryan et al., 1997). Photosynthesis of branches and bark also is an important feature of certain arid-zone species (Gibson, 1983), including *Gutierrezia sarothae* (Depuit and Caldwell, 1975) and *Eriogonum inflatum* (Osmond et al., 1987). Some desert shrubs that shed their leaves during the dry season have persistent, photosynthetically active stems and branches. For example, palo verde is leafless during most of the year and its green branches may produce up to 40% of the total annual photosynthate (Adams and Strain, 1969).

CHLOROPLAST DEVELOPMENT AND STRUCTURE

Energy capture and CO_2 fixation occur in the chloroplasts of higher plants (Fig. 5.1; see also Fig. 5.11). These organelles are noted most commonly in leaves, but also may develop in other organs including cotyledons, buds (Larcher and Nagele, 1992), the

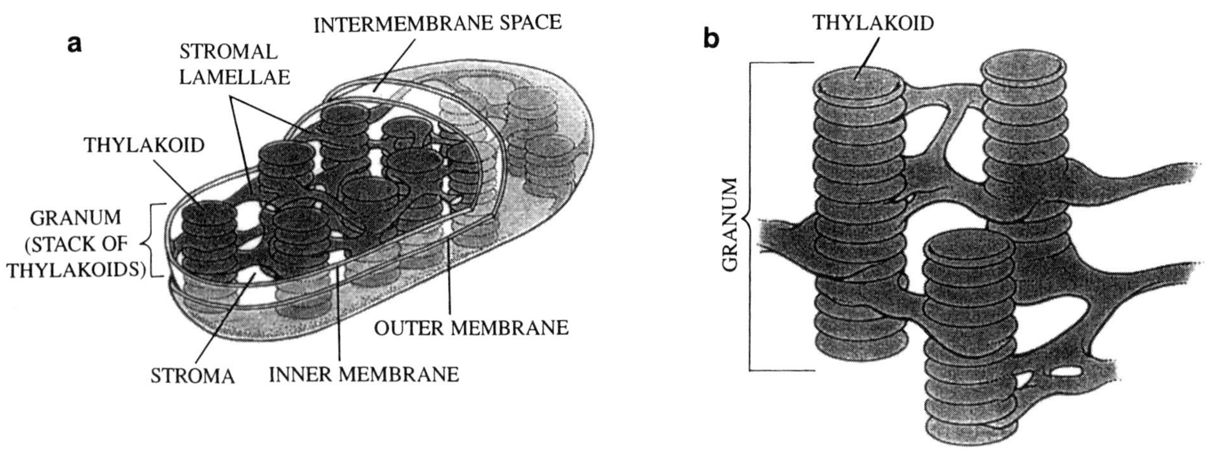

FIGURE 5.1. Schematic illustrating three-dimensional structure (a) and arrangement of thylakoid membranes (b) of chloroplasts of higher plant cells. The stacks of disklike thylakoids, the grana, are connected by thylakoids (stroma thylakoids, or lamellae) that span the stroma. From Raven et al. (1992).

bark of stems and branches (Schaedle, 1975; Coe and McLaughlin, 1980; Pfanz et al., 2002), flowers (Dueker and Arditti, 1968), fruits (Blanke and Lenze, 1989), and cones (Linder and Troeng, 1981a). Chloroplasts generally pass between generations and do not arise from less organized cellular contents (Leech, 1984). Inheritance is primarily maternal (through the egg) in angiopserms and paternal (through the sperm) in gymnosperms (Ahuja, 2001). Within dividing cells of meristematic tissues, proplastid progenitors of chloroplasts divide repeatedly, and the ultimate complement of chloroplasts in a mature leaf cell may depend on the intensity of chloroplast division.

Mature chloroplasts arise during tissue ontogeny from small (1 μm) roughly spherical proplastids that have a double membrane envelope and rudimentary internal membrane structure (Leech, 1984). Young plastids contain protochlorophyllide, a precursor of chlorophyll. In angiosperms, exposure to light activates conversion, through a phytochrome-mediated mechanism, of protochlorophyllide to chlorophyll. In cotyledons of conifers, light is not required for accumulation of chlorophyll *a* and *b* (Lewandowska and Öquist, 1980). In developing chloroplasts there is rapid accumulation of membrane systems and synthesis of proteins, some of which are coded by chloroplast DNA (cpDNA) and translated on chloroplast ribosomes. Other chloroplast proteins (e.g., the small subunit of ribulose bisphosphate carboxylase-oxygenase, Rubisco, and light harvesting chlorophyll complex protein) are coded by nuclear DNA. These proteins are synthesized on cytoplasmic ribosomes and transported into the chloroplast, where they often undergo final conversion to a functional form (Dyer, 1984).

It is now widely accepted that chloroplasts (and mitochondria) have their origin in procaryotic bacteria that originally associated symbiotically with protoeucaryotes. Eventually, the bacteria entered the cells via endocytosis and over evolutionary time became intimately integrated into cell metabolism. Structural, functional, and DNA evidence support this hypothesis (Blankenship, 2002).

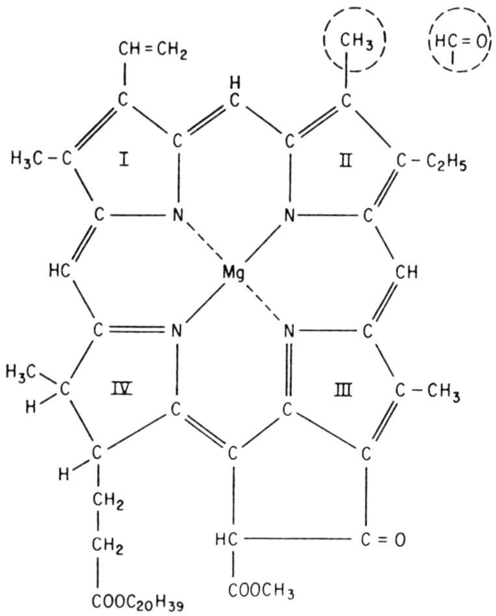

FIGURE 5.2. Structural formula for chlorophyll *a*. The formula for chlorophyll *b* is the same except for the substitution of an aldehyde for a methyl group (indicated by dashed circles). Rearrangement of double bonds can occur with attendant shift in linkage to Mg (dashed lines).

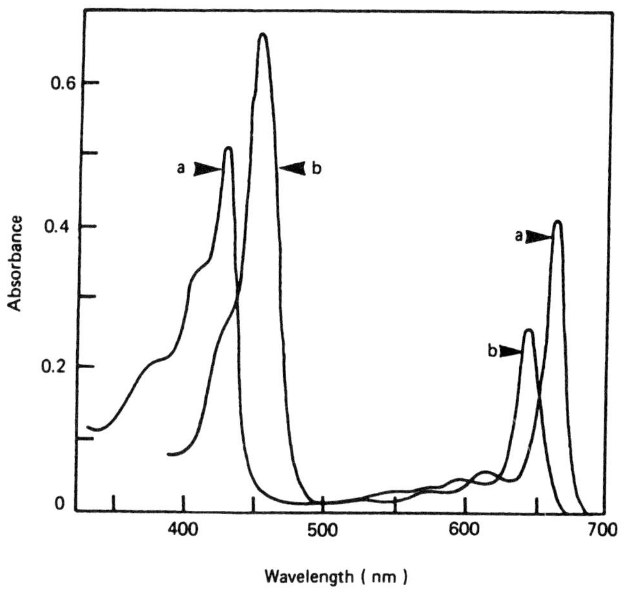

FIGURE 5.3. Absorption spectra of 80% acetone solutions of chlorophylls *a* and *b*. From Gregory, R. P. F. (1989b). *Photosynthesis.* Figure 31a, p. 47. Chapman & Hall, New York.

Pigments

As for abundance and functional significance, the chlorophylls are the most important pigments found in chloroplasts. Two forms of chlorophyll, *a* and *b*, are present in higher plants (Fig. 5.2) along with several "accessory" pigments. The latter include carotenes such as β-carotene and xanthophylls that function in energy transfer to chlorophyll or in protective and regulatory roles to be discussed later. The absorption spectra of the chlorophylls indicate two peaks of absorption in the visible region, one in the blue and one in red wavelengths (Fig. 5.3). The exact maxima of absorption depend upon the molecular environment provided to the chlorophyll molecules. Chlorophyll *a* dissolved in organic solvents exhibits absorption maxima at 450 and 660 nm, whereas maxima of chlorophyll *b* are drawn in slightly (about 20 nm) toward intermediate wavelengths. The relative absence of absorption in the green region clearly illustrates why these pigments give plants and much of the world a greenish color.

Leaves of a few varieties of trees such as copper beech and Crimson King Norway and Japanese maples are red or purple because of the presence of anthocyanin pigments that occur in the cell sap of the vacuole rather than in the chloroplasts. Many other trees produce anthocyanins in the autumn as discussed in Chapter 7.

Proteins

Once proplastids are exposed to light they rapidly accumulate proteins (Lawlor, 1987). Over a period of about four days there is about a 400% increase in thylakoid proteins, with similar rates of synthesis of stromal proteins associated with carbon fixation and metabolism. Several dozen proteins are involved. Some protein synthesis occurs both in the chloroplasts and cytoplasm, where chloroplast proteins subsequently are transported through the envelope membranes (Fig. 5.1). Up to half the protein in chloroplasts consists of Rubisco and it is the most abundant protein in leaves of most plants (Gray, 1984). It probably also is the most abundant protein on earth.

Membrane Systems

As happens with mitochondria, chloroplasts are delimited from other cell compartments by a double membrane envelope (Fig. 5.1). Suspended in the internal stromal matrix is another membrane system containing a spatially complex and well-ordered arrangement of pigment, protein, and lipid molecules. Membrane proteins function as enzymes, associate with pigment molecules, or they may function with prosthetic groups as electron carriers. The three-dimensional hydrophobic–hydrophilic character of a membrane protein determines its location and orientation relative to adjacent bilayered lipid, pigment, and protein molecules. In this membrane system the energy of photons of light entering the chloroplast is captured and converted to a metabolically useful form.

THE PHOTOSYNTHETIC MECHANISM

In simple terms photosynthesis consists of reduction of atmospheric CO_2 to carbohydrate by use of light energy, with an associated release of oxygen from water. This reaction can be summarized by the following generalized equation:

$6CO_2$	+	$12H_2O$	$\rightarrow$	$C_6H_{12}O_6$	+	$6O_2$	+	$6H_2O$
Carbon Dioxide		Water		Glucose		Oxygen		Water

Like many physiological processes, photosynthesis consists of several sequential steps. Because of their complexities, a thorough discussion lies beyond the scope of this book and for more details I refer you to the reviews of various aspects of photosynthesis by Flore and Lakso (1989), Ghanotakis and Yocum (1990), Golbeck (1992), Hartman and Harpel (1994), Long et al. (1994), and Ort (2001), and books such as that edited by Baker and Bowyer (1994), and those by Ke (2001), Lawlor (2001), Blankenship (2002), and the *Advances in Photosynthesis* series edited by Govindjee. However, the enormous importance of photosynthesis requires that the main features of the mechanism of the process be discussed briefly. Some knowledge of the process also is necessary to understand how it is affected by environmental factors.

Photosynthesis can be broken down into the following sequential events:

1. Trapping of light energy by chloroplasts.
2. Splitting of water and release of high-energy electrons and O_2.
3. Electron transfer leading to generation of chemical energy as ATP and reducing power as $NADPH_2$.
4. Terminal steps involving expenditure of energy of ATP and the reducing power of $NADPH_2$ to fix CO_2 molecules in phosphoglyceric acid (PGA) and reduce it to phosphoglyceraldehyde, and finally to convert this compound into more complex carbohydrates such as glucose.

Discussions of photosynthesis naturally place primary emphasis on carbohydrates as the principal products. However, large quantities of the primary products rapidly are converted into other compounds, such as lipids, organic acids, and amino acids, which are equally important in plant metabolism. Some pathways followed in these conversions are discussed in later chapters.

Light Reactions

The light reactions involve events that capture light energy and result in production of ATP and $NADPH_2$. Although there are many individual steps in the light reactions (Fig. 5.4), they proceed very rapidly, usually taking less than a millisecond (10^{-3} s) and often occur-

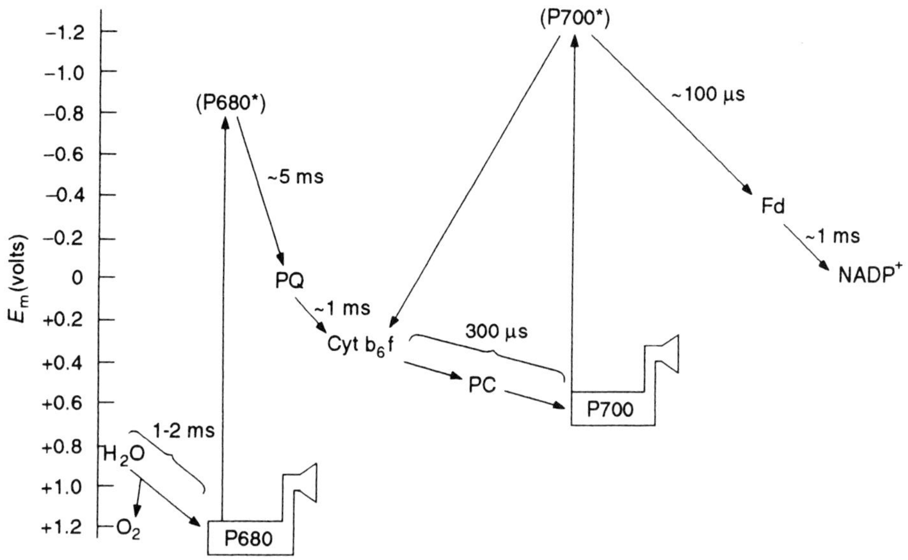

FIGURE 5.4. Summary diagram of the light reactions showing electron flow in the photochemical stage of photosynthesis as related to reduction-oxidation potentials. Light trapped by light harvesting chlorophyll and reaction center (P680) of PS II causes photolysis of water, release of O_2 and H^+, and creation of an ion pair, the reduced member of which loses an electron to lipid-soluble plastoquinone (PQ). Plastoquinone presumably diffuses within the membrane to reduce the cytochrome b_6f complex. Plastocyanin (PC), a water soluble Cu-containing protein, is reduced by this complex and subsequently reduces oxidized reaction centers of PS I (P700). $NADP^+$ is reduced by ferredoxin (Fd) after its reduction by products of the PS I reaction center. Characteristic reaction times for the various stages are shown. Several transient intermediates are not shown. Modified from Gregory, R. P. F. (1989a). *Biochemistry of Photosynthesis, 3rd ed.* Copyright 1989 John Wiley & Sons. Reprinted by permission of John Wiley & Sons, Ltd.

ring in picoseconds (10^{-12} s). The products of the light reactions are utilized in a variety of ways, but principally in fixation of carbon dioxide and reduction of the resultant product to sugars.

Photochemistry

Photosynthesis begins with a very brief (10^{-15} s) encounter between a photon of light and a molecule of pigment, during which an electron in the pigment moves to a higher energy state. The pigment may be a carotenoid, but more often is chlorophyll associated with protein molecules and arranged into an identifiable light harvesting (LH) complex. The LH complex of one photosystem (PS II, see later) is composed of 12 chlorophyll molecules and two carotenoid molecules embedded in the thylakoid membrane with an intimately associated polypeptide (Kühlbrandt and Wang, 1994).

After excitation, the pigment molecule may fluoresce, transfer its energy to another pigment molecule, or lose energy as heat. Carotenoid pigments may transfer energy to chlorophyll molecules in the LH complex and serve in a photoprotective role by preventing singlet oxygen from forming from oxygen and excited chlorophyll *a* molecules. Chlorophylls may undergo numerous transfers of excitation from one pigment molecule to another until a chlorophyll molecule in a specific configuration with certain proteins receives the energy, triggering a photochemical reaction. The pigment-protein complexes involved in photochemical reactions are known as reaction centers. Two reaction centers of green plants (P700 and P680) can be identified and, considered with their associated LH complexes, are described as Photosystems I and II (PS I and PS II), respectively. Charge separation occurs within the reaction centers, resulting in the formation of an ion-pair (Gregory, 1989a):

$$[CX] \rightarrow [C^*X] \rightarrow [C^+X^-] \quad (5.2)$$

This ion pair consists of an oxidized, positively charged chlorophyll dimer (C^+) and a reduced electron acceptor (X^-). This event marks the point at which light energy is converted to chemical energy (Blankenship, 2002). Once the pair is formed the electron may drop back to the donor in a process called recombination; this would be the favored process unless the electron acceptor can be quickly shuttled to another acceptor and the charge further separated spatially from the original donor. The latter process occurs through a series of electron transfers.

Electron Transport

The charge separation described earlier is a key event in the flow of electrons through a series of molecules acting as electron donors and receptors. From an oxidation-reduction perspective, the complete mechanism of electron transport in the light reactions resembles a prone letter "Z" (hence the commonly applied name "Z scheme" for these processes). Electrons are passed from donors of higher (more negative) to lower redox potential. At the two photosystem reaction centers the redox state is elevated in the highly reduced member of the photoinduced ionized pair.

Much progress has been made in recent years in understanding the structure and functioning of the electron transport chain, but uncertainty about some steps remains. The ultimate donor to PS II is water with molecular oxygen as a product (Fig. 5.4), but the exact mechanism is still unclear. The ultimate reduced product of the chain is $NADPH_2$. Some components of the chain are localized to specific areas of the membrane. For example granal stacks are rich in PS II, whereas PS I complexes are found only on the external surfaces of grana and on stromal membranes (Fig. 5.1). The cytochrome b_6f complex (cyt b_6f), which is reduced from the reducing side of PS II by plastoquinone (PQ) and gives up its electrons to plastocyanin, is found nearly equally distributed between grana disks and stromal thylakoids. Plastoquinone is lipid soluble and free to diffuse within the membrane, whereas plastocyanin may diffuse into the aqueous lumen space of thylakoids.

The diffusibility of these compounds is quite significant because it overcomes the apparent hindrance of a spatial barrier between photosystems. However, the need for diffusion of one or the other carrier does provide some limitation of overall electron transport. The relatively slow rate of reduction of the PQ pool and its subsequent reoxidation apparently limit whole-chain electron transport under most conditions. Plastocyanin must travel about the same distance as PQ between source and destination, but apparently moves much faster (Gregory, 1989b) and does not substantially limit whole-chain electron flow. A cyclic path for electron transport also exists between PS I and the cyt b_6f complex. Transport within this cycle results in no net production of $NADPH_2$, but does allow synthesis of ATP.

When electrons move through the electron transport chain a pH gradient builds across the thylakoid membranes as PQ molecules, which are reduced to the quinol form at the interior membrane surface, and release hydrogen ions upon oxidation at the exterior thylakoid surface.

$NADP^+$ Reduction

Ferredoxin, an iron-sulfur protein, is reduced after electron transport through PS I (Fig. 5.4). Ferredoxin, in turn, is oxidized by an enzyme, ferredoxin:$NADP^+$ reductase, which is found on the stroma-facing surfaces of thylakoids. This enzyme reduces $NADP^+$ to $NADPH_2$, providing the reducing power by which fixed carbon dioxide is converted to simple carbohydrates.

Photophosphorylation

Formation of ATP in chloroplasts is catalyzed by ATP synthase, an enzyme found on stromal and external granal surfaces (Gregory, 1989b). The chloroplast ATP synthase consists of two components (CF_0 and CF_1). The CF_0 subunit, which consists of several polypeptides, is hydrophobic and embedded in the thylakoid membrane (Fig. 5.5). The CF_1 subunit is water-soluble, also consists of several polypeptides, and rests on the CF_0 complex on the membrane exterior. A recent structural model of the enzyme proposes two groups of repeating polypeptide units connected by additional polypeptides on one side of the complex (McCarty et al., 2000). The CF_1 subunit has a six-member symmetric ring consisting of equal numbers of two different polypeptides (α and β). The CF_0 subunit has a barrel-like multi-unit member consisting of nine to 12 identical polypeptides (c). On the outside

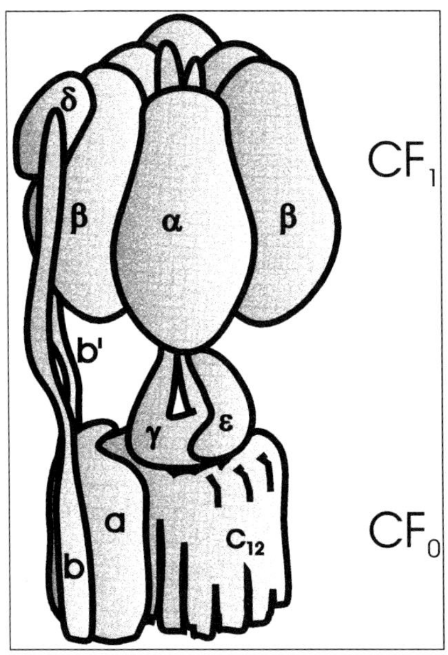

FIGURE 5.5. A current structural model of ATP synthase. The C_{12}, γ, and ε subunits rotate as a unit as protons flow through the CF_o complex. The b, b′, and δ units serve as the stator that holds the α/β heterohexamer. From Groth and Strotmann (1999).

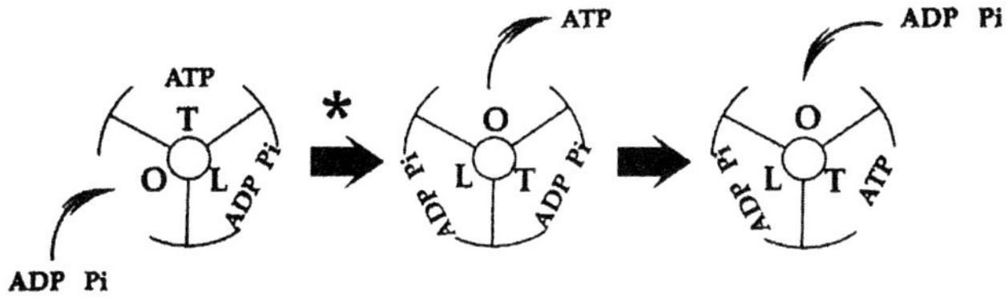

FIGURE 5.6. The alternating sites mechanism of ATP synthase. Three sites interact cooperatively to alternate between tight (T), loose (L), and open (O) states. The energy (*) required for the synthesis of ATP is used for substrate binding and product release. From McCarty et al. (2000). Reprinted, with permission, from the *Annual Review of Plant Physiology and Plant Molecular Biology*, Volume 51 © 2000 by Annual Reviews www.annualreviews.org

are two spanning (b and b′) and two other associated polypetides (a, δ). In the intervening space between repeating units one polypeptide (γ) is linked with the bottom structure, but extends into the central region of the CF_1 α,β ring. Also in this space is another polypeptide (ε) associated with the γ member.

Formation of ATP is linked to passage of H^+ through the CF_0 component, but the exact mechanism is still unknown. Based on several types of evidence an "alternating sites mechanism" has been proposed that posits three binding sites for nucleotide pairs on the CF_1 complex, one for each αβ unit (Fig. 5.6). The three sites alternate among open, loose, and tightly bound configurations with respect to nucleotides and P_i. Transition may be linked to energy-requiring opening and closing of catalytic CF_1 sites, which in turn are driven by rotation of the asymmetric γ polypeptide within the CF_1 complex driven by proton flow through the protein complex of CF_o (Fig 5.5). The ATP synthase thus behaves mechanically like a motor with the c-complex serving as a rotor that spins the γ protein through connection with ε while the b and b′ arms and the a subunit hold the CF_1 complex rigid as in the stator of a motor (Groth and Strotmann, 1999).

Rotation of the c barrel has been proposed to arise from protonation of an aspartic acid of a c subunit in a well at the interface between c and a subunits on the acidic side of the membrane (Fig. 5.7). Loss of charge in this amino acid promotes diffusion of the c subunit into the lipid bilayer. Back diffusion is deterred by the adjacent presence of a positively charged arginine residue on the a subunit. Each net protonation event thus "ratchets" the c barrel in one direction. Each such move of the barrel also exposes a previously protonated residue to an exit well on the other (basic) side of the membrane and on the opposite side of the charged arginine, promoting dissociation of the proton. Mitichondrial ATP synthase operates similarly, and the mechanism can also (by reversal of proton gradients) explain the capacity for ATP hydrolysis and proton pumping seen in some organisms.

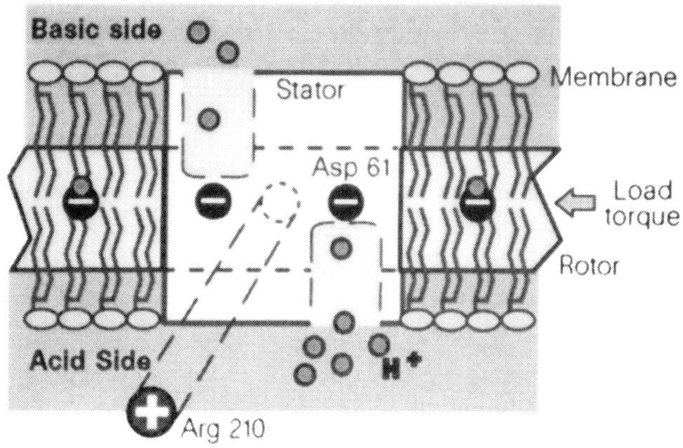

FIGURE 5.7. Charge geometry in ATP synthase showing relative locations of the stator and rotor charges and proton channels. The two proton reservoirs are connected by offset half-channels that make the assembly nonsymmetric. Arg (arginine) 210 is located on a subunit slightly offset from the plane of the rotor. Reprinted by permission from McMillan Publishers Ltd.: Nature Elston, T., Wang, H., and Oster, G. (1998). Energy transduction in ATP synthase. *Nature* **391**, 510–513." © 1998.

Photoinhibition

When plants are exposed to light intensity characteristic of open habitats, the photosynthetic apparatus often shows substantial inhibition (Kozlowski, 1957; Kozlowski et al., 1991). Both shade intolerant and shade tolerant plants may show such photoinhibition. Shade tolerant species and plants developed in the shade are particularly prone to photoinhibition (Watling et al., 1997). Photoinhibition is reversible within minutes to hours on exposure to low light intensity, and it is now thought to be associated with the engagement of photoprotective mechanisms in the leaf. When the limits of these mechanisms are reached, long-term injury to the photosynthetic apparatus generally follows (Layne and Flore, 1993; Choudhury and Behera, 2001).

There are many examples of photoinhibition and only a few will be given here. Photoinhibition was evident in attached leaves of kiwifruit grown in natural light (not exceeding PPFD of 300 μmol m^{-2} s^{-1}) (Greer et al., 1988). When willow leaves that had developed in the shade were exposed to full sunlight, they showed more photoinhibition than leaves developed in the light (Ögren, 1988). Photoinhibition was more pronounced in leaves of cotton and ivy plants grown at a low PPFD than in those grown at high PPFD. The rate of recovery also was higher in plants grown in high PPFD (Demmig and Björkman, 1987). Photoinhibition was demonstrated in seedlings of several species of tropical trees when they were transferred from shade to bright sun (Langenheim et al., 1984; Oberbauer and Strain, 1986). When cacao leaves were continually exposed to light intensities higher than half of that in which instantaneous maximum photosynthesis occurred, the rate of photosynthesis began to decline after four hours (Raja Harun and Hardwick, 1987). The rate of decline increased with further increases in light intensity. At light intensities higher than 100% of saturating photosynthetic intensity, the decline began almost immediately.

Because photoinhibition is accentuated by stresses other than high light intensity, some degree of photoinhibition commonly occurs. Photoinhibition has been demonstrated, often under even moderate light intensity, in plants exposed to temperature extremes (Verhoeven et al., 1999; Close and Beadle, 2003; Hendricksen et al., 2003), drought (Valladares and Pearcy, 1997), nutrient stress (Close et al., 2000), and UV-B (Krause et al., 2003). Gamon and Pearcy (1990) showed that both high and low temperatures can cause photoinhibition, as photosynthesis of California wild grape leaves under high PPFD was inhibited more at high and low than at intermediate temperatures. In the field, however, at high PPFD photoinhibition often occurs even when other stress factors are not severe (Ögren, 1988; Valladares and Pearcy, 1997).

Recovery from photoinhibition is temperature dependent, with the rate of recovery in weak light increasing with elevated temperature. Leaves of shade-grown kiwifruit recovered from photoinhibition when high-light stress was alleviated. However, recovery depended on temperature, with maximum recovery at temperatures of 25 to 35°C, slow recovery below 20°C, and no apparent recovery at 10°C (Greer and Laing, 1988).

Photoinhibition is characterized by reduction in quantum efficiency (moles of CO_2 fixed per mole photons absorbed) and in maximum capacity for photosynthesis (Osmond, 1987; Long et al., 1994). Although the molecular nature of severe photoinhibitory damage is not completely understood, an increasing body of evidence suggests that it results primarily from overexcitation of PS II reaction centers from bright light resulting in inactivation of the primary photochemistry (Critchley, 1988; Kyle, 1987; Cleland, 1988; Long et al., 1994; Ort, 2001). Persistent excitation of chlorophyll molecules can lead to production of singlet oxygen molecules ($^1O_2$) and promote prevalence of the oxidized form of the P680 chlorophyll dimer in the PS II reaction center (Niyogi, 1999). Both of these species can damage lipids and proteins in the reaction center vicinity, including the D1 protein of the PS II reaction center. This protein possesses a binding site for PQ and thus is vital for electron transfer in the light reactions. It also is the site that binds the herbicides diuron and atrazine, displacing PQ and preventing electron transport. In photoinhibition, the D1 protein is exposed to a protease enzyme that removes it from the PS II complex. Although photoinhibitory damage occurs, repair processes at the molecular level also can occur, as new D1 protein is continuously synthesized. In fact, although D1 protein makes up only about 1% of the protein content of the chloroplast, its turnover rate is similar to that of Rubisco, which may constitute 50% of chloroplast protein (Singh, 2000). Hence the development of photoinhibition depends upon both the rate of degradation and synthesis.

Although less studied, PS I components also respond to high light intensity. For example, excessive reduction of the electron acceptor side of P700 by high light may result in reduction of O_2 to superoxide anion radical (O_2^-), which can be metabolized to H_2O_2 and the extremely reactive hydroxyl radical (OH·) (Niyogi, 1999). These reactive molecules are very destructive to many biomolecules, especially membrane lipids.

Apart from repair mechanisms noted earlier, there are several other adaptive responses of plants to excess radiation intensity. Under moderate conditions, photosynthesis and photorespiration place a demand on the electron transport system that prevents the accumulation of excess excitation energy in pigment arrays and reaction centers. Under high excitation pressure provided by intense light, the capacity to divert excess excitation energy from the photosystems by thermal dissipation appears to be quite important. Ort (2001) described two alternative states for the photosynthetic electron transport system. At low-to-moderate light intensity (Fig. 5.8) excited chlorophylls transfer energy to the PS I and PS II reaction centers with little loss, and electron flow through the system is uninhibited. At high light an increased pH gradient across the thylakoid membranes induces a shift to a photoprotected state.

With respect to PS II, the increased ΔpH induces conformational changes in certain proteins of the reaction center and activates violaxanthin epoxidase, an enzyme that converts violaxanthin (V), a xanthophyll pigment that is bound to the PS II light harvesting complex, to zeaxanthin (Z) and antheroxanthin (A). These changes promote the diversion of excitation energy away from the electron transport chain and to thermal dissipation. The pool of xanthophylls often is increased in leaves exposed to high light, indicating acclimation of plants to light regime. For example, Verhoeven et al. (1999) observed significantly higher levels of V + A + Z pigment per unit chlorophyll in *Vinca minor* plants grown outdoors compared with those grown at low light in the laboratory. Additionally, photoinhibition of low-light acclimated plants induced increases in the V + A + Z pool.

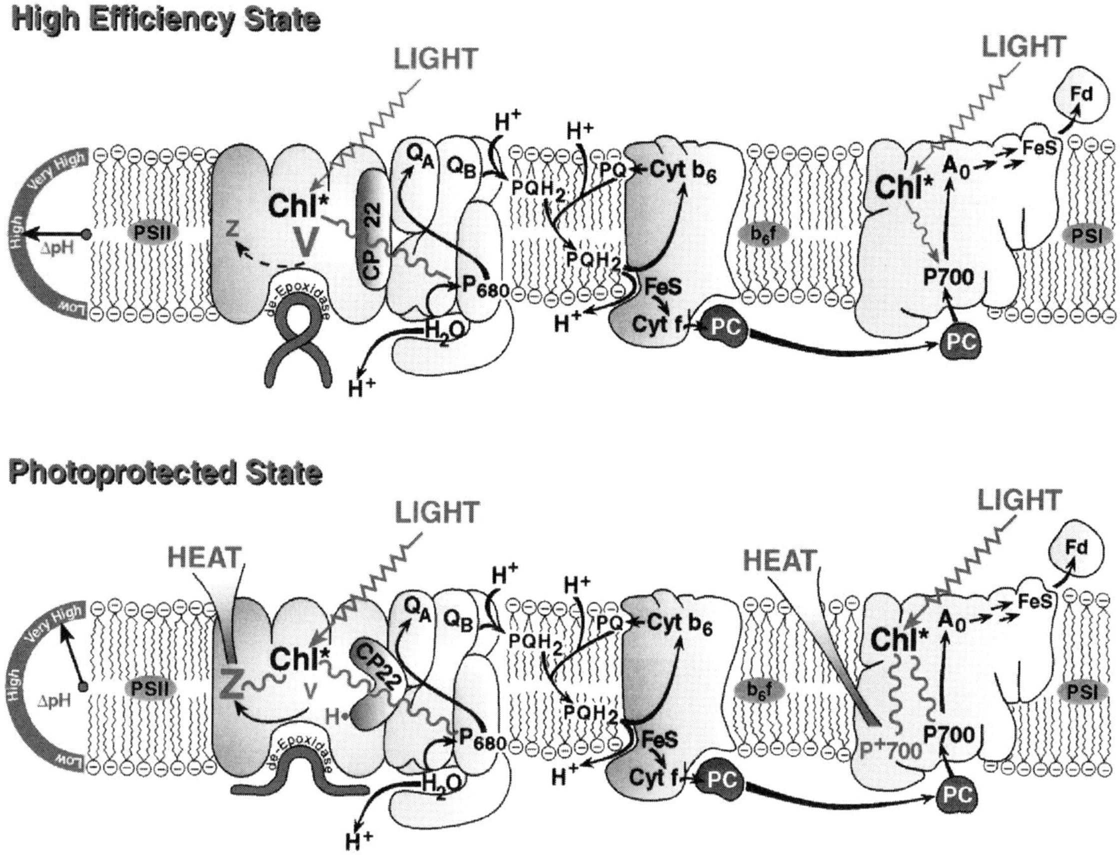

FIGURE 5.8. Model depicting the conversion of the thylakoid membrane at excess light from the high efficiency state (top) to the photoprotected state (bottom). The excess light condition is sensed by a very large ΔpH that initiates the nonphotosynthetic thermal dissipation of absorbed light. PS II; PS I, Photosystem I complex; b_6f, cytochrome b_6f complex; P_{680}, reaction center chlorophyll of PS II; Q_a and Q_b, quinine receptors of PS II; PQ and PQH_2, plastoquinone and reduced plastoquinone; Cyt, cytochrome; FeS, Rieske iron sulfur protein; PC, plastocyanin; P_{700} and P_{700}^+, reduced and oxidized forms of the reaction center chlorophyll of PS I; A_o, primary acceptor of PS I; FeS, bound iron sulfur acceptors of PS I; Fd, soluble ferredoxin; Chl*, excited chlorophyll molecule; Z, zeaxanthin; V, violaxanthin; CP22, minor PS II protein required for regulated thermal every dissipation and believed to function in protonation-dependent reorganization in LHCII. Used with permission of the American Society of Plant Biologists, from Ort, D. R. (2001). When there is too much light. *Plant Physiol.* **125**, 29–32; permission conveyed through Copyright Clearance Center, Inc.

With respect to PS I, excess light intensity likely results in accumulation of the oxidized primary donor of PS I (P700+), which is an effective quencher and thermal dissipater of excited chlorophyll in the PS I antenna complex (Ort, 2001). Thus, PS I photoprotection may be regulated by the electron flow from PS II when the photosynthetic apparatus is in the photoprotected state.

Avoidance of damage from reactive oxygen species (ROS) molecules produced by the light reactions is afforded by several scavenging enzyme systems. For example, O_2^- produced by PS I at high light can be converted to water by the action of superoxide dismutase and ascorbate peroxidase (Chapter 6). This process has been referred to as the water–water cycle because water is both the substrate (via PS II) and product. This cycle may be a significant pathway to dissipate excess photons through electron transport when CO_2 reduction and photorespiration do not provide adequate sinks for electrons. Other antioxidant compounds such as carotenoid molecules, α-tocopherol, ascorbate, and glutathione can quench or chemically scavenge excited chlorophylls and ROS (Niyogi, 1999).

Changes in light regime result in acclimation responses in light harvesting capacity. For example, both mRNA and proteins associated with light harvesting chlorophyll-binding proteins of PS II decreased with an increase in light intensity applied during growth (Maxwell et al., 1995). This response appeared to be controlled by the redox state of the first stable quinone receptor of PS II, Q_a. Finally, avoidance of photoinhibition may be related to reduced light absorption by acclimation of leaf orientation or wilting (Ludlow and Björkman, 1984; Gamon and Pearcy, 1989). For example, Valladares and Percy (1997) observed increased photoinhibition in horizontally restrained leaves of the sclerophyll shrub *Heteromeles arbutifolia* compared with normally more vertically oriented vertical leaves in sun-acclimated plants.

Dark Reactions

CO_2-fixing Enzymes

Most herbaceous crop plants and nearly all woody plants are termed C_3 plants because after CO_2 combines with the five-carbon sugar, ribulose bisphosphate (RuBP), two molecules of the three-carbon compound, phosphoglyceric acid (PGA), are produced (Fig. 5.9A). Rubisco is the carboxylating enzyme and also functions as an oxygenase. As a result, O_2 is a competitive inhibitor of CO_2 fixation. Given its central role in the physiology of plants, it is not surprising that Rubisco is a highly regulated enzyme. Several types of regulation of the activity of Rubisco may exist, including that associated with binding of Mg^{2+} and CO_2, influences of RuBP concentration and pH, activation by light and activase enzymes, and inhibition associated with certain metabolites. Hartman and Harpel (1994) provided a detailed discussion of these properties in a more extensive review of regulation of the photosynthetic process at the enzyme level.

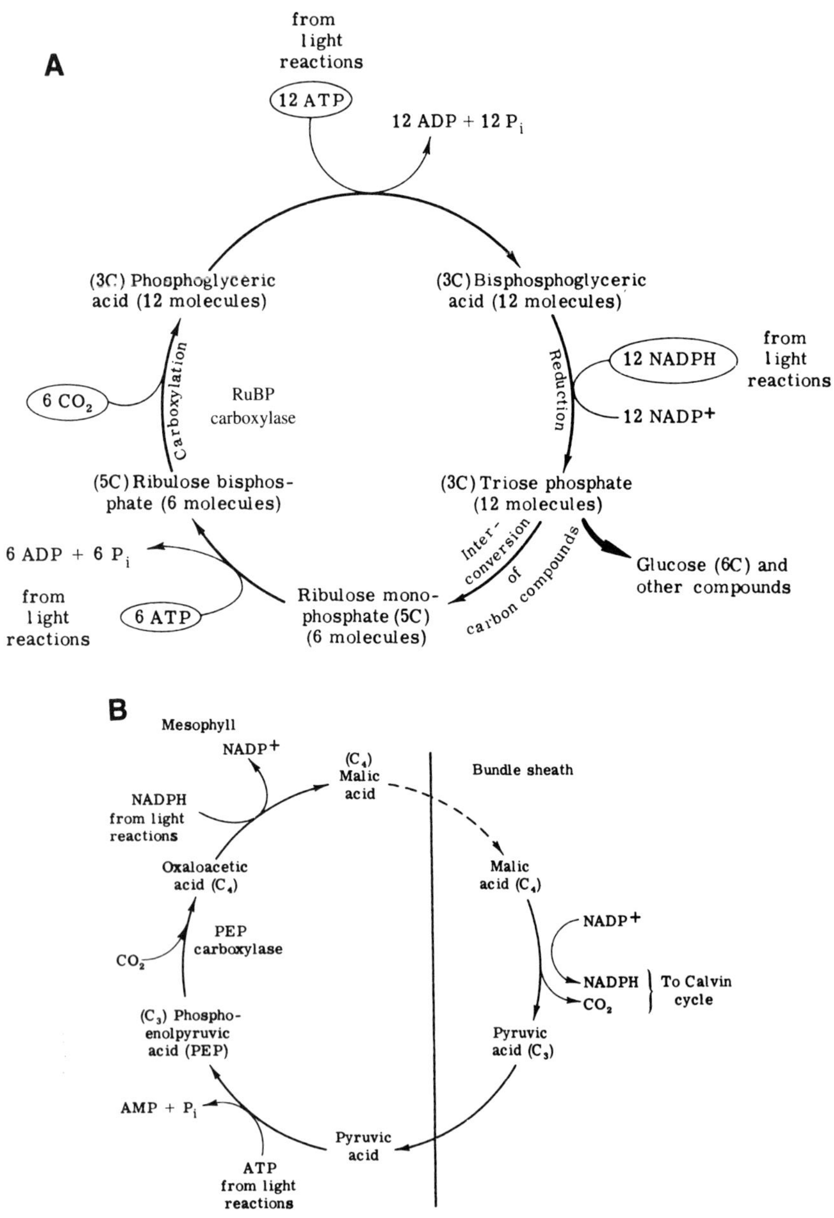

FIGURE 5.9. A. The path of carbon fixation by the C_3 or Calvin-Benson cycle. B. The C_4 or Hatch-Slack pathway for carbon fixation. After Govindjee and Govindjee (1975).

In some species of plants such as maize and sugar cane the first detectable products of photosynthesis are four-carbon compounds, chiefly aspartic, malic, and oxaloacetic acids. In this system, known as the C_4 or Hatch-Slack pathway (Fig. 5.9B), phosphoenolpyruvic acid (C_3) is the CO_2 acceptor and phosphoenolpyruvate carboxylase (PEP carboxylase) is the carboxylating enzyme. This enzyme has a very high affinity for CO_2 and is not inhibited by O_2, as is Rubisco. Among woody plants, only a few species of *Euphorbia* (Pearcy and Troughton, 1975; Pearcy and Calkins, 1983) and certain shrubs in arid central Asia (Winter, 1981; Pyankov et al., 2000) have been reported to conduct C_4 photosynthesis. The absence of C_4 species among forest trees has been attributed to "adaptive troughs" characterized by low biochemical efficiencies and quantum yields of photosynthesis that may be associated with species evolving to the C_4 mode. Such troughs would be very maladaptive in the shaded environments of forests (Monson, 1989, 1999).

Another hypothesis offered by Monson (2003) posits that the long generation times of trees do not favor the rapid appearance of new mutations in duplicated genes that would facilitate evolution of C_4 photosynthesis. The C_4 mechanism has apparently evolved numerous times in both monocots and dicots with multiple biochemical and anatomical variations (Sage, 2004). Sage (2004) presented evidence showing that time of appearance of C_4 photosynthesis was associated with the Oligocene Period, 24 to 33 million years B.P., when global environmental conditions were increasingly unfavorable for C_3 photosynthesis (decreasing CO_2 and increasing aridity).

In most C_4 plants the initial fixation of carbon in C_4 organic acids occurs in a wreath-like ring of mesophyll cells; the acids then are transported into large bundle sheath cells where decarboxylation occurs, releasing pyruvic acid (which reenters the Hatch-Slack system) and CO_2 that is then fixed in the Calvin cycle as shown in Fig. 5.9A (Sage, 2004). This pattern of anatomical differentiation has been termed "Kranz" anatomy, and is characteristic of most C_4 plants. This transport process concentrates CO_2 where Rubisco can more efficiently fix it into PGA because of a higher CO_2/O_2 ratio. It should be emphasized that the Hatch-Slack carbon pathway is not an alternative to the Calvin-Benson cycle, but an added system that increases its efficiency under some circumstances (Moore, 1974). Only the Calvin cycle is truly autocatalytic in the sense that it can generate more CO_2 acceptor than originally was present and therefore produces a net increase in fixed carbon (Kelly et al., 1976).

Besides the difference in the first product of photosynthesis and a higher CO_2 compensation point, C_3 plants become light saturated at a lower intensity than C_4 plants, and their rate of photosynthesis is increased if the oxygen concentration is lowered. Furthermore, they lack the large-bundle sheath cells characteristic of leaves of C_4 plants where CO_2 can be concentrated. The high CO_2 compensation point of C_3 plants is at least partly attributable to the fact that the carboxylase activity of Rubisco is competitively inhibited by oxygen. This also accounts for the observation that a decrease in O_2 concentration results in a large increase in CO_2 fixation in C_3 plants. Increasing the CO_2 concentration reduces the oxygenase activity of Rubisco, favoring increased net CO_2 fixation. This difference in sensitivity to CO_2 suggests that rising atmospheric CO_2 levels associated with burning of fossil fuels by humans will stimulate photosynthesis in C_3 plants more than in C_4 species (Kozlowski et al., 1991). Although many fewer studies under elevated CO_2 have been performed on C_4 than C_3 plants, likely because of these theoretical considerations, the few data that have been assembled suggest that photosynthesis of C_4 species is in fact less responsive to elevated CO_2 (e.g., 15% stimulation of light-saturated photosynthetic rate in C_4 plants vs. 34% stimulation for C_3 plants; Ainsworth and Long, 2005).

An important characteristic of plants having the C_4 pathway is the very low rate of photorespiration as compared with C_3 plants. Photorespiration refers to light-dependent production of CO_2 by photosynthetic tissue and is in no way related to the basic respiration discussed in Chapter 6 that involves a cytochrome system. The substrate is a recent product of photosynthesis, probably glycolic acid derived from RuBP by the oxygenase activity of Rubisco. The measurable rate appears to be negligible in C_4 plants, probably because nearly all the CO_2 released by photorespiration is recycled by the more efficient combination of carboxylases in the combined Hatch-Slack and Calvin-Benson cycles than in the Calvin cycle alone.

Because of its apparent wasteful expenditure of photosynthate, photorespiration frequently has been viewed as an unalloyed disadvantage. However, as noted earlier, nearly all woody species possess the C_3 pathway and exhibit photorespiration, yet they are very successful in nature. Although the C_4 carbon pathway is more efficient than the C_3 pathway at the leaf level and appears to have evolved as a mechanism that promotes reduced photorespiration (Ehleringer and Monson, 1993), C_4 metabolism often seems to provide little advantage at the level of plant stands (Gifford, 1974; Snaydon, 1991). Further, Gregory (1989a) suggested that photorespiration serves a synthetic role, as the amino acid serine is produced via the photorespiratory pathway. Additionally, photorespira-

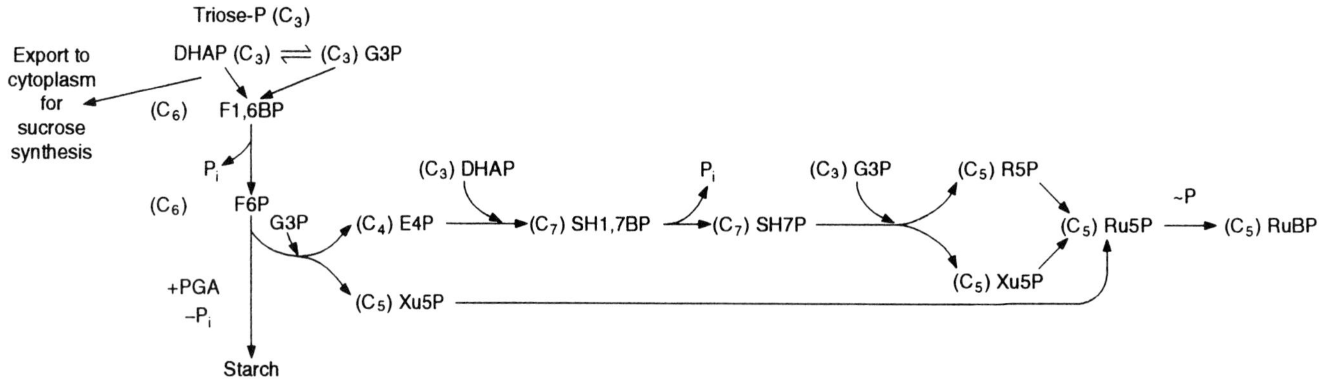

FIGURE 5.10. Alternative pathways in the photosynthetic carbon reductive cycle. Fixed carbon can be: (1) moved to the cytoplasm (primarily as DHAP) for cell metabolism and export as sucrose, (2) converted to starch in the chloroplast, or (3) regenerated as RuBP. Abbreviations: G3P, glyceraldehyde 3 phosphate; DHAP, dihydroxyacetone phosphate; F1,6BP, fructose 1,6 bisphosphate; F6P, fructose 6 phosphate; PGA, phosphoglyceric acid; E4P, erythrose 4 phosphate; Xu5P, xylulose 5 phosphate; SH1,7BP, sedoheptulose 1,7 bisphosphate; SH7P, sedoheptulose 7 phosphate; R5P, ribose 5 phosphate; Ru5P, ribulose 5 phosphate; RuBP, ribulose 1,5 bisphosphate. After Gregory, R. P. F. (1989a). *Biochemistry of Photosynthesis, 3rd ed.* Copyright 1989 John Wiley & Sons. Reprinted by permission of John Wiley & Sons, Ltd.

tion, through internal production of CO_2 that may be refixed, protects from damage a highly illuminated photosynthetic mechanism deprived of external CO_2. Whereas absence of an electron acceptor would result in production of injurious oxygen free radicals and superoxide, photorespiratory CO_2 production provides continuous consumption of ATP and $NADPH_2$.

Another variation of carbon fixation, known as Crassulacean acid metabolism (CAM), is found in many succulents and a few other plants that can fix large quantities of CO_2 in darkness. Crassulacean acid metabolism carboxylation reactions lead to the formation of oxaloacetic and malic acids in the dark through the activity of PEP carboxylase. In the light malic acid is decarboxylated, yielding pyruvic acid and CO_2, which is used in photosynthesis by way of the Calvin-Benson cycle. In such plants the acidity becomes very high at night, but decreases during the day while the sugar content increases. In many succulents the stomata are open at night, allowing absorption of CO_2, but they are mostly closed during the day, preventing excessive loss of water when evaporative demands are highest. Whereas C_4 plants avoid the effects of photorespiration through a spatial separation of CO_2 capture and fixation, CAM plants employ a temporal solution to capture and eventually fix CO_2 under harsh conditions (high vapor pressure deficit, VPD). This arrangement is very efficient for plants growing in dry habitats, but it has been observed in only one genus of trees (*Clusia* spp.) (Tinoco-Ojanguren and Vasquez-Yanes, 1983; Popp et al., 1987).

Photosynthetic Carbon Metabolism

After reduction of phosphoglyceric acid to triose phosphate, fixed carbon can follow several courses (Fig. 5.10). First, triose phosphate may move to the cytoplasm in exchange for phosphate by means of a translocator located in the chloroplast envelope membranes. In the cytoplasm triose phosphate is converted into sucrose, much of which is exported from the cell. Within the chloroplast the triose phosphate also may enter a complex sequence of interconversions by which RuBP, consumed as CO_2 as fixed by Rubisco, is regenerated. Finally, within this cycle fructose-6-phosphate may be diverted to starch synthesis in the chloroplast.

Flow of fixed carbon to carbohydrates, particularly starch and sucrose, is highly regulated (Geiger and Servaites, 1994). In the presence of strong sinks for photosynthate within the plant, much newly fixed triose-P moves to the cytoplasm for synthesis and export of sucrose. The released inorganic P subsequently is cycled back into the chloroplast via the phosphate translocator. If sink demand is small, phosphate is depleted and PGA and triose phosphates accumulate within the chloroplast. As starch synthesis is stimulated by low phosphate and high PGA levels, fixed carbon consequently is diverted to starch in chloroplasts (Trethewey and Smith, 2000). Electron micrographs of chloroplasts (Fig. 5.11) frequently reveal large starch grains, which change in size diurnally and with other influences on sink strength and photosyn-

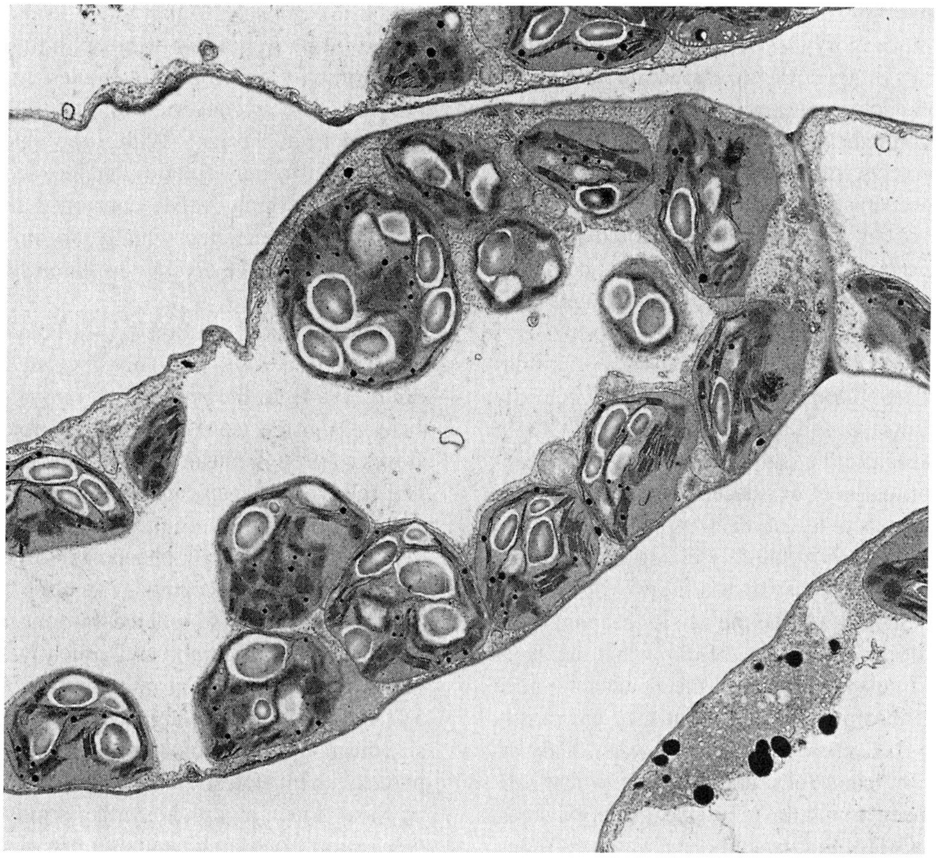

FIGURE 5.11. Electron micrograph of mesophyll cells showing chloroplast structure and arrangement of chloroplasts within a cell. Courtesy of J. White, Univ. of Missouri.

thetic rates. Accumulation of starch has obvious physical limits and photosynthesis must be reduced if neither export nor starch synthesis can accommodate existing photosynthetic rates. Such reductions in photosynthesis may result from P_i-limited photophosphorylation (and hence reduced CO_2 fixation) (Sharkey, 1985; Sage, 1994) and from feedback inhibition of photosynthetic gene expression by accumulating sugars (van Oosten et al., 1994; Sheen, 1994).

CARBON DIOXIDE UPTAKE BY PHOTOSYNTHETIC TISSUES

In the gas phase external to the leaf and within the substomatal cavity, the path of diffusion of CO_2 during photosynthesis is roughly similar to that of water vapor, but in the opposite direction. Gas phase movement of CO_2 within the leaf differs somewhat from that of water vapor in that whereas most water vapor moving through the stomata originates from cells lining the substomatal cavity, inward-moving CO_2 diffuses farther into the intercellular air spaces of the leaf mesophyll (Parkhurst, 1994). Concentration of CO_2 progressively decreases along this path through dissolution into wet mesophyll cell walls. Movement of the two molecules also differs markedly in that CO_2 must move in the liquid phase to the chloroplasts, a route that requires diffusion through different compartments, including wetted cell walls, plasmalemma and chloroplast membranes, and the matrix of the cytoplasm and stroma (Fig. 5.12). Some researchers have suggested that photosynthesis might be increased by the wind because it can reduce boundary layer thickness and promote mass flow of CO_2 through amphisomatous, fluttering leaves. Most evidence indicates that this effect is measurable but slight (Shive and Brown, 1978; Roden and Pearcy, 1993a).

Carbon dioxide diffuses at least 10,000 times as fast in air as in water, suggesting that movement in the liquid phase might severely limit flux of photosynthetic substrate. However, certain anatomical features of leaves reduce the distance that CO_2 must move in liquid. Mesophyll cells are relatively small, thus having a large surface area to volume ratio. In a well-hydrated leaf, turgor pressure forces adjacent cells to have relatively small areas of contact (Levitt, 1980b; Fellows and Boyer, 1978). In this way, mesophyll cells may have

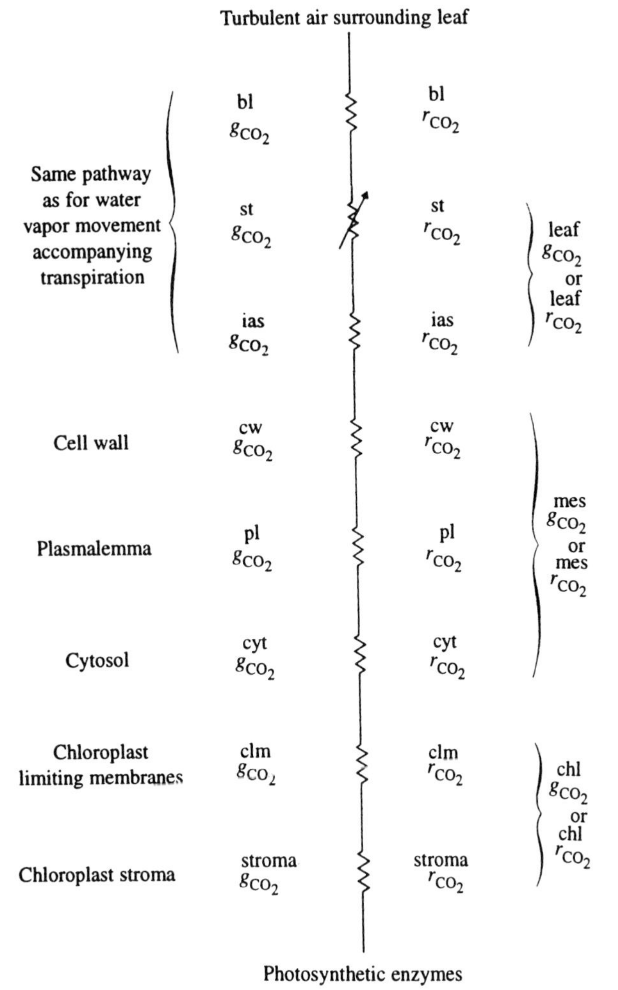

FIGURE 5.12. Major diffusion conductances (g_{CO2}) and resistances (r_{CO_2}) encountered by CO_2 moving from turbulent air around the leaf to the point of fixation by photosynthetic enzymes in the chloroplast. bl, Boundary layer; st, stomata; ias, intercellular air space. After Nobel (1991).

areas (A_{mes}) that far exceed the corresponding leaf area (A), and in turgid leaves most of this surface is in contact with intercellular air spaces. Nobel (1991) reported A_{mes}/A ratios between 10 and 40 for mesophytic species and 20 to 50 for xerophytic species. Additionally, chloroplasts are positioned very close to the plasmalemma (Fig. 5.11). This combination of features makes the actual diffusion distance for CO_2 in the liquid phase relatively small (<1 µm) for a large proportion of the chloroplasts.

Comparative estimates of diffusion resistances (and their reciprocal, conductances) (Table 5.1) indicate that the liquid phase constitutes a small percentage (5 to 22%) of total diffusion resistance in trees, but is relatively greater (23–60%) in herbaceous crop plants that generally have lower leaf resistances than do woody plants. In the liquid phase, molecules of CO_2 must pass at least three membranes (the plasmalemma and the double membrane of the chloroplast) for entry to the stroma. Plant membrane resistance to CO_2 is unknown, but estimates based on red blood cell and artificial membranes suggest a resistance of about 50% of the total liquid resistance (Evans and Loreto, 2000). Once inside the chloroplast, CO_2 diffusion is accelerated by carbonic anhydrase (CA), an enzyme that facilitates rapid interconversion between CO_2 and HCO_3^- in solution. As the HCO_3^-/CO_2 ratio is about 45 at typical chloroplast pH (~8), CO_2 removed by Rubisco at the site of fixation is rapidly replenished by CA from abundant local HCO_3^-. It should be remembered that actual net CO_2 fixation rates depend not only on these diffusion characteristics of leaves, but also upon biochemical and photochemical capacities of the chloroplasts to fix CO_2 and simultaneous respiratory processes.

A useful formulation of the flux of CO_2 into the leaf can be described by the following equation:

$$P_{net} = [CO_{2a}] - [CO_{2chlpst}]/r_{CO_{2a}\rightarrow chlplst} \qquad (5.3)$$

where P_{net} = net flux of CO_2 (e.g., µmol m^{-2} sec^{-1}), $[CO_{2a}]$ = concentration of CO_2 in air beyond the leaf boundary layer, $[CO_{2chlpst}]$ = CO_2 concentration in the chloroplast, and $r_{CO2a\rightarrow chlplst}$ = integrated diffusion resistance to CO_2 from the air to the chloroplast. At steady state, this equation can be partitioned into two equations describing gaseous pathway and leaf mesophyll segments:

$$\begin{aligned} P_{net} &= [CO_{2a}] - [CO_{2i}]/r_{CO_{2a}\rightarrow i} \\ &= [CO_{2i}] - [CO_{2chlpst}]/r_{CO_{2i}\rightarrow chlplst} \end{aligned} \qquad (5.4)$$

where $[CO_{2i}]$ = CO_2 concentration in the intercellular air spaces of the leaf, $r_{CO2a\rightarrow i}$ = diffusion resistance from bulk air to intercellular spaces of the leaf, and $r_{CO2i\rightarrow chlplst}$ = resistance to CO_2 from the intercellular air spaces to the chloroplast. Estimates of CO_2 resistance can be obtained from measurements of water vapor flux outward from the leaf (Chapter 12), accounting for the greater molecular weight of CO_2 compared with water vapor (44 g/mole vs 18 g/mole, respectively). In this way $[CO_{2i}]$, hereafter called c_i, can be computed as:

$$c_i = [CO_{2a}] - r_{CO_{2a}\rightarrow i} \cdot P_{net} \qquad (5.5)$$

The response of c_i under various conditions provides valuable insight into the relative importance of stomatal aperture and mesophyll components in control of photosynthesis. For example under water stress, if both P_{net} and c_i decline significantly, it indicates that stomatal closure is more responsible for photosynthetic decline than mesophyll aspects of photosynthesis because the decline in c_i reflects starva-

TABLE 5.1. **Representative Values of Conductances and Resistances for CO_2 Diffusing into Leaves[a]**

	Conductance		Resistance	
Component	**mm s^{-1}**	**mmol m^{-2} s^{-1}**	**s m^{-1}**	**m^2 s mol^{-1}**
Leaf (lower surface), gas phase				
Crops, open stomata	1.2–6	50–250	160–800	4–20
Trees, open stomata	0.3–2	12–75	500–2500	13–80
Cell wall	30	1200	30	0.8
Plasmalemma	10	400	100	2.5
Cytosol	100	4000	10	0.25
Mesophyll				
Estimation	7	300	140	3.5
Measurements, mesophytes	2.5–25	100–1000	40–400	1–10
Chloroplast				
Estimation	10	400	100	2.5
Measurements	>5	>200	<200	<5

[a]From Nobel (1991).

tion of a competent photosynthetic apparatus. On the other hand, increased c_i indicates that liquid-phase transport and/or photosynthetic carbon fixation and metabolism cannot keep pace with the capacity for stomata to supply CO_2 to the mesophyll. This analysis, although somewhat oversimplified as will be discussed later, provides a valuable method of analysis of stomatal versus nonstomatal limitations on photosynthesis.

Because of the additional resistances associated with CO_2 flux into leaves compared to water vapor, partial closure of stomata increases the resistance to water movement relatively more than the resistance to CO_2 movement and should reduce transpiration more than it reduces photosynthesis. This theory forms the basis for use of metabolic antitranspirants, which cause closure of stomata to reduce transpiration (see Chapter 12).

CARBON AND OXYGEN ISOTOPE DISCRIMINATION DURING PHOTOSYNTHESIS

The rate of CO_2 fixation varies if CO_2 molecules contain different natural isotopes of carbon (^{12}C and ^{13}C being the most common at 98.9 and 1.1% of atmospheric CO_2, respectively). First, $^{13}CO_2$ diffuses somewhat more slowly across the boundary layer and through stomatal pores, tending to make the interior of the leaf slightly depleted in $^{13}CO_2$ compared to that which would otherwise be the case. Second, the primary carboxylating enzyme in C_3 plants, Rubisco, preferentially fixes $^{12}CO_2$. As a result, the ratio of carbon isotopes ($^{13}C/^{12}C$) in C_3 plants tends to be lower than that in the atmosphere. In contrast, carboxylating enzymes of the C_4 photosynthetic pathway (and CAM plants fixing the bulk of daily CO_2 at night) show much less discrimination against $^{13}CO_2$. The extent of depletion ^{13}C in plant dry matter, termed carbon isotope discrimination and represented by the symbol Δ, is described by the following equation (Ehleringer, 1993, cited in Smith and Griffiths, 1993):

$$\Delta = \frac{\delta_a - \delta_p}{1 + \delta_p} \tag{5.6}$$

where δ_a represents the deviation in carbon isotope composition of the atmosphere from that in a standard (a sample of belemnite from the Pee Dee formation) and δ_p represents a similar deviation of plant carbon isotope composition from the standard (δ_p is frequently abbreviated as $\delta^{13}C$). Reference of plant carbon isotope composition to atmospheric composition often is needed to eliminate the effects of variation in source air isotopic composition on the isotope ratios found in plant materials. For example, in urban areas burning of fossil fuels can change the $^{13}C/^{12}C$ ratio of the air considerably. Additionally, the carbon isotope ratio of the air in a forest tends to vary with height. This pattern is found because the flux of CO_2 into the canopy is downward and the progressive removal of greater amounts of ^{12}C by the upper canopy tends to enrich ^{13}C at lower positions. However, respiratory contributions from roots and soil decomposition processes create an atmosphere near the surface that is depleted in ^{13}C because of the relative scarcity of this isotope in plant matter. These differences are most noticeable if air mixing is restricted, as in many tropical forests.

Farquhar et al. (1982) developed a commonly employed predictive model relating Δ to diffusion and carboxylation isotope effects and the ratio of internal to external concentrations of CO_2:

$$\Delta = a + (b - a)\frac{c_i}{c_a} \qquad (5.7)$$

Here a is the fractionation that is associated with diffusion (4.4‰) and b is the net fractionation that occurs with carboxylation (about 27‰ for carboxylation via Rubisco). The discrimination against ^{13}C of PEP carboxylase in C_4 photosynthesis is only about 5.71‰. As c_a is relatively stable, Δ values essentially provide an estimate of the internal CO_2 concentration of the leaf (Ehleringer, 1993). This equation predicts that environmental or genotypic factors that tend to reduce c_i/c_a (e.g., stomatal closure or genetically based increases in mesophyll capacity for photosynthesis) will tend to decrease Δ (and increase $\delta^{13}C$). As noted earlier, low c_i/c_a values often indicate that flux rates of CO_2 into the leaf interior are restricted compared to the capacity of the mesophyll to fix the available CO_2. As the fixation process exhibits a preference for $^{12}CO_2$, the residual air in the mesophyll interior becomes enriched in $^{13}CO_2$ as the influx of $^{12}CO_2$ that would dilute the concentration of $^{13}CO_2$ is reduced by stomatal restrictions. Consequently, relatively more $^{13}CO_2$ molecules will be fixed by Rubisco despite its greater affinity for $^{12}CO_2$, resulting in reduced discrimination. In contrast, conditions that tend to elevate c_i/c_a (e.g., low light levels, decreased mesophyll conductance) will increase Δ.

Carbon isotope discrimination was widely employed in studies of C_3 and C_4 photosynthesis as a convenient method of determining a plant's primary photosynthetic pathway. It also has become apparent that Δ varies among and within species of plants from a photosynthetic group. For example, in desert C_3 plants there appears to be a negative relationship between Δ and longevity of species (Ehleringer and Cooper, 1988, cited in Ehleringer, 1993).

As values of $\delta^{13}C$ and Δ are closely related to c_i/c_a, they can provide an integrated estimate of the relative limitation of photosynthesis by stomata and mesophyll capacity. However, in comparative studies an increase in $\delta^{13}C$ (and hence lower inferred c_i/c_a ratio) may reflect either lower stomatal conductance or increased mesophyll capacity for photosynthesis. This problem can be circumvented by providing identical and stable environmental conditions to plants to be compared, but at the cost of relevance to the dynamic and variable conditions in the field. For example, when plants to be compared come from environments known to differentially affect stomatal aperture (e.g., high and low VPD, see Chapter 12), it is impossible to employ $\delta^{13}C$ values to compare photosynthetic attributes.

Recently, researchers have employed tandem measurements of $\delta^{13}C$ and $\delta^{18}O$ in leaf organic matter to distinguish stomatal from photosynthetic capacity effects (Scheidegger et al., 2000; Dawson et al., 2002; Sullivan and Welker, 2007). This strategy capitalizes on the fact that leaf water becomes enriched in the natural isotope ^{18}O during transpiration (being more massive that ^{16}O it evaporates more slowly). Carbonic anhydrase in the chloroplast catalyzes exchange of oxygen atoms from ^{18}O-enriched water to CO_2 resulting in higher ^{18}O contents in substrate CO_2 (Gillon and Yakir, 2000). Subsequent fixation of ^{18}O-enriched CO_2 into carbohydrates provides a signal of the level of enrichment of this atom in leaf water. As ^{18}O enrichment of leaf water is a function of integrated evaporation rate, organic matter ^{18}O content is a proxy indicator for stomatal aperture. In this way, the effects of photosynthetic capacity on c_i/c_a can be separated from stomatal effects. If $\delta^{13}C$ and $\delta^{18}O$ increase in plant dry matter, there is evidence that the increase in $\delta^{13}C$ arises primarily from decreased stomatal conductance. This conclusion follows because an increase in $\delta^{18}O$ is associated with increased evaporation rate due to higher VPD, and as noted elsewhere in this chapter, stomatal closure is common as VPD increases (Scheidegger et al., 2000).

Analysis of carbon isotope ratios also has been employed in water relations research. As one impact of moderate water stress is stomatal closure in the absence of substantial mesophyll inhibition, Δ values frequently have been used to estimate the impact of drought on plants. Plants growing under chronic drought conditions tend to have lower Δ values than those that are well watered. A distinct advantage of this approach is the integrated nature of the measurement, as plant dry matter accumulated over periods ranging from days to months can be analyzed for Δ. In contrast, conventional gas exchange measurements provide only instantaneous estimates of photosynthesis and gas diffusion resistances. Additionally, as stomatal closure tends to reduce photosynthesis more than it does transpiration, values can sometimes be effectively related to seasonal water use efficiency (see Chapter 12).

VARIATIONS IN RATES OF PHOTOSYNTHESIS

The rate of photosynthesis of woody plants varies widely and is influenced by interactions of many environmental and plant factors. It sometimes is difficult

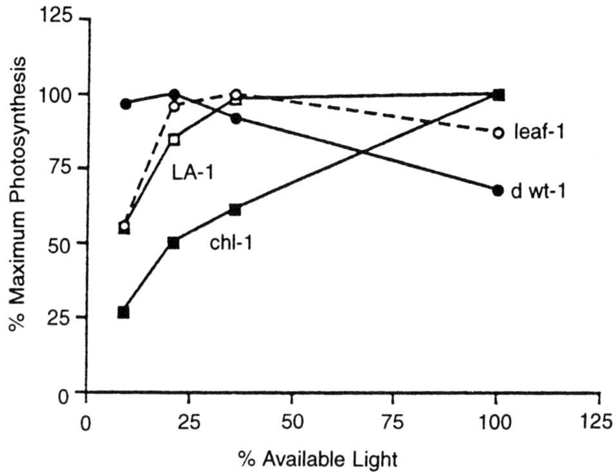

FIGURE 5.13. Photosynthesis of peach leaves at different light intensities when the rate (% of maximum) was expressed per unit of leaf area (□), per leaf (○), per mg of chlorophyll (■), and per unit of leaf dry weight (●). From Flore and Lakso (1989).

to make meaningful comparisons of rates as determined by different investigators because of variations in measurement techniques and methods of expressing rates of photosynthesis. For example, widely different rates of photosynthesis have been reported for apple trees. Avery (1977) attributed such differences to variations in methods of measuring rates and preconditioning effects of environment on leaf anatomy. Photosynthesis of peach expressed on a leaf area or whole leaf basis was constant with decreasing light intensity until the available light decreased below 20% of full sunlight (Fig. 5.13). When expressed per unit of chlorophyll, however, photosynthesis decreased linearly with increasing shade. When photosynthesis was expressed on a dry weight basis, it increased with increasing shade.

Photosynthetic capacity sometimes is estimated from net dry weight increment of plants (see Chapter 3). Such estimates usually are meaningful because the weight increment represents an average increase over a considerable time in an environmental regime that includes the usual periodic environmental stresses. In contrast, short-term measurements of CO_2 uptake often have been made in a constant, controlled environment rather than a fluctuating one. Such an environment may be favorable to all or only some of the species or genotypes compared. Photosynthesis of a number of species of angiosperms is as efficient at low light intensities as it is at high ones, whereas photosynthesis of many gymnosperms is much more efficient at high intensities. Comparisons of species of these two groups at low or at high light intensities often give different impressions of their comparative photosynthetic capacity. Furthermore, evergreens often accumulate some dry weight during the dormant season, whereas deciduous angiosperms lose some through respiration. Therefore, an evergreen tree with a slightly lower rate of photosynthesis than a deciduous tree at some time during the growing season may accumulate as much or more dry matter over a year because of the much longer seasonal duration of photosynthetic activity (Lassoie et al., 1983).

Species and Genetic Variations

Photosynthetic capacity often varies appreciably among species and genotypes. Such variations usually are related to basic differences in metabolism and/or leaf anatomy. In addition, both species and genotypes differ in crown development and architecture, and greater leaf production or a longer growing season often compensates for a low rate of photosynthesis per unit of leaf area or dry weight.

Species Variations

In general, the leaves of deciduous species have higher rates of light-saturated photosynthesis than do leaves of evergreens when the rates are expressed on a leaf dry weight basis. A few examples of differences among species will be given. It should be remembered, however, that rates of photosynthesis vary widely in different leaves on the same tree and comparisons among species may be misleading when sampling as well as measurement techniques and environmental conditions under which the plants developed were different.

In forest trees, high rates of photosynthesis have been measured in several angiosperms, including poplar, apple, ash, and eucalyptus, and in such gymnosperms as Douglas-fir, larch, and metasequoia (Larcher, 1969). The reasons for these inherent differences are discussed later in this chapter.

Kramer and Decker (1944) found that CO_2 absorption was much higher per unit of leaf area in northern red and white oaks than in flowering dogwood or loblolly pine seedlings. Polster (1955) measured high rates of photosynthesis in Douglas-fir, intermediate rates in white pine, and low rates in Norway spruce. The rate of photosynthesis was higher in well-watered post oak and white oak than in sugar maple or black walnut seedlings (Ni and Pallardy, 1991). The rate of photosynthesis of mature leaves of oil palm is among the highest reported for trees (Ceulemans and Saugier, 1991). Photosynthetic rates of even closely related species may differ significantly. For example, Mooney et al. (1978) found wide variations in photosynthesis of several species of *Eucalyptus*.

Variations in rates of photosynthesis also have been reported for fruit and nut trees. Flore and Lakso (1989) found that among deciduous temperate fruit crops sour cherry, nectarine, and sweet cherry had the highest rates; apple, peach, blueberry, and almond had intermediate rates; and apricot had a low rate. Citrus trees generally had lower rates than apple trees, with orange and grapefruit having the highest rates of the citrus trees and lemon the lowest. These investigators advised, however, that these rankings should be viewed cautiously because the values were based on different numbers of observations. De Jong (1983) found the mean maximum rates of peach, plum, and cherry to be similar and intermediate, and that of almond was highest and apricot lowest when expressed on a leaf area basis (Fig. 5.14). Other examples of variations among species in photosynthetic capacity are shown in Figures 5.15, 5.16, 5.17, and 5.23.

Intraspecific and Hybrid Variation

There are many examples of genotypic variation in rates of photosynthesis. Differences have been reported among cultivars of *Rhododendron simsii* (Ceulemans et al., 1984) as well as clones of rubber (Samsuddin and Impens, 1978, 1979), obeche (Lapido et al., 1984), fig (Otteson, 1990), and poplars (Fig. 5.17). Variations in photosynthesis among families of black locust were reported by Mebrahtu and Hanover (1991).

Populations of balsam fir (Fryer and Ledig, 1972) and cabbage gum (Slatyer, 1977) from various altitudes had different temperature optima for CO_2 uptake. Photosynthetic capacity was higher for sugar maple progenies from high altitudes than those from low altitudes (Ledig and Kurbobo, 1983). Variations in rates of photosynthesis among Norway spruce, Scotch pine, and black spruce provenances also have been reported (Pelkonen and Luukkanen, 1974; Zelawski and Goral, 1966; Johnsen et al., 1996). The optimum temperature for photosynthesis varied among cultivars of olive in accordance with the climate of their origin (Bongi et al., 1987).

Genotypic variations in photosynthesis per plant often reflect differences in rates of leaf production and retention. The higher photosynthetic production of loblolly pine seedlings from Florida over those from Georgia was attributed to variations in amounts of foliage, rather than to higher rates per unit of foliage (McGregor et al., 1961). Much higher photosynthetic production of loblolly pine seedlings from Florida than from Arkansas, Oklahoma, or Texas was attributed to differences in leaf area as well as in net photosynthesis per unit of leaf area (Boltz et al., 1986). Late-season growth and photosynthesis of the Florida source increased the provenance differences that were established early in the growing season. Prolonged retention of green leaves in the autumn by some poplar hybrids accounted for appreciable production and storage of late-season photosynthate (Nelson et al., 1982; Nelson and Isebrands, 1983).

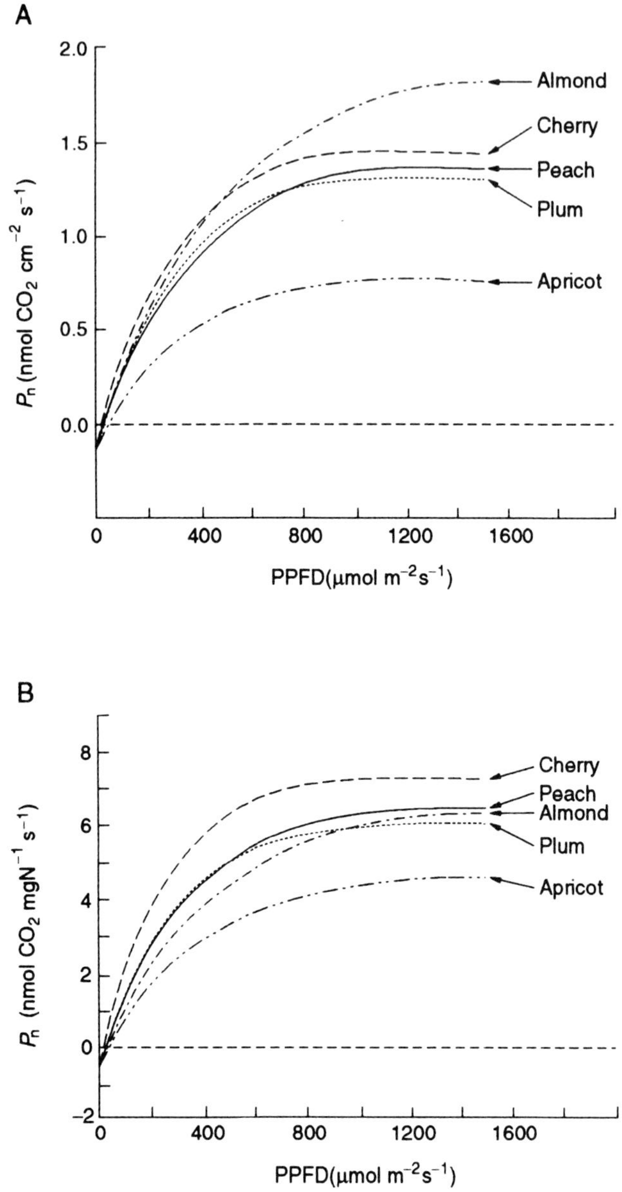

FIGURE 5.14. Effect of photosynthetic photon flux density (PPFD) on net photosynthesis (P_n) of five species of fruit trees on a leaf area basis (A) and a leaf N basis (B). From De Jong (1986).

Photosynthesis and Productivity

Much interest has been shown in using of rates of photosynthesis as indices of growth potential of tree species and genotypes. However, both high and low

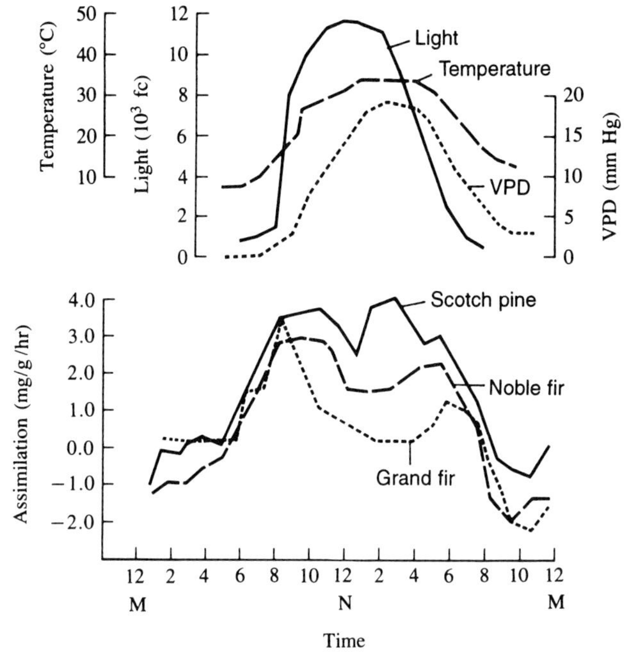

FIGURE 5.15. Diurnal variation in photosynthesis of Scotch pine, noble fir, and grand fir on a clear day during the summer. VPD, vapor pressure deficit. From Hodges (1967).

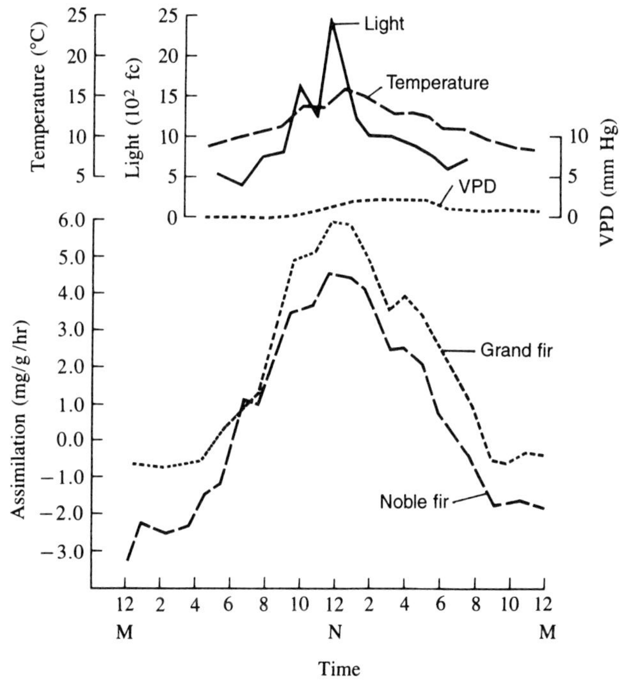

FIGURE 5.16. Diurnal variation in photosynthesis of grand fir and noble fir on an overcast day during the summer. VPD, vapor pressure deficit. From Hodges (1967).

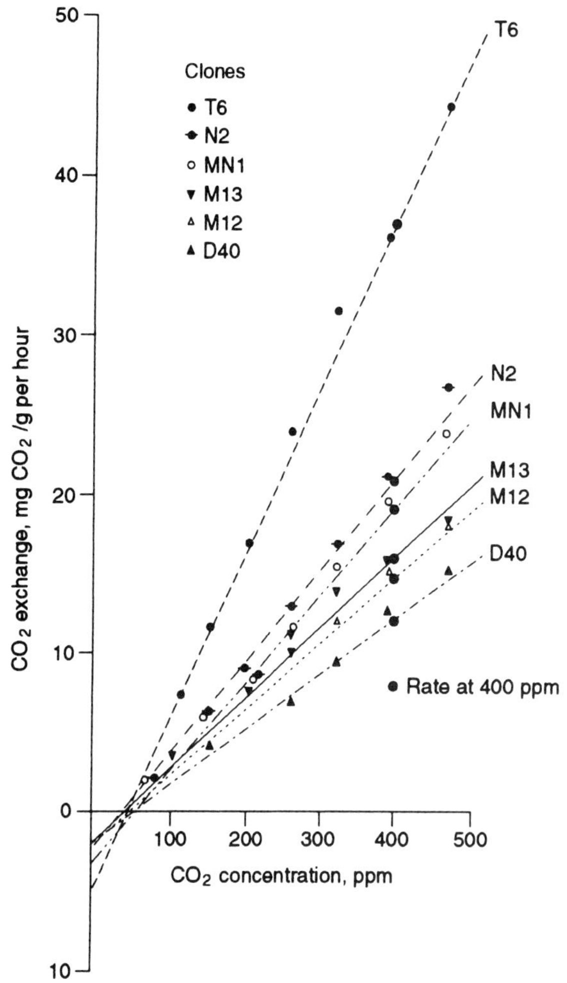

FIGURE 5.17. Effect of ambient CO_2 concentration on net photosynthesis of six poplar clones at 25°C. From Luukkanen and Kozlowski (1972).

and even negative correlations between photosynthetic capacity and growth of trees have been demonstrated (Kramer and Kozlowski, 1979). Short-term measurements of photosynthetic capacity often are not reliable for estimating growth potential because, in addition to photosynthetic rates alone, at least four other important physiological characteristics determine growth. These include the seasonal pattern of photosynthesis, the relation of photosynthesis to respiration, partitioning of photosynthate within the tree, and the amount of foliage produced.

A plant with a high rate of photosynthesis at one stage of its seasonal cycle may have a low rate at another stage. Johnsen et al. (1996) attributed at least part of the poor growth rates of northern provenances of black spruce to low rates of photosynthesis late in the growing season. Hence, prediction of dry matter increment from measurements of photosynthesis should be based on both rates and rate-duration aspects of the process.

Despite the difficulties of relating photosynthetic capacity and growth (biomass production), measurement of photosynthesis may be useful in genetic improvement of certain trees. Isebrands et al. (1988) showed that clonal differences in leaf photosynthesis and integrated whole-tree photosynthesis (the sum of maximum photosynthesis of all the leaves of the tree over time) exist throughout the genus *Populus*. There also are clonal differences in such traits as stomatal frequency and structure as well as leaf area and shape that are closely related to photosynthetic rates. Hence, clonal differences in integrated whole tree photosynthesis may be useful in identifying the growth potential of poplar clones.

Diurnal Variations

The rate of photosynthesis generally changes greatly during the day. The rate commonly is low early in the morning of a clear day and is associated with low light intensity and sometimes with low temperature, despite a high leaf water potential (Ψ) and high CO_2 concentration in the intercellular spaces of leaves. As the light intensity increases and the air warms, the stomata open and net photosynthesis begins to increase rapidly and may reach a maximum before or near noon. Sometimes the maximum rate is followed by a midday decrease, which may be slight or severe, and often is followed by another increase in the late afternoon. A final subsidence in photosynthesis generally follows the late afternoon and early evening decrease in light intensity and temperature (Figs. 5.16, 5.18). Double-peaked daily patterns of photosynthesis have been reported for temperate-zone plants (Figs. 5.15, 5.18, 5.19), arid-zone plants (Pearcy et al., 1974), Mediterranean-climate plants (Fig. 5.20) (Tenhunen et al., 1981), and arctic-zone plants (Tieszen, 1978; Kauhanen, 1986).

Because of variations in environmental conditions from day to day, and within the same day, any particular diurnal pattern often deviates considerably from the trends just described. For example, diurnal patterns of photosynthesis of white oak differed appreciably at various times in the growing season and were associated with changes in leaf temperature and leaf conductance (Fig. 5.18). Diurnal patterns of photosynthesis of Tasmanian blue gum varied greatly between sunny and cloudy days (Fig. 5.20).

Except for the midday decreases, diurnal changes in photosynthesis often are reasonably well correlated with changes in light intensity. For example, in an open area the peak rates of photosynthesis of three species of angiosperms occurred at midday and corresponded to peaks in solar radiation (Flore and Lakso, 1989). Under a forest canopy, the rate of daily photosynthesis fluctuated considerably, with the highest rates occurring during sunflecks. The daily pattern of photosynthesis of gymnosperms was very different on cloudy and sunny days (Figs. 5.15 and 5.16). On overcast or cloudy days, net photosynthesis of grand fir and noble fir increased to a maximum about noon, then either decreased or remained stable for an hour or two, and finally decreased. By comparison, on bright sunny days, photosynthesis normally increased rapidly, reached a peak between 9 A.M. and 12 P.M., then decreased until late afternoon when it increased again and reached a second but much lower peak.

Causes of Diurnal Variations

Daily variations in photosynthesis may have different causes, including environmental influences (e.g., light and VPD effects on stomatal aperture; temperature and light effects on mesophyll photosynthetic capacity) and endogenous factors affecting only the stomata or mesophyll photosynthetic capacity (Küppers et al., 1986; Flore and Lakso, 1989).

During the first minute or two of full illumination, RuBP regeneration is the major limiting factor in photosynthesis. After approximately two minutes, stomatal opening and activation of Rubisco by light are the primary factors that regulate the rate of photosynthesis. The degree to which each factor limits photosynthesis depends on their response dynamics and state at the beginning of photosynthetic induction (Kirschbaum and Pearcy, 1988; Pearcy, 1990; Tinoco-Ojangunen and Pearcy 1993).

Midday depressions in photosynthesis often are observed on sunny days despite adequate soil water supply (e.g., Figs. 5.19, 5.20). The causes of this phenomenon are several and vary from species to species, although not all show midday depressions. For example, even under the intense solar radiation and high VPD of the Brazilian cerrado photosynthesis and g_s of trees of *Miconia fallax* closely followed diurnal patterns of PPFD (Franco and Lüttge, 2002). In contrast, *Qualea grandiflora* exhibited accentuated midday suppression of gas exchange under the same conditions. Plants under high PPFD and VPD potentially are subject to both excessive excitation energy and leaf water deficits (the latter even in moist soil because of great evaporative demand and internal resistances to water flow within the plant).

In some plants stomatal closure is induced by high VPD (see also Chapter 12) as in *Populus deltoides* and *Prosopis juliflora* (Pathre et al., 1998), whereas in others,

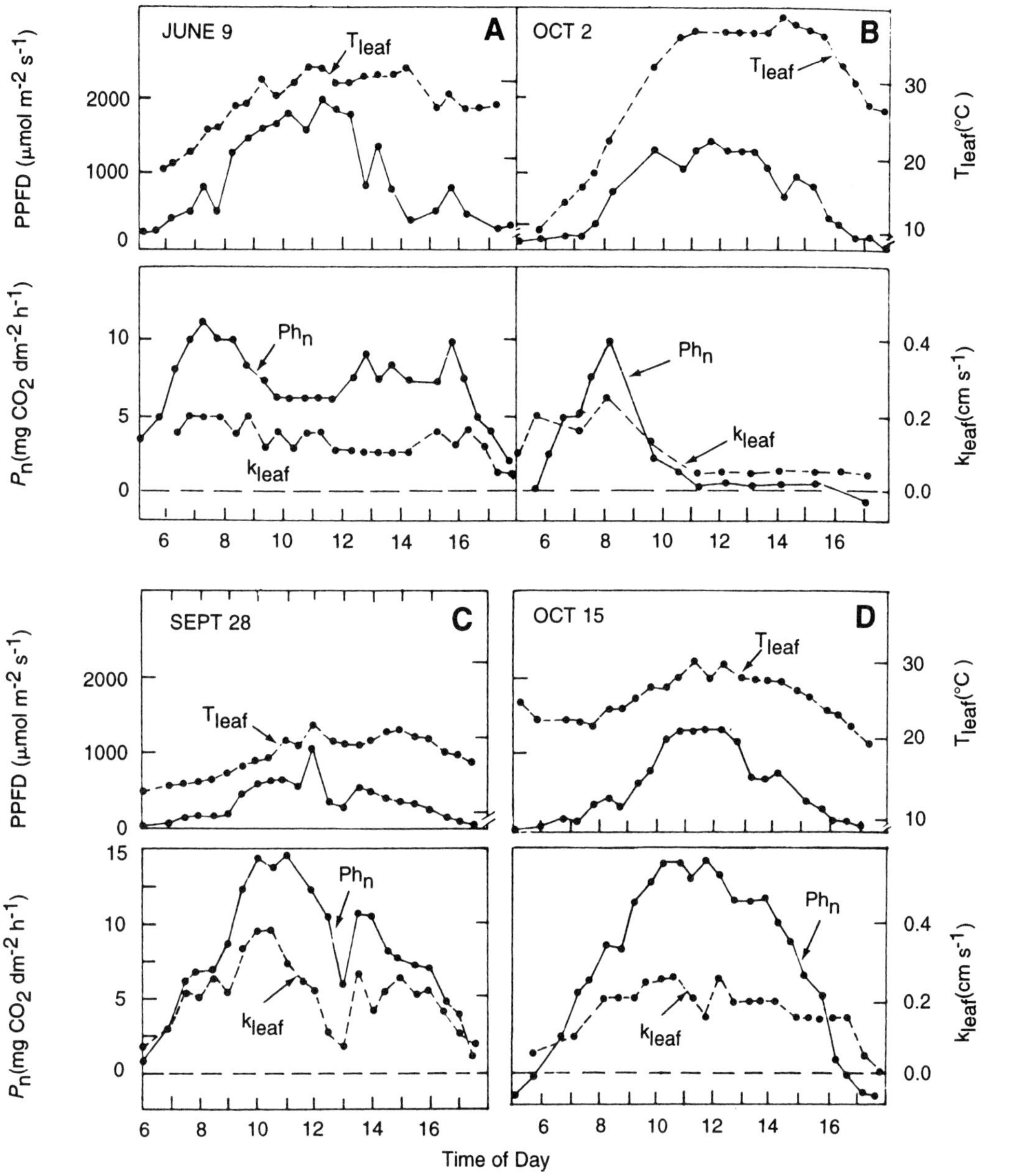

FIGURE 5.18. Diurnal variations in net photosynthesis (P_n), photosynthetic photon flux density (PPFD), leaf temperature (T_{leaf}), and leaf conductance (k_{leaf}) in white oak at various times of the year: (A) June 9, (B) October 2, (C) September 28, (D) October 15. From Dougherty and Hinckley (1981).

leaf water deficits may reduce stomatal aperture (e.g., kiwifruit, Gucci et al., 1996; *Quercus coccifera* and *Juniperus phoenicia*, Martinez-Ferri et al., 2000). Stomatal closure imposes potentially injurious conditions for the photosynthetic apparatus, as the availability of CO_2 as a sink for the light-driven electron transport system becomes limited. Thus, the photosynthetic apparatus becomes susceptible to damage from excess excitation energy, as described earlier in this chapter.

Several possible adaptive responses may function to prevent permanent damage. In some species, photorespiration is elevated at midday, providing internally produced CO_2 as a sink for photosynthetic electron transport (e.g., *Macaranga conifera*, Ishida et al., 1999; *P. deltoides*, *P. juliflora*, *Acacia auriculiformis*, Pathre et al., 1998; several cerrado species, Franco and Lüttge, 2002). In many species midday depressions of photosynthesis are associated with transient declines in quantum yield of PS II and increased thermal dissipation processes via activity of the xanthophyll cycle (Faria et al., 1996; Düring, 1999), and ΔpH-induced shifts in conformational state of the reaction center chlorophylls (Ort,

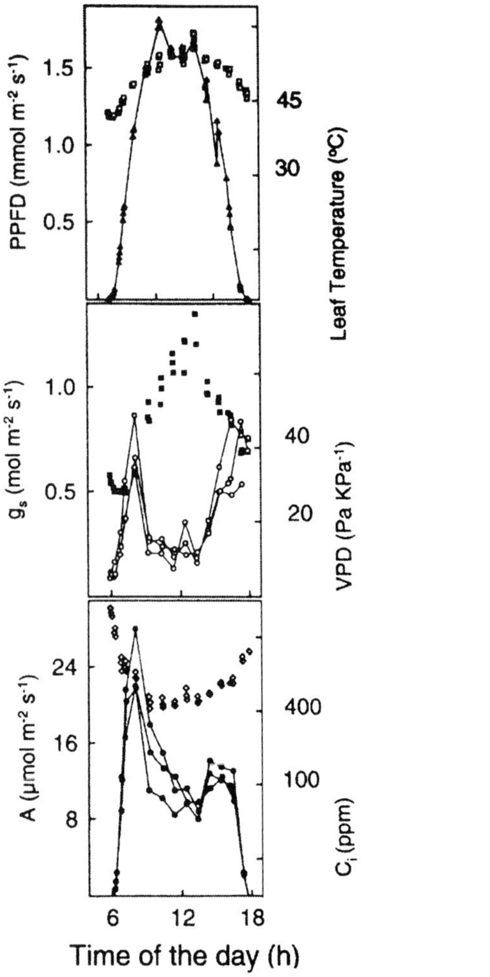

FIGURE 5.19. Diurnal variation of irradiance (PPFD, ▲), leaf temperature (□), leaf-to-air vapor pressure deficit (■), net photosynthesis (A, ●), stomatal conductance (g_s, ○), and internal CO_2 concentration (Ci, ◇) measured on three *Populus deltoides* trees in autumn. After *Trees*, Factors determining the midday depression of photosynthesis in trees under monsoon climate, Pathre, U., Sinha, A. K., Shirke, P. A., and Sane, P. V., **12**, 472–481, Figure 1. © 1998 with kind permission from Springer Science and Business Media.

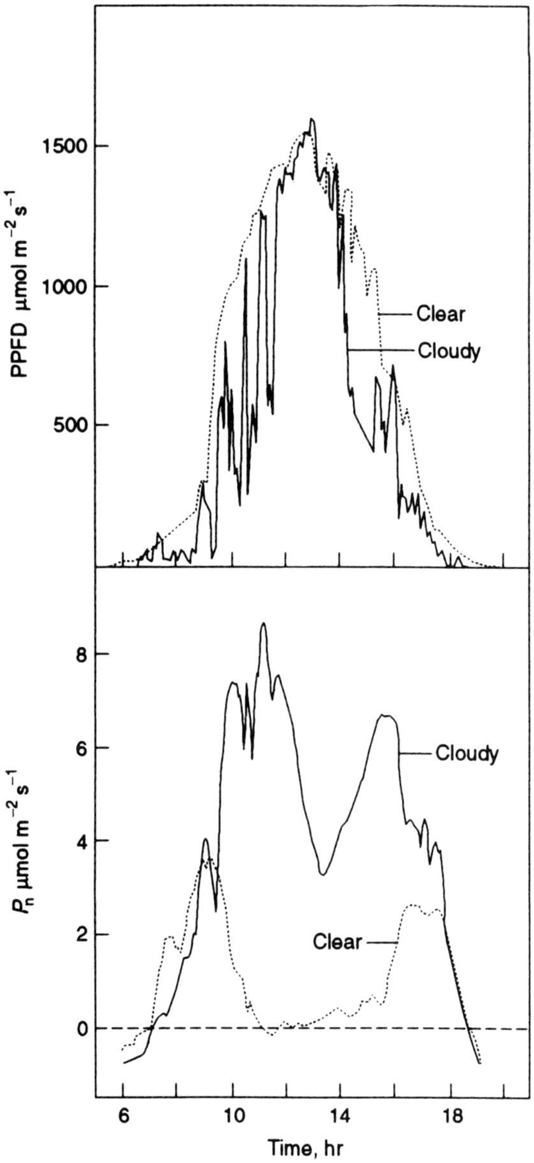

FIGURE 5.20. Diurnal variations in net photosynthesis (P_n) of Tasmanian blue gum trees in Portugal and photosynthetic photon flux density (PPFD) on a clear day (September 13) and a cloudy day (September 16). From Pereira et al. (1986).

2001). Zhang and Gao (1999) observed such a reciprocal pattern of response in distribution of light energy between photosynthetic electron transport or thermal dissipation in *Populus* clones (Fig. 5.21). This "dynamic" photoinhibition appears to be adaptive because it disappears or is much reduced by the next morning, and causes no long-term degradation of photosynthetic electron transport (Fig. 5.22) (Faria et al., 1996; Pathre et al., 1998; Martizes-Ferri, 2000; Senevirathna et al., 2003).

Seasonal Variations

It is important to distinguish between seasonal variations in the photosynthetic capacity of trees associated with leaf ontogeny and actual rates in the field that are determined by dynamic changes in both photosynthetic capacity and effects of superimposed environmental conditions. Actual rates of photosynthesis in the field show substantial fluctuation from day to day because of environmental conditions.

Gymnosperms

In the temperate zone, seasonal changes in photosynthetic capacity occur more gradually in evergreen gymnosperms than in deciduous angiosperms. As the

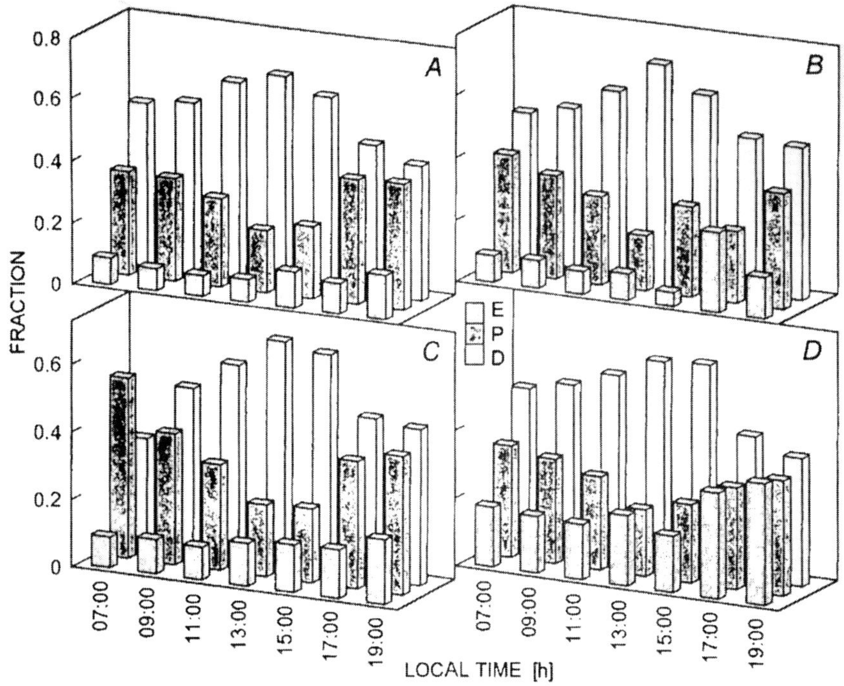

FIGURE 5.21. Diurnal course of allocation of light absorbed by PS II antenna pigments in hybrid *Populus* clones B346 (A), B342 (B), B11 (C), and ZH6 (D). Light energy was used in photosynthetic electron transport (P) or thermal dissipation (D); E indicates excessive excitation energy retained in the PS II reaction center. From Zhang and Gao (1999).

temperature increases in the spring and night frosts become less frequent, the photosynthetic capacity of gymnosperms increases gradually. The rate also declines gradually in the autumn. The wide distribution and dominance of gymnosperms in the northwestern part of the United States have been related to their high potential for photosynthesis outside the growing season (with 50–70% annual photosynthesis of Douglas-fir occurring outside the summer months), water stress limitations on net photosynthesis of all plants during the growing season, and high nutrient use efficiency (dry matter production per unit of nutrient) (Lassoie et al., 1985).

In Sweden, net photosynthesis of Scotch pine continued for approximately eight months (April to November) (Linder and Troeng, 1980). However, the seasonal duration of photosynthesis differed by up to a month in different years because of variations in weather. During early spring, the photosynthetic apparatus, which had been partly inactivated by low winter temperature, was reactivated but net photosynthesis was not recorded until the soil water was unfrozen. More than two months were required to reestablish maximum photosynthetic capacity following the effects of winter temperatures on the photosynthetic apparatus. Photosynthetic rates were high during the summer and were largely controlled by light intensity, air temperature, and water supply. In the autumn, the decreasing rate of photosynthesis was associated with shortening days and low irradiance.

Seasonal changes in photosynthetic capacity of loblolly pine and eastern white pine seedlings in North Carolina were studied by McGregor and Kramer (1963). The seedlings were kept outdoors but brought into the laboratory where measurements of CO_2 uptake were made periodically at 25°C and saturating light. Beginning in February, the rate of photosynthesis per unit of fascicle length for both species increased slowly until April, then accelerated rapidly, and subsequently declined during the autumn and winter (Fig. 5.23). The maximum rate per seedling for loblolly pine was reached in mid-September, after which the autumn decline was rapid (Fig. 5.24). The maximum rate for eastern white pine occurred between July 15 and September 15, and the autumn decline was more gradual. The higher and later peak of photosynthesis per seedling of loblolly pine was largely due to the fact that the seedlings made three flushes of shoot growth, adding new needles until late summer. The eastern white pine seedlings, however, made only one flush of shoot growth that occurred early in the season.

Some of the increase in the rate of photosynthesis after April 9 for each species was attributed to increasing amounts of foliage. However, the significant increase in photosynthesis from February 14 to April 9 could not be explained on this basis because no new foliage had expanded by April 9, but must have resulted from recovery of photosynthetic capacity of the needles already present. Similarly, the decrease after the mid-season maximum in both species was not

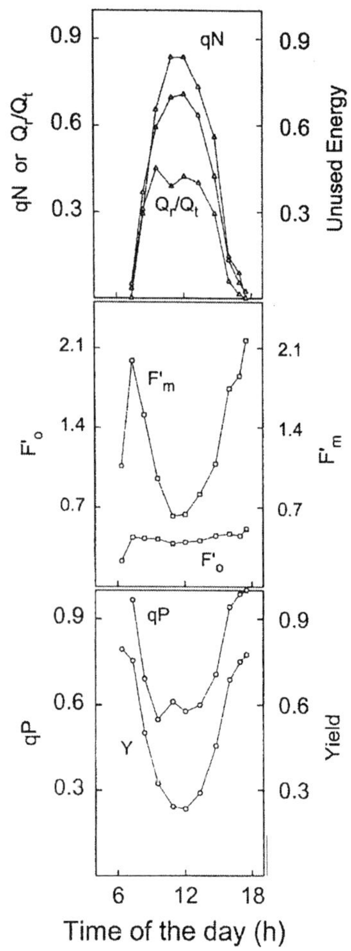

FIGURE 5.22. Diurnal course of fluorescence parameters of overall quantum yield (Y), initial fluorescence (F'_o), maximum fluorescence (F'_m), photochemical (qP), and nonphotochemical quenching (qN) of *Populus deltoides*. Fluorescence yields were used to calculate qN, reduction state of Q (Q_r/Q_t) and the fraction dissipated (Unused Energy, unlabeled line in top panel). After *Trees*, Factors determining the midday depression of photosynthesis in trees under monsoon climate, Pathre, U., Sinha, A. K., Shirke, P. A., and Sane, P. V., **12**, 472–481, Figure 6. © 1998 with kind permission from Springer Science and Business Media.

caused by loss of needles but rather from a decreased photosynthetic capacity of existing needles. This conclusion is supported by evidence of changes in photosynthetic capacity during the life of loblolly pine needles (Strain et al., 1976).

The low rates of photosynthesis during the establishment phase of banana plants were attributed to leaves not fully adapted to high radiation, high VPD, moderate temperatures, and an undeveloped root system (Eckstein and Robinson, 1995a,b; Eckstein et al., 1995). Major factors contributing to high rates of photosynthesis during sunny summer days included high average temperatures, high PPFD, monthly renewals of foliage, high soil Ψ, and a functional root system capable of coping with high evaporative demand. Low rates of photosynthesis during the winter were associated with low minimum temperatures and a root system unable to cope with any significant evaporative demand.

Winter Photosynthesis

Photosynthesis of evergreens during the winter varies with climatic regions but may be substantial even in the temperate zone. In Mediterranean regions the photosynthetic capacity of evergreens is reduced only slightly during the winter. In the southeastern United States, evergreens often increase in dry weight during the winter, indicating photosynthetic activity. Helms (1965) reported appreciable winter photosynthesis of 35-year-old Douglas-fir trees near Seattle, Washington as well as large variations in rates between years. During the wet, relatively warm winter of 1961, net photosynthesis, which was twice that during the 1962 winter, amounted to 25% of the total for the entire year. Schaberg et al. (1995) noted that rates of photosynthesis of red spruce in Vermont generally were low during the winter but increased substantially during thaws, and on some days the rates of individual trees were as high as those during the growing season. In regions with severe winters the rate of photosynthesis becomes negligible when night freezes are prolonged (Larcher and Bauer, 1981).

The rate of net photosynthesis of Norway spruce and Swiss stone pine in Austria was appreciable until late autumn (Pisek and Winkler, 1958). Thereafter, variations of a few degrees below and above freezing caused CO_2 uptake to fluctuate. As soon as the temperature dropped below −4 to −5°C, net photosynthesis stopped, and if freezes recurred for several nights thereafter, photosynthesis was inhibited during the day, even when the temperature rose above freezing. Strand et al. (2002) and Schwarz et al. (1997) noted similar inhibition of photosynthesis in Scotch pine and red spruce after a single nighttime freeze in Autumn. After a freeze of −6 to −8°C, net photosynthesis of Norway spruce and Swiss stone pine ceased and several mild days in succession were required for restoration of the capacity for positive net photosynthesis (Pisek and Winkler, 1958). Complete recovery of photosynthesis did not occur before the temperature increased in the spring. When temperatures fluctuated in the spring, the rate of photosynthesis did also. Hence, the photosynthetic apparatus remained functional only as long as the winter was without severe freezes or prolonged moderate freezes. At timberline, temperatures were so low for four to five months that photosynthesis was essentially eliminated.

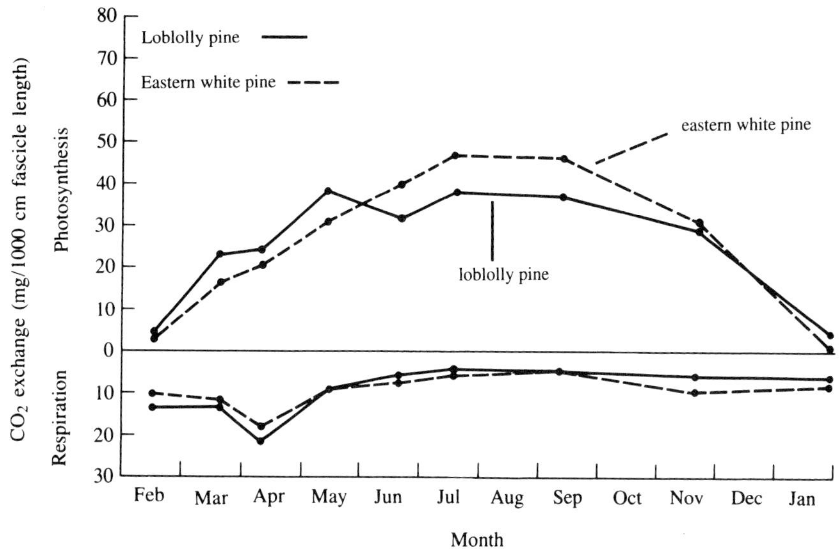

FIGURE 5.23. Seasonal changes in net photosynthesis and respiration per unit of fascicle length of loblolly pine (solid line) and eastern white pine (dashed line) seedlings. The seedlings were kept outdoors, but photosynthesis was measured indoors at 25°C and at a light intensity of 43,000 lux. From McGregor and Kramer (1963).

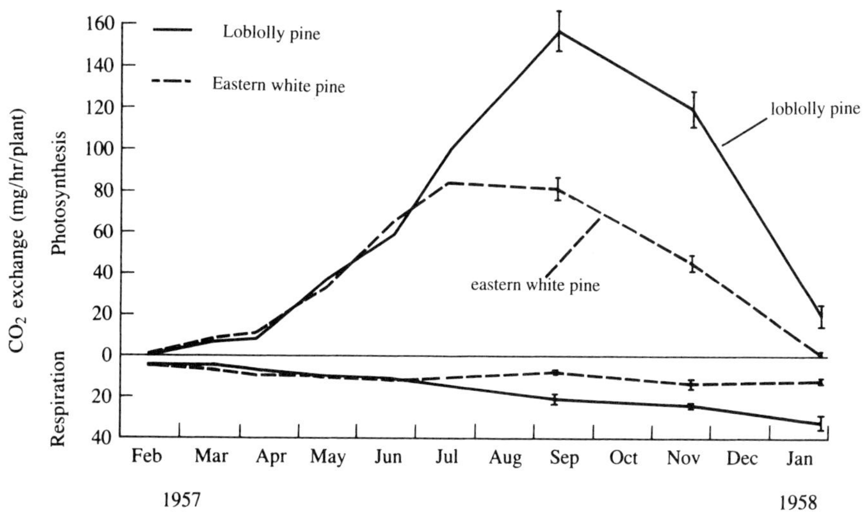

FIGURE 5.24. Seasonal changes in net photosynthesis and respiration per plant of loblolly pine (solid line) and eastern white pine (dashed line) seedlings. From Planta, Water-storage capacity of Thuja, Tsuga and Acer stems measured by dehydration isotherms-The contribution of capillary water and cavitation. (1990). **182**, 420–426, Figure 1 © 1998 with kind permission from Springer Science and Business Media.

The inhibition of photosynthesis subsequent to freezing nights has been attributed to both stomatal and mesophyll-level effects. Strand et al. (2002) observed that both stomatal conductance and c_i of Scotch pine leaves were reduced the day after night frost, a result that suggests that stomatal limitation of photosynthesis was more important than mesophyll inhibition. However, this pattern is not always observed (Schwarz et al., 1997) and other studies have reported that c_i (and hence mesophyll limitations) after summer frost are elevated (Lundmark et al., 1990). These results suggest that frost may influence both CO_2 diffusion capacity and mesophyll components of photosynthesis depending on its severity and duration, and the physiological condition of the affected plants.

Angiosperms

In many deciduous angiosperms, the rate of photosynthesis typically accelerates rapidly in the spring as trees refoliate, remains high during the summer, and declines rapidly in late summer or early autumn as the leaves senesce before abscising (e.g., Wilson et al., 2000a,b).

The seasonal capacity for whole-plant photosynthesis in deciduous angiosperms varies among species that have different patterns of leaf development. Species with shoots fully preformed in the winter buds (Chapter 3) achieve maximum leaf area early in the season, whereas other species continue to add foliage during much of the summer, either gradually or in flushes. Hence, total photosynthetic capacity may be expected to vary as the leaf surface area changes. Retention of green foliage late into the autumn appears to be an important factor contributing to the rapid growth of some broadleaved trees. For example, in Wisconsin, fast growing poplar clones retained their green leaves for a long time and had appreciable rates of photosynthesis at least until the first hard frost (Nelson et al., 1982). In Illinois black alder retained its leaves and continued to photosynthesize until mid- November, a month longer than white basswood (Neave et al., 1989). Pin cherry retained green leaves with a high rate of photosynthesis longer into the autumn than American beech or sugar maple (Amthor, 1990).

In apple trees the photosynthetic rate of spur shoot leaves (which expand rapidly) remains rather constant for several weeks after the leaves are fully expanded and decreases when these leaves senesce. The rate of photosynthesis of leaves of extension shoots stays higher than the rate of spur leaves. In August and September the rate may be three times as high as in spur leaves (Palmer, 1986).

ENVIRONMENTAL FACTORS

Many environmental factors influence photosynthesis, including light, temperature, CO_2 concentration of the air, water supply, vapor pressure difference between leaves and the air, soil fertility, salinity, pollutants, applied chemicals, insects, diseases, and various interactions among these. Photosynthesis also is responsive to cultural practices such as thinning of stands, pruning, fertilization, and irrigation that alter the environmental regimes of plants. Environmental conditions influence photosynthesis in the short term (days to weeks) by regulating stomatal conductance and mesophyll photosynthetic capacity. In the longer term, photosynthesis also is environmentally regulated through changes in leaf area.

Light Intensity

The rate of photosynthesis varies greatly with light intensity. Photosynthetic responses to light are important to growers because the light microclimate can be modified by thinning of stands, as well as pruning, and spreading of branches. To produce the maximum amount of high-quality fruit it generally is desirable to expose as great a proportion of the tree crown as possible to light during the entire season (Chapter 6, Kozlowski and Pallardy, 1997).

Both stomatal aperture and nonstomatal photosynthetic processes are affected by light intensity. Stomata may respond to light directly or through photosynthetic depletion of CO_2 in the mesophyll, or both may occur concurrently (Sharkey and Ogawa, 1987; Roelfsma and Hedrich, 2005). In addition to light providing the energy for photosynthesis, exposure to light for some critical length of time (induction) is necessary for attainment of full photosynthetic capacity (Perchorowitz et al., 1981), probably through regulation of important enzymes, especially Rubisco, in the photosynthetic process (Portis et al., 1986; Salvucci et al., 1986). The effects of light intensity on photosynthesis are modified by interactions with other environmental factors. For example, injury to the photosynthetic apparatus by very high light intensity (photoinhibition) may be increased by extreme drought or temperature (Powles, 1984).

Light-Response Curves

In darkness there is no photosynthesis; therefore, CO_2 produced in respiration is released from leaves. With increasing light intensity the rate of photosynthesis increases until a compensation point is reached, at which photosynthetic uptake of CO_2 and its release in respiration are equal, hence there is no net gas exchange between the leaves and the atmosphere. The light compensation point varies with species and genotype, leaf type (shade leaves have lower light compensation points than sun leaves), leaf age (young leaves have higher light compensation points than mature leaves), CO_2 concentration of the air, and temperature (Kozlowski et al., 1991). Since respiration increases faster than photosynthesis with rising temperature, the light compensation point also increases, and reaches very high values at temperatures above 30°C (Larcher, 1983).

With additional light intensity above the compensation point, the rate of photosynthesis increases linearly. However the rate often departs from linearity before light saturation occurs. The efficiency of the photosynthetic apparatus under light limiting conditions is indicated by the quantum efficiency, the initial slope of the absorbed light versus photosynthesis curve. With further increase in light intensity, decreased utilization efficiency becomes evident and the increase in photosynthesis is progressively less than proportional to the increase in light. Eventually, light saturation occurs and the rate of photosynthesis becomes more or less constant (Figs. 5.25 and 5.26). In some species very high light intensities may even cause a decline in photosynthesis (Kozlowski, 1957), especially if leaves are shade-adapted (Fig. 5.26).

Within-Leaf Light Variation

Development of fiber optic microprobes in the 1980s has allowed exploration of the light environment

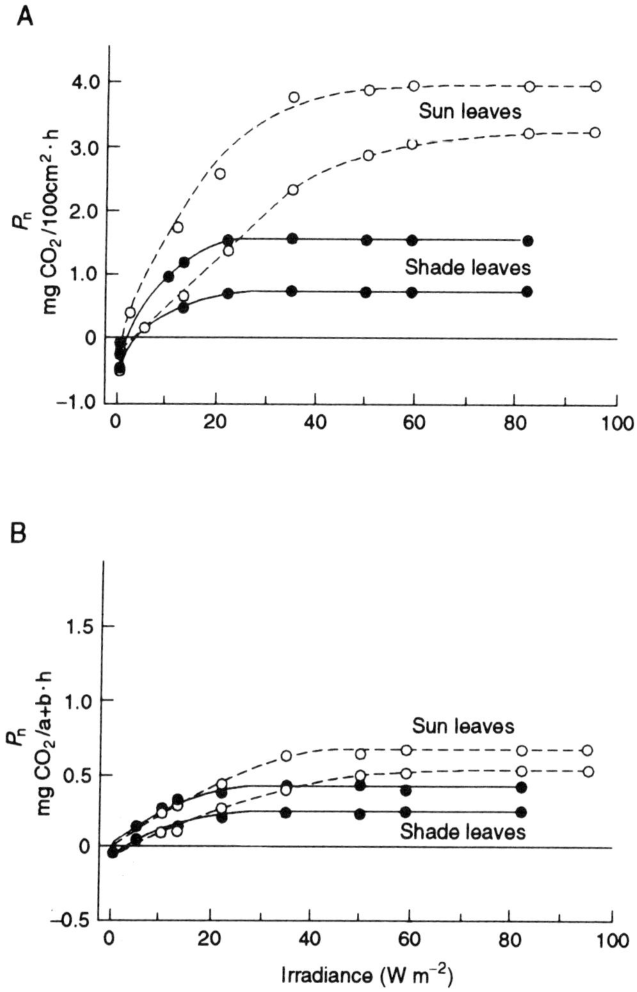

FIGURE 5.25. Response of photosynthesis (P_n) of sun leaves (○) and shade leaves (●) of European beech to light intensity. Rates of photosynthesis are given on a leaf area basis (A) and a chlorophyll content basis (B). The dual curves for both sun and shade leaves are for upper and lower values attained with nine different sun and shade leaves. From Lichtenthaler et al. (1981). Reprinted by permission of Kluwer Academic Publishers.

within the typically thin angiosperm leaf (Vogelmann et al., 1988). As light enters a leaf from the adaxial surface, complex patterns of concentration (lensing), absorption, and scattering ensue. Except for cells of the stomatal complex, epidermal cells are transparent and have a planoconvex shape, which tends to focus light. This geometry results in chloroplasts that may be exposed to three times the normal maximum PPFD incident to the leaf (~6,000 μmol m^{-2} s^{-1}) (Vogelmann et al., 1996). Palisade cells act as "light pipes," especially to collimated light, allowing deeper penetration of direct beam radiation (Vogelmann, 1993; Terashima and Hikosaka, 1995). It is interesting to note that shade leaves have reduced or no palisade layer development and hence no capacity for light piping, but such is not necessary to moderate the light gradient in a shaded leaf under diffuse light.

The abundant air spaces in the spongy mesophyll provide many air water interfaces that serve to scatter radiation that penetrates through the palisade layers. Some of this radiation is scattered backward to be reabsorbed in the leaf interior and some subsequently is reflected backward by the abaxial epidermal cell layer. Mean within-leaf pathlengths for photons can exceed leaf thickness. For example, Fukshansky et al. (1993) calculated that mean pathlength for a 460 μm-thick *Catalpa bignonioides* leaf was between 1 and 2 mm and pathlengths of up to 7 mm were possible. From the perspective of a bilateral leaf with a developed palisade, most absorption occurs in this tissue with relatively little in the spongy parenchyma (Fig. 5.27). As green light is only weakly absorbed in the palisade layer, there is greater relative absorption by the spongy parenchyma.

The convexity of the light response of photosynthesis is largely determined by the pattern of light absorption and local acclimation of chloroplast photosynthesis within leaves (Kull, 2002). The shape of the curve can change in a matter of days with a change in irradiation regime. Chloroplast orientation within cells changes with light regime, becoming more parallel to direct radiation and chloroplast properties such as saturated photosynthetic rate per unit chlorophyll, thylakoid volume compared to stroma volume, and number of thylakoids per granum all respond to the local light regime within leaves (Terashima and Hikosaka, 1995).

Crown Depth and Photosynthesis

The PPFD decreases rapidly with increasing depth of tree crown. Also, the rate of photosynthesis of open-grown trees typically is much higher in leaves in the outer crown than in those in the inner crown. In apple trees, the average net assimilation rate was 16.7 μmol CO_2 m^{-2} s^{-1} in the high-light-intensity zone of the outer crown and 4.7 μmol CO_2 m^{-2} s^{-1} in the low-intensity zone in the inner crown (Heinicke, 1966). As the leaves of apple trees became fully expanded in May, the rate of photosynthesis was high in various parts of the crown. However, between the end of June and early August, the rate of photosynthesis declined rapidly in the shaded crown interior (Porpiglia and Barden, 1980).

The pattern of distribution of light in tree crowns often varies between peach and apple trees because of differences in training of their crowns. In apple trees, which often are trained to a central leader, the decreases in PPFD are exponential from the crown periphery

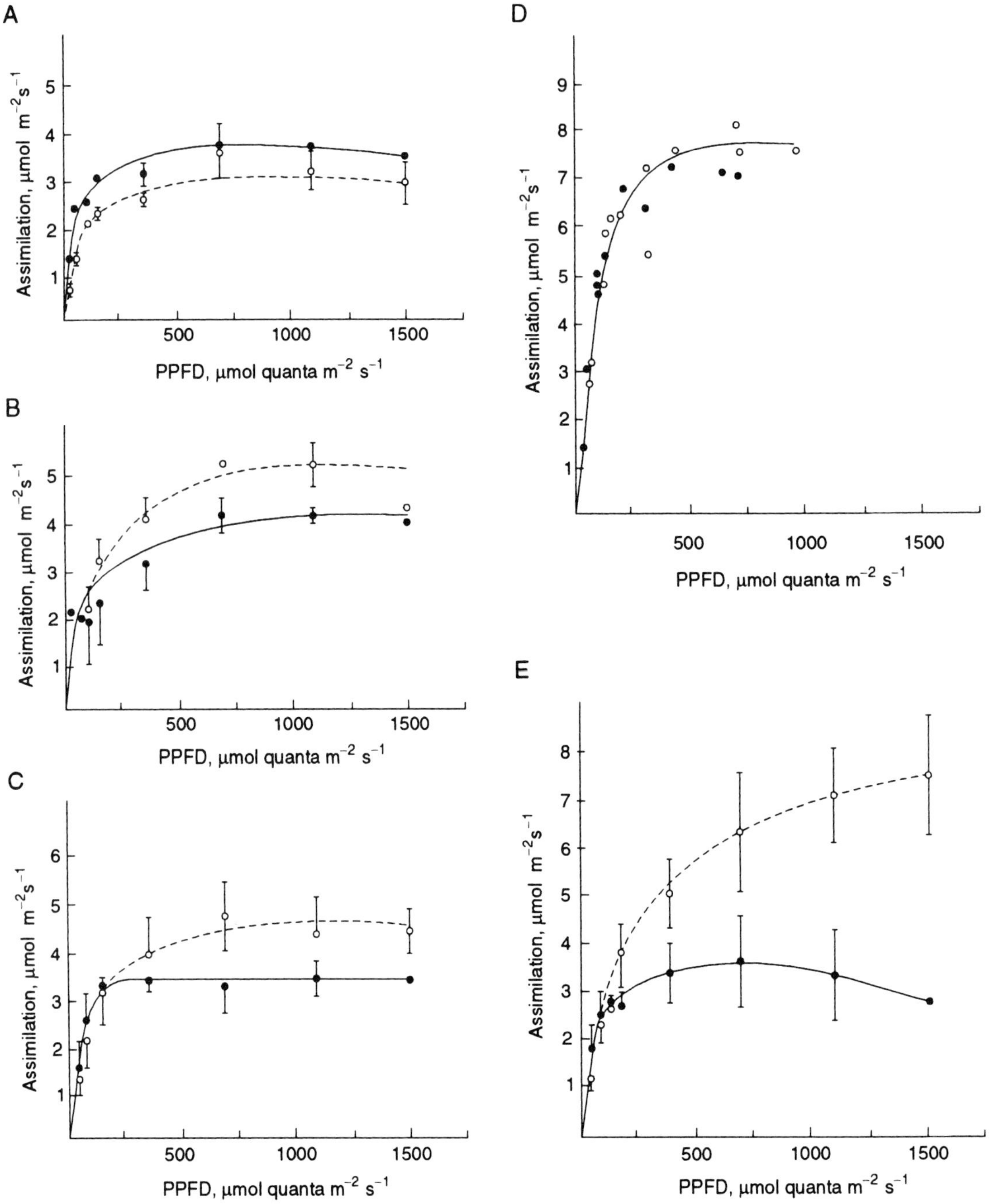

FIGURE 5.26. Response of photosynthesis to photosynthetic photon flux density (PPFD) in leaves of rain forest tree seedlings grown in full sunlight (○) and 6% of full sunlight (●): (A) *Hymenaea parvifolia*, (B) *Hymenaea courbaril*, (C) *Agathis macrostachya*, (D) *Copifera venezuelana*, (E) *Agathis robusta*. Bars indicate ± 1 SE. From *Oecologia*, Photosynthetic responses to light in seedlings of selected Amazonian and Australian rainforest tree species, Langenheim, J. H., Osmond, C. B., Brooks, A., and Ferrar, P. J., **63**, 215–224, Figure 2. © 1984, Springer-Verlag.

inward. By comparison, peach trees commonly are trained to an open center. In such trees the rate of photosynthesis is greatest for peripheral leaves, lowest for leaves midway into the crown, and intermediate for leaves in the center of the crown (Marini and Marini, 1983).

Canopy Height and Photosynthesis

Generally the leaves near the top of the canopy have higher rates of photosynthesis and become saturated at higher light intensities than those near the bottom of the canopy. Such differences are correlated with pro-

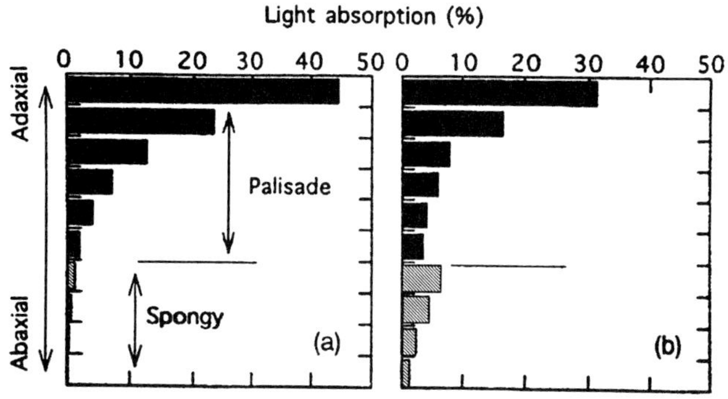

FIGURE 5.27. Light absorption in mesophyll layers in a leaf of *Camellia japonica* at red (680 nm, panel (a)) and green (550 nm, panel (b)) wavelengths. From Terashima, I. and Hikosaka, K. (1995). Comparative ecophysiology of leaf and canopy photosynthesis. *Plant Cell Environ.* **18**, 1111–1128. Blackwell Publishing.

gressive decreases in stomatal conductance and mesophyll photosynthetic capacity from the top toward the bottom of the canopy. Although CO_2 concentration, temperature, and humidity differ with canopy height, the largest differences by far are found in the light intensity regime (Jarvis et al., 1976).

Total photosynthetic output in both gymnosperms and angiosperms varies greatly with tree height, largely as a result of differences in the amounts of foliage and in shading. The inefficient lower branches, with relatively few and heavily shaded leaves, often do not contribute any carbohydrates for growth of the main stem. In a 38-year-old Douglas-fir tree with 18 whorls of branches, maximum photosynthesis occurred in current-year needles around whorl seven from the top in the zone between full light and full shade (Fig. 5.28). The rate then decreased progressively toward the base of the crown. Despite diurnal and seasonal variations in photosynthesis at all crown heights, the crown surface could be stratified into zones of photosynthetic efficiency (Woodman, 1971). The photosynthetic capacity of a 7-m-tall Monterey pine tree enclosed in a controlled environment room also varied with crown height. The upper and middle third of the crown, which had 14 and 60% of the needles, contributed 19 and 60%, respectively, of the total photosynthate (Rook and Corson, 1978). In Norway spruce the upper sunlit portion of the crown, which represented approximately two-thirds of the total leaf weight, accounted for 71% of the total annual photosynthesis. The lower shaded portions of the crown accounted for 6.4% of the total leaf weight and 3.4% of total annual photosynthesis (Schulze et al., 1977).

In a mixed turkey oak-European hornbeam forest the rate of photosynthesis of leaves of the upper canopy layer was higher than that of the lower canopy layer

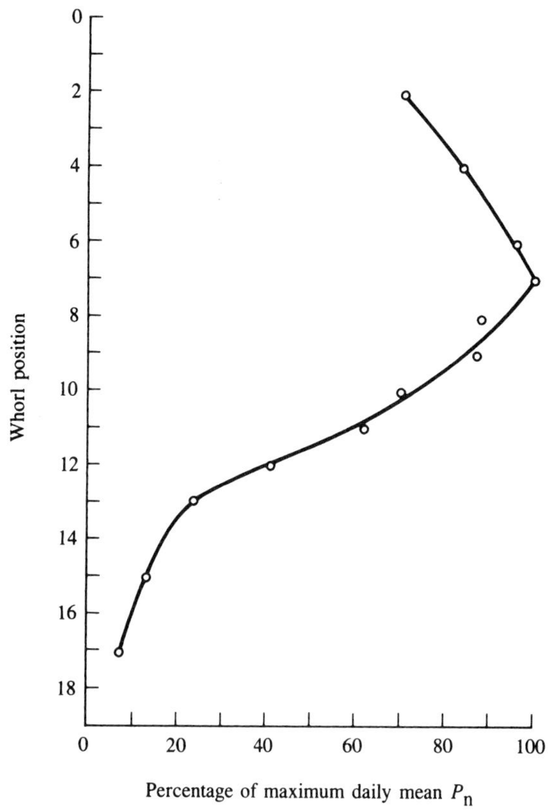

FIGURE 5.28. Variations in photosynthesis in different parts of the crown of a 38-year-old Douglas-fir tree. From Woodman (1971).

(Fig. 5.29). In fact, the photosynthetic contribution of lower canopy leaves to productivity was significant only during midday (Marek et al., 1989). In a 12-year-old Australian forest dominated by *Eucalyptus*, at midday the foliage in the layers located seven to ten, four to seven, and one to four meters from the ground contributed 35, 44, and 20.7%, respectively, to the total C assimilated by the canopy. Although the middle level (four to seven meters) provided most of the photosynthate, the rate of photosynthesis per unit of leaf area decreased with canopy height (Wong and Dunin, 1987).

In an 18-m-tall closed-canopy sugar maple forest, the upper 10% of the canopy contributed 30 to 40% of the entire canopy photosynthesis (Ellsworth and Reich, 1993). Both leaf mass and maximum photosynthesis per unit area decreased two-fold from the upper to the lower canopy. Whereas N per unit leaf area decreased with canopy height, N per unit leaf dry weight did not (Fig. 5.30). The differences in leaf traits along the canopy gradient were largely structural rather than biochemical (change in chlorophyll content was an exception). The increases in photosynthetic capacity in the upper canopy were attributed to an increased investment in N and leaf mass in this part of the crown.

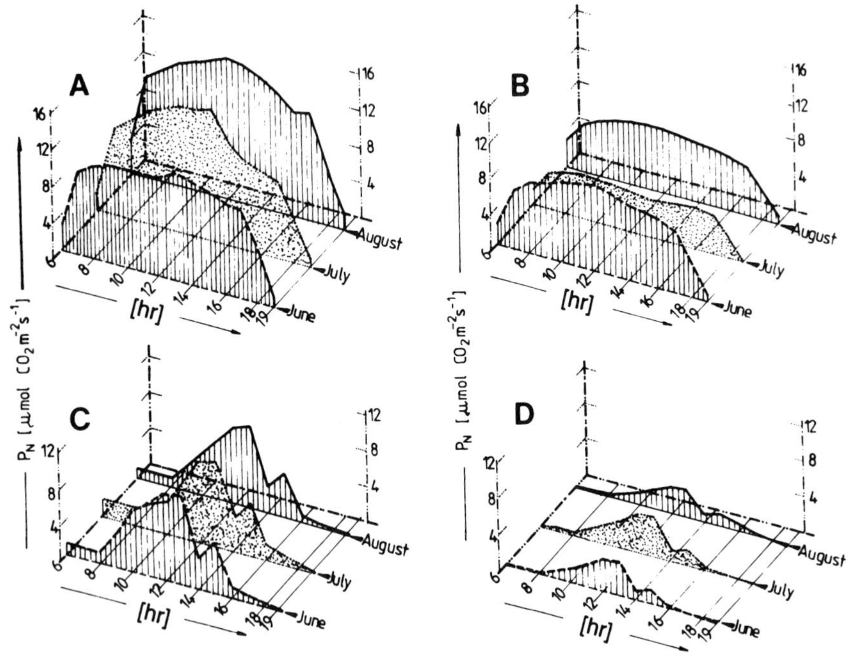

FIGURE 5.29. Diurnal and seasonal variations in net photosynthesis (P_n) of turkey oak (A-upper canopy, C-lower canopy) and hornbeam (B-upper canopy, D-lower canopy). From *Trees*, Stand microclimate and physiological activity of tree leaves in an oak-hornbeam forest. II. Leaf photosynthetic activity. Marek, M., Masarovicova, E., Kratchilova, I., Elias, P., and Janous, D., 3(4), 234–240, Figure 2A–D. © 1989, Springer-Verlag.

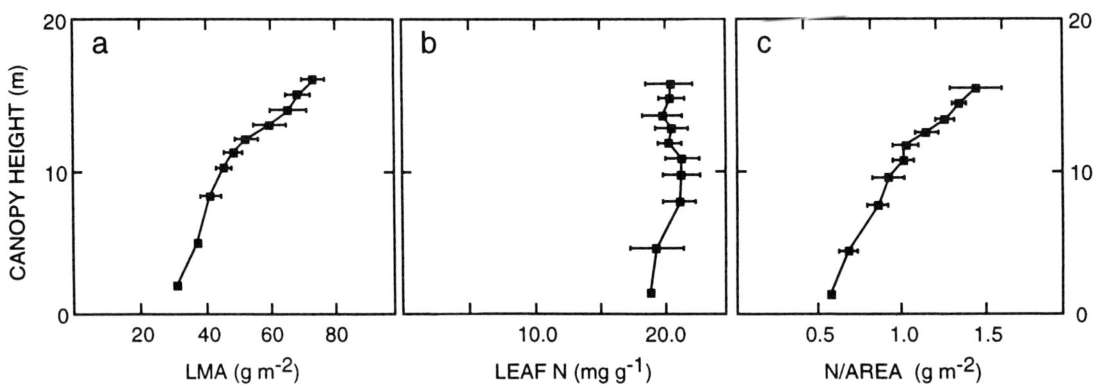

FIGURE 5.30. Vertical variation in leaf mass per unit area (LMA), mass-based leaf N concentration (LEAF N), and leaf N content per unit leaf area (N/AREA) with canopy height of sugar maple in a closed canopy forest. Error bars represent variation (±1 SE) among 6 different locations within the stand. From *Oecologia*, Canopy structure and vertical patterns of photosynthesis and related traits in a deciduous forest. Ellsworth, D. S. and Reich, P. B., **96**, 169–178, Fig. 3a–c. © 1993 Springer-Verlag.

Similar structural changes were found in sugar maple seedlings growing along a light intensity gradient among different habitats (Ellsworth and Reich, 1992). Reich et al. (1990) divided the forest canopy of an oak-maple forest into four horizontal layers and developed predictive models for progressively decreasing photosynthesis in each layer. These models were useful for comparing canopy net CO_2 exchange capacity with other measures of stand productivity and resource availability.

Light use efficiency of plant canopies is higher under diffuse than direct beam radiation because proportionally more of the leaf surface functions below the light saturation point (Gu et al., 2002). Thus the reduction in canopy photosynthesis during cloudy days, which are characterized by a greater fraction of diffuse radiation, will not be proportionally reduced based on total incident PPFD. Events that increase diffuse radiation on a global scale have been shown to increase total photosynthesis of the earth's terrestrial ecosystems. For example, Gu et al. (2003) observed that the eruption of Mt. Pinatubo in the Philippines in 1991, which caused a global stratospheric SO_2 aerosol layer to form that persisted for several years, was associated with an elevation of community photosynthesis (net ecosystem exchange) of a Massachusetts temperate deciduous forest that persisted for two years. Similar influences would occur for climatic changes that alter cloud development as noted earlier.

Sunflecks and Photosynthesis

The understory plants of dense forests are exposed to sunflecks (short periods of direct irradiance) as well as to persistent, dim, diffuse background irradiance. The duration of exposure to intermittent irradiance varies greatly among forest types. For example, in closed canopy understory sites within tropical forests most sunflecks are less than two minutes long. By comparison, in open coniferous forests "sunpatches" may last for an hour or more. Sunflecks, which may comprise up to 80% of the total irradiance at the forest floor, are essential for survival of many understory plants (Chazdon and Pearcy, 1991).

Growth of understory plants is greatly influenced by sunflecks (Pearcy, 1988). Up to 60% of the C gain of understory plants was attributed to utilization of sunflecks (Pearcy et al., 1994). The relative growth rate of *Euphorbia forbesii* and claoxylon in the understory of a Hawaiian evergreen forest was linearly correlated with minutes of exposure to sunflecks (Pearcy, 1983). Height growth of *Lecythis ampla*, a shade-tolerant rain forest tree, was correlated with the photon flux density contributed by sunflecks (Oberbauer et al., 1988). On bright microsites the potential stimulation of growth by sunflecks may be negated by drought, high leaf temperatures, or competition for nutrients (Pearcy et al., 1994).

The stimulatory effects of sunflecks on plant growth are mediated by their influence on photosynthesis. Short sunflecks significantly increased photosynthesis of understory plants of Hawaiian and Australian rain forests (Pearcy et al., 1985; Chazdon and Pearcy, 1986). Significant increases in photosynthesis in European ash, European filbert, and Wych elm occurred during fairly long sunflecks (three minutes or longer). Photosynthesis of shade leaves of European beech and English holly, which were almost light-saturated at their site irradiances, responded less to sunflecks (Harbison and Woodward, 1984).

There is a strong preconditioning effect of light intensity on photosynthetic responses to sunflecks. For example, total CO_2 uptake of claoxylon was greater and the photosynthetic response was faster after the leaves had been exposed to a high light intensity than when exposed to low light intensity for the previous two hours (Pearcy et al., 1985). Once photosynthetic induction occurs, there is a carryover stimulatory effect of sunflecks on photosynthesis. Utilization of sunflecks apparently depends on induction of a balance in chloroplast components, possibly pools of electron transport carriers and photosynthetic intermediates that permit a fast response of electron transport to increased light, and subsequent post-illumination consumption of its products (Anderson and Osmond, 1987). During sunflecks, shade grown alocasia plants accumulated substrates for post-illumination CO_2 fixation (Sharkey et al., 1986). The initial rise in photosynthesis in response to sunflecks was attributed to an increase in RuBP concentration resulting from accelerated electron transport. The subsequent post-illumination decrease in photosynthesis was related to depletion of RuBP by Rubisco and reductions in regeneration of RuBP because of slower electron transport (Pearcy et al., 1994). Photosynthesis of shade-adapted plants appears to adapt rapidly to short periods of irradiation by exhibiting a low rate of loss of photosynthetic induction, high electron transport capacity relative to carboxylation capacity, and stomatal opening at low photon flux density (Chazdon and Pearcy, 1991; Küppers and Schneider, 1993).

The fluttering of upper canopy leaves of trembling aspen influences the amount of light available to leaves of the lower canopy. As poplar leaves fluttered at the top of the canopy, the number of sunflecks increased in the lower canopy, hence exposing the lower leaves to more light (Roden and Pearcy, 1993b). The greater light penetration to the lower canopy was correlated with decreased light interception by the fluttering leaves of the upper canopy. However, this decreased interception likely would not substantially decrease C gain for the upper leaves because they often were light-saturated.

Stomatal opening and closing in leaves of understory plants in a deciduous forest fluctuated greatly during the day. In the morning, when the leaves were frequently exposed to moving sunflecks, stomatal conductance was maximal. As the intensity of sunflecks decreased in the afternoon, stomatal conductance fell to a very low value, reflecting stomatal closure. Changes in leaf water relations were correlated with stomatal aperture. Because the frequency of sunflecks, stomatal conductance, and transpiration rate were high in the morning, and leaf water deficits increased rapidly and were greatest in the late afternoon, by which time the rate of transpiration had decreased (Elias, 1983). Woods and Turner (1971) reported that the stomata of shade-tolerant plants opened faster during sunflecks than the stomata of intolerant species, allowing the former to carry on photosynthesis during very short periods of exposure to high light intensities. However, Pereira and Kozlowski (1976) did not find a close relation between shade tolerance and stomatal responsiveness to light. Unless a sunfleck is more than

5 to 10 min. long, stomata may open too slowly to benefit plants by utilization of the sunfleck. However, short sunflecks may increase photosynthesis during subsequent sunflecks.

Acclimation to Low Light

Curves of response of photosynthesis over a range of light intensities have been widely used to show differences of shade- and sun-grown plants to light intensity (Figs. 5.25 and 5.26). Ramos and Grace (1990) showed considerable variation among four tropical tree species to shading. These species differed more in their growth rate than in rates of photosynthesis, emphasizing the importance of partitioning of carbohydrates to growth. Maximum photosynthetic rates at high light intensities are higher for plants that were grown in high light intensities than for the same plants grown at low light intensities. McMillen and McClendon (1983) compared light-saturated photosynthesis of 10 species of trees grown in the field or under a canopy that transmitted approximately 18% of full light. Photosynthetic capacity was consistently higher for the sun-grown leaves than for those grown in the shade.

Photosynthetic responses to light may vary among plants grown out of doors and those grown in greenhouses or in controlled-environment chambers. For example, light-saturated net photosynthesis per unit of leaf area was 1.6 to 2.1 times higher in mature leaves of poplar grown in the field than in plants grown in a greenhouse or a controlled-environment growth room (Nelson and Ehlers, 1984). The differences in rates were associated with the thicker leaves of the field-grown trees and the thin leaves of greenhouse- and growth-chamber-grown plants, as is characteristic of sun-grown and shade-grown leaves (Chapter 3). The higher rates of the field-grown leaves were attributed to more photosynthesizing tissue per unit of leaf area.

Consistent with the trends noted earlier, maximum photosynthetic capacity decreases with depth in the canopy and is most closely associated with relative irradiance regime (Kull, 2002; Niinemets et al., 2004). The decline in capacity is linked with reduced carboxylation and electron transport capacity (area-based V_{cmax} and J_{max}, respectively) (Niinemets et al., 1998). The decline is determined primarily by the decrease in leaf mass per unit area in lower canopy leaves, as the photosynthetic capacity per unit leaf mass remains relatively stable with height.

At low light intensities, photosynthesis of shade-grown leaves of many species is more efficient than that of sun-grown leaves. Shade leaves typically have increased quantum yield, an estimate of quantum use efficiency during CO_2 fixation. Because most leaves in a tree canopy are shaded to various degrees, a higher quantum yield of shade leaves permits efficient utilization of the existing microenvironment (Mbah et al., 1983). Greater investment in chlorophyll and the proportion of leaf nitrogen in light harvesting constituents were shown in both shade-tolerant (*Tilia cordata*, *Corylus avellana*) and shade-intolerant (*Populus tremula*) species as irradiance in the canopy declined. The increase was especially marked in shade-tolerant species at very low light levels (Fig. 5.31). Higher rates of photosynthesis (leaf area basis) at low light intensity have been shown for the shade leaves of such temperate-zone species as beech, sugar maple (Logan and Krotkov, 1968), black walnut (Dean et al., 1982), red maple, northern red oak, and yellow-poplar (Loach, 1967), and such tropical species as coffee (Friend, 1984) and Venezuelan copal tree (Langenheim et al., 1984). Shade-grown leaves of weeping fig had a photosynthetic advantage over sun-grown leaves at low levels of PPFD, whereas at high levels the reverse was true. Shade-grown leaves were light saturated at 200 $\mu mol\ m^{-2}\ s^{-1}$. The daily photosynthetic rate of shade-grown leaves was not affected until 1600 hr, whereas the rate of sun-grown leaves began to decline around noon. By late afternoon the rates of sun-grown leaves were only about a third of the daily maximum rates (Fails et al., 1982).

Adaptation to Shade

Trees vary widely in their capacity to grow in the shade of other trees and this difference often is decisive in the success of certain species in competitive situations. Much of the difference in shade tolerance is related to variations among species in adaptation of the photosynthetic apparatus to low light intensity, although other attributes such as a conservative growth habit and investment in herbivore and decay defenses also may be important in some species (Kitajima, 1994). Photosynthetic adaptations are associated with effects of shade on both leaf production and on photosynthetic capacity. Successful growth of plants at low light intensity requires capacity to efficiently trap the available light and convert it into chemical energy, maintain a low rate of respiration, and partition a large fraction of the carbohydrate pool into leaf growth.

Shade-tolerant species generally have lower dark respiration rates and hence lower light compensation points (Loach, 1967; Field, 1988; Givnish et al., 2004), and lower light saturation points for photosynthesis than do shade-intolerant species. The leaves of shade-tolerant species also usually contain lower levels of

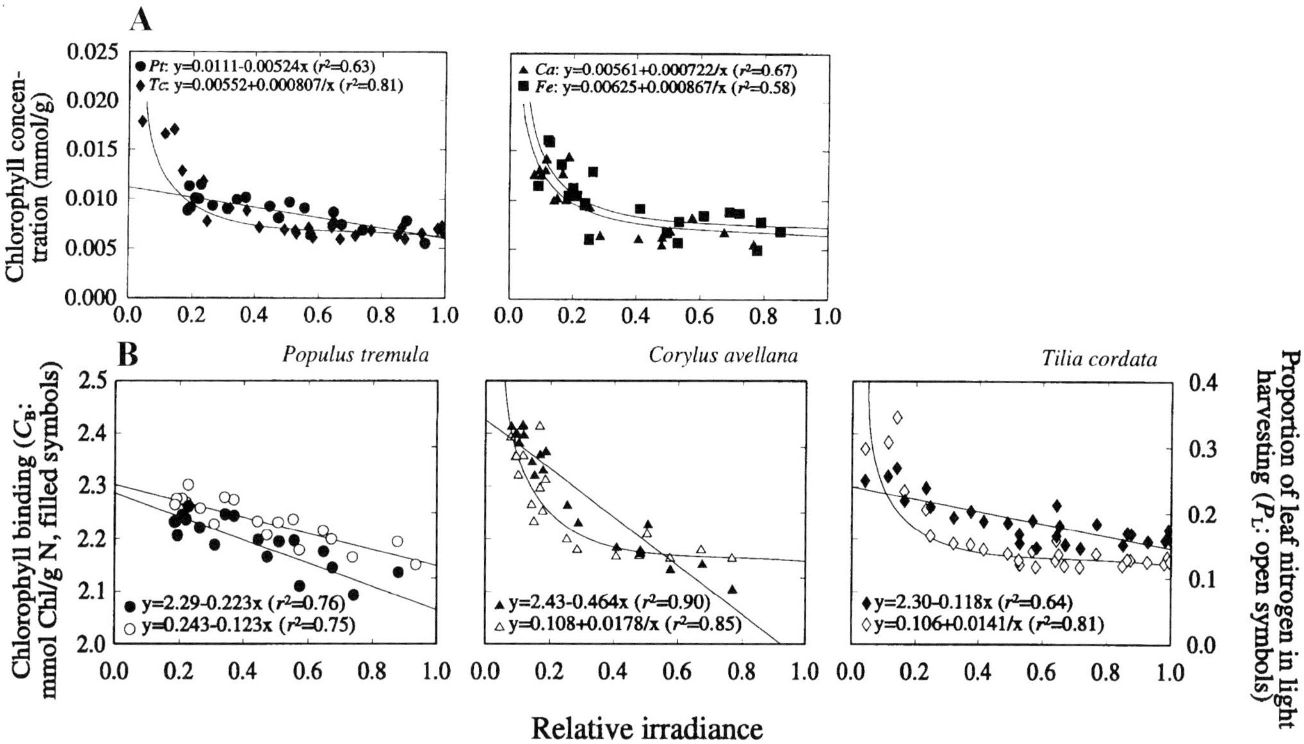

FIGURE 5.31. Effects of relative irradiance on (A) leaf chlorophyll (a + b) concentration, and (B) chlorophyll binding (C_B, ratio of leaf chlorophyll to leaf nitrogen in light harvesting) and fractional investment of leaf nitrogen in light harvesting (P_L). Shade tolerant species: *Corylus avellana* (Ca), *Tilia cordata* (Tc); shade intolerant species: *Populus tremula* (Pt), *Fraxinus excelsior* (Fe). Relative irradiance was determined from hemispherical photographs. From Niinemets et al. (1998).

Rubisco, ATP synthase, and electron carrier per unit of leaf surface, which is reflected in lower values of V_{cmax} and J_{max} in shade species growing under low light. The reduced amounts of these constituents are consistent with a lowered capacity for electron transport and carbon fixation (Lewandowska et al., 1976; Syvertsen, 1984; Givnish et al., 2004). Niinemets et al. (1998) found a close relationship between daytime dark respiration and maximum carboxylation rate (V_{cmax}) in both shade-tolerant and shade-intolerant species (Fig. 5.32), a result that strongly suggests high carboxylation capacity must come at the expense of greater respiratory expenditures. Further, shade-intolerant *Populus tremula* maintained higher R_d and V_{cmax} than did shade-tolerant *Tilia cordata* and *Corylus avellana*.

Shade-intolerant species tend to respond to high light regimes with much increased photosynthetic capacity. For example, the rate of light-saturated photosynthesis of seedlings of such shade intolerant species as idigbo, afara, and obeche grown at high light intensities was approximately twice as high as that of shaded seedlings. By comparison, seedlings of the shade-tolerant kuka showed little change in photosynthesis whether grown at high or low light intensities.

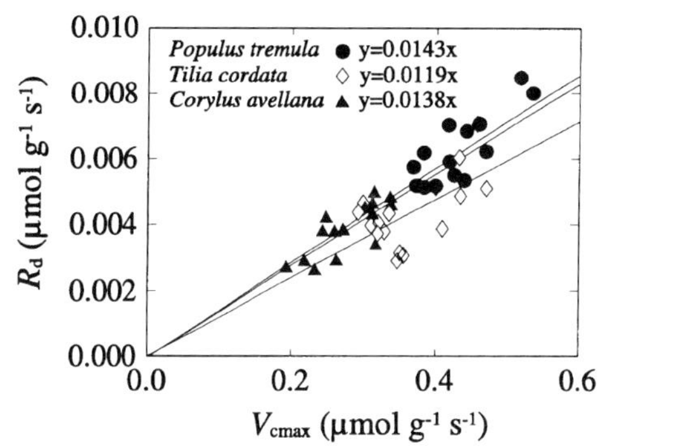

FIGURE 5.32. Relationships between leaf day respiration rate (R_d) and maximum carboxylation capacity (V_{cmax}) for three species of forest trees. Regression lines are forced through the origin and all are highly statistically significant ($p < 0.001$). From Niinemets et al. (1998).

In a particularly notable example of apparent evolutionary adaptation to habitat light levels, Givnish et al. (2004) found that maximum photosynthetic rates decreased monotonically in Hawaiian lobeliad species descended from a single island colonization and now naturally distributed across habitats of decreasing light.

Other shade-tolerant species (e.g., iripilbark tree) show moderate (Oberbauer and Strain, 1986) or large (e.g., coffee; Friend, 1984) increases in light-saturated photosynthesis under a high light regime during development. Capacity to adjust maximum photosynthesis to utilize the light regime in which they were grown was demonstrated for 14 early-, middle-, and late-successional species grown in full sunlight and in the shade (Bazzaz and Carlson, 1982).

Air Temperature

Photosynthesis of woody plants occurs over a wide temperature range from near freezing to over 40°C, the specific range depending on species and genotype, plant age, plant origin, and season. With increasing CO_2 supply and high light intensity, net photosynthesis usually increases with rising air temperature up to some critical temperature above which it begins to decline rapidly. In most temperate-zone species, the rate of photosynthesis increases from near freezing until it attains a maximum at a temperature between 15 and 25°C. The effect of air temperature usually is modified by light intensity, CO_2 availability, soil temperature, water supply, and preconditioning effects of environmental factors.

In tropical trees the rate of photosynthesis often is reduced by temperatures below about 15°C but above freezing (Fig. 5.33). For example, after coffee trees were exposed to 4°C at night, the rate of photosynthesis was reduced by more than half. Exposure to 0.5°C induced leaf necrosis and the plants no longer absorbed CO_2. If the leaves were not lethally injured by chilling temperatures, photosynthesis recovered completely within two to six days. Chilling on successive nights at 4 to 6°C reduced CO_2 absorption progressively on each day. After 10 nights of exposure to such chilling temperatures the rate of photosynthesis declined to less than 10% of the initial rate.

Woody plants often show both photosynthetic acclimation and adaptation to the temperature regime in which they are grown. For example, the optimum temperature for photosynthesis of apricot shifted from 24 to near 38°C in mid-August, followed by a drop to near 27°C by the end of September (Lange et al., 1974). The optimum temperature also varies with the altitudinal temperature gradient. Balsam fir seedlings showed a clinal pattern of adaptation, the optimum temperature for photosynthesis decreasing 2.7°C for each 300 m of elevation (Fryer and Ledig, 1972).

The time required for plants to acclimate the temperature of photosynthesis in response to temperature change varies with species, ontogeny, and nutritional status. Whereas photosynthesis of California encelia acclimated to a new temperature regime within 24 hours (Mooney and Shropshire, 1967), seven to 30 days were required for acclimation of different altitudinal populations of cabbage gum (Slatyer and Ferrar, 1977).

Subjecting plants to either a high or low temperature affects the subsequent rate of photosynthesis at another temperature (Pearcy, 1977). As mentioned, subfreezing temperatures inhibit the photosynthetic mechanism, but the damage generally is reversible with time at temperatures above freezing. The inhibitory effect of high-temperature preconditioning may

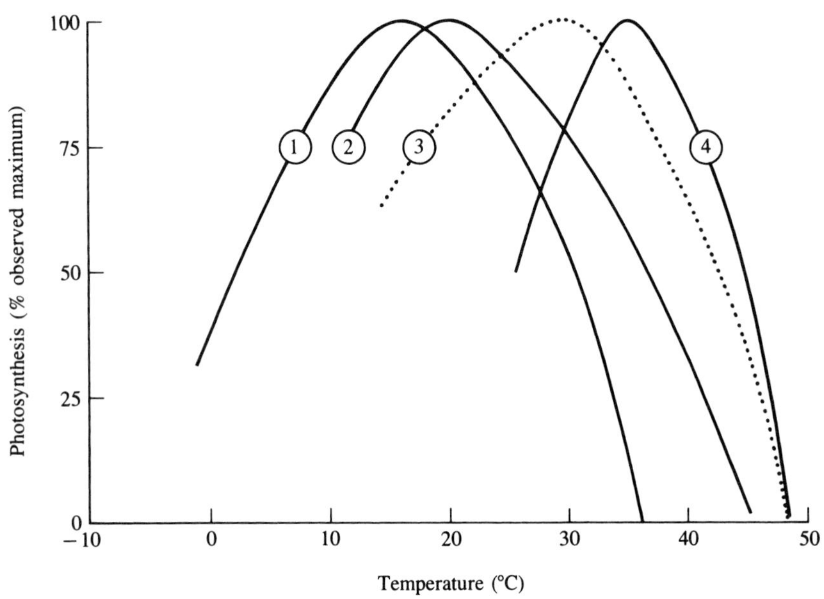

FIGURE 5.33. Effect of temperature on photosynthesis of temperate-zone species (curve 1, Swiss stone pine; curve 2, European beech) and tropical species (curve 3, India laurel; curve 4, acacia). After Larcher (1969); from Kozlowski et al. (1991).

last for many days, with the aftereffect much greater on photosynthesis than on respiration. For example, when European silver fir and sycamore maple leaves were exposed to nearly lethal heat for 60 min., CO_2 uptake at moderate temperatures was inhibited for many days whereas respiration rates were not appreciably altered (Larcher, 1969).

Mechanisms of Photosynthetic Inhibition

The mechanisms by which photosynthesis is reduced by low or high temperatures are complex and involve both stomatal and nonstomatal inhibition.

Low Temperature Inhibition

Induction of winter dormancy or frost hardening by short days and low temperatures above freezing often decreases the rate of photosynthesis (Öquist, 1983). Reduction in photosynthesis at very low temperatures involves multiple effects on the photosynthetic mechanism. Alterations in the photosynthetic machinery of conifers during the winter include changes in chloroplast structure, reduction in chlorophyll content, changes in activity of photosynthetic enzymes, disruptions in photosynthetic electron transport, and stomatal closure (Berry and Björkman, 1980; Larcher and Bauer, 1981; Schaberg et al., 1995).

Öquist and Huner (2003) reviewed the literature concerning winter photosynthesis and concluded that the photosynthetic apparatus in overwintering evergreen plants experiences potentially injurious effects of high excitation pressure from a combination of continued light harvesting and low-temperature induced inhibition of capacity for consumption of reducing power and ATP in CO_2 fixation and the photosynthetic carbon reduction cycle. As noted previously, severe photoinhibition and photo-oxidation of chlorophyll may ensue under such conditions, which are reflected in reductions in the quantum yield of photosynthesis, fluorescence yields of Photosystem II, loss of leaf chlorophyll and the D1 protein of the PS II reaction center (Fig. 5.34). The plastoquinone pool on the downstream side of PS II becomes reduced and reoxidation is inhibited. Photosystem I appears more resistant to inhibition by freezing temperatures. For example, Ivanov et al. (2001) demonstrated that photochemistry capacity of the PS I reaction center was inhibited only 20% during the winter in Scots pine as compared to 60% for PS II.

Acclimation mechanisms of evergreen species to low temperature include reduction in light-harvesting chlorophyll antenna size and partial loss of PS II reactions centers to reduce downstream excitation pressure, sustained nonphotochemical quenching via xanthophyll cycle interconversions (Fig. 5.34) (Öquist

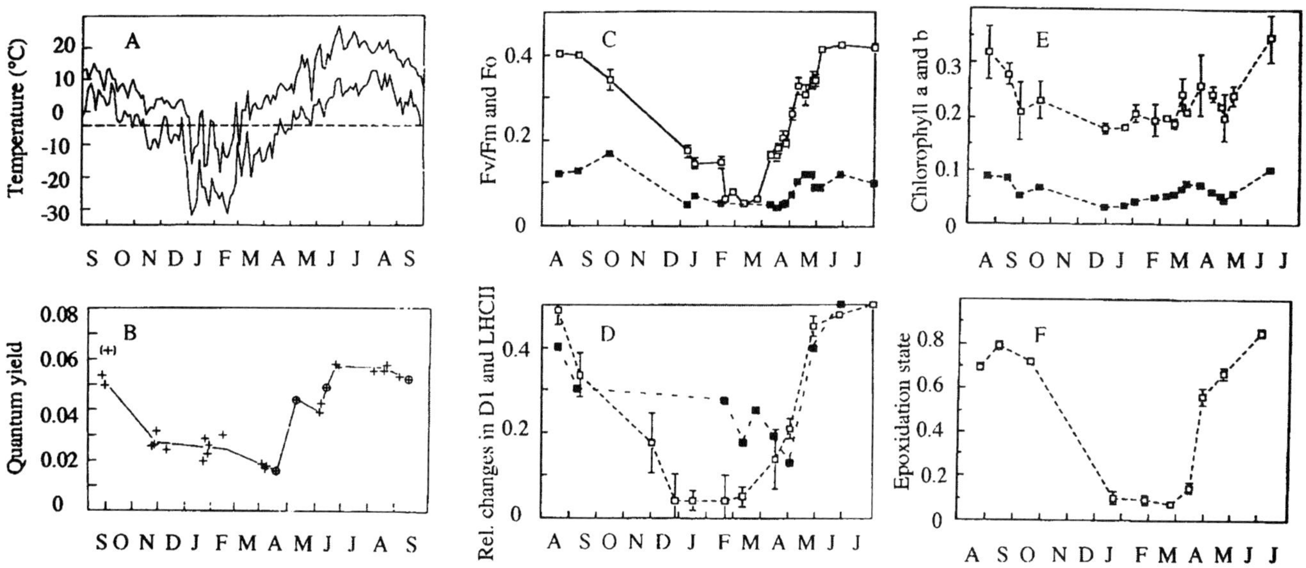

FIGURE 5.34. Seasonal variations in temperature and photosynthetic parameters of Scotch pine growing in an open stand in northern Scandinavia. (A) minimum and maximum temperatures for 3-day periods, (B) quantum yield for CO_2 uptake measured at 25°C, (C) photochemical efficiency of PS II (Fv/Fm, open symbols) and intrinsic fluorescence yield (Fo; closed symbols), (D) changes in PS II reaction center D1 protein (open symbols) and LHCII chlorophyll-protein complex of PS II (closed symbols), (E) chlorophyll a (open symbols) and b (closed symbols) in g m^{-2}, (F) epoxidation state of xanthophyll-cycle pigments (low values indicate an increased capacity for nonphotochemical dissipation of absorbed radiation). From Öquist and Huner (2003). Reprinted, with permission, from the *Annual Review of Plant Biology*, Volume 54 © 2003 by Annual Reviews www.annualreviews.org

and Huner, 2003), and protection of the intersystem electron transport system from oxidative damage by maintenance of a highly reduced condition. Photosystem I capacity is protected in winter because of increased activity of reactive oxygen scavenging systems, particularly the ascorbate peroxidase system (Chapter 6) and perhaps by cyclic electron flow that dissipates excitation pressure at PS I.

High Temperature Inhibition

Inhibition of net photosynthesis by heat often occurs because respiration continues to increase above a critical high temperature at which photosynthesis begins to decrease. Reduction of photosynthesis by heat does not appear to be caused primarily by stomatal closure, even though the stomata often close progressively at high temperatures. Such closure usually results from a stomatal response to the increased leaf-to-air VPD when the leaf temperature is raised. However, several studies showed that stomatal conductance remained high when the leaves were exposed to temperatures that injured the photosynthetic mechanism. Furthermore, stomata may open in the dark when the temperature is raised to between 35 and 40°C (Berry and Björkman, 1980).

Impairment of the photosynthetic mechanism at high temperatures largely reflects direct inhibition of chloroplast function. For a given species, inactivation of chloroplast activity begins at approximately the same temperature at which irreversible inhibition of light-saturated CO_2 uptake occurs. Reduction in photosynthesis at high temperatures is associated with changes in properties of thylakoid membranes, inactivation of enzymes of photosynthetic carbon metabolism, and decrease in the amount of soluble leaf proteins as a result of denaturation and precipitation (Berry and Björkman, 1980). For example, there is evidence that the amount of active Rubisco in heat-stressed plants is reduced by enhanced formation of abnormal catalytic products (e.g., xylulose bisphosphate and 3-ketoarabinitol bisphosphate) after binding of RuBP. These "dead-end" products then bind tightly to the catalytic site (Salvucci and Crafts-Brander, 2004). Rubisco activase appears to be quite important in freeing Rubisco active sites of inhibitory bound sugar phosphates (Portis, 2003) and the activase enzyme appears to be quite sensitive to heat denaturation (Crafts-Brandner and Salvucci, 2000).

Many organisms respond to heat stress with synthesis of certain proteins (heat shock proteins). These proteins may be synthesized only in response to stress or they may be present in unstressed plants but increase greatly upon exposure to heat. Researchers are just beginning to understand the functions of heat shock proteins, but one role they may play is stabilizing other protein molecules, thereby preventing heat denaturation (Vierling, 1990).

Injury to the photosynthetic apparatus by high temperature can be repaired at a lower temperature, provided the membranes associated with compartmentation in cells (including mitochondrial, chloroplastic, nuclear, and vacuolar membranes) have not been severely injured. The extent of recovery often depends on the severity of the heat stress and time for recovery. For example, English ivy leaves exposed for 30 minutes to a temperature of 44°C lost about half of their photosynthetic capacity but recovered it within a week after subsequent exposure to 20°C. However, after exposure to 48°C, photosynthetic capacity was reduced by nearly three-fourths and recovery at 20°C required eight weeks. Changes in properties of chloroplast membrane lipids play a major role in acclimation of photosynthesis to high temperatures by increasing the heat stability of membranes (Berry and Björkman, 1980).

Soil Temperature

Soil temperature as well as air temperature affects photosynthesis, with the rate of CO_2 uptake decreasing at low temperature (Figs. 5.35 and 5.36). Low rates of photosynthesis of Engelmann spruce were directly related to low (night) air and soil temperatures, but at different times during the early summer growth period. After cessation of freezing nights, low soil temperature was the dominant limiting factor for photosynthesis (DeLucia and Smith, 1987). Similar results were reported for Norway spruce in a soil warming study (Bergh and Linder, 1999). Artificial warming of the soil by electrical cables in a 31-year-old spruce plantation accelerated snowmelt and elevated soil temperatures; however, elevated soil temperature was associated with enhanced photosynthesis only in late May when air temperatures moderated and there was a cessation of freezing nights.

The extent of inhibition of photosynthesis at low soil temperature depends on how much the soil is chilled and how well the plants have hardened to frost. For example, the rate of photosynthesis of frost-hardened Sitka spruce plants recovered fully overnight at soil temperatures above –4°C, but recovered to only 58% of the initial rate at the same time from a soil temperature of –8°C. By comparison, photosynthesis of unhardened plants recovered fully overnight when the soil temperature was above –0.5°C, but did not recover fully in 17 days when exposed to lower temperatures (Turner and Jarvis, 1975).

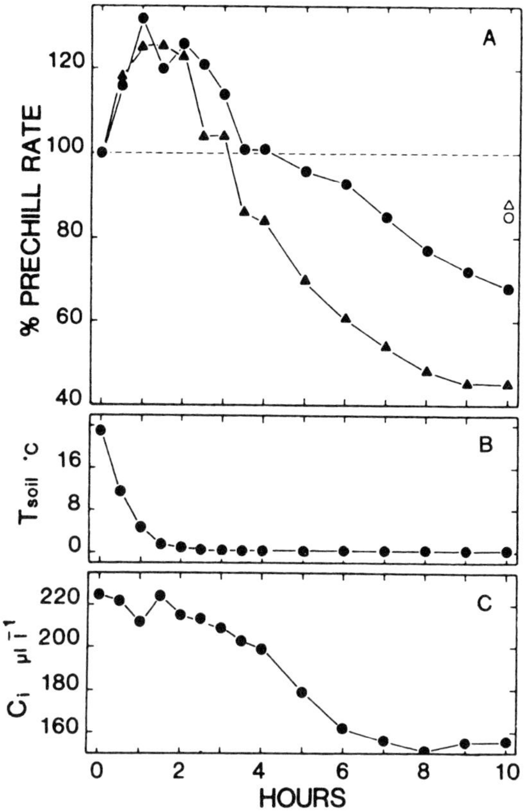

FIGURE 5.35. Short-term response of net photosynthesis (●) and stomatal conductance (▲) to chilling of roots in Engelmann spruce seedlings (A). Changes in soil temperature and intercellular CO_2 concentration are shown in (B) and (C), respectively. Open symbols indicate changes in gas exchange parameters of unchilled plants measured after 12 hours in the measurement cuvette. From DeLucia (1986).

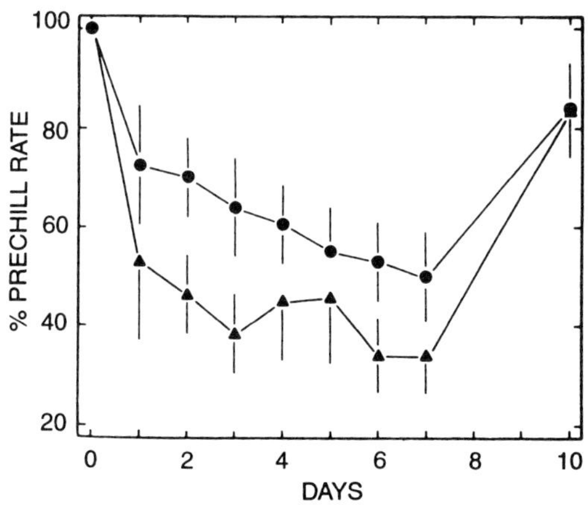

FIGURE 5.36. Long-term response of net photosynthesis (●) and stomatal conductance (▲) to chilling of roots (0.7°C) of Engelmann spruce seedlings. Chilling of roots was initiated on day 0 and terminated on day seven. Bars indicate ±1 SE. From DeLucia (1986).

Reduction of photosynthesis in response to low soil temperatures often involves both stomatal and nonstomatal inhibition. Stomatal closure due to decreased absorption of water at low soil temperature accounted only partly for a decline in photosynthesis of Sitka spruce seedlings as shown by substantial reduction of mesophyll conductance (Turner and Jarvis, 1975). In contrast, Bassirad et al. (1993) found that cooling roots systems of the shrub *Artemesia tridentata* to 5°C while maintaining the shoots at 20°C reduced stomatal conductance by 50% and reduced c_i by 10%, a result which suggests predominant stomatal limitation in this species. Reductions in c_i at soil temperatures below 15°C also were shown for four boreal species (*Populus tremuloides, Pinus banksiana, Picea mariana, P. glauca*) (Dang and Cheng, 2004).

Soil temperatures between 10 and 20°C did not affect photosynthesis of Engelmann spruce needles. However, both photosynthesis and leaf conductance declined sharply when the soil temperature was below 8°C. Photosynthesis and stomatal conductance decreased by 50 and 66%, respectively, after seven days at a soil temperature of 0.7°C. The decline during the first few hours was associated with a decrease in the CO_2 concentration in the leaf interior, suggesting primarily stomatal effects in photosynthetic inhibition (Fig. 5.35). However, in the longer term, the internal CO_2 concentration was high and the decrease in photosynthesis was attributed largely to lowered carboxylation efficiency and reduced apparent quantum yield (DeLucia, 1986).

The mechanism behind stomatal closure at low soil temperature is not well resolved. Low soil temperature has induced lower shoot water potentials in some studies (*Populus tremuloides*, Wan et al., 1999), but not in others (*P. tremuloides*, Wan et al., 2004). Nonhydraulic control of stomatal aperture in cold soil has also been studied (see also Chapter 12). Abscisic acid concentration of xylem sap of sugar maple trees was elevated in the spring, if soil had been allowed to freeze more than normal during the winter by snow cover removal (Bertrand et al., 1994). Reduced stomatal conductance was associated with elevated root and shoot ABA concentration in arctic tussock grasses exposed to reduced root temperatures between 10 and 0°C (Starr et al., 2004). Wan et al. (2004) found that ABA concentration in the xylem sap of *Populus tremuloides* increased when plants were subjected to 5°C root-zone temperatures, but that this increase was preceded by increased sap pH. The latter increase was better correlated with declining stomatal conductance and suggested that pH could serve as a signal for stomatal closure by stimulating release of sequestered ABA from chloroplasts.

Very high root temperatures may lead to reduction in photosynthesis. After three weeks the rate of photosynthesis of holly plants grown with soil temperatures of 38 or 42°C was lower than that of plants in soil at 30 or 34°C (Ruter and Ingram, 1992). Leaf chlorophyll and carotenoid contents decreased, whereas leaf soluble proteins increased as the soil temperature was raised. Rubisco activity per unit of leaf fresh weight increased linearly in response to increasing root zone temperature, but when Rubisco activity was expressed per unit of protein or chlorophyll there was a reduction in activity at both high (42°C) and low (30°C) root temperature. The reduction in photosynthesis at high root temperature was modest (13%) and the authors suggested that holly plants adjusted their metabolism or redistributed photosynthetic products to maintain photosynthetic rates at high root temperatures.

Carbon Dioxide

Photosynthesis of woody plants that are well watered and exposed to sunlight is limited chiefly by the low CO_2 concentration of the air, which is only approximately 0.0375% by volume (375 ppm or $\mu mol\ mol^{-1}$). As noted earlier, the availability of CO_2 to photosynthetic cells is strongly limited by resistances in its inward diffusion path, including boundary layer or air, cuticular, stomatal, and mesophyll air space and liquid diffusion resistance. The boundary layer of most conifer needles is small and increases with leaf size and decreases with wind speed. Because the cuticular surfaces of most leaves are relatively impermeable to CO_2, stomatal conductance becomes very important in regulating CO_2 uptake. Total mesophyll conductance is determined both by biochemical and diffusional characteristics and also is related to the concentrations of the photosynthetic carboxylating enzyme and photochemistry capability of leaves. This measure of mesophyll conductance tends to be low in comparison to stomatal conductance. Nevertheless, plants with comparable stomatal and boundary layer conductances may have different photosynthetic rates because of dissimilarities in mesophyll conductance.

Locally the CO_2 concentration may rise far above the world-wide average because of industrial activity, or it may fall below it because of depletion by photosynthesis. In the absence of wind, the CO_2 content of the air fluctuates diurnally with a minimum in the afternoon. The CO_2 concentration near the ground often is high because of root respiration and release by decay of organic matter, and the concentration in the plant canopy sometimes is decreased by use in photosynthesis. By midday, the CO_2 concentration in forest stands may decrease by a fourth or more, presumably because of removal of CO_2 by photosynthesis (Miller and Rüsch, 1960). Increased photosynthesis may occur on foggy days if light is not limiting because the CO_2 content of the air may be higher on such days than on clear days (Wilson, 1948).

Much interest has been shown in potential effects of rising CO_2 concentration of the air on photosynthesis. The increase in CO_2 is caused largely by accelerated use of fossil fuels in transportation, industrial processes, and possibly by extensive destruction of forests. Analyses of air bubbles trapped in ancient ice indicate that CO_2 in the air has been increasing for many years. The concentration increased from near 260 ppm in the middle of the nineteenth century to 300 ppm early in this century. In 2003 the CO_2 concentration at Mauna Loa observatory in Hawaii rose to 375 ppm (Fig. 5.37) and a concentration of 500 ppm by the year 2050 and 700 ppm by 2100 has been predicted (Prentice et al., 2001, cited in Long et al., 2004).

Much research on the effects of rising atmospheric CO_2 on plants has been conducted over the past 25 years. Plants are both affected by and affect atmospheric CO_2 concentrations. Annual exchange of CO_2 between terrestrial ecosystems and the atmosphere (120–125 Gt) dwarfs anthropogenic contributions (6–7 Gt), and changes in the terrestrial fluxes have the potential both to exacerbate or ameliorate human inputs (Field et al., 1998; Worrell et al., 2001; Gonzalez-Meler et al., 2004). Much early CO_2 research employed short-term and some longer-term exposures of potted seedlings to elevated CO_2 (200–400 ppm above ambient) (Ceulemans et al., 1999; Norby et al., 1999). In greenhouses and controlled-environment chambers, with favorable water and mineral supplies, an increase in CO_2 concentration typically is accompanied by at least a temporary rise in the rate of photosynthesis and in dry weight increase of plants (Table 5.2; see Chapter 5, Kozlowski and Pallardy, 1997). However, many early investigators reported that the increased rate of photosynthesis associated with CO_2 increase gradually decreases (DeLucia et al., 1985). As noted earlier, most experiments that showed such inhibition were conducted with plants grown in containers. Such conditions can retard both photosynthesis and growth of plants by inhibiting root growth and nutrient uptake (Jarvis, 1989).

Subsequently, experiments in open-topped chambers (OTC) around plants rooted in large soil volumes or in the field were developed that overcame many of the deficiencies of pot experiments. Both yellow-poplar and white oak seedlings and saplings grown in OTC showed increased rates of photosynthesis for three consecutive years in response to CO_2 enrichment of the atmosphere (Gunderson et al., 1993). These increases

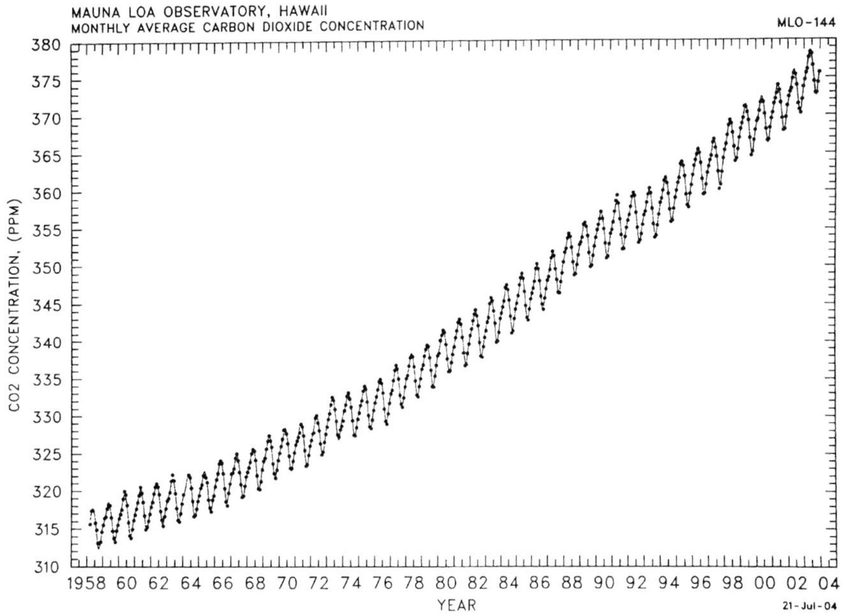

FIGURE 5.37. Increase in atmospheric concentration of CO_2 at the Mauna Loa observatory in Hawaii and the seasonal cycle in concentration each year since 1958. Source: Dave Keeling and Tim Whorf. Scripps Institution of Oceanography. Data provided by the Carbon Dioxide Information Analaysis Center, Oak Ridge National Laboratory.

TABLE 5.2. Effects of Atmospheric CO_2 Enrichment on Growth of Sweetgum and Loblolly Pine at 32 Weeks after Planting (End of First Growing Season)[a,b]

CO_2 concentration (ppm)	Stem weight (g)	Leaf weight (g)	Root weight (g) sweetgum	Root–shoot ratio	Total weight (g)	Percentage increase in total weight
350	2.1a	1.5a	6.3a	1.7ab	9.9a	
500	3.4b	1.6ab	9.2b	1.8b	14.2b	43.4
650	3.7b	2.0b	8.6ab	1.5a	14.3b	44.4
			Loblolly pine			
350	0.4a	1.4a	2.1a	1.2a	3.9a	
500	1.0b	2.3b	2.9b	0.9a	6.2b	58.9
650	0.8b	2.1b	3.2b	1.1a	6.1b	56.4

[a]From data of Sionit et al. (1985).
[b]Means followed by the same letter within a column for each species are not significantly different at the 5% level of probability. Values are means of measurements on 15 seedlings.

were similar to those reported for these species in short-term, controlled-environment experiments in which nutrients and physical restriction were not limiting (Norby and O'Neill, 1989, 1991; Gunderson et al., 1993). Photosynthesis was stimulated by CO_2 enrichment even without supplemental irrigation and fertilization, and despite decreasing leaf N and chlorophyll concentrations in yellow-poplar (Norby et al., 1992). In reviews by Gunderson and Wullschleger (1994) and Ceulemans and Mousseau (1994), mean enhancements in photosynthesis of between 40 and 61% were reported in angiosperm and coniferous tree species when photosynthesis was measured at the CO_2 levels in which they were grown.

Although less subject to artifacts, open-topped chamber experiments themselves are hindered by space limitations, microclimatic influences of the chambers, and edge effects on experimental plants. They cannot realistically simulate natural or managed ecosystems. Most recently, field-based experiments employing free air CO_2 enhancement (FACE) installations have allowed exposure of larger stands of plants (up to 700 m^2) to elevated CO_2. FACE studies employ a ring of release tubes for tank CO_2 with computer

FIGURE 5.38. Aerial view of a young sweetgum plantation near Oak Ridge, Tennessee, in which Free Air CO_2 Enhancement (FACE) rings have been installed around experimental plots. Each plot is surrounded by 24 vent pipes spaced 3.3 m apart suspended from 12 aluminum towers. Rings are 25 m in diameter. Photo courtesy of Oak Ridge National Laboratory.

control of release pattern and rate (Fig. 5.38). On-site weather monitoring instruments monitor wind direction and velocity measurements to direct release of the appropriate amounts of CO_2 into the downwind direction. FACE experiments have been undertaken with crop, shrub, and tree species, and multiyear data sets are now emerging.

A sufficient number of FACE studies have been conducted to allow meta-analysis (e.g., Long et al., 2004; Nowak et al., 2004; Ainsworth and Long, 2005). The results indicate that an elevated CO_2 level induces long-term stimulations of light-saturated photosynthetic rates (A_{sat}) and daily photosynthetic gains (A′) (Fig. 5.39), with trees and shrubs showing intermediate-to-high A_{sat} and A′ responses compared to herbaceous plants. Apparent quantum yield at low light levels increases under elevated CO_2, presumably because high CO_2 curbs the diversion of ATP and $NADPH_2$ needed in photorespiratory metabolism by suppressing the oxygenase activity of Rubisco (Long et al., 2004). Starch content of leaves increases substantially and sucrose concentration increases somewhat under elevated CO_2.

There often is long-term down-regulation of some photosynthetic parameters under elevated CO_2 as manifested largely in a decrease in Rubisco content and activity (reduced V_{cmax}) (Fig. 5.40). Reduction in Rubisco generally is associated with a corresponding decline in leaf N on a mass basis. Reduction in the capacity for electron capacity (J_{max}) is less affected than is V_{cmax}. Trees exhibit somewhat less response than herbaceous plants in these parameters. On the other hand Leaf Area Index increases under elevated CO_2 and more so in trees than in C_3 grasses.

The closing response of stomata in elevated CO_2 generally results in decreased g_s and canopy conduc-

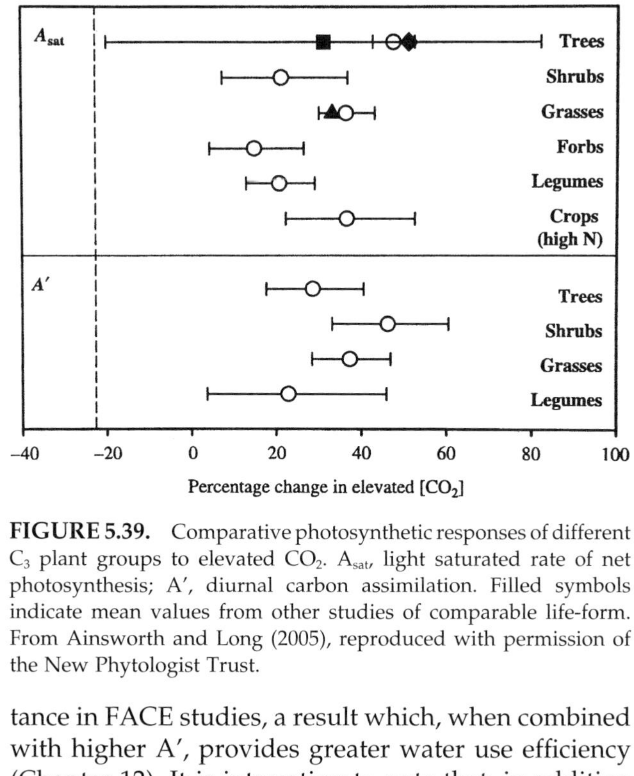

FIGURE 5.39. Comparative photosynthetic responses of different C_3 plant groups to elevated CO_2. A_{sat}, light saturated rate of net photosynthesis; A′, diurnal carbon assimilation. Filled symbols indicate mean values from other studies of comparable life-form. From Ainsworth and Long (2005), reproduced with permission of the New Phytologist Trust.

tance in FACE studies, a result which, when combined with higher A′, provides greater water use efficiency (Chapter 12). It is interesting to note that, in addition to stomatal closing responses, stomatal density also declines under elevated CO_2 (Woodward and Kelly, 1995), a trend that would also tend to elevate WUE. However, it is not clear that soil water content will be higher under future elevated CO_2 conditions because of correspondingly greater LAI developed by forests under these conditions and uncertainty in the frequency and severity of droughts in a climate-altered world.

Thus, it must be cautioned that even though increased concentrations of CO_2 increase the rate of

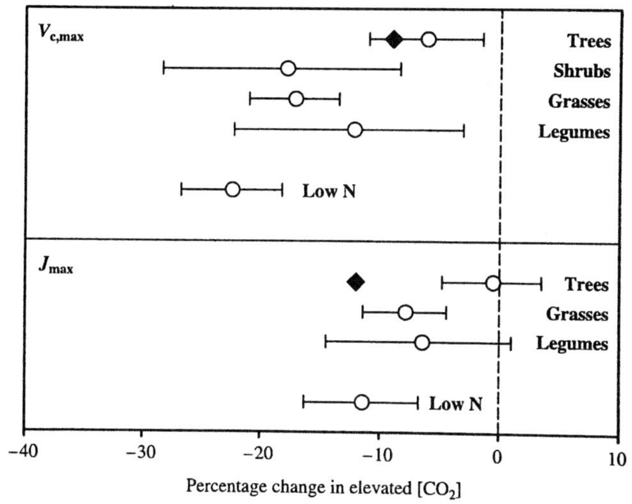

FIGURE 5.40. Comparative acclimation responses of maximum carboxylation velocity (V_{cmax}) and maximum rate of electron transport (J_{max}) of different C_3 plant groups to elevated CO_2. Filled symbols indicate mean values from other studies of comparable life-form. From Ainsworth and Long (2005), reproduced with permission of the New Phytologist Trust.

photosynthesis and growth of plants in experimental studies in the field, greenhouses, and controlled environments, there is no assurance that similar increases will occur on a global scale where other stresses such as deficiencies of water and nitrogen already inhibit photosynthesis (Kramer, 1981). Further, Körner (2003) suggested that elevation of atmospheric CO_2 concentrations already experienced since the beginning of the Industrial Age have stimulated photosynthesis to the point where plants are no longer carbon-limited in growth. Under conditions where one would expect depleted carbohydrate and lipid contents of tree tissues (e.g., at timberline in high elevation ecosystems, Hoch and Körner, 2003; under drought in temperate deciduous forests, Hoch et al., 2003) measurements indicate no depletion compared to nearby trees not subjected to extreme environments. Hence, limitations in sink strength in a high-CO_2 world may restrict further increases in photosynthesis and growth.

WATER SUPPLY

Photosynthesis is very responsive to availability of water, with the rate decreased by both drying and flooding of soil.

Soil Drying and Photosynthesis

Water deficits reduce photosynthesis by closing stomata, decreasing the efficiency of the carbon fixation process, suppressing leaf formation and expansion, and inducing shedding of leaves. With a high soil moisture supply, daily net photosynthesis of white oak in Missouri averaged more than 5.1 μmol CO_2 m^{-2} s^{-1}, and maximum rates were greater than 8.8 μmol m^{-2} s^{-1}. During a severe drought, however, average and daily maximum net photosynthesis decreased to less than 1.0 and 5.1 μmol CO_2 m^{-2} s^{-1}, respectively (Dougherty and Hinckley, 1981).

There has been considerable controversy over the years about the soil moisture content at which photosynthesis is first reduced. Some investigators reported that photosynthesis decreased when irrigated soil dried only slightly. Others claimed that photosynthesis did not decrease appreciably until much of the available soil water was depleted (for review see Kozlowski, 1982a). Gollan et al. (1985) reported that photosynthesis of oleander decreased when about half the available soil water was depleted. The early controversy about the critical soil moisture content at which photosynthesis decreases arose at least partially because some investigators who measured both soil moisture content and photosynthesis assumed that soil water deficits and leaf water deficits were closely correlated. However, soil water retention curves show that during drying, water potential of a soil changes only a small amount until soil water is substantially depleted (Chapter 11). Also, leaves of trees growing in dry soil may not develop severe water deficits if the relative humidity of the air is high. Conversely, when the relative humidity is low, even trees growing in well-watered soil tend to dehydrate. Leaf water deficits depend on relative rates of absorption and transpiration, and not on absorption alone (Kozlowski et al., 1991). Hence, although leaf water deficits may reduce the rate of photosynthesis, slight decreases in soil moisture content do not always do so.

Evidence shows that stomatal inhibition of photosynthesis of plants in dry soil is not entirely traceable to leaf dehydration (see also Chapter 12). Photosynthesis of water-stressed black walnut seedlings and canopy trees was more closely related to soil water status than to leaf Ψ, suggesting that stomatal closure was directly influenced by soil water status (Parker and Pallardy, 1991; Loewenstein and Pallardy, 1998a,b). Plants sometimes close their stomata before leaf turgor shows measurable change (Schulze, 1986). This response is explained by "root sensing" of soil water deficits and consequent transport of a signal, possibly hormonal, in the xylem from roots to leaves, leading to stomatal closure. A good candidate for the signal in many species is ABA (Zhang and Davies, 1989, 1990a; Zhang et al., 1987; Dodd, 2003). When the xylem sap extracted from maize plants was applied to maize foliage, stomatal closure followed and transpiration was reduced as much as by application of ABA solutions of equivalent concentration (Zhang and Davies, 1991). When ABA was removed from extracted xylem sap, the antitranspirant effect disappeared. Loewenstein and Pallardy (1998a,b) found that xylem sap ABA concentration of black walnut was elevated and stomatal conductance reduced during drought in both potted seedlings in the greenhouse and canopy trees in the field. These responses occurred in the absence of leaf water deficits. Feeding excised leaves of black walnut artificial sap solutions containing ABA also induced stomatal closure, supporting the conclusion that xylem ABA produced by roots in drying soil was the primary stimulus in stomatal closure in this species. Similar results were obtained for *Acacia confusa* and *Leucaena leucocephala* (Liang et al., 1997).

Other types of chemical signals imprinted on xylem sap may also influence stomatal aperture, including sap pH and ion contents (Wilkinson and Davies, 2002). For example, drought may increase xylem sap pH with a corresponding influence on leaf ABA partitioning that favors release of chloroplast-sequestered ABA into the apoplast where it can move to the guard cells (Sobeih et al., 2004). It is interesting to note that woody plants do not show increases in xylem sap pH during drought as often as do herbaceous plants (Wilkinson and Davies, 2002; Loewenstein and Pallardy, 1998a), apparently relying more on ABA as a xylem signal. It should be emphasized that chemical signals likely provide only one of numerous sources of stomatal control, and may also interact with other factors (e.g., VPD, leaf Ψ, prior water stress, temperature) in regulating stomatal aperture (see Fig. 12.7, Tardieu and Simonneau, 1998; Loewenstein and Pallardy, 2002; Wilkinson and Davies, 2002). The influence of root signals on stomatal aperture is discussed further in Chapter 12.

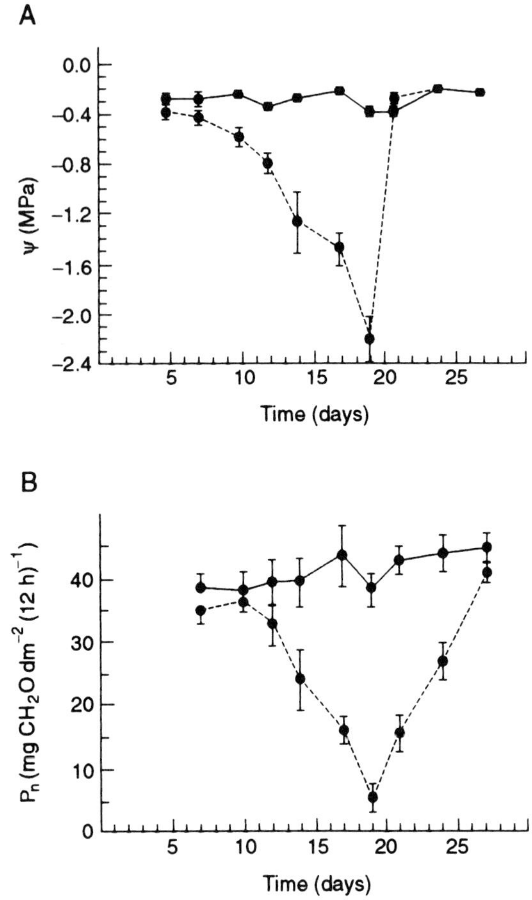

FIGURE 5.41. Time course of water potential (Ψ) and net photosynthesis (P_n) in mature leaves of irrigated (——) and nonirrigated (----) cacao seedlings during a period of soil drying and recovery. Water was withheld from nonirrigated seedlings from day 0 through day 19, at which time the plants were rewatered. From Deng et al. (1989).

Leaf Water Potential (Ψ) and Photosynthesis

As leaves dehydrate, their Ψ becomes more negative and the rate of photosynthesis is reduced (Fig. 5.41). Several investigators attempted to identify a critical leaf Ψ at which photosynthesis of various species growing in drying soil begins to decline (for review see Kozlowski, 1982a, pp. 70–71). However, it is difficult to establish a precise leaf Ψ at which photosynthesis is first reduced because this value differs with species, genotype, habitat, past treatment of the plant, and prevailing environmental conditions. Additionally, as noted earlier, leaf Ψ may interact with other factors such as hormones translocated in the transpiration stream and VPD in control of stomatal aperture (Tardieu and Davies, 1992).

Woody plants of arid regions often maintain photosynthesis at lower leaf Ψ values than do more mesic plants. For example, the rate of photosynthesis of alder (*Alnus oblongifolia*) and green ash decreased when the leaf Ψ dropped to −1.0 MPa; that of creosote bush, a desert shrub, dropped at a leaf Ψ of −2.0 MPa (Chabot and Bunce, 1979). Photosynthesis of boxelder from a streamside habitat stopped at shoot Ψ values 1.0 to 1.5 MPa higher than that of Gambel oak and big sagebrush of dry areas (Dina and Klikoff, 1973). Photosynthesis of post oak, a xeric species, was appreciable at leaf Ψ down to nearly −3.0 MPa (Ni and Pallardy, 1991). Photosynthesis of the mesic black walnut showed much greater sensitivity to leaf Ψ.

Causes of Reduction in Photosynthesis

Some of the short-term reduction in photosynthesis during drought has been attributed to increased resistance to diffusion of CO_2 to the chloroplasts and some to metabolic inhibition.

Much emphasis has been placed on the importance of stomatal closure in reducing photosynthesis. Some investigators found close correlations between stomatal aperture, transpiration, and photosynthetic rates (Regehr et al., 1975). Lakso (1979) reported a linear relationship between stomatal aperture and photosynthesis of apple leaves. Water stress appreciably increased the relative stomatal compared to nonstomatal limitation of photosynthesis of post oak (Ni and Pallardy, 1992) and black spruce seedlings (Stewart et al., 1995). Temporary midday reductions in photosynthesis occur commonly and often have been associated with stomatal closure, which limits absorption of CO_2 by leaves (Tenhunen et al., 1982).

Reduction in photosynthesis during a drought may also be traceable to inhibition of the photosynthetic apparatus as well as to stomatal closure. For example, decreased rates of photosynthesis of *Encelia farinosa* (Ehleringer and Cook, 1984), *Encelia frutescens* (Comstock and Ehleringer, 1984), apple (Swietlik et al., 1983), white oak, post oak, sugar maple, black walnut (Ni and Pallardy, 1992), and loblolly pine (Teskey et al., 1986) during soil drying were attributed to both stomatal and nonstomatal components. Drawing upon much experimental evidence, Lawlor and Cornic (2002) presented two models of photosynthetic response to water deficits (Fig. 5.42). In plants displaying a Type 1 response, inhibition of photosynthesis by reduced RWC can initially be completely overcome by elevating the CO_2 to which the leaf is exposed, indicating a lack of effect on photosynthetic metabolic capacity per se. Hence in this region stomatal closure contributes the primary limitation to photosynthesis. As RWC declines further, photosynthesis can no longer be completely restored regardless of the CO_2 concentration, indicating eventual mesophyll-level inhibition under severe water stress. In plants that display the Type II response, maximal photosynthesis cannot be restored by elevated CO_2 even under mild water deficits, but rather the mesophyll limitation is induced in a progressive fashion as RWC declines.

With regard to component processes that determine metabolic limitation, photosynthetic electron transport appears to be relatively resistant to inhibition under water stress (Keck and Boyer, 1974; Epron and Dreyer, 1992, 1993; Epron et al., 1992; Havaux, 1992). In contrast, photophosphorylation and photosynthetic carbon metabolism seem to be more sensitive to dehydration. Photophosphorylation may be particularly sensitive to the toxic effects of high concentrations of Mg^{2+} that accompany removal of water from chloroplasts in dehydrating leaves (Boyer and Younis, 1983; Rao et al., 1987; Santakumari and Berkowitz, 1991; Kramer and Boyer, 1995). Tezara et al. (1999) noted both decreased ATP and ATP synthase in water-stressed sunflower leaves. The primary influence of water stress on carbon metabolism does not appear to involve Rubisco, as numerous studies have shown that the activity of this enzyme is at most only slightly reduced under water stress (e.g., Huffaker et al., 1970; Björkman et al., 1980; Gimenez et al., 1992; Tezara et al., 1999). Gimenez et al. (1992) found that the regeneration of one of the *substrates* of Rubisco, RuBP, was progressively inhibited in sunflower plants as Ψ decreased. As the activities of photosynthetic carbon reduction cycle enzymes and concentrations of other metabolites did not vary much with the imposition of stress, the authors concluded that RuBP regeneration was an important limiting factor in water-stressed plants. Sharkey and Seemann (1989) observed similar patterns of response in severely water-stressed bean plants. The reduced RuBP levels in water-stressed plants appear to reflect reduced ATP and not decreased activity of enzymes of the photosynthetic carbon reduction cycle (Lawlor and Cornic, 2002).

Another possible source of mesophyll limitation to photosynthetic capacity is increased diffusional resistance to CO_2 movement from intercellular spaces to the chloroplasts. As early as 1978, Fellows and Boyer noted

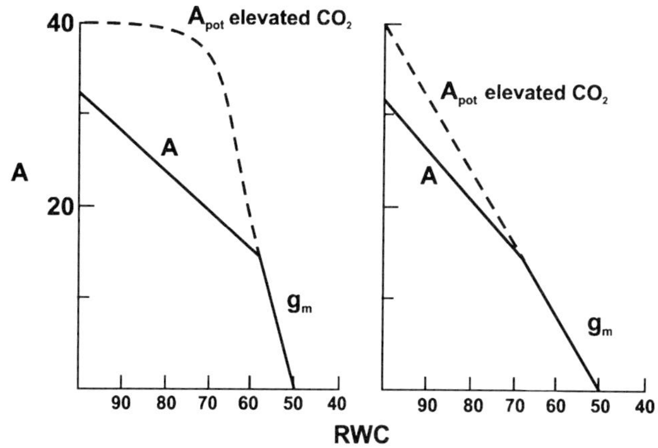

FIGURE 5.42. Theoretical relationships between net photosynthesis (A) and Relative Water Content (RWC). In the Type 1 response (left) potential photosynthesis (A_{pot}) is stable as RWC declines from 100 to 75%. A measured at ambient CO_2 falls over this RWC range, but elevation of CO_2 provided to the leaf can raise A to A_{pot}. Below 75% RWC A_{pot} falls as mesophyll limitation of photosynthesis occurs and elevation of CO_2 no longer restores A to the well-hydrated level of A_{pot}. In Type 2 species (right) A and A_{pot} decrease progressively as well-hydrated leaves lose water. In both types initial regulation of A is by stomatal closure, but at low RWC mesophyll-level effects predominate. From Lawlor, D. W. and Cornic, G. (2002). Photosynthetic carbon assimilation and associated metabolism in relation to water deficits in higher plants. *Plant. Cell Environ.* **25**, 275–294. Blackwell Publishing.

that water-stressed sunflower leaves exhibited internal alterations in cell-to-cell contact that could increase the more high-resistance liquid path flow distance to the chloroplast. Other more recent studies have suggested that not all the decrease in mesophyll photosynthetic capacity under mild and moderate water stress might arise from inhibition of metabolic activity such as ATP synthesis, but rather that liquid-phase mesophyll conductance is decreased (Flexas et al., 2004). The mechanism(s) responsible for decreased mesophyll diffusional conductance are not known, but may involve already-mentioned anatomical changes (Fellows and Boyer, 1978) that lengthen liquid diffusion paths and perhaps alteration in activity of aquaporins in membranes that are known to pass CO_2 molecules in addition to water (Uehlein et al., 2003).

The relative importance of stomatal and nonstomatal inhibition of photosynthesis during drought varies with the drought tolerance of tree species. Ni and Pallardy (1992) found that relative stomatal limitation increased under water stress in xeric post oak but it decreased in mesic sugar maple. Similarly, Kubiske and Abrams (1993) found that stomatal limitation of photosynthesis increased during drought in xeric and mesic species, whereas in wet-mesic species mesophyll limitation was at least as important as stomatal limitation in limiting photosynthesis. There also was a general trend toward decreased relative mesophyll limitation in xeric species, suggesting a capacity for high carbon fixation in drought-tolerant species. These results suggest the more drought-tolerant species tend to possess photosynthetic attributes of the Type I category of Lawlor and Cornic (2002) (Fig. 5.42).

Brodribb (1996) and Brodribb and Hill (1998) monitored c_i/c_a ratios in leaves of several Southern Hemisphere conifer species for which plants were allowed to gradually deplete soil moisture. In all species, initial inhibition of photosynthesis was associated with declining c_i/c_a, indicating predominant stomatal limitation. As water stress became more severe, c_i/c_a ratio increased substantially, indicating a reversal of the pattern of limitation with mesophyll then predominating, a result that was supported by independent chlorophyll fluorescence measurements of photoinhibition. Interestingly, both the level of water stress at which this reversal occurred and minimum c_i/c_a reached were correlated with field drought tolerance evaluations of the species (Fig. 5.43). These results support the tendency for more drought-tolerant species to have a more robust Type I photosynthetic apparatus in the face of water stress (e.g., Ni and Pallardy, 1992).

Many claims for nonstomatal, short-term inhibition of photosynthesis are based on reports that the internal CO_2 concentration remains high in water-stressed

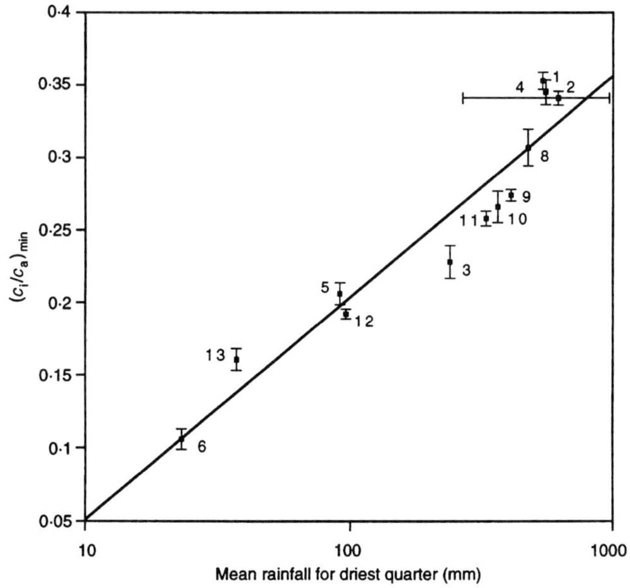

FIGURE 5.43. Relationship between minimum c_i/c_a during drought [(c_i/c_a) min] of 12 Southern Hemisphere conifer species and precipitation during the driest consecutive four months. 1, *Acmopyle pancheri*; 2, *Dacrycarpus dacrydioides*; 3, *Podocarpus lawrencii*; 4, *D. compactus*; 5, *Callitris rhomboidaea*; 6, *Actinostrobos acuminatus*; 8, *Lagarostrobos franklinii*; 9, *Microstrobos niphophilis*; 10, *Athrotaxis selaginoides*; 11, *Diselma archerii*; 12, *Podocarpus drouynianus*; 13, *Widdringtonia cedarbergensis*. From Brodribb. T. and Hill, R. S. (1998). The photosynthetic drought physiology of a diverse group of southern hemisphere conifer species is correlated with minimum seasonal rainfall. *Funct. Ecol.* **12**, 465–47. Blackwell Publishing.

leaves after the stomata close. Such reports have been interpreted cautiously because of evidence of patchy closure of stomata across the leaves of heterobaric species (McClendon, 1992) and the potential it carries for causing artifacts in calculations of internal CO_2 concentrations (Düring, 1992; Ni and Pallardy, 1992; Beyschlag et al., 1994; Pospisilova and Santrucek, 1994). Downton et al. (1988) found that ABA application induced closure of groups of stomata rather than causing stomatal closure uniformly across the leaf. These investigators concluded that stomatal closure can fully account for many examples of previously assumed nonstomatal inhibition of photosynthesis. Sharkey and Seemann (1989) examined Rubisco activity and concentrations of photosynthetic carbon reduction cycle intermediates under water stress conditions in bean and concluded that reduction in photosynthesis in mildly water-stressed plants was largely attributable to stomatal closure.

After-Effects of Water Deficits on Photosynthesis

Unless drought has been so severe as to cause hydraulic disconnection of shoots from roots by cavitation (Chapter 11), leaf Ψ recovers in a matter of hours.

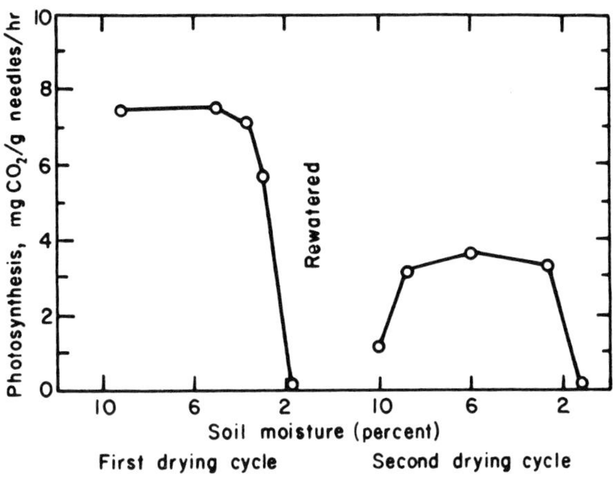

FIGURE 5.44. Effect of two soil drying cycles on net photosynthesis of Douglas-fir seedlings. From Zavitkovski and Ferrell (1970).

For example, leaf Ψ of seedlings of *Populus deltoides*, *Salix nigra*, *Quercus alba*, and *Q. velutina* subjected to moderate water stress recovered within six hours of rewatering (Loewenstein and Pallardy, 2002). However, the rate of photosynthesis may or may not return to predrought levels, depending on plant species, severity and duration of the drought, and dryness of the air. The rate of photosynthesis of droughted Japanese red pine seedlings recovered to the rate of control plants within a day (Negisi and Satoo, 1954); and of cacao within eight days (Fig. 5.41). By comparison, the photosynthetic rate of Douglas-fir seedlings subjected to drought did not recover to predrought levels after irrigation (Fig. 5.44). After irrigation of droughted seedlings of five species of broadleaved trees, the increase in photosynthesis several days later varied from 50 to 90% of the predrought values (Davies and Kozlowski, 1977). After five days at high soil moisture following a drought, photosynthesis of post oak and white oak seedlings had recovered to the predrought levels whereas it had not fully recovered in sugar maple and black walnut (Ni and Pallardy, 1992).

The harmful effects of drought on the photosynthetic process sometimes last for weeks to months. The failure of water-stressed plants to recover full photosynthetic capacity may be associated with failure of stomata to reopen fully as well as to injury to the photosynthetic apparatus (Kozlowski, 1982a). Of course, total photosynthesis of severely droughted plants also is lowered by a reduced leaf area (Parker and Pallardy, 1985; Pallardy and Loewenstein, 2004).

Recovery of photosynthesis of cabbage gum from a water deficit occurred in two separate stages. After rewatering of droughted plants no recovery of photosynthesis took place for 50 to 60 minutes (Kirschbaum, 1988). Thereafter, photosynthesis recovered rapidly to about half of the final recovery rate, with completion between 30 minutes and 3 hours after rewatering. Once recovery began it was related to concurrent increases in stomatal conductance. This stage of rapid recovery was followed by a constant or decreasing rate of photosynthesis for the remainder of the light period. A second stage of recovery occurred and was completed during the night following rewatering. The second stage probably was not associated with a change in leaf water status (recovery of leaf Ψ was largely completed within a few hours after rewatering). Photosynthetic recovery in the second stage probably required repair of photosynthetic components that were damaged during water stress. Protein synthesis was necessary for reversal of injury to the photosynthetic apparatus.

Humidity

Exposure to dry air is commonly followed by stomatal closure in many woody temperate zone plants (Grace et al., 1975; Turner et al., 1984; Meinzer et al., 1997) and tropical plants (Meinzer et al., 1984; Sena Gomes et al., 1987; Clough and Sim, 1989; Thomas et al., 2000) (see Chapter 12). For example, the stomata of cacao were more open, and the rate of photosynthesis was higher at high than at low relative humidity (Fig. 5.45). Increasing the relative humidity from 28 to 86% almost doubled the rate of photosynthesis of orange trees (Ono et al., 1978). Stomata of many gymnosperms also close as the leaf-to-air vapor pressure difference increases (Whitehead and Jarvis, 1981). The response of stomata to humidity change can occur within seconds (Fanjul and Jones, 1982). Such stomatal closure likely reflects responses of epidermal turgor to water vapor content of the air that arise from changes in transpiration rather than responses to changes in bulk leaf Ψ (Pallardy and Kozlowski, 1979a; Appleby and Davies, 1983; Sena Gomes et al., 1987; Schulze, 1986, 1993; Mott and Parkhurst, 1991; Monteith, 1995).

Exposure of plants to rain sometimes causes rapid suppression of photosynthesis by inducing stomatal closure and injury to the photosynthetic mechanism. When bean plants were exposed to misty rain the stomata closed completely within two minutes of continuous treatment and opened to half their original aperture within one hour. The rate of photosynthesis of these wetted leaves changed with stomatal aperture and decreased by 30 to 40% within one hour. The rate of photosynthesis did not recover to pretreatment levels even after three days of treatment. Analysis of relationships between photosynthesis and c_i indicated that misty rain caused nonstomatal and well as stomatal inhibition of photosynthesis. Dry weight increase of plants exposed to rain for seven days was only half

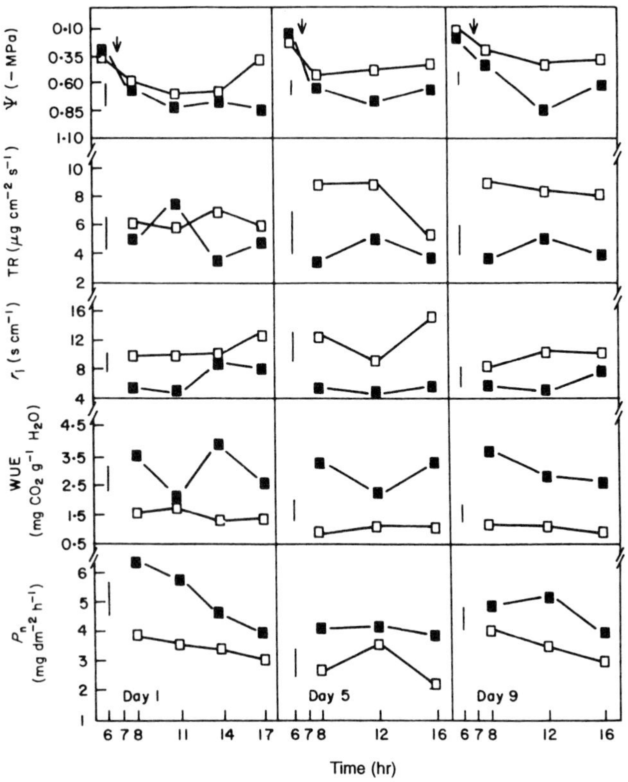

FIGURE 5.45. Effects of high (■) and low (□) relative humidity on leaf water potential (Ψ), rate of transpiration (TR), stomatal diffusive resistance (r_l), water-use efficiency (WUE), and rate of photosynthesis (P_n) of cacao seedlings. Bars indicate LSD at $p = 0.05$. From Sena Gomes et al. (1987).

that in control plants (Ishibashi and Terashima, 1995). However, it is obvious that persistent drought will cause far greater injury to plants than will wetting of leaves associated with rain.

Flooding

Soil inundation typically is followed by rapid reduction in the rate of photosynthesis (Kozlowski and Pallardy, 1984; Kozlowski et al., 1991). For example, the rate of photosynthesis of citrus (Phung and Knipling, 1976), apple (Childers and White, 1942), pecan (Loustalot, 1945), eastern cottonwood (Regehr et al., 1975), silver maple (Bazzaz and Peterson, 1984), sweetgum (Pezeshki and Chambers, 1985a), cherrybark oak (Pezeshki and Chambers, 1985b), blueberry (Davies and Flore, 1986), and Douglas-fir (Zaerr, 1983) declined drastically within a day or two after the soil was flooded.

Early reduction in the rate of photosynthesis of flooded plants is associated with stomatal closure, resulting in reduced CO_2 absorption by leaves (Pereira and Kozlowski, 1977a; Sena Gomes and Kozlowski, 1980, 1986; Kozlowski, 1982b; Pezeshki and Chambers, 1986). The stomata of cacao began to close within two hours after the soil was flooded (Sena Gomes and Kozlowski, 1986). Using gas exchange and chlorophyll fluorescence techniques, Mielke et al. (2003) concluded that reductions in photosynthesis during a 63-day period of flooding of seedlings of *Genipa americana*, a tropical fruit-tree species, were mostly attributable to stomatal closure and some reduction in leaf area. Li et al. (2004) observed similar stomatal limitation of photosynthesis in rooted *Salix nigra* plants subjected to continuous flooding for eight weeks.

After the flood water drains away following a short period of flooding, the stomata often reopen slowly and the rate of photosynthesis increases accordingly. For example, the stomata of rabbiteye blueberry plants closed within a few days after flooding, but when flooding was discontinued after 18 days the stomata reopened (Davies and Flore, 1986). Similarly the closed stomata of flooded mango trees reopened after the flood water drained away (Larson et al., 1989). The closed stomata of flooded pecan trees also reopened when flooding for eight days was discontinued, but not when flooding was ended after 15 days (Smith and Ager, 1988).

The photosynthetic response during prolonged flooding varies with species differences in flood tolerance. Flood-sensitive species apparently lack a mechanism to reopen stomata that close under soil hypoxia. For example, the stomata of paper birch remained closed during two weeks of flooding (Tang and Kozlowski, 1982). By comparison, in the flood tolerant baldcypress, flooding induced early stomatal closure and a decrease in photosynthesis but the stomata reopened and photosynthesis recovered to 92% of the initial rate within 14 days of flooding (Pezeshki, 1993). Pereira and Kozlowski (1977a) also noted that early stomatal closure of black willow was followed by stomatal reopening during prolonged flooding.

In the longer term, photosynthesis of flooded plants also is inhibited by adverse effects on photosynthetic capacity (Bradford, 1983a,b), which may be associated with changes in carboxylation enzymes, reduced chlorophyll content of leaves, and a reduced leaf area (the result of inhibition of leaf formation and expansion, injury, and abscission) (Kozlowski, 1982b; Mielke et al., 2003).

Mineral Nutrition

Deficiencies of essential macronutrients and micronutrients, as well as nutrient imbalances, may lower the rate of photosynthesis. However, the concentration of each element can vary over a fairly wide range in leaves without significantly altering the rate of photosynthesis. Although chlorosis and necrosis of leaf tissues may accompany the decreased photosynthetic capacity of mineral-deficient leaves, photosynthesis commonly is reduced even when such visible symptoms are not evident.

The effects of mineral nutrients on photosynthesis are complex and may be both direct and indirect. In mineral-deficient leaves the rate of net photosynthesis may be reduced by depressed chlorophyll synthesis, decreased capacity for photosynthetic electron transport, lowered activity of carboxylating and other enzymes, decreased stomatal conductance, and increased respiration. In the long term, total photosynthesis of mineral-deficient plants is greatly reduced by a decrease in leaf area.

Increased rates of photosynthesis often, but not always, follow additions of fertilizer to woody plants. The response to fertilizer varies with tree vigor; species; tree age; amount, timing, and composition of the fertilizer; stand density; soil moisture content; soil fertility; temperature; and light conditions (Linder and Rook, 1984).

Macronutrients

Nitrogen deficiency occurs commonly and usually decreases photosynthesis more than deficiencies of other macronutrients (Chapter 9). The most important long-term effect of N deficiency on photosynthesis is a decrease in leaf growth, resulting in reduced total production of photosynthate (Brix, 1983). Application of N fertilizer to young N-deficient loblolly pines was followed by a 50% increase in leaf area (Vose, 1988; Vose and Allen, 1988). Similarly, increased N availability to both well-watered and droughted American elm seedlings resulted in greater leaf area, together with increased production of photosynthate (Walters and Reich, 1989). The increased production of carbohydrates following application of N fertilizer may reflect an increase in the number, size, and longevity of leaves as well as a longer time during which they remain photosynthetically active (Linder and Rook, 1984). Addition of N fertilizer also may increase the amount of palisade tissue, thus increasing the potential for high rates of photosynthesis per unit of leaf area (Kozlowski and Keller, 1966; Kozlowski, 1992).

A relationship between photosynthetic capacity and N content of foliage may be expected because the soluble proteins of the Calvin cycle and thylakoid membranes contain most of the leaf N (Evans, 1989) (Chapter 9). Correlations between the rate of photosynthesis and leaf N content have been demonstrated in a wide variety of woody plants. Examples are peach and other stone fruit species (De Jong, 1982, 1983), several deciduous tree species of forests in eastern North America (Abrams and Mostoller, 1995), and conifer species (Tan and Hogan, 1995). Photosynthesis on a dry weight basis of a number of *Eucalyptus* species from several sites also was highly correlated with leaf N content (Fig. 5.46). This relationship extends in a robust fashion across regional and global samples that include deciduous and evergreen species (e.g., Reich et al., 1999; Wright et al., 2004) and community-level pooled estimates of photosynthetic capacity and mass-based N (Fig. 5.47).

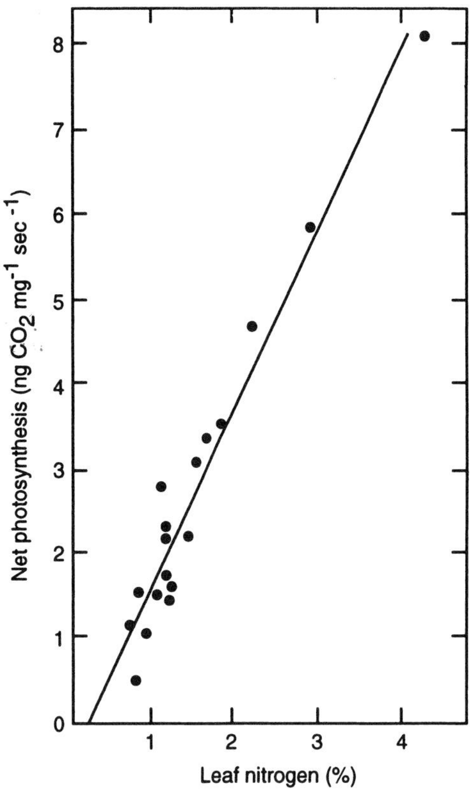

FIGURE 5.46. Relation between N content of *Eucalyptus* leaves and the rate of net photosynthesis. From *Oecologia*, Photosynthetic capacity and carbon allocation patterns in diverse growth forms of *Eucalyptus*, Mooney, H. A., Ferrar, P. J., and Slatyer, R. O., **36**, 103–111, Figure 3. © 1978 Springer-Verlag.

Photosynthesis is reduced by N deficiency through its effects on chlorophyll synthesis, level and activity of carboxylating and other photosynthetic enzymes, and stomatal conductance to CO_2 transfer (Natr, 1975). Of the nonstomatal nitrogenous limits, those imposed by Rubisco activity are best documented. Reduced carboxylation efficiency, and presumably lower Rubisco activity, appeared to be the primary limitations to photosynthesis in N-deficient jack pine seedlings (Tan and Hogan, 1995). Mitchell and Hinckley (1993) found a

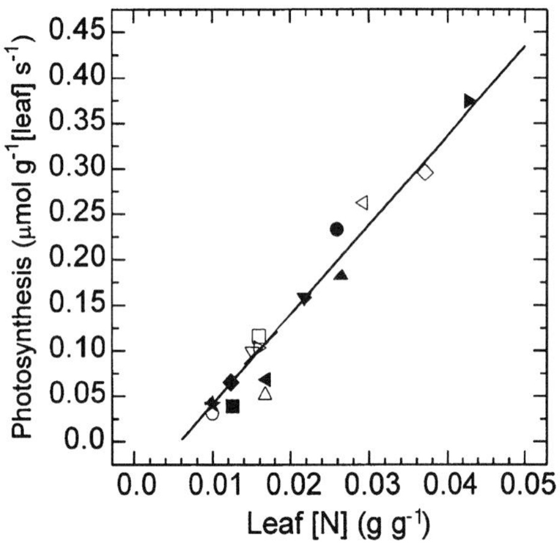

FIGURE 5.47. Relationship between mass-based net photosynthesis and leaf N concentration for pooled species of a variety of plant community types. Points represent the mean for each community or vegetation type. Line represents points fitted by regression. Communities represented span temperate and tropical regions, early to late succession, herbaceous species, deciduous trees, and pines. After Peterson et al. (1999) "The photosynthesis leaf nitrogen relationship at ambient and elevated atmospheric carbon dioxide: A meta-analysis." *Global Change Biol.* **5**, 331–346. Blackwell Publishing.

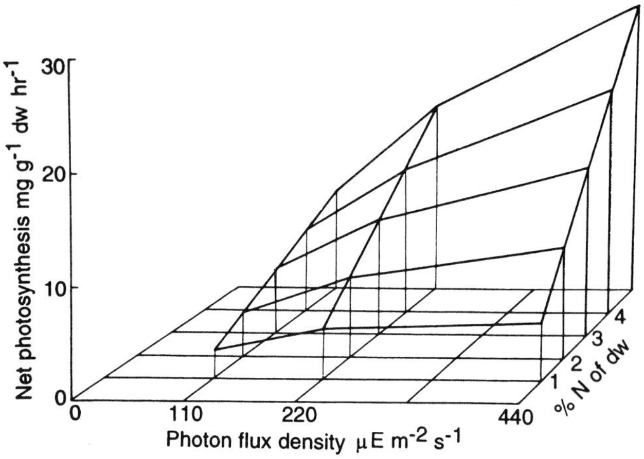

FIGURE 5.48. Effects of N concentration of leaves and irradiance at which plants were grown on net photosynthesis of silver birch seedlings. Measurements were made at the irradiance at which the plants were grown. From Linder et al. (1981).

strong positive correlation between foliar N concentration and photosynthesis in Douglas-fir. High N shoots had greater mesophyll conductance than low-N shoots. Increased photosynthesis also was reported by von Caemmerer and Farquhar (1981) and attributed to increased allocation of foliar N to Rubisco in high-N plants.

The rate of photosynthesis often increases after N fertilizers are applied (Linder and Troeng, 1980; Brix, 1983; Kozlowski et al., 1991). Photosynthetic responses sometimes vary with the form in which N is supplied. For example, photosynthesis of Scotch pine seedlings was increased more by N fertilizers supplied as ammonium chloride than by nitrate or ammonium nitrate (Lotocki and Zelawski, 1973; Zajaczkowska, 1973).

The effect of N on photosynthesis is modified by environmental conditions. In silver birch seedlings grown with suboptimal N supply and three levels of light, the rate of photosynthesis (leaf dry weight basis) was linearly related to leaf N levels (Fig. 5.48). When photosynthesis was measured at the light intensities at which the plants were grown, the rates at the same concentration of leaf N differed appreciably. However, when photosynthesis was measured at light saturation, the rates were similar because of changes in leaf morphology with light and nutrient status (Linder et al., 1981). Roberntz (2001) found that photosynthetic capacity of Norway spruce leaves under both ambient and elevated CO_2 concentrations showed essentially no relationship to N concentration at 8°C, but an

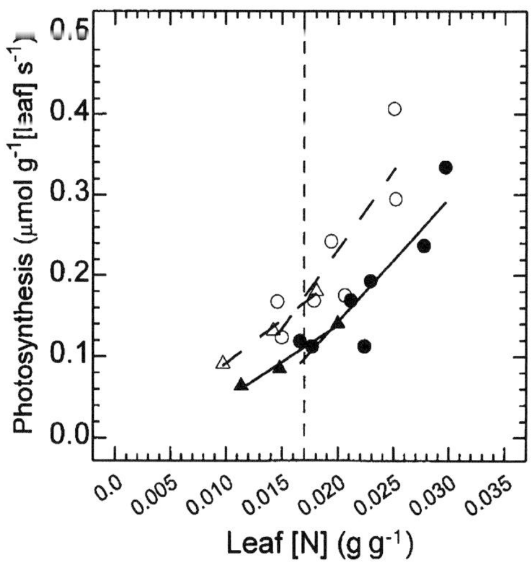

FIGURE 5.49. The effect of elevated CO_2 (650 or 700 ppm) on the relationship between net photosynthesis and leaf N concentration for (▲) pines and (●) deciduous trees on a mass basis. Open symbols = elevated CO_2; closed symbols = ambient CO_2. "From Peterson et al. (1999) "The photosynthesis leaf nitrogen relationship at ambient and elevated atmospheric carbon dioxide: A meta-analysis." *Global Change Biol.* **5**, 331–346. Blackwell Publishing.

increasingly strong positive response as leaf temperatures increased to 24°C. There also is an enhancement of photosynthesis rate at a given N content by elevated CO_2 (Peterson et al., 1999) (Fig. 5.49).

Water stress often greatly reduces or eliminates the stimulatory effect of N on photosynthesis (Jose et al., 2003; Runion et al., 1999). Water stress can influence the relation between leaf N and photosynthesis by inhibiting expression of photosynthetic capacity through decreased stomatal or mesophyll conductance to CO_2 fixation, or both. For example, increased water stress in American elm seedlings resulted in a greater relative decline in net photosynthesis and leaf conductance in high-N than in low-N plants (Walters and Reich, 1989). In well-watered American elm seedlings, biochemical rather than stomatal limitations appeared to account for reduction in photosynthesis with decreasing leaf nitrogen. In contrast, in water-stressed plants, stomatal conductance was more important in reducing photosynthesis (Reich et al., 1989).

In addition to N deficiency, low levels of other macronutrients often inhibit photosynthesis. In eastern white and Monterey pines, photosynthetic rate was highly correlated with total foliar P (Reich and Schoettle, 1988; Sheriff, 1995). In peach trees, CO_2 assimilation, mesophyll conductance, and leaf conductance to water vapor were linearly related to leaf P (De Jong, 1982). Phosphorus fertilization increased leaf area of *Liquidambar styraciflua* seedlings, especially at high N fertilization rates (Chang, 2003), thus increasing the capacity for whole-plant photosynthesis. A shortage of K, like that of P deficiency, may impede photosynthetic energy transfer and increase respiration rate (Pirson, 1958), thus lowering the rate of net photosynthesis. Leaf potassium concentrations less than 0.5–0.6% appeared to limit leaf CO_2 exchange rate in field-grown almond trees (Basile et al., 2003). Stomatal conductance of almond leaves was unaffected by leaf K content, and thus K limitation appeared to be associated with mesophyll components of photosynthesis. On the other hand, the level of K also may affect photosynthesis through regulation of stomatal aperture. Stomatal opening, associated with high K availability, resulted in increased gas exchange of silver maple seedlings (Noland and Kozlowski, 1979).

Micronutrients

Several micronutrients have direct or indirect effects on photosynthesis, but the amounts needed vary over a considerable range for each element.

Iron influences photosynthesis because it is necessary for chlorophyll synthesis and affects enzymatic activity. Iron occurs in ferredoxin and the cytochromes, which are essential components of electron transport systems in both photosynthesis and respiration. Deficiencies of Fe are a common cause of chlorosis and low rates of photosynthesis (Keller and Koch, 1962, 1964). Chlorophyll contents of leaves of fertilized poplar trees were highly correlated with Fe content. The rate of photosynthesis was consistently higher in poplar plants fertilized with Fe chelate than in unfertilized plants over a range of light intensities (Table 5.3).

Because Mn is an essential cofactor for release of oxygen during photosynthesis and may also serve as an activator of enzyme systems, it can influence photosynthesis if present in too small amounts. A deficiency of Mn in Norway spruce depressed photosynthesis, but the rate increased after application of the deficient element (Kreutzer, 1972). Manganese deficiency inhibited CO_2 uptake by tung leaves more by decreasing leaf area than by reducing photosynthetic efficiency per unit of leaf area. Partially chlorotic leaves were not much less efficient than leaves that had recovered a normal green color after an application of manganese sulphate (Reuther and Burrows, 1942). Severe deficiencies of Cu and Zn lowered the rate of photosynthesis of tung leaves by 30 to 55%, respectively, often without inducing chlorosis or necrosis (Loustalot et al., 1945; Gilbert et al., 1946).

TABLE 5.3. Effect on Photosynthesis of Fertilizing Hybrid Poplar Plants with Iron Chelate[a,b]

Light intensity (lux)	Photosynthesis ($mg\ CO_2\ dm^{-2}\ hr^{-1}$)		Increase (%)
	Control	Fertilized	
500	0.96	1.26	31
2,500	1.98	2.94	48
5,000	3.60	5.03	40
10,000	5.86	9.34	59
20,000	7.92	13.57	71
40,000	10.09	14.72	46

[a]After Keller and Koch (1964); from Kramer and Kozlowski (1979).

[b]The potted plants were watered in late June with 500 ml of 0.2% solution of iron chelate. Photosynthesis was measured in late August and early September.

Salinity

Deicing salts and salt spray along sea coasts adversely influence photosynthesis (Gibbs and Burdekin, 1980; Kozlowski, 1986a). Progressive decreases in rates of photosynthesis with increasing salinity have been demonstrated for many species of plants including green ash (Table 5.4), littleleaf linden, sycamore maple, Scotch pine (Cornelius, 1980), citrus (Walker et al., 1982; Lloyd et al., 1987, 1990), grape (Downton, 1977; Walker et al., 1981), poplars (Fung et al., 1998), and ponderosa pine (Bedunah and Trlica, 1979).

Salinity influences photosynthesis both directly and indirectly, with the mechanism of short-term effects often different from that of long-term effects. Most photosynthetic reduction by salinity has been attributed to nonstomatal effects (Long and Baker, 1986; Pezeshki et al., 1987; Ziska et al., 1990). When seedlings of green ash were irrigated with low-salt solutions, increasing leaf dehydration caused partial stomatal

TABLE 5.4. Effect of Salt Applications on Stomatal Conductance (g_w), Transpiration (T_r), Net Photosynethesis (P_n), and Plant Water Potential (Ψ) of Green Ash[a]

Variable	Treatment[b] I	II	III	IV	V
g_w (cm s^{-1})	0.33 ± 0.02	0.21 ± 0.02	0.16 ± 0.01	0.11 ± 0.01	0.08 ± 0.02
T_r (μg H_2O cm^{-2} s^{-1})	3.30 ± 0.2	2.20 ± 0.3	1.80 ± 0.1	1.20 ± 0.2	0.80 ± 0.2
P_n (mg CO_2 dm^{-2} hr^{-1})	8.30 ± 1.01	4.47 ± 0.78	3.61 ± 0.11	2.62 ± 0.62	0.96 ± 0.23
Ψ (MPa), midday	−0.67 ± 0.04	−0.81 ± 0.04	−0.86 ± 0.04	−0.52 ± 0.03	−0.43 ± 0.03

[a]From Pezeshki and Chambers (1986).
[b]Total amounts of NaCl added to each pot were as follows: I, 0.0 g; II, 7.3 g; III, 14.6 g; IV 36.5 g; V, 73.0 g.

closure and reduced CO_2 absorption. When salinity levels were higher, ion toxicity, membrane disruption, and complete stomatal closure were dominant factors in reducing photosynthesis (Pezeshki and Chambers, 1986). For a few days after treatment, NaCl decreased stomatal conductance in fig but photosynthesis was influenced only slightly (Colombck and Ludders, 1993). This short-term response to salinity contrasted with the marked inhibition of photosynthesis by longer-term salinity treatment. In the long term, salinity effects are particularly complex because they involve direct effects on photosynthetic functioning as well as developmental modifications of the photosynthetic apparatus. Because salinity suppresses both leaf initiation and expansion, it reduces the amount of photosynthetic surface produced (Long and Baker, 1986; Chapter 5, Kozlowski and Pallardy, 1997).

Pollution

Much attention has been given to the effects of air pollutants on photosynthesis because the rate of photosynthesis often is correlated with biomass production of woody plants. A major impact of certain pollutants, such as O_3, at ambient concentrations appears to be exerted on photosynthesis (Reich, 1987).

Reduction in the rate of photosynthesis would be expected when leaves are injured or shed by exposure to environmental pollutants, but photosynthesis often is inhibited long before visible injury or growth reduction occurs. The amount of reduction in photosynthesis varies with specific pollutants and dosage (Table 5.5) and with species, clones, cultivars, and environmental conditions (Kozlowski and Constantinidou, 1986b).

Most studies show that SO_2 at high dosages rapidly and substantially reduces the rate of photosynthesis, but the response varies for different species and genotypes (Keller, 1983). For example, the amount of reduction of photosynthesis by SO_2 varied in the following order: sycamore maple > English oak, horse-chestnut, European ash > European white birch (Piskornik, 1969). Sulfur dioxide at five pphm for two hours decreased photosynthesis of sensitive eastern white pine clones by 27% and that of tolerant clones by 10% (Eckert and Houston, 1980). Variations in photosynthetic responses to SO_2 also were shown among Scotch pine clones and provenances (Lorenc-Plucinska, 1982; Oleksyn and Bialobok, 1986). The superior resistance to SO_2 injury of *Pinus pinaster* compared with *P. radiata* was associated with greater activity of detoxifying enzymes (superoxide dismutase, ascorbate peroxidases, guaiacol peroxidases) in the apoplastic fluid and cell walls of leaves (Durán-Carril and Buján, 1998).

Many examples are available of reduction of photosynthesis by relatively low levels of ozone (O_3) (Kozlowski and Constantinidou, 1986a). The rates of photosynthesis of plum, apricot, almond, prune, apple, and pear trees were lower in air with twice ambient O_3 partial pressures than in charcoal-filtered air; photosynthesis of peach, nectarine, and cherry was not affected (Retzlaff et al., 1991). Reich (1983) emphasized that chronic exposure of hybrid poplar leaves to low levels of O_3 decreased photosynthesis and chlorophyll contents while increasing dark respiration. During the first seven days, photosynthesis of hybrid poplar leaves chronically exposed to 0.125 $\mu l\ l^{-1}$ O_3 differed only slightly from that of control leaves. However, once the leaves were fully expanded, the rate of photosynthesis was greatly reduced (Fig. 5.50).

Accelerated leaf aging by O_3 was partly responsible for decreased photosynthetic capacity. The control leaves lived about 10 to 15% longer than those exposed to O_3. Reich and Amundson (1985) reported that exposure to O_3 concentrations representative of those found in clean ambient air and in mildly to moderately polluted ambient air appreciably reduced photosynthesis of eastern white pine, northern red oak, sugar maple, and poplar seedlings. The reductions were linear with

TABLE 5.5. Threshold Pollutant Doses for Suppression of Photosynthesis of Forest Tree Seedlings and Saplings[a]

Pollutant	Concentration	Time	Experiment Duration	Species	Reference
SO_2	660 pphm (15.7×10^3 µg m^{-3})	4–6 hr	Single treatment	Red maple	Roberts et al. (1971)
	100 pphm (2620 µg m^{-3})	2–4 hr	Single treatment	Quaking aspen, white ash	Jensen and Kozlowski (1974)
	10 pphm (262 µg m^{-3})	Continuous	2 weeks	White fir	Keller (1977b)
	20 pphm (524 µg m^{-3})	Continuous	2 weeks	Norway spruce, Scotch pine	Keller (1977b)
O_3	30 pphm (588 µg m^{-3})	9 hr day^{-1}	10 days	Ponderosa pine	Miller et al. (1969)
	15 pphm (294 µg m^{-3})	Continuous	19 days	Eastern white pine	Barnes (1972)
	15 pphm (294 µg m^{-3})	Continuous	84 days	Slash pine, pond pine, loblolly pine	Barnes (1972)
F	30 µg g^{-1} dry weight basis foliar tissue			Pines (various)	Keller (1977a)
Pb	<10 µg g^{-1} dry weight basis foliar tissue			American sycamore	Carlson and Bazzaz (1977)
Cd	<10 µg g^{-1} dry weight basis foliar tissue			American sycamore	Carlson and Bazzaz (1977)
SO_2	100 pphm (2,620 µg m^{-3})	30 min	Single treatment	Silver maple (excised leaves)	Lamoreaux and Chaney (1978b)
	50 pphm (1,310 µg m^{-3})	7–11 hr	1–2 days	Black oak	
	50 pphm (1,310 µg m^{-3})	7–11 hr	1–2 days	Sugar maple	Carlson (1979)
	50 pphm (1,310 µg m^{-3})	7–11 hr	1–2 days	White ash	
O_3	50 pphm (980 µg m^{-3})	4 hr	Single treatment	Eastern white pine	Botkin et al. (1972)
	50 pphm (980 µg m^{-3})	7–11 hr	1–2 days	Black oak	Carlson (1979)
	50 pphm (980 µg m^{-3})	7–11 hr	1–2 days	Sugar maple	Carlson (1979)
Cd	~100 µg g^{-1} dry weight	45 hr	Single treatment	Silver maple (excised leaves)	Lamoreaux and Chaney (1978a)

[a]From Smith (1981) with permission of Springer-Verlag.

respect to O_3 concentrations, and no visible injury was detected. Photosynthesis of species with high stomatal conductances, and hence higher potential for O_3 uptake, was reduced more than in species with low stomatal conductances. Commonly, photosynthetic inhibition caused by low-to-moderate concentrations of O_3 is most apparent in older leaves (e.g., trembling aspen, Noormets et al., 2001), and there may even be some stimulation of photosynthesis in younger leaves exposed to O_3, presumably because of greater N availability as this nutrient is remobilized from O_3-induced accelerated leaf senescence (Dizengremel, 2001).

In conifers the effect of O_3 on photosynthesis often varies for needles of different age classes. In August and September the current-year needles of O_3-treated ponderosa pine plants had higher photosynthetic capacity than the older needles, a result of photosynthetic stimulation in plants that had shed older, O_3-injured needles (Beyers et al., 1992). As with angiosperms, this photosynthetic compensatory response in the current-year needles was correlated with their high N content, a response associated with accelerated loss of the older needles.

Immediately after exposure to NO_2 (0.5, 1.0, or 2.0 cm^3 m^{-3}), photosynthesis of seedling progenies of Scotch pine was reduced; the amount of reduction was greatest at the highest dosage (Lorenc-Plucinska, 1988). Inhibition of photosynthesis by F is common near smelting, aluminum, and fertilizer plants. Reduction of photosynthesis near an F-emitting aluminum smelter varied widely among species, with Scotch pine much more sensitive than Norway spruce or Douglas-fir (Keller, 1973b).

Particulates such as cement kiln dusts, some fluorides, soot, magnesium oxide, iron oxide, foundry dusts, and sulfuric acid aerosols often inhibit photosynthesis (Keller, 1973a; Auclair, 1976, 1977). Several heavy metals reduce photosynthesis by both stomatal and nonstomatal inhibition (Lamoreaux and Chaney, 1977, 1978a,b).

Combined Pollutants

Air pollutants rarely exist singly. Rather, the environment contains a complex mixture of gaseous and particulate air pollutants, and it often is difficult to

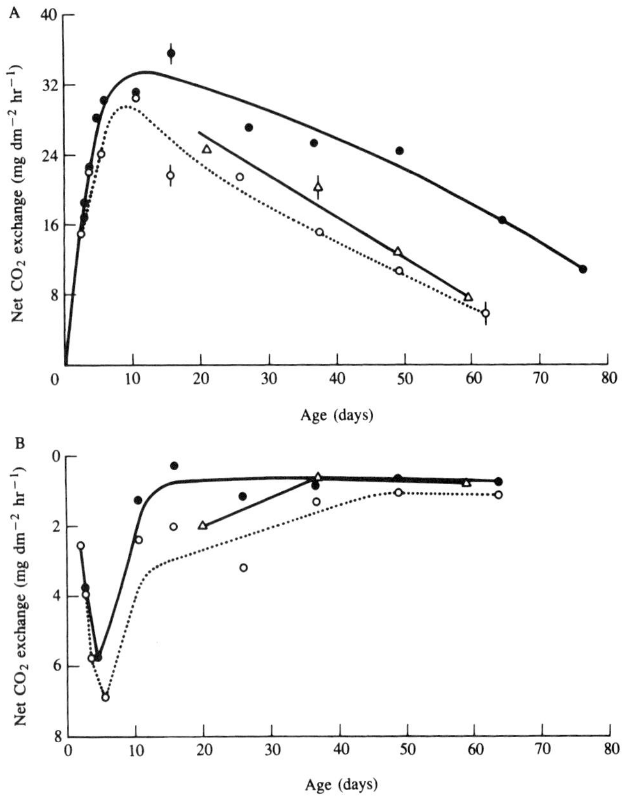

FIGURE 5.50. Net photosynthesis (A) and dark respiration (B) of variously aged poplar leaves (●, control) and leaves chronically exposed to (Δ) 0.085 or 0.125 μl l^{-1} (○) ozone. After Reich (1983); from Kozlowski et al. (1991).

identify those involved in reducing photosynthesis. Often, tolerable levels of a pollutant reduce photosynthesis when present with another pollutant at the same concentration. The effects of mixtures of gaseous pollutants vary with the concentrations of each gas in the mixture, the relative proportions of the gases, whether the combined pollutant stress is applied simultaneously or intermittently, and the age and physiological condition of the exposed plant tissues (Reinert et al., 1975).

Most studies of effects of pollutant mixtures have been conducted with SO_2 and O_3. For example, reduction of photosynthesis by SO_2 plus O_3 was reported for white ash (Carlson, 1979) and silver maple (Jensen, 1983). Photosynthesis also was reduced by SO_2 plus Cd (Lamoreaux and Chaney, 1978b) and SO_2 plus F (Keller, 1980). Several other examples are given by Smith (1990).

Mechanisms of Photosynthetic Inhibition

Environmental pollutants may alter the rate of photosynthesis by several mechanisms:

- Clogging of stomatal pores
- Altering optical properties of leaves by changing reflectance, and by decreasing light intensity that reaches the leaf interior
- Altering the heat balance of leaves
- Inhibiting the photosynthetic process through breakdown of chlorophyll as well as changing activity of carbon fixing enzymes, phosphorylation rate, and pH buffering capacity
- Disrupting the integrity of membranes and ultrastructure of organelles
- Inducing changes in leaf anatomy

Ozone exposure has multiple effects on photosynthetic processes. Reduction in the rate of photosynthesis by O_3 has been attributed to lowered maximum photosynthetic capacity as well as reduced carboxylation efficiency and quantum yield (Matyssek et al., 1995). Rubisco activity and quantity are both reduced in older, more sensitive leaves compared to leaves of similar age not exposed (Rebbeck and Loats, 1997; Dizengremel, 2001). Rubisco activase also declines, as well as chlorophyll and Calvin cycle enzymes. $NADPH_2$ is diverted to the ascorbate peroxidases-glutathione pathway that is associated with detoxification of O_3.

In the long term, total photosynthesis per plant is lowered by decreased leaf formation and expansion, and by necrosis, and abscission.

Applied Chemicals

A number of applied chemicals may adversely affect photosynthesis, especially when used at higher than recommended dosages (Ayers and Barden, 1975; Kozlowski and Constantinidou, 1986a). Such chemicals include insecticides, fungicides (Kramer and Kozlowski, 1979), herbicides (Sasaki and Kozlowski, 1967; Kramer and Kozlowski, 1979), antitranspirants (Davies and Kozlowski, 1974a, 1975a,b; Olofinboba et al., 1974), and salts used in deicing of roads (Kozlowski, 1986a). The effects vary with specific chemicals and dosage, species and genotype (Akinyemiju and Dickmann, 1982), age of plants (Kozlowski, 1976b), and method of application. Photosynthetic reduction by chemicals often is associated with injury to leaves (Kozlowski and Clausen, 1966a; Kozlowski et al., 1991).

PLANT FACTORS

The leaf factors that may regulate photosynthesis include

- Capacity of enzymatic steps of carbon fixation and metabolism

- Capacity for photosynthetic electron transport and phosphorylation
- Conductance to CO_2 diffusion from outside the plant to the chloroplasts
- Leaf age

The first two factors may influence photosynthesis both at saturating and normal CO_2 partial pressures; the third factor can affect photosynthesis only when CO_2 partial pressures are limiting. Changes in any of these factors commonly are associated with changes in leaf structure (Björkman, 1981).

The reflective nature of leaf surfaces may influence the rate of photosynthesis, especially at low light intensities. The pubescent leaves of *Encelia* absorbed only 30% of the solar radiation, whereas glabrous leaves with the same chlorophyll content absorbed 84% (Ehleringer et al., 1976). At low light intensities, the quantum yield for photosynthesis was proportional to absorptance, but at saturating light intensity, the rate of photosynthesis of pubescent and glabrous leaves did not differ (Ehleringer and Björkman, 1978).

Stomatal Characteristics and Capacity of Photosynthetic Partial Processes

The resistance offered by stomata to CO_2 uptake by leaves often provides a major limitation for photosynthesis. Variations in stomatal size, stomatal frequency (number per unit area), control of stomatal aperture, and stomatal occlusion by waxes influence stomatal conductance and the rate of photosynthesis (Siwecki and Kozlowski, 1973). When CO_2 partial pressure is insufficient to saturate photosynthesis, stomatal conductance influences photosynthesis through its effects on the CO_2 partial pressure in the intercellular spaces. In several gymnosperms the presence of a stomatal antechamber increases the length of the diffusion path for CO_2. Furthermore, considerable wax accumulates in the antechamber (Davies et al., 1974a), making the pathway more tortuous and decreasing the cross-sectional area available for diffusion. Jeffree et al. (1971) estimated that the wax in the antechamber of stomatal pores of Sitka spruce reduced the rate of photosynthesis by approximately half when the stomata were fully open.

As emphasized by Kramer and Boyer (1995), stomatal closure may or may not inhibit photosynthesis, depending on the need for CO_2 by photosynthetic metabolism. If the rate of metabolism remains high, the CO_2 partial pressures decrease within the leaf during stomatal closure and such decreases can inhibit photosynthesis. However, if the rate of metabolism decreases, the CO_2 requirement also decreases and the partial pressures of CO_2 may rise as the stomata close. When this occurs the photosynthetic inhibition is traceable to the lowered metabolism (CO_2 has become *more* available). Hence, the degree of stomatal inhibition of photosynthesis cannot always be determined solely from stomatal closure.

There are large variations in mesophyll capacity for carbon fixation and reduction. Wullschleger (1993) reported that both the maximum capacities for carbon fixation by Rubisco (V_{cmax}) and for electron transport (J_{max}) varied about 20-fold for 109 C_3 woody and herbaceous species from a variety of plant taxa (Table 5.6). There was substantial variation in both parameters within a particular group of plants. For example, some woody species exhibited photosynthetic attributes characteristic of herbaceous crop plants. Other investigators also have made the point that photosynthetic rates of trees may be comparable to those of annual crop plants (Nelson, 1984). However, these examples tended to represent the high extremes of woody plant response patterns and average representatives of annual crop species. Hence there also are distinctive differences in photosynthetic attributes among plant groups. Despite the fact there are no systematic differences in V_{cmax} and J_{max} between monocot and dicot C_3 plants, annuals have higher carboxylation and electron transport capacities than do perennials (Table 5.6), although these differences may be exaggerated somewhat because of assumptions the author made with respect to the concentrations of CO_2 at the chloroplast (Epron et al., 1995; Manter and Kerrigan, 2004). Similarly, deciduous angiosperm trees generally have higher values of V_{cmax} and J_{max} compared with conifers. There was a high correlation between V_{cmax} and J_{max} within species (Fig. 5.51), emphasizing that there apparently is close coordination in development of the constituent processes of photosynthesis during leaf growth (Wullschleger, 1993).

Maximum stomatal conductance varies widely among plant taxa and often is correlated with mesophyll capacity for photosynthesis (Nobel, 1991). However, mesophyll limitation of photosynthesis in unstressed plants substantially exceeds limitation by stomata in a wide variety of C_3 plants, including many woody angiosperm and gymnosperm species (Teskey et al., 1986; Briggs et al., 1986; Ni and Pallardy, 1992; Kubiske and Abrams, 1993; Stewart et al., 1995).

Photosynthetic capacity of leaves with an abnormally light green color often is lower than in leaves with a healthy, dark green color. Under controlled conditions the leaf chlorophyll content and CO_2 uptake often are highly correlated. In the field, however, the rate of photosynthesis may not vary much over a considerable range of leaf color, indicating that chloro-

TABLE 5.6. Estimates for Maximum Rate of Carboxylation (V_{cmax}) and Maximum Rate of Electron Transport (J_{max}) as Calculated for Several Broad Plant Categories[a]

	V_{cmax} (μmol m^{-2} s^{-1})		J_{max} (μmol m^{-2} s^{-1})	
Plant categories	**Mean[b]**	**Range**	**Mean[b]**	**Range**
Agricultural crops				
Dicots (n = 40)	90 ± 40	29–194	171 ± 57	87–329
Monocots (n = 12)	68 ± 21	35–108	157 ± 43	87–229
Horticultural crops				
Fruit trees (n = 6)	37 ± 23	11–69	82 ± 40	29–148
Vegetables (n = 17)	59 ± 29	15–97	137 ± 77	40–290
Temperate forests				
Hardwood (n = 19)	47 ± 33	11–119	104 ± 64	29–237
Conifers (n = 10)	25 ± 12	6–46	40 ± 32	17–121
Tropical forests (n = 22)	51 ± 31	9–126	107 ± 53	30–222
Understory herbs and forbs (n = 10)	66 ± 49	11–148	149 ± 92	31–269
Desert annuals and perennials (n = 3)	153 ± 54	91–186	306 ± 58	264–372
Sclerophyllous shrubs (n = 7)	53 ± 15	35–71	122 ± 31	94–167
Orthogonal contrasts ($p > F$)				
Dicots versus monocots	0.1460		0.4526	
Hardwoods versus conifers	0.0292		0.0115	
Annuals versus perennials	0.0001		0.0001	

[a]From Wullschleger, S. D. (1993). *J Exp Bot* **44**, 907–920, by permission of Oxford University Press.
[b]Mean ± 1 standard deviation.

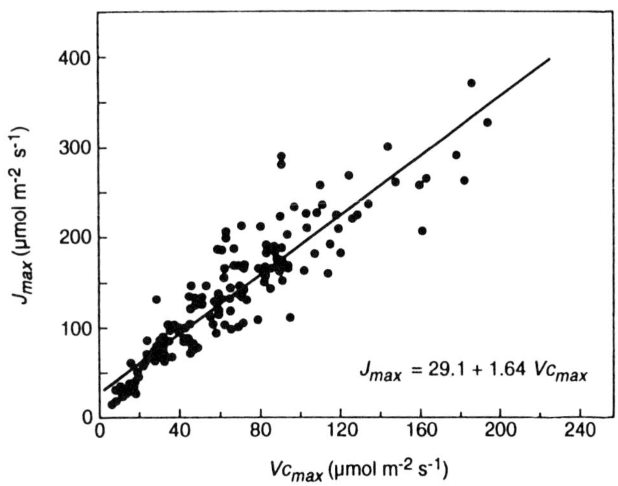

FIGURE 5.51. The correlation between the maximum carboxylation rate (V_{cmax}) and maximum rate of electron transport (J_{max}) estimated from the relationship between photosynthesis and internal CO_2 concentration for 109 C_3 plant species. From Wullschleger, S. D. (1993). *J Exp Bot* **44**, 907–920 by permission of Oxford University Press.

phyll content often is less important than other factors in controlling photosynthesis. The organization of the chlorophyll in terms of number and size of photosynthetic units may be as important as the amount (Alberte et al., 1976). Nevertheless, severe chlorosis from whatever cause invariably is correlated with reduced photosynthesis. The winter decline in photosynthesis often is associated with disorganization of chloroplasts and breakdown of chlorophyll (Kozlowski et al., 1991).

Source-Sink Relations

As mentioned previously, photosynthesis is influenced by the rate of translocation of photosynthetic products from sources to sinks (Kozlowski, 1992). Hence, a variety of cultural practices such as thinning of stands, pruning of branches and roots, fertilization, application of growth regulators, irrigation, and failure to protect plants against pests may be expected to influence photosynthesis directly or indirectly by affecting some type of sink activity (Flore and Lakso, 1989; Kozlowski and Pallardy, 1997).

Strong vegetative and reproductive sink strengths often have been associated with high photosynthetic rates. The rate of photosynthesis of second-flush leaves of northern red oak seedlings increased when the third-flush leaves began to expand and became strong carbohydrate sinks (Hanson et al., 1988). Photosynthesis also has been altered by changing the ratio of carbohydrate sinks to sources. If trees are partially defoliated, the rate of photosynthesis increases in the remaining leaves as they supply a proportionally larger carbohydrate sink. For example, the rate of photosynthesis of the residual leaves of partially defoliated poplar plants was higher than in corresponding leaves of intact plants. Photosynthesis was stimulated within 24 hours after defoliation and the increase was measurable for up to five weeks (Bassman and Dickmann, 1982). There also was a 30 to 60% increase in the rate of photosynthesis of the remaining leaves of partially defoliated young red maple and northern red oak trees. The photosynthetic stimulation was correlated with increased leaf conductance (Heichel and Turner, 1983).

Removal of 30% of the leaves of sour cherry reduced photosynthesis within one to three weeks (Layne and Flore, 1992). Removal of less than 30% of the leaf area was compensated by higher carboxylation efficiency and higher RuBP regeneration capacity. The threshold level of leaf area removal based on photosynthesis of individual leaves was 20%. Dry weight increase of whole plants was reduced at each of three levels of leaf area removal (10, 20, or 30%), but a disproportionally large decrease in dry weight increment occurred when the amount of leaf area removed increased from 20 to 30%.

The rate of photosynthesis sometimes is increased by root pruning, but the response may take considerable time. Root pruning may stimulate production of new roots, which become strong carbohydrate sinks (Kozlowski et al., 1991). Translocation of photosynthetic products to the roots of Monterey pine seedlings approximately tripled within a month after the roots were pruned (Rook, 1971).

The strong sink strength of growing reproductive structures often has been associated with high photosynthetic rates. The rate of photosynthesis of leaves of bearing apple trees was 45 to 60% higher than in leaves on trees without fruits (Avery, 1977). Tartachynk and Blanke (2004) compared photosynthetic characteristics of apple trees allowed to hold their fruit past the normal commercial harvest date with those harvested normally. Photosynthetic rates, as well as Rubisco activity, electron transport capacity, and chlorophyll content were higher in leaves of trees with retained fruit. Photosynthesis of stone fruits often increases appreciably during late stages of fruit expansion when the sink strength is greatest (Fujii and Kennedy, 1985). Increased rates of photosynthesis of bearing over nonbearing apple trees corresponded to a 30% increase in leaf conductance and only minor changes in mesophyll conductance or leaf photosynthetic capacity as indicated by leaf N content (De Jong, 1986).

The effect of sink strength of growing fruits on photosynthesis was not apparent in certain studies. Flore and Lakso (1989) attributed inconsistent effects of fruits on photosynthesis to differences in time of measurement, method of measurement, environmental stress, and location and position of sources and sinks. Hence, to establish the effects of fruits on photosynthesis, care should be taken that they not be masked by competing sinks (Herold, 1980).

Regulation of photosynthetic capacity by sinks is quite complex, involving numerous signaling molecules and pathways (Paul and Foyer, 2001). In the short term, as mentioned earlier, the dynamic balance of PS I and PS II capabilities is modulated by the oxidation-reduction state of plastoquinone. Highly reduced PQ promotes expression of genes coding for PS I proteins within minutes, whereas oxidized PQ stimulates the transcription of PS II-related genes (Paul and Foyer, 2001). Also as noted earlier, sucrose and starch metabolism can regulate photosynthetic rate via influence on P_i availability within the chloroplasts.

Leaf carbohydrate concentrations, plant hormones (especially cytokinins), and plant nitrogen status all appear to be involved in long-term regulation of photosynthetic capacity. One line of investigation implicates a direct feedback mechanism by which the rate of photosynthesis changes to meet the requirements of meristematic tissues for assimilates. The studies of Bassman and Dickmann (1982), Heichel and Turner (1983), and Hanson et al. (1988) tend to support such a mechanism. When sinks such as fruits are removed, soluble carbohydrates accumulate in leaves (e.g., apple; Tartachynk and Blanke, 2004), setting into motion downregulation of photosynthetic genes by sugar signaling (Sheen, 1994). Similarly, preventing rapid withdrawal of photosynthate from leaves is followed by a decrease in the rate of photosynthesis. For example, photosynthesis was reduced by approximately 30% following girdling of grapevines (Harrell and Williams, 1987).

There also is evidence that hormonal signals alter the rate of photosynthesis as growth rates change. Various hormones possess the necessary properties of a messenger (Chapter 13). Their synthesis often is correlated with changes in sink activity; they are translocated over long distances at rates commensurate with the period between a change in sink activity and photosynthetic response; and specific hormonal effects on photosynthesis have been demonstrated (Herold, 1980). During development of cacao leaves, concentrations of auxins and cytokinins were high. The concentrations declined after leaf maturity, but cytokinins increased again just before renewal of elongation of apical buds (Orchard et al., 1981). Cytokinins also increased before a second flush of shoot growth in English oak (Smith and Schwabe, 1980). At the molecular level cytokinins have been shown to increase tran-

TABLE 5.7. Physiological and Anatomical Features of Juvenile and Adult Leaves of English Ivy[a,b]

Parameter	Juvenile		Adult	
	L	LH	L	LH
Net photosynthesis (μmol CO_2 m^{-2} s^{-1})	4.2 ± 0.6	6.0 ± 0.9**	8.4 ± 0.3	10.2 ± 0.66**
Stomatal conductance (cm s^{-1})	0.297 ± 0.068	0.268 ± 0.077	0.349 ± 0.017	0.268 ± 0.016**
Intercellular CO_2 (μl $liter^{-1}$)	264 ± 3	244 ± 8**	238 ± 4	206 ± 8**
Quantum yield (mmol CO_2 mol^{-1} photons)	63 ± 9	59 ± 4	62 ± 5	61 ± 4
Saturating light (μmol photons m^{-2} s^{-1})	100	200	200	400
Carboxylation efficiency (cm s^{-1})	0.047 ± 0.004	0.064 ± 0.005**	0.100 ± 0.003	0.139 ± 0.016**
RuBP carboxylase activity (μmol CO_2 m^{-2} s^{-1})	6.7 ± 1.0	9.2 ± 0.6**	11.8 ± 1.1	15.9 ± 1.3**
Soluble protein (g m^{-2})	4.5 ± 1.2	5.4 ± 0.4	6.5 ± 0.8	15.1 ± 0.8**
Leaf thickness (μm)	198 ± 9	303 ± 35*	254 ± 50	373 ± 21**
Palisade parenchyma thickness (μm)	71 ± 5	143 ± 34*	92 ± 19	177 ± 21**
Spongy parenchyma thickness (μm)	95 ± 6	141 ± 5**	130 ± 31	173 ± 15*

[a]From Bauer and Thoni (1988).
[b]The leaves were allowed to expand fully in low light (L) and were then transferred to moderately high light (LH). Asterisks, * and **, denote values of LH significantly different from L at the 5 and 1% probability levels, respectively.

scription of genes for the small subunit of Rubisco and the chlorophyll a/b binding protein (Axelos et al., 1987; Ohya and Suzuki, 1991). Hence hormonal influences on photosynthetic rates are likely exerted both through effects on translocation of carbohydrates through influence on sink activity (Farnsworth, 2004) and on photosynthetic protein synthesis.

Nitrogen status of the plant also regulates photosynthetic capacity. Nitrogen deficiency promotes degradation of Rubisco, which is often synthesized abundantly under high N conditions, to allocate the N elsewhere in the plant. However, this response occurs only if leaf sugar levels are high (Paul and Foyer, 2001). Abundant N availability to roots stimulates cytokinin synthesis and translocation to shoots where this hormone may thus increase photosynthetic capacity as noted earlier. Hence both direct and indirect effects of N can be shown, and there is a growing recognition that source-sink control of photosynthetic capacity must be considered in terms of whole-plant carbon-nitrogen balance. Hormonal, nutrient, and sugar influences on photosynthesis at the leaf level are transduced through complex, intertwined signaling pathways.

Age of Leaves and of the Plant

The photosynthetic capacity of leaves varies greatly during their development. Differences occur in rates of photosynthesis of juvenile and adult leaves and in adult leaves of different ages. Photosynthesis on a leaf area basis in both low and high-light regimes was higher in adult leaves than in juvenile leaves of English ivy (Table 5.7). The higher photosynthetic capacity of the adult leaves was associated with a greater number of chloroplasts per cell, thicker leaves, a thicker palisade layer, higher carboxylation efficiency, and higher activity of Rubisco (Bauer and Thoni, 1988). In general, juvenile leaves possessed characteristics of shade leaves. The rate of photosynthesis typically is low in very young leaves, increases to a maximum as leaves expand (usually to near full size), and declines as leaves senesce (e.g., Wilson et al., 2000a). Photosynthesis of eastern cottonwood leaves was negative (CO_2 was released) when the leaves were very small, but the rate became positive by the time the leaves expanded to 1/20 of their full size. The rate then increased progressively until the leaves were fully expanded. Thereafter, the rate stabilized and finally decreased as the leaves senesced (Dickmann, 1971). Peak photosynthesis of grape leaves was reached when they became fully expanded and then declined gradually (Fig. 5.52). Photosynthesis of sour cherry leaves reached a maximum by the time the leaves were 80% expanded, remained relatively stable for two to four weeks, and then gradually declined (Sams and Flore, 1982).

In some fast-growing species such as poplars, which produce leaves during much of the growing season (Chapter 3), the photosynthetic capacity of individual leaves declines rapidly after the maximum rate is reached near the time of full leaf expansion (Dickmann et al., 1975; Reich, 1984b). By comparison, in species that produce their full complement of leaves during a

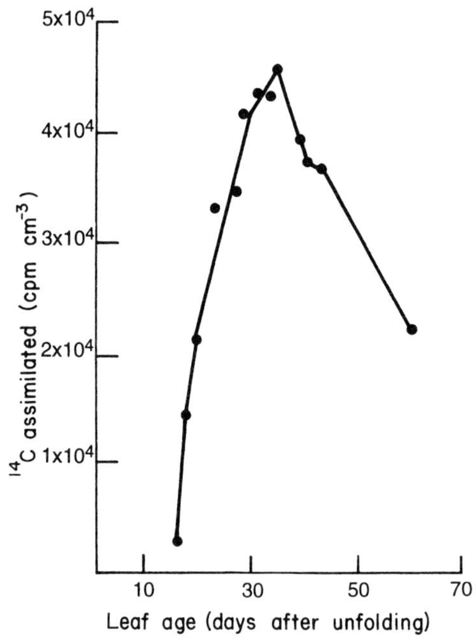

FIGURE 5.52. Effect of leaf age on photosynthesis ($^{14}CO_2$ assimilated) of grape leaves. From Kriedemann et al. (1970).

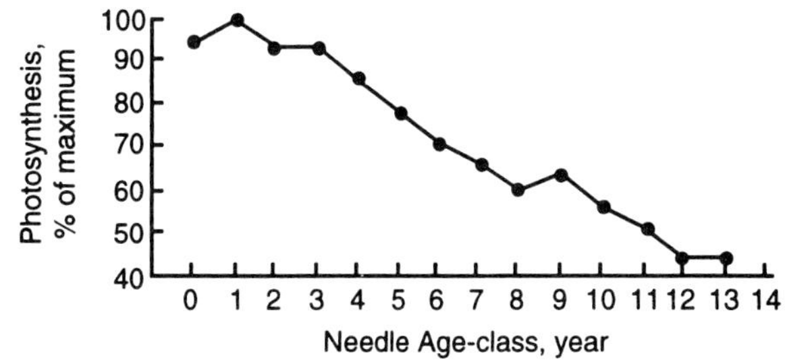

FIGURE 5.53. The photosynthetic capacity of different age classes of needles of black spruce expressed as a percentage of the maximum rate. From Hom and Oechel (1983).

single early-season flush of growth, individual leaves may maintain high photosynthetic capacity for a long time after they are fully expanded. For example, the rate of photosynthesis of sugar maple, red maple, and American beech leaves increased to a maximum by the time the leaves were fully grown in June. The rate then remained high until the middle of September and declined rapidly as the leaves senesced. The leaves of northern red oak, which were not shed as early as leaves of the maples, maintained high rates of photosynthesis into October (Jurik, 1986).

Photosynthetic rates of pine needles usually increase until near full size is attained and then decrease. In needles more than one year old, the rate decreases progressively each year. Net photosynthesis was highest in one-year-old needles of black spruce (Fig. 5.53). Fully expanded current-year needles had slightly lower rates. Needles up to four years old had high rates and rates of older needles declined progressively with age. In 13-year-old needles, the rate of photosynthesis was approximately 40% of the maximum rate (Hom and Oechel, 1983). Current-year needles of Pacific silver fir developed slowly. They began to expand by June 22, but were not fully grown until early August, and the highest rates of photosynthesis did not occur until September and October. The highest rates in one-year-old needles were recorded in July and coincided with the period of major expansion of current-year needles. Net photosynthesis decreased progressively with age of needles. However, the rate of seven-year-old needles was only 48% less than that of one-year-old needles, indicating that the old needles contributed appreciably to growth (Teskey et al., 1984a).

Variations in photosynthetic capacity of gymnosperm leaves of different ages have important implications in source-sink relations of growing tissues. For example, the rapidly expanding needles of red pine had high carbohydrate requirements and could not supply enough photosynthate for their own growth. Early in the season they obtained large amounts of photosynthate from the one-year-old needles, which had higher rates of photosynthesis and replaced the old needles as major exporters of carbohydrates (Dickmann and Kozlowski, 1968).

The pattern of decline in photosynthesis with leaf aging varies among species with different leaf longevities. For example, Martin et al. (1994) found that Costa Rican evergreen tree species had less reduction in photosynthetic rate and stomatal conductance between young and old leaves than did deciduous species. In general, in early successional species with short leaf life spans, the rate of photosynthesis decreases faster than it does in late successional species with longer-lived leaves. However, when the rate of photosynthesis is expressed as a proportion of the leaf life span, it does not vary greatly among species with different leaf life spans (Reich et al., 1995).

Changes in photosynthesis with leaf age are associated with various anatomical and physiological alterations. Increases during leaf expansion are related to development of internal leaf tissues and stomata, synthesis of chlorophyll, and increases in stomatal conductance, capacity for photosynthetic electron transport and phosphorylation, protein synthesis, and Rubisco activity; and an abrupt decrease in mitochondrial respiration. The largest increases in photosynthesis of developing leaves of apple coincided with the period of their greatest expansion and chlorophyll synthesis as well as increased stomatal conductance (Kennedy and Johnson, 1981). The gradual decline in photosynthesis after full leaf expansion has been correlated with decreases in stomatal conductance, and reduced photophosphorylation and amounts of Rubisco. For example, Rubisco activity as indicated by measurements of V_{cmax} declined over a period of about 100 days from maxima in mid-August in white and chestnut oak, sugar and red maple, and black gum trees growing in eastern Tennessee (Wilson et al., 2000b). Furthermore, relatively high levels of photorespiration develop and large decreases in mitochondrial respiration occur in old leaves (Dickmann et al., 1975).

Many tropical woody plants show delayed greening during leaf development. The young leaves may be white, red, blue, or very light green, and typically have very low amounts of chlorophyll and Rubisco. Development of the light harvesting system is not completed until well after the leaves are fully expanded. Hence, photosynthetic capacity is lower than in plants with normal green leaves. Three tropical woody species with delayed greening, *Ouratea lucens*, *Xylopia micrantha*, and *Connarus panamensis*, did not reach the photosynthetic compensation point under saturating light until the leaves were fully expanded. Thirty additional days were required for maturation of the photosynthetic apparatus. Kursar and Coley (1992a) estimated that these species synthesize 50 to 90% of their Rubisco after leaf expansion is completed. In addition, the chlorophyll binding and coupling factor proteins and Calvin cycle enzymes accumulate late in leaf development.

Delayed greening is more important on some sites than on others. Under the high light intensities of treefall gaps, the lost photosynthetic potential of species with delayed greening may be appreciable. In contrast, in the understory of a tropical forest, where the light intensity may be less than 1% of full sun, the photosynthetic advantage of young green leaves over nongreen leaves may be small (Kursar and Coley, 1992b).

Comparisons of photosynthesis among trees of the same species at different life stages sometimes, but not always, show declines in net photosynthetic rate (and stomatal conductance) with age (Bond, 2000). Various hypotheses have been advanced to explain these differences, including increased whole-plant xylem transport resistance that depresses leaf Ψ, reduced photosynthetic capacity that is inherently associated with aging, changes in leaf anatomy, greater sensitivity of stomata to VPD in older trees, and reduced N availability to older trees because of immobilization of available N in biomass. Comparisons among life stages often are confounded by differences in plant size, light, and nutrient regimes, thus complicating sound inferences from field studies. Careful comparisons of fast-growing *Eucalyptus saligna* plantations in Hawaii suggested that neither hydraulic limitation nor photosynthetic capacity or photosynthetic rates varied between one-year-old (7 m tall) and five-year-old (26 m tall) trees (Barnard and Ryan, 2003). More such carefully controlled studies utilizing a variety of species and ages will be needed to assess the occurrence and generality of age-related photosynthetic decline and its possible causes.

SUMMARY

Photosynthesis is the process by which light energy is used to synthesize reduced carbon compounds in green plants. It provides the primary substances and energy by which ecosystems are supported in addition to a host of products important to humans. Most photosynthesis occurs in foliage leaves, but there are exceptions in which other tissues, such as cotyledons, buds, stems, flowers, and fruits may conduct some photosynthesis.

In higher plants, photosynthesis occurs in chloroplasts, double membrane-bounded organelles that possess an additional, complex internal membrane system containing pigment and lipid molecules as well as proteins. The aqueous matrix of the chloroplasts is rich in proteins, especially the primary carboxylating enzyme ribulose bisphosphate carboxylase-oxygenase (Rubisco). In the process of photosynthesis light energy is captured by pigment molecules and transferred to reaction centers where photochemical reactions occur that drive electron flow through a series of carriers. Photochemistry of the photosynthetic apparatus causes splitting of water with the evolution of molecular oxygen and creates high energy intermediates that ultimately result in production of ATP and $NADPH_2$. These products are consumed when Rubisco catalyzes a reaction in which CO_2 is combined with ribulose bisphosphate and subsequently reduced to 3-carbon sugars. Sugars produced may be utilized within the cell, exported (as sucrose, primarily), or diverted to starch synthesis within the chloroplast itself.

Some variations in this C_3 photosynthetic process have been reported. The C_4 plants initially fix CO_2 into 4-carbon compounds that subsequently are transported to other cells, decarboxylated, and the CO_2 released is then fixed by Rubisco. This spatial separation of photosynthesis overcomes a number of inefficiencies of the process, especially the largely wasteful proclivity of Rubisco to utilize O_2 as well as CO_2 as a substrate. In another type of plant (CAM), initial fixation of carbon into a 4-carbon compound (malic acid) also occurs. However, in C_4 photosynthesis decarboxylation occurs after transport to another location, but in CAM plants it is temporally separated. In an effective adaptation to arid conditions, CAM plants absorb CO_2 through open stomata at night when evaporative demands are minimal. During the following day when stomata are closed, CO_2 is released internally and fixed by Rubisco. Only a few tree species possessing C_4 and CAM photosynthesis have been identified.

The rate of photosynthesis is influenced by interactions among hereditary, environmental, and plant factors. Variations in rates are traceable to restriction of CO_2 uptake by stomatal closure and to nonstomatal inhibition usually associated with increased liquid diffusion resistance, reduced chlorophyll content, decreased photosynthetic electron transport, low activity of photosynthetic enzymes, and changes in properties of membranes. The relative amounts of stomatal and nonstomatal inhibition of photosynthesis vary with time and with different environmental stresses. In the long term total photosynthesis is reduced by environmental stresses because of inhibition of leaf formation, expansion, and premature abscission.

Comparisons of photosynthetic rates as determined by different investigators often are difficult because of variations in measurement techniques and methods of expressing rates of photosynthesis. Photosynthetic rates (as CO_2 absorption of leaves) have been variously expressed (e.g., per unit of leaf fresh weight, dry weight, leaf area, stomata-bearing leaf surface area, leaf volume, chlorophyll content, and leaf N content). Differences in rates of photosynthesis between species and genotypes are related to variations in metabolism and/or leaf anatomy. Total photosynthesis also varies among species and genotypes because of differences in leaf production. Usually, rates of light-saturated photosynthesis are higher in broadleaved deciduous species than in evergreens when expressed on a leaf dry weight basis.

The rate of photosynthesis has been used as an index of growth potential of various species and genotypes. However, short-term measurements of photosynthetic capacity often are not reliable indices of growth potential because, in addition to photosynthetic rate, growth is determined by the seasonal pattern of photosynthesis, the relation of photosynthesis to respiration, partitioning of photosynthate within the tree, and the amount of foliage produced.

The rate of photosynthesis changes diurnally and seasonally. The rate generally is low early in the morning and increases to a maximum before or near noon. The maximum rate may be followed by a midday decrease and a subsequent increase before a late afternoon decline as the light intensity decreases. Seasonal changes in photosynthesis occur more gradually in conifers than in broadleaved deciduous trees.

The major environmental factors that regulate photosynthesis include light intensity, temperature, CO_2, drought, soil flooding, humidity, soil fertility, salinity, pollution, applied chemicals, and various interactions among them. Photosynthesis also is influenced by cultural practices such as thinning of stands, pruning of branches and roots, application of fertilizers, and irrigation.

In forests and orchards the amount of light available to the lower canopy is low because of shading by neighboring trees. The light intensity also decreases rapidly with increasing depth of tree crowns. In darkness there is no photosynthesis and CO_2 produced in respiration is released by leaves. With increasing light intensity a compensation point finally is reached at which absorption of CO_2 by leaves and its release in respiration are equal. The light compensation point varies with plant species, genotype, leaf type, leaf age, CO_2 concentration of the air, and temperature. As light intensity increases above the compensation point, the rate of photosynthesis increases linearly until light saturation occurs and the rate of photosynthesis either stabilizes or declines. At very high light intensities, excessive excitation of the photosynthetic reaction centers may occur, especially under environmental conditions that limit CO_2 fixation capacity (e.g., low temperature, drought). Inhibition of photosynthetic capacity, photoinhibition, often follows exposure to high light and may be reversible or persistent. Metabolic adjustment involving enhanced thermal dissipation through operation of the xanthophyll cycle pigments is a common acclimation response to excess light.

The rate of photosynthesis typically is much higher in leaves in the outer crown than in those in the inner crown. Also the leaves near the top of the canopy have higher rates of photosynthesis and become saturated at higher light intensities than those in the lower canopy. At low light intensities the rate of photosynthesis is higher in shade-tolerant than in shade-intolerant species. Much of the difference in shade tolerance of trees is related to variations in adaptation of the photosynthetic apparatus to shade.

Photosynthesis occurs over a temperature range from near freezing to more than 40°C. In most temperate zone species photosynthesis increases from near-freezing temperatures and attains a maximum between 15 and 25°C. In tropical species photosynthesis is detectable at temperatures several degrees above freezing and becomes maximal at temperatures greater than 25°C. The effect of air temperature on photosynthesis is modified by light intensity, CO_2, soil temperature, water supply, and environmental preconditioning of plants. Inhibition of net photosynthesis by very high temperature often occurs because respiration continues to increase above the critical temperature at which photosynthesis begins to decrease.

Photosynthesis of healthy, well-watered plants exposed to light is limited largely by the low CO_2 concentration of the air. Availability of CO_2 to leaf mesophyll cells is limited by resistances to diffusion, including boundary layer or air, cuticular, stomatal, and mesophyll air space diffusion resistances. Increase in CO_2 of the air above ambient values usually is accompanied by increases in the rate of photosynthesis and dry weight increment of plants.

The rate of photosynthesis is decreased by both drying and flooding of soil. When droughted plants are irrigated the rate of photosynthesis may or may not recover to predrought levels depending on plant species, severity and duration of the drought, and relative humidity of the air. The harmful effects of drought often are of long duration (weeks to months). Failure of droughted plants to recover photosynthetic capacity following irrigation often is associated with failure of stomata to reopen and injury to the photosynthetic apparatus.

Stomata of many plants are good humidity sensors. Exposure to dry air often is followed by stomatal closure and reduced CO_2 absorption by leaves. Under most conditions, the stomatal response to dry air appears most closely linked with changes in transpiration rate, and consequently epidermal turgor, and significant responses can be observed with little or no change in bulk leaf water potential.

Flooding of soil is followed by reduction in the rate of photosynthesis. Early reduction in photosynthesis is associated with stomatal closure; later reduction by adverse effects on the photosynthetic process and on reduced leaf area (the result of inhibition of leaf formation and expansion as well as leaf injury and abscission).

Deficiencies of both macro- and micronutrients, as well as nutrient imbalances, often decrease photosynthesis. Nitrogen deficiency decreases the rate more than deficiency of other macronutrients. Chlorosis and necrosis of leaves commonly are associated with reduced photosynthesis but the rate often is lowered even where visible symptoms of mineral deficiency do not occur.

Pollutants reduce the rate of photosynthesis, often before visible injury and growth reduction occur. The amount of photosynthetic inhibition varies with the type of pollutant or pollutants and dosage, species, clones, cultivars, and environmental conditions. Because the environment contains a mixture of variable amounts of gaseous and particulate pollutants, it often is difficult to quantify the effects of individual pollutants. A tolerable level of a given pollutant may lower the rate of photosynthesis when present with another pollutant at the same concentration. The effects of combined pollutants vary with the concentration of each in the mixture, their relative proportions, whether combined pollutants are applied simultaneously or intermittently, and the age and physiological condition of plant tissues.

At high dosages several applied agricultural chemicals may decrease the rate of photosynthesis. Such chemicals include some insecticides, fungicides, herbicides, and antitranspirants.

Several cultural practices (e.g., thinning of stands, pruning of branches and roots, and application of fertilizers) may influence photosynthesis directly or indirectly.

The rate of photosynthesis is influenced by several plant factors, including inherent differences in mesophyll CO_2 fixation capacity, leaf anatomy, leaf and sometimes tree age, stomatal size, stomatal frequency, and control of stomatal aperture.

General References

Ainsworth, E. A. and Long, S. P. (2005). What have we learned from 15 years of free-air CO_2 enrichment (FACE)? A meta-analytic review of the responses of photosynthesis, canopy properties and plant production to rising CO_2. *New Phytol* **165**, 351–371.

Baker, N. R. and Bowyer, J. R., eds. (1994). *Photoinhibition of Photosynthesis: From Molecular Mechanisms to the Field.* Bios Scientific, Oxford.

Blankenship, R. E. (2002). *Molecular Mechanisms of Photosynthesis.* Blackwell Science Ltd., Oxford.

Ceulemans, R. and Mousseau, M. (1994). Effects of elevated atmospheric CO_2 on woody plants. *New Phytol* **127**, 425–446.

Chazdon, R. L. and Pearcy, R. W. (1991). The importance of sunflecks for forest understory plants. *BioScience* **41**, 760–766.

Dawson, T. E., Mambelli, S., Plamboeck, A. H., Templer, P. H., and Tu, K. P. (2002). Stable isotopes in plant ecology. *Annu Rev Ecol Syst* **33**, 507–559.

Ehleringer, J. R. and Monson, R. K. (1993). Evolutionary and ecological aspects of photosynthetic pathway variation. *Annu Rev Ecol Syst* **24**, 411–439.

Flore, J. A. and Lakso, A. N. (1989). Environmental and physiological regulation of photosynthesis in fruit crops. *Hortic Rev* **11**, 111–157.

Geiger, D. R. and Servaites, J. C. (1994). Diurnal regulation of photosynthetic carbon metabolism in C_3 plants. *Annu Rev Plant Physiol Plant Mol Biol* **45**, 235–256.

Holbrook, N. M. and Lund, C. P. (1995). Photosynthesis in forest canopies. In *Forest Canopies*, M. D. Lowman and N. D. Nadkarni, eds., pp. 411–430. Academic Press, San Diego.

Jarvis, A. J., Mansfield, T. A., and Davies, W. J. (1999). Stomatal behaviour, photosynthesis and transpiration under rising CO_2. *Plant, Cell Environ* **22**, 639–648.

Ke, B. (2001). *Photosynthesis: Photobiochemstry and Photobiophysics. Advances in Photosynthesis*, Vol. 10. Kluwer Academic Publishers, Dordrecht.

Kozlowski, T. T. (1992). Carbohydrate sources and sinks in woody plants. *Bot Rev* **58**, 107–222.

Lawlor, D. W. (2001). *Photosynthesis, 3rd edition*. BIOS Sci. Publ., Oxford.

Leegood, R. C., Sharkey, T. D., and von Caemmerer, S., eds. (2000). *Photosynthesis: Physiology and Metabolism. Advances in Photosynthesis*, Vol. 9. Kluwer Academic Publishers, Dordrecht.

Long, S. P., Humphries, S., and Falkowski, P. G. (1994). Photoinhibition of photosynthesis in nature. *Annu Rev Plant Physiol Plant Mol Biol* **45**, 633–662.

Ort, D. R. (2001). When there is too much light. *Plant Physiol* **125**, 29–32.

Ort, D. R. and Yocum, C. F., eds. (1996). *Oxygenic Photosynthesis: The Light Reactions. Advances in Photosynthesis*, Vol. 4. Kluwer Academic Publishers, Dordrecht.

Parkhurst, D. F. (1994). Diffusion of CO_2 and other gases inside leaves. *New Phytol* **126**, 449–479.

Paul, M. J. and Foyer, C. H. (2001). Sink regulation of photosynthesis. *J Exp Bot* **52**, 1383–1400.

Pearcy, R. W. and Sims, D. A. (1994). Photosynthetic acclimation to changing light environments. In *Exploitation of Environmental Heterogeneity by Plants*, M. M. Caldwell and R. W. Pearcy, eds., pp. 145–174. Academic Press, San Diego.

Pearcy, R. W., Chazdon, R. L., Gross, L. J., and Mott, K. A. (1994). Photosynthetic utilization of sunflecks: A temporally patchy resource on a time scale of seconds to minutes. In *Exploitation of Environmental Heterogeneity by Plants*, M. M. Caldwell and R. W. Pearcy, eds., pp. 175–208. Academic Press, San Diego.

Portis, A. R. (2003). Rubisco activase—Rubisco's catalytic chaperone. *Photosyn Res* **75**, 11–27.

Sage, R. F. and Monson, R. K., eds. (1999). *C_4 Plant Biology*. Academic Press, San Diego.

Saxe, H., Ellsworth, D. S., and Heath, J. (1998). Tree and forest functioning in an enriched CO_2 atmosphere. *New Phytol* **139**, 395–436.

Schulze, E.-D. and Caldwell, M. M. (1993). *Ecophysiology of Photosynthesis*. Springer-Verlag, New York and Berlin.

C H A P T E R

6

Enzymes, Energetics, and Respiration

INTRODUCTION

Among the important processes occurring in living organisms are the release by respiration of the chemical energy stored in foods and its use to transform carbohydrates, fats, and proteins into new protoplasm and new tissue (assimilation). An understanding of how these complex processes occur requires at least an elementary knowledge of enzyme activity and energy transfer. First enzymes and energy transfer will be discussed in general terms and then an overview of the biochemistry of respiration will be presented. Respiration rates in whole trees and various organs and tissues and the factors that influence respiration rates will then be addressed.

ENZYMES AND ENERGETICS

Enzymes

One of the most important characteristics of living cells is the high rate at which chemical reactions occur within them at temperatures of 5 to 40°C. The same reactions occur very slowly if at all in the laboratory, at those temperatures. Even glucose must be heated to a high temperature to burn (oxidize) in air, but it is readily oxidized at 5 to 10°C in living cells. This is because most chemical reactions, even those that release energy, do not occur spontaneously but require addition of a certain amount of energy, the energy of activation, to begin.

Enzymes are organic catalysts that lower the energy of activation to a point where reactions can occur at ordinary temperatures. This usually is accomplished by momentarily binding substrate molecules on the surfaces of the enzyme molecules and thereby increasing the probability of a reaction. Most enzymes are very specific and catalyze only one reaction or one type of reaction. This is because enzymes are basically protein molecules with specific structural configurations that permit combination only with substrate molecules having a certain molecular structure. While the substrate molecules are temporarily bound on the surfaces of the enzyme there is a rearrangement of atoms and chemical bonds, resulting in production of different molecules, often with a different free energy. Many enzymes catalyze reversible reactions that can proceed in either direction, depending on the concentration of reactants, pH, and other factors.

There are thousands of different kinds of enzymes in living cells, many of them carefully compartmentalized in various organelles. For example, the enzymes involved in the Krebs cycle are found in mitochondria, enzymes involved in electron transport occur in both mitochondria and chloroplasts, and those involved in glycolysis and the pentose shunt occur principally in the cytoplasm. Some extracellular enzymes even occur on the external surfaces of cells or diffuse out into the surrounding medium.

The basic structure of an enzyme is a protein molecule. Enzymes vary because of differences in the sequence of amino acids in their proteins. Some enzymes such as urease and papain consist only of protein molecules, but many require a nonprotein constituent, often termed a cofactor or coenzyme, closely associated with or bound to the protein molecule. Cofactors that are integral parts of an enzyme, such as the copper in tyrosinase and ascorbic acid oxidase and the iron in catalase, are known as prosthetic groups. Many enzymes are active only in the presence of ions such as Mg^{2+}, Mn^{2+}, Ca^{2+}, and K^+. These are known as metal activators. The most important function of the micronutrient elements in plants and animals is as prosthetic groups or metal activators of enzymes. Some enzymes are active only in the presence of complex organic molecules, which if tightly bound are called prosthetic groups, but if loosely bound, are termed cofactors or coenzymes.

Several vitamins, especially those of the B complex, play important roles as enzyme cofactors. Pyridine nucleotides are of paramount importance as cofactors of enzymes involved in metabolic energy transfers. NAD^+ (nicotinamide adenine dinucleotide) and $NADP^+$ (nicotinamide adenine dinucleotide phosphate) are essential coenzymes of the enzymes involved in oxidation-reduction systems of living cells, both in respiration and in photosynthesis. In their reduced form, NADH or NADPH, they are high-energy compounds that supply reducing power in such processes as electron transport and the reduction of carbon in the process of photosynthesis. Two other important enzymes involved in oxidation-reduction reactions are the flavin nucleotides FMN (flavin mononucleotide) and FAD (flavin adenine dinucleotide), which are derived from one of the B_2 vitamins, riboflavin.

Thiamine pyrophosphate is derived from vitamin B_1 (thiamine) and serves as a coenzyme for various decarboxylases, oxidases, and transketolases. Pyridoxal, pyridoxine, and pyridoxamine constitute the vitamin B_6 complex, from which is derived pyridoxal phosphate, an important enzyme in reactions in amino acid synthesis. Another vitamin, pantothenic acid, is a precursor of coenzyme A, which plays an important role in metabolism (see Figs. 6.1 and 6.22). Thus several vitamins are essential because of their roles as coenzymes in important metabolic reactions.

The complement of enzymes produced in a plant is determined primarily by its genotype, and occasionally metabolic disorders are caused by gene mutations that eliminate specific enzymes. Some genetic control also exists outside the nucleus, especially in the chloroplasts. Among the most conspicuous mutations in plants are those that produce defects in chlorophyll development, but many others also occur. These often result in death of seedlings and go unnoticed. Study of mutations provides much information about the role of enzymes in metabolism. The regulation of enzyme activity is complex and not fully understood, but it involves feedback by accumulation of end products, activation by metabolites, and energy charge regulation.

In addition to genetic controls, enzyme activity is affected by such factors as temperature; hydrogen ion concentration; concentration of enzyme, substrate, and end products; hydration; and various growth regulators. In both microorganisms and seed plants, formation of enzymes often is induced by the presence of substrate. For instance, formation of nitrate reductase seems to be induced by the presence of nitrate (see Chapter 9). Various substances also inhibit enzyme action and much has been learned about metabolic processes by use of selective inhibitors. There are two general classes of inhibitors, competitive and noncompetitive. Competitive inhibitors are compounds so similar in structure to the substrate molecule that they partially replace it in reaction sites on the enzyme and interfere with normal enzyme action. An example is the blocking of conversion of succinic acid to fumaric acid by addition of malonic acid, which resembles succinic acid in structure. Noncompetitive inhibitors such as fluoride, cyanide, azide, copper, and mercury form permanent combinations with enzyme molecules, rendering them inactive.

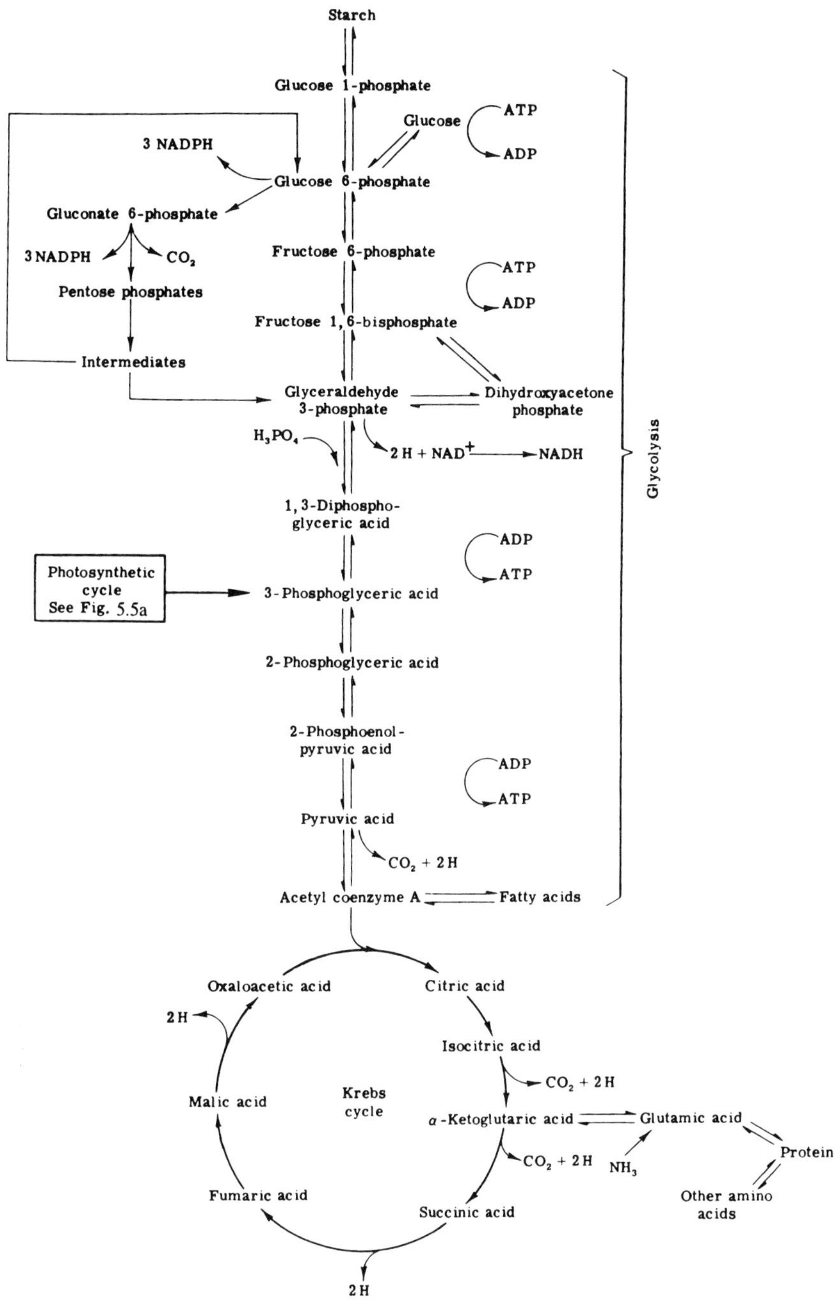

FIGURE 6.1. Outline of the principal steps in glycolysis and the Krebs cycle. Each pair of hydrogens released yields 3 ATP and a molecule of water when it passes through the terminal electron transport system shown in Figure 6.2. Figure 6.22 shows the relationship among various products of these reactions and other important compounds found in plants. The processes in the linear portion called glycolysis occur in the cytoplasm, but those of the Krebs cycle occur in the mitochondria.

Enzyme Classification

Enzymes are classified on the basis of the reactions they catalyze. The following major types are recognized:

- Oxidoreductases: Catalyze oxidation-reduction reactions (e.g., oxidases, dehydrogenases)
- Transferases: Catalyze transfer of a chemical group from a donor compound to an acceptor compound (e.g., transaminase)
- Hydrolases: Catalyze hydrolytic cleavage of C-O, C-N, C-C, and some other bonds (e.g., sucrase)
- Lyases: Catalyze removal of chemical groups from substrates by nonhydrolytic means; these enzymes cleave C-C, C-O, C-N, and other bonds by elimination, leaving double bonds or adding groups to double bonds (e.g., decarboxylases)
- Isomerases: Catalyze conversion of a compound into some of its isomers (e.g., triose phosphate isomerase)
- Ligases (synthetases): Catalyze linking of two molecules together with hydrolysis of a pyrophosphate bond in ATP (e.g., thiokinases)

Isozymes

Enzymes in woody plants exist in several molecular forms that act on the same substrate. These are known as isozymes or allozymes. Isozymes may form by various mechanisms. They may arise through the binding of a single polypeptide to various numbers of coenzyme molecules or other prosthetic groups such as divalent metals. They also may result from conjugation or deletion of molecules with reactive groups such as amino, carboxyl, or hydroxyl groups of the amino acid residues of the polypeptide chain (Scandalios, 1974).

Studies with isozymes have become routine in forest genetics and tree improvement research. Several applications of isozyme analysis in tree improvement programs were described by Cheliak et al. (1987). These included certification of pedigrees of plant families and clones, increased accuracy in describing relative species purity in cases where hybridization occurs between two parental taxa, and, in seed orchards, estimation of mating systems, elucidation of mating patterns among orchard clones, determination of relative fertility, determination of proportions of contamination, and quantitative assessments of pollen flow.

Starch gel electrophoresis of isozymes has been extensively used to elucidate patterns of population variation within communities of woody plants. Plants are not distributed randomly within communities but tend to be clustered in patches. Genetic variation in plant populations also is distributed nonrandomly. Genes and genotypes tend to be clumped, with genetic differences occurring over short distances. Linked isozyme loci have been reported in a large number of species of forest trees. Examples include balsam fir (Neale and Adams, 1981), giant sequoia (Fins and Libby, 1982), white spruce (King and Dancik, 1983), western white pine (Steinhoff et al., 1983), black spruce (Boyle and Morgenstern, 1985), tamarack (Cheliak and Pitel, 1985), incense-cedar (Harry, 1986), and Table Mountain pine (Gibson and Hamrick, 1991). Gymnosperms are particularly suited for isozyme studies because the seeds contain haploid female tissue, which is identical with the female contribution in the enclosed embryo. This information on inheritance can be obtained without a necessity for breeding (King and Dancik, 1983).

As genetic markers isozymes may serve primarily as labels that can be used to certify identity of seed lots, determine validity of controlled crossing, study genetic efficiency of seed orchards, determine the effectiveness of supplemental mass pollination, and aid in selection of economically important traits. Isozymes also may provide information useful for gene conservation (Adams, 1983). However, the use of isozymes in research has some limitations. As Stebbins (1989) cautioned, enzymes extracted and subjected to electrophoresis represent only a very small nonrepresentative sample of all the proteins present in a plant. Furthermore, electrophoretic differences are small and represent only one kind of difference that exists between related proteins. Hence, Stebbins emphasized caution in using isozyme evidence to generalize about major evolutionary and genetic problems. Molecular methods such as restriction fragment length polymorphism (RFLP), cleaved, amplified polymorphic sequences (CAPS), amplified fragment length polymorphism (AFLP), simple sequence repeats (SSR), and random amplified polymorphic DNA (RAPD) techniques also are quite useful in genetic analyses and offer some advantages over isozyme analysis, particularly with respect to the number of loci available for analysis and amount of material needed for analysis (Neale and Williams, 1991; Newton et al., 1999) (see Chapter 9, Kozlowski and Pallardy, 1997).

Energetics

Living organisms depend on a continuous supply of energy for use in synthesis of new protoplasm, maintenance of the structure of organelles and membranes, and mechanical activity such as cytoplasmic streaming. Transport of ions into root cells is driven by respiration. The respiratory energy is essential to overcome electropotential gradients across membranes, counteract diffusion gradients, and expel excess ions (Ryan, 1991). In motile organisms much energy is used for locomotion, and in warm-blooded organisms for maintenance of body temperature. The immediate source of this energy is food—that is, carbohydrates, fats, and proteins accumulated in the organism—but the ultimate source varies. Autotrophic organisms manufacture their own food, either by photosynthesis (Chapter 5) or chemosynthesis. Examples of chemosynthetic organisms are bacteria that obtain energy to synthesize carbohydrates by oxidizing NH_3 to NO_2^-, NO_2^- to NO_3^-, H_2S to SO_4^{2-}, or Fe^{2+} to Fe^{3+}.

Heterotrophic organisms, which include animals and nongreen plants, depend on green plants for the food that supplies their energy. The success of plants depends on their capacity to acquire, store, and release energy as needed. The acquisition and storage of energy in green plants by the process of photosynthesis were discussed in Chapter 5. Here we are concerned with how energy is used, stored, and released in various tissues.

A brief discussion of the energetics of oxidation-reduction reactions is needed at this point. A compound is said to be oxidized if it loses electrons or hydrogen atoms and it is reduced if it gains electrons or hydrogen atoms. Obviously, when one substance is reduced another is oxidized. In the following equation compound A is oxidized and compound B is reduced as the reaction proceeds to the right:

$$AH_2 + B \xrightarrow{\text{dehydrogenases}} A + BH_2 \quad (6.1)$$

The oxidation of glucose can be summarized as follows:

$$C_6H_{12}O_6 + 6O_2 \rightarrow 6CO_2 + 6H_2O + 686{,}000 \text{ calories/mole} \quad (6.2)$$

The carbon in the glucose is oxidized to CO_2 by removing hydrogen atoms and combining them with oxygen to form water. During the rearrangement of atomic bonds a large amount of energy is released. If this process occurs as ordinary combustion in the laboratory the energy is released as heat, but in living cells about two-thirds of the energy is captured and stored in ATP, from which it later can be released to do chemical work. In living cells this process requires about 25 enzymes and numerous steps, as shown in Figures 6.1 and 6.2. The reverse is the reduction of the carbon in CO_2 to glucose, with input of light energy that is used to split H_2O and release H^+. However, because of the inherent inefficiency of the photosynthetic system, far more than 686,000 calories are needed to fix one mole of CO_2 into glucose. The H^+ combines with NADP to produce the reducing compound NADPH, which ultimately supplies the hydrogen required to reduce the carbon. Light energy also is used to produce ATP, which supplies the energy for other steps in the carbon cycle. Thus, in summary, as described in Chapter 5, carbon is reduced, NADPH is oxidized, and energy is stored in the products of photosynthesis:

$$CO_2 + 2NADPH + 2H^+ \rightarrow CH_2O + 2NADP^+ + H_2O \quad (6.3)$$

However, this also occurs in many steps and requires nearly 20 enzymes. Thus reduction is accompanied by an increase in free energy, and oxidation by the release of free energy. For example, the reduced forms of NAD^+ and $NADP^+$, NADH and NADPH, have much higher free energies than the oxidized forms and, because they can supply hydrogen atoms, they are said to have reducing power. The same is true of FMN and FAD and their reduced forms, $FMNH_2$ and $FADH_2$. An example of the role of NAD is the conversion of malic acid to oxaloacetic acid, in which malic acid loses two hydrogen atoms and NAD^+ is reduced to NADH.

$$\underset{\text{malic acid}}{\begin{matrix} CH_2\text{-}COOH \\ | \\ HO\text{-}CH\text{-}COOH \end{matrix}} + NAD^+ \xrightarrow[\text{dehydrogenase}]{\text{malate}} \underset{\text{oxaloacetic acid}}{\begin{matrix} CH_2\text{-}COOH \\ | \\ O{=}C\text{-}COOH \end{matrix}} + NADH + H^+ \quad (6.4)$$

Oxidation and reduction of cytochrome oxidase are accomplished by change in valence of the iron, which shifts from Fe^{3+} (oxidized) to Fe^{2+} (reduced). The role of ATP as a medium for energy transfer is discussed later.

RESPIRATION

Respiration can be defined as the oxidation of food (substrate) in living cells, bringing about the release of energy. The energy released is stored as chemical energy in the substrate molecules. Products of respiration include energy and metabolic intermediates that provide carbon skeletons for cell constituents. Both products are required for growth and maintenance of tissues, absorption of mineral nutrients, and translocation of organic and inorganic materials. Strong correlations between respiration and growth of woody plants have been demonstrated (Anekonda et al., 1993, 1994). The rates of shoot and root respiration per unit of shoot dry weight and root dry weight, respectively, vary systematically such that fast-growing species have higher rates than slow-growing species (Poorter et al., 1991). In white spruce seedlings, root respiration varied seasonally and was related to the number of white roots present (Johnson-Flanagan and Owens, 1986). Sometimes respiration is rapid in maturing fruits, with much of the energy released as heat that seems to serve no useful purpose. The storage life of fruits and seeds can be prolonged significantly by storing them under environmental conditions that keep the rate of respiration low (Chapter 8, Kozlowski and Pallardy, 1997). On the other hand, reduction of the respiration rate in growing tissues by low temperature or low oxygen concentration is undesirable because it reduces the rate of growth.

The control of respiration is complex, involving both environmental and internal factors. With respect to internal controls, Cannell and Thornley (2000) felt that respiration should be viewed as colimited by both the supply of carbon substrates and by demand for high-energy and reduced compounds (ATP and NAD(P)H).

Respiration is much more complex and occurs in many more steps than is indicated by the summary equation given in Eq. 6.2, which does not explain how the oxygen is used, how the carbon dioxide and

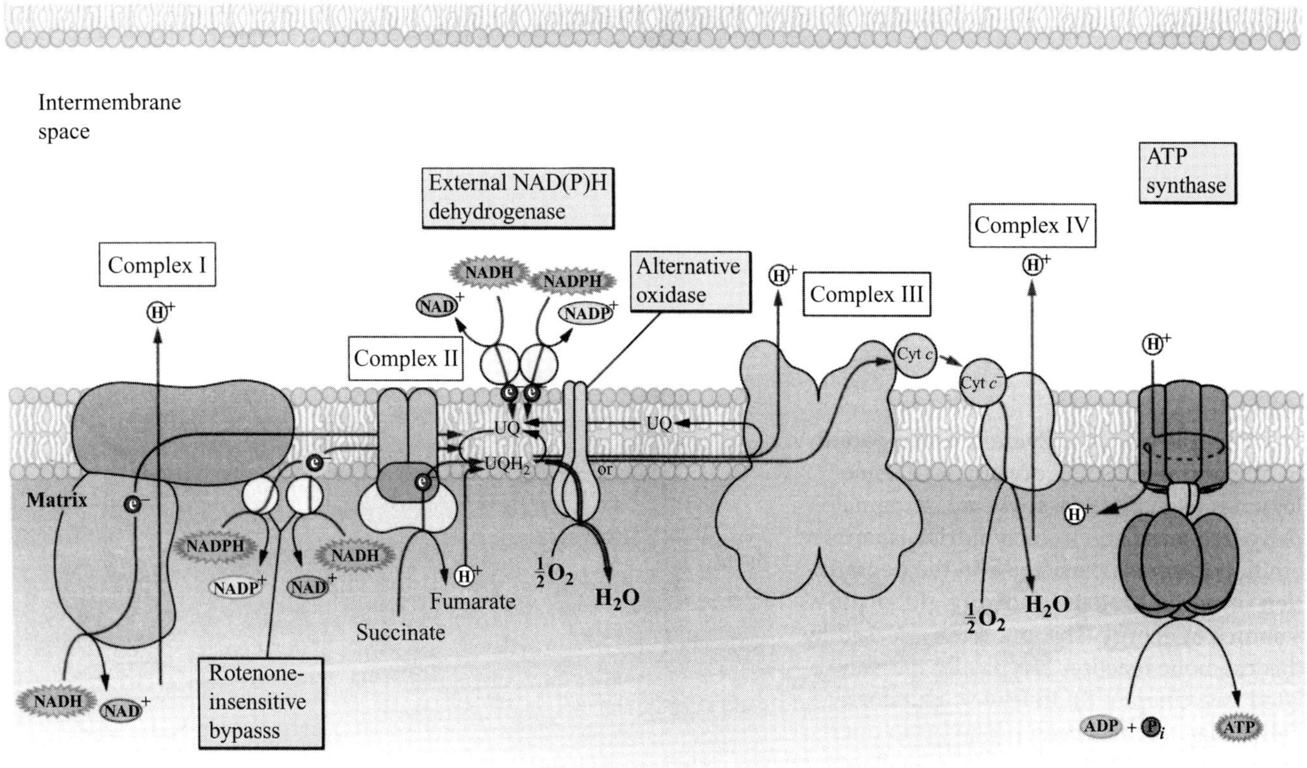

FIGURE 6.2. Organization of the plant electron transport chain in the inner mitochondrial membrane. Electron transfer Complexes I–IV, ATP synthase, four rotenone-resistant NAD(P)H dehydrogenases, and the alternative oxidase are shown. Oxidized ubiquinone (UQ) and reduced ubiquinol (UQH_2) diffuse freely within the membrane and transfer electrons derived from dehydrogenases to either Complex III or the alternative oxidase. Movement of protons across the membrane drives synthesis of ATP via the ATP synthase. Used with permission of the American Society of Plant Biologists, from Siedow, J. N., and Day, D. A. (2000). Respiration and photorespiration. In *Biochemistry and Molecular Biology of Plants* (B. Buchanan, B., W. Gruissem, and R. L. Jones, eds.), pp. 676–728.; permission conveyed through Copyright Clearance Center, Inc.

water are formed, or how energy is released in usable form.

Biological Oxidations

As mentioned earlier, oxidations usually involve removal of hydrogen atoms or transfer of electrons from substrates to an acceptor. In cells hydrogen atoms (proton plus an electron) are split off from the substrate by enzymes known as dehydrogenases and transferred to an acceptor or oxidizing agent. This acceptor usually is NAD^+ (nicotinamide adenine dinucleotide) or $NADP^+$ (nicotinamide adenine dinucleotide phosphate), which become NADH or NADPH when reduced. The hydrogens are then transferred from NADH or NADPH through a series of acceptors in a terminal oxidase system containing cytochromes and combined with oxygen to form water (see Fig. 6.2). At the same time energy is released, part of which is used to form the high-energy compound ATP (adenosine triphosphate) by attaching a phosphate to ADP (adenosine diphosphate).

ATP

As ATP is the most important high-energy compound in plants, it deserves description. Adenosine is formed from the purine, adenine, and the 5-carbon sugar, ribose. When a phosphate group is added to adenosine by an ordinary ester bond, AMP (adenosine monophosphate) results. If a second phosphate is added by a pyrophosphate bond, ADP results, and addition of a third phosphate group by another pyrophosphate bond produces ATP. This is termed a high-energy compound because, when the terminal pyrophosphate bond is broken by hydrolysis at pH 7.0 and 25°C, nearly 7,300 calories/mole of energy are released, in contrast to only about 3,200 calories/mole when the ester bond of AMP is broken. It should be emphasized that the high energy is not in the bond, but refers to the difference in energy content between the original compound and its products. Energy is released because there is a new arrangement of electrons after the terminal phosphate is split off, with a much lower total energy than existed in ATP. The actual amount of energy released by the reaction ATP + $H_2O \rightarrow$ ADP + P_i (inorganic phosphate) varies somewhat with the pH and other factors, but often is given as 7,300 or 7,600 calories/mole, though under the conditions existing in cells it may approach 12,000 calories/mole.

Metabolically active tissue contains large amounts of adenosine phosphates. For example, Ching and Ching (1972) reported that embryos in germinating seedlings of ponderosa pine contained 80 times as much adenosine phosphate as those of nongerminating seeds. The ratio of ATP and ADP to total adenosine phosphate is called the adenylate energy charge and it is considerably higher in growing tissue than in nongrowing tissue.

Other High-Energy Compounds

There are other phosphate compounds such as phosphoenolpyruvate, 1,3-diphosphoglycerate, and acetyl phosphate, that have even higher standard free energies of hydrolysis than ATP. Acetyl coenzyme A (acetyl CoA) possesses a high free energy and the pyridine nucleotides in their reduced forms (NADH and NADPH) also have high free energies. However, ATP occupies a unique position because it is intermediate in the free energy scale and can transfer phosphate groups from compounds with a high free energy to those with a lower free energy such as glucose and glycerol, resulting in glucose-6-phosphate and glycerol-3-phosphate. Thus, ATP and ADP are involved in nearly all enzymatic phosphate transfer reactions in cells and thereby control the flow of energy (Amthor, 1989; Becker and Deamer, 1991).

Glycolysis and the Krebs Cycle

The oxidation of glucose occurs in a series of steps, which fall into two groups, a sequence of reactions called glycolysis (splitting of sugar), in which glucose is converted to pyruvic acid, followed by the Krebs or tricarboxylic acid (TCA) cycle, in which carbon dioxide and water are produced. These steps are shown in Figure 6.1. Glycolysis, an anaerobic process, also called the Embden-Meyerhof pathway, or the EMP (Embden, Meyerhof, and Parnas) pathway occurs in the cytoplasm, whereas the reactions of the Krebs cycle and oxidative phosphorylation occur in the mitochondria.

Glycolysis

Glycolytic metabolites presumably arrive at enzyme sites by diffusion within the aqueous phase of the cytosol (matrix in which cytoplasmic organelles are suspended). The metabolites are converted and then dissociate from the enzyme and reenter the aqueous metabolite pool (Amthor, 1989).

Becker and Deamer (1991) divided glycolysis into three sequential phases including splitting of sugar, an oxidative event that drives the entire pathway, and two steps at which the reaction sequence is correlated with production of ATP. Glycolysis can be summarized by the following equation:

$$\text{glucose} + 2\text{NAD}^+ + 2\text{ADP} + 2\text{P}_i \rightarrow 2\ \text{pyruvate} + 2\text{NADH} + 2\text{H}^+ + 2\text{H}_2\text{O} + 2\text{ATP} \qquad (6.5)$$

Thus one glucose molecule is converted to two molecules of pyruvate. The energy yield for each glucose molecule is two molecules of ATP and two molecules of NADH.

Krebs Cycle

The Krebs cycle has two important functions. One is production of intermediate compounds im-portant in the synthesis of substances as amino and fatty acids. The other is formation of large quantities of ATP, which provides energy for various synthetic processes.

In the presence of oxygen a molecule of CO_2 and two hydrogen atoms are split from pyruvic acid, and acetyl-coenzyme A (acetyl CoA) is formed in a reaction catalyzed by the pyruvate dehydrogenase complex. Acetyl CoA combines with oxaloacetic acid in the Krebs cycle to form the six-carbon atom citric acid. Citric acid is successively converted to five- and four-carbon acids as carbon dioxide molecules and hydrogen atoms are split off, and the cycle finally returns to oxaloacetic acid (Fig. 6.1). Becker and Deamer (1991) summarized the Krebs cycle as follows:

$$\text{Acetyl CoA} + 3\text{H}_2\text{O} + 3\text{NAD}^+ + \text{FAD} + \text{ADP} + \text{P}_i \rightarrow 2\text{CO}_2 + 3\text{NADH} + 3\text{H}^+ + \text{FADH}_2 + \text{CoA-SH} + \text{ATP} + \text{H}_2\text{O} \qquad (6.6)$$

When this equation is adjusted for the two cycles needed to metabolize both of the acetyl CoA molecules derived from one molecule of glucose, the overall expression of glycolysis through pyruvate, and oxidative decarboxylation of pyruvate to acetyl CoA, the summary equation from glucose for the Krebs cycle becomes (Becker and Deamer, 1991):

$$\text{glucose} + 6\text{H}_2\text{O} + 10\text{NAD}^+ + 2\text{FAD} + 4\text{ADP} + 4\text{P}_i \rightarrow 6\text{CO}_2 + 10\text{NADH} + 10\text{H}^+ + 2\text{FADH}_2 + 4\text{ATP} + 4\text{H}_2\text{O} \qquad (6.7)$$

The total energy yield from one molecule of glucose (including the yields of glycolysis, pyruvate to acetyl-CoA conversion, and the Krebs cycle) is 36 ATP (Raven et al., 1992).

Electron Transfer and Oxidative Phosphorylation

The conversion of hydrogen to water, called oxidative phosphorylation, is a complex and a very important process because it is accompanied by production of many ATP molecules. The general scheme is shown in Figure 6.2. Electron transport in plant mitochondria differs from that in animals in important ways that will be detailed after the conventional pattern common to both groups is described.

In the conventional sequence of reactions, NADH produced via the Krebs cycle transfers electrons first to FMN (flavin mononucleotide), then to a bound ubiquinone in Complex I (an NADH dehydrogenase). The proton separated from this electron is extruded into the intermembrane space creating a proton gradient across the inner membrane. Bound, reduced ubiquinone in turn reduces free oxidized ubiquinone within the membrane. Complex II, which contains an enzyme of the Krebs cycle (succinate dehydrogenase) that oxidizes succinate to fumarate, also transfers electrons to this free ubiquinone pool. Complex III (cytochrome *bc*$_1$) oxidizes reduced ubiquinone and transfers electrons to cytochrome *c*, which in turn is oxidized by Complex IV (cytochrome *c* oxidase). Complexes III and IV also extrude protons during electron flow and the action of Complex IV transfers electrons to oxygen as a terminal acceptor, producing H_2O. The resulting proton gradient drives ATP formation by ATP synthase located in the inner mitochondrial membrane (see Chapter 5). One ATP is produced for each pair of hydrogens. Thus, the oxygen is reduced to produce water in the final stage of respiration, and a major part of the ATP is produced. In contrast to this oxidative phosphorylation, the production of ATP in glycolysis is termed substrate-level phosphorylation because the phosphate added to ADP to form ATP comes chiefly from the substrate, rather than from inorganic phosphate as in oxidative phosphorylation and in the photophosphorylation associated with photosynthesis.

In addition to the linear system shared with animals, plant mitochondria have alternative electron transport pathways that create a highly branched respiratory chain (Rasmusson et al., 2004). There are four additional NAD(P) dehydrogenases present in plant mitchondria, two on the outer and two on the inner surface of the inner mitochondrial membrane (Fig. 6.2). These enzymes are rotenone-insensitive and reduce the quinone pool, but do not result in proton extrusion across the inner mitochondrial membrane. The NADH dehydrogenase activity of these enzymes has a lower affinity for NADH compared to Complex I, and is appreciable only when NADH levels are high. In plants, electrons from reduced ubiquinone may bypass the cytochrome pathway via an alternative oxidase enzyme located on the inner surface of the inner mitochondrial membrane. Flow through the alternative oxidase is not sensitive to cyanide inhibition as is the activity of Complex IV (cytochrome *c* oxidase). Its activity consumes O_2 and produces H_2O and heat, but does not result in proton extrusion or ATP synthesis.

The possible functions of these alternative pathways remain the subject of debate (Møller, 2001; Millenaar and Lambers, 2003; Rasmusson et al., 2004). The alternative oxidase pathway has a documented role in heating (10–25°C above ambient) of the upper spadix of flowers of plants of the Araceae. The resultant temperature elevation favors volatilization of carrion-like odors that attract the pollinators of flowers of these plants. Other potential roles proposed for the alternative oxidase and nonproton-pumping dehydrogenases include:

1. Consumption of excess carbohydrates and reductants.
2. Maintenance of activity of the Krebs cycle and its production of synthetic precursor molecules under conditions when the linear electron transport chain activity is low (e.g., low ADP concentrations).
3. Maintenance of the ubiquinone pool at an oxidation level that prevents formation of injurious levels of reactive oxygen species (ROS).

Metabolic Decomposition of Respiration

Total respiration can be decomposed into functional components in a variety of ways. Much of the impetus for such separation derives from the needs of modelers for mechanistic and realistic predictive models of plant respiration (Thornley and Cannell, 2000). One approach is a division into growth respiration (also called synthesis or constructive respiration), and maintenance respiration (also called dormant or basal respiration). Traditionally, growth respiration has been viewed as that required for synthesis of new tissues, maintenance respiration for the energy needed to keep existing tissues healthy. The energy produced by maintenance respiration is used in (1) resynthesis of compounds that undergo renewal in metabolic processes, (2) maintenance of gradients of ions and metabolites; and (3) processes involved in physiological adaptation to stressful environments (Penning de Vries, 1975a,b). Protein turnover is the maintenance process that uses the greatest amount of respiratory products and maintenance respiration is consequently closely correlated with tissue nitrogen concentration (Ryan, 1991; Amthor, 1984, 1994).

Environmental stresses may affect growth respiration and maintenance respiration differently. Both growth and maintenance respiration are variously reduced by mild and severe stress. However, maintenance respiration may be increased, decreased, or may remain stable depending on stress severity. Slowly developing stresses (e.g., drought, shading) often inhibit growth without injuring plant tissues. Following acclimation to such mild stresses, maintenance respiration may show little change. However, when plant tissues are injured by severe stresses, maintenance expenditure will be higher because of the cost of repair (Amthor, 1994).

Appreciable amounts of carbohydrates are consumed in maintenance respiration during the dormant season (Kozlowski, 1992). Gordon and Larson (1970) exposed young red pine trees to $^{14}CO_2$ late in the growing season and followed its redistribution during the next season's height growth in trees brought into a greenhouse in January. More than 85% of the CO_2 originally absorbed was lost, probably through respiration before growth began.

Large amounts of carbohydrates are depleted by respiration of nursery seedlings that are not shipped in the autumn but kept in cold storage until spring (Ronco, 1973). After 100 days of storage in sealed bags at 4.5°C, the dry weight of white spruce and red pine seedlings decreased by 4.0 to 4.5% as a result of maintenance respiration (van den Driessche, 1979). Carbohydrates were depleted by respiration from Sitka spruce and Douglas-fir seedlings in cold storage from September to April at a rate of 0.4 to 0.6 mg g^{-1} day^{-1} (Cannell et al., 1990).

The proportion of the carbohydrate pool that is used in maintenance respiration is quite variable and differs among species, and with stand density and season. Maintenance respiration was lower than growth respiration in Scotch pine (Linder and Troeng, 1981b), about the same as growth respiration in Pacific silver fir (Sprugel, 1990), and higher than growth respiration in loblolly pine (Kinerson et al., 1977) and Douglas-fir (Korol et al., 1991). Over a three-year period the annual contribution to total respiratory consumption of the above-ground parts of a hinoki cypress tree was 21% for growth respiration and 79% for maintenance respiration (Fig. 6.3). Amthor and Baldocchi (2001) reported

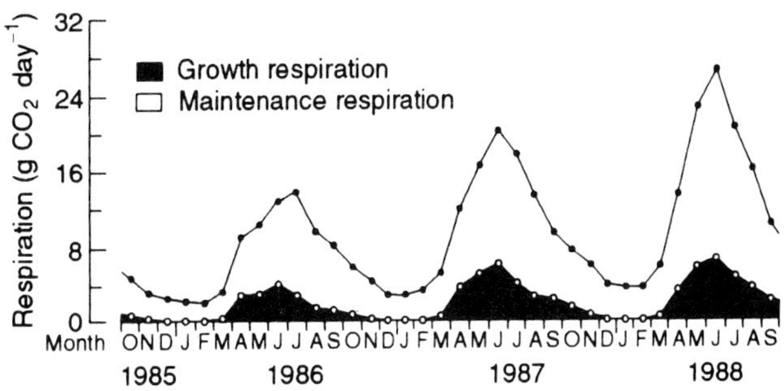

FIGURE 6.3. Seasonal variations in growth respiration and maintenance respiration of a hinoki cypress tree over 3 years. From Paembonan et al. (1992).

roughly equal proportions of growth and maintenance respiration in crop plants, whereas mixed deciduous and oak-pine forests were estimated to have greater proportions of maintenance respiration (88 and 78% of total respiration, respectively) (Ryan, 1991; Amthor and Baldocchi, 2001). Additionally, crops tend to show a lower total respiration to photosynthesis ratio (Amthor, 2000). These attributes are associated with inherently high yield of biomass per unit of food invested in the reproductive structure (growth yield) and a relatively low maintenance respiration requirement in this organ. As crops have been selected for greater relative allocation to reproductive parts (i.e., harvest index), there may have been indirect selection for reduced respiration compared to photosynthesis.

Given the growing recognition of the alternative oxidase and dehydrogenase systems in plant mitochondria, a term for "wastage" respiration may be added to growth and maintenance respiration to account for electron flow that does not result in ATP production (Amthor, 2000; Cannell and Thornley, 2000) (Fig. 6.4). Certain processes supported by respiration may contribute both to growth and maintenance (e.g., ion transport, phloem loading), and hence don't neatly fit into a dichotomous growth and maintenance paradigm. Based on these issues, Amthor (2000) described a general paradigm that relates respiration to processes that it supports.

Other Oxidases

There are a number of other oxidase systems in plants in addition to the cytochromes. Catalase, an iron-containing enzyme, is very abundant. It splits hydrogen peroxide into water and molecular oxygen, whereas peroxidase transfers hydrogen from a donor to peroxide, producing two molecules of water. NADH and NADPH also are oxidized by peroxidase; this may have a regulatory effect on cell metabolism. Peroxidase

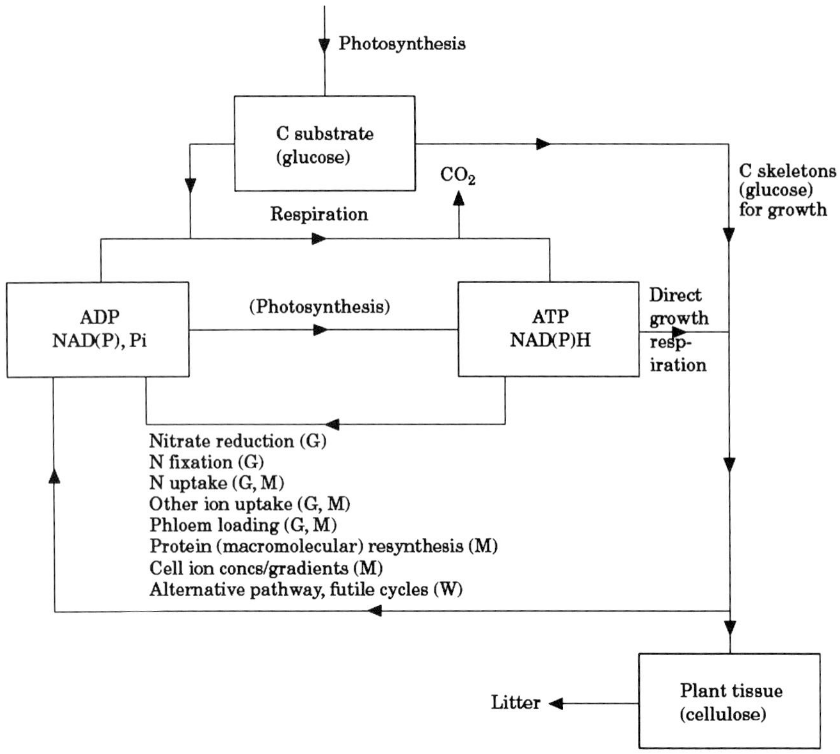

FIGURE 6.4. Simplified scheme of the carbon biochemistry of growth and respiration. Arrows indicate fluxes. G, growth; M, maintenance; W, wastage respiration. From Cannell, M. G. R. and Thornley, J. H. M. "Modelling the components of plant respiration: Some guiding principles." (2000). *Ann. Bot.* **85(1)**, 45–54, by permission of Oxford University Press.

and phenol oxidase appear to be involved in synthesis of lignin. The phenol oxidases are copper-containing enzymes responsible for the darkening of cut surfaces of plant tissue such as apple or potato. They apparently are compartmentalized in intact tissue, but are released and cause discoloration when cells are damaged. Another important oxidase system, glycolic acid oxidase, occurs in stems and leaves, but not in roots; it converts glycolic acid to glyoxylic acid, an important reaction in photorespiration. These are termed soluble oxidases because they occur in the cytoplasm, in contrast to the cytochromes, which occur on membranes within the mitochondria.

The Pentose Shunt

An alternate pathway for oxidation of glucose in plants is the pentose shunt, or hexose monophosphate pathway. Glucose-6-phosphate is oxidized to gluconate-6-phosphate and NADP is reduced to NADPH. This is followed by another oxidation yielding the pentose sugar ribulose-5-bisphosphate, carbon dioxide, and another NADPH. Thus, the pentose shunt produces considerable reducing power in the form of NADPH, which is used in reactions such as fatty acid synthesis. The pentose sugar undergoes a series of transformations similar to those in the Calvin-Benson cycle, and produces compounds needed for synthesis of nucleic acids, adenine and pyridine nucleotides, and other substances. Further rearrangements of the pentose sugar lead to glyceraldehyde-3-phosphate and pyruvic acid or back to glucose-6-phosphate. Although this pathway yields less ATP than glycolysis and the Krebs cycle do, it is important because it produces reducing power and molecules needed for other synthetic processes. According to some investigators it is a major metabolic pathway for hexose metabolism in germinating seeds.

Anaerobic Respiration

Under conditions of limited O_2 availability (hypoxia), plant metabolism is characterized by concurrent aerobic respiration and some degree of anaerobic respiration. Appreciable anaerobic respiration often occurs in internal tissues of seeds, buds, stems, roots, fruits, and also seedlings in cold storage. In maize roots exposed to low internal O_2 concentrations, the stele was exposed to hypoxia but the cortex was not (Thomson and Greenway, 1991). There are wide differences among species in capacity to function under hypoxia or complete lack of O_2 (anoxia) (Chapter 5, Kozlowski and Pallardy, 1997).

In the absence of oxygen the terminal electron transport system cannot operate, oxidative phosphorylation does not occur, and the only energy released is that made available during glycolysis. The general situation is shown in the following diagram:

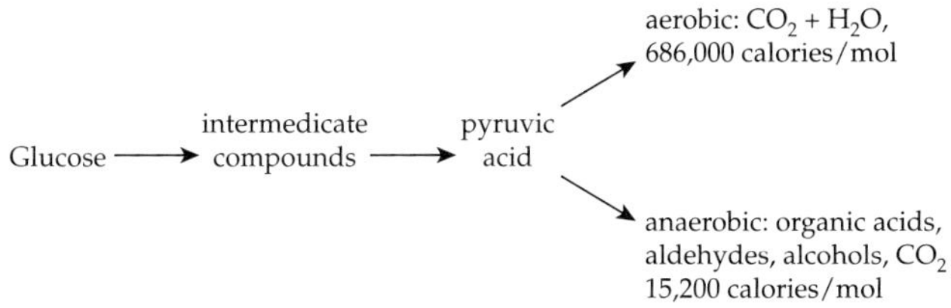

(6.8)

Thus, anaerobic respiration is very inefficient and does not supply enough energy to support rapid growth. Furthermore, accumulation of incompletely oxidized compounds may be injurious to plants. Under anaerobic soil conditions, changes in permeability of root cell membranes result in loss of ions by leaching (Rosen and Carlson, 1984).

Survival of different tissues under anoxia varies appreciably. Roots of dryland species generally are injured or die when exposed to anaerobic conditions for a few hours but germinating seeds of some species may survive for days (Chapter 2, Kozlowski and Pallardy, 1997). Survival depends on availability of a fermentable substrate. Cells may die before such substrates are depleted, suggesting that toxic products are involved. Ethanol alone does not appear to be very toxic (Jackson et al., 1982), but acetaldehyde plus ethanol is toxic.

Regulation of cytosolic pH seems to be important for survival of tissues under anaerobic conditions. Lactic acid participates by producing overacidification of the cytoplasm (Ricard et al., 1994). Maintenance of carbohydrate supply and capacity to convert this supply to ATP at low O_2 also appear to be importance in tolerance of anoxia (Armstrong et al., 1994; Crawford and Braendle, 1996). Flood-sensitive plants suffer massive membrane damage from active oxygen species and post-anoxic metabolism of ethanol. Some flood-tolerant plants exhibit increased production of free-radical-scavenging enzymes such as superoxide dismutase and enzymes associated with metabolism of antioxidant compounds such as ascorbic acid and glutathione (Kozlowski and Pallardy, 2002).

Respiratory Quotient

The nature of the substrate respired markedly affects the ratio of CO_2 produced to O_2 used, the respiratory quotient or RQ. This is illustrated by the following summary equations. The complete oxidation of carbohydrate has an RQ of 1:

$$C_6H_{12}O_6 + 6O_2 \rightarrow 6CO_2 + 6H_2O \; (6CO_2/6O_2) = 1.0 \quad (6.9)$$

The complete oxidation of highly reduced compounds such as proteins and fats gives an RQ considerably less than 1. For example, oxidation of tripalmitin, a common fat, gives an RQ of 0.7:

$$C_{51}H_{98}O_6 + 72.5O_2 \rightarrow 51CO_2 + 49H_2O$$
$$(51CO_2/72.5O_2) = 0.7 \quad (6.10)$$

In plants such as the succulents that oxidize organic acids, the RQ may be considerably more than 1. For malic acid it is 1.33:

$$C_4H_6O_5 + 3O_2 \rightarrow 4CO_2 + 3H_2O$$
$$(4CO_2/3O_2) = 1.33 \quad (6.11)$$

If sugars are converted to organic acids, oxygen may be used without production of CO_2.

Because several different substrates may be oxidized at once and several reactions may occur simultaneously in plant tissue, the respiratory quotient is not necessarily a good indicator of the predominant type of reaction.

Photorespiration

The Krebs cycle and mitochondrial electron transport system likely function in the light, but at lower activities than in the dark (Krömer, 1995). However, many investigators have shown that CO_2 production of woody plants having the C_3 carbon pathway is much higher in the light than in darkness. The process of photorespiration is quite different from dark respiration because it responds differently to inhibitors and oxygen.

Photorespiration is influenced by temperature as well as light intensity, with high temperatures and high light intensity usually accelerating formation of glycolate and flow through the photorespiratory pathway. The temperature effect appears to be partly due to a differential effect on the kinetic properties of Rubisco and differential solubility responses of O_2 and CO_2 to temperature. Stimulation of photorespiration by light sometimes is associated with excessive heating of leaves. Alternatively, the response to light may occur because the absolute amount of phosphoglycolate produced is proportional to the availability of RuBP (ribulose 1,5-bisphosphate) (Artus et al., 1986). The C_3 plants use 20 to 50% of the CO_2 fixed by photosynthesis in photorespiration.

The major events associated with photosynthetic and photorespiratory C metabolism in C_3 plants are shown in Figure 6.5. As summarized by Artus et al. (1986), photosynthetic and photorespiratory C metabolisms consist of two interlocking cycles initiated by carboxylation or oxygenation of RuBP. Both reactions are catalyzed by Rubisco. Carboxylation of RuBP produces two molecules of PGA that are then utilized in the reactions of the Calvin cycle to regenerate RuBP. Excess fixed carbon is stored in the chloroplasts as starch or transported to the cytoplasm as triose phosphate. The primary reaction of photorespiration is oxygenation of RuBP to one molecule of PGA and one of 2-phosphoglycolate. Photorespiration may have some function in protecting the photosynthetic apparatus from high light injury (Osmond and Chow, 1988; Foyer and Noctor, 2000). For more details on the biochemistry of photorespiration, refer to Artus et al. (1986).

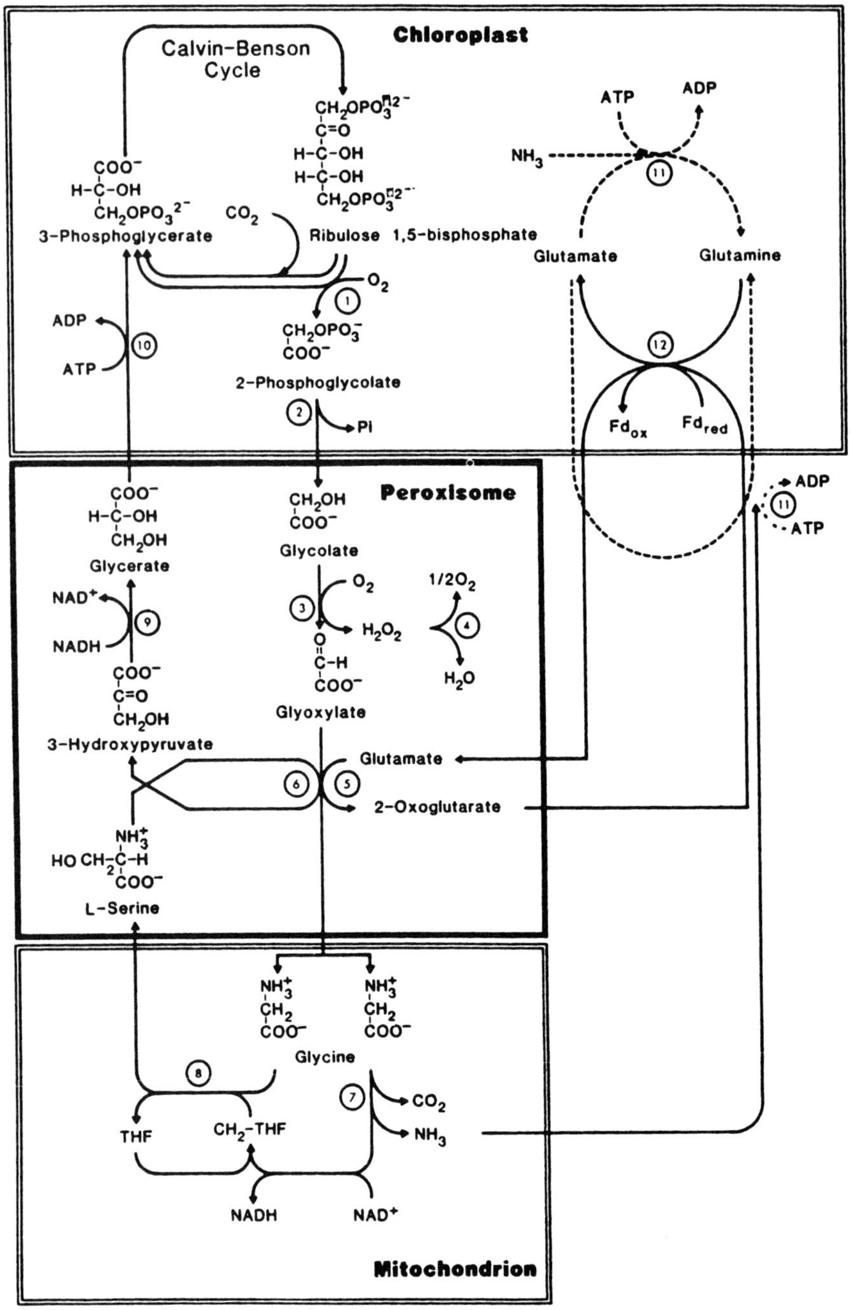

FIGURE 6.5. The pathway of photorespiratory metabolism. Because of the two complementary routes of glycine metabolism in the mitochondrion, two molecules of 2-phosphoglycolate must enter the pathway for each molecule of serine, CO_2, and NH_3 produced. The encircled numbers correspond to the following enzymes: (1) D-ribulose-1,5-bisphosphate carboxylase/oxygenase, (2) phosphoglycolate phosphatase, (3) glycolate oxidase, (4) catalase, (5) glutamate:glyoxylate aminotransferase, (6) serine:glyoxylate aminotransferase, (7) glycine decarboxylase, (8) serine transhydroxymethylase, (9) hydroxypyruvate reductase, (10) glycerate kinase, (11) glutamine synthetase, and (12) glutamate synthase. THF, Tetrahydrofolic acid; CH_2-THF,N^5,N^{10}-methylene tetrahydrofolic acid; Fd_{ox}, Fd_{red}, oxidized and reduced ferredoxin, respectively. From Artus, N. N., Somerville, S. C., and Somerville, C. R. (1986). The biochemistry and cell biology of photorespiration. *Crit Rev Plant Sci* **4**, 121–147. Copyright CRC Press, Boca Raton, Florida.

RESPIRATION OF PLANTS AND PLANT PARTS

Amount of Food Used in Respiration

The total amount of food used in respiration by plants is of interest because it affects how much is available for use in the assimilation processes associated with vegetative growth and the amount accumulated in fruits and seeds. In most plants practically all the carbohydrates are synthesized in the leaves (Chapter 5), but they are consumed by respiration in every living cell. The total amount of food depleted by respiration of leaves, twigs, and living cells of stems, roots, and reproductive structures is a very large fraction of the total available amount.

Respiration of Entire Trees

Respiration commonly depletes from 30 to more than 60% of the daily production of photosynthate (Kozlowski, 1992). Landsberg (1986) estimated that respiratory depletion of carbohydrates by branches, stems, and roots of temperate-zone trees ranged from 25 to 50%. Similarly, Gifford (2003) reported that the ratio of respiration to photosynthesis measured in numerous studies ranged from 0.35 to 0.6. Carbohydrate depletion by respiration of 14-year-old loblolly pine trees in North Carolina was estimated to be about 58% (Kinerson, 1975). In the tropics where night temperatures are high, carbohydrate losses by respiration of woody plants may amount to 65% or more (Kira, 1975; Sprugel and Benecke, 1991).

Respiration of Various Plant Parts

The rates of respiration in different organs vary greatly, largely because of differences in their proportions of physiologically active tissues.

Buds

The respiration rates of individual buds on a tree vary widely because of large differences in their size and the amount of metabolically active tissue they contain. Many individual buds fail to open or die (Chapter 3), and hence they have low or negligible respiration rates (Kozlowski, 1992).

Although buds constitute a very small part of the mass of a tree, during the growing season they are organs of high physiological activity. Apical buds of kiwifruit vines maintained low, stable respiration rates during the dormant season in New Zealand (Fig. 6.6). As buds swelled in the spring, specific respiration rates increased manyfold. Interestingly, attempts to force buds to open in the winter by exposing them to 20°C for a week suppressed rather than stimulated respiration rates (McPherson et al., 1997).

Several studies showed that bud scales hinder the entrance of oxygen, and the respiration rate (oxygen uptake) of intact Norway maple buds was only about half as high as that of buds from which the scales have been removed (Pollock, 1953). The effect of removal of the scales on respiration of eastern white pine buds is shown in Figure 6.7.

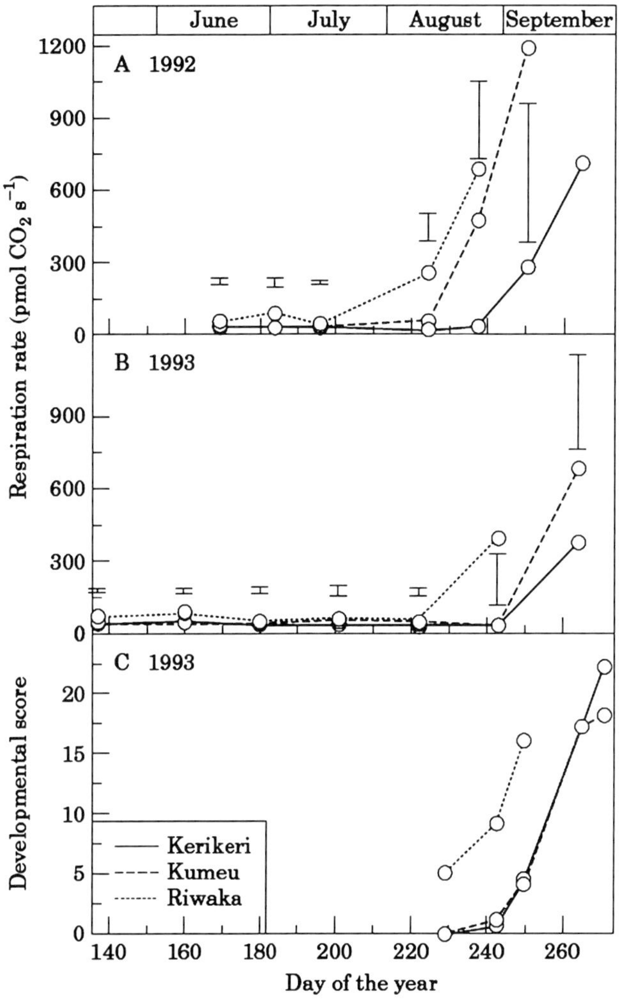

FIGURE 6.6. Changes in dark respiration of dormant apical buds of kiwi in 1992 (A) and 1993 (B). Canes were excised from vines grown in three contrasting regions and measured at 20°C the following day. Vertical bars indicate least significant difference ($p = 0.05$). C, changes in state of development of reproductive buds where 0 = undeveloped primordia to visible petals and sepals at the highest numerical values. From McPherson, H. G., Snelgar, W. P., Manson, P. J., and Snowball, A. M. "Bud respiration and dormancy of kiwifruit (*Actinidia deliciosa*)." *Ann. Bot.* (1997), **80(4)**, 411–418, by permission of Oxford University Press.

Leaves

Leaves comprise a small part of the mass of a tree but they have high rates of respiration because they contain a very large percentage of living tissue (see Fig. 6.16). They have been estimated to account for 50% of the total respiration in a 60-year-old beech forest, 60% in a tropical rain forest (Larcher, 1975), and 32% in a young loblolly pine stand (Kinerson et al., 1977). The daytime respiration rate of leaves of lemon and orange trees was 15 to 20% of the rate of photosynthesis. If these rates were maintained during the entire 24 hours, they would use 30 to 40% of the carbohydrates manufactured (Wedding et al., 1952). However, the lower temperature at night should reduce the night rate materially below that during the day.

The rate of respiration of leaves varies greatly with their age, season, and location in the crown. Such variations are not surprising because respiration is closely coupled with photosynthesis, which is regulated by light intensity (Chapter 5) and the age of leaves. Respiration in mature tissues depends on protein turnover and active transport to maintain ionic gradients. The rates of both processes decline as photosynthesis decreases. Light influences respiration in the short

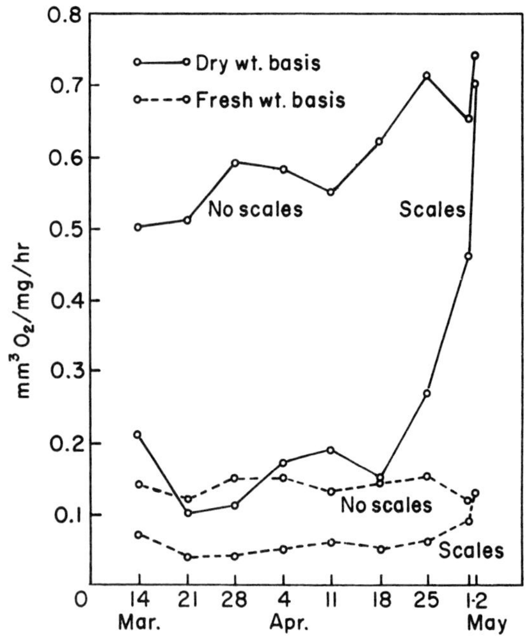

FIGURE 6.7. The effect of removal of bud scales on respiration (oxygen uptake) of buds of eastern white pine at various stages in development, calculated on dry and fresh weight basis. From Kozlowski and Gentile (1958).

term (up to an hour) by affecting metabolism and carbohydrate availability and over periods of days to years by altering leaf morphology and photosynthetic capacity. Furthermore, a decline in light intensity over a long time, as occurs in developing forest stands, reduces the size of the pool of photosynthetic enzymes, which further reduces the rate of respiration (Brooks et al., 1991). Respiration is very high during early stages of leaf growth, when the rate of synthesis of chlorophyll, proteins, and structural compounds is high. As the photosynthetic system becomes active, the rate of respiration decreases. Dark respiration rates were high for newly formed cottonwood leaves but rapidly declined as leaves matured (Dickmann, 1971). In white oak saplings, specific respiration rates of expanding leaves on May 3 were higher (66.5 nmol CO_2 g^{-1} s^{-1}), slightly lower on May 5, and much lower by May 18 (25.5 nmol CO_2 g^{-1} s^{-1}) (Wullschleger and Norby, 1992). The pattern of decreasing respiration with increasing leaf age is consistent with data for other species (Koch and Keller, 1961).

Leaf respiration rates are high in upper exposed parts of the canopy and decrease with increasing crown depth. In Pacific silver fir, respiration of all age classes of needles decreased with depth in the crown (Brooks et al., 1991). Respiration differed more with location of needles in the crown than with needle age. The rate of needle respiration at a tree height of 4.7 m was approximately six times the rate of needles of the same age at 2.6 m. In Pacific silver fir the pattern of needle respiration within the crown was similar to that for photosynthesis (Teskey et al., 1984b) and other conifers (Beadle et al., 1985). Higher respiration rates in sun leaves than in shade leaves have been reported for both deciduous trees and evergreens (Negisi, 1977; Kozlowski et al., 1991). For example, sun-grown leaves of fig trees had higher respiration rates than shade-grown leaves (Fails et al., 1982).

Considered across biomes, mass-based dark respiration rate of leaves was correlated with leaf mass-based N, specific leaf area (SLA), and leaf lifespan (Reich et al., 1997, 1998) (Fig. 6.8). Respiration itself varied from 3.1 to 65 nmol CO_2 g^{-1} s^{-1} and averaged 27.3, 14.4, 11.4, and 4.9 nmol CO_2 g^{-1} s^{-1} for forbs, broad-leaved shrubs, broadleaved trees, and needle-leaved trees, respectively. Leaf respiration rate increased with N concentration and SLA and decreased with leaf lifespan. The high leaf respiration rates of forbs were associated with higher SLA and shorter leaf lifespan in this group, so that at a given N concentration forbs had higher leaf respiration. Low leaf respiration rate in needle-leaved species was linked with low SLA, low N concentration, and long leaf lifespan characteristic of this group. Leaf respiration also was positively correlated with maximum photosynthetic rates, likely because of the maintenance respiration needs of protein turnover in the photosynthetic apparatus. It should be noted also that studies spanning a more limited set of species and climatic gradients may not show strong dark respiration leaf-massed based N relationships (e.g., Meir et al., 2001; Mitchell et al., 1999) because of limitations in the functional groups present and localized environmental influences (canopy light variability, elevation).

Branches and Stems

The rate of respiration varies appreciably with the size of branches and is higher in current-year twigs than in larger branches or stems (Fig. 6.9). The aggregate respiration of all the branches on a tree may be high. For example, Kinerson (1975) attributed approximately half the total autotrophic respiration of a loblolly pine plantation to branch respiration.

In tree stems and large branches most of the respiration occurs in the new phloem and xylem adjacent to the cambium (Fig. 6.10). Although living ray and axial parenchyma cells absorb oxygen, their number is so small that their total respiration is low. In one experiment the heartwood absorbed a small amount of oxygen, but this probably resulted from oxidation of organic compounds in dead tissue rather than from respiration, because boiled blocks of heartwood absorbed nearly as much oxygen as unboiled wood (Goodwin and Goddard, 1940).

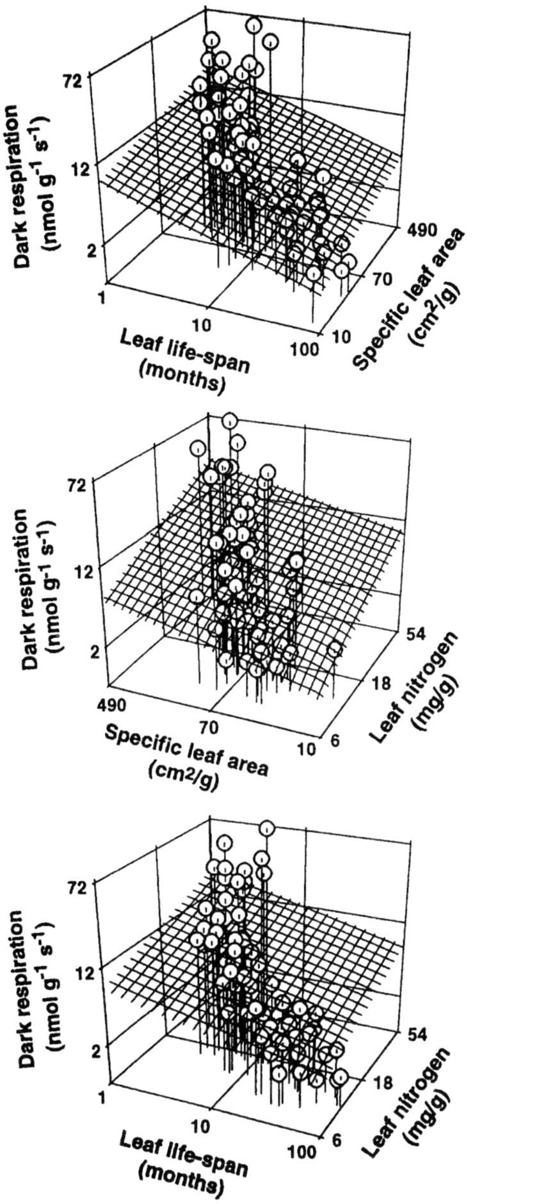

FIGURE 6.8. Leaf dark respiration related to combinations of specific leaf area (SLA) and leaf lifespan (top), SLA and leaf N concentration (middle), and leaf lifespan and leaf N concentration (bottom). From *Oecologia*, Relationships of leaf dark respiration to leaf nitrogen, specific leaf area and leaf life-span—a test across biomes and functional groups. Reich, P. B., Walters, M.. B., Ellsworth, D. S., Vose, J. M., Volin, J. C., Gresham, C., and Bowman, W. D. (1998). **114**, 471–482, Figure 5. © 1998 with kind permission from Springer Science and Business Media.

The rate of respiration decreases from the outer to the inner sapwood at a rate that varies among species. For example, respiration decreased faster from the cambium inward in oak than in beech stems because of differences in distribution of living cells (Möller, 1946). In lodgepole pine and Engelmann spruce trees, stem maintenance respiration was linearly related to the volume of living cells (Ryan, 1990).

The rates of stem respiration per unit of stem surface often vary greatly among trees in the same stand because of differences in their rates of cambial growth and various other factors. The average rate of respiration of a thinned balsam fir stand was greater than that of an unthinned stand (Lavigne, 1988). Nevertheless, the rate of stem respiration per hectare was nine times greater in the unthinned stand. This was because the stem surface area of the unthinned stand was 15 times greater than that of the thinned stand.

Roots

Respiration of roots is variable but often very high. Under some conditions root respiration may account for more than half of the CO_2 evolution from forest soils (Ewel et al., 1987). Much root respiration occurs in the fine roots. For example, more than 95% of the root respiration of pine and birch stands occurred in the fine roots (Mamaev, 1984).

Mycorrhizal fungi contribute materially to root respiration. It has been estimated that mycorrhizal fungi, which may comprise no more than 5% of the root system, may account for as much as 25% of the respiration of mycorrhizal root systems (Harley, 1973; Phillipson et al., 1975). The high respiration rates of mycorrhizal roots are associated with direct contributions of mycorrhizal fungi to respiration, increased mineral absorption, translocation of carbohydrates to roots, and changes in hormone balance (Reid et al., 1983). Below-ground respiration of ectomycorrhizal ponderosa pine seedlings was more than twice as high as that of nonmycorrhizal seedlings (Rygiewicz and Andersen, 1994). The higher rate of mycorrhizal seedlings was attributed to three factors: (1) the fungal hyphae had the highest respiration rate of any seedling parts, (2) the fungus-colonized roots had higher respiration rates than the noninoculated roots, and (3) mycorrhizal roots had a higher percentage of fine roots than the roots of noninoculated seedlings.

In some plants root respiration is increased by root nodules, with more carbohydrates used in nodule respiration than in growth of nodules (Schwintzer, 1983). For example, red alder seedlings used three times as much photosynthate in nodule respiration as in growth of nodules (Tjepkema, 1985).

Pneumatophores

Trees of swamp habitats or those subject to tidal flooding, such as mangroves, often have specialized root systems, called *pneumatophores* (Chapter 2), which often are involved in gas exchange. Mangroves of the type represented by *Avicennia nitida* may produce thousands of air roots or pneumatophores, which protrude from the mud around the base of the tree.

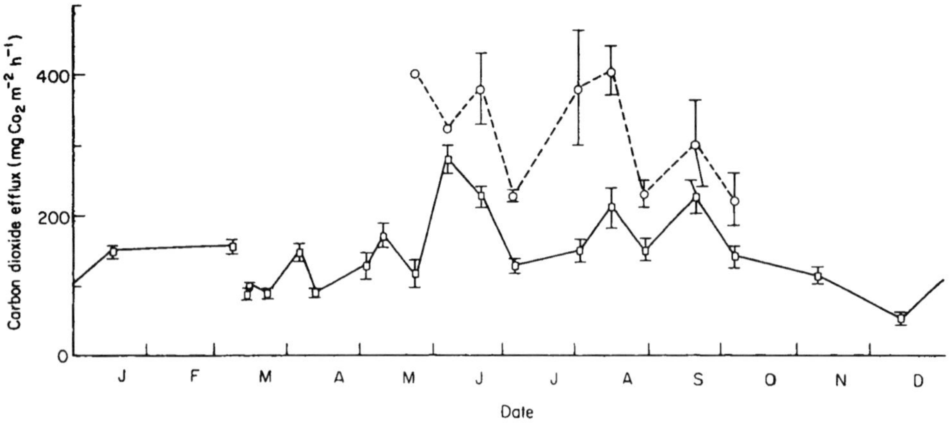

FIGURE 6.9. Seasonal changes in respiration (CO_2 efflux) of current-year twigs (○) and older branches (>1.0 cm in diameter) (□) of tulip poplar. The bars represent standard errors. From McLaughlin et al. (1978).

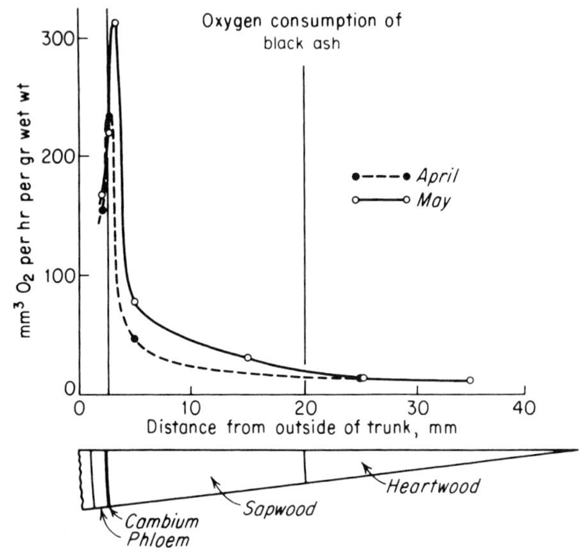

FIGURE 6.10. Rates of respiration of various parts of a trunk of a black ash tree before (April) and after (May) bud opening, measured as oxygen uptake. From Goodwin and Goddard (1940).

Scholander et al. (1955) reported that air is sucked in through lenticels of vertical pneumatophores in *Avicennia* when the tide falls and is forced out when the tide rises. The stilt roots of *Rhizophora mangle* have lenticels on the surface, which are connected by air spaces to roots buried in the mud. Plugging the lenticels with grease caused the O_2 content of the roots buried in mud to decrease, indicating that the stilt roots serve as aerating mechanisms for the submerged roots (Scholander et al., 1955).

Although claims have been made that cypress knees (Chapter 2) serve as aerating organs and supply O_2 to submerged roots, the evidence that they play such a role is not convincing. If O_2 transfer commonly occurs through the knees to the root system, then when knees are detached the amount of O_2 they absorb should be immediately reduced. Kramer et al. (1952) found, however, that respiration in detached knees was higher than in attached knees for two days but then decreased with time. Because of the large amount of active cambial tissue in knees, it appears that most of the oxygen is utilized locally and that cypress knees are not important as aerating organs.

Fruits and Cones

Respiration rates vary appreciably with the type of fruit and stage of fruit development. Fruit respiration consumed 16 to 23% of total carbohydrates of developing sweet cherry fruits (Loescher et al., 1986), 31% in sour cherry fruits (Kappes and Flore, 1986), and 16 to 21% in peach fruits (DeJong and Walton, 1989). Respiration accounted for 18 to 38% of the carbohydrate costs of producing fruits of 15 temperate deciduous tree species (Bazzaz et al., 1979).

The rate of respiration of fruits differs greatly during their development. For example, respiration of apple fruits is highest immediately after fruits are set and decreases rapidly during the early summer and slowly during late summer, with a rise called the "climacteric" just after picking (Fig. 6.11). Respiration of bananas is highest early in their development, then falls rapidly and remains steady thereafter (Thomas et al., 1983). The rate of respiration of grapes on a dry weight basis was high early in their development and then rapidly declined on a single berry basis, but respiratory peaks occurred in later developmental stages (Niimi and Torikata, 1979). Respiration of pistachio fruits increased progressively during seed growth and development and gradually declined after seed growth was completed (Toumadje et al., 1980). Respiratory losses of carbohydrates of sour cherry fruits were much higher in a mid stage than in early or late stages of development. The high rates in the mid stage were associated with lignification and lipid synthesis during pit hardening and embryo development (Kappes and Flore, 1986).

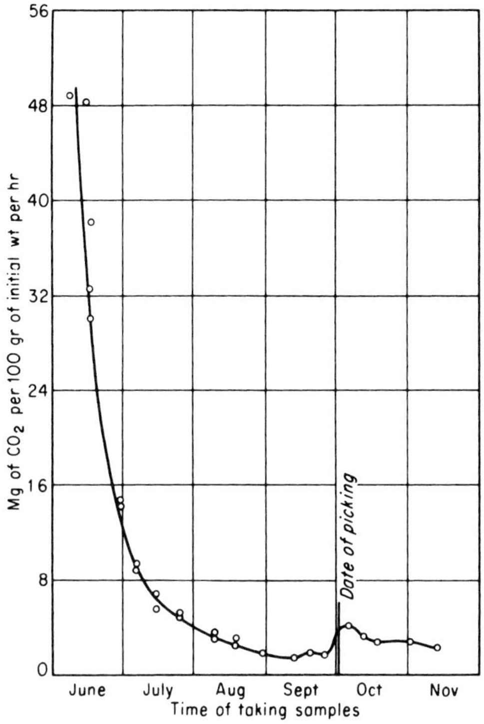

FIGURE 6.11. Change in rate of respiration during development of an apple fruit. From Krotkov (1941). © American Society of Plant Physiologists.

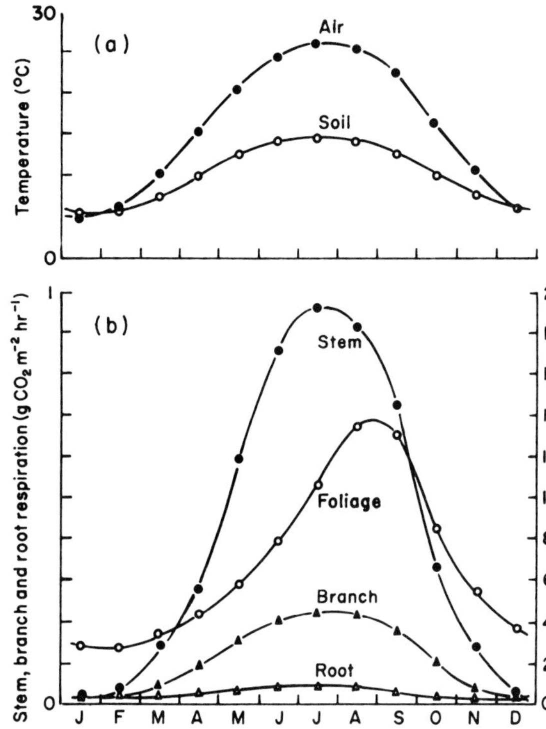

FIGURE 6.12. (a) Mean monthly air (●) and soil (○) temperatures during the investigation (b) seasonal patterns of CO_2 evolution per unit area of ground from stem (●), branch (▲), root (△), and foliage (○) of a loblolly pine plantation. Note the change of scale for foliage respiration. From Kinerson (1975), Blackwell Science Ltd.

The rate of respiration (based on dry weight) of first-year conelets of pines is considerably higher than that of second-year cones. However, total respiratory losses are higher in the much larger second-year cones (Han and Kim, 1988). First-year conelets of red pine in Wisconsin were only about one-fortieth the weight of mature second-year cones (Dickmann and Kozlowski, 1969a). Respiration of Douglas-fir conelets was high during the pollen receptive stage and declined after pollination. More carbohydrates were consumed in respiration by the developing seeds than by the cone scales. Respiration of both seeds and scales decreased with maturity (Ching and Ching, 1962; Ching and Fang, 1963).

Seeds

Respiration rates are high in early stages of seed development but decrease in mature seeds as they dehydrate. Respiration in mature seeds increases greatly as they imbibe water, as discussed in Chapter 2 in Kozlowski and Pallardy (1997).

Seasonal Variations

There are marked seasonal variations in the rate of respiration as shown for whole trees, buds, leaves, branches and stems (Figs 6.9 and 6.12), and reproductive structures (Kozlowski, 1992). In Alabama, respiratory losses of ^{14}C by loblolly pine seedlings varied from 22 to 87% at various times of the year (Kuhns and Gjerstad, 1991). Seasonal changes in respiration rates reflect effects of warming and cooling on the growth cycle as well as direct effects of temperature on the rate of respiration. Hence, stem respiration is increased more by high temperature after cambial activity begins than while the cambium is inactive. Respiration of apple trees was low when the trees were dormant and rose rapidly to a peak in spring (before full bloom) and then declined steadily during the summer (Butler and Landsberg, 1981).

Scaling of Respiration to the Ecosystem Level

Plant physiological ecologists need to scale respiration rates of individual trees and various tissues to entire forest stands. Ecosystem respiration is such a large component of annual carbon balance that it commonly exceeds net primary productivity (Amthor and Baldocchi, 2001). There is much interest in assessing the balance between photosynthesis and respiration on a global scale because each of these CO_2 fluxes far exceeds anthropogenic emissions of CO_2 (Goulden et al., 1996; Cao and Woodward, 1998; Valentini et al., 2000); therefore, small changes in either component caused by natural causes or human influences can ameliorate or exacerbate the buildup of CO_2 in the atmosphere.

Scaling of respiration of plant parts to the stand level is beset with challenges (Sprugel et al., 1995; Hanson et al., 2000; Edwards and Hanson, 2003). Particularly challenging is the need for year-round, frequent (preferably continuous) sampling of multiple trees and locations. Accurate estimation of the volume of living tissue in various organs also is difficult. Although measurement of leaf respiration theoretically is not difficult, it is not easy to create models with proper weighting of canopy variation in respiration (Bolstad et al., 2004), and that account for seasonal acclimation of respiration to temperature (Bolstad et al., 1999, 2003), and possibly adjust for reduced respiration in the light (Curtis et al., 2005). Stem and branch respiration estimates by conventional gas exchange techniques are not easy to model because of the difficulty in accurately estimating the amount of sapwood volume in trees and the number of living cells per unit sapwood volume (Edwards and Hanson, 2003). Although estimates of efflux of CO_2 from the soil are now routinely measured with chambers (Fig. 6.13), partitioning of root, mycorrhizal, and soil components of the total flux are difficult to obtain. A general deficiency of quantitative data on fine root biomass for different forest ecosystems, on respiration rates of fine roots within ecosystems, and on factors influencing respiration, have impeded scaling of root respiration to whole stands.

Despite these difficulties, results of several studies (Table 6.1) indicate that soil respiration constitutes the largest contribution to forest ecosystem respiration, with leaf and stem respiration contributing a much smaller and variable proportion. There is some evidence that total ecosystem and soil respiration are dependent on gross primary productivity of a forest ecosystem, as was shown for European forest ecosystems from a broad latitudinal and climatic gradient (Janssens et al., 2001). In a methods critique and meta-analysis of 37 studies of forest ecosystems, Hanson et al. (2000) reported a mean value of total soil CO_2 efflux that could be attributed to roots of 48%, with most values between 40 and 70%. However, these data represented only mid-summer time periods, and winter root contributions were substantially lower. Amthor and Baldocchi (2001), in an assemblage of numerous studies from forest ecosystems from all parts of the world, conifer and angiosperm, managed and unmanaged, presented ratios of autotrophic respi-

FIGURE 6.13. Automated soil respiration chamber after the design of Edwards and Riggs (2003). Soil respiration rate, conducted during periodic lid closure, is obtained by measurement of inlet and outlet CO_2 concentrations of air drawn through the chamber at a measured rate of flow. Photo by the author.

TABLE 6.1. Respiration Components for a Number of Representative Temperate Forest Ecosystems

Ecosystem	Ponderosa pine[a]	Northern hardwood[b]	Mature aspen[b]	Intermediate aspen[b]	Boreal aspen[c]	Northern hardwood[d]
Respiration component (Mg C ha^{-1} yr^{-1})						
Leaves	1.57	0.57	1.10	0.99	1.98	0.94
Wood/stem	0.54	2.33	1.54	0.18	1.55	1.66
Soil	6.83	8.94	11.16	8.93	9.62	10.03
Total	8.94	11.83	13.80	10.10	13.15	12.63

[a]From Law et al. (1999).
[b]From Bolstad et al. (2004).
[c]From Griffis et al. (2004).
[d]From Curtis et al. (2005).

ration to gross photosynthesis that had mean value of 0.59 and ranged between 0.39 and 0.93. Clearly, more research is needed to find reliable and accurate ways of estimating respiration of entire tree stands.

Respiration of Harvested Fruits

Much attention has been given to the biochemistry of harvested fruits in attempts to find ways of prolonging their life in storage. Fruits continue to consume carbohydrates in respiration after they are harvested and many show a marked climacteric increase in CO_2 production before they senesce and disintegrate. The climacteric rise appears to be a response to an increased requirement for metabolic energy (Lambers, 1985). The climacteric burst of CO_2 production is shown by many fruits and not by others (Table 6.2). It occurs in avocado fruits after they are picked but not in fruits attached to the tree. Differences in the magnitude of CO_2 production during the climacteric rise are shown in Figure 6.14.

Ripening and the climacteric respiratory burst are associated with increased ethylene production, and exogenous ethylene induces both ripening and the climacteric burst. The effects of ethylene on respiration of climacteric and nonclimacteric fruits differ. Treatment of climacteric fruits with low concentrations of ethylene shifts the time of onset of the rise in respiration without necessarily changing the shape of the curve of the climacteric cycle. Ethylene is effective only if applied during the preclimacteric stage prior to a burst of ethylene production by the fruit. By comparison, in nonclimacteric fruits the rise of respiration is stimulated by exposure to ethylene throughout the post-harvest life of the fruit (Biale and Young, 1981).

The climacteric finally terminates in deterioration of fruits, which then become increasingly susceptible to destruction by microorganisms. The rapid deterioration of fruits during senescence suggests that control over enzyme activity is lost, perhaps because cellular compartmentation is breaking down, releasing enzymes that normally are compartmentalized.

TABLE 6.2. Examples of Fruit of Woody Plants Showing a Climacteric or a Nonclimacteric Pattern of Respiration during Ripening

Climacteric	Nonclimacteric
Apple	Blueberry
Apricot	Cacao
Banana	Sweet cherry
Fig	Grape
Mango	Grapefruit
Papaya	Lemon
Peach	Lychee
Pear	Olive
Persimmon	Orange
Plum	Strawberry

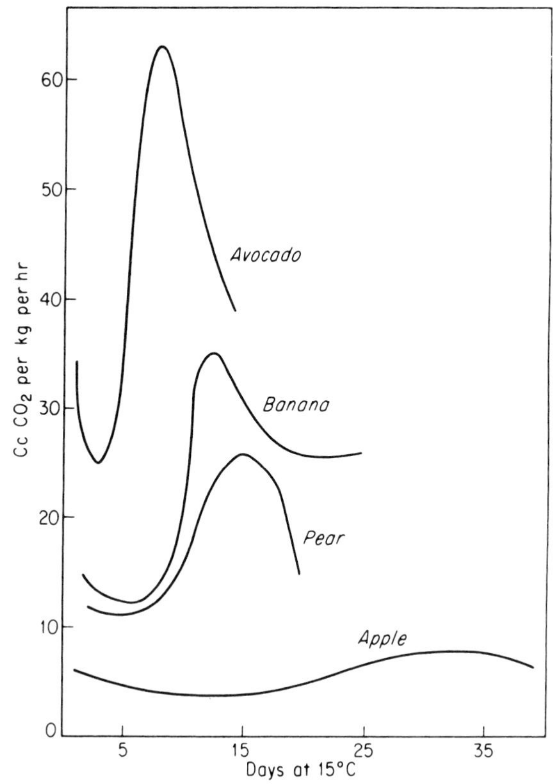

FIGURE 6.14. Climacteric rise in respiration of different fruits. From Biale (1950). Reproduced with permission from the *Annual Review of Plant Physiology*, Vol. 1. © 1950 by Annual Reviews Inc.

Fruit Storage

Successful storage of fruits and vegetables is closely related to control of respiration, and storage at low temperatures is the simplest method of reducing the rate of respiration. However, storage of apples, pears, and other fruits and certain vegetables in refrigerated spaces sometimes is accompanied by pitting, internal browning, and other evidences of physiological breakdown. Cooling does not reduce the rates of all processes to the same degree, and injury may occur from accumulation of products of anaerobic respiration, phenol oxidase activity, and other abnormal physiological processes. Storage life in refrigeration of many plant materials can be greatly increased by lowering the O_2 concentration to 2 or 3% and increasing the CO_2 concentration to 2 to 6% or even to 20% or more in some instances. The best temperature, O_2, and CO_2 levels for storage vary with the type of fruit and cultivar (Dewey, 1977; Smock, 1979; Weichmann, 1986). The high concentration of CO_2 not only inhibits anaerobic respiration, but also reduces ethylene injury and sometimes increases shelf life when fruit is removed from cold storage. Accumulation of ethylene increases the rate of deterioration and is sometimes removed by absorption in $KMnO_4$. Storage of harvested fruits is discussed further in Chapter 8 in Kozlowski and Pallardy (1997).

FACTORS AFFECTING RESPIRATION

In simplest terms the success of trees and other plants depends on the relative rates of respiration and photosynthesis. The rate of respiration is influenced by several internal and environmental factors that often interact. Among the important internal factors are age and physiological condition of tissues, amount of oxidizable substrate, and tissue hydration. Environmental factors include soil and air temperature; gaseous composition of the soil; available soil moisture; light; injury and mechanical disturbances; and chemicals such as herbicides, fungicides, insecticides, fertilizers, and environmental pollutants.

Age and Physiological Condition of Tissues

Young tissues with a high proportion of protoplasm to cell wall material and few dead cells have higher mass-based respiration rates than mature tissues that contain less physiologically active mass. For example, respiration of small twigs is more rapid per unit of dry weight than respiration of branches, and respiration rates of young leaves are higher than those of older leaves. It was mentioned earlier that respiration of buds increases several fold when growth begins and decreases rapidly when growth ceases. High tissue N concentration often is associated with elevated respiration, presumably because of the greater maintenance respiration demands of tissue protein metabolism (e.g., *Pinus sylvestris*, Zha et al., 2002; *Liquidambar styraciflua*, Tissue et al., 2002; *Populus deltoides*, Griffin et al., 2002; *Quercus alba*, *Q. rubra*, and *Acer rubrum*, Lee et al., 2005).

Available Substrate

As noted previously, the law of mass action applies to respiration; hence an increase in the amount of oxidizable substrate usually results in a higher rate of respiration. This is very noticeable in ripening fruits in which conversion of starch to sugar is accompanied by an increase in the rate of respiration. The high carbohydrate concentration of the youngest sapwood may be a factor in its high rate of respiration (see Fig. 6.10). Respiration rates in upper-canopy leaves are often higher than those in lower canopy leaves, and this pattern is correlated with higher soluble sugar content concentrations in leaves of the upper canopy (e.g., Griffin et al., 2002).

Light

Aside from photorespiration discussed previously, the level of "dark respiration" in the light is also of interest. In the light, mitochondrial O_2 consumption may increase, decrease, or remain stable depending on how much photosynthetic carbohydrate and reducing equivalents arrive at the mitochondria from chloroplasts and how much photorespiratory NADH is consumed there (Atkin et al., 2005). However, nonphotorespiratory release of CO_2 from mitochondria is frequently and often substantially inhibited in the light. Although the mechanism responsible for this is unknown, it is known that key enzymes of the mitochondrial respiratory metabolism, pyruvate decarboxylase, and possibly malic enzyme, are inactivated by light. Electron transport and phosphorylation of ADP proceed in the mitochondria in the light and supply the ATP needed for sucrose synthesis in the cytoplasm (Krömer, 1995).

Hydration

Within limits the rate of respiration is correlated with the water content of tissues. This is particularly conspicuous for dry seeds, in which the rate of respiration decreases as the seeds mature and become dry, but increases almost immediately when the seeds are wetted (Chapter 2, Kozlowski and Pallardy, 1997). Parker (1952) reported that if conifer twigs and needles were severely dehydrated there was a temporary increase in respiration followed by a decrease. Brix (1962) reported a similar phenomenon in loblolly pine. In general, total respiration appears to be reduced by moderate-to-severe water stress in vegetative tissues such as leaves (Ribas-Carbo et al., 2005) and roots (e.g., in *Citrus volkameriana*, Bryla et al., 1997; *Acer saccharum*, Burton et al., 1998; "Concord" grape, Huang et al., 2005). This response may reflect the net effects of down-regulation of the cytochrome pathway and somewhat enhanced electron flow through the alternative pathway (Ribas-Carbo et al., 2005). The relative effect of water stress on photosynthesis is much greater than that on respiration (e.g., Xu and Baldocchi, 2003; Ribas-Carbo et al., 2005).

Temperature

Respiration rate (especially maintenance and "wastage" respiration) is greatly influenced by temperature and therefore varies with changes in soil and air temperatures, as was indicated in Figure 6.12. Respiration rates typically increase exponentially with a rise in temperature but only over a rather narrow temperature range, usually between 10 and 25°C. Below 10°C the typical response to temperature is approximately linear. At temperatures above 35°C respiration often declines (Brooks et al., 1991).

Growth respiration is affected indirectly by temperature through changes in the rate of production of new tissues. However, the rates of respiration per unit of tissue dry weight should not change. By comparison and as noted earlier, maintenance respiration is very sensitive to temperature (Sprugel et al., 1995).

Foote and Schaedle (1976) found measurable respiration in stems of trembling aspen at −11°C, the lowest temperature imposed, but no photosynthesis below −3°C. These results are similar to those reported for conifers, in which CO_2 production began when the stems were warmed to about 3°C, possibly because the cortical cells thawed at that temperature. The seasonal course of respiration and photosynthesis for aspen twigs is shown in Figure 6.15. The effects of temperature on respiration are mediated by other factors such as tissue water content and amount of available substrate (Lavigne, 1987).

The effects of temperature on plant processes often are indicated by their Q_{10} value, which refers to the ratio of the rate of a process at temperature T to the rate at temperature T +10°C. If the rate doubles the Q_{10} is 2. The dependence of the Q_{10} value on temperature has been attributed to a shift in the activation energy of enzymes. The Q_{10} values for tropical trees are higher than for temperate zone trees (Sprugel and Benecke, 1991).

Kinerson (1975) calculated the Q_{10} for loblolly pine stems to be 2.9. Linder and Troeng (1981a,b) reported a Q_{10} of approximately 2.0 for Scotch pine at various times of the year (except for the high values during the winter when the stems were frozen). Hagihara and Hazumi (1991) noted that the Q_{10} of aboveground parts of hinoki cypress trees varied seasonally from 1.4 to 3.4 and was highest in the winter and lowest in the summer. The greater Q_{10} in winter seems to be a general response of trees (Atkin et al., 2005). Roots tend to exhibit lower values of Q_{10} than do mature leaves under the same conditions (Loveys et al., 2003), and upper canopy leaves may have different Q_{10} values than do lower canopy leaves (Griffin et al., 2002; Turnbull et al., 2003).

The effect of temperature on respiration of various parts of apple trees is shown in Figure 6.16. Increasing

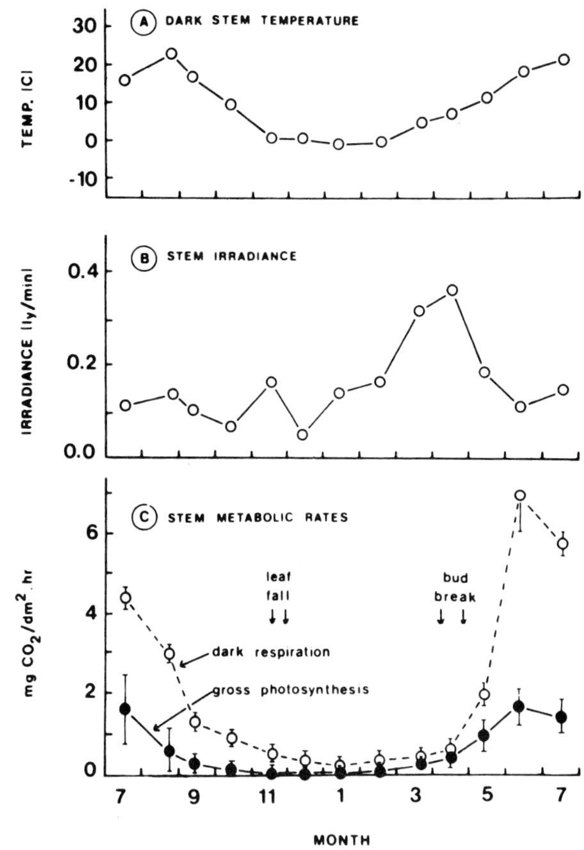

FIGURE 6.15. Seasonal course of respiration and gross photosynthesis of stems of trembling aspen. Individual measurements indicated that respiration occurred down to −11°C, the lowest temperature during the study. No measurable photosynthesis occurred below −3°C. From Foote and Schaedle (1976). © American Society of Plant Physiologists.

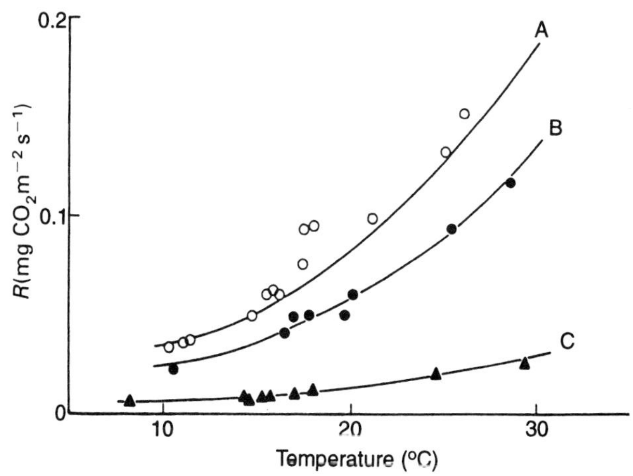

FIGURE 6.16. Effect of temperature on respiration (R) of young apple trees: (A) whole trees, (B) trees without fruit, and (C) trees without fruit and leaves. From Butler and Landsberg (1981).

soil temperatures from 5 to 25°C increased total root respiration from three to five times in seedlings of paper birch, balsam poplar, trembling aspen, and green alder. Both total and maintenance respiration increased exponentially with soil temperature (Lawrence and Oechel, 1983).

Respiration of leaves and roots shows substantial acclimation to shifts in prevailing temperature. For example when roots of *Citrus volkmeriana* plants were exposed to 35°C after growth at 25°C, respiration rates initially increased but thereafter declined over a period of seven days to nearly the level exhibited by roots kept at 25°C (Bryla et al., 2001). However, roots lowered to 15°C, after showing an initial decrease in respiration, did not show similar recovery to the level of plants maintained at 25°C, suggesting a limit to the capacity to acclimate to lower temperature in this subtropical species. Lee et al. (2005) measured leaf respiration at 24°C in *Quercus alba*, *Q. rubra*, and *Acer rubrum* seedlings before and after passage of summer weather fronts that resulted in warming of the air by 7 to 10.5°C. Within three days after warming, dark respiration rates, measured at 24°C, were up to 62% lower than those during cool weather, indicating substantial short-term acclimation. This acclimation was correlated with changes in leaf N content, SLA, and soluble sugar concentrations that could possibly explain some of the acclimation response, but there also appeared to be an additional direct effect of temperature. Although the seedlings in this study were from nearly range-wide collections of seed sources, there was little apparent genotypic variation in acclimation response. Atkin et al. (2005) reviewed the effects of temperature on plant respiration.

Composition of the Atmosphere

Some organs such as leaves are well adapted for rapid gas exchange. Because they have a high surface-to-volume ratio and are constantly exposed to bulk air they are unlikely to experience hypoxia. The inner phloem probably is well supplied with O_2 by inward diffusion through lenticels and cracks in the outer bark. In contrast, the internal tissues of some organs often become deprived of oxygen. Meristems are especially likely to undergo O_2 deficits because their cells are compactly arranged, diffusion is limited by lack of intercellular spaces, and rates of O_2 consumption during metabolism and growth are very high (Harry and Kimmerer, 1991). Hypoxia in the vascular cambium is associated with high activity of alcohol dehydrogenase (ADH), which functions in anaerobic metabolism to catalyze reduction of acetaldehyde to ethanol. Substantial amounts of ethanol occur in tree stems (Kimmerer and Stringer, 1988; MacDonald and Kimmerer, 1991).

Pollock (1953) showed that a higher than normal O_2 concentration did not accelerate respiration of dormant buds, but increased it in growing buds (Fig. 6.17). Respiration of roots and soil organisms reduces the soil O_2 concentration and increases the CO_2 concentration, and the deviation from normal usually increases with soil depth and is greater in the summer than in the winter.

Although elevated CO_2 has demonstrable inhibitory effects on respiratory enzymes (e.g., cytochrome *c* oxidase, succinate dehydrogenase) of isolated mitochondria (Gonzalez-Meler, 1996), there is no convincing evidence that elevated CO_2 inhibits respiration at

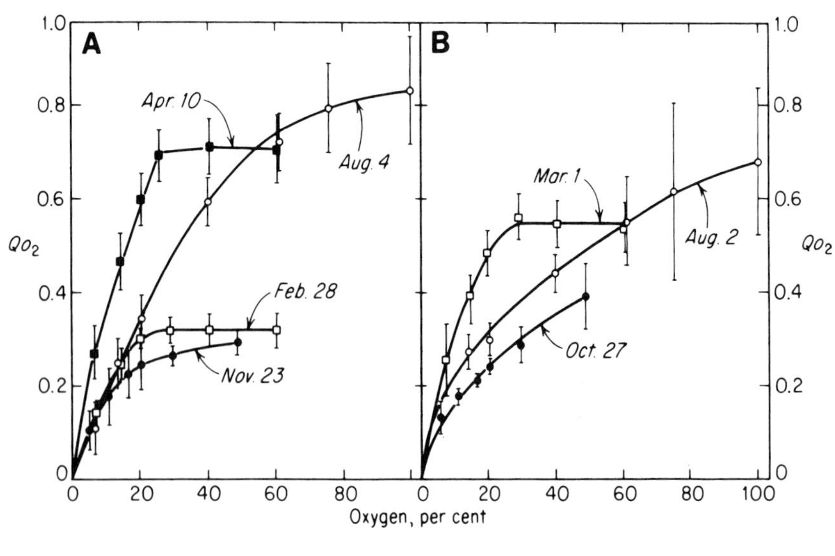

FIGURE 6.17. Effects of season and external oxygen concentration on oxygen uptake by Norway maple (A) and sugar maple (B) buds at 25°C. Oxygen uptake (QO_2) is expressed in microliters of oxygen per milligram dry weight per hour. From Pollock (1953).

the tissue level. This apparent paradox may be attributable to elevated activity of the alternative respiratory pathway at high CO_2, which could compensate for the decline in the cytochrome pathway (Gonzalez-Meler et al., 2004).

Soil Aeration

The soil O_2 that normally is consumed by root respiration is replaced by diffusion from the aboveground atmosphere. However, root respiration is reduced when gas exchange between the air above ground and that in the soil is impeded by soil compaction, a high or perched water table, impermeable layers such as hardpans and pavements, and flooding of soil (Fig. 6.18). Low soil O_2 contents also characterize many heavy-textured soils (Kozlowski, 1984b, 1985a, 1986b).

The rate of root respiration commonly is low in soils compacted by pedestrian traffic, heavy machinery, and grazing animals. Soil compaction is characteristic of many campsites, parks, golf courses, and timber harvesting areas. Forest soils are especially prone to compaction. Soil compaction decreases the number and size of macropores and increases the proportion of micropore space. Such changes inhibit water drainage as well as diffusion of O_2 into and diffusion of CO_2 out of the soil (Chapter 5, Kozlowski and Pallardy, 1997).

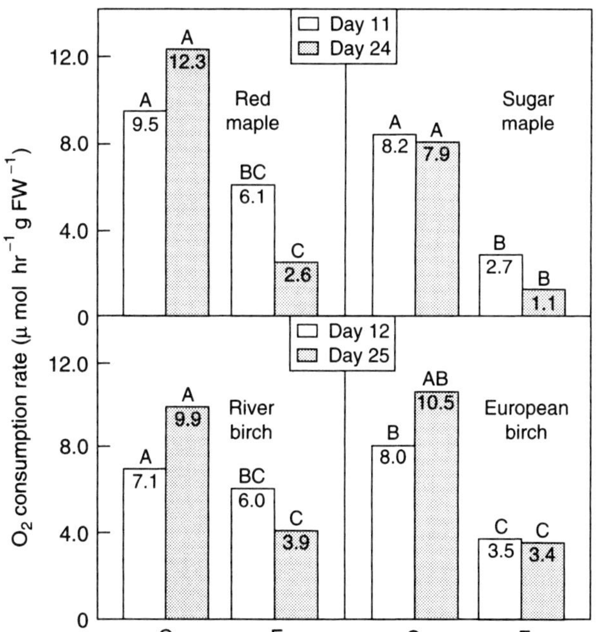

FIGURE 6.18. Effect of flooding of soil on respiration (O_2 consumption) of excised root tips of four species of woody plants. C, control; F, flooded. Uppercase letters indicate differences between means as determined by Tukey's HSD (honestly significant difference) test at the 1% level. Modified from Tripepi and Mitchell (1984).

A deficiency of soil O_2 also may result from placement of fill around shade trees. In an area where clay fill was placed around trees, the soil O_2 concentration declined to near 1% and CO_2 concentration increased to 20%. By comparison, the soil O_2 content in an adjacent undisturbed forest was at least 18% and CO_2 content did not exceed 2.5% (Fig. 6.19). Arborists sometimes install wells, tiles, and gravel fills to increase soil aeration (Harris, 1992). Some species are more tolerant than others of soil fills, presumably because their roots are more tolerant of low O_2 concentrations.

When a soil is flooded, the water occupies the soil pores, causing almost immediate deficiency of soil O_2. The small amounts of remaining O_2 in the soil are consumed by roots and microorganisms within a few hours (Kozlowski, 1984a,b).

A number of wetland species are morphologically adapted to poor soil aeration. Some species absorb O_2 through stomatal pores or lenticels from which it moves downward and diffuses out of the roots to the rhizosphere. Such O_2 transport benefits plants by oxidizing reduced soil compounds such as toxic ferrous and manganous ions (Opik, 1980). Entry of O_2 through leaves is well known in willows (Armstrong, 1968) and lodgepole pine (Philipson and Coutts, 1978, 1980) and through lenticels of twigs, stems, and roots of several species of woody plants (Hook, 1984).

Other adaptations of flood-tolerant species include formation of hypertrophied lenticels on submerged portions of stems and on roots as well as formation of aerenchyma tissue with large intercellular spaces through which O_2 is easily transported (Chapter 5, Kozlowski and Pallardy, 1997).

Inadequate aeration of roots produces a series of physiological disturbances that lead to a reduction in growth and often to death of trees. This is discussed further in Chapter 5 of Kozlowski and Pallardy (1997).

Mechanical Stimuli and Injuries

Handling, rubbing, and bending of leaves often cause large increases in the rate of respiration, as shown in Fig. 6.20. This suggests that care should be taken to avoid rough handling of plant tissues before measuring respiration and perhaps some other processes. Wounding, such as slicing fruits, severing twigs, or cutting out a block of bark or wood, usually is accompanied by an increase in respiration. For this reason many measurements of respiration of severed twigs are higher than the rates in twigs of intact plants.

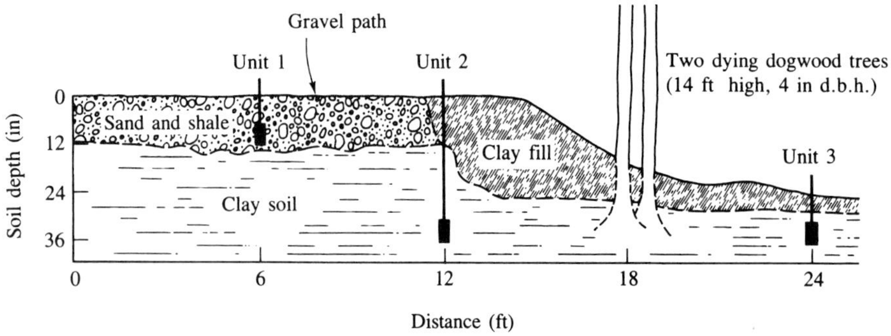

Soil air content (%)

Date	6 ft O_2	6 ft CO_2	12 ft O_2	12 ft CO_2	24 ft O_2	24 ft CO_2
8/10/62	Installed		Installed		Installed	
8/17/62	18.0	2.5	4.0	>20.0	8.0	12.5
10/13/62	19.0	1.5	4.0	>20.0	4.5	15.5
11/16/62	19.0	1.0	4.0	>20.0	6.0	15.0
3/31/63	—	—	1.0	>20.0	—	—

FIGURE 6.19. Change in oxygen and carbon dioxide concentrations beneath soil fills of sand and clay. From Yelenosky (1964).

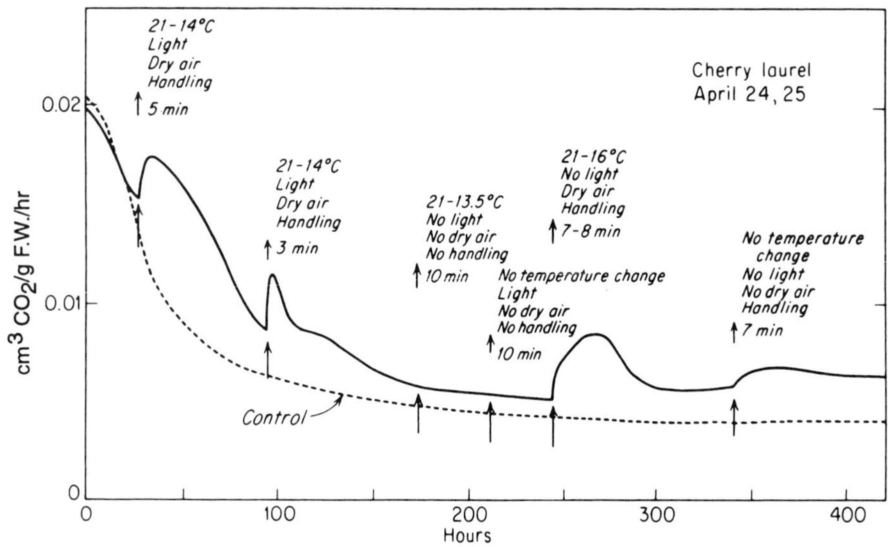

FIGURE 6.20. Effects of mechanical disturbance on respiration of cherry laurel leaves. The controls were disturbed as little as possible; the other group was subjected to some handling during measurements. From Godwin (1935).

Acceleration of respiration following injury is associated with loss of integrity of subcellular organelles, increased availability of O_2, and initiation of repair processes by the infected plant (McLaughlin and Shriner, 1980).

An increase in respiration of plants follows invasion by pathogens that interfere with feedback controls that regulate respiration (Daly, 1976). The high respiration rates of diseased plants reflect increased metabolic activity of the host, the pathogen, or both (Kozlowski, 1992).

Chemicals

Respiration is sensitive to a variety of chemicals that inhibit various stages of the overall process. For example, fluoride blocks conversion of phosphoglycerate to phosphoenolpyruvate, and specific steps in the

Krebs cycle are blocked by fluoroacetate and malonate. Antimycin A blocks between cytochromes *b* and *c*, and inhibitors such as cyanide and carbon monoxide block the final stage of electron transport. Cyanide-resistant respiration also increases in some storage tissues when incubated aerobically (Ikuma, 1972; Solomis, 1977).

Air Pollutants

Effects of pollutants on respiration are complex, with the rate increased or decreased depending on plant species and genotype, the specific pollutant and dosage, plant nutrient balance, developmental stage, extent of injury, time after exposure, and environmental conditions (particularly, light, temperature, and humidity). Stimulation of respiration by pollutants often reflects use of energy in repair processes and hence may be a consequence rather than a cause of cellular injury. Pollutants also affect respiration directly as shown by changes in the rate even in the absence of visible injury.

Fluorides

Exposure to fluoride may stimulate or decrease the rate of respiration, depending to a great extent on available pools of respiratory and photosynthetic intermediates, the relative activity of various respiratory pathways, and concentration of F in plant tissue (Weinstein, 1977; Black, 1984). An increase in respiration usually accompanies early stages of F injury. If the injury is severe the respiratory increase is followed by a decrease. In the absence of visible injury, respiration often is stimulated by low F concentrations and inhibited by high concentrations.

Exposure of eastern white pine and loblolly pine seedlings to low concentrations of F increased respiration (McLaughlin and Barnes, 1975). However, respiration was inhibited in tissues with high concentrations of F. Current-year needles were more sensitive than one-year-old needles. The low respiration rates of tissues with high concentrations of F often have been attributed to inhibition of activity of oxidative enzymes, including enolase, hexokinase, phosphoglucomutase, and succinate dehydrogenase.

Ozone

Ozone (O_3) is a very reactive pollutant. Several investigators reported that low concentration of O_3 stimulated respiration in the absence of or before visible injury occurred. For example, respiration of Valencia orange leaves was stimulated by O_3 when no injury was evident (Todd and Garber, 1958). Barnes (1972) demonstrated increases in respiration of as much as 90% when four species of pine seedlings were exposed to concentrations of O_3 of 100 to 300 $\mu g\ m^{-3}$. McLaughlin et al. (1982) also reported increases in respiration following chronic exposure to oxidants. In contrast, Edwards (1991) reported that annual root respiration rates were 12% lower in loblolly pine seedlings exposed to twice ambient O_3 than in seedlings exposed to subambient O_3. The lower rates of respiration may have been associated with a reduced supply of photosynthate to the roots of plants exposed to elevated O_3. Reduced allocation of photosynthate to roots may be associated with an increased respiratory requirement for maintenance and repair of leaf tissues injured by O_3. Because O_3 has a primary impact on leaves by damaging membranes, a respiratory cost for repair of cells may be expected after a pollution episode (Amthor, 1989).

The effects of O_3 on leaf respiration vary appreciably with the age of leaves. Dark respiration increased rapidly in *Populus deltoides* × *P. trichocarpa* leaves exposed to 0.125 ml l^{-1} O_3 and then rapidly to low levels in the next 10 days (Reich, 1983). Respiration rates of treated leaves declined progressively with leaf age. In leaves four to 35 days old, respiration was higher than in control leaves. After 35 days, respiration of treated leaves was still higher than in control leaves but the differences were small and the rates were low.

Sulfur Dioxide

In the absence of injury, respiration rates may be increased or decreased by SO_2, often depending on dosage. Respiration rates usually return to control values if the SO_2 concentration is low and the duration of exposure short. This change in response may be associated with capacity of plants to detoxify sulphite or repair SO_2 injury (Black, 1984).

Several investigators reported stimulation of respiration following exposure of woody plants to SO_2. Examples are Scotch pine (Oleksyn, 1984; Katainen et al., 1987) and Carolina poplar (Van Hove et al., 1991). Others reported a decrease. Photorespiration was inhibited much more by SO_2 in a susceptible clone of Scotch pine than in a tolerant one (Lorenc-Plucinska, 1989).

Nitrogen Oxides

Only a few studies are available of the effects of NO_2 on respiration of woody plants. However, work with herbaceous plants showed that both respiration and photorespiration were inhibited by NO_2 with the amount of decrease increasing with higher NO_2 dosage, temperature, and duration of exposure (Srivastava et al., 1975). Both photorespiration and dark respiration of seedlings of both tolerant and susceptible Scotch pine trees were inhibited within 30 minutes after fumigation and stimulated or unaffected after 24 and 48 hours (Fig. 6.21).

ASSIMILATION

The term assimilation as used here refers to the conversion of food—that is, carbohydrates, fats, and proteins—into new tissue. Not only does this require large amounts of energy supplied by respiration, but it also uses compounds synthesized in various parts of the respiratory cycle. Reference to Figure 6.22 indicates that nucleic acids and nucleotides, amino acids, fatty acids, and other substances important in metabolism of plants originate in different parts of the respiratory cycle. Many of the metabolic pathways in the synthesis and degradation of these substances pass through acetyl coenzyme A, as mentioned earlier, which forms the crossroads for a variety of important metabolic reactions.

Assimilation is an integral part of growth, and it therefore is most conspicuous in meristematic regions such as the cambia and root and stem tips. The simple carbohydrates translocated to these meristematic regions are converted into cellulose, pectic compounds, and lignin in the cell walls, and the amino acids and amides are transformed into the protein framework and enzymes of new protoplasm. The existing protoplasm not only produces new protoplasm and new cell walls, but a wide variety of other substances such as organic N compounds; chlorophyll; the carotenoids and other pigments; lipids and isoprene derivatives such as essential oils, oleoresins, and rubber; sterols; tannins; alkaloids (Chapter 8); hormones; and numerous other compounds. Most of these play important roles in plant metabolism, but some, such as the alkaloids and rubber, seem to have no known essential functions in plants. The origin of some of these compounds is shown in Figure 6.22.

Table 6.3 shows the approximate chemical composition of the shoots (needles and adjacent twig) of loblolly pine, expressed as percentage of total dry weight. The high percentage of phenolics is surprising in view of the high energy requirement for their synthesis and the uncertainty concerning their utility. The relative amounts of substrate used and CO_2 produced per gram of each of the major constituents is shown in Table 6.4. The amounts of substrate required and CO_2 released increase with the degree to which a compound is reduced, because more energy is required to produce highly reduced compounds such as lipids and pheno-

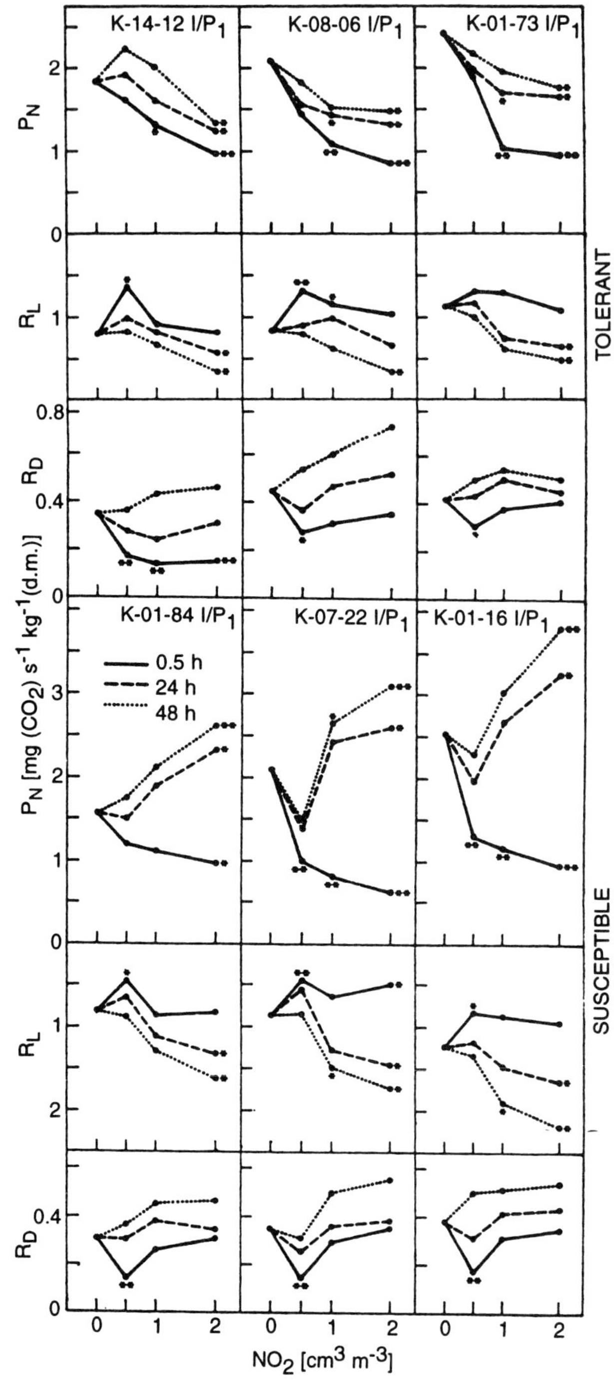

FIGURE 6.21. Rates of net photosynthesis, (P_N) photorespiration (R_L), and dark respiration (R_D) of tolerant (top) and susceptible (bottom) Scotch pine seedlings treated with 0.5, 1.0, and 2.0 cm^3 NO_2 m^{-3} for 3 days, 6 hrs per day. Times after NO_2 fumigation were: 0.5, 24, and 48 hrs. Asterisks *, ** indicate values significantly different from control (0 cm^3 NO_2 m^{-3}) at the 0.05 and 0.01 levels, respectively. From Lorenc-Plucinska (1988).

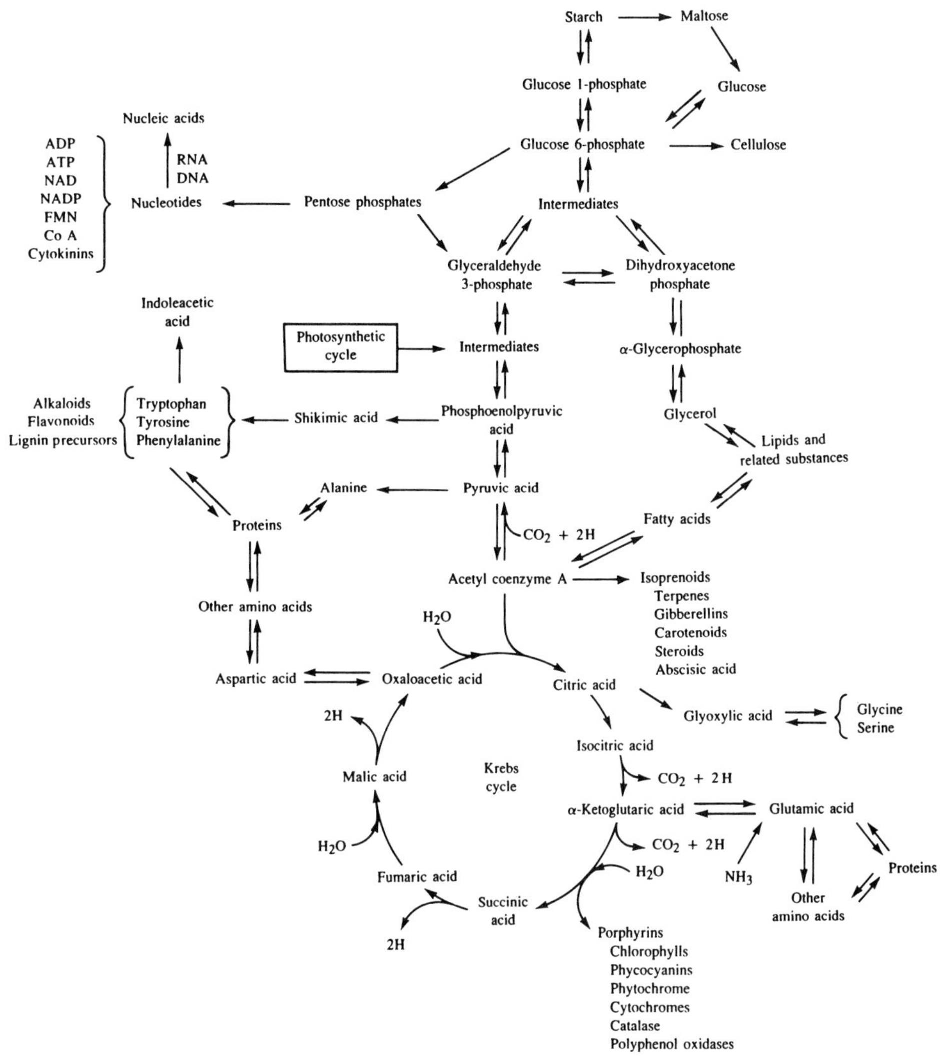

FIGURE 6.22. Some metabolic pathways in plants.

lics. For example, synthesis of 1 g of carbohydrate requires about 1.2 g of glucose and 0.11 g of O_2 with 0.149 g of CO_2 being released, but synthesis of 1 g of highly reduced lipid requires 3.0 g of glucose and 0.3 g of O_2 with 1.5 g of CO_2 being released. Synthesis of lignin also is expensive in terms of the amount of carbohydrate used.

The course of assimilation and the kinds of substances produced are regulated by the enzymes present. These in turn are controlled by heredity, although the amounts and kinds of enzymes present may be modified by the environment. Various plant families produce characteristic chemical compounds. For example, each species of pine produces its own characteristic oleoresins, and there sometimes are differences among geographic races (Mirov, 1954). Various attempts have been made to correlate plant classification with chemical composition (e.g., Gibbs, 1958, 1974; Fairbrothers et al., 1975).

SUMMARY

The course of assimilation and the kinds of substances produced in woody plants are regulated by many enzymes, which are classified on the basis of the reactions they catalyze. Most chemical reactions require addition of enzymes that lower the energy of activation sufficiently for chemical reactions to occur at room temperatures. Isozymes (enzymes that exist in several molecular forms and act on the same substrate) are important in forest genetics and tree improvement research. Enzyme activity is influenced by temperature, pH, concentration of enzyme and substrate, as well as end products, tissue hydration, and hormonal growth regulators.

Growth and development of woody plants depend on respiration for a continuous supply of carbon skeletons and energy for use in synthesis of new protoplasm, maintenance of the structure of organelles and

TABLE 6.3. Chemical Composition of Shoots of Loblolly Pine as Percentages of Oven Dry Weight[a]

Constituents	Percent of dry weight
Nitrogenous compounds	8.4
Amino acids	(7.2)
Protein	(90.7)
Nucleic acids	(2.1)
Carbohydrates	38.0
Reducing sugars	(5.1)
Sucrose	(8.2)
Cellulose	(56.0)
Hemicelluloses	(25.7)
Pectin	(5.0)
Lipids	5.3
Fatty acids and resin acids	(74.8)
Glycerols	(6.0)
Unsaponifiable	(19.2)
Lignin	23.3
Organic acids	3.5
Shikimic	(48.7)
Quinic	(51.3)
Phenolics	20.0
Minerals	1.5

[a]From Chung and Barnes (1977). Numbers in parentheses are subfraction percentages of major fractions.

TABLE 6.4. Amounts of Substrate Used and CO_2 Produced during Synthesis of the Principal Constituents of the Shoots of Loblolly Pine[a,b]

	Substrate used		Byproduct	
Constituents	Glucose	O_2	CO_2	H_2O
Nitrogenous compounds	1.58	0.28	0.40	0.65
Carbohydrates	1.18	0.11	0.14	0.16
Lipids	3.02	0.30	1.50	0.82
Lignin	1.90	0.04	0.27	0.66
Organic acids	1.48	0.35	0.48	0.35
Phenolics	1.92	0.37	0.56	0.73

[a]Amounts are in grams per gram of constituent.
[b]From Chung and Barnes (1977).

membranes, and active transport of ions and molecules across membranes. Assimilation (conversion of foods into new tissues) uses materials synthesized in the respiratory cycle. For example, nucleic acids, nucleotides, amino acids, fatty acids, and other compounds important in plant metabolism are synthesized from various respiratory intermediates.

Appreciable amounts of carbohydrates are consumed by woody plants in maintenance respiration during the dormant season and in seedlings kept in cold storage.

The oxidation of glucose can be summarized as follows:

$$C_6H_{12}O_6 + 6O_2 \rightarrow 6CO_2 + 6H_2O + 686{,}000 \text{ calories/mole} \quad (6.12)$$

Oxidation of glucose occurs in several steps, which fall into two general groups: (1) glycolysis, by which glucose is converted to pyruvic acid, followed by the Krebs (tricarboxylic acid) cycle in which large amounts of ATP are formed, and (2) the pentose shunt. Reducing power (NADH) produced in the Krebs (tricarboxylic acid) cycle supports electron transport in the cytochrome pathway, leading to proton pumping at ATP formation. Additionally, plants possess membrane-bound dehydrogenase activity that does not lead to proton pumping, and an alternative oxidase that reduces oxygen to water but does not cause ATP formation from proton extrusion. The possible functions of these alternative pathways in the plant's highly branched respiratory system are not known.

Under O_2 deficiency, metabolism is characterized by concurrent aerobic respiration and some anaerobic respiration. The latter is inefficient and does not supply enough energy to support rapid growth. Under anaerobic conditions carbohydrates may not be completely oxidized to CO_2 and water, and intermediate compounds accumulate. Incompletely oxidized compounds may injure plants.

The amount of food depleted by respiration of various organs and tissues may be high, sometimes exceeding 60% of the daily production of photosynthate. The rates of respiration and food consumption in different organs vary greatly, largely because of differences among them in the proportions of physiologically active tissues. Seasonal variations in respiration reflect effects of warming and cooling on the growth cycle as well as the direct effects of temperature on respiration. From an ecosystem perspective, a combination of autotrophic (root) and heterotrophic respiration from the soil dominates total respiration.

The rate of respiration is influenced by both internal and environmental factors. Important internal factors are age and physiological condition of tissues, the amount of oxidizable substrate, and tissue hydration. The important environmental factors include air and soil temperature, light, soil aeration, soil moisture, mechanical stimuli and injury, and various chemicals such as herbicides, fungicides, insecticides, fertilizer, and environmental pollutants.

General References

Amthor, J. S. (1989). *Respiration and Crop Productivity*. Springer-Verlag, New York and Berlin.

Amthor, J. S. (1994). Plant respiratory responses to the environment and their effects on the carbon balance. In *Plant-Environment Interactions*, R. E. Wilkinson, ed., pp. 501–554. Marcel Dekker, New York.

Amthor, J. S. (2000). The McCree-de Wit-Penning de Vries-Thornley respiration paradigms—30 years later. *Ann Bot* **86**, 1–20.

Amthor, J. S. and Baldocchi, D. D. (2001). Terrestrial higher plant respiration and net primary production. In *Terrestrial Global Productivity*, J. Roy, B. Saugier, and H. A. Mooney, eds., pp. 33–59. Academic Press, San Diego.

Artus, N. N., Somerville, S. C., and Somerville, C. R. (1986). The biochemistry and cell biology of photorespiration. *Crit Rev Plant Sci* **4**, 121–147.

Becker, W. M. and Deamer, D. W. (1991). *The World of the Cell, 2nd edition*. Benjamin Cummings, Redwood City, California.

Brady, C. J. (1987). Fruit ripening. *Annu Rev Plant Physiol* **38**, 155–178.

Cannell, M. G. R. and Thornley, J. H. M. (2000). Modeling the components of plant respiration: Some guiding principles. *Ann Bot* **85**, 45–54.

Fersht, A. (1985). *Enzyme Structure and Mechanism*. Freeman, New York.

Hagihara, A. and Hozumi, K. (1991). Respiration. In *Physiology of Trees*, A. S. Raghavendra, ed., pp. 87–110. Wiley, New York.

Hook, D. D. and Crawford, R. M., eds. (1978). *Plant Life in Anaerobic Environments*. Ann Arbor Science, Ann Arbor, Michigan.

Kozlowski, T. T. and Pallardy, S. G. (2002). Acclimation and adaptive responses of woody plants to environmental stresses. *Bot Rev* **68**, 270–334.

Millenaar, F. F. and Lambers, H. (2003). The alternative oxidase: In vivo regulation and function. *Plant Biol* **5**, 2–15.

Lassoie, J. P. and Hinckley, T. M., eds. (1991). *Techniques and Approaches in Forest Tree Ecophysiology*. CRC Press, Boca Raton, Florida.

Lehninger, A. L. (1973). *Bioenergetics*. Benjamin, Menlo Park, California.

Page, M. I. (1984). *The Chemistry of Enzyme Action*. Elsevier, New York.

Palmer, T. (1985). *Understanding Enzymes, 2nd edition*. Wiley, New York.

Price, N. C. and Stevens, L. (1989). *Fundamentals of Enzymology, 2nd edition*. Oxford University Press, Oxford and New York.

Rasmusson, A. G., Soole, K. L., and Elthon, T. E. (2004). Alternative NAD(P)H dehydrogenases of plant mitochondria. *Annu Rev Plant Biol* **55**, 23–39.

Schulte-Hostede, S., Darrall, M., Blank, L. W., and Wellburn, A. R., eds. (1988). *Air Pollution and Plant Metabolism*. Elsevier Applied Science, London and New York.

Siedow, J. N. and Day, D. A. (2001). Respiration and photorespiration. In *Biochemistry and Molecular Biology of Plants*, B. Buchanan, W. Gruissem, and R. L. Jones, eds., pp. 676–728. *Amer Soc Plant Physiol*, Rockville, Maryland.

Soltis, P. E. and Soltis, P. S., eds. (1989). *Isozymes in Plant Biology*. Dioscorides Press, Portland, Oregon.

Sprugel, D. G., Ryan, M. G., Brooks, J. R., Vogt, K. A., and Martin, T. A. (1995). Respiration from the organ level to the stand. In *Resource Physiology of Conifers: Acquisition, Allocation, and Utilization*, W. K. Smith and T. M. Hinckley, eds., pp. 255–299. Academic Press, San Diego.

CHAPTER

7

Carbohydrates

INTRODUCTION

Carbohydrates are of special importance because they are direct products of photosynthesis and are therefore the primary energy storage compounds and the basic organic substances from which most other organic compounds found in plants are synthesized. Carbohydrates also are chief constituents of cell walls; they are the starting point for the synthesis of fats and proteins; large amounts are oxidized in respiration (Chapter 6); another fraction is accumulated as reserve foods; and still another portion is variously lost from plants. Soluble carbohydrates decrease the osmotic potential of the cell sap, and such carbohydrates as the pentosans, pectic compounds, gums, and mucilages increase the water-holding capacity of tissues. Quantitatively, carbohydrates are the most important constituents of woody plants, comprising up to three-fourths of their dry weight. This chapter deals with the kinds of carbohydrates found in woody plants, their transformations, uses, losses, and accumulation. The importance of carbohydrates in seed germination, vegetative growth, and reproductive growth is discussed in more detail in Chapters 2, 4, and 6 of Kozlowski and Pallardy (1997).

KINDS OF CARBOHYDRATES

Carbohydrates are made up of carbon, hydrogen, and oxygen approximating the empirical formula $(CH_2O)_n$. Many carbohydrates also contain other elements such as phosphorus or nitrogen. Carbohydrates can be classified in three main groups: monosaccharides, oligosaccharides, and polysaccharides. Figure 7.1 shows the classification of the more important carbohydrates.

Monosaccharides

The monosaccharides include simple sugars and their derivatives. They are the basic carbohydrate units from which more complex compounds are formed. Monosaccharides consist of carbon atoms to which are attached hydrogen atoms, at least one hydroxyl group, and either an aldehyde (RCHO) or ketone (RCOR) group. The number of carbon atoms in monosaccharides varies from three to eight, but the most common number is five (e.g., pentoses, $C_5H_{10}O_5$) or six (e.g., hexoses, $C_6H_{12}O_6$). Monosaccharides do not yield smaller molecular weight sugars on hydrolysis.

Many simple sugars occur in woody plants, usually in very small amounts, probably because of their rapid

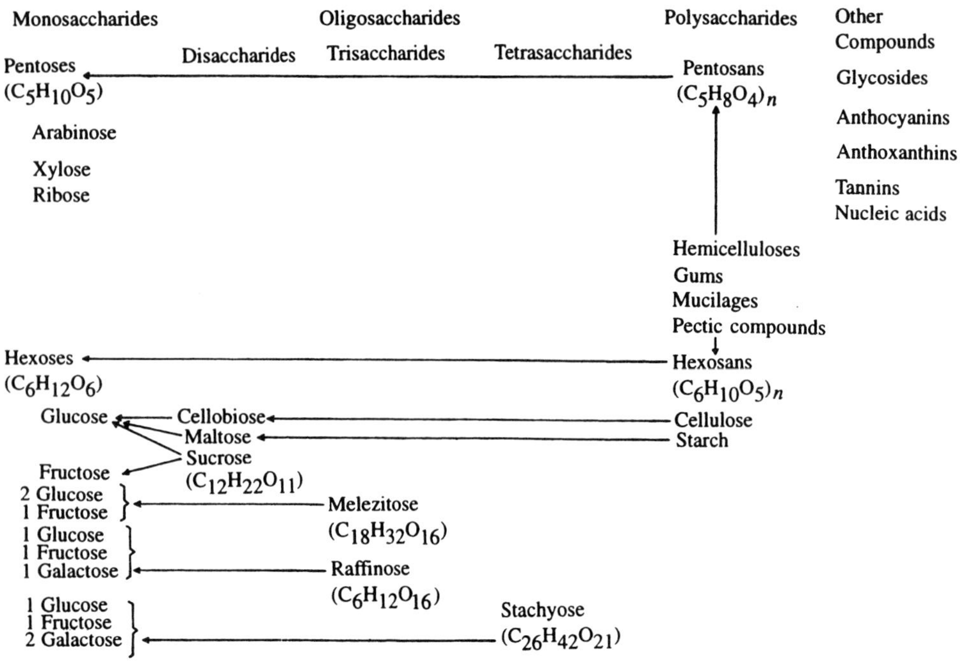

FIGURE 7.1. Relationships among some important carbohydrates and products of their hydrolysis.

incorporation in polysaccharides. Exceptions are the six-carbon sugars, glucose, and fructose. Glucose is present in large amounts, especially in certain fruits, and probably occurs in every living cell. Fructose also is common and abundant, although its concentration usually is lower than that of glucose. Glucose and fructose occur not only in living cells, but also in the xylem sap of certain trees such as maples and birches (Chapter 5, Kozlowski and Pallardy, 1997). Derivatives of glucose and fructose that have been phosphorylated—that is, have had phosphate groups attached to them—form the starting point for many metabolic transformations of carbohydrates (Fig. 6.22). Some of the hexose sugars occur chiefly as polymers, for example, galactose and mannose as galactans and mannans.

Although only traces of pentose sugars are found free in plants, their condensation products, the pentosans, are important constituents of cell walls. The pentose sugars, arabinose and xylose, rarely occur free but often are present as parts of cell wall polymers, the arabans and xylans. Ribose, another pentose sugar, also occurs in a combined form as a constituent of such nucleotide coenzymes as ATP, NAD^+, $NADP^+$, FAD, and coenzyme A. Ribose also is found as a part of ribonucleic acid (RNA). Deoxyribose occurs in the nucleotides that constitute deoxyribonucleic acid (DNA).

Many of the monosaccharides are associated with the Calvin-Benson cycle of photosynthesis and in the alternate pentose shunt of respiration (Chapter 6). Glucose is the principal compound produced in the former and serves as substrate for the latter. The pentose, ribulose 1-5 bisphosphate (RuBP), reacts with CO_2 in photosynthesis. Both cycles, but especially the pentose shunt, are the source of pentoses and many other monosaccharides found as parts of more complex molecules present in plants.

Oligosaccharides

The oligosaccharides consist of linkages of two or more molecules of monosaccharides. The major oligosaccharides include disaccharides (e.g., sucrose, maltose), trisaccharides (e.g., raffinose, melezitose), and tetrasaccharides (e.g., stachyose). The disaccharide sucrose is considered the most important oligosaccharide in plants because of its high concentration in cells, wide distribution, and metabolic importance. Together with starch, sucrose is a major reserve carbohydrate. In many plants sucrose represents over 95% of the dry weight of the material that is translocated in the sieve tubes of the phloem. Maltose also is common, but it usually occurs in lower concentrations than sucrose.

Sugars other than sucrose and maltose often are found in variable amounts. For example, small amounts of the higher oligosaccharides of the raffinose family (raffinose, stachyose, and verbascose) are found in sieve tubes of certain plants. These sugars are related and consist of sucrose with variable numbers of galactose units attached (Fig. 7.2). Whereas sugars of the raffinose family are relatively unimportant in most

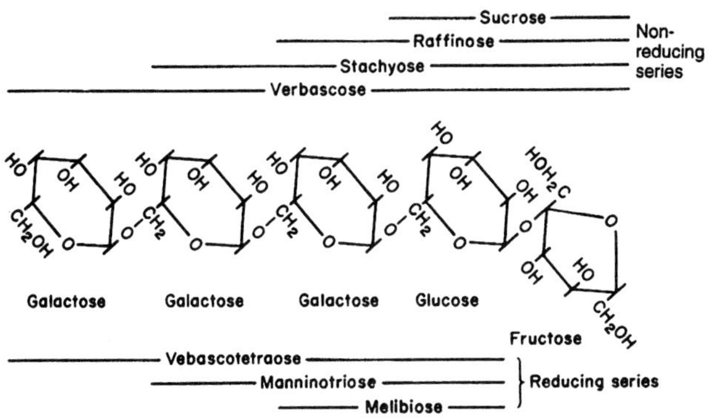

FIGURE 7.2. The raffinose family of oligosaccharides. From Zimmermann and Brown (1971), with permission of Springer-Verlag.

plants, they are of considerable importance in a few plant families, including the Bignoniaceae, Celastraceae, Combretaceae, Myrtaceae, Oleaceae, and Verbenaceae (Zimmermann, 1957).

The disaccharide trehalose is a nonreducing sugar consisting of glucose molecules linked by an α-1,1 oxygen-ether bridge. Trehalose serves as a storage carbohydrate like sucrose in bacteria, fungi and insects, and also may confer protection against desiccation, freezing, and heat stress (Müller et al., 2001; Schluepmann et al., 2003). Until recently only desiccation tolerant resurrection plants (Chapter 12) were known to accumulate appreciable amounts of trehalose (Drennan et al., 1993). However, genes for metabolism of trehalose are widely distributed in plants (Paul et al., 2001) and experiments with validamycin A, an inhibitor of trehalase—the enzyme that degrades trehalose—resulted in detectable accumulation of trehalose in plants. These results indicated a functioning trehalose biosynthetic pathway in higher plants. Current research suggests that trehalose-6-phosphate (T-6-P) may function as a signaling molecule by modulating the activity of hexokinase, a key enzyme in glycolysis (Schluepmann et al., 2003). High T-6-P levels may inhibit hexokinase activity and render the photosynthetic apparatus insensitive to sugar feedback regulation (Chapter 5). Paul et al. (2001) cited evidence that tobacco plants transformed to elevate expression of T-6-P synthase activity had significant elevations in light saturated photosynthetic rates. Hence manipulation of trehalose metabolism may aid in efforts to increase photosynthesis in plants.

Polysaccharides

The most important polysaccharides in woody plants are cellulose and starch. Cellulose is the most abundant organic compound. It has been estimated

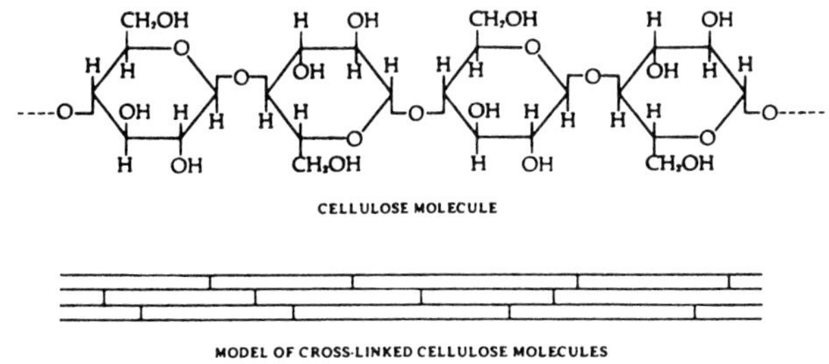

FIGURE 7.3. Structure of cellulose. The OH groups that project from both sides of the chain form hydrogen bonds with neighboring H groups, resulting in bundles of cross-linked parallel chains.

that the global standing crop contains 9.2×10^{11} tons of cellulose, produced at a rate of 0.85×10^{11} tons per year (Duchesne and Larson, 1989). Cellulose is the chief constituent of the cell walls that form the framework of woody plants. Whereas the expanding primary cell walls contain 20 to 40% cellulose (dry weight basis), secondary walls contain 40 to 60%.

Each cellulose molecule consists of at least 3,000 glucose residues linked together by oxygen-ether bridges between the 1 and 4 carbon atoms of adjacent molecules to form long straight, unbranched chains (Fig. 7.3). The chains are organized into microfibrils. The bundles of cellulose molecules in the microfibrils give the cell wall high tensile strength (Duchesne and Larson, 1989). The β1-4 linkage of cellulose results in a stiff molecule capable of forming fibrils by hydrogen bonding. The spaces in pure cellulose walls, such as those of cotton fibers, are occupied by water but become partly filled with lignin in woody tissue and with pectin compounds, cutin, or suberin if these substances are present. Among the notable characteristics of cellulose are its insolubility in water and organic solvents as well as its high resistance to both chemical and enzymatic degradation.

A multi-subunit enzyme complex, cellulose synthase, is responsible for cellulose synthesis. This complex, which is assembled in the Golgi apparatus and then transported to the plasmalemma, consists of six subunits, each of which contains six polypeptides (Doblin et al., 2002). Each polypeptide produces a single β-(1,4) glucan chain. Six subunits may associate to form a 36-polypeptide rosette (Fig. 7.4). Cellulose microfibrils arise from the glucan chains produced by a rosette. Molecular evidence suggests that the genes responsible for cellulose synthase proteins (*Ces* genes) are likely derived from an endosymbiotic transfer from cyanobacteria to plants (Nobles et al., 2001). Among woody plants, *Ces* genes have been identified in poplar and loblolly pine (Wu et al., 2000; Nairn and Haselkorn, 2005). The enzymatic breakdown of cellulose to glucose

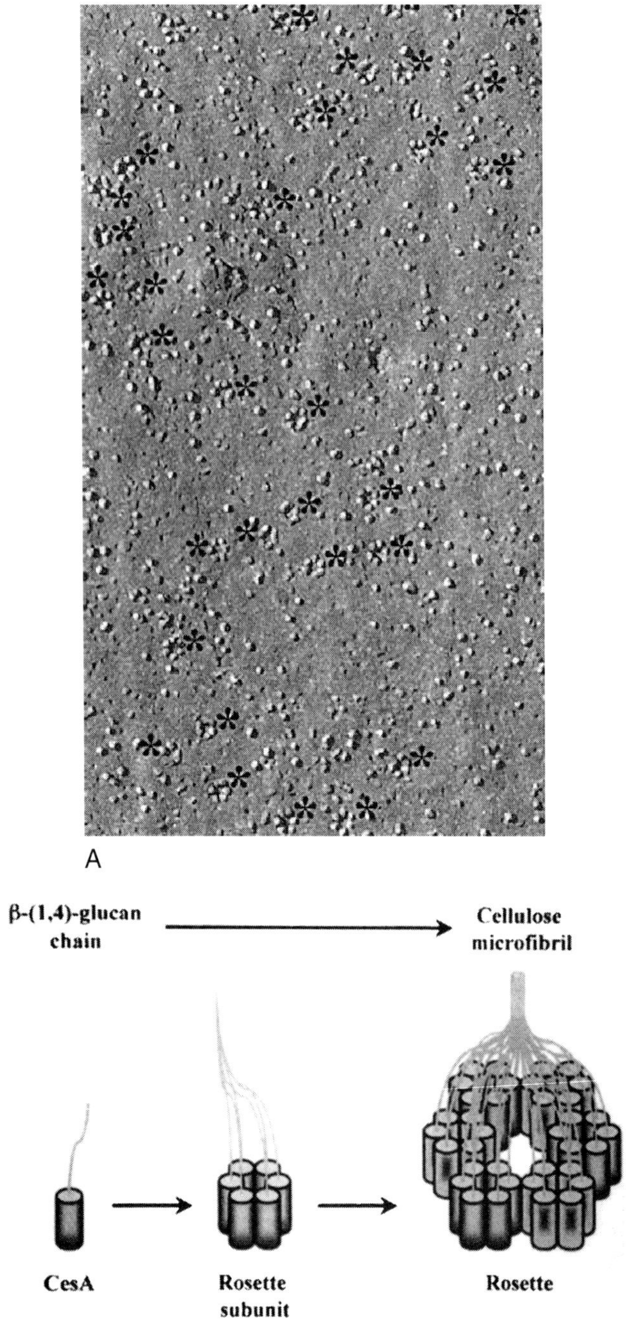

FIGURE 7.4. Structure of the cellulose synthase complex. (A) Freeze-fracture electron micrograph of the inner face of the plasma membrane of *Zinnia elegans* cells. Asterisks are located adjacent to intact cellulose synthase "rosettes." (B) A model of rosette structure. (A) From *Planta*, Characterization of a novel cellulose synthesis inhibitor. Kiedaisch, B. M., Blanton, C. R. L., and Haigler, C. H. **217**, 922–930. Figure 4a © 2003 with kind permission from Springer Science and Business Media. (B) From Doblin, M. S., Kurek, I., Jacob-Wilk, D., and Delmer, D. P. Cellulose biosynthesis in plants: From genes to rosettes. *Plant Cell Physiol.* (2002). **43(12)**, 1407–1420, by permission of Oxford University Press.

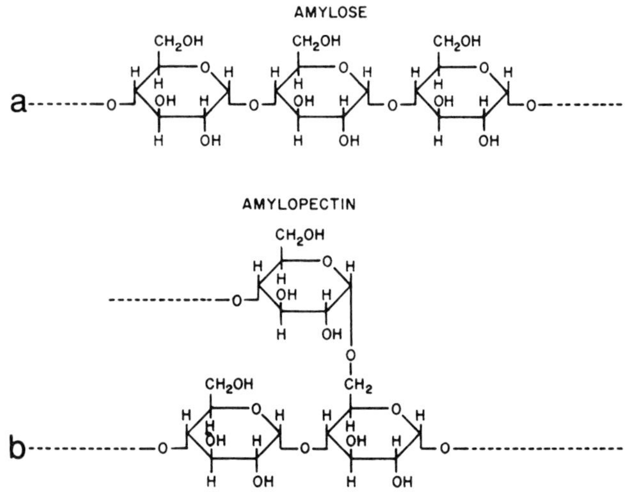

FIGURE 7.5. Structure of starch: (a) amylose; (b) amylopectin.

requires two enzymes: (1) cellulase, which catalyzes the formation of cellobiose; and (2) cellobiase, which carries the digestion further to glucose.

Genes similar to *Ces*, called cellulose synthase-like (*Csl*) genes because of base-pair similarities, may code for noncellulosic polysaccharides. For example, when *Drosophila* cells engineered to express *Csl* genes from plants were provided GDP-mannose they produced β-linked mannan polymers, and when provided with GDP-mannose and GDP-glucose produced mixed β-linked glucomannan polymers (Liepman et al., 2005).

Starch is the most abundant reserve carbohydrate in woody plants. Starch grains cannot pass from cell to cell; hence starch must have been synthesized in the tissues in which it is found. It is formed by condensation of hundreds of glucose molecules into long, often spiraled chains. As in cellulose, the glucose residues are linked together by oxygen-ether bridges between the C-1 and C-4 atoms, but starch has an α linkage and cellulose has a β linkage (see Figs 7.3 and 7.5). Once cellulose is formed glucose cannot be recovered and reused because plants lack the enzymes necessary to degrade cellulose. In contrast, starch is readily degraded enzymatically in plants and thus becomes important in metabolic processes and growth of plants. Starch has two forms that differ in some physical properties. The most abundant component of most starch is amylopectin, which consists of very long molecules with numerous branched side chains. The other component is amylose, which consists of unbranched chains containing 300 to 1,000 residues (Fig. 7.5). Amylose gives a deeper blue color with iodine and is more water soluble and more viscous than amylopectin.

Starch accumulates in grains formed of many layers, which give the grains a laminated appearance. Starch grains often occur in living cells (axial and ray parenchyma) of the sapwood of woody plants. They also occur in the living phloem cells of the inner bark. The amount of starch in the woody structure of trees varies seasonally. Starch grains also occur in large numbers in the chloroplasts of cells of almost all leaves (Fig. 5.11). The structure of starch grains is a useful diagnostic feature in identifying sources of different plant materials, including spices, drugs, and some archeological samples. Because the size, shape, and structure of starch grains in a plant vary within defined limits, the starch grains of different species often can be identified (Cortelli and Pochettino, 1994).

Hemicelluloses are matrix polysaccharides that hydrogen bond to cellulose microfibrils. They are found in all woody tissues and include arabans, xylans, galactans, and mannans. Unlike cellulose, the constituents of hemicellulose differ among plants. In the wood of angiosperms the primary hemicellulose is a xylan (a polymer of xylose), whereas in gymnosperm wood the primary hemicellulose is a glucomannan. However, some of each of these polysaccharides occurs in both types of wood. Hemicelluloses are not forms of cellulose and differ from cellulose primarily in three ways (Nevell and Zeronian, 1985):

1. They contain several different sugars instead of only glucose.
2. They show variable chain branching (whereas cellulose is a linear polymer).
3. The extent of polymerization of cellulose is many (up to 100) times greater than that of most hemicelluloses.

Hemicelluloses occur in some seeds, including those of persimmon and certain palms, and they are metabolized during germination (see Chapter 2 of Kozlowski and Pallardy, 1997). Although hemicelluloses of xylem cell walls generally do not function as reserve foods, a few exceptions have been cited. It has been claimed, for example, that hemicelluloses of the cell walls of white oak xylem may serve as reserve foods (McLaughlin et al., 1980). The hemicellulose in *Eucalyptus obliqua* and apple also were reported to function as reserves and support root growth (Kite, 1981; Stassen, 1984).

Pectic compounds are hydrophilic substances that occur in the middle lamella and in the primary walls of cells, especially in fruits, but they are not present in woody tissues in large amounts. Gums and mucilages are complex carbohydrates of high molecular weight that somewhat resemble pectic compounds. An example is the well-known gum arabic, also known as gum acacia, that is produced by *Acacia senegal*. Although gums vary appreciably in their composition and are specific for genotypes, they are mainly polysaccharides based on glucuronic acid with associated hexose and pentose sugars. They also generally contain phenolic substances. The chemistry of gums and mucilages is similar, but some mucilages also contain proteins.

Gum formation, which appears to be a natural phenomenon in many species, is greatly accelerated by injury or invasion by pathogens. Excessive gum formation or gummosis in response to wounding is well known in sweetgum, acacia, citrus, peach, prune, apricot, plum, sweet cherry, sour cherry, and almond.

Gum-producing ducts often develop in response to fungal and virus diseases. For example, when citrus trees were infected with *Phytophthora citrophthora*, the causal agent of brown rot gummosis, gum ducts developed schizogenously by separation of cells (Fahn, 1988a,b). Some investigators attributed gum formation to decomposition of cell walls of specialized parenchyma cells that differentiate in the cambium and later disintegrate and form both the duct and the gum (Stösser, 1979). However, many studies show that gums commonly are synthesized in epithelial cells surrounding gum ducts (Fig. 7.6) (Gedalovich and Fahn, 1985a; Morrison and Polito, 1985).

Ethylene appears to play a role in formation of gum ducts during cambial activity. This is inferred by formation of gum ducts following application of ethephon (which forms ethylene) to a variety of plants (Stösser, 1979; Gedalovich and Fahn, 1985b). Ethylene may act by inducing synthesis of hydrolytic enzymes in xylem mother cells. The enzymes dissolve the middle lamella between the cells, leading to formation of a duct lumen.

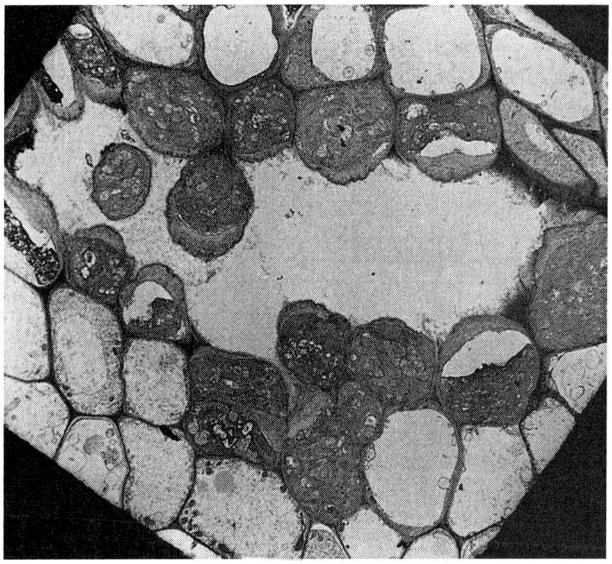

FIGURE 7.6. Young gum duct in citrus, showing the epithelial secretory cells. Photo courtesy of A. Fahn, Hebrew University.

Another growth regulator, methyl jasmonate, also may induce the accumulation of gums alone or synergistically with ethylene (Saniewski et al., 2004; Skrzypek et al., 2005).

Vascular wilt diseases such as oak wilt, Dutch elm disease, and verticillium wilt of elm and maple are associated with occlusion of xylem vessels by gums and tyloses, possibly leading to dehydration of leaves by preventing water transport from the roots (Kozlowski, 1962; Shah and Babu, 1986). This is discussed further in Chapter 11.

Examples of mucilages are the slimy substances from the inner bark of slippery elm and the sticky substances found in the seed pods of carob and honeylocust. Mucilages may play a role as food sources, adhesives in dispersal of seeds, regulation of seed dormancy, lubrication of growing root tips, and influences on root-microorganism interactions (Fahn, 1988a,b).

CARBOHYDRATE TRANSFORMATIONS

Many of the carbohydrates found in plants are continually undergoing conversion from one form to another or are being transformed into compounds used in respiration or synthesis of fats, proteins, and other noncarbohydrates (see Fig. 6.20).

Starch-sugar conversions in both vegetative and reproductive tissues occur commonly. In developing seeds, for example, there is a period during which sugars, principally sucrose, are converted to starch. In cottonwood leaves starch was degraded during the night and free sugars were converted to sucrose (Dickson, 1987). In a number of ripening fruits starches are converted to sugars. Starch concentrations in developing apples increased early in the season and decreased near fruit maturation, at which time sucrose concentrations increased (Pavel and De Jong, 1995). Starch apparently was converted to sucrose at that time as indicated by increasing activities of sucrose synthetase and sucrose phosphate synthase (Moriguchi et al., 1992).

Phosphorylation

The first step in many carbohydrate transformations is phosphorylation, a priming process in which monosaccharide sugars react with ATP (adenosine triphosphate) to form phosphate esters, while ATP is converted to ADP (adenosine diphosphate). A sugar, for example, may be converted to another sugar by the following general scheme:

$$\text{Sugar A} + \text{ATP} \xrightarrow{\text{enzyme A}} \text{sugar A-phosphate} + \text{ADP}$$

$$\text{Sugar A-phosphate} \xrightarrow{\text{enzyme B}} \text{sugar B-phosphate.}$$

A specific example is the conversion of glucose to glucose-6-phosphate and then to fructose 6-phosphate:

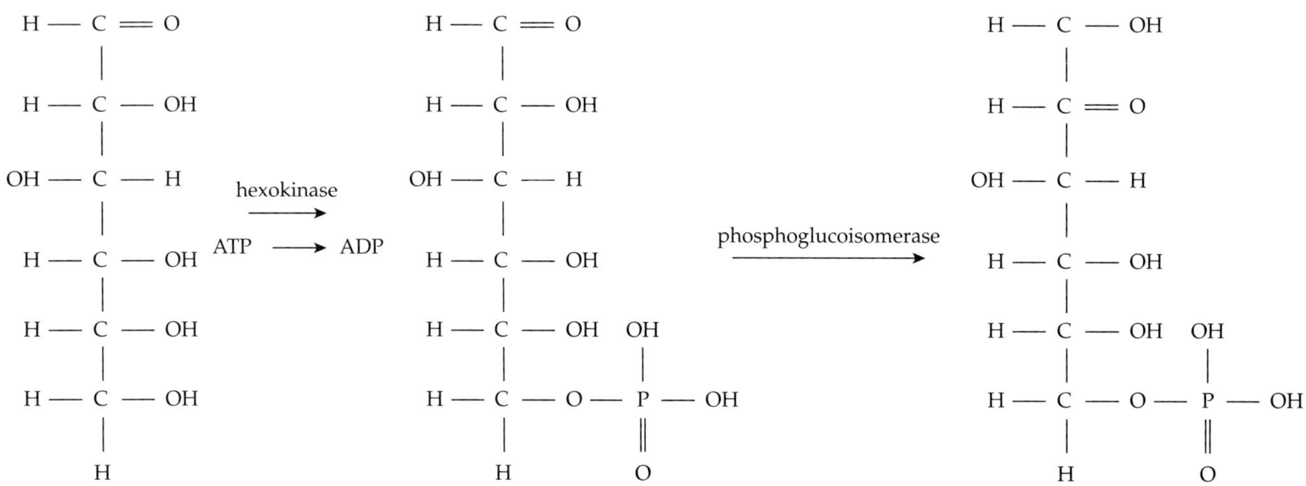

Glucose Glucose-6-phosphate Fructose-6-phosphate

Phosphorylated monosaccharides are among the primary products of photosynthesis. They are directly involved in chemical reactions or converted to translocated and accumulated forms. Starches accumulate whenever a high level of sugars occurs and are transformed to sugars when sugar concentrations are low. At low temperatures the equilibrium is shifted in favor of sugars.

Phosphate esters are particularly important since they are intermediates in synthesis and degradation of starch and sucrose. They also are substrates in glycolysis, fermentation of sugars, photosynthetic CO_2 fixation, and a number of oxidative processes. In addition, phosphate esters are constituents of nucleic acids and coenzymes.

The important carbohydrate transformations are not limited to the sugars. For example, sugars may be converted to alcohols, as in the formation of sorbitol and mannitol from glucose and mannose. In woody plants of the Rosaceae, sorbitol is the main form of carbon translocated and the principal reserve carbohydrate in nonphotosynthesizing cells (Oliveira and Priestley, 1988).

Sucrose

Formation of sucrose from glucose and fructose may take place when one of the sugar units occurs as a sugar-nucleotide complex. For example, glucose in the form of uridine diphosphoglucose (UDPG) may react with fructose to form sucrose:

$$\text{UDP–D-glucose} + \text{D-fructose} \xrightleftharpoons{\text{sucrose synthetase}} \text{sucrose} + \text{UDP}$$

In addition, UDPG may also react with fructose-6-phosphate to form sucrose phosphate, which in turn is hydrolyzed by a phosphatase, resulting in formation of free sucrose (Hassid, 1969).

$$\text{UDP–D-glucose} + \text{D-fructose 6-phosphate} \xrightleftharpoons{\text{sucrose-phosphate synthetase}} \text{sucrose phosphate} + \text{UDP}$$

Sucrose yields glucose and fructose when hydrolyzed. The reaction is catalyzed by sucrase and is not reversible.

$$\underset{\text{Sucrose}}{C_{12}H_{22}O_{11}} + H_2O \xrightarrow{\text{sucrase}} \underset{\text{glucose}}{C_6H_{12}O_6} + \underset{\text{fructose}}{C_6H_{12}O_6}$$

Starch

The synthesis of starch occurs in several ways. Perhaps the most important are the following:

1. Phosphorylase reaction, which involves the joining together of glucose-1-phosphate units until a starch molecule is formed according to the following scheme (Meyer et al., 1973):

$$\text{Glucose 1-phosphate} + \text{glucose chain (n units)} \xrightleftharpoons{\text{phosphorylase}} \text{glucose chain (n+1 units)} + \text{phosphate}$$

2. Uridine diphosphoglucose (UDPG) pathway:

$$\text{Glucose 1-phosphate} + \text{UTP} \xrightleftharpoons{\text{pyrophosphorylase}} \text{UDPG} + \text{pyrophosphate}$$

$$\text{UDPG} + \text{glucose chain (n units)} \xrightleftharpoons{\text{transglucosylase}} \text{UDP} + \text{glucose chain (n+1 units)}$$

3. Adenosine diphosphoglucose pathway (ADPG) pathway:

$$\text{Glucose 1-phosphate} + \text{ATP} \xrightleftharpoons{\text{pyrophosphorylase}} \text{ADPG} + \text{pyrophosphate}$$

$$\text{ADPG} + \text{glucose chain (n units)} \xrightleftharpoons{\text{transglucosylase}} \text{ADP} + \text{glucose chain (n+1 units)}$$

Starch is degraded to sugar by two separate reactions involving phosphorylases or hydrolases. The reaction involving phosphorylase predominates in tissues that do not have a major food storage role. Hydrolases are found in highest concentrations in tissues having a major food storage function. The hydrolase reaction involves starch digestion by means of the breaking of bonds together with incorporation of water. The maltase reaction involves hydrolysis of starch to glucose via intermediate products:

$$\text{Starch} \rightarrow \text{Dextrins} \rightarrow \text{Maltose} \rightarrow \text{Glucose}$$

Starch is degraded to maltose by α-amylase and β-amylase, both of which act on amylose and amylopectin. The final conversion of maltose to glucose is accomplished by the enzyme maltase.

USES OF CARBOHYDRATES

The carbohydrates formed by photosynthesis have several fates. The largest fraction is oxidized in respiration, releasing the energy needed in the synthetic processes associated with growth (Chapter 6). A very large portion is used in growth, being translocated to the stem and root tips, the cambium, and reproductive structures, where it is converted into new protoplasm and cell walls. Another fraction is accumulated as reserve food and eventually used in metabolism and growth. Some carbohydrates are diverted for production of defensive chemicals, and small amounts are lost by leaching, exudation, translocation through root grafts to other plants, and losses to parasites such as mistletoes and other sap feeders (Kozlowski, 1992).

Respiration

Losses by maintenance and growth respiration amount to between 30 and approximately 60% of the daily production of photosynthate (Chapter 6). Respiratory consumption of carbohydrates is especially high in diseased plants, reflecting increased metabolic activity of the host, pathogen, or both. Respiration is discussed in more detail in Chapter 6.

Growth

Woody plants use both stored and currently produced carbohydrates, often concurrently, for growth. The proportions of the carbohydrate pool that are used by various vegetative and reproductive tissues vary greatly with species, genotype, age of plants, and growing conditions (Cannell, 1985, 1989; see Chapter 3, Kozlowski and Pallardy, 1997).

Annual use of carbohydrates varies among species in accordance with their inherent patterns of shoot growth. Species exhibiting fixed growth (Chapter 3), which complete shoot expansion in a small proportion of the frost-free season, generally use lower amounts of carbohydrates for shoot growth than species that exhibit free or recurrently flushing growth and thus show shoot elongation during a large part of the summer (Kozlowski et al., 1991). Some tropical pines use very large amounts of carbohydrates for shoot growth because their shoots grow rapidly and more or less continuously throughout the year (Kozlowski and Greathouse, 1970). Tropical angiosperm trees had the lowest levels of starch and soluble sugars in woody tissues during periods of rapid canopy growth and the highest levels when growth was constrained (e.g., during drought) (Newell et al., 2002; Würth et al., 2005), suggesting that sink activity has a major influence on use of carbohydrate reserves. Wide genotypic differences in use of carbohydrates for shoot growth also are well known. These are traceable to variations in time of bud opening, rates of shoot growth, and seasonal duration of shoot growth (Kozlowski, 1992).

Large amounts of carbohydrates are used in production of xylem and phloem mother cells, their division and differentiation into xylem and phloem cells, and expansion of the cambial sheath. Small amounts are used for production of phellem (cork) and phelloderm by the phellogen (cork cambium), as discussed in Chapter 5 of Kozlowski and Pallardy (1997).

Carbohydrates are used during initiation, elongation, and thickening of roots as well as in growth of mycorrhizae and root nodules (Kozlowski, 1992). Both the large perennial roots and short-lived fine roots consume carbohydrates during growth. Use of carbohydrates in thickening of the large perennial roots is much more irregular than it is in stems. In perennial roots, carbohydrates are used early in the season for xylem production near the soil surface and later in deeper parts of roots. The use of carbohydrates for xylem production around the root circumference is uniform in young perennial roots but becomes very uneven within a few years (Kozlowski, 1971a).

During periods of heavy fruiting and seed production, a very large proportion of the available carbohydrate pool is diverted from vegetative to reproductive growth (Kozlowski et al., 1991). When a heavy fruit or seed crop is produced, shoot growth, cambial growth, and root growth are reduced during the same year or the following year. The inhibitory effects of reproductive growth on leaf growth of 10 European white birch trees are shown in Figure 7.7. Cone formation in lodgepole pine trees was associated with a 27 to 50% reduction in the number of needles per cone-bearing branch (Dick et al., 1990a,b). Small xylem increments were produced during years of heavy cone production in Douglas-fir, grand fir, and western white pine (Eis et al., 1965). During good seed years the width of annual xylem rings of European beech was about half

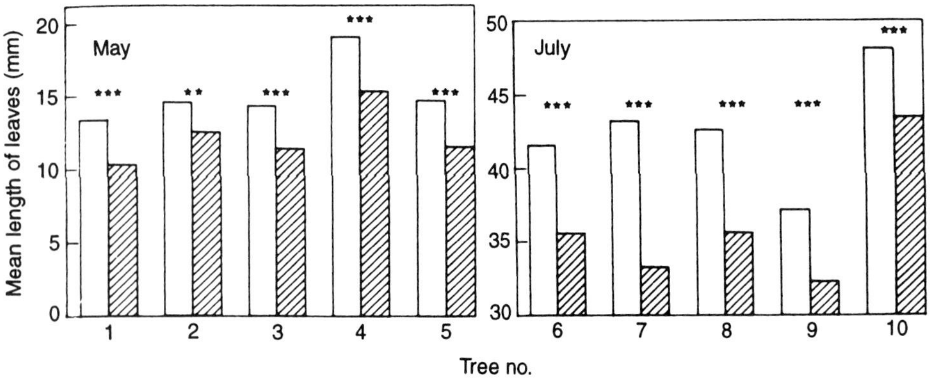

FIGURE 7.7. Leaf growth of European white birch on reproductive (hatched bars) and nonreproductive (open bars) shoots of 10 individual trees, Asterisks indicate that all differences between shoot types were significant ($P \leq 0.05$). From Toumi et al. (1982).

of that in years of low seed production. Heavy seed production reduced xylem increment of European beech for two years (Holmsgaard, 1962). The use of carbohydrates in growth of woody plants is discussed further in Chapters 3 and 4 of Kozlowski and Pallardy (1997).

Defense

Woody plants produce a variety of secondary compounds from carbohydrates, amino acids, and lipids that provide protection against disease-causing lower plants, herbivores, and competing higher plants (Kozlowski, 1992).

Chemicals that protect plants against disease-causing organisms include simple phenols, coumarins, tannins, and lignins, with the phenols most important. Important roles of gums and resins are in prevention of insect attack and as antifungal compounds (Babu and Menon, 1990). Some plants are protected because large amounts of defensive chemicals are maintained throughout much of the host plant's lifecycle. In other plants, certain defensive chemicals (called phytoalexins) increase rapidly only in response to infection (Creasy, 1985).

Each species of woody plants produces a unique array of defensive chemicals that deters attack by herbivores. Some of these compounds have a direct toxic effect and others reduce digestibility of plant tissues (Rhoades, 1985). There also is some evidence of production of defensive chemicals in trees that are not attacked by insects but are near trees undergoing defoliation by insects (Rhoades, 1983; Baldwin and Schultz, 1983).

Individual monoterpenes, triterpenes, and phenols provide constitutive defenses and deter feeding by snowshoe hares. Inducible chemical defenses against mammals are uncommon. Constitutive chemical defenses in the bark and phloem provide some defense against bark beetles. However, chemicals that accumulate during defensive reactions provide additional defense against these insects. Secondary metabolites that dominate induced defenses include several terpenoids, resin acids, and phenolics (Bryant and Raffa, 1995). The amounts and types of defensive chemicals formed by woody plants vary with species and their growth characteristics. For example, the concentration of defensive compounds in leaves of fast-growing trees on good sites may only be half as high as in slow-growing species on poor sites (Feeny, 1976; Coley, 1983).

Many woody plants allocate carbohydrates for synthesis of certain defensive chemicals (allelochems) that may arrest seed germination and growth of competing higher plants. Such defensive chemicals include organic acids, alcohols, aliphatic aldehydes, ketones, unsaturated lactones, fatty acids, polyacetylenes, naphthoquinones, anthroquinones, complex quinones, phenols, benzoic acids, cinnamic acid, coumarins, flavonoids, tannins, terpenoids, steroids, amino acids, polypeptides, alkaloids, cyanohydrins, sulfides, mustard oil glycosides, purines, and nucleosides (Rice, 1974, 1984). Synthesis of defensive chemicals involves the direct carbon cost of constructing molecules and maintaining the processes involved in such construction, as well as the indirect cost of subsequent reduction in plant growth because of the diversion of carbon to defense. The direct costs vary over time because defensive chemicals form at different rates as plants age (Gulmon and Mooney, 1986).

Leaching

Woody plants lose both organic and inorganic compounds by leaching, primarily from leaves but also from branches and stems. Compounds lost by leaching may include sugars, sugar alcohols, organic acids, pectic compounds, minerals, growth hormones, alkaloids, and phenolic compounds. However, most of the leached compounds are carbohydrates, primarily sugars (Tukey, 1970a,b, 1980; Parker, 1983). For example, fructose was the major carbohydrate leached from leaves of trembling aspen, but glucose, galactose, inositol, sucrose, maltose, and raffinose also were present in the leachate (Wildman and Parkinson, 1981).

The amounts and types of compounds leached from plants vary with species, cultivar, site, and environmental conditions. Evergreen leaves generally are less susceptible than deciduous leaves to leaching. However, evergreen leaves are leached for a longer time each year, so annual losses of leached substances from evergreens may exceed those from deciduous species (Thomas and Grigal, 1976). The leaves of apple and pear are less resistant to leaching than those of black currant (Tukey, 1971). Leaves with a waxy surface (e.g., citrus) are not easily leached. Because of cuticle degradation senescing leaves are more easily leached than young, rapidly growing leaves (Tukey, 1970a).

Exposure of plants to air pollutants may affect leaching by preventing wax formation on leaves and degrading surface waxes. For example, exposure of yellow birch foliage to acid mist accelerated leaching of carbohydrates (Scherbatskoy and Klein, 1983). Similarly, exposure of Norway spruce needles to acidic mist increased leaching of sucrose, galactose, and fructose. However, the amounts of carbohydrates leached amounted to less than 1% of the nonstructural carbohydrates present in the needles (Mengel et al., 1988, 1990).

Exudation

The roots of woody plants release a wide variety of compounds to the soil, including small amounts of carbohydrates as well as amino acids, organic acids, flavonones, mineral elements, and enzymes to the soil. Root exudates may leak from or between epidermal cells or be actively excreted (Uren and Reisenauer, 1986).

The amounts of carbohydrates exuded are small and rarely exceed 0.4% of the amount of carbon fixed (Rovira, 1969). Shortleaf pine roots lost less than 0.1% of the fixed carbon by exudation (Norby et al., 1987). Loss of carbohydrates by exudation is not restricted to roots. For example, grape berries exude a variety of compounds, including sugars, through the cuticle and epicuticular wax (Padgett and Morrison, 1990).

ACCUMULATION OF CARBOHYDRATES

Accumulation of carbohydrates during the growing season is essential for survival of plants. Stored carbohydrates play an important role in metabolism, growth, defense, development of cold hardiness, and postponement or prevention of mortality (Kozlowski, 1992). Seed germination, early growth, and survival of seedlings are influenced by amounts of stored carbohydrates. Heavy seeds with large carbohydrate reserves usually germinate faster and the young seedlings grow faster than those emerging from light seeds (see Chapter 2 of Kozlowski and Pallardy, 1997). Early-season shoot growth of many broadleaved species depends largely on stored foods, as in *Fraxinus ornus* (Boscaglia, 1983). Evergreens also use some reserve carbohydrates for early-season growth. Reserve carbohydrates are particularly important for regrowth following pruning of shoots, insect defoliation of angiosperm trees, or killing of young leaves by early-season frosts (Kozlowski et al., 1991). Cold hardiness has been associated with accumulation of sugars in the autumn. Increased susceptibility of pecan to winter cold was coupled with depletion of carbohydrate reserves by heavy fruiting (Wood, 1986). When tree vigor declines as a result of environmental stresses such as drought or mineral deficiency, outbreaks of stem cankers, diebacks, declines, and insect attacks often follow (Kozlowski et al., 1991). Seedlings and mature trees with low carbohydrate reserves undergo high risk of mortality. In contrast, vigorous trees generally accumulate enough carbohydrates to heal injuries, synthesize defensive chemicals, and maintain physiological processes at levels necessary to sustain life when exposed to environmental stresses (Waring, 1987).

Carbohydrate Distribution

Carbohydrate reserves are accumulated largely in parenchyma cells and death of cells is preceded by withdrawal of reserves (Ziegler, 1964). Many studies have been made of the accumulation of starch and sugars in different tissues and organs. As might be expected, there are marked variations in the amounts of carbohydrates in various parts of woody plants and there also are large seasonal differences in the amounts and kinds of carbohydrates present. There also are differences between deciduous and evergreen species and between temperate and tropical species in seasonal carbohydrate accumulation (see Chapter 3, Kozlowski and Pallardy, 1997).

The types of stored carbohydrates vary in different organs and tissues and their contents vary seasonally as well. Starch is considered the most important reserve carbohydrate and often has been used as the sole indicator of the carbohydrate status of plants. Starch accumulates whenever a high level of sugar builds up; it is converted to sugars when sugar contents are low and at low temperature (Kozlowski and Keller, 1966; Kozlowski, 1992). Among the soluble carbohydrates, sucrose is the primary transportable and storage carbohydrate. In addition to starch and sucrose, a variety of other compounds accumulate in woody plants. They vary from simple molecules (e.g., various sugars, polyols, oligosaccharides, amino acids) to complex compounds (e.g., polysaccharides and lipids). Sorbitol, a six-carbon alcohol, occurs commonly in most temperate-zone fruit trees. Sorbitol comprises up to 85% of the carbohydrates in the phloem of apple trees and also is found in large amounts in apricot, plum, and peach. Mannitol is found in coffee and olive. Soluble carbohydrates found in small amounts in both roots and shoots include inositol, xylose, rhamnose, maltose, trehalose, arabinose, ribose, mannose, raffinose, and stachyose (Loescher et al., 1990).

Storage Sites

Carbohydrates accumulate in a variety of tissues and organs, including buds, leaves, branches, stems, roots, seeds, fruits, and strobili. Some carbohydrates also accumulate in the vascular sap.

It is important to distinguish between the total amount and concentration of carbohydrates in differ-

TABLE 7.1. Distribution of Starch and Sugars in Grimes Apple Trees in Mid-October[a]

Plant part	Dry weight (kg)	Starch and sugar kg	Starch and sugar %[b]	Starch kg	Starch %[b]	Sugar kg	Sugar %[b]
Leaves	14.92	1.40	9.41	0.42	2.80	0.99	6.61
Spurs	2.88	0.31	10.90	0.17	5.85	0.15	5.05
Wood							
1 year	2.70	0.30	10.94	0.16	5.89	0.14	5.05
2 years	3.24	0.33	10.25	0.21	6.32	0.13	3.93
3 years	3.97	0.39	9.73	0.23	5.84	0.15	3.89
4–6 years	21.03	1.13	5.36	0.60	2.85	0.53	2.51
7–10 years	56.36	3.27	5.81	2.03	3.60	1.25	2.21
11–18 years	46.00	2.70	5.88	1.82	3.96	0.88	1.92
Main stem	31.00	3.44	11.10	2.79	9.00	0.65	2.20
Total above ground	182.10	13.28	7.29[c]	8.42	4.63[c]	4.85	2.67[c]
Root stump	28.55	4.52	15.83	3.14	11.00	1.38	4.83
Roots							
18–14 years	21.32	3.74	17.53	2.88	13.52	0.85	4.01
13–7 years	10.24	2.51	24.51	1.80	17.63	0.71	6.88
6–1 years	2.45	0.50	20.37	0.39	15.93	0.11	4.44
Total soil	62.55	11.26	18.01[c]	8.22	13.14[c]	3.05	4.87[c]
Total tree	244.65	24.54	10.03[c]	16.64	6.75[c]	7.90	3.23[c]

[a]Adapted from Murneek (1942).
[b]Percent of oven dry weight.
[c]Mean.

ent parts of plants. Carbohydrate distribution often is expressed in percent of dry weight of various tissues. This may be misleading because high concentrations of carbohydrates often occur in tissues that comprise a low proportion of the total dry weight of a plant. For example, in both young and old trees the concentration of carbohydrates usually is higher in the roots than in the shoots. Nevertheless, in adult trees the aboveground parts often are the primary carbohydrate reservoirs because the stem, branches, and leaves make up more of the total dry weight of a tree than do the roots. The apple trees studied by Murneek (1933, 1942) had a higher carbohydrate concentration in the roots than in the stems, yet the aboveground parts, which were about three times as massive as the roots, contained more total carbohydrates (Table 7.1). Although carbohydrate concentrations were low in the stem wood of Monterey pine trees, the total amounts were higher in the stem wood than in other parts (Cranswick et al., 1987). The proportional distribution of carbohydrates above and below ground also is a function of tree age because root-shoot ratios decrease progressively with increasing age. In one-year-old apple trees carbohydrate reserves were about equally distributed above and below ground (Priestley, 1962).

Buds

Both vegetative and flower buds accumulate carbohydrates during the latter stages of their development (Lasheen and Chaplin, 1971; Nelson and Dickson, 1981). The kinds of carbohydrates stored in buds vary among species. The inactive buds of Monterey pine had high concentrations of soluble carbohydrates but little starch (Cranswick et al., 1987), whereas the principal carbohydrates of the dormant buds of pear were sucrose and sorbitol (Watts and De Villiers, 1980). Carbohydrates in rhododendron buds consisted primarily of glucose, fructose, and starch. Small amounts of maltose and raffinose also were present (Wright and Aung, 1975). Deep dormancy of buds of *Populus trichocarpa* × *P. deltoides* cv. Raspalje was associated with low levels of carbohydrates, which continued to decline slightly. A subsequent, short transitional phase was characterized by a decrease in starch and increase in soluble sugars. Most variation in carbohydrates occurred after bud dormancy was broken. Together with sucrose, raffinose was the major sugar present. Raffinose fluctuated rapidly during postdormancy and rapidly disappeared before bud break (Bonicel and Medeiros Raposo, 1990).

Leaves

Although leaves often have a high concentration of carbohydrates, they usually contain only a small proportion of the total amount present in a woody plant (Kramer and Kozlowski, 1979). The carbohydrate concentration of starch plus sugar of apple leaves was 9%, a value higher than reported for other tissues. However, the total amount of carbohydrates in the leaves was only about 5% of the carbohydrate reserve of the whole tree (Murneek, 1942). Leaves possessed only 3% of total tree nonstructural carbohydrates in a semideciduous tropical forest in Panama (Würth et al., 2005).

The leaves of evergreens accumulate carbohydrates that subsequently are used in metabolism and growth, as demonstrated for jack pine seedlings (Glerum and Balatinecz, 1980; Glerum, 1980). The ^{14}C-photosynthate that was stored in red pine needles later was used in growth of new shoots (Gordon and Larson, 1970). This pattern was consistent with the rapid decrease in dry weight of one-year-old needles of red pine as the new shoots expanded (Kozlowski and Clausen, 1965; Clausen and Kozlowski, 1967). The amounts of carbohydrates stored in needles of evergreen species often vary with their shoot growth patterns. For example, the much lower accumulation of carbohydrates in needles of Monterey pine than in those of Scotch pine was attributed to use of carbohydrates in growth of Monterey pine throughout the year (Rook, 1985).

Citrus leaves accumulated carbohydrates at a rate of 8 mg g dry wt^{-1} day^{-1}, or approximately twice the rate of accumulation in stems (Yelenosky and Guy, 1977). Leaves of olive also store significant amounts of carbohydrates (Priestley, 1977). Leaves of some deciduous trees also may accumulate some carbohydrates late in the growing season (Tschaplinski and Blake, 1989).

Stems and Branches

The amount of stored carbohydrates varies appreciably in different parts of the tree stem and branches. For example, starch concentrations in plane trees always were higher at the stem base than in the middle or upper stem. Large amounts of starch also accumulated at the stem-branch junctions. The large branches in the upper part of the crown stored more starch than the large branches in the lower part of the crown (Haddad et al., 1995). Carbohydrates are stored in both wood and bark tissues. The starch in the xylem of adult trees is stored mainly in ray parenchyma and axial parenchyma cells. In seedlings pith cells also are important in starch storage (Glerum, 1980).

The amount and concentration of reserve carbohydrates in the xylem vary with the age of tissues. Starch grains are abundant in ray cells near the cambium and they decrease toward the inner sapwood. Ray cells in the heartwood contain only negligible amounts of starch or none at all.

In October the amounts of reserve carbohydrates (starch, glucose, fructose, and sucrose) were highest in the outer sapwood of Scotch pine trees, and they decreased gradually toward the inner sapwood and heartwood (Saranpää and Höll, 1989; Saranpää, 1990). Healthy American elm trees stored starch in 12 to 18 xylem rings, whereas diseased trees stored starch in the outer ring only (Shigo et al., 1986). In sugar maple, carbohydrate reserves were about twice as high in the outer sapwood as in the inner sapwood (Murneek, 1942).

In the bark, starch is deposited in parenchyma and albuminous cells, which are distributed in a variety of patterns. The concentration of carbohydrates in the bark often is very high but in many species the total amount in the bark is less than in the wood. In mature European beech trees in winter the inner living phloem of branches, main stems, and roots often had two to three times as high a concentration of carbohydrate reserves as the wood. Nevertheless, the total amount was greater in the wood than in the bark (Gäumann, 1935). Wenger (1953) also found a higher concentration of carbohydrates in stem bark than in stem wood of sweet gum but the wood had much more total carbohydrate. The root bark of shortleaf pine had a much higher concentration of carbohydrates than the root wood (Hepting, 1945).

Roots

Large amounts of carbohydrates are stored in the large perennial roots (Fig. 7.8) and in the fine roots. For example, both the large and fine roots of white oak accumulated starch to a maximum in late autumn (McLaughlin et al., 1980).

The amounts of stored carbohydrates in perennial roots vary with root size. In northern red oak and white oak, the smaller branch roots had higher starch contents than the large roots from which they emerged. In the smaller roots the starch-storing xylem rays were closer together and the proportion of ray tissue to woody tissue was greater than in the large-diameter roots, resulting in a higher starch concentration in the former (Wargo, 1976).

Appreciable amounts of carbohydrates may accumulate in the fine roots. In Scotch pine trees starch increased in the fine roots to a maximum near 30% (dry wt. basis) (Ericcson and Persson, 1980). Nguyen et al. (1990) demonstrated much loading of starch and sugars in the fine roots of two poplar cultivars in the autumn.

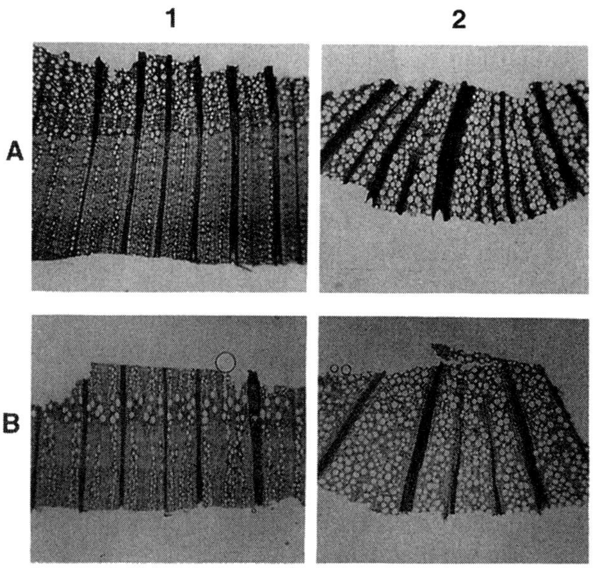

FIGURE 7.8. Cross sections of wedges of root tissues stained with I_2KI, showing the number and width of rays and large amounts of starch stored in the ray tissues of (A) northern red oak and (B) white oak (1) at the root collar and (2) 1.05 m from root collar. From Wargo (1976).

Starch concentrations in the fine roots more than doubled in cv. Tristis and increased in cv. Eugenei by 75 times. During the same time the sugar concentrations in the fine roots doubled.

Xylem Sap

The xylem sap of woody plants contains sugars, mostly sucrose, in addition to very small amounts of organic acids, nitrogen compounds, inorganic salts, growth hormones, and enzymes. The sugar concentration of the xylem sap varies with species and genotype, season, time of day, age of plants, and nutritional status (Ferguson, 1980). The xylem sap of sugar maple may contain 2 to 10% sugar (Chapter 11); of willow, 3 to 5%; and that of American beech, only negligible amounts (Taylor, 1956; Sauter, 1980).

Taylor (1956) reported significant differences in the sugar content of maple sap from different trees and stands of trees. This finding indicated that selection of trees for high yields of sugar might be possible. Morselli et al. (1978) reported that high yielding maple trees have more and larger rays than low yielding trees. The sugar concentration of maple sap typically is low early in the season, rises quickly to a maximum, then gradually decreases later in the season. In addition to sucrose and a small amount of glucose, maple sap contains small amounts of inorganic salts, nitrogenous compounds such as peptides and amino acids, amylases, and unidentified organic constituents (Taylor, 1956). The characteristic taste of maple sap is attributed to certain amino acids and is developed by heating (Pollard and Sproston, 1954). The sugar comes from starch accumulated during the preceding summer that is converted to sucrose in the late autumn and early winter. The activity of the enzymes involved in this conversion seems to be increased by low temperatures. According to Sauter et al. (1973) the sucrose is secreted into the xylem, causing a high concentration of sugar in the xylem sap (see also Chapter 11). Apparently the loss of sugar by tapping is not injurious because many trees have been tapped for decades without apparent harm.

Fruits

During their development fruits accumulate carbohydrates, generally as starch, sucrose, or hexose sugars (see Chapter 4 of Kozlowski and Pallardy, 1997). A distinction often is made between reproductive sinks (which store carbohydrates in fruits and seeds) and utilization sinks in leaves, stems, and roots (which use much of the imported carbohydrates for growth (Giaquinta, 1980; Ho, 1988; Cannell and Dewar, 1994). Fruits that store starch (e.g., banana) sweeten after harvest because of conversion of starch to soluble sugars. Fruits that lack stored carbohydrates (starch) must remain attached to the plant for accumulation of soluble sugars to take place (Hubbard et al., 1991).

AUTUMN COLORATION

Autumn coloration will be discussed at this time because the anthocyanin pigments responsible for the pink, red, and purple colors are related to the carbohydrates and carbohydrate accumulation favors their formation. Anthocyanins are glycosides formed by reactions between various sugars and complex cyclic compounds called anthocyanidins (Fig. 7.9A). They are water soluble and usually occur in the cell sap of the vacuole. Anthocyanins usually are red in acid solution and may become purplish to blue as the pH is increased. The amount of anthocyanin pigments depends primarily on the possession of certain hereditary potentialities for their production, but environmental factors also have an influence.

With declining autumn temperatures, the leaves of trees stop producing chlorophyll, and at the same time certain species that contain large amounts of carbohydrates and the hereditary potential to do so begin to form anthocyanins in their leaves. As chlorophyll synthesis stops, the chlorophyll already present begins to

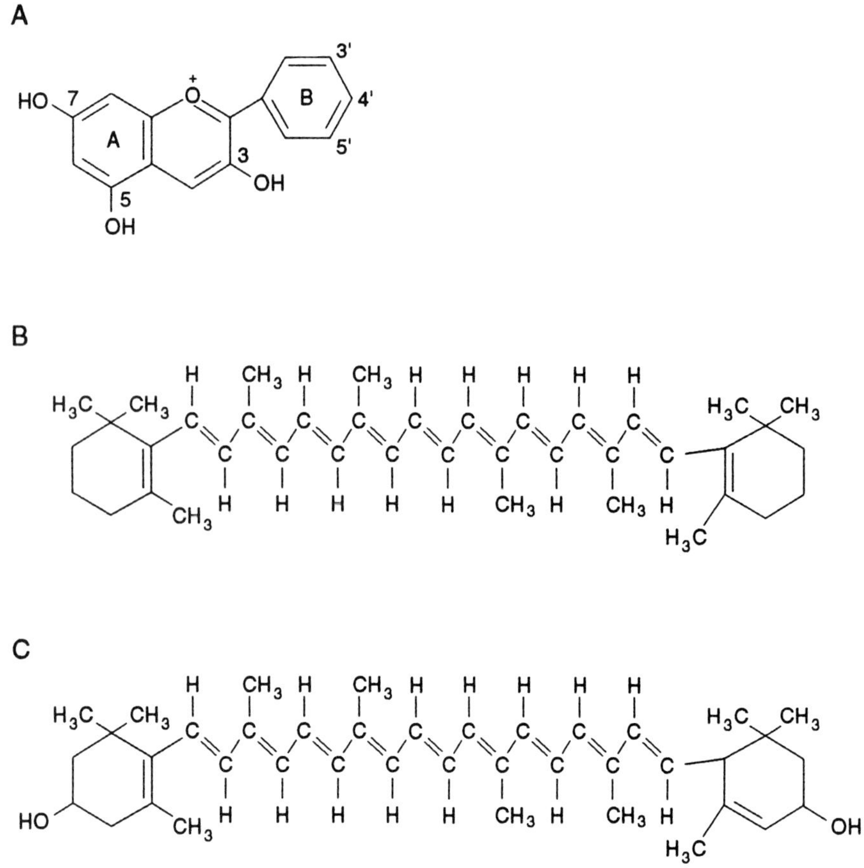

FIGURE 7.9. Structure of pigments that become prominent during autumn coloration of leaves of woody plants: (A) anthocyanidin; (B) β-carotene, a yellow carotenoid pigment with the formula $C_{40}H_{56}$; and (C) β-lutein, a yellow xanthophyll pigment with the formula $C_{40}H_{56}O_2$.

decompose chemically and the newly formed anthocyanins are unmasked. In those species that do not form anthocyanin pigments, the autumn breakdown of chlorophyll unmasks the relatively more stable yellow carotene and xanthophyll pigments (Fig. 7.9B,C), resulting in clear-yellow colored leaves, as in ginkgo, yellow-poplar, and hickory; or there may be an admixture of red anthocyanin pigment with yellow carotene to give a bright orange color, as in some species of maple. In other species both chlorophyll and carotenoids disintegrate simultaneously and new carotenoids are synthesized. Thus, by disintegration of green pigments and the unmasking of yellow ones, the formation of red pigments, or all three, the leaves may assume various shades of yellow, orange, crimson, purple, or red.

Trees such as alders and blue ash show little autumnal color change. In contrast, leaves of a large group of trees, including black walnut, catalpa, elm, basswood, and sycamore turn to a mixture of rusty green and yellow. Leaves of poplars, honeylocust, hickory, ginkgo, beech, green ash, and most species of birch change to yellow of various shades. But by far the most dazzling displays are the shades of orange, red, and purple seen in red and sugar maples, dogwood, sassafras, sumac, white oak, scarlet oak, shadbush, tupelo gum, sweetgum, zelkova, and winged euonymus, which form large amounts of anthocyanin pigments. Various species and cultivars of maple show much gradation from yellow to deep red (Santamour and McCardle, 1982). Variability in both the degree and seasonal duration of red coloration has been the major source of selection among red maple cultivars (Sibley et al., 1995). Red maple seedlings from more northern locations had better autumn colors than progenies from 45 other locations (Townsend, 1977).

Trees of the same species growing together often show much difference in color because of variations among individual trees in amounts of soluble carbohydrates. Some reach their peak of color later than others. Species of oaks color late in the autumn, usually after the best maple color has developed and disappeared. The yellow-brown colors of beech and some species of oak are caused by the presence of tannins in leaves, in addition to yellow carotenoids.

Variations among species in rate of autumnal color change often reflect wide differences in rates and amounts of chlorophyll breakdown. As senescence of trembling aspen progressed the chlorophyll content decreased by 99% (Dean et al., 1993). Wieckowski

TABLE 7.2. **Chlorophyll Content of Green and Yellow Autumn Leaves of Forest Trees**[a]

Species	Total chlorophyll (mg/g)		Chlorophyll *a* (%)		Reduction in total chlorophyll (%)
	Green leaves	Yellow leaves	Green leaves	Yellow leaves	
Liriodendron tulipifera	2.19	0.29	67.5	40.1	86.8
Populus nigra var. *italica*	1.79	0.27	73.6	63.4	85.2
Magnolia grandiflora	1.74	0.14	75.4	47.9	91.9
Cercis canadensis	1.55	0.12	71.8	43.0	92.2
Acer saccharum	1.38	0.19	69.4	56.5	86.5
Liquidambar styraciflua	1.23	0.05	70.1	57.4	96.2
Acer saccharinum	1.19	0.26	62.5	42.7	78.1
Juglans nigra	1.10	0.26	65.4	49.4	76.4
Celtis occidentalis	1.06	0.32	71.5	65.1	69.7
Cornus florida	0.97	0.18	64.9	53.6	81.4
Ulmus americana	0.93	0.15	70.4	57.9	83.6
Quercus macrocarpa	0.92	0.11	68.0	47.4	87.6
Fagus grandifolia	0.90	0.22	64.4	57.4	75.3
Quercus palustris	0.87	0.17	71.3	54.2	80.6
Carya sp.	0.76	0.16	70.7	67.1	78.9

[a]From Wolf (1956).

(1958) reported that, whereas rapid chlorophyll disintegration occurred in a species of magnolia (35 days), slow breakdown took place in white mulberry (more than 60 days). Before they abscised, the leaves of sycamore maple and beech lost practically all their chlorophyll, whereas those of lilac lost only 40%. Wolf (1956) demonstrated wide variations in the chlorophyll content of leaves and in the rate of chlorophyll breakdown during the autumn (Table 7.2). Chlorophyll *a* was destroyed more rapidly than chlorophyll *b* in many species.

Goodwin (1958) followed changes in both chlorophyll and carotenoid pigments from June to November in red plum, English oak, and sycamore maple. In oak and maple both chlorophylls and carotenoids decreased almost to zero. In oak these were depleted simultaneously, whereas in maple the decline in chlorophyll preceded that in carotenoids. In red plum the carotenoids tended to disappear first, but carotenoids and chlorophylls decreased by only about half. Eichenberger and Grob (1962) found that when maple leaves began to change color, a carotenoid pigment was formed that was different from the carotenoids present in the summer and the total amount of carotenoids decreased.

In angiosperm trees the rate of autumn coloration within tree crowns may be expected to vary somewhat with differences among species in hereditary patterns of leaf production. Koike (1990) found that leaves of angiosperm trees with neoformed (free) growth patterns began to develop autumn color from the inner part of the crown whereas trees with preformed (fixed) leaf production begin color development from the outer part of the crown. These changes were associated with differences in leaf senescence.

Any environmental factor that influences the synthesis of carbohydrates or the conversion of insoluble to soluble carbohydrates will favor anthocyanin formation and bright autumn colors in species that have the genetic potential to synthesize these pigments. Among the most important factors controlling autumn coloration are temperature, light, and water supply. The lowering of temperature above the freezing point favors anthocyanin formation. However, severe early frosts actually make red autumn colors less brilliant than they otherwise would be. Bright light also favors red colors, and anthocyanin pigments usually develop only in leaves that are exposed to light. If one leaf is covered by another during the period when red anthocyanin pigments are forming, the lower leaf usually does not form the red pigment at all. Water supply also affects formation of anthocyanin pigments, with mild drought favoring bright red colors. Rainy days without much light occurring near the time of peak coloration actually will decrease the intensity of autumn colors. In summary, the best autumn colors occur under conditions of clear, dry, and cool but not freezing weather.

The potential adaptive value of anthocyanin pigment synthesis has been debated for many years. The most widely held view was that these compounds were simply dead end products of secondary metabolic pathways and had no demonstrable function in plants (Matile, 2000). An alternative perspective, the light screen hypothesis, was offered long ago (and reviewed early by Wheldale, 1916), and was recently revisited and extended by Hoch et al. (2001). The masking hypothesis asserts that anthocyanin pigments, with their intense absorption of solar radiation in the photosynthetically active blue wavelengths, protect the photosynthetic apparatus from oxidative damage during disassembly and recovery of mineral nutrients, particularly nitrogen. Recovery of nitrogen requires phloem translocation, which is driven by carbohydrate loading and unloading at source (i.e., the leaf) and sink sites, respectively. Carbohydrate availability for this process is inadequate unless photosynthetic competence is maintained in leaves.

Feild et al. (2001) showed that yellow leaves of red-osier dogwood endured more inhibition of photosynthetic electron transport and slower recovery from inhibition compared with highly red-colored leaves. This protection may be critical in the autumn as lower temperatures combined with high light intensities are particularly conducive to photoinhibitory damage (Chapter 5), and these environmental conditions are known to intensify anthocyanin synthesis in many species. Hoch et al. (2001) also noted greater abundance of species that synthesize anthocyanins in regions of continental climate that frequently have early autumn cold spells. Species of the same genera from more moderate maritime regions often lacked this capacity.

During the autumn the needles and stems of some conifer seedlings turn purple and this color persists throughout the winter (Toivonen et al., 1991). In Scotch pine seedlings the purple coloration was attributed to two anthocyanin pigments, cyanidin-3-glucoside and delphinindin-3-glucoside. In Finland the seedlings changed color as they hardened to frost. Northern seedlings turned purple and hardened earlier than southern seedlings (Toivonen et al., 1991). Presumably, the accumulation of anthocyanins by evergreen species in winter would have the same protective effects on the photosynthetic apparatus at low temperature noted previously.

SUMMARY

Carbohydrates are direct products of photosynthesis, and the basic organic substances from which most other organic compounds in woody plants are synthesized. The important carbohydrates include monosaccharides (pentoses, hexoses), oligosaccharides (sucrose, maltose, raffinose), and polysaccharides (cellulose, starch). Many carbohydrates continuously undergo transformation, with starch-sucrose interconversions occurring commonly. However, once glucose has been incorporated into cellulose by the cellulose synthase enzyme complex, it remains in this compound because plants lack the required enzymes for cellulose degradation.

Although large amounts of carbohydrates are used in growth of woody plants, most are lost by maintenance and growth respiration and lesser amounts by leaching, exudation, translocation to other plants, losses to parasites, and production of defensive chemicals. Respiratory losses may amount to 60% or more of the daily production of photosynthate.

Both stored and currently produced carbohydrates are used for growth. Utilization of carbohydrates in shoot growth varies greatly among species in accordance with their inherent patterns of shoot growth. Use of carbohydrates for cambial growth varies among species, genotype, and environmental conditions. It also varies with stem height and around the stem circumference. Carbohydrates are used in initiation, elongation, and thickening of the roots, as well as in growth of mycorrhizae and root nodules. Fruiting and seed production preferentially use a large portion of the available carbohydrate pool, resulting in inhibition of vegetative growth.

A variety of secondary compounds, produced from carbohydrates, amino acids, and lipids protect plants against disease-causing lower plants, herbivores, and competing higher plants. Compounds that protect plants against disease-causing organisms include simple phenols, coumarins, tannins, and lignins, with phenols most important. Gums and resins are important in defense against insects and as antifungal compounds. Some defensive compounds have a direct toxic effect on herbivores and others render plant tissues less digestible. Some carbohydrates are lost by leaching, mostly from leaves, and some are lost to the soil by exudation from roots. The amounts and types of compounds leached from plants vary with species and genotype, site, and environmental conditions. Deciduous leaves are more susceptible than evergreen leaves to leaching, but evergreen leaves are leached for a longer time each year. Air pollutants may increase leaching by eroding leaf waxes.

Carbohydrates that are not used in metabolism and growth or lost, accumulate in a variety of vegetative and reproductive tissues and organs and in the vascular sap. Starch is the most important reserve carbohydrate. Other important reserves include sucrose, other sugars, alcohols, fats and oils, and nitrogen compounds.

The bright red autumn colors of trees such as red maple, sumac, and sassafras are associated with disintegration of chlorophyll and formation of anthocyanin pigments. In trees that do not form anthocyanin pigments (e.g., birches, poplars) breakdown of chlorophyll in the autumn unmasks the more stable yellow carotene and xanthophyll pigments. Among the important environmental factors that control autumn coloration are temperature, light, and water supply. Formation of anthocyanin pigments may protect the photosynthetic apparatus from damage during recovery in the autumn of nitrogen from senescing leaves.

General References

Brett, C. and Waldron, K. (1990). *Physiology and Biochemistry of Plant Cell Walls*. Unwin Hyman, London.

Bryant, J. A., Burrell, M. M., and Kruger, N. J., eds. (1999). *Plant Carbohydrate Biochemistry*. BIOS Scientific, Oxford.

Day, P. M. and Dixon, R. A. (1985). *Biochemistry of Storage Carbohydrates in Green Plants*. Academic Press, London.

Duchesne, L. C. and Larson, D. W. (1989). Cellulose and the evolution of plant life. *BioScience* **39**, 238–241.

Duffis, C. M. and Duffis, J. H. (1984). *Carbohydrate Metabolism in Plants*. Longman, London.

Gupta, A. K. and Kaur, N., eds. (2000). *Carbohydrate Reserves in Plants: Synthesis and Regulation*. Elsevier, Amsterdam and New York.

Haigler, C. H. and Weimer, P. J., eds. (1991). *Biosynthesis and Biodegradation of Cellulose*. Marcel Dekker, New York.

Kozlowski, T. T. (1992). Carbohydrate sources and sinks in woody plants. *Bot Rev* **58**, 107–222.

Levin, D. A. (1976). The chemical defenses of plants to pathogens and herbivores. *Annu Rev Ecol Syst* **7**, 121–159.

Lewis, D. H., ed. (1984). *Storage Carbohydrates in Vascular Plants*. Cambridge Univ. Press, Cambridge.

Lewis, N. G. and Price, M. G., eds. (1989). *Plant Cell Wall Polymers: Biogenesis and Biodegradation*. American Chemical Society, Washington, D.C.

Preiss, J., ed. (1988). *The Biochemistry of Plants. Vol. 14. Carbohydrates*. Academic Press, San Diego.

CHAPTER

8

Lipids, Terpenes, and Related Substances

INTRODUCTION

The compounds dealt with in this chapter are a heterogeneous group that have little in common except their low solubility in water and high solubility in organic solvents such as acetone, benzene, and ether. Included are simple lipids, cutin, suberin, waxes, and compounds composed of substances in addition to glycerides (e.g., phospholipids and glycolipids). Another large group of compounds discussed in this chapter are the isoprenoids or terpenoids, which are derived from isoprene. This group includes essential oils, resins, carotenoids, and rubber. The relationships of the major groups are shown in Figure 8.1 and the terpenoids or isoprenoids are shown in more detail in Figure 8.6.

The fats and other lipids of woody plants are both physiologically and economically important. They are physiologically important because fats and phospholipids are essential constituents of protoplasm and occur in all living cells. Fats are important storage forms of food in seeds (Chapter 2, Kozlowski and Pallardy, 1997) and are found in small amounts in the leaves, stems, and roots of most woody plants. Membrane lipids (phospholipids, glycolipids, sterols) play an important role in maintaining the integrity and function of biological membranes. Lipids (mainly fatty acid derivatives) are important volatile components of flowers that act to stimulate and guide pollinators (Knudsen et al., 1993). In the form of cutin, wax, and suberin, lipids form protective coverings over the outer surfaces of leaves, fruits, and stems. Internal deposits of cutin or related lipids occur in certain tissues (Esau, 1965, pp. 155–157).

Lipids also play a significant role as second messengers in plants, participating in signaling networks by which plants respond to biotic and abiotic stresses. A number of lipid kinase and lipase enzymes act on lipid molecules to produce these signaling compounds, which then activate enzymes or attract proteins with specific lipid-binding domains to localized membrane regions (Meijer and Munnink, 2003; Wang, 2004). Additionally, jasmonic acid and related compounds (Chapter 13), which are synthesized from α-linolenic acid, are potent growth regulators involved in inhibition of growth, pollen and seed germination, regulation of storage protein accumulation in seeds, fruit ripening, insect and disease responses, and wounding (Crozier et al., 2000; Farmer et al., 2003; Pieterse et al., 2004).

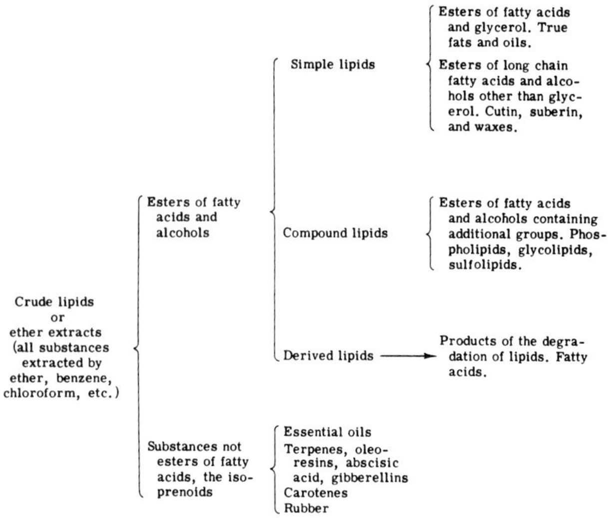

FIGURE 8.1. Relationships among principal substances found in ether extracts of plant tissue.

Some lipids are of great commercial value. Examples are oil from palm, olive, and tung trees. So-called essential oils are extracted from a variety of trees and used for flavoring.

Terpenoids, the largest group of organic substances in the plant kingdom, are physiologically, ecologically, and commercially important. They act as hormones, components of membranes, photoprotective pigments, and membrane-bound sugar carriers in synthesis of glycoproteins and polysaccharides. Terpenoid metabolites protect plants against herbivores and pathogens and participate in allelopathic interactions, nutrient cycling, and attraction of pollinators (Gershenzon, 1994). The terpenes obtained from pines are important sources of commercial rosin and turpentine.

LIPIDS

The term lipid includes the simple triglycerides found in common oils and fats; various compound lipids such as phospho-, sulfo-, and glycolipids; and other compounds such as cutin, suberin, and the waxes.

Simple Lipids

The simplest and most common lipids are the triglycerides. They are esters of glycerol and various fatty acids that form ordinary oils and fats. If the ester is a liquid at ordinary temperatures it is called an oil, but if it is a solid it is a fat. A generalized formula for an oil or fat is:

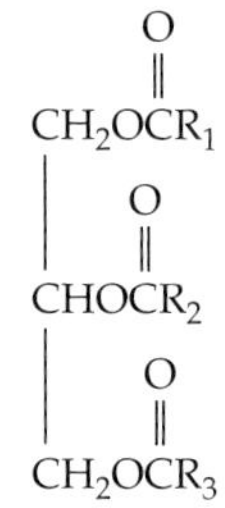

R_1, R_2, and R_3 represent the carbon chains of the same or different fatty acids. In simplified form, the synthesis of oils and fats involves a reaction between the three -OH groups of glycerol and three molecules of fatty acid. The fatty acids linked to a given glycerol molecule often are different, producing mixed glycerides. The essential features of the reaction, with the enzyme lipase as the catalyst, can be shown as follows:

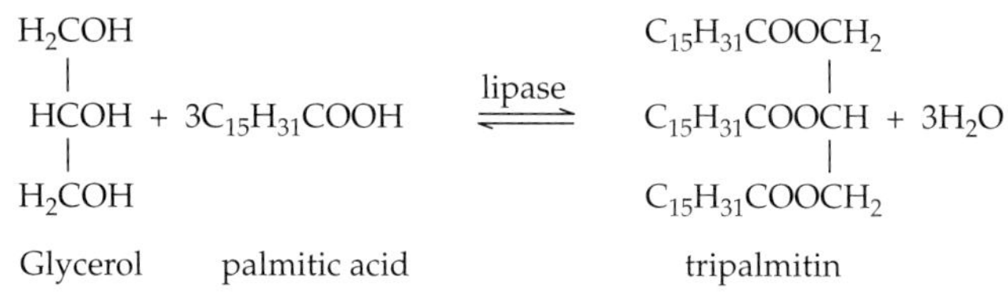

The actual starting materials probably are α-glycerophosphate and fatty acids built up of acetate units derived from acetyl coenzyme A. The place of fat synthesis in the general metabolic scheme is shown in Figure 6.22. Readers are referred to plant biochemistry books (see "General References," later) for more detailed descriptions of the synthesis of glycerol and fatty acids and their reaction to form fats. In leaves, synthesis of many lipids appears to occur in the chloroplasts, and over 50% of the lamellar membranes is composed of lipid material. Lipid synthesis obviously occurs in the cells where they are found because their insolubility in water makes translocation from cell to cell impossible.

Fatty Acids

The fatty acids are classified as saturated if no double bonds occur or unsaturated if double bonds occur between carbon atoms. The most important fatty acids found in woody plants are shown in Table 8.1. Palmitic acid is the most widely distributed fatty acid in woody plants, but most fatty acids in woody plants are unsaturated, oleic and linoleic being most common. The oils containing large amounts of unsaturated fatty acids combine with oxygen when exposed to the air and form the hard films characteristic of drying oils, such as those obtained from tung and flax. The unusual

TABLE 8.1. Important Fatty Acids Found in Plant Lipids

Saturated acids		
Lauric	$C_{12}H_{24}O_2$	$CH_3(CH_2)_{10}COOH$
Myristic	$C_{14}H_{28}O_2$	$CH_3(CH_2)_{12}COOH$
Palmitic	$C_{16}H_{32}O_2$	$CH_3(CH_2)_{14}COOH$
Stearic	$C_{18}H_{36}O_2$	$CH_3(CH_2)_{16}COOH$
Unsaturated acids		
Oleic	$C_{18}H_{34}O_2$	$CH_3(CH_2)_7CH{=}CH(CH_2)_7COOH$
Linoleic	$C_{12}H_{32}O_2$	$CH_3(CH_2)_4CH{=}CHCH_2CH{=}CH(CH_2)_7COOH$
Linolenic	$C_{18}H_{30}O_2$	$CH_3CH_2CH{=}CHCH_2CH{=}CHCH_2CH{=}CH(CH_2)_7COOH$

property of cocoa butter of remaining a solid to just below body temperature, then suddenly melting, is attributed to the special arrangement of fatty acids in its triglycerides (Wolff, 1966).

The major acids found in liquid triglycerides (oils) are unsaturated, such as oleic, linoleic, and linolenic, whereas the major fatty acids in solid triglycerides (fats) are saturated, such as palmitic and stearic acids. In the plant kingdom as a whole, oleic and palmitic acids probably are the most abundant fatty acids. The leaf tissues of most kinds of plants contain similar fatty acids, chiefly palmitic, linoleic, and linolenic, and considerable linolenic acid occurs in chloroplasts. There is much more variation among species with respect to the fatty acids found in seeds than in leaves, and those in seeds usually differ from those found in the vegetative structures of the same plant. According to Hitchcock (1975), the fruits of oil palm contain chiefly palmitic, oleic, and linoleic acids, but the seed contains considerable lauric, myristic, and oleic acid in addition to palmitic acid.

It seems that plants growing in cool climates usually produce more unsaturated fatty acids such as linoleic and linolenic than plants growing in warm climates (Lyons, 1973). It also appears that species with a wide climatic distribution contain more unsaturated fatty acids in the cooler parts of their range, although Mirov (1967) pointed out some exceptions in pines. Lipid metabolism is discussed in detail in books by Bonner and Varner (1976), Tevini and Lichtenthaler (1977), Thomson et al. (1983), Harwood and Russell (1984), Fuller and Nes (1987), Stumpf and Conn (1980–1987), Quinn and Harwood (1990), and Moore (1993). In the digestion of oil and fat the fatty acids are degraded to acetyl CoA and then can either be transferred to the Krebs cycle and oxidized with the release of energy or enter the glyoxylate cycle and be transformed into sugar.

Lipid Distribution

Lipids are widely distributed throughout woody plants and may be found in leaves, stems, roots, fruits, flowers, and seeds. Lipid contents vary among species and genotypes and in different parts of plants. By far the highest lipid concentrations are found in reproductive structures of certain species. Some seeds may contain over 70% fat on a dry weight basis but seeds of many species contain very little fat (Chapter 2, Kozlowski and Pallardy, 1997). Vegetative tissues rarely contain more than 4% fat and often have much less.

Lipid contents also vary seasonally. In littleleaf linden triglycerides in stems were hydrolyzed well before bud break. The amounts of triglycerides were high in December, low in March, and increased in May. The perennial roots contained considerable triglyceride, with much less present in the outer wood than in the inner wood. The roots contained less triglyceride than the stem in December but more than the stem in March. In May there was little difference in the amounts of triglycerides in the root wood and stem wood (Höll and Priebe, 1985).

In one-year-old Scotch pine needles lipids decreased dramatically during expansion of new shoots, but the pattern varied among specific lipids (Fig. 8.2). By comparison, seasonal changes in lipid contents of stems were negligible (Fig. 8.3).

Fats contain more energy per unit of weight than do carbohydrates or proteins. The heat of combustion of 1 g dry wt fat is 38.9 kJ; of proteins 23.9 kJ; and of carbohydrates 21.3 kJ. However, fats accumulate in relatively small amounts in vegetative tissues, and hence they are less important than carbohydrates as reserve foods. Triglycerides, the major storage lipids, increased from 1 to 3% of residual dry weight in cottonwood stems as the plants became dormant, and total non-

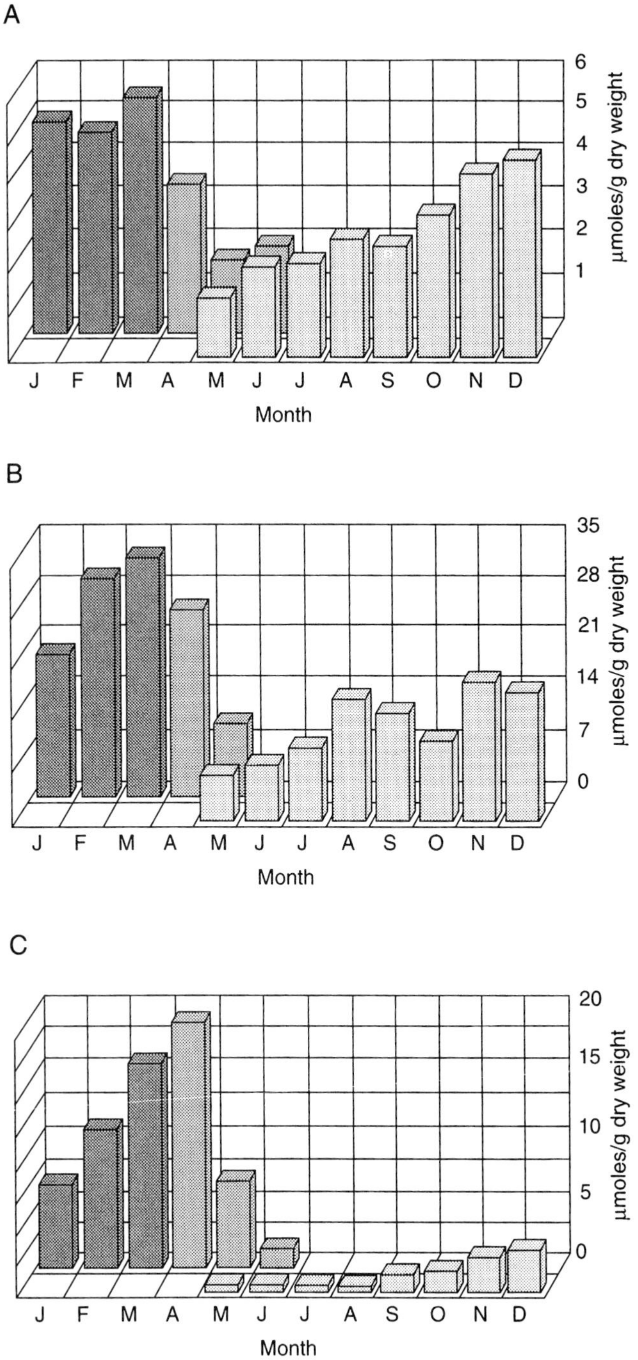

FIGURE 8.2. Seasonal changes in three classes of lipids in Scotch pine needles: (A) diacylglycerols, (B) free fatty acids, and (C) triacylglycerols. Budbreak occurred at the end of April and needle maturation in August. May and June needles initiated in different years as shown in different rows. From Fischer and Höll (1991).

structural carbohydrates increased to more than 35% (Nelson and Dickson, 1981). It has been estimated that the amount of stored lipids in trees would need to increase by nearly 10 times to equal the energy found in stored carbohydrates (Dickson, 1991).

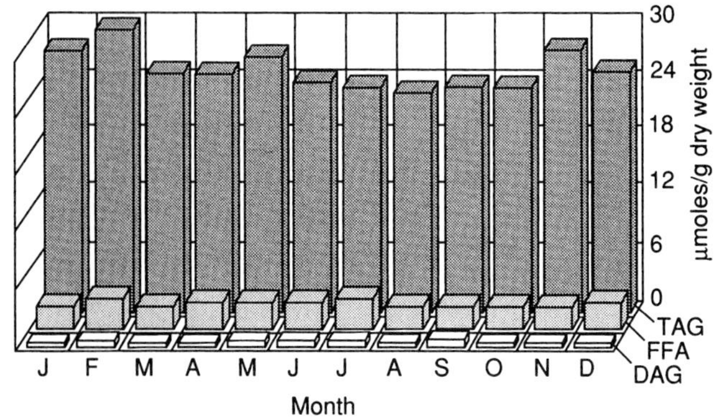

FIGURE 8.3. Seasonal changes in three classes of lipids in Scotch pine sapwood. TAG, triacylglycerols; FFA, free fatty acids; DAG, diacylglycerols. From Fischer and Höll (1992).

WAXES, CUTIN, AND SUBERIN

Cuticle

The outer surfaces of herbaceous stems, leaves, fruits, and even flower petals usually are covered by a relatively waterproof layer, the cuticle. As mentioned in Chapter 2, the cuticle is composed of wax and cutin and is anchored to the epidermal cells by a layer of pectin.

Waxes

The waxes are esters of long-chain monohydric alcohols and longer-chain fatty acids than those found in simple lipids, that is, with carbon chains containing more than 20 carbon atoms. Waxes also contain alkanes with odd numbers of carbon atoms, primary alcohols, and very long-chain free fatty acids.

There are two kinds of leaf waxes, epicuticular and intracuticular. The epicuticular waxes comprise the outer part of the cuticle; intracuticular waxes are embedded in cutin (Stammitti et al., 1995). Wax synthesis occurs in the epidermal cells of apple fruits and several kinds of leaves, and it must occur near the site where it is deposited because of the difficulty of transporting such an insoluble material. Waxes probably are generally synthesized in the epidermal cells as droplets, pass out through the cell walls, and form layers on the outer surfaces. Some wax is pushed out through the cutin-wax layer, forming a deposit on the cuticle and producing the bloom characteristic of some leaf and fruit surfaces (see Fig. 8.4). Waxes also occur in suberin-rich barks (Martin and Juniper, 1970). Apparently wax generally accumulates on the external surfaces of plants, in contrast to suberin, which accumulates in cell walls, and to cutin, which sometimes accumulates on internal as well as external surfaces. An excep-

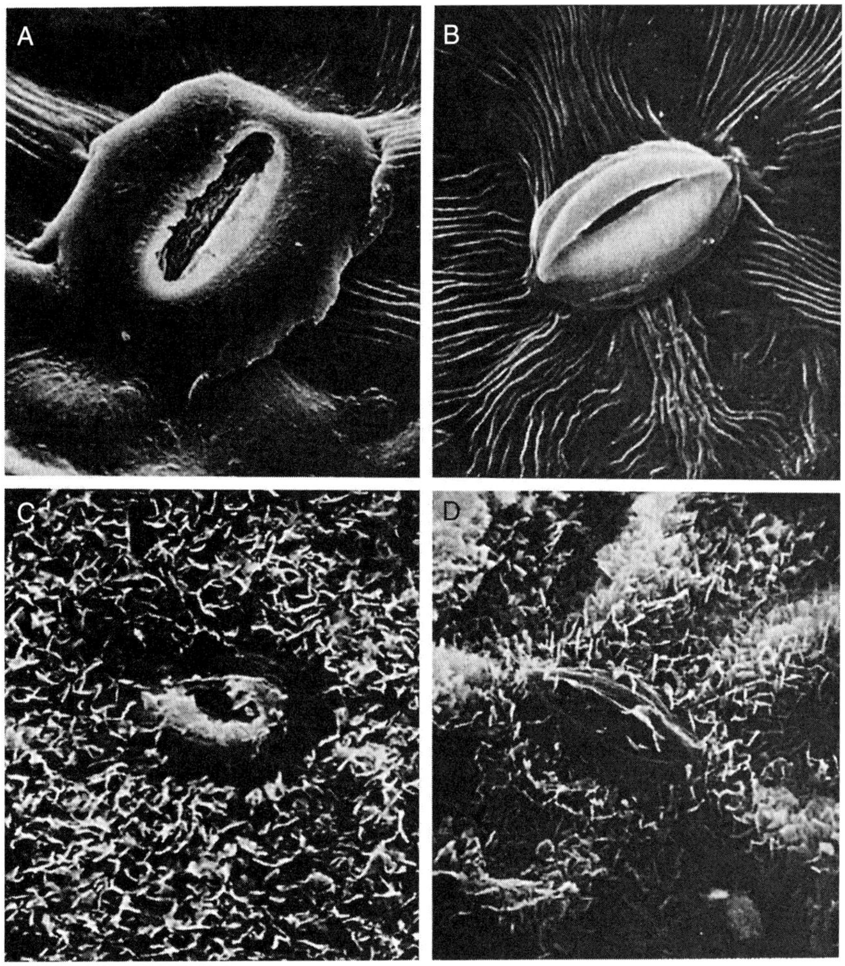

FIGURE 8.4. Variations in leaf waxes of angiosperm trees. (A) American elm (×2,000), (B) white ash (×2,000), (C) sugar maple (×2,000), and (D) eastern redbud (×2,000). Photos by W. J. Davies.

tion is the accumulation of liquid wax in the seeds of jojoba.

Epicuticular waxes are physiologically important because they restrict transpirational water loss, contribute to control of gas exchange, reduce leaching of nutrients, provide a barrier to air pollutants, and influence entry of agricultural chemicals into leaves, fruits, and stems. When present in irregular masses, waxes make leaf surfaces difficult to wet, hence a wetting agent or "spreader" added to spray materials often ensures even coverage. Some of the chemicals in cuticular wax inhibit growth of pathogenic organisms (Martin and Juniper, 1970). In some cases, however, components of leaf waxes stimulate spore germination and development of pollen tubes, thus promoting pathogenesis (Schuck, 1972).

Deposition of wax on leaves is an important adaptation to drought. Transpiration rates of drought tolerant plants with closed stomata commonly vary from 2 to 20% of the rates when the stomata are open. By comparison, mesophytic plants with thinner layers of leaf waxes generally lose from 20 to 50% as much water with closed stomata as they do with open stomata (Levitt, 1980b). The permeability coefficient for diffusion of water vapor through the cuticle increased by 300 to 500 times following extraction of the cuticular wax (Schönherr, 1976), emphasizing the importance of leaf waxes in desiccation avoidance by plants.

In some species the occlusion of stomatal pores with wax greatly reduces water loss and photosynthesis (Chapters 5 and 12). Leaf waxes in stomatal pores also increase resistance to penetration by some fungal pathogens (Patton and Johnson, 1970; Franich et al., 1977).

Some waxes are of considerable commercial importance. Among the best known is carnauba wax, obtained from the leaves of a palm, *Copernicia cerifera*, found in Brazil. It contains about 80% alkyl esters of long-chain fatty acids and 10% free monohydric alcohols. Palm wax occurs on the trunk of the wax palm (*Ceroxylon andicola*) in layers up to 2 or 3 cm in thickness. It consists of about one-third true wax, the remainder being resin. Other commercial palm waxes are ouricuri wax, obtained from the Attalea palm

(*Attalea excelsa*), and raffia wax, obtained from the dried leaves of the Madagascar raffia palm (Deuel, 1951). *Eucalyptus gunnii* var. *acervula* of Tasmania, and the leaves of white sandalwood also yield wax. The leaves of *Myrica carolinensis* supply the fragrant wax used in bayberry candles.

Leaf waxes have been classified into two major types: (1) flat deposits (including wax granules, rods and filaments, plates, and scales) and (2) localized deposits (including layers and crusts as well as liquid or soft coatings). The amount and structure of wax often differ between the two surfaces of the same leaf and even between different locations on the same leaf surface. For example, in *Eucalyptus polyanthemos* the wax was plate-like over most of the leaf blade but tubular over the midrib (Hallam, 1967). The structure of leaf wax has been used as a taxonomic character to separate species of *Eucalyptus* and *Cupressus* (Hallam and Chambers, 1970; Dyson and Herbin, 1970).

The amount of wax on leaves varies from a trace to as much as 15% of the dry weight of the leaf, and differs with plant species, genotype, leaf age, and environmental conditions. White ash leaves had thin leaf waxes; sugar maple leaves not only had thick deposits of wax but many of the stomatal pores were occluded with wax (Kozlowski et al., 1974; Davies and Kozlowski, 1974b). Genetic variations in wax deposition have been reported in *Eucalyptus* and *Hevea* (Barber and Jackson, 1957; Rao et al., 1988).

The amount of leaf wax that forms is favored by high light intensity, low relative humidity, and drought (Baker, 1974; Weete et al., 1978). In some species changes in leaf waxes occur in response to selection by environmental factors. In Tasmania, for example, nonglaucous (green) phenotypes of *Eucalyptus* were present in sheltered habitats and glaucous phenotypes in exposed sites. At elevations of 2,000 ft. the leaves of *Eucalyptus urnigera* were nonglaucous and had predominantly flaky wax; at 2,300 ft. the leaf waxes consisted of flakes and rods; and at 3,200 ft. the leaves were glaucous and their waxes consisted of masses of rodlets (Hall et al., 1965).

Waxes are produced largely during early stages of leaf expansion. Fully expanded leaves generally have lost capacity to produce large amounts of wax. Hence, old leaves with their thin layers of wax often have high transpiration rates, lose large amounts of minerals by leaching, and have low resistance to pathogens (Romberger et al., 1993).

The structure of epicuticular waxes changes during leaf development. In Douglas-fir fusion of crystalline wax rods into amorphous (solid) wax began several weeks after bud break (Thijsse and Baas, 1990). Increase in the amount of solid wax occurred similarly but more slowly in one- and two-year-old needles. Very young Scotch pine needles had more amorphous wax than older needles. This observation, together with the presence of wax rodlets on top of amorphous wax crusts, indicated that wax was recrystallized (Bacic et al., 1994).

The structure of leaf waxes is influenced by mineral nutrition of plants. Proportionally more tubular wax and less scalelike wax were produced by Douglas-fir trees that were fertilized with N and K than by unfertilized trees (Chiu et al., 1992). The deteriorating effects of unbalanced mineral nutrition on coverage and structure of wax were evident in the stomatal furrows of Scotch pine needles within a year and in the epistomatal chambers a year later (Ylimartino et al., 1994). Deficiencies of Ca and Mg decreased wax coverage in both the stomatal furrows and epistomatal chambers. Coverage in the epistomatal chambers also was decreased by K deficiency and N excess (and hence N/K ratios). Waxes in both the stomatal furrows and epistomatal chambers changed from tubelike to more fused and netlike structures as a result of deficiencies of K, Mg, and Ca (hence increased N/K, N/Mg, and N/Ca ratios).

Cutin and Suberin

The polymeric compounds cutin and suberin comprise barriers that prevent diffusion of water and other molecules, largely because of the waxes that are deposited with the polymers. Cutin, which is composed of hydroxy and epoxy fatty acids, occurs on almost all aerial parts of plants, including stems (except the bark), leaves, flower parts, fruits, and seed coats. Cutin impedes penetration by germinating fungal spores unless they are induced by contact with plant surfaces to produce cutinase, which digests cutin to form pathways for penetration of germ tubes (Kolattukudy et al., 1987; Podila et al., 1988). Many microorganisms, including plant pathogens, can grow on cutin as their only C source (Kolattukudy, 1980).

Suberization occurs in belowground tissues including epidermis, exodermis, root and tuber phellem (cork), as well as in the cork cells of stem bark (Bernards, 2002). Suberized and lignified cells are associated with different meristems, as in roots where meristem initials that give rise to suberized cells such as those found in the endodermis are different from those that produce lignified cells in the stele (Bernards, 2002). Suberin composition differs in a number of ways from that of lignin, with suberin having hydroxycinnamic acids and fewer monolginols than lignin. Suberin structure (Fig. 8.5) is dimorphic, appearing to consist of (1) a polyphenolic-dominated region in the

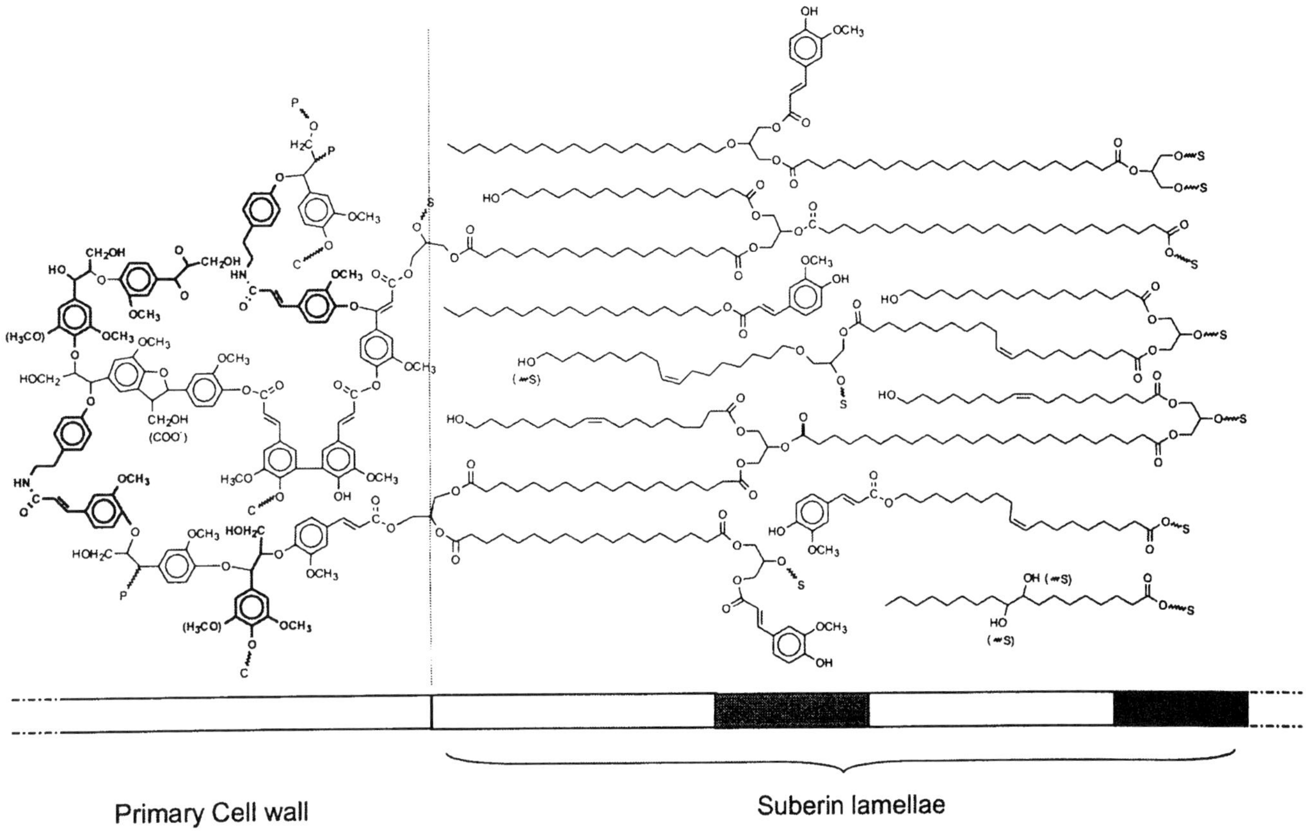

FIGURE 8.5. Model for the structure of suberin of potato. The suberized primary cell wall possesses a polyphenolic-dominated region whereas the adjacent, covalently bound lamellar region has polyaliphatic constituents. From Bernards (2002).

primary cell wall and (2) an adjacent, covalently bound lamellar region of polyaliphatic constituents. The latter consist of shorter-chain (most C_{18}) fatty acids that are highly oxidized and long-chain (C_{24} to C_{32}) alcohols and oxidized fatty acids. The phenolic and aliphatic components are cross-linked by esterification to glycerol.

Suberin restricts the movement of water within and from the plant. The suberin in the Casparian bands/strips of the endodermis and exodermis reduces the apoplastic transport of water and solutes. Loss of water from roots in dry soil is reduced by suberin in the exodermis (Hose et al., 2001). Suberized cells also may protect plants from attack by bacterial and fungal pathogens (Kolattukudy, 1980, 1981, 1984; Lulai and Corsini, 1998; Bernards, 2002). Suberization also appears to be involved in wound healing of plant tissues. Wounding triggers induction of enzymes involved in synthesis of suberin (Dean and Kolattukudy, 1976). Polyphenolic constituents appear first in response to wounding followed a few days later by aliphatic components (Lulai and Corsini, 1998).

INTERNAL LIPIDS

The surfaces of cell walls that are exposed to intercellular spaces often are covered with a hydrophobic lipid layer that increases the resistance to evaporation of water from mesophyll cell surfaces. In this connection readers are reminded that although cutin and wax layers are relatively impermeable to water when dry, they are more permeable when moist, as is the case with roots in moist soil and in the interior of leaves. Also, when leaf surfaces are wetted, substances in solution penetrate more readily through the cuticle.

Phospholipids

The phospholipids are the major lipids in a variety of membranes including those of mitochondria, nuclear membranes, tonoplast, plasmalemma, and endoplasmic reticulum. Phospholipids are diesters of phosphoric acid with diacylglycerol and various alcohols. They contain both a hydrophobic, water-insoluble portion (fatty acids) and a hydrophilic, water-soluble portion (inositol, choline, ethanolamine, serine). The biosynthesis and turnover of phospholipids are discussed by Mudd (1980) and Mazliak and Kader (1980).

Glycolipids

In glycolipids, a terminal hydroxy group of the glycerol moiety is attached to a sugar, either galactose or glucose. Mono- and digalactosyl diglycerides are the major glycerolipids in chloroplasts where they are esterified with linolenic and linoleic acids. The glycolipid containing sulfo-quinovose is called sulfolipid. It likewise is most abundant in chloroplasts, but also occurs in nonphotosynthetic tissues.

Membrane Lipids

The basic framework of all cell membranes is a double layer of lipid molecules. Lipids comprise 20 to 40% by weight of biological membranes. The same lipids are found in nonphotosynthetic and photosynthetic tissues, but in different proportions. In nonphotosynthetic tissues the most abundant membrane phospholipids are phosphatidylcholine (PC) and phosphatidylethanolamine (PE), together comprising more than 70% of the total phospholipid weight (Mazliak and Kader, 1980). In leaves the amount of glycolipids greatly exceeds the amount of phospholipids. Total lipids of most photosynthetic tissues consist of about 20% phospholipids and 40% glycolipids. The most abundant leaf phospholipid is PC, but PE, phosphatidylglycerol (PG), and phosphatidylinositol (PI) also are major components.

There is considerable evidence of lipid exchange between different membranes of plant cells. Newly synthesized lipids accumulated sequentially in microsomes, mitochondria, and nuclei, suggesting lipid transfer among organelles (Mazliak et al., 1977). The turnover of a given phospholipid in one organelle may be different from turnover in another.

Functions of biological membranes may be influenced by the lipid-protein ratio, molecular species of lipids, and perturbation of the lipid bilayer or biochemical properties of membranes. Changes in lipid fluidity and phase properties of membranes of senescing cells influence membrane permeability as shown by leakage of pigments and metabolites from senescing tissues. A decrease in membrane fluidity during senescence may inhibit the activity of membrane-associated enzymes and receptors. The change in state of lipids in cell membranes from a liquid crystalline to a solid gel state at chilling temperatures has been associated with chilling injury (Levitt, 1980a; Wang, 1982). It has been claimed that this change induces contraction, which causes cracks or channels in membranes, hence increasing their permeability. However, phase transition of membranes probably is less important in chilling injury than originally proposed. For example, phase transitions in membranes of chilling-sensitive soybeans were not evident in the temperature range in which chilling injury occurred (O'Neill and Leopold, 1982). Although Kenrick and Bishop (1986) found wide variations in the chilling sensitivity of 27 species of plants, their contents of high-melting-point fatty acids were unrelated to sensitivity to chilling.

ISOPRENOIDS OR TERPENOIDS

The isoprenoids or terpenoids are of both biochemical and economic interest. They are hydrocarbons built up of varying numbers of isoprene (C_5H_8) units and include essential oils, resins, carotenoids, and rubber (Fig. 8.6). All plants can synthesize carotenoids and steroids, but the capacity to synthesize other terpenes is scattered very irregularly through the plant kingdom. For many years, it was assumed that isoprenoid compounds in plants were synthesized from precursors derived via the mevalonate pathway with acetate as a starting point (Eisenreich et al., 2001, 2004). However, an alternative isoprenoid synthetic pathway in the chloroplast (alternately named the methyl 4-phosphate pathway, MEP, or the deoxyxylulose pathway) has been identified for which pyruvate and glyceraldehye 3-phosphate are precursors (Rohmer et al., 1996; Eisenreich et al., 2004). Sterols are synthesized primarily from acetate via the mevalonate pathway. The phytol side chain of chlorophyll, carotenoids, and numerous natural plant products are derived entirely or at least partly via the deoxyxylulose pathway.

The construction costs of terpenoids per gram are higher than those of most other metabolites. In addition, the enzyme costs of synthesizing terpenoids are high because the terpenoid-synthesizing enzymes are not involved in other pathways. Because terpenoids are sequestered in multicellular secretory structures, the costs of storing them also are high. However, maintenance of terpenoid pools is inexpensive because large amounts of terpenoids are not lost by metabolic turnover, volatilization, or leaching (Gershenzon, 1994).

Essential Oils

The essential oils are straight chain or cyclic compounds and may be mono-, sesqui-, or diterpenes. Their varying characteristics are determined by the

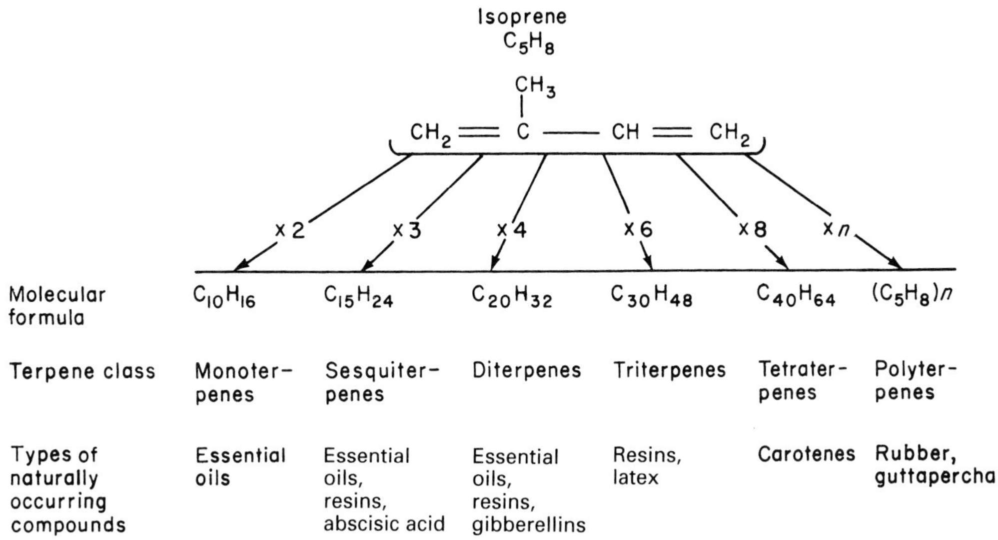

FIGURE 8.6. Relationships among isoprenoid compounds. After Bonner (1950).

chemical groups associated with them. Essential oils are the source of most of the odors found in the flowers, fruit, or wood of many plants. They are most common in species of the Pinaceae, Apiaceae, Myrtaceae, Lauraceae, Rutaceae, Lamiaceae, and Asteraceae. All organs of plants, including the bark, wood, and leaves, may contain essential oils. They often are produced in groups of glandular cells or in glandular hairs on flowers, leaves, and stems, and sometimes they are secreted into specialized ducts in leaves and stems. The essential oils have no known important functions in plant metabolism although they may be useful in attracting pollinators or repelling herbivores. Many are volatile and evaporate into the air, especially on warm days, producing the typical odors of various flowers and coniferous forests. Most loss of essential oils occurs through stomatal pores but some loss occurs through the cuticle (Schmid and Ziegler, 1992).

Rasmussen and Went (1965) estimated that the vegetation of the world releases 438×10^6 tons of volatile material annually. More is released on warm sunny days than on rainy days, and large amounts are released from dying vegetation. The latter is the source of the odor of recently mown grass and hay and of recently fallen autumn leaves. Photooxidation of this volatile material is believed to produce the blue haze characteristic of the Blue Ridge Mountains of Virginia and other heavily vegetated areas.

The release of volatile hydrocarbons from vegetation is an important source of air pollution. For example, it is reported that from 0.1 to 3.0% of the carbon fixed in photosynthesis by live oak leaves is lost as isoprene (Tingey and Ratsch, 1978). Leaves of numerous species synthesize isoprene, primarily via the MEP pathway in the chloroplast, with the loss greatest in bright light at high temperatures (Rasmussen, 1970; Monson and Fall, 1989; Loreto and Sharkey, 1990; Sharkey and Singass, 1995; Sharkey 1996; Karl et al., 2002; Wolfertz et al., 2003; Sanadze, 2004). Some potential adaptive roles for isoprene production have been proposed, including increased thermal tolerance of the photosynthetic apparatus, especially during heatflecks (short-term spikes in leaf temperature) (Sharkey et al., 2001; Wolfertz et al., 2004), dissipation of excess reducing power under light stress when CO_2 fixation is restricted by stomatal closure (Sanadze, 2004), and detoxification of destructive reactive oxygen species in leaves (Loreto et al., 2001).

Essential oils are extracted commercially on a small scale by steam distillation from leaves of pines, eastern arborvitae, black spruce, balsam fir, and eastern hemlock, and from wood of cedar, sweet birch bark, and sassafras roots and buds. Turpentine is the most economically important essential oil obtained from trees.

The large quantities of essential oils in the leaves of some shrubs make them very flammable and greatly increase the speed with which fire spreads. An example is chamise, an important shrub of the California chaparral (Kozlowski and Ahlgren, 1974, p. 338). Some species of sagebrush and eucalypts also are very flammable.

Resins

Resins are a heterogeneous mixture of resin acids ($C_{20}H_{30}O_2$), fatty acids, esters of these acids, sterols, alcohols, waxes, and resenes (mixtures of neutral alkali-resistant compounds containing carbon, hydrogen, and oxygen). Both conifers and broadleaved trees synthesize resins, but conifers usually produce much larger amounts. Resin yields of 0.8 to 25% have been reported for coniferous woods as compared to only 0.7 to 3% for woods of angiosperm trees (Wise and Jahn, 1952). Most resin used commercially comes from trees of the Pinaceae, legume families, and Dipterocarpaceae. Copals are a group of resins extracted from leguminous forest trees and are known for their hardness and high melting point. Trees of the Dipterocarpaceae produce a resin called dammar in commerce. Another commercially important resin is kauri gum, obtained from the kauri tree of New Zealand (Howes, 1949). Amber is a fossil resin. Secretory structures associated with resins are discussed in Chapter 2.

Oleoresins

The most important commercial resins are the oleoresins obtained from pines. Oleoresins consist of approximately 66% resin acids, 25% turpentine (an essential oil), 7% nonvolatile neutral material, and 2% water (Wise and Jahn, 1952).

Oleoresins are produced by the living epithelial cells that line the resin ducts and that are especially active in the outer sapwood. Scarifying the tree trunk exposes the resin ducts and oleoresin oozes out. Oleoresins are obtained commercially from pines by chipping or cutting through the bark and exposing the surface of the sapwood, as shown in Figure 8.7. Oleoresin yield varies not only among tree species but also among different trees of the same species. Bourdeau and Schopmeyer (1958) found that the amount of resin flow in slash pine was controlled by the number and size of resin ducts, resin pressure, and viscosity of the exudate. Resin pressures vary diurnally, the highest pressures occurring about dawn and the lowest in the afternoon when the water content of the trunk is lowest. In Monterey pine trees diurnal changes in resin pressures were in phase with changes in stem diameters (Neher, 1993). Lorio and Hodges (1968) reported that diurnal variations in oleoresin exudation pressure of loblolly pine are related to soil and atmospheric moisture conditions. In loblolly pine trees resin flow increased from May to August (Tisdale and Nebeker, 1992). The seasonal increase was consistent with Lorio's (1986) growth differentiation hypothesis, which states that early in the season the rapidly growing tissues are preferential sinks for photosynthate, hence resin production is low. As the rate of growth decreases later in the season, more photosynthate is available for resin production. Seasonal variations in resin flow also are

FIGURE 8.7. Slash pine chipped to produce oleoresin. Photo courtesy of U.S. Forest Service.

appreciably influenced by soil moisture content and temperature (Blanche et al., 1992).

Stimulation of Flow of Oleoresins

Epithelial cells progressively lose their secretory function as they age. However, they can be induced to increase resin production by wounding, infestation by insects and parasitic fungi, and by treatment with certain chemicals such as paraquat.

In maritime pine the resin content of cortical woody tissues increased near wounds as a consequence of reactivation of epithelial cells lining the resin ducts (Walter et al., 1989). Wounding also commonly alters the composition of terpenes produced. In maritime pine, for example, wounding was followed by large increases in α- and β-pinene whereas other terpenes increased only slightly. The increases resulted from reactivation of resin duct secretory cells of primary origin (Marpeau et al., 1989). In addition to reactivating preexisting resin ducts, wounding induces the cambium to produce proportionally more resin ducts on both sides of the cambial zone (Cheviclet, 1987), further contributing to increased resin production in injured pines.

Injection of the herbicide paraquat (1,1′-dimethyl-4,4′-bipyridilium dichloride; methyl viologen) into pine stems or stumps greatly increases production of oleoresins in a strip of wood often extending several meters above the point of injection (Roberts and Peters,

1977; Kossuth and Koch, 1989). The oleoresin is deposited in the sapwood, producing the resin-soaked wood known as lightwood, which is an important source of naval stores such as rosin and turpentine. Treatment with paraquat was effective on several species of pine but not on Douglas-fir, balsam fir, hemlock, tamarack, or Norway spruce (Rowe et al., 1976). Lightwood formation also is stimulated by mechanical injury and invasion by pathogens.

The increase in oleoresin content of the sapwood following treatment with paraquat reflects preferential repartitioning of the tree's fixed C resources to synthesis of oleoresin. The resin-loading process is associated with several events such as (1) an increase in ethylene production and respiration, (2) rapid mobilization of reserve carbohydrates, (3) increase in lipid peroxidation, and (4) injury to membranes and subcellular organelles (Schwarz, 1983). Because injection of ethrel increases resin formation in conifers, it has been suggested that the increased resin production following wounding may be caused by ethylene released from the wounded tissue (Wolter, 1977).

Unfortunately, paraquat kills the cambium to a considerable height above the point of treatment, producing a strip of dead wood and bark. This limits the amount of stem circumference that can be treated. Treated trees also are more susceptible to attack by bark beetles and the total effect of treatment of a tree population is a considerable increase in mortality.

The composition of turpentine from southeastern and western pines of the United States varies considerably. Turpentine of southeastern pines is of relatively simple composition and consists essentially of two monoterpenes, α- and β-pinene. Turpentines of western pines are more complex and contain, in addition to the pinenes, some aliphatic hydrocarbons, aliphatic aldehydes, 3-carene, and sesquiterpenes. Turpentine of southeastern pines resembles that of European species; that of trees growing in western regions resembles turpentine from pines of southeast Asia (Mirov, 1954). Franklin (1976) reported variations in the composition of oleoresins of slash pine from the base of the tree to the crown as well as among trees. He suggested that the base-to-crown variation may be controlled by growth regulators transported downward from the crown.

After the volatile turpentine has been removed from the oleoresin by distillation, the remaining substance is a hard resin called rosin, which varies in color from amber to almost black. Its chief constituent is abietic acid.

Other commercially important oleoresins are Canada balsam and Oregon balsam. The former, obtained from balsam fir, is secreted in resin canals formed by the separation of cells in the bark and occurs in small blisters under the bark. Oregon balsam, obtained from Douglas-fir, is found in cavities in trees that were produced by wind shake. Venetian turpentine, used in the arts, is obtained from European larch.

Monoterpenes

In recent decades much attention has focused on monoterpenes, largely because of their importance in studies of genetic variation and geographical distinction of plants, identification of origins of conifers in commercial plantations, seed certification, and identification of seed sources of plantations of unknown origin (Fady et al., 1992).

Terpene composition and tree growth are controlled genetically as shown for white and blue spruce (Von Rudloff, 1972, 1975), Norway spruce (Esteban et al., 1976), black spruce (Chang and Hanover, 1991), western white pine (Hanover, 1966), slash pine (Squillace, 1971), loblolly pine (Squillace and Swindel, 1986), Scotch pine (Forrest, 1980; Yazdani et al., 1982), and lodgepole pine (White and Nilsson, 1984). McRae and Thor (1982) found variations in monoterpene composition of 12 loblolly pine provenances in Tennessee. An east-to-west gradient was found in contents of limonene, myrcene, and α-pinene, whereas a high β-phellandrene content was more frequent in western than in eastern provenances. No clear trend was evident in β-pinene contents.

In chemosynthetic studies with conifers, information on seasonal variations in terpenes is important. Von Rudloff (1972, 1975) showed that large seasonal changes in terpenes of white spruce and blue spruce occurred only in the buds and young leaves after bud burst and continued until midsummer. The composition of the mature leaves and twigs changed little during the same period (Fig. 8.8). In blue spruce there also were major differences in synthesis of monoterpenes in the leaves when compared with the twigs and buds. The leaves contained large amounts of the closely related santene, tricyclene, camphene hydrate, borneol, and bornyl acetate. Only small amounts of these compounds were present in twigs and buds. In buds 3-carene predominated and the relative amounts of β-pinene, sabinene, terpinolene, and 4-terpinenol were significant.

The emission of terpenes to the atmosphere is regulated by water supply and correlated with foliage water content. During a prolonged drought the water content of Italian cypress (*Cupressus sempervirens*) foliage changed in three sequential steps (Yani et al., 1993):

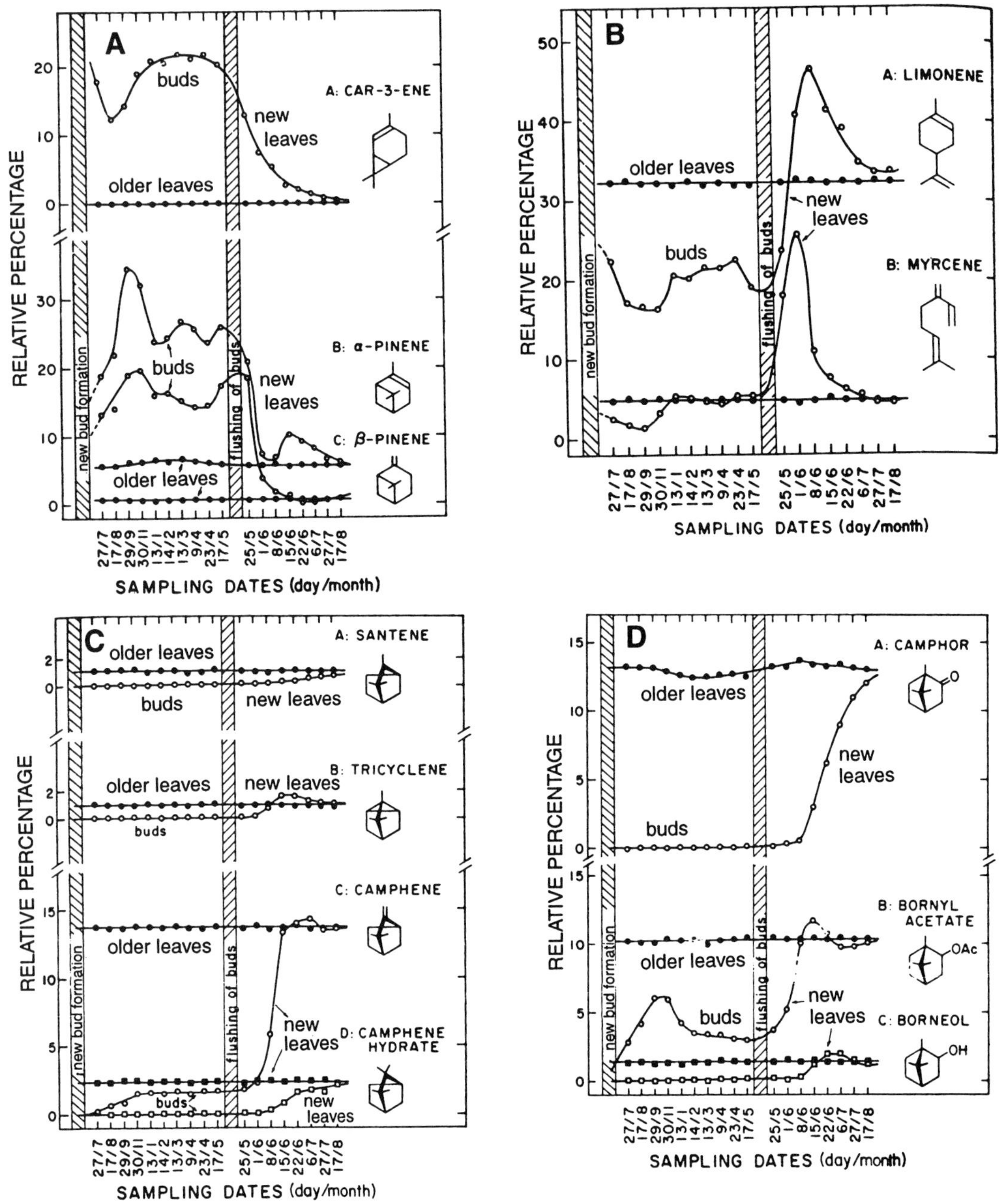

FIGURE 8.8. Seasonal changes in relative percentages of monoterpenes in the volatile oil of mature leaves, buds, and young leaves of blue spruce. (A) 3-carene, α-pinene, and β-pinene; (B) limonene and myrcene; (C) santene, tricyclene, camphene, and camphene hydrate; and (D) camphor, bornyl acetate, and borneol. From von Rudloff (1975).

1. During the first 20 days there was no large loss of water.
2. After two months of drought severe dehydration of foliage was evident.
3. Finally the rate of water loss declined greatly and the water content stabilized at about 300 mg g^{-1} fresh weight.

Significant amounts of terpenes, mostly monoterpenes, were released to the atmosphere only during step 1 and the first part of step 2. Thereafter terpene emission decreased greatly until no more terpenes were emitted. The decline in terpene emission was attributed to stomatal closure or metabolism of terpenes, reducing the amounts of terpenes available for emission.

Hansted et al. (1994) identified 11 floral volatiles of black currant including monoterpenes, hydrocarbons, and monoterpene ethers. These were emitted in a rhythmic manner, with a maximum in the middle of the photoperiod. The period of maximum emission coincided with the flight activity of important pollinating insects.

Carotenoids

The only naturally occurring tetraterpenes are the carotenoid pigments, which have the formula $C_{40}H_{56}$. The carotenes are pure hydrocarbons, which include red, orange, and yellow pigments and may occur in all organs of plants. They are involved in light trapping in photosynthesis. The xanthophylls are yellow or brownish pigments that occur commonly in leaves and protect chlorophyll from photooxidation (Chapter 5). They contain a small amount of oxygen and generally have the formula $C_{40}H_{56}O_2$.

Carotenoids, which form by condensation of eight isoprene units, can be derived from the basic C skeleton of lycopene by hydrogenation, dehydrogenation, cyclization, oxidation, and combinations of these. Synthesis of carotenoids is influenced by light, nutrition, temperature, pH, and O_2. Because these factors influence several aspects of cell metabolism, their effects are rather nonspecific. However, light plays a predominant regulatory role in synthesis of carotenoids, with the effect mediated by phytochrome (Goodwin, 1980; Jones and Porter, 1985). The synthesis of carotenoids generally is stimulated by red light and can be prevented by a flash of far-red light (Rau, 1983).

Rubber

Rubber is a polyterpene (*cis*-1,4-polyisoprene) composed of 500 to 5,000 isoprene units joined linearly in the following pattern:

$$
\begin{array}{c}
\mathrm{CH_3} \qquad\qquad\qquad\qquad \mathrm{CH_3} \\
| \qquad\qquad\qquad\qquad\quad | \\
\mathrm{—CH_2—C{=}CH—CH_2—CH_2—C{=}CH—CH_2—}
\end{array}
$$

Biosynthesis of rubber occurs sequentially by (1) generation of acetyl CoA, (2) conversion of acetyl CoA to isopentenyl pyrophosphate (IPP) via mevalonic acid, and (3) polymerization of IPP into rubber (Backhaus, 1985; Archer and Audley, 1987). Rubber particles from India rubber, rubber trees, and guayule plants contain similar proteins that may share common functions in rubber synthesis and/or rubber particle structure (Siler and Cornish, 1993).

Rubber is formed by about 2,000 species of plants, including herbs, shrubs, trees, and vines. It is formed only in dicotyledonous angiosperms and is not synthesized by monocotyledons, gymnosperms, or lower plants. Especially well represented with rubber-producing species are the families Euphorbiaceae, Moraceae, Apocynaceae, Asclepiadaceae, and Asteraceae. Most rubber-producing woody plants are tropical, and guayule is said to be the only temperate zone woody plant that produces enough rubber for commercial extraction. The chief source of natural rubber is the tropical tree, *Hevea brasiliensis*, which is in the family Euphorbiaceae. The trans isomer of rubber, gutta-percha, is obtained chiefly from *Palaquium gutta*, which is in the family Sapotaceae.

Rubber is occasionally found in parenchyma cells, as in guayule. More often it occurs as suspended globules in latex, a complex liquid system containing a variety of substances in solution or suspension. Among these are terpene derivatives, sugars, starch grains, organic acids, sterols, and enzymes. The exact composition of latex varies widely among species and even among individual plants of the same species. Starch grains occur in latex of *Euphorbia*, but not in *Hevea* latex. In addition to rubber, *Hevea* latex contains vacuolar components called lutoids, which are bound by a single membrane, and Frey-Wyssling particles (organelles with double membranes containing carotenoid pigments) (Backhaus, 1985). *Ficus* latex is high in protein; the latex of *Papaver somniferum* is high in the alkaloids of opium; and the latex of *Carica papaya* is the commercial source of the enzyme papain. The chicle traditionally used in chewing gum is obtained from latex of a tropical tree, *Achras zapota*, which grows in Mexico, Central America, and Venezuela.

Rubber does not occur in the latex of all plants and, when present, usually is found in very low concentrations. It occurs in commercially useful quantities in only a few species, notably in *Hevea brasiliensis*. The rubber content of *Hevea* is about 25% of the dry weight per volume of tapped latex, which accounts for about 2% of the dry weight of the plant. In contrast, guayule plants can accumulate up to 22% of their dry weight as rubber. This difference emphasizes that relatively few cells, the latex vessels, synthesize rubber in *Hevea*, whereas essentially all the parenchyma cells of guayule may produce rubber. Nevertheless, total yields of rubber are much higher from *Hevea* than from guayule (Leong et al., 1982). Neither latex nor rubber is used as a reserve food by plants, even though large amounts of resources are allocated for rubber synthesis. Once formed, rubber remains in the plant because plants lack the enzymes capable of degrading it (Backhaus, 1985).

The physiological role of latex in plants is conjectural. Suggestions have been made that latex may

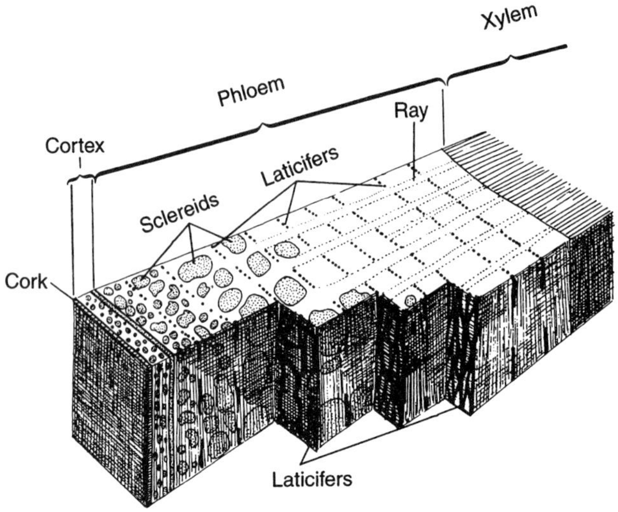

FIGURE 8.9. Bark of *Hevea brasiliensis*, showing arrangement of laticifers in the secondary phloem. Adapted from Vischer (1923); from Fahn (1990).

FIGURE 8.10. Tapping of a rubber tree to produce latex. Photo courtesy of Rubber Research Institute, Kuala Lumpur, Malaysia.

serve as a water-regulating system or as a reserve food in stressed plants. However, strong evidence to support these hypotheses is lacking. Polhamous (1962) concluded that rubber is an end product and is not reused in metabolism. It is unlikely that latex functions as a reserve food because the rubber content in stressed plants is not reduced after as much as 60% of the reserve carbohydrates have been depleted (Hunter, 1994). After a prolonged light-starvation period, the starch grains in *Euphorbia esula* latex did not function as utilizable carbohydrates (Nissan and Foley, 1986). There is considerable evidence that latex does play a role in defense of plants against herbivores (Farrell et al., 1991).

As mentioned, in some species latex is distributed throughout the plant body; in others it is confined to cells and tubes (laticifers) that may be branched or unbranched. Laticifers arise in two ways: they are laid down in the embryo or seedling, elongate, and often branch at the apices, or they form by cambial activity by a method similar to that by which vessels are initiated. In young *Hevea* plants, the tubes develop from a longitudinal series of cells. The end walls between the cells become disorganized and each series of cells is converted to a tube (Metcalf, 1967). The latex vessels of *Hevea* are located in concentric layers in the bark of the stem, branches, and roots (Fig. 8.9) as tubes up to several meters long. The development of laticifers is discussed in detail by Fahn (1988a,b, 1990).

Rubber trees usually are tapped by cutting a spiral groove in the bark halfway or more around the stem at an angle of 25° to 30° from the horizontal (Fig. 8.10). The latex flows down the groove from the opened latex vessels and is collected at the bottom. Because the nuclei and mitochondria in laticifers are concentrated in the parietal cytoplasm, most are not expelled during tapping. The initial flow is caused by elastic contraction of the latex vessels, but later there is osmotic movement of water into the vessels and the viscosity and rubber concentration of the latex decrease. Flow stops after a few hours because the latex coagulates when exposed to air and every second day a thin slice is removed from the bottom of the groove, causing renewed flow. Wounding stimulates metabolic activity in the phloem, and the ribosomes, mitochondria, enzymes, and rubber particles lost in the outflow quickly are regenerated. The tapping of *Hevea* stems is not deep enough to injure the cambium so the bark is regenerated in a few years and the process can be started over again.

Latex yield varies greatly among *Hevea* clones and with tree vigor, season, stand density, age of trees, site, and cultural practices. Stems of trees with high rubber yield sometimes are grafted onto disease-resistant rootstocks. Tapping of rubber trees generally decreases their growth, presumably because regeneration of latex consumes carbohydrates that otherwise would be used in growth.

The pressure and rate of flow of latex are correlated with changes in environmental factors that control turgor in the latex vessel system (Raghavendra, 1991). The rate of flow usually is greater in the morning, when turgor is high, than in the afternoon and is reduced during dry weather (Buttery and Boatman, 1976). The low yields of latex during the dry season are associated with a tendency for plugging of laticifers (Devakumar et al., 1988).

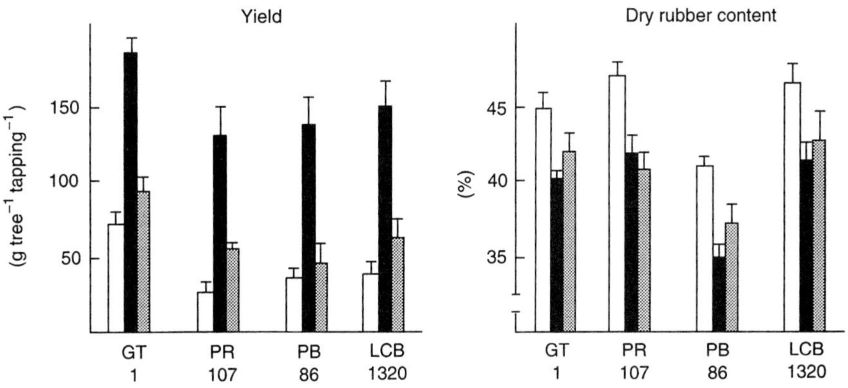

FIGURE 8.11. Effects of application of ethephon on yield and rubber content of latex of four *Hevea brasiliensis* clones (GT1, PR107, PB86, and LCB1320): open bars, before treatment; solid bars, one week after treatment; gray bars, four weeks after treatment. From Eschbach et al. (1984). *Physiol Veg*, Gauthier-Villars Publishers.

The flow of latex in *Hevea* can be greatly increased by applying certain chemicals. For a long time injections of Cu and B at the tapping cut or application of 2,4-dichlorophenoxyacetic acid (2,4-D) or α-naphthaleneacetic acid (NAA) were used to stimulate flow. These compounds largely have been replaced by ethephon (Ethrel), which greatly stimulates both the rate and duration of latex flow (Fig. 8.11). It is generally believed that the lutoids in latex are disrupted during tapping and induce coagulation and plugging of latex vessels. Ethephon appears to lessen disruption of the lutoids and also may increase thickening of the walls of latex vessels, making them less likely to contract during tapping. Ethephon treatment also increases the pH of latex, which regulates the activity of latex invertase, the enzyme controlling the use of sucrose in latex metabolism (Eschbach et al., 1984, 1986).

Related Compounds

Several important compounds are derived from the terpenoids:

Abscisic Acid. This important plant growth regulator, which is derived from a sesquiterpene, is discussed in Chapters 12 and 13.

Gibberellins. The gibberellins are another important group of plant growth regulators derived from diterpenes, and are also discussed in more detail in Chapter 13.

Steroids. The steroids or sterols are an important group of compounds, derived from isoprenoid compounds synthesized via the mevalonic acid pathway in the cytoplasm (Eisenreich et al., 2004) and found in both plants and animals. The triterpene squalene is a precursor of cholesterol, which in turn is the precursor of other steroids. Those produced in plants often are termed phytosterols. Examples are stigmasterol and ergosterol.

Most plant sterols occur in cellular organelles and the plasmalemma. The interaction of sterols with phospholipids stabilizes membranes and likely regulates their permeability (Grunwald, 1980). Sterols also may function in signal transduction in plant metabolic pathways (Piironen et al., 2000). The sterol content is higher in the heartwood than in the sapwood. Sterols apparently are decomposed or transformed during heartwood formation because the change in composition of free sterols occurs in the transition zone between the sapwood and heartwood (Saranpää and Nyberg, 1987).

Terpenoid Glycosides. Some terpenoids, especially sterols, exist as glycosides, including the saponins and cardiac glycosides such as those obtained from *Digitalis*.

Phytol. Phytol, an alcohol which is a component of chlorophyll, is derived from a diterpene.

SUMMARY

Lipids are physiologically important as constituents of protoplasm, storage forms of foods in seeds, in maintenance of the integrity and function of biological membranes, and as protective coverings on leaves, fruits, and stems. The simplest and most common lipids are the triglycerides. They are esters of glycerol and fatty acids. Palmitic acid, a saturated fatty acid, is the most widely distributed fatty acid in woody plants, but most fatty acids are unsaturated, oleic and linoleic being most common.

Lipids are widely distributed in woody plants and may be found in leaves, stems, roots, flowers, fruits, and seeds. The amount of lipids varies among species and genotypes as well as in different parts of plants. They also vary seasonally.

Waxes are esters of long-chain monohydric alcohols and fatty acids that contain more than 20 carbon atoms.

Epicuticular waxes of leaves comprise the outer part of the cuticle; the intracuticular waxes are embedded in cutin. Leaf waxes are of two major types: (1) flat deposits (including wax granules, rods, filaments, plates, and scales) and (2) localized deposits (layers, crusts, and liquid or soft coatings). The amount of wax on leaves varies with species and genotype, leaf age, and environmental conditions. Epicuticular waxes control water loss and may influence gas exchange, reduce leaching of nutrients from leaves, provide a barrier to air pollutants, and affect entry of agricultural chemicals into leaves, fruits, and stems. Commercially important waxes include carnauba wax, palm wax, ouricuri wax, and raffia wax.

Cutin and suberin are barriers to diffusion of moisture. Cutin, composed of polymers of hydroxy and epoxy fatty acids, occurs in almost all aerial parts of plants, including stems (except bark), leaves, flower parts, fruits, and seed coats. Suberin, which contains polymers of long-chain (C_{16}-C_{26}) hydroxy and dicarboxylic acids, is attached to the cell walls of periderms, the endodermis, and seed coats. Suberization is a common response to wounding of plants.

Compound lipids, especially phospholipids and glycolipids, are important components of membranes of chloroplasts, mitochondria, nuclear membranes, plasmalemma, and endoplasmic reticulum. The functions of membranes are influenced by the lipid-protein ratio, molecular species of lipids, and perturbation of the lipid bilayer or biochemical properties of membranes.

Terpenoids (isoprenoids) include essential oils, resins, carotenoids, and rubber. Essential oils account for odors of flowers, fruits, and wood of many plants. Essential oils are extracted commercially from the leaves of several conifers and wood of cedar, bark of sweet birch, and roots and buds of sassafras. Turpentine is the most important essential oil obtained from trees.

Resins are a mixture of resin acids, fatty acids, esters of these acids, sterols, alcohols, waxes, and resins. The most important commercial resins are the oleoresins obtained from pine trees.

Monoterpenes are important in studies of genetic variation of plants, identification of origins of conifers in plantations, seed certification, and identification of seed sources of plantations of unknown origin. Monoterpenes also play a role in defense of plants. They may repel or attract insects and their associated fungi, depending on the rate of flow and monoterpene composition as well as the specific insect and associated fungus. Abundant resin flow often repels bark beetles. Some monoterpenes also attract insects. For example, limonene repels western pine beetles but α-pinene attracts the same insect.

Rubber, a polyterpene comprised of 500 to 5,000 isoprene units, is obtained primarily from the latex of *Hevea brasiliensis* trees. Rubber also occurs in parenchyma cells, as in guayule. Total yields of rubber are much higher from *Hevea* than from guayule plants. Latex yield varies greatly among *Hevea* clones and with tree vigor, stand density, age of trees, site, and cultural practices. The pressure of latex vessels and latex flow are correlated with changes in environmental factors that control turgor in the latex vessel system. Both the rate and duration of flow of latex can be greatly increased by application of ethephon (Ethrel) to the tapping cut.

General References

Bernards, M. A. (2002). Demystifying suberin. *Can J Bot* **80**, 227–240.

Browse, J. (1991). Glycerolipid synthesis. *Annu Rev Plant Physiol Mol Biol* **42**, 467–506.

Brydson, J. A. (1978). *Rubber Chemistry*. Applied Science Publishers, London.

Charlwood, B. V. and Benthorpe, B. V., eds. (1991). *Terpenoids*. Academic Press, London.

Fahn, A. (1979). *Secretory Tissues in Plants*. Academic Press, London.

Frank, H. A., Young, A., Britton, G., and Cogdell, R. J., eds. (1999). *The Photochemistry of Carotenoids*. Advances in Photosynthesis, Vol. 8. Kluwer Academic Publishers, Dordrecht, the Netherlands and Boston.

Fuller, G. and Nes, W. D., eds. (1987). *Ecology and Metabolism of Plant Lipids*. Am. Chemical Society, Symposium Series 325, Washington, D.C.

Harborne, J. B. and Tomas-Barbevan, F. A., eds. (1991). *Ecological Chemistry and Biochemistry of Plant Terpenoids*. Oxford, New York.

Harwood, J. L. and Russell, N. V. (1984). *Lipids in Plants and Microbes*. Allen and Unwin, London.

Heredia, A. (2003). Biophysical and biochemical characteristics of cutin, a plant barrier biopolymer. *Biochim Biophys Acta* **1620**, 1–7.

Kerstiens, G. (1996). *Plant Cuticles: An Integrated Functional Approach*. BIOS Scientific Publishers, Oxford, England.

Mahlberg, P. G. (1993). Laticifers: An historical perspective. *Bot Rev* **59**, 1–23.

Moore, T. S., Jr., ed. (1993). *Lipid Metabolism in Plants*. CRC Press, Boca Raton, Florida.

Nes, W. D., Fuller, G., and Tsai, L.-S., eds. (1984). *Isoprenoids in Plants*. Dekker, New York.

Quinn, P. J. and Harwood, J. S., eds. (1990). *Plant Lipid Biochemistry: Structure and Utilization*. Portland Press, London.

Siegenthaler, P.-A. and Murata, N., eds. (1998). *Lipids in Photosynthesis: Structure, Function, and Genetics*. Kluwer Academic Publishers, Dordrecht the Netherlands and Boston.

Stumpf, P. K. and Conn, E. E., eds. (1987). *The Biochemistry of Plants*. Vol. 9. *Lipids, Structure and Function*. Academic Press, New York.

C H A P T E R

9

Nitrogen Metabolism

INTRODUCTION

Compounds containing nitrogen make up only a few percent of the dry weight of woody plants but they are extremely important physiologically. The N concentration of the foliage may exceed 2% of the dry weight in some tree species, whereas that of the wood is much lower (Fig. 9.1, Table 9.1). The small amounts of N-containing compounds occur in living cells, where they have essential roles in biochemical and physiological processes. Among these compounds are the structural proteins that form the protoplasm, as well as enzymes that catalyze the biochemical processes of plants. A large part of the N in leaves occurs as enzymes in the chloroplasts and mitochondria. Large amounts of protein also accumulate in the seeds of some plants (Chapter 2, Kozlowski and Pallardy, 1997). Significant amounts of N occur in chlorophyll, amides, amino acids, nucleic acids, nucleotides and other nitrogenous bases, hormones, vitamins, and alkaloids (see Fig. 6.22). Most of these substances are physiologically important, but the alkaloids, although economically important, seem simply to be byproducts of metabolism that at most provide modest protection against attacks by pests.

From the seedling stage to maturity of trees, nitrogen is required for growth and N deficiency is, after water stress, the most common limitation to growth. The demand for N is closely related to the amount of plant growth. Trees completing a large part of their annual growth early in the season use large amounts of N then. Orchardists commonly supply N to fruit trees to ensure an adequate supply during this critical period of growth. Most forests undergo some degree of N deficiency (see Anonymous, 1968), and foresters are now finding that they can afford to apply fertilizers under at least some conditions. However, N deficiency is less serious for trees with their long growing season, in which absorption can occur over a long period, than for annual crop plants, which make the major part of their growth and require most of their N within a few weeks. As N is very mobile within plants, it is commonly translocated from inactive to active tissues, making a deficiency somewhat less obvious than for a less mobile element.

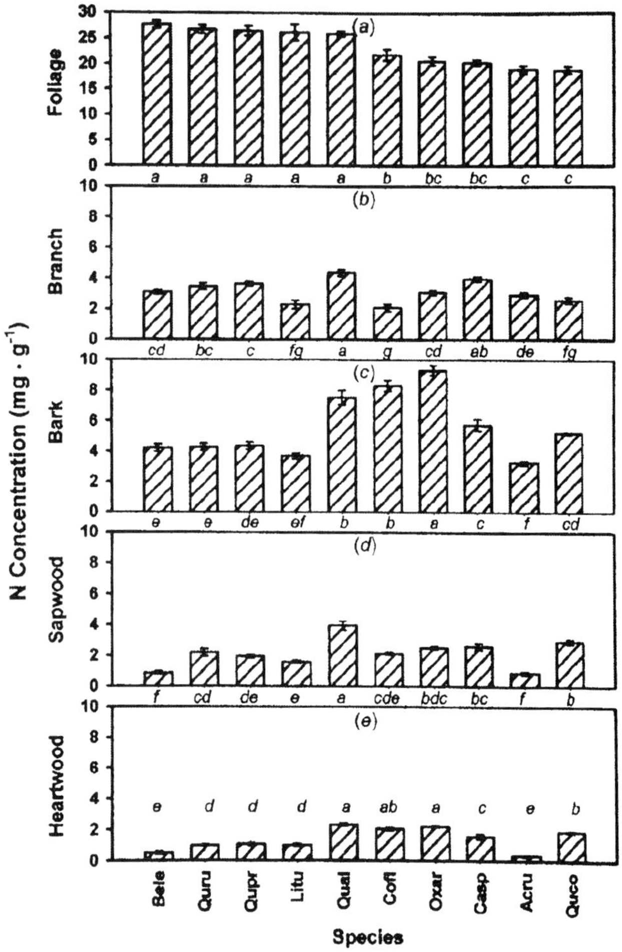

FIGURE 9.1. Tissue nitrogen concentrations for (a) foliage, (b) branches, (c) bark, (d) the last two years of sapwood, and (e) heartwood for 10 deciduous species. Bars are ordered by foliage concentrations. Species with the same letter below were not significantly different ($p < 0.05$). Species codes: Bele = *Betula lenta*, Quru = *Quercus rubra*, Qupr = *Q. prinus*, Litu = *Liriodendron tulipifera*, Cofl = *Cornus florida*, Oxar = *Oxydendrum arboreum*, Casp = *Carya* spp., Acru = *Acer rubrum*, Quco = *Q. coccinea*. From Martin et al. (1998).

TABLE 9.1. Amounts of Nitrogen in Various Parts of Trees in a 16-Year-Old Loblolly Pine Plantation[a]

Tree part	Nitrogen (kg/ha)	Percentage
Needles, current	55	17.1
Needles, total	82	25.5
Branches, living	34	10.6
Branches, dead	26	8.0
Stem wood	79	24.6
Stem bark	36	11.2
Aboveground total	257	80.0
Roots	64	19.9
Total	321	

[a]From Wells et al. (1975).

DISTRIBUTION AND SEASONAL FLUCTUATIONS OF NITROGEN

Because of the physiological importance of nitrogen, numerous studies have been made of fluctuations in the tissue N concentrations of trees. The amount of N present varies with the tissue, the age or stage of development, and the season. The highest concentrations of N are found in tissues composed chiefly of physiologically active cells, including leaves and meristematic tissues such as cambia and root and stem tips. Seeds also often are high in N, but the N in seeds occurs chiefly as a reserve and is relatively inactive physiologically. Some data on seed protein contents are given in Chapter 2 of Kozlowski and Pallardy (1997).

Concentration in Various Tissues

Figure 9.1 shows the distribution of nitrogen in various parts of 10 southern Appalachian (U.S.) tree species. The concentration in the leaves was higher than in any other part of the tree. Essentially the same situation apparently exists in broadleaved evergreen trees such as citrus (Cameron and Compton, 1945); nearly 50% of the total N present in bearing orange trees was in the leaves, about 10% in twigs, and 25% in the branches and trunk. Less than 20% of the total was in the roots. The distribution of N in various parts of 16-year-old loblolly pine trees is shown in Table 9.1. As in angiosperm species, the concentration of N is higher in the leaves of pine than in any other part of trees. The primary carboxylating enzyme, Rubisco (Chapter 5), contains a large fraction of leaf N, and some researchers believe Rubisco may serve as a storage form of N. Nitrogen deficiency promotes degradation of Rubisco, which is often synthesized abundantly under high N conditions, to allocate the N elsewhere in the plant. However, this response occurs only if leaf sugar levels are high (Paul and Foyer, 2001).

The nitrogen concentration of the heartwood usually is lower than that of the sapwood, as shown in Figure 9.1. This decrease in N concentration from sapwood to heartwood is associated with the death of parenchyma cells and movement of their N to growing regions (Chapter 3). The pith and adjacent annual ring contain more N than the bulk of the heartwood does and sometimes more than the sapwood. Similar patterns of distribution were found in conifer and hardwood stems by Merrill and Cowling (1966), who reported that a high N concentration favors the activity of wood-rotting fungi. Allsopp and Misra (1940) found that the newly formed xylem of ash and elm contained about

TABLE 9.2. Proportions of Various Nitrogen Compounds Present in Apple Xylem Sap during the Growing Season in New Zealand[a,b]

Compound	Date					
	Oct. 12	Oct. 19	Oct. 26	Nov. 29	Dec. 29	Jan. 26
Aspartic acid	12	10	22	31	58	41
Asparagine	65	59	52	65	34	48
Glutamic acid	<1	<1	<1	1	2	3
Glutamine	20	27	24	4	6	7
Serine	<1	1	<1	<1	<1	2
Threonine	2	1	1	<1	<1	<1
Methionine + valine	1	1	1	<1	<1	1
Leucine	1	1	<1	<1	<1	<1
Total nitrogen (μg/ml)	41	117	131	48	22	12
Percentage of total nitrogen contributed by aspartic acid + asparagine	77	69	74	96	92	89

[a]From Bollard (1958).
[b]Results are expressed as percent N by each compound.

5% N, but the older sapwood contained only 1.3 and 1.7%, respectively. Considerable amounts of soluble N compounds, mostly amides and amino acids, also occur in the xylem sap (Table 9.2) (Barnes, 1963a,b; Kato, 1981; Vogelmann et al., 1985; Sauter and van Cleve, 1992; Avery, 1993).

The phloem ("bark") contains considerably more N than the wood (Fig. 9.1) and is an important source of N for growth (see later). In orange trees, a decrease in the N concentration of branch bark accompanies each flush of shoot growth (Cameron and Appleman, 1933). Seeds often contain considerable N because in some species the principal reserve food is protein (Chapter 2, Kozlowski and Pallardy, 1997).

Seasonal Changes in Nitrogen Concentration

There is strong interest in seasonal changes in the concentration of N and other constituents in leaves and woody tissue of perennial plants because they provide the compounds required for the first flush of growth in the spring. There also is interest in learning when these compounds accumulate, because this information may show the best time to fertilize trees and aids in predicting the severity of injury resulting from defoliation by insects, pathogenic organisms, or storms. In both deciduous and evergreen trees the concentration of N in the woody tissue tends to increase during the autumn and winter, decrease when growth begins, and then to increase again as growth slows and ceases. Figure 9.2 shows seasonal changes in the N concentration of various parts of 15-year-old Stayman Winesap apple trees over an entire year. The total amount of N in the wood decreased when growth was rapid and increased when growth ceased. Apparently much of the N in the leaves was translocated back to the spurs in the early autumn before leaf fall occurred. Figure 9.3 shows seasonal variations in several forms of N in apple leaves.

Specific bark storage proteins have been identified in apple, beech, ash, basswood, birch, oak, poplar, willow, maple, and elderberry (O'Kennedy and Titus, 1979; Nsimba-Lubaki and Peumans, 1986; Wetzel et al., 1989a; Wetzel and Greenwood, 1991). Similar proteins also occur in ray parenchyma cells of the xylem (Sauter et al., 1988; Sauter and van Cleve, 1990; Wetzel et al., 1989b). In the winter, protein is localized in protein bodies of storage vacuoles dispersed through the cytoplasm (Fig. 9.4). By mid-summer, there is no sign of protein bodies.

The nature and accumulation patterns of these proteins have been studied most extensively in poplars (Cooke and Weih, 2005). In late summer there is an accumulation of both the messenger RNA (mRNA) coding for the poplar storage protein and the 32 kDa protein itself (Clausen and Apel, 1991; Coleman et al., 1991, 1992; Langheinrich and Tischner, 1991). The poplar bark storage protein is particularly rich in serine, leucine, phenylalanine, and lysine. It is not known for certain whether accumulation of bark storage protein is a direct phytochrome-mediated

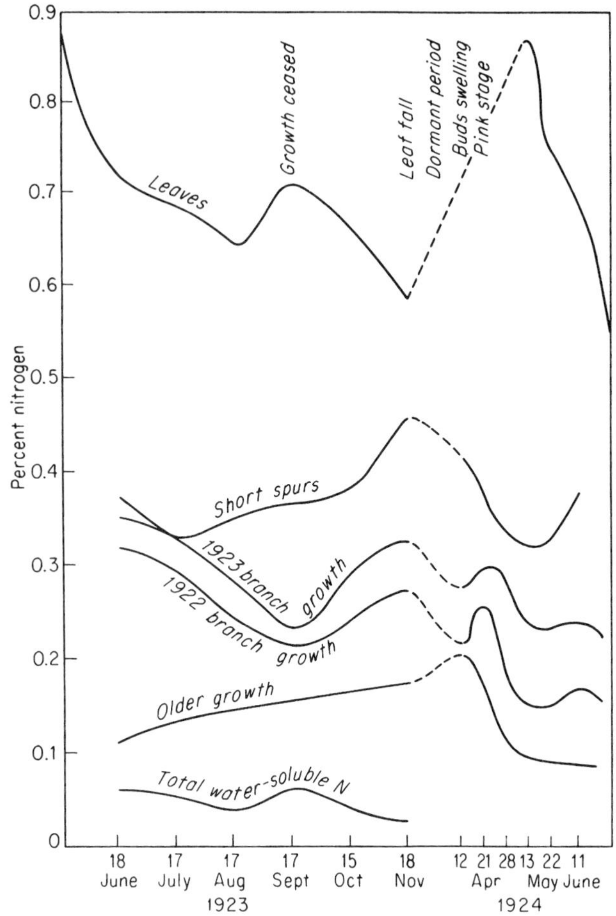

FIGURE 9.2. Seasonal changes in the total N concentration of leaves, short spurs, and 1-year, 2-year, and older branches of apple trees as percentages of fresh weight. From Thomas (1927). © American Society of Plant Physiologists.

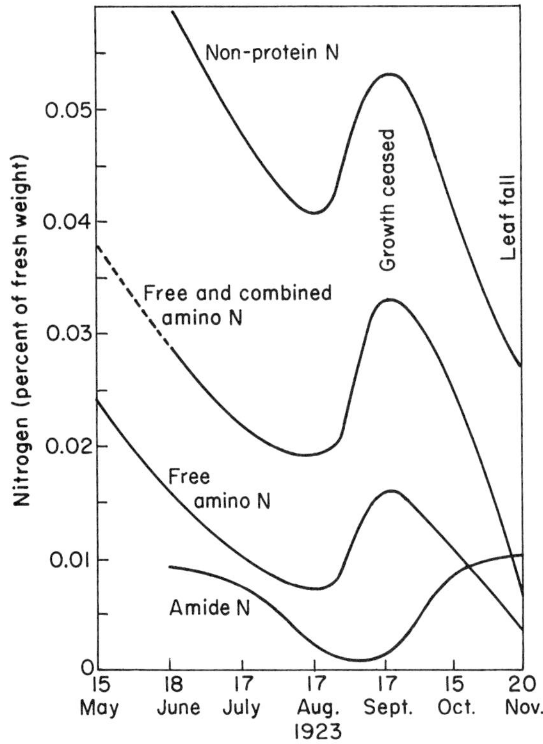

FIGURE 9.3. Seasonal changes in the forms of N present in apple leaves, expressed as percentages of fresh weight. Dashed lines indicate estimates for which no data were available. From Thomas (1927). © American Society of Plant Physiologists.

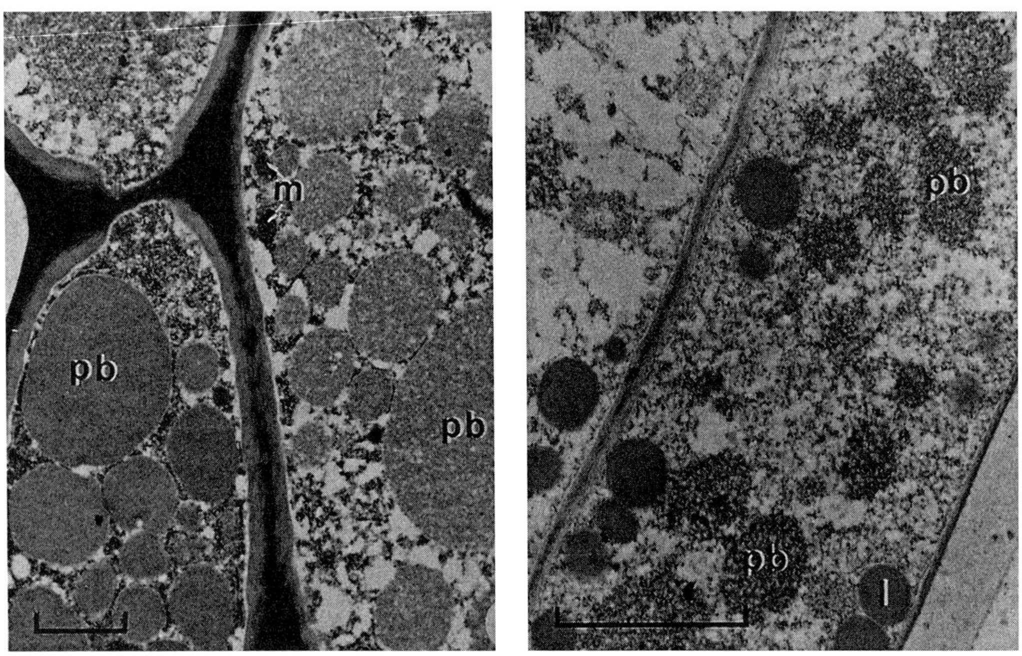

FIGURE 9.4. Electron micrographs of phloem parenchyma cells of *Tilia* in winter (left) and of the cambial region of *Betula* (right). Both plates show protein bodies (pb); lipid bodies (l) and a mitochondrion (m) also are found. Samples were collected on January 15, 1988. Bars: 2 μm. From Wetzel et al. (1989b).

response or an indirect response to changes in N source-sink relations associated with growth cessation or senescence (Coleman et al., 1991, 1992; Sauter and Neumann, 1994). However, synthesis of the storage protein also may be induced even under long days by treating plants with ammonium nitrate or glutamine or exposing them to low temperatures (van Cleve and Appel, 1993; Coleman et al., 1994; Zhu and Coleman, 2001). Further, Zhu and Coleman (2001) demonstrated that the activity of a promoter of gene transcription for a bark storage protein in poplar was greatly increased by a combination of sucrose and glutamine compared to that provided by glutamine alone. Hence, it is more likely that bark storage proteins are synthesized as an indirect response to changes in N and C availability, such as would occur during growth cessation and proteolysis associated with leaf senescence, rather than directly via a phytochrome-mediated mechanism. Winter storage of N in roots also is substantial either as free amino acids (e.g., *Malus domestica*, Tromp, 1983), or storage proteins (Millard and Proe, 1991; Wisniewski et al., 2004).

The decrease in the N concentration of stem wood and bark and roots in the spring is associated with transport to developing buds and new shoots. Release from bud dormancy and the presence of intact buds appear to be essential for degradation of the molecular weight 32,000 bark storage protein in poplar (Coleman et al., 1992). Nitrogen stored in the bark and wood is more available for new growth than N supplied externally, as shown for apple (Oland, 1963; Tromp, 1970), peach (Taylor and May, 1967), and grape (Possingham, 1970). This conclusion is further supported by direct observations that trees supplied with a ^{15}N-enriched N source for a year showed large fractional recovery of stored ^{15}N in growing shoot tissues the following year whatever the current N supply (Millard and Proe, 1991; Millard, 1996; Malaguti et al., 2001; Grassi et al., 2003; Guak et al., 2003). When seasonal growth starts, nearly as much stored N is removed from the tissues of fertilized as from those of unfertilized plants (Millard and Thomson, 1989). Levi and Cowling (1968) found that as leaves developed there was a marked decrease in the N concentration of the sapwood of southern red oak. Similarly, each flush of growth in orange trees produces a decrease in the N concentration of the adjacent woody tissue (see Fig. 9.2). Roots provided much of the stored N remobilized in *Prunus avium* and *Juglans nigra* × *regia* (Grassi et al., 2003; Frak et al., 2002). In a few species, remobilization of stored N and current root N uptake both supply growing shoots after budburst (e.g., *Juglans nigra* × *regia*, Frak et al., 2002, *Pinus sylvestris*, Millard et al., 2001, and perhaps *Betula pendula*, Millard et al., 1998, Millard et al., 2001).

The data on seasonal changes in N in conifers are limited but suggest that they are similar to those in deciduous species. Study of loblolly pine by Nelson et al. (1970) in the mild climate of Mississippi showed that both dry matter and N are accumulated by trees throughout the year, although the rate becomes quite low in the autumn and winter. There was evidence of translocation of N out of the bark and wood during rapid stem elongation and an increase during late summer and autumn to a winter maximum, followed by a decrease in late winter (Fig. 9.5). Translocation out of wood and bark in late winter began before stem elongation resumed, but the N presumably was used in the first flush of growth. Most retranslocation of stored N in conifers appears to occur from existing, young leaves to growing shoots (Millard, 1996). Some of this N may derive from catabolism of Rubisco protein (Chapter 5).

Autumn Movement from Leaves

As mentioned earlier, leaves may contain over 40% of the total N concentration of trees. Fortunately, a

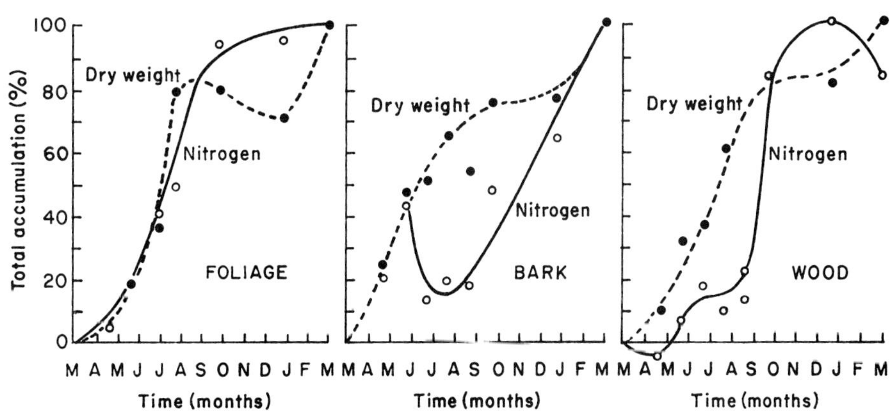

FIGURE 9.5. Relative dry matter and N accumulation in foliage, bark, and wood during the fifth year of development of a loblolly pine stand. From Nelson et al. (1970).

considerable part of the N and other mineral nutrients in the leaves usually is translocated back into the twigs and branches before leaf abscission occurs. This movement is quite important—otherwise a large fraction of the N in the plant would be lost, at least temporarily, by leaf fall.

Autumn movement of N out of leaves has been observed in many deciduous species, including aspen, birch, beech, elm, forsythia, cherry, horse-chestnut, maple, oak, pear, poplar, larch, plum, and willow (Fig. 9.6) (Oland, 1963; Grigal et al., 1976; Chapin and Kedrowski, 1983; Titus and Kang, 1982; Côté and Dawson, 1986; Millard and Thomson, 1989; Millard and Proe, 1991; Gower et al., 2000; Kurdali, 2000; Norby et al., 2000; von Fircks et al., 2001). Translocation out of leaves ranges from one-fourth to two-thirds of their N concentration. Over 80% of the retranslocated soluble protein N in apple leaves appears to arise from a preferential degradation of Rubisco protein (Millard and Thomson, 1989). Leaching of N from leaves of healthy plants appears to be a minor factor in changes in leaf N status. Chapin and Kedrowski (1983) found that leaves of birch, alder, larch, and spruce that were submerged in water and shaken for 12 minutes lost a maximum of 0.3% of their organic and inorganic N. Further shaking for 2 hours resulted in even smaller additional losses.

Chapin (1991) reported that leaching could account for 15% of the return to the soil of above ground N, but this figure included N deposited on external surfaces of plants. Autumn frosts that kill still-green leaves reduce retranslocation of N (Norby et al., 2000). Drought may also influence the efficiency of recovery of N from leaves. Minolettii and Boerner (1994) observed that early leaf litterfall after summer drought in a beech-maple forest in Ohio, consisted of green unsenesced leaves that exhibited higher N concentrations and lower resorption efficiency than senesced leaves.

In contrast to most seasonal data on deciduous species, Cameron and Appleman (1933) reported no decrease in the N concentration of orange leaves before abscission. However, Milla et al. (2005) observed retranslocation of N from leaves before abscission in eight Mediterranean evergreen angiosperm species, and both new tissue growth and senescence were associated with N retranslocation from older leaves of *Eucalyptus globulus* (Saur et al., 2000). There are fewer data for evergreen conifers, but Wells (1968) and Chapin and Kedrowski (1983) reported no changes in the N concentration of loblolly pine and black spruce needles during the autumn and early winter in the southeastern United States and Alaska, respectively (Figure 9.6).

As noted previously, nutrient retranslocation from nonsenescent leaves to growing regions does occur in evergreen conifers, and senescence of needles is preceded by nutrient withdrawal (Nambiar and Fife, 1991; Gower et al., 2000). Additionally, as could be deduced from the variability in N retranslocation rates mentioned previously, not all deciduous species exhibit efficient recovery of N from leaves. For example, Côté and Dawson (1986) reported that leaves of eastern cottonwood and white basswood exhibited declines in leaf N as the autumn progressed, but those of black alder did not. These differences apparently resulted from a lack of capacity of the alder to break down salt-insoluble leaf proteins. Kurdali (2000) also reported that N concentration of *Alnus orientalis* leaves had declined only 8 to 16% after abscission, indicating much reduced capacity for N retranslocation.

Changes in Distribution with Age

Changes in N concentration associated with aging of leaves are confounded with the effects of season, especially in leaves of deciduous plants. For example, as leaves grow older the proportion of cell wall material increases and this causes an apparent decrease in N concentration as a percentage of dry weight (Tromp, 1970). Madgwick (1970) reported that the N concentration of needles of Virginia pine decreased from 1.2% in first-year needles to 1.0% in third-year needles, but some of this apparent decrease possibly was caused by an increase in dry weight of needles. In California most of the N moves into leaves of Gambel oak and California black oak early in their development, but leaf expansion continues, resulting in a gradual decrease in N per unit of leaf area during the summer. As mentioned previously, there often is a rapid decrease in N concentration in the autumn as senescence occurs and N compounds are translocated back to the stem before abscission occurs (Sampson and Samisch, 1935).

The decrease in N concentration with increasing organism age is particularly noticeable in the woody parts of trees. In general, N appears to move out of cells as they become senescent and the N concentration of old tissue typically is lower than that of young tissue. Table 9.3 shows little difference in the N concentration of new growth, leaves, or fruit of apple trees of various ages, but there is a decrease in the woody parts of older trees. In loblolly pine the N concentration of the woody parts decreases with age, but that of the current leaves of older trees remains high, as shown in Table 9.4. Thus, conifers and hardwoods seem to show similar trends. It was mentioned earlier that the sapwood contains more N than the heartwood does and the youngest sapwood has the highest

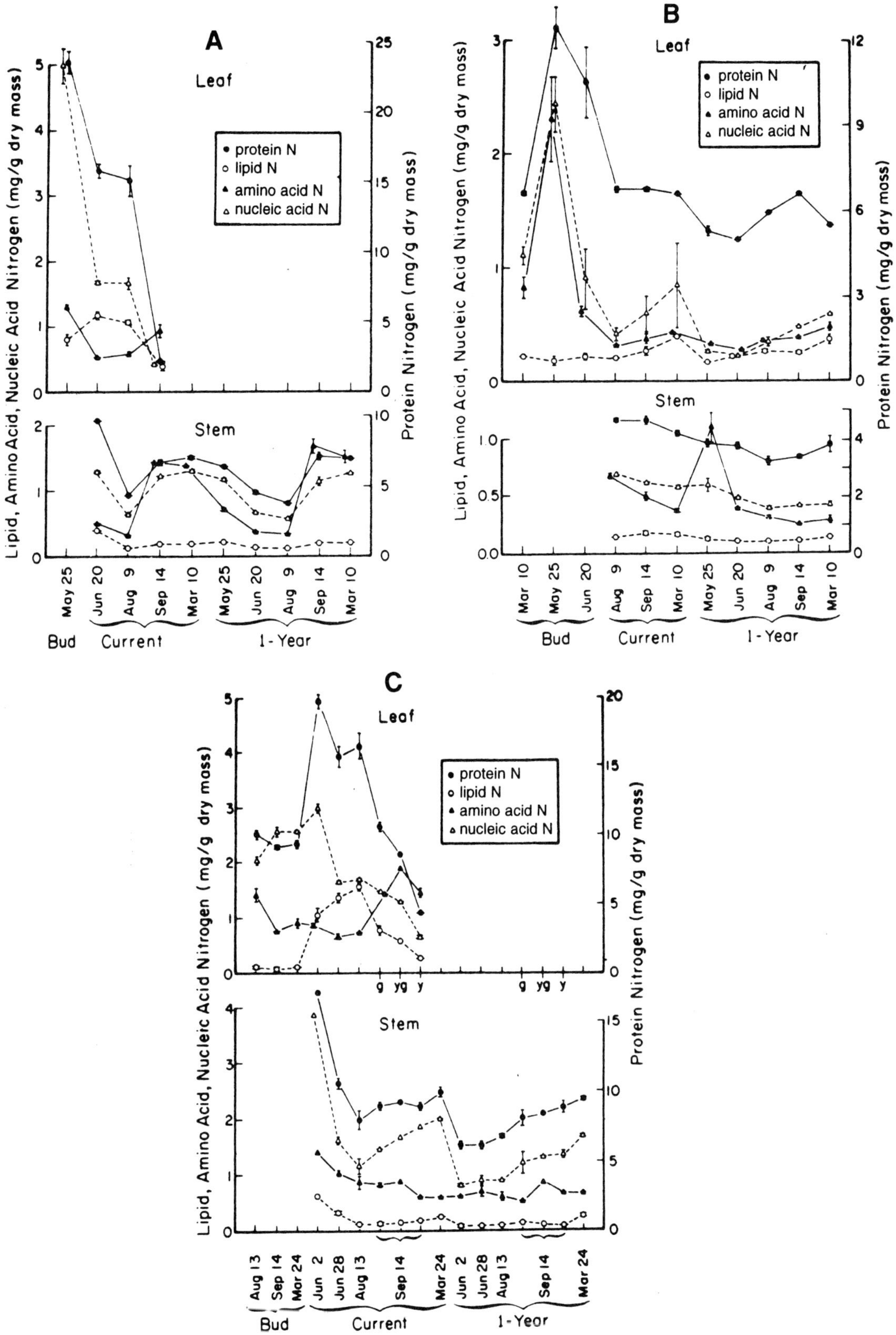

FIGURE 9.6. Seasonal patterns of N concentration in major chemical fractions of leaves and current terminal stems of eastern larch (A), needles and stems of black spruce (B), and leaves and current terminal stems of paper birch (C). For paper birch, on September 14 leaves and stems were separated into shoots for which leaves were green (g), mixed yellow and green (yg), or yellow (y). Data are given as means ± SE. From Chapin and Kedrowski (1983).

TABLE 9.3. Effect of Age on Nitrogen Content of Leaves, New Growth, Trunks and Branches, Roots, and Fruit of Apple Trees[a]

	Leaves		New growth		Trunks and branches		Roots		Fruit	
Age (years)	Oven dry weight (%)	Nitrogen (g)	Oven dry weight (%)	Nitrogen (g)	Oven dry weight (%)	Nitrogen (g)	Oven dry weight (%)	Nitrogen (g)	Oven dry weight (%)	Nitrogen (g)
1	1.71	0.44			0.30	0.29	0.39	0.20		
2	2.09	1.51			0.57	1.36	0.88	1.14		
5	1.76	7.84	0.89	1.93	0.48	17.20	0.64	9.85		
9	1.70	61.50	0.82	9.08	0.35	85.50	0.58	81.00	0.31	10.55
30	2.09	394.00	0.95	13.60					0.31	258.00
100	1.04	435.00	1.04	390.00	0.27	2,863.00	0.22	417.00		

[a]From Gardner et al. (1952). Copyright 1952; used with permission of the McGraw-Hill Book Company.

TABLE 9.4. Changes in Percentage of Nitrogen with Increasing Age in Various Tissues of Loblolly Pine Growing on Good Sites[a]

Tree age (years)	Current foliage	Older branches	Stem bark	Stem wood
4	1.00	0.41	0.42	0.16
8	0.95	0.24	0.24	0.06
18	1.08	0.23	0.23	0.06
30	1.22	0.22	0.19	0.04
56	1.16	0.21	0.17	0.03

[a]From Switzer et al. (1968).

concentration, presumably because it contains the most living parenchyma cells.

IMPORTANT NITROGEN COMPOUNDS

Having discussed the distribution of N in trees, I will consider some compounds in which N occurs. Among the most important of these are amino acids, amides, nucleic acids, nucleosides and nucleotides, proteins, and alkaloids (Fig. 6.22). Only the outlines of protein metabolism can be mentioned, and readers are referred to Hewitt and Cutting (1979) and Marcus (1981) for more detailed discussions.

Amino Acids

The amino acids are the basic building blocks of protoplasmic proteins. Most amino acids have the basic formula $RCHNH_2COOH$ and have properties of both bases and acids, because each amino acid has an amino group (NH_2) and a carboxyl group (COOH). In the simplest amino acid, glycine, R is represented by a hydrogen atom (CH_2NH_2COOH). In others, R may be very complex and may contain additional amino or carboxyl groups. Certain amino acids also contain hydroxyl (—OH) groups, sulfur (—CH_2SH, —CH_2SCH_3, or disulfide —CH_2SSCH_2— bridges), additional N groups, and cyclic C or C—N rings. Some 20 amino acids commonly are considered components of plant proteins and there are additional naturally occurring amino acids, such as ornithine and citrulline, that are not found in proteins (Miflin, 1981).

Amino Acid Synthesis

Amino acids can be produced in several ways, including through the assimilation of ammonia, transamination, chemical transformation of acid amides or other N compounds, and hydrolysis of proteins by enzymes. The first two methods probably are the most important.

Nitrate Reduction

Nitrate is a source of N for plants, and reduction of nitrate to ammonia is an important step in N metabolism. Nitrogen in plant litter is released as ammonium during mineralization, and it subsequently may either be absorbed by roots, absorbed and effectively immobilized in the microbial biomass, or converted to nitrate by nitrifying bacteria.

Nitrate is absorbed by trees and usually is quickly reduced, although it may accumulate if the carbohydrate supply and level of metabolic activity are low. A two-step reaction is involved (Fig. 9.7) in which nitrate in the cytoplasm is first reduced to nitrite (NO_2^-) by nitrate reductase with NADH supplying the

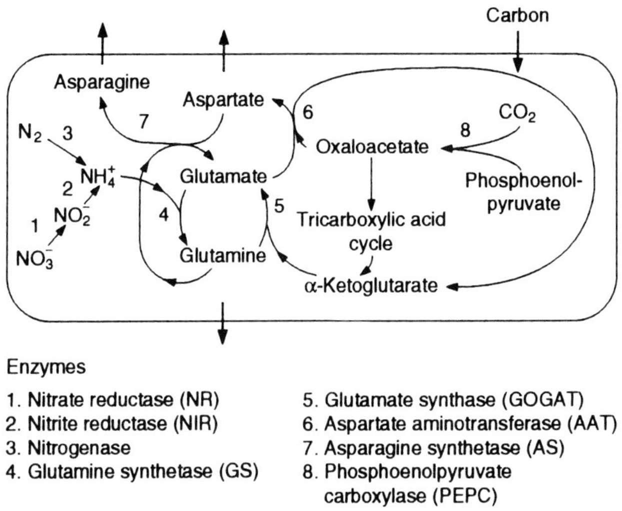

FIGURE 9.7. General scheme of N assimilation in higher plants and the enzymes involved. Glutamine, asparagine, and aspartate are the primary amino acids transported to other cells and plant organs. Carbon skeletons for amino acid biosynthesis are produced during photosynthate metabolism in the tricarboxylic acid (TCA) cycle. Carbon skeletons also may be derived from nonphotosynthetic CO_2 fixation of respired and/or atmospheric CO_2. From Dennis and Turpin (1990), with permission from Longman Group Limited.

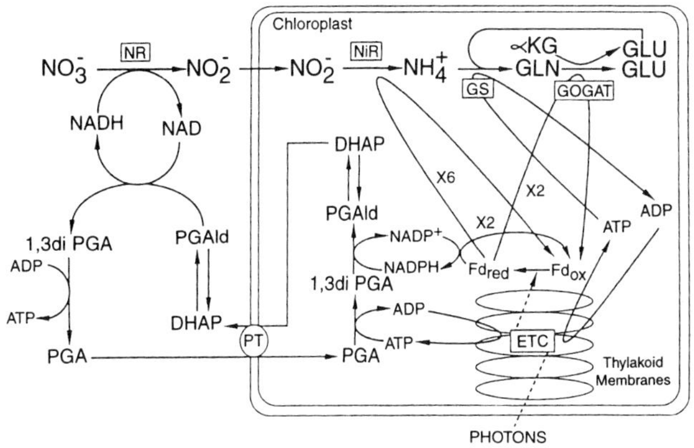

FIGURE 9.8. Pathway of NO_3^- reduction to glutamate in a photosynthetic, eukaryotic cell in the light. The photosynthetic light reactions provide the reductant and ATP for both the cytosolic nitrate reductase (NR) and the chloroplastic nitrite reductase (NiR), glutamine synthetase (GS), and glutamate synthase (GOGAT) activities. PT, Phosphate translocator; PGA, 3-phosphoglycerate; 1,3diPGA, 1,3-diphosphoglycerate; PGAld, phosphoglyceraldehyde; DHAP, dihydroxyacetone phosphate; ETC electron transport chain; Fd_{red} and Fd_{ox}, reduced and oxidized ferredoxin, respectively; GLN, glutamine; αKG, α-ketoglutarate; GLU, glutamate. From Dennis and Turpin (1990), with permission from Longman Group Limited.

reducing power. Nitrite subsequently is reduced to NH_4^+ in the chloroplasts of leaves or plastids of roots by nitrite reductase, usually with reduced ferredoxin as reductant.

The energy for nitrate reduction is derived from oxidation of carbohydrates, or more directly from products of photosynthetic light reactions (Fig 9.8), and the ammonia produced is finally combined with organic acids to form amino acids, as will be shown later. Nitrite is toxic in moderate concentrations, but it ordinarily does not accumulate in sufficient quantities to cause injury.

Nitrate reduction may occur in roots or leaves or in both organs. The energy costs of nitrate reduction in roots are greater than those in leaves, as carbohydrates must be transported in the phloem the length of the plant and oxidized to provide reducing power in roots. In contrast, excess NADPH and ATP produced in the light reactions of photosynthesis can be utilized in leaf nitrate reduction (Fig. 9.8). Up to 25% of the reducing power produced in photosynthesis may be consumed in nitrate assimilation (Chapin et al., 1987). The need to maintain pH levels against the production of OH^- that attends NO_3^- reduction may offset somewhat the advantages of reduction in the leaves.

It has long been assumed that, at least in well-fertilized crop plants, nitrate reduction occurs chiefly in the leaves (Beevers and Hageman, 1969), the nitrate being translocated from roots to leaves in the xylem sap (Shaner and Boyer, 1976). However, in woody plants nitrate was thought to be reduced in the roots because the available information indicated that organic forms of N predominated in the xylem sap (Bollard, 1956, 1957, 1958, 1960; Pate, 1980). This simple categorization has been succeeded by recognition that the location of nitrate reduction within woody plants varies both with species and nitrate availability. In a study of the capacity for nitrate reduction and the inducibility of nitrate reductase in over 500 species of woody plants, Smirnoff et al. (1984) noted that several plant taxa (i.e., gymnosperms, members of the Ericaceae and Proteaceae) had inherently low capacities for nitrate reduction in leaves. However, nearly 75% of the species of other woody taxa showed substantial nitrate reductase activity in leaves. Further, only members of the Ericaceae, which are characteristic of very acid soils in which the availability of nitrate usually is low, failed to show increased nitrate reductase activity in leaves when nitrate supplies to plants were increased.

Similarly, Fredeen et al. (1991) noted that nitrate reductase activity of foliage of woody *Piper* species that were growing in sunny forest gaps was highly inducible whereas that of broadly distributed species or those restricted to shady spots was not. Root nitrate reductase activities were an order of magnitude lower than those of leaves. Hence, the location of nitrate reduction appears to depend on nitrate supply and often on species adaptations to environmental factors that characterize habitat, especially light availability and nitrification capacity of the soil (Andrews, 1986; Yandow and Klein, 1986; Stadler and Gebauer, 1992).

When ammonium ions predominate in the soil solution, they are absorbed by roots and directly incorporated into organic compounds, usually amino acids (see later). For example, whereas $^{15}NO_3^-$ constituted 97% of the labeled N in the xylem sap of four-year-old citrus trees supplied with $K^{15}NO_3^-$, xylem sap amino acids (arginine, asparagine, and proline) contained 79% of labeled N in trees supplied $(^{15}NH_4)_2SO_4$ (Kato, 1981). Barnes (1963b) found 17 amino acids and ureides in the xylem sap of 60 species of North Carolina trees analyzed in June and July. Citrulline and glutamine comprised 73 to 88% of the organic N in the seven species of pine that were studied, and Bollard (1957) found the same compounds predominating in the xylem sap of Monterey pine, sampled in New Zealand.

Ammonia Assimilation

Most evidence supports the existence of a pathway for incorporation of ammonia into amino acids by means of a cyclic reaction sequence involving two enzymes (Inokuchi et al., 2002; Lea and Miflin, 2003; Suzuki and Knaff, 2005). The initial entry of NH_4^+ into an organic form occurs via the activity of glutamine synthetase, with NH_4^+, glutamate, and ATP as substrates and glutamine being the primary product (Fig. 9.7). Subsequently glutamate is produced from glutamine and α-ketoglutaric acid via glutamate synthase, resulting in a net gain of one glutamate molecule and consumption of one molecule of α-ketoglutaric acid for each "turn" of the cycle. Another possible reaction by which NH_4^+ might be assimilated involves glutamate dehydrogenase, NADH or NADPH as coenzymes, NH_4^+, and α-ketoglutaric acid with glutamate as the primary product. Although for many years this latter reaction was thought to be the primary means of NH_4^+ assimilation, it is now generally accepted that glutamate dehydrogenase is active in the oxidation of glutamate rather than in its formation unless NH_4^+ concentrations are high (Oaks and Hirel, 1985; Coruzzi and Last, 2000).

Transamination

Transamination involves transfer of an amino group from one molecule to another. This is exemplified by the reaction between glutamic and oxaloacetic acid to produce α-ketoglutaric acid and aspartic acid. An aminotransferase enzyme is involved.

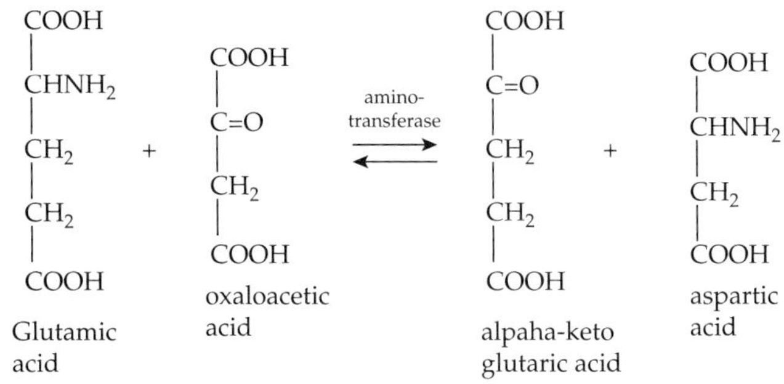

Another group of amino acids, tyrosine, phenylalanine, and tryptophan, originates from shikimic acid (see Fig. 6.22). Phenylalanine and tyrosine may be incorporated directly into protein or may serve as precursors of alkaloids or components of lignin. Tryptophan also gives rise to alkaloids but is best known as the precursor of the plant hormone indoleacetic acid.

Peptides

Peptides, like proteins, consist of amino acids joined by peptide linkages; that is, by the bonding of the carboxyl group of one amino acid molecule with the amino group of another molecule. If the molecular weight of the resulting compound is less than 6,000 it is arbitrarily designated a peptide. Some smaller peptides are of great interest to human pathology. Penicillin is a tripeptide, pollen allergens also are peptides, and other peptides such as insulin are pharmacologically active.

Amides

Among the common N compounds found in plants are the amides glutamine and asparagine. Glutamine is formed from glutamic acid by reaction with ammonia as noted earlier. It would be expected that asparagine would be formed from aspartic acid in the same manner, but in most situations glutamine is the substrate in a reaction that also consumes ATP (Dennis and Turpin, 1990, p. 408):

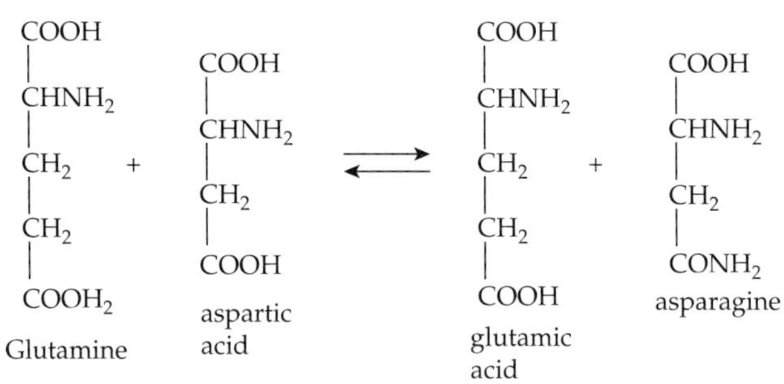

It is supposed that synthesis of these amides prevents accumulation of injurious concentrations of ammonia in plants. Amides also appear in seedlings when storage proteins are used in growth. If severe carbohydrate deficiency develops amides can be oxidized, but this may be accompanied by the release of injurious amounts of ammonia. Glutamine also is important metabolically, as it is the donor of the N group in the synthesis of carbamyl phosphate and asparagine unless cellular NH_4^+ concentrations are very high (Massiere and Badetdenisot, 1998).

Proteins

Proteins are the principal organic constituents of protoplasm. They are exceedingly complex nitrogenous substances of high molecular weight that differ in shape, size, surface properties, and function. They all, however, have in common the fact that they are built up from amino acids, are amphoteric (and therefore possess the properties of both an acid and base), and have colloidal properties. Proteins possess both basic NH_2 and acidic COOH groups. They are positively charged at pH values below the critical "neutral" value, known as the isoelectric point, and negatively charged at pH values above the isoelectric point.

Although the molecular weights of proteins always are high, there is considerable variation in weights of the various types. Some seed or storage proteins have molecular weights of 200,000 to 400,000. Amandin, a protein of almond, has a molecular weight of 329,000, and hippocastanum, a protein found in horse-chestnut seeds, has a molecular weight of 430,000. Some enzymes also are very large; urease has a molecular weight of 400,000 and catalase, 500,000. On the other hand, some protein molecules are quite small, with molecular weights of 10,000 to 50,000. On a dry-weight basis, proteins usually contain 50 to 55% carbon, 6 to 7% hydrogen, 20 to 23% oxygen, and 12 to 19% N. All plant proteins contain small amounts of sulfur and some also contain phosphorus.

The structural proteins include those in protoplasm and its components, such as chloroplasts. Storage proteins are particularly abundant in seeds and are an important source of food for humans and animals as well as for germinating seedlings. As noted earlier, storage proteins also occur seasonally in the bark and xylem ray parenchyma cells of many temperate deciduous tree species. Some enzymes such as urease and papain function alone. They are simply protein molecules and their catalytic properties are determined by the arrangement of the amino acid residues of which they are composed. Others require that a cofactor or prosthetic group be associated with the protein molecule. Sometimes the cofactor is a metal ion such as copper or iron. Still other enzymes require the presence of nonprotein prosthetic groups or more loosely associated cofactors such as the pyridine nucleotides NAD^+ and $NADP^+$. When decomposed by acid or alkali, proteins produce a mixture of amino acids, but during hydrolysis several products of intermediate complexity are formed as follows:

$$\text{proteins} \rightarrow \text{proteoses} \rightarrow \text{peptones} \rightarrow \text{polypeptides} \rightarrow \text{dipeptides} \rightarrow \text{amino acids}$$

Enzymatic degradation of proteins is more specific, the enzymes usually attacking specific chemical bonds and splitting off specific amino acids or groups of amino acids, rather than the large, poorly defined groups or compounds split off by acid or alkali hydrolysis.

A protein molecule consists of a long chain of amino acids brought together by peptide linkages or bonds in which the carboxyl group of an amino acid unites with the amino group of another amino acid, with water being split off in the reaction. An example of a peptide linkage is the union of two molecules of glycine (CH_2NH_2COOH) in a condensation reaction:

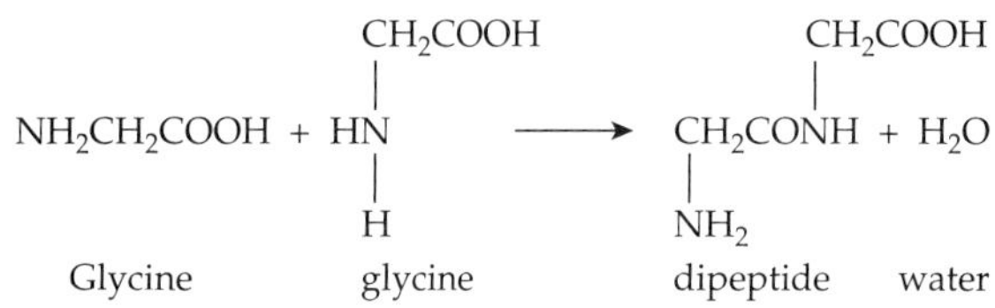

Inspection of the dipeptide formed in this reaction shows a free carboxyl and a free amino group available for possible linkage to other amino acids. No matter how many additional amino acids are linked to such a dipeptide, there always are free amino and carboxyl groups in the resulting complex molecule. Thus, with the union of several hundred amino acids in peptide linkages, a protein is formed. The skeleton of a protein molecule might be pictured as follows:

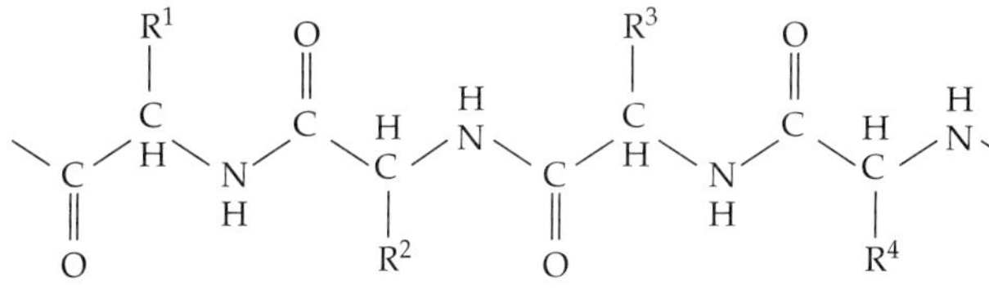

where R^1, R^2, R^3, and R^4 represent residues of different amino acids.

In a living plant, proteins are in a dynamic state and are constantly being broken down and reformed, but the total amount of protein may remain constant over considerable periods because degradation is balanced by synthesis. Various proteins differ in stability and have characteristic rates of turnover. Protein synthesis is controlled by a multitude of internal and external factors. Abundant synthesis of proteins occurs in cells of meristematic regions where cell division is occurring and in storage organs where protein is accumulated, as in endosperms of seeds. Induction of synthesis of specific proteins in response to stresses such as drought (Chapter 12) and heat (Chapter 5 of Kozlowski and Pallardy, 1997) also is common. Many of these stress-induced proteins have hypothesized protective or repair functions.

Synthesis takes place on ribosomes, organelles composed of protein and RNA in three locations within the cell (Spermulli, 2000). About three-quarters of cell protein is made in the cytoplasm where the sequence of amino acids, and therefore the kinds of protein formed, is controlled genetically by mRNA from the cell nucleus. Units of mRNA are produced under the control of DNA strands in the nucleus and migrate to the ribosomes where they function as templates or models for formation of protein molecules. Transfer RNA also is involved in the movement of amino acids to ribosomes. In photosynthetically active cells about 20% of cell protein is similarly synthesized on chlorplast ribosomes from internally produced mRNA templates and a further small amount (2–5%) is produced within the mitochondria from mitochondrial mRNA transcription. Refer to Chapter 9 of Kozlowski and Pallardy (1997) for a more complete discussion of this important and complex process.

Nucleic Acids and Related Compounds

Substituted purine and pyrimidine bases are constituents of many extremely important compounds. These include nucleic acids (RNA and DNA); nucleosides such as adenosine, guanosine, uridine, and cytidine; nucleotides such as AMP, ADP, and ATP; the nicotinamide nucleotides (NAD^+ and $NADP^+$); thiamine; coenzyme A; and cytokinins (see Fig. 6.22). Nucleotides are phosphate esters of nucleosides and the nucleic acids (DNA and RNA) are high molecular weight polymers formed from long chains of four kinds of nucleotide units derived from adenine, guanine, thymine, and cytosine. The genetic material in the nucleus is DNA, each molecule consisting of two polynucleotide chains arranged in a double helix. Small amounts of DNA occur outside the nucleus in mitochondria and chloroplasts.

RNA molecules consist of single strands. The sugar associated with RNA is ribose, and that associated with DNA is deoxyribose. Thymine is replaced by uracil in RNA molecules. Three kinds of RNA occur in ribosomes, nuclei, and other organelles: mRNA, transfer RNA (tRNA), and ribosomal RNA (rRNA). Their role in protein synthesis is discussed more extensively in Chapter 9 of Kozlowski and Pallardy (1997).

Alkaloids

Alkaloids are a large and complex group of cyclic compounds that contain N. About 2,000 different alkaloids have been isolated, some of which are of pharmacological interest. Important alkaloids include morphine, strychnine, atropine, colchicine, ephedrine, quinine, and nicotine. They are most common in herbaceous plants, but some occur in woody plants, chiefly tropical species.

Alkaloids commonly are concentrated in particular organs such as the leaves, bark, or roots. For example, although nicotine is synthesized in the roots, 85% of that in a tobacco plant occurs in the leaves, and the cinchona alkaloids are obtained from the bark. Alkaloids also sometimes occur in wood, and the wood of some species of the families Anacardiaceae, Apocynaceae, Euphorbiaceae, the legume families, Rutaceae, and Rubiaceae contains so much alkaloid that it produces dermatitis (Garratt, 1922). Among alkaloids derived from trees, the cinchona alkaloids are best known because of their use in treatment of malaria. They occur in the Andean genera *Cinchona* and *Remijia* of the family Rubiaceae.

In spite of their wide occurrence in plants, no essential physiological role has been found for alkaloids. However, it is possible that in some instances they discourage fungal, bacterial, or insect attacks. They may be byproducts of N metabolism, which ordinarily cause neither injury nor benefit to the plants that produce them, and the amount of N diverted into them apparently is too small to be of selective importance in plant competition. The biochemistry of secondary plant products, including alkaloid compounds, was thoroughly reviewed in the volume edited by Conn (1981).

NITROGEN REQUIREMENTS

Many estimates have been made of the N requirements of forest and fruit trees. Table 9.5 shows the categorized annual N requirements of individual 20- and 25-year-old apple trees. Nearly one-third of the total annual requirement of the younger trees and about one-fifth of that of the older trees accumulates in the fruit and is removed by harvesting it. As trees grow older, proportionally less of the total N is used in fruits and a higher proportion is used in root and top growth. Five to 6% of the N is temporarily lost by the shedding of flowers and young fruits in the spring and 30 to 40% by abscission of leaves in the autumn.

TABLE 9.5. Estimated Annual Nitrogen Requirements of Apple Trees[a,b]

	20-year-old trees[a]		25-year-old trees[b]	
	Grams	Percent of total	Grams	Percent of total
For fruit crop	180	30.53	150	21.57
Loss (temporary) from abscised blossoms and fruit	30	4.59	40	5.88
Loss (temporary) from abscised leaves	180	30.53	270	39.22
For top and root growth	160	26.72	230	33.30
Removed by pruning	50	7.63	230	
Total	590	100.00	690	100.00

[a]From Murneek (1942).
[b]From Magness and Regeimbal (1938).

It is reported that in years when a heavy fruit crop develops, little N is stored and existing reserves are seriously or completely depleted (Murneek, 1930). This results in decreased vegetative growth and inhibition of flower bud formation during the current and even the following year. Sometimes more than one year is required for complete recovery. These effects are similar to the effects of heavy fruiting on carbohydrate reserves, discussed in Chapter 7.

Estimates of annual N uptake by deciduous forests range from 30 to 70 kg ha^{-1} (Baker, 1950; Cole and Rapp, 1981). According to Switzer et al. (1968), the trees of a 20-year-old stand of loblolly pine contain over 300 kg ha^{-1} of N and use about 70 kg ha^{-1} annually for the production of new tissues. Of this, 38 kg comes from the soil and the remainder comes from within the trees, chiefly from the leaves. The distribution of N in young pine trees was shown in Table 9.1. Finzi et al. (2002) reported annual N requirements of young (14–17-year-old) loblolly pine stands ranged from about 45 to 61 kg ha^{-1} yr^{-1} over four years with between 48 and 66% recovered from senescing leaves and the balance derived from soil N uptake. Bormann et al. (1977) stated that a 55-year-old beech, maple, and birch forest in New England contained 351 kg ha^{-1} of N in the aboveground biomass and 180 kg ha^{-1} in the belowground biomass. About 120 kg were used in growth, of which one-third was withdrawn from storage in the plant tissue. About 20 kg ha^{-1} of N were added to the system each year, of which about 14 kg were supplied by fixation in the soil and about 6 kg by precipitation. Very little N was lost from this forest.

These data indicate that considerably less N is required for tree growth than for cultivated crops. Corn, for example, absorbs over 175 kg ha^{-1} of N and alfalfa may use over 200 kg ha^{-1} in a growing season, most of which must be absorbed from the soil in a short time. According to Hardy and Havelka (1976), soybeans use nearly 300 kg ha^{-1} of N during the growing season, of which about 25% is fixed in root nodules. Fruit trees typically have high annual N requirements, ranging from 60 to 175 kg ha^{-1} depending on the number and size of the trees. Forest trees obtain most of their N from the decay of litter and from atmospheric inputs, but fruit trees and tree seedlings in nurseries often must be fertilized to maintain yields.

There is evidence that tree species differ in their N requirements, because some species occur only on fertile soil, whereas others can grow on infertile soil. Mitchell and Chandler (1939) divided 12 common deciduous species of forest trees into three categories with respect to N requirements. Red, white, and chestnut oak, trembling aspen, and red maple were most tolerant of low N; pignut hickory, sugar maple, beech, and black gum were intermediate; and white ash, yellow-poplar, and basswood had high N requirements. Cole (1986) presented evidence of large differences among various ecosystems in aboveground production per kg of N uptake (Table 9.6). Whereas boreal conifer forests were most efficient in use of N for new biomass production, boreal deciduous forest and Mediterranean ecosystems were only one-third as efficient.

Physiological processes and growth of plants frequently show good correlations with the supply of available N. For example, the maximum rate of photosynthesis often shows some relationship with leaf N concentration (Fig. 5.46), primarily because of the high N content of constituents of the photosynthetic apparatus including chlorophyll, thylakoid proteins, and the soluble enzymes involved in carbon fixation and photosynthetic carbon metabolism (Chapter 5) (Field and Mooney, 1983, 1986; Seemann et al., 1987; Reich et al., 1991b). Dry matter production also may be

TABLE 9.6. Aboveground Biomass Production per Unit of Nitrogen Uptake[a]

Forest region (number of sites)	Average aboveground production per kg of N uptake (kg ha^{-1} $year^{-1}$)
Boreal coniferous (3)	295
Boreal deciduous (1)	92
Temperate coniferous (13)	179
Temperate deciduous (14)	103
Mediterranean (1)	92
Tropical (7)	120

[a]From Cole (1986). Reprinted by permission of Kluwer Academic Publishers.

closely linked with N status. These correlative relationships have led some researchers to analyze N relations of plants from the perspective of "N use efficiency" similar to water use efficiency (Chapter 12).

Evans (1989) noted that within species there are strong correlations between N and important determinants of photosynthetic capacity such as Rubisco and chlorophyll, and mass-based relationships between leaf N and net photosynthesis are robust across plant functional groups and climatic regimes (e.g., Wright et al., 2004). However, based on leaf area the photosynthetic capacity per unit of leaf N may differ substantially among species. Field (1991) noted a two- to three-fold range in this relationship for species from a variety of habitats, but excluding evergreen sclerophyll species that have low specific leaf areas. When considered on a leaf mass basis, values for sclerophyll species fell within the stated range. These differences reflect variation in N allocation to different leaf pools, and differences among species in electron transport capacity and specific activities of Rubisco. Even within a species, as leaf N per unit leaf area increases, the proportion of N found in thylakoid proteins is stable, whereas that in soluble proteins increases. Differences in light intensity during development of leaves result in substantially altered allocation of leaf N to chlorophyll and thylakoid proteins.

Roberds et al. (1976) reported considerable variation in response to N fertilization among different families of loblolly pine. Li et al. (1991a) observed that traits in loblolly pine families relating to both the efficiency with which N was absorbed (N taken up per unit of N applied) and utilized (stem biomass produced per unit of N uptake) were moderately to highly heritable when soil N levels were low (5 ppm). At high N levels (50 ppm) only utilization efficiency varied significantly among families. They further reported that N level had a complex relationship to genetic makeup and allocation of biomass (Li et al., 1991b). At low soil N, seedlings were smaller and allocated proportionally more biomass to roots at the expense of needles and/or stems. Family differences in biomass allocation were observed at low but not high soil N levels. More research on the N requirements of tree species and genotypes would be useful, as existing research suggests a substantial potential for economically increasing the N-limited productivity of forest stands.

SOURCES OF NITROGEN

Trees can use N in the form of nitrates, nitrites, ammonium salts, and organic N compounds such as urea, but whatever its initial form, most N probably is absorbed in the form of nitrate or ammonium (Hauck, 1968). An interesting exception to this generalization is the absorption of amino acids (e.g., glycine, aspartic acid, glutamic acid) by roots of herbaceous plants, and deciduous and evergreen shrubs native to arctic regions (Chapin et al., 1993; Kielland, 1994). Based on laboratory experiments in which uptake rates were compared, Kielland (1994) estimated that amino acid absorption may account for between 10 and 82% of the total N absorbed by naturally occurring arctic plants, with deciduous shrub species exhibiting the greatest potential for absorption of amino acids. Näsholm et al. (1998) and Nordin et al. (2001) also demonstrated that boreal region plants (e.g., *Pinus sylvestris*, *Picea abies Vaccinium myrtillus*) also can accumulate organic N (glycine), thereby bypassing the need for nitrogen mineralization. In these cold environments release of N by mineralization is very slow (Table 9.7), and species apparently have evolved unusual mechanisms for N absorption.

Woody plants show a general preference for ammonium ion uptake over that of nitrate, although there are exceptions to this generalization and it should be remembered that some nitrate uptake usually will occur as well. Preference for NH_4^+ was shown by *Eucalyptus nitens* (Garnett and Smithhurst, 1999; Garnett et al., 2003), *Pinus sylvestris* (Nordin et al., 2001), pecan (Kim et al., 2002), five of six tree dipterocarps (Norisada and Kojima, 2005), and by sugar maple and eastern hemlock (Templer and Dawson, 2004). Adams and Attiwill (1986a,b) reported that the fraction of the N taken up as nitrate in Australian eucalypt forests never exceeded one-third, even under strongly nitrifying soil conditions. Kronzeker et al. (1997) reported that in white spruce, uptake of NH_4^+ was up to 20 times that of NO_3^-. In contrast, American beech preferentially absorbed NO_3^- and northern red oak showed no statistically significant preference in N form (Templer

TABLE 9.7. **Mean Residence Period in Years for the Forest Floor and Nutrient Constituents as Found in Major Forest Regions of the World**[a]

Region	Organic matter	N	K	Ca	Mg	P
Boreal coniferous	350	230	94	150	455	324
Boreal deciduous	26	27	10	14	14	15
Subalpine coniferous	18	37	9	12	10	21
Temperate coniferous	17	18	2	6	13	15
Temperate deciduous	4	6	1	3	3	6
Mediterranean	3	4	<1	4	2	1
Tropical	0.7	0.6	0.2	0.3	—	0.6

[a]From Cole (1986). Reprinted by permission of Kluwer Academic Publishers.

and Dawson, 2004). Typically, forest soils tend to be acidic, a property that favors NH_4^+ availability (Garnett and Smethhurst, 1999) and supports NH_4^+ preference as generally adaptive. Similarly, forest tree species often are rare in highly disturbed habitats where soil conditions promote nitrification, and herbaceous plants that are more flexible in absorption of different forms of N are likely favored (Crawford and Glass, 1998).

The principal sources of the N used by forest trees are that fixed in the soil by microorganisms, that washed out of the atmosphere by rain and snow, and that released by decay of litter in the forest floor. Commercial fertilizers are the most important source of N for orchard trees and ornamental shrubs and are beginning to be used on forest trees. Some N is lost to the atmosphere during ammonification and denitrification, and some is lost by leaching during heavy rainfall.

Nitrogen Fixation

Although the atmosphere consists of about 80% N_2, atmospheric nitrogen is very inert, and this potential source can be used by trees only after it is "fixed" or combined with other elements. Nitrogen fixation by microorganisms, humans, and lightning replaces N lost from the soil by leaching, fire, and absorption by plants and prevents its ultimate exhaustion. The biological aspects of N fixation were discussed by Sprent and Sprent (1990), Stacey et al. (1992), and Postgate (1998).

Nitrogen fixation occurs by the same process in both free living and symbiotic organisms. Nitrogenase, the prokaryotic enzyme that accomplishes this reaction, consists of two component proteins. A Mo-Fe protein with a molecular weight of 200,000 to 250,000 contains the active site of the enzyme. This protein consists of two pairs of distinct subunits, and each "half" of the enzyme apparently functions independently. Another protein, the iron protein, has a molecular weight of about 60,000 and dissociates into two similar subunits. The ferredoxin-reduced Fe-protein repeatedly feeds electrons to the Mo-Fe protein until sufficient reducing power is present to convert N_2 to ammonia. In an evidently wasteful side-reaction, a molecule of hydrogen gas (H_2, from H^+) also is produced in fixation of one molecule of N_2. Molecular hydrogen is produced at the point where the Mo-Fe protein attains a state where it has received between three and four electrons and actually binds N_2. Sixteen molecules of ATP are hydrolyzed to ADP and P_i in producing two molecules of ammonia and one molecule of H_2. Some ATP ultimately may be recovered by other enzymes, called uptake hydrogenases, that oxidize H_2.

Plant hemoglobin found in root nodules seems to play an important role in N fixation by controlling oxygen concentration and flux to the nodule interior. Nitrogenase has contradictory requirements for relatively low O_2 tension (it is inactivated at high O_2) and a large supply of ATP derived from O_2-supported respiration. The presence of hemoglobin in nodules apparently provides the capacity to maintain low free O_2 *concentrations* and for a high *rate of flux* of O_2 to support respiration. For example, whereas free O_2 concentration in soybean nodules typically is 11 nM, the concentration of oxygen bound up in hemoglobin is about 55,000 times greater (600 μm) (Appleby, 1985). Consumed O_2 is rapidly replaced from the hemoglobin pool, while a steep gradient in dissolved O_2 between the nodule and external O_2 supply, which supports rapid diffusion, can be maintained.

Nitrogenase can reduce various other substrates, including, as noted earlier, hydrogen ions, as well as acetylene ($CH{\equiv}CH$). Reduction of acetylene leads to production of ethylene ($CH_2{=}CH_2$), and this reaction is widely used to measure nitrogenase activity because it is easy to measure ethylene production by gas chromatography. The nitrogenase complex also has the capacity to reduce other compounds beside N_2, and acetylene that have multiple bonds, including nitrous oxide, cyanide, and azide (Crawford et al., 2000). Free-living bacteria fix less N per unit of protoplasm than those in nodules because they use more energy in growth (Mulder, 1975).

Nonsymbiotic Nitrogen Fixation

Nitrogen fixation in the soil occurs largely as a result of activity by saprophytic bacteria of the genera

Azotobacter and *Clostridium*. These bacteria are mostly free-living in the soil but a few species have been found that are restricted to the rhizosphere of certain plants. Some blue-green algae also fix N and are effective colonizers of raw soil and other extreme habitats. Anaerobic forms are said to be more common than aerobic forms in forest soils, probably because the high acidity common in forest soils is unfavorable for the aerobic N-fixing bacteria. According to Russell (1973, p. 353) the number of N-fixing bacteria is so low in many soils that it is doubtful that they fix a very large quantity of N.

Symbiotic Fixation by Legumes

Bacteria of the genera *Rhizobium, Sinorhizobium, Azorhizobium, Mesorhizobium,* and *Bradyrhizobium* penetrate the roots of many species of legumes, producing root nodules in which N fixation occurs Vessey et al. (2004). Symbiotic N fixation is most important in forestry where trees associate with wild, herbaceous legumes, as in the southeastern United States. Chapman (1935) reported increased height and diameter growth of several species of hardwoods planted beside black locust trees. He also found that total N in the soil was greatest near the black locust trees and concluded that the improved growth was the result of N fixation by the locust.

Nitrogen fixation by bacterial nodules on roots can be affected in several ways by water stress. Infection and nodule formation can be reduced and N fixation can be decreased by loss of water from nodules in drying soil. An effect of water stress on nodules appears to be a reduction in oxygen diffusion rate to the nodule interior that is attributable to physical alterations in the outer portion of the shrinking nodule (Sprent and Sprent, 1990). Reductions in carbohydrate supplies and excessive accumulation of NH_4^+ in nodules because of reduced photosynthesis and water transport, respectively, also may be involved in inhibition of N fixation under water stress. Serraj et al. (1999) suggested that inhibition of phloem transport by plant water deficits might be a key mechanism of reduced nitrogen fixation in nodules, because it could cause reduced fixed carbon supply as well as N accumulation and feedback inhibition of N fixation. Additionally, reduced water supply to nodules could alter diffusion rates of O_2 through the nodule surface because of shrinkage.

Fixation in Nonlegumes

Nitrogen-fixing root nodules usually are identified with the legume families. However, root nodules occur on some nonleguminous dicotyledonous plants. In one unusual exception, *Bradyrhizobium* forms N-fixing nodules on a woody species of the Ulmaceae native to Java, anggrung (*Parasponia parviflora*) (Akkermans et al., 1978). Other cases of root nodule formation on woody plants have been associated with actinomycetes of the genus *Frankia*, and many actinorhizal host plant species of trees and shrubs in various families have been identified, including members of the Betulaceae, Elaeagnaceae, Myricaceae, Rhamnaceae, Casuarinaceae, Coriaricaceae, Rosaceae, and Datiscaceae (Baker and Mullin, 1992). The nodules on nonleguminous plants are composed of much-branched lateral roots, whereas those on legumes usually are developed from cortical cells (Torrey, 1976) (Fig. 9.9). Most plants in these families are adapted to grow on poor or disturbed sites (Baker and Mullin, 1992).

In certain nonleguminous species aboveground nodules are important in N fixation. For example, nodulated aboveground adventitious roots of red alder fixed atmospheric N at comparable rates to belowground roots in wet forests of the western United States (Coxson and Nadkarni, 1995).

Nitrogen fixation by nodulated nonlegumes appears to be of considerable ecological significance in some places. For example, Crocker and Major (1955) noted that at Glacier Bay, Alaska, an average of 61.6 kg ha^{-1} of N accumulated under alder thickets, creating a favorable site for Sitka spruce, which succeeded alder. There is increasing interest in the N-fixing capacity of red alder in forests of the Pacific Northwest (Harrington, 1990). The value of interplanting alder in conifer plantations to improve growth of conifers has long been recognized by Europeans and also is practiced in Japan (Chapter 7, Kozlowski and Pallardy, 1997). The beneficial effects undoubtedly are due to greater N availability. Virtanen (1957) showed that when spruce was planted beside alder, it obtained N fixed in the root nodules of the alder. He calculated that, in a grove of alders about 2.5 m high and with 10,000 trees per hectare, the leaf fall and roots remaining in the soil would add about 200 kg ha^{-1} of N. Nitrogen losses were not considered in his calculations. For comparison, Hardy and Havelka (1976) state that soybeans use about 300 kg ha^{-1} of N, of which about 25% is fixed in root nodules. Actinomycete nodules on the roots of California chaparral also fix significant amounts of N (Kummerow et al., 1978).

Generally, blue-green algae are said to form symbiotic relationships with and supply N to mosses, lichens, and some seed plants, including woody cycads (Rai et al., 2000; Vessey et al., 2004). Cross-sections of coralloid roots of woody cycads show internal rings composed of blue-green cells of *Nostoc* sp., and there is evidence that the root provides a microaerobic environment to the algal cells as well as fixed carbon in

FIGURE 9.9. Actinorrhizal root nodules of European black alder. From Becking (1972).

return for N transfer in the form of amino acids (most likely glutamine and citrulline). Nitrogen-fixing bacteria form loose associations with root systems of several economically important tropical grasses. For example, N-fixing populations of *Azotobacter paspali* inhabit mucilaginous sheaths on outer root surfaces of paspalum (*Paspalum notatum*). The ecological significance of such associations remains debated, but their existence and demonstrated capacity to enhance the N status of the plant are now generally accepted (Elmerich et al., 1992).

It has been reported that some bacteria form nodules and fix N in the leaves of several kinds of plants, including species of *Psychotria*, *Pavetta*, *Ardisia*, and *Dioscorea*. However, van Hove (1976) concluded from acetylene reduction and growth tests that if N reduction occurs in these leaf nodules it is too limited in amount to be important in the N economy of the plants. It is possible, however, that the bacteria produce growth regulators such as cytokinin or other substances that are beneficial to the plants. It is also claimed that bacteria and blue-green algae living on leaf surfaces (the phyllosphere) can fix N (Ruinen, 1965; Jones, 1970). For example, Jones claimed that bacteria living on the surfaces of Douglas-fir needles fix measurable amounts of N, and Bentley and Carpenter (1984) showed that N fixed by blue-green algae on the frond surfaces of a palm (*Welfia georgii*) accounted for between 10 and 25% of frond N concentration.

There is wide interest in finding methods for increasing the amount of N fixed by vegetation. This includes searching for methods of increasing N fixation in those plants in which it already occurs and possibly inducing microbial N fixation in the rhizosphere of species other than the tropical grasses in which it has been observed (Rai et al., 2000). A more exotic approach involves genetic manipulation by use of recombinant DNA techniques to introduce the N-fixing gene into plant species where it does not exist (Chapter 9, Kozlowski and Pallardy, 1997). Another possibility is fusion of protoplasts to transfer the gene or genes that make legumes good hosts for *Rhizobium* to other plants. However, because of the complex structural and biochemical requirements involving closely coordinated plant and bacterial gene expression, much further research must be undertaken before the desired types of plants can be produced by these techniques.

Atmospheric Nitrogen Fixation

Measurable amounts of N are returned to the soil in rain and snow. Precipitation brings down ammonia and N oxides fixed by electrical storms, released by volcanic and industrial activity, and a small amount leached from tree canopies (Chapter 10). The amounts range widely, from less than 2 kg ha^{-1} yr^{-1} where the influences of industrial activity are negligible to greater than 60 kg ha^{-1} yr^{-1} in the mountainous regions of the northeastern United States (Aber et al., 1989).

Release from Litter

Some N absorbed by trees is returned to the soil in fallen litter. Maintenance of forest soil fertility is partly dependent on the return of N and mineral nutrients by decay of litter. Leaves and twigs that are shed annually add up to several thousand kilograms of organic material possessing approximately 1% N-containing compounds. Cole (1986) reported annual litterfall of 5,400 kg ha^{-1} yr^{-1} for temperate deciduous and 4,380 kg ha^{-1} yr^{-1} for temperate coniferous forests. Wide variations occur, however, and values as low as 500 kg ha^{-1} have been measured on poor beech sites, whereas the best European beech stands return as much as 6,700 kg ha^{-1} yr^{-1} of leaf and twig debris (Baker, 1950).

The amount of N in litter varies greatly with species. Chandler (1941) found that the N concentration of leaf litter of hardwoods in central New York State varied from 0.43 to 1.04%, with an average of 0.65%, and Coile (1937) found values ranging from 0.50 to 1.25% in conifers and hardwoods in the Piedmont of North Carolina. Hardwood leaves and litter generally have higher average N concentrations than do coniferous leaves. Conifer leaves that have been shed contain about 0.6 to 1.0% N, and fallen hardwood leaves generally contain from 0.8 to 2.0% N (Baker, 1950). However, values considerably lower than 0.8% have been reported for several hardwoods (Coile, 1937; Chandler, 1941; Alway et al., 1933).

With an average addition of 3,400 kg ha^{-1} of litter by forest trees and an average N concentration of 0.6 to 2.0%, the return of N is approximately 20 to 70 kg ha^{-1}. Larcher (1975, p. 126) reported an N loss in leaf fall of 61 kg ha^{-1} from a mixed deciduous forest in Belgium and 33 kg ha^{-1} yr^{-1} for an evergreen oak forest in southern France. This is approximately 70% of the N absorbed.

The rate of decomposition of litter varies with species, nutrient conditions of the soil, aeration, moisture conditions, and temperature. In general, hardwood leaves decompose more rapidly than do coniferous leaves. Decomposition is slow in northern latitudes and most rapid in tropical areas (Chapter 10). Summarizing data from several sources, Cole (1986) noted that in northern conifer forests mean residence time for N in litter can exceed 200 years, and serious mineral deficiencies can develop because of the slow rate of decay (Table 9.7). The increased rate of growth following thinning of overstocked young stands is caused at least in part by the release of N and other nutrients from the decay of the slash (Tamm, 1964, pp. 148–149). In tropical forests decay is very rapid and the turnover is correspondingly fast. In shallow, acid, infertile tropical soils it is believed that the fungal mat often found in the surface layer plays an important role in speeding up release of nutrients (Went and Stark, 1968).

THE NITROGEN CYCLE

Given the common limitation of plant growth by N, there has long been a keen interest in the rates at which N is absorbed and lost during cycling in various kinds of plant stands and ecosystems.

Larcher (1975, pp. 91–100) and Curlin (1970) summarized considerable information on this subject. According to Switzer et al. (1968) the 20-year-old pine stand studied by them contained 2,300 kg ha^{-1} of N, including about 1,900 kg ha^{-1} occurring in the surface soil, but only used 70 kg ha^{-1} yr^{-1}, of which 38 kg came from the soil and the remainder from foliage and twigs before they abscised. The falling foliage and other plant parts were estimated to return about 30 kg ha^{-1} to the soil, leaving 8 kg to be supplied from other sources. This much probably could be supplied by atmospheric deposition. A simple diagram of the N cycle of this pine stand is shown in Figure 9.10. Mitchell et al. (1992) presented the N budget for a 300-year-old sugar maple-dominated forest in central Ontario (Fig. 9.11). This system seems to contain considerably more N than either pine stand shown in Table 9.7, but it is considerably older. About 87% of the N occurs in the soil and less than 10% in the trees. Bormann et al. (1977) observed similar patterns in a New England hardwood forest, where about 90% of the N was in soil organic matter, 9.5% in the vegeta-

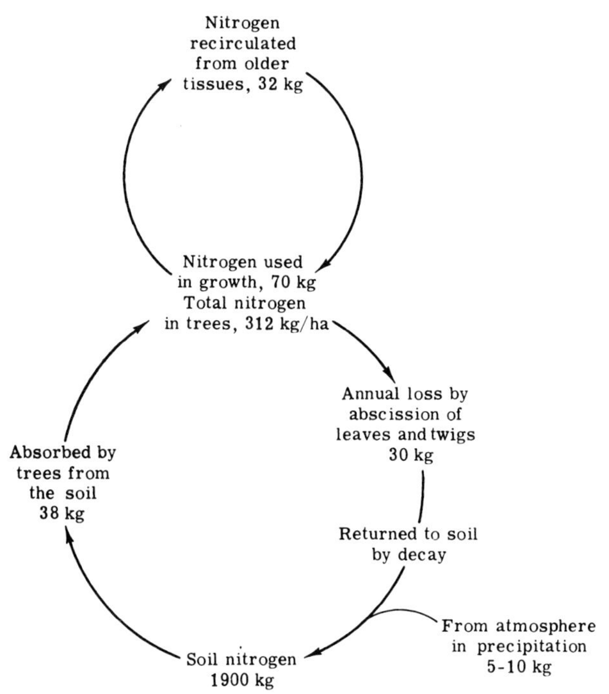

FIGURE 9.10. Simplified diagram of the major components of the N cycle in a 20-year-old loblolly pine stand, based on data of Switzer et al. (1968).

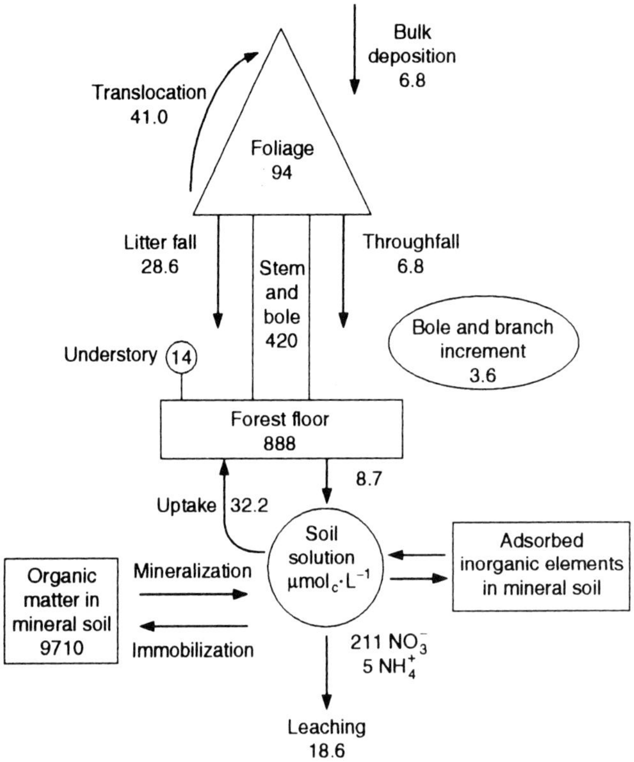

FIGURE 9.11. Major pools (kg ha^{-1}) and fluxes (kg ha^{-1} yr^{-1}) of N in a 300-year-old hardwood forest dominated by sugar maple. Adapted from Mitchell et al. (1992).

tion, and 0.5% as available N in the soil. In contrast, nearly 30% of the N in the surface soil, and trees of a tropical forest ecosystem, is in the trees (Sanchez, 1973).

A generalized diagram of the global N cycle is shown in Figure 9.12. The input of N comes from fixation by symbiotic and nonsymbiotic bacteria and other organisms, atmospheric fixation by lightning, that escaping from volcanoes and industrial processes, and that supplied by fertilization. Conversion of organic-N to NH_4^+ in the process of mineralization is an important process leading to reentry of N into the biotic portion of an ecosystem. The rate of release of N by N mineralization may limit productivity in forest ecosystems, most often because low temperatures (especially in boreal forests, Table 9.7), deficient soil moisture or soil oxygen availability, and litter nutrient concentration and chemistry reduce the rate of organic matter decomposition (Melillo et al., 1982; Adams and Attiwill, 1986a,b, White and Gosz, 1987; White et al., 1988; Zak and Pregitzer, 1990; Attawill and Adams, 1993; Updegraff et al., 1995).

In an extreme example, Ehleringer et al. (1992) demonstrated that there was essentially no return of N to the soil in litter in the mesquite-dominated Atacama Desert in Chile. In this very dry ecosystem, N concentrations and C:N ratios of recently deposited litter were virtually the same as that identified to be at least 40 years old. Further, a thick impenetrable surface crust of carbonate minerals prevents root growth into the litter layer. Mesquite trees derived all their N from nitrogen fixation in subsurface nodules and water and other mineral nutrients were necessarily absorbed from groundwater. Once released as NH_4, N may quickly be absorbed and effectively immobilized by the microbial biomass of soil and litter, especially if the supply of fixed carbon available to the microflora and fauna is not limiting.

The losses of N are caused chiefly by the action of denitrifying bacteria that reduce nitrates to molecular N, loss of ammonia during decay of plant and animal residues, fire, and leaching. Forest fires, set either by lightning or humans, have occurred from the earliest times and greatly affect the types of tree stands. For example, the forests of the southeastern United States are composed chiefly of fire-resistant species. Prescribed burning has become a standard silvicultural tool to control diseases, reduce hardwood reproduction, and decrease the damage from wildfires. This has produced considerable interest in the effects of fire on the mineral nutrition of trees.

Considerable N is lost when the litter and organic matter on the forest floor are burned, but other elements are released and become immediately available so that plant growth actually may be increased. Viro (1974, p. 39) concluded that in northern Europe the gains from increased mineralization of N after fire greatly outweigh the effect of the losses during fires. This also may be true in California chaparral, where mineralization is slow and nitrate N increases in the soil after burning (Christensen, 1973). Also in the southeastern United States pine stands have been burned repeatedly over many years without significantly reducing the rate of growth. Effects of fire on site quality also are discussed in Chapter 5 of Kozlowski and Pallardy (1997).

Losses of N and other nutrients by leaching are negligible in undisturbed forests (Bormann et al., 1977; Cole, 1986) and ordinarily are not seriously increased by prescribed burning. If all vegetation is destroyed, as in one Hubbard Brook experiment (Likens et al., 1970), losses of N may be heavy. However, there usually is rapid regrowth of vegetation after clear cutting or burning, and most of the released nutrients are recaptured. Vitousek and Melillo (1979) reviewed the influences of disturbance on and the process of recovery from N losses from forest ecosystems.

Because of human activities, particularly the burning of fossil fuels, forest ecosystems in many parts of the

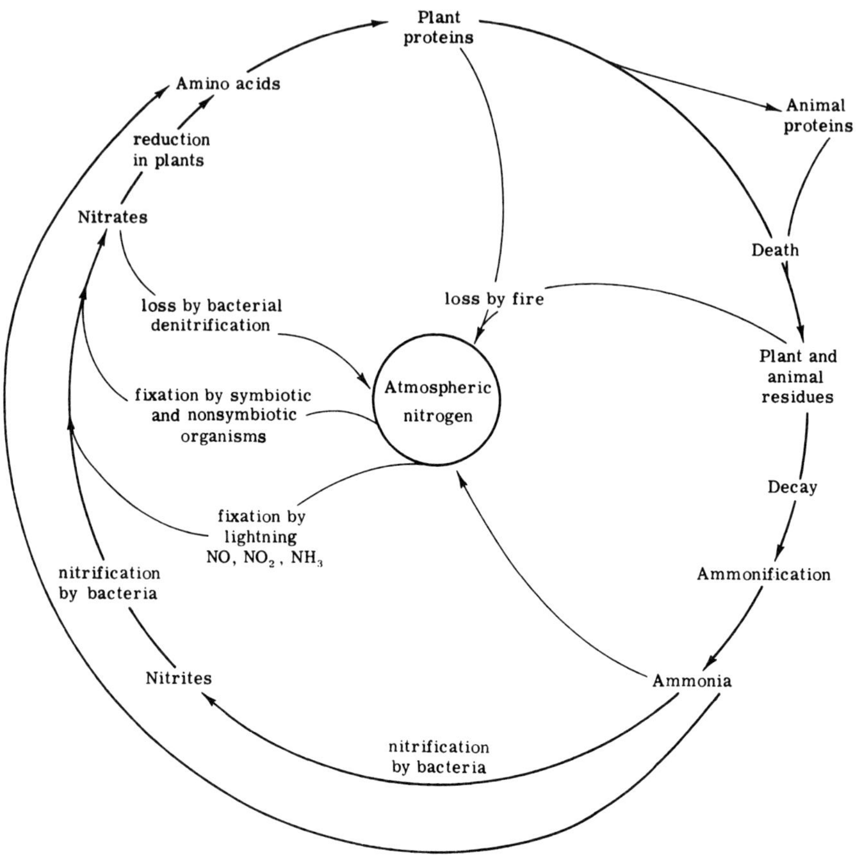

FIGURE 9.12. A simple, generalized N cycle. From Kramer and Kozlowski (1979).

world now receive continuously elevated levels of N as atmospheric deposition. Little previous attention has been directed toward studying the effects of *excess* N on forests, as N deficiency was considered a dominant problem in forest nutrition research (Ågren and Bosatta, 1988; Skeffington and Wilson, 1988). However, interest and concern are now growing about the long-term impacts of chronic, low-level N additions on forest productivity and ecosystem stability. Nitrogen saturation occurs when N addition to a forest ecosystem exceeds that amount necessary to meet plant and microbial needs (Aber et al., 1989). The effects of chronic inputs were distinguished from those associated with fertilization by differences in N concentration (2 to 40 kg ha^{-1} yr^{-1} in chronic additions vs. 100 to 400 kg ha^{-1} yr^{-1} with fertilization) and the short-term impact of fertilization. The primary effects of fertilization usually are increases in foliage N concentration and leaf biomass. Subsequently, leaf N concentration returns to prefertilization levels and added N moves to inactive pools in the plant stem and soil.

Chronic addition of N to a forest ecosystem may lead to an initial increase in foliar biomass and net primary productivity, but also to long-term declines in the latter (Fig. 9.13). When an ecosystem reaches N saturation, foliage N concentrations may become permanently elevated, investment in root biomass declines, and cold hardiness may be reduced (Friedland et al., 1984). Soil processes associated with N cycling and transformations may be drastically affected under N-saturating conditions. Nitrification may be substantially accelerated, even at low pH, and this will greatly increase the likelihood of nitrate leaching from an ecosystem. As a result, forest ecosystems may be transformed to N sources rather than sinks. Production of N_2O may be stimulated during nitrification and by elevated denitrification. Ultimately, imbalances in nutrient concentration and biomass allocation may predispose trees to environmental stresses, leading to forest decline and mortality, or conversion to rapidly growing fast-N cycling forest types via successional processes (Fig. 9.13). Additionally, the combination of elevated N and CO_2 likely to characterize future environments may lengthen the length of time NPP may be sustained because of the greater N demand of high CO_2-exposed plants (Finzi et al., 2002; Luo et al., 2004).

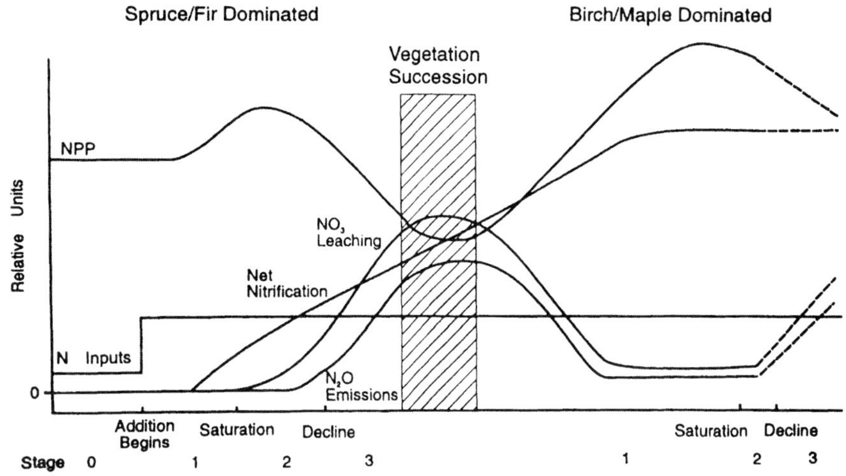

FIGURE 9.13. Time course of conceptual model of forest ecosystem responses to chronic N addition as proposed by Aber et al. (1989, 1995) based on data from forests in New England and in recent fertilizations studies (McNulty et al., 1996). Dashed portions of lines indicate unobtainable trends near and beyond stand decline. However, nitrate leaching and trace gas emissions from soil may increase further even with higher N deposition. Line positions along the y-axis are diagrammatic and not intended to accurately convey relative rates of one process to another. From Fenn et al. (1998).

SUMMARY

Although nitrogen makes up only a small fraction of plant dry weight (often less than 1%), N compounds are extremely important physiologically. Structural and storage proteins, enzymes, amino acids and amides, nucleic acids, and plant hormones all contain N. Nitrogen concentration is high in tissues and organs in which physiological activity is greatest, as in leaves and developing fruits and seeds, and low in inactive tissues such as heartwood. Seasonally, considerable N is retranslocated within the plant before parts are shed, a process that results in adequate N for new growth. Nitrogen may be absorbed either as nitrate or ammonium ions or, rarely, in organic form. If nitrate is absorbed it is reduced to ammonia in the roots or in the leaves, or in both organs. Once reduced, ammonia is assimilated into glutamine and subsequently into other amino acids.

Among the most important N compounds in plants are amino acids joined by peptide linkages to form proteins that may serve in structural, storage, and catalytic roles in the plant. The substituted purine and pyrimidine bases that constitute nucleic acids (DNA, RNA), nucleotides such as AMP, ADP, and ATP, and nicotinamide nucleotides (NAD and NADP) also contain N. Alkaloids, another class of N-containing compounds, appear to have no essential physiological role but are found in great variety in many plants. However, they may serve as biological deterrents to attacking organisms, and certain alkaloid compounds have been widely and very successfully employed in human medicine.

Fixed N enters the biosphere primarily by natural atmospheric and biological fixation of molecular N, and by deliberate and incidental fixation by human activities. Biological fixation is accomplished by many prokaryotic organisms, often in association with plants, particularly legumes and species of several families of woody plants that form nodules with actinomycetes. Nitrogen fixation is accomplished by the O_2-labile enzyme nitrogenase in an environment in which oxygen levels are tightly controlled and buffered. Such control is needed to reduce the chances of enzyme inactivation and simultaneously assure the respiratory rates needed to support the energy-demanding reduction of N_2 to NH_3.

Nitrogen deficiency limits growth of plants more often than any other mineral element and much research has been done to quantify the inputs, internal fluxes and storage pools, and losses of N from ecosystems. Most N in forest ecosystems is found in soil organic matter, with lesser relative amounts in the litter, stems, branches, and foliage. Nitrogen inputs to forest ecosystems often exceed losses, unless disturbance has induced accelerated removals attributable to erosive losses and excessive rates of mineralization and nitrification. Recent evidence suggests that high anthropogenic inputs of N from the atmosphere have the potential to disrupt ecosystem function after a period of growth stimulation.

General References

Amâncio, S. and Stulen, I., eds. (2004). *Nitrogen Acquisition and Assimilation in Higher Plants*. Kluwer Academic Publishers, Dordrecht and Boston.

Andrews, M. (1986). The partitioning of nitrate assimilation between root and shoots of higher plants: Mini-review. *Plant Cell Environ* **9**, 511–519.

Attiwill, P. M. and Adams, M. A. (1993). Nutrient cycling in forests. *New Phytol* **124**, 561–582.

Binkley, D. (1986). *Forest Nutrition Management*. Wiley, New York.

Cassob, G. I. and Varner, J. E. (1988). Cell wall proteins. *Annu Rev Plant Physiol Plant Mol Biol* **39**, 321–353.

Gresshoff, P. M. (1990). *Molecular Biology of Symbiotic Nitrogen Fixation*. CRC Press, Boca Raton, Florida.

Haynes, R. J. (1986). *Mineral Nitrogen in the Plant-Soil System*. Academic Press, New York.

Lea, P. J. and Morot-Goudry, J.-F., eds. (2001). *Plant Nitrogen*. Springer, Berlin and New York.

Mengel, K. and Pilbeam, D. J. (1992). *Nitrogen Metabolism in Plants*. Oxford Univ. Press (Clarendon), Oxford.

Postgate, J. R. (1998). *Nitrogen Fixation, 3rd edition*. Cambridge University Press, Cambridge and New York.

Shewry, P. R. (1995). Plant storage proteins. *Biol Rev* **70,** 375–426.

Singh, B. K., Flores, H. E., and Shannon, J. C., eds. (1992). *Biosynthesis and Molecular Regulation of Amino Acids in Plants*. American Society of Plant Physiologists, Bethesda, Maryland.

Solomonson, L. P. and Barber, M. J. (1990). Assimilatory nitrate reductase: Functional properties and regulation. *Annu Rev Plant Physiol Plant Mol Biol* **41**, 225–253.

Stacey, G., Burris, R. H., and Evans, H. J., eds. (1992). *Biological Nitrogen Fixation*. Chapman & Hall, New York.

Suzuki, A. and Knaff, D. B. (2005). Glutamate synthase: Structural, mechanistic and regulatory properties, and role in the amino acid metabolism. *Photosynth Res* **83**, 191–217.

Tamm, C. O. (1991). *Nitrogen in Terrestrial Ecosystems: Questions of Productivity, Vegetational Changes and Ecosystem Stability*. Ecological Studies, Vol. 81. Springer-Verlag, Berlin.

Titus, J. S. and Kang, M. (1982). Nitrogen metabolism, translocation and recycling in apple trees. *Hortic Rev* **4**, 204–206.

Vessey, J. K., Pawlowski, K., and Bergman, B. (2004). Root-based N_2-fixing symbioses: Legumes, actinorhizal plants, *Parasponia* sp. and cycads. *Plant Soil* **266**, 205–230.

CHAPTER

10

Mineral Nutrition

INTRODUCTION

A supply of mineral nutrients is essential for plant growth. Unfortunately, mineral deficiencies are common and often limit the growth of woody plants. Nitrogen deficiency is particularly well documented throughout the world and deficiency of phosphorus occurs in Australia, New Zealand, the southeastern United States, and many tropical countries. In Norway, deficiencies of nitrogen, phosphorus, potassium, and frequently boron limit productivity of forest stands on drained sites (Braekke, 1990). Growth reduction and shoot dieback in forest plantations in the Philippines are associated with unbalanced mineral nutrition (Zech, 1990). Mineral deficiencies in forests usually are chronic rather than catastrophic and hence their impacts often are not obvious.

More than half the elements in the periodic table have been found in woody plants, but not all are essential. Parker (1956) found platinum, tin, and silver in leaves of ponderosa pine, and considerable quantities of aluminum, silicon, and sodium occur in plants; however, none of these elements is regarded as essential. A mineral element is considered essential if it is part of a molecule of some essential plant constituent and/or the plant can be so deprived of the element that it manifests abnormal growth, development, or development compared with a plant not so deprived (Epstein & Bloom, 2005). The essentiality of an element can be determined under only the most carefully controlled conditions, which exclude the possibility of contamination with the element under study from the salts, the water, the containers in which the plants are grown, and even from dust in the air. The minimum amounts of various elements necessary for growth can be determined most readily by using soil, sand, or water cultures, or by field fertilization experiments. The adequacy of the supply of various elements in the field also can be studied by analysis of soil and plant tissues (foliar diagnosis) and by observing the effects of supplying various elements to the soil or directly to the foliage (Walker, 1991).

The elements required by plants in fairly large quantities, usually at least 1,000 ppm, are nitrogen (N), phosphorus (P), potassium (K), calcium (Ca), magnesium (Mg), and sulfur (S); these sometimes are called the major elements or macronutrients. Elements required in much smaller quantities include iron (Fe), manganese (Mn), zinc (Zn), copper (Cu), boron (B), molybdenum (Mo), chlorine (Cl), and nickel (Ni). Those elements required in very small quantities often are called the minor elements, trace elements, or micronutrients.

FUNCTIONS OF MINERAL NUTRIENTS AND EFFECTS OF DEFICIENCIES

Mineral nutrients have many functions in plants; they are, for example, constituents of plant tissues, catalysts in various reactions, osmotic regulators, constituents of buffer systems, and regulators of membrane permeability. Several elements including Fe, Cu, and Zn, although required in very small quantities, are essential because they are prosthetic groups or coenzymes of certain enzyme systems. Other minerals function as activators or inhibitors of enzyme systems. Some elements, such as B, Cu, and Zn, which are required in extremely small quantities in enzyme systems, are very toxic if present in larger quantities. Toxicity of these and other ions such as silver and mercury probably is related chiefly to their injurious effects on enzyme systems.

Although much of the osmotic potential of cell sap is attributable to soluble carbohydrates, a measurable fraction results from the presence of mineral salts, and salts often are the major sources of the low osmotic potential of halophytes. Phosphates form one of the important plant buffer systems, and elements such as Ca, Mg, and K constitute the cations of cellular organic acid buffer systems. The kinds of ions present often affect the hydration of protoplasm and permeability of cell membranes, with di- and trivalent cations usually decreasing and monovalent cations increasing permeability. Certain ions tend to counterbalance the effect of others. For example, a low concentration of Ca is required to balance Na and prevent the injury that occurs by a solution of NaCl alone.

Deficiencies of any of the essential elements alter physiological processes and reduce plant growth, often before visible symptoms appear. Deficiencies also produce morphological changes and injuries. For example, micronutrient deficiencies have been associated with twisting of stems and branches of Monterey pine as well as prostrate tree growth (Turvey, 1984; Turvey et al., 1992). Leaves and stem and root tips are particularly sensitive to mineral deficiency. Mineral-deficient plants tend to be small, chlorotic, and sometimes have dead areas at the tips and margins or between the veins (Fig. 10.1) (see Chapter 5 of Kozlowski & Pallardy, 1997). Sometimes they develop in tufts or rosettes and have other abnormalities that enable experienced observers to diagnose their cause. Other symptoms of mineral deficiency include dieback of stem tips and twigs, bark lesions, and excessive gum formation.

Nitrogen

The essential role of N as a constituent of amino acids, the building blocks of proteins, is well known (Chapter 9). It occurs in a variety of other compounds such as purines and alkaloids, enzymes, vitamins, hormones, nucleic acids, and nucleotides. Both leaf area development and photosynthesis depend greatly on N supply (Chapter 5). Nitrogen deficiency is accompanied by failure to synthesize normal amounts of chlorophyll, resulting in chlorosis of older leaves and, when deficiency is severe, of young leaves also. In fruit and nut trees, N deficiency may be associated with leaf abscission, decreased fruit set, poorly developed fruit buds, and small and early maturing fruits (Shear and Faust, 1980).

Phosphorus

The element phosphorous is a constituent of nucleoproteins and phospholipids, and the high energy bonds associated with phosphate groups constitute the chief medium for energy transfer in plants. Phosphorus occurs in both organic and inorganic forms and is translocated readily, probably in both forms.

Deficiency of P often causes severe stunting of young forest trees in the absence of other visible symptoms. In fruit trees the major P deficiency problems involve the fruit (P deficiency symptoms essentially do not occur on leaves in orchards). Both flowering and fruiting are reduced by P deficiency. Stone fruits may ripen early and often are soft and of low quality.

Potassium

Although large amounts of K are required by plants, it is not known to occur in organic forms. Potassium, which is highly mobile in plants, is involved in enzyme

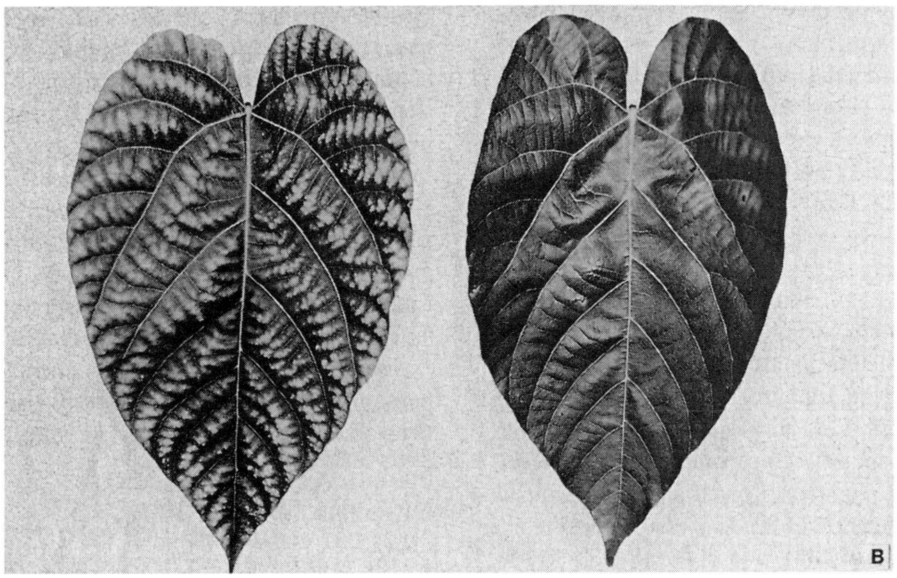

FIGURE 10.1. Visual symptoms of mineral deficiencies. (A) Apple leaves developing Mg deficiency. (B) Manganese-deficient (left) and normal tung leaves (right). (A) Courtesy of Crops Research Division, U.S. Department of Agriculture; (B) Courtesy of R.B. Dickey, Florida Agricultural Experiment Station.

activation, protein synthesis, osmoregulation, stomatal opening and closing, photosynthesis, and cell expansion. It is interesting to note that plant cells distinguish between K and Na, and the latter cannot be completely substituted for the former.

Potassium deficiency often is characterized by marginal scorching of old leaves, with chlorosis often preceding the scorching (Fig. 10.2). In nut trees the high sink strength of reproductive tissues for K sometimes is associated with the premature leaf shedding.

Sulfur

Sulfur is a constituent of the amino acids cysteine and methionine as well as coenzymes, ferredoxin, biotin, and thiamine. It also is a component of sulfolipids and hence affects biological membranes (Marschner, 1995). Deficiency of sulfur causes chlorosis and failure to synthesize proteins, resulting in accumulation of amino acids. In S-deficient trees the young leaves show general yellowing similar to that of N-deficient leaves. In older leaves both interveinal chlorosis and necrotic areas may be found. Other deficiency symptoms include marginal chlorosis and rosettes of small lateral shoots near terminals.

Calcium

Calcium occurs in considerable quantities in cell walls as calcium pectate and apparently influences cell wall elasticity. At low concentrations of Ca, wall deposition of Norway spruce needles was inhibited, mainly as a result of reduced deposition of lignin and noncellulosic polysaccharides (Eklund and Eliasson, 1990). Calcium also is involved in some manner in N metabolism and is an activator of several enzymes, including

FIGURE 10.2. Potassium deficiency with marginal leaf scorching in raspberry. Courtesy of East Malling Research Station.

amylase. It is relatively immobile, and a deficiency results in injury to meristematic regions, especially root tips. By acting as a second messenger, Ca often modifies the functions of various growth hormones (Chapter 13). Surplus Ca often accumulates as Ca oxalate crystals in cell vacuoles in leaves and woody tissue. McLaughlin and Wimmer (1999) reviewed the role of Ca in physiological processes and in terrestrial ecosystem processes.

Symptoms of Ca deficiency include chlorosis and necrosis of leaves as well as decreased root growth. Calcium deficiency of fruits shortens their storage life. A number of physiological fruit disorders also are associated with Ca deficiency, often when the level of Ca is high enough for normal vegetative growth. In pome fruits, disorders associated with low levels of Ca include bitter pit, cork spot, sunburn, lenticel breakdown, watercore, internal breakdown, and low temperature breakdown (Shear and Faust, 1980). Development of Ca-related disorders depends on the tissue N concentration. When the N:Ca ratio (based on element mass) was 10, metabolic disorders did not develop; when the ratio was 30, metabolic disorders were common (Faust, 1989).

Calcium probably is the most important element that affects the quality of fruits. It is particularly important in apples and pears because they are stored for long periods. The effects of Ca on storage quality cannot be replaced by other factors (Faust, 1989). Whereas N, P, and K applied to the soil reach the absorbing roots rapidly, Ca only slowly becomes available to roots because of slow dissolution and transport within the soil solution to root surfaces. Hence Ca must be applied before planting and mixed into the soil. Faust (1989) has an excellent discussion of the role of Ca in fruit nutrition.

Magnesium

The element magnesium is a constituent of the chlorophyll molecule and also is involved in the action of several enzyme systems as well. Magnesium also is involved in maintaining the integrity of ribosomes, which disintegrate in its absence. It is translocated readily in most plants. A deficiency of Mg usually induces chlorosis; severe deficiency also may induce marginal scorching. Initially the apices of older leaves become chlorotic and chlorosis spreads interveinally toward the leaf base and midrib, resulting in a herringbone pattern. Developing fruits have a high Mg requirement, and Mg is translocated from neighboring leaves to fruits, culminating in severe deficiency symptoms in leaves and early leaf shedding (Shear and Faust, 1980).

Iron

Much of the iron in leaves occurs in the chloroplasts, where it plays a role in synthesis of chloroplast proteins. It also occurs in respiratory enzymes such as peroxidases, catalase, ferredoxin, and cytochrome oxidase. Iron is relatively immobile, and deficiencies usually develop in new tissues because Fe is not translocated out of older tissues.

Deficiency of Fe is one of the most common and conspicuous micronutrient deficiencies, occurring chiefly in trees growing in alkaline and calcareous soils in which a high pH prevents its absorption of Fe. An early symptom of Fe deficiency is chlorosis of very young leaves. The interveinal tissues become chlorotic while the veins remain dark green. Dieback of shoots often is associated with Fe deficiency.

The total amount of Fe seldom is deficient in soils but it exists in two valence states, ferric (Fe^{3+}) and ferrous (Fe^{2+}), which are not equally available to plants. Many plants absorb and use ferrous Fe better than ferric Fe, but soil Fe occurs predominantly in the ferric form. Some plants can respond to Fe deficiency stress by inducing plant reactions that make Fe available in a useful form. Such "iron efficiency" is enhanced by release from roots of Fe chelating compounds, hydrogen ions, and reductants as well as by reduction of Fe^{3+} to Fe^{2+} by roots, and an increase in organic acids (especially citrate) by roots (Brown and Jolley, 1989). Both Hagstrom (1984) and Korcak (1987) presented good discussions of corrective measures for Fe deficiency.

Manganese

The element manganese is essential for chlorophyll synthesis. Manganese deficiency results in reduced contents of chlorophyll and constituents of chloroplast membranes (e.g., phospholipids, glycolipids). Manganese plays a role in the O_2 evolution step of photosynthesis. Its principal function probably is the activation of enzyme systems, and it also affects the availability of Fe. A deficiency often causes malformation of leaves and development of chlorotic or dead areas between the major veins. The symptoms differ from those caused by Fe deficiency. Chlorosis does not occur on very young leaves of Mn-deficient plants, nor do the small veins remain green. Symptoms of Mn deficiency often develop in leaves shortly after they become fully expanded. Excessive Mn uptake may lead to chlorosis, early leaf shedding, inhibition of flower bud formation, reduced growth, and bark necrosis.

Zinc

Zinc acts as a metal component of enzymes and as an enzyme cofactor. Because it is a cofactor for RNA polymerase, Zn influences protein synthesis. In Zn-deficient plants protein synthesis is inhibited while amino acids and amides accumulate (Faust, 1989). In several species of trees, Zn deficiency produces leaf malformations resembling virus diseases, possibly because it is involved in synthesis of tryptophan, a precursor of indoleacetic acid. Symptoms of Zn deficiency include interveinal chlorosis, development of small leaves, and reduced elongation of internodes. These symptoms often have been characterized as "little-leaf" and "rosette." The purplish color of the lower leaf surface of Zn-deficient trees is called bronzing. In New Zealand, Zn deficiency in Monterey pine is associated with lack of apical dominance, stunted growth, and terminal rosettes of buds (Thorn and Robertson, 1987). When Zn deficiency is severe, necrotic areas appear on the leaf surface (Shear and Faust, 1980). Resin exudation around buds of Monterey pine in Australia also is associated with Zn deficiency. Phosphorus-induced Zn deficiency sometimes occurs in intensive cropping systems (e.g., nurseries, hybrid poplar plantations). Symptoms, including rosetting of small leaves at ends of dwarf shoots, follow application of high dosages of P fertilizers (Teng and Timmer, 1990a,b, 1993).

Copper

Copper is a constituent of certain enzymes, including ascorbic acid oxidase and tyrosinase. Very small quantities are needed by plants and too much is toxic. A deficiency of Cu occasionally occurs in some forest trees but not often in fruit trees. Copper deficiency causes various degrees of stem deformity (Turnbull et al., 1994), interveinal chlorosis, leaf mottling, defoliation, and dieback of terminal shoots.

Boron

Boron is another element required in very small quantities, the specific requirements varying from 5 to 15 ppm, depending on the species. In plants most B exists as boric acid ($B(OH_3)$) with a small amount of borate anion $B(OH)_4^-$. Both forms readily form complexes with sugars and other compounds that have *cis*-hydroxyl groups, and the best evidence for a functional role for B in plants is in stabilizing the cell wall pectic network and regulating cell wall pore size (Blevins and Lukaszewski, 1998; Brown et al., 2002; Bolaños et al., 2004; Kakegawa et al., 2005). In B-deficient tissues, excessive phenolics accumulate that adversely affect membrane permeability. Plants deficient in B contain more sugars and pentosans and have lower rates of water absorption and transpiration than normal plants. Boron deficiency often occurs in orchard trees, and is one of the most common micronutrient deficiencies in forest plantations all over the world. For example, B deficiency is an important cause of poor stem form and leaf malformation in several species of *Eucalyptus* in China (Dell and Malajczuk, 1994).

Although symptoms vary somewhat among species, in general B deficiency is characterized by breakdown of meristematic tissues and walls of parenchyma cells as well as by weak development of vascular tissues, particularly the phloem. Boron deficiency often is associated with scorched leaf margins, dead shoot tips, and weak lateral shoots, giving the tree a bushy appearance. The dark green, thick, and brittle leaves are shed early. Symptoms of B deficiency generally appear on fruits before they are evident on vegetative tissues. Mild B deficiency may result in small and deformed fruits, and more severe deficiency in internal and external cork formation.

Unfortunately, the B concentration for optimal growth closely approaches the toxic concentration in some species. Boron toxicity is shown in early maturation of fruits, premature fruit drop, shortened storage life, and senescence breakdown in storage (Faust, 1989).

Molybdenum

The element molybdenum is required in the lowest concentration of any essential element, with less than 1 ppm sufficing for most plants. Molybdenum is involved in the nitrate-reducing enzyme system. Deficiency symptoms include uniform chlorosis of young leaves, burning of tips and margins of older leaves, and leaf abscission. Molybdenum deficiency is not common in orchard or forest trees.

Chlorine

It appears that chlorine is essential for plants, and it may be involved in the water splitting step of photosynthesis. However, there probably is no serious chlorine deficiency in forest trees or fruit trees.

Nickel

In recent years the essentiality of nickel has been increasingly accepted by researchers, although Ni deficiency is very rarely encountered in the field. Nickel is a required cofactor of urease, an enzyme required for ureide metabolism (Gerendás et al., 1999). As certain species translocate assimilated N from roots as ureides, particularly legumes, Ni deficiency can disrupt nitrogen metabolism. "Mouse-ear," a growth disorder of pecan that is occasionally present in orchard trees, can be reversed by foliar applications of nickel sulphate solutions (Wood et al., 2004). Bai et al. (2006) reported alterations of carbon and nitrogen metabolism in Ni-deficient pecan foliage that resulted in abnormal accumulation of certain amino acids (especially glycine) and organic acids (oxalic and lactic acids). The toxic accumulation of these two organic acids in rapidly growing tips and margins of leaflets was suggested as the potential cause of morphological symptoms in the syndrome.

Other Mineral Elements

Silicon (Si), sodium (Na), and aluminum (Al) may occur in large quantities in some plants, but although these elements sometimes increase plant growth and may be required for certain plants, they generally are not regarded as essential. Silicon has been regarded as an essential element only for diatoms and *Equisetum* (Epstein, 1999), although it accumulates in large amounts in some plants, particularly monocots such as rice and sugarcane where it stimulates growth (Ma, 2005). However, its biochemical role(s) is not known. Silicon affords protection against some diseases (Epstein, 1999) and may reduce Al toxicity because it stimulates the release of phenolic compounds from roots that promote formation of stable complexes with toxic aluminum ions (Kidd et al., 2001). It also forms precipitates with this element and excess Zn (Richmond and Sussman, 2003). Sodium appears essential for certain C_4 plants such as some species of *Atriplex*, *Kochia*, *Panicum*, and *Distichum*, where it appears to play an important role in regeneration of phosphoenolpyruvate phosphate (PEP) in mesophyll sheath cells during photosynthesis (Subbarao et al., 2003). Sodium can stimulate growth in plants, particularly in certain taxa such as the Chenopodiaceae, which includes sugar beet, red beet, and spinach. This stimulation occurs even in the presence of adequate K, a nutrient with which Na can often be interchanged.

Aluminum is very toxic in low concentrations, especially if the pH is less than 4.7. Aluminum toxicity is the major factor limiting crop yield on acid soils (Kochian, 1995; Kochian et al., 2005). Toxic levels of Al have been associated with leaf symptoms such as those of Ca deficiency, root growth inhibition and malformation, and eventual death of plants (Shear and Faust, 1980; Kochian et al., 2005). Excess Al affects many physiological processes. The visible expression of physiological changes may vary with plant species and vigor as well as with environmental conditions. Aluminum toxicity commonly is considered a complex rather than a simple abiotic disease with a single mode of action. Aluminum toxicity influences energy transformations, cell division, and membrane function. Because Al reduces accumulation of Mg and Ca in both the roots and shoots, as well as translocation of P to shoots, high levels of Al in plants often lead to dysfunctions associated with deficiencies of Ca, Mg, and P (Sucoff et al., 1990; see also Chapter 3, Kozlowski and Pallardy, 1997).

There are many complicated interactions among various mineral nutrients, with one element modifying absorption and utilization of others. For discussions of these interactions, I refer you to books by Mengel and Kirkby (1982), Tinker and Läuchli (1986), Wild (1989), Marschner (1995), and Epstein and Bloom (2005).

ACCUMULATION AND DISTRIBUTION OF MINERAL NUTRIENTS

Young seedlings obtain nearly all their nutrients from the soil. Beyond the seedling stage, increasing proportions of nutrients come from internal redistribution and from nutrients released from decomposing litter and roots (Attiwill, 1995). The amounts of mineral nutrients in plants vary with species and genotype, age of plants, site, season, and in different organs and tissues of the same plant. *Total nutrient contents* of plants of different ecosystems vary in the following order: tropical > temperate broadleaf > temperate coniferous > boreal (Table 10.1). Generally, N, P, K, Ca, and Mg contents of tropical forests are three to five times higher than those of temperate broadleaved forests. Nutrient *concentrations* fall in the order: tropical > boreal > temperate broadleaf > temperate coniferous forests (Marion, 1979).

In temperate regions there are major differences in nutrient accumulation by deciduous and evergreen trees (Table 10.2). For example, trees in a white oak stand in North Carolina may contain twice as much N, P, and K, and 15 times as much Ca as loblolly pine trees in a stand of equal basal area (Ralston and Prince, 1965). Differences among species and genotypes in accumulation of minerals are discussed further in the section on factors affecting absorption of nutrients.

The distribution of minerals varies appreciably in different tissues and organs of the same tree. Partitioning of nutrients within trees depends on distribution of biomass and nutrient concentrations in different organs and tissues. As trees increase in size the proportion of their biomass in foliage decreases, whereas the proportions in the stem and bark increase (van den Driessche, 1984).

In general the concentration of minerals as a percentage of plant dry weight varies as follows: leaves > small branches > large branches > stems. This relationship is shown in Figure 10.3 for loblolly pine and three species of broadleaved trees. The relative amounts of mineral nutrients in various parts of several species of trees are shown in Tables 10.2, 10.3, and 10.4.

TABLE 10.1. Median and Range of Total Biomass and Nutrient Content (kg ha^{-1}) for Major Forest Types[a]

	Biomass		N		P	
Forest type	Median	Range	Median	Range	Median	Range
Boreal	129,000	37,000–336,000	447	174–1,915	50	33–67
Temperate coniferous	291,000	274,000–604,000	664	375–1,327	47	—
Temperate broadleaf	338,000	147,000–504,000	1,085	406–1,608	73	36–99
Tropical	378,000	276,000–1,189,000	4,260	3,280–5,290	241	26–1,212
	K		**Ca**		**Mg**	
	Median	Range	Median	Range	Median	Range
Boreal	291	133–449	488	243–732	108	74–142
Temperate coniferous	263	—	717	—	—	—
Temperate broadleaf	463	286–531	1,142	644–1,334	115	93–123
Tropical	2,157	1,606–10,395	4,005	1,891–13,472	437	374–1,452
	S		**Fe**		**Mn**	
	Median	Range	Median	Range	Median	Range
Boreal	58	24–91	46	16–76	28	27–28
Temperate coniferous	—	—	—	—	—	—
Temperate broadleaf	76	70–82	27	—	125	—
Tropical	—	—	—	—	—	—

[a]From Marion (1979). Reprinted with permission of the State University of New York, College of Environmental Science and Forestry (ESF), Syracuse, NY.

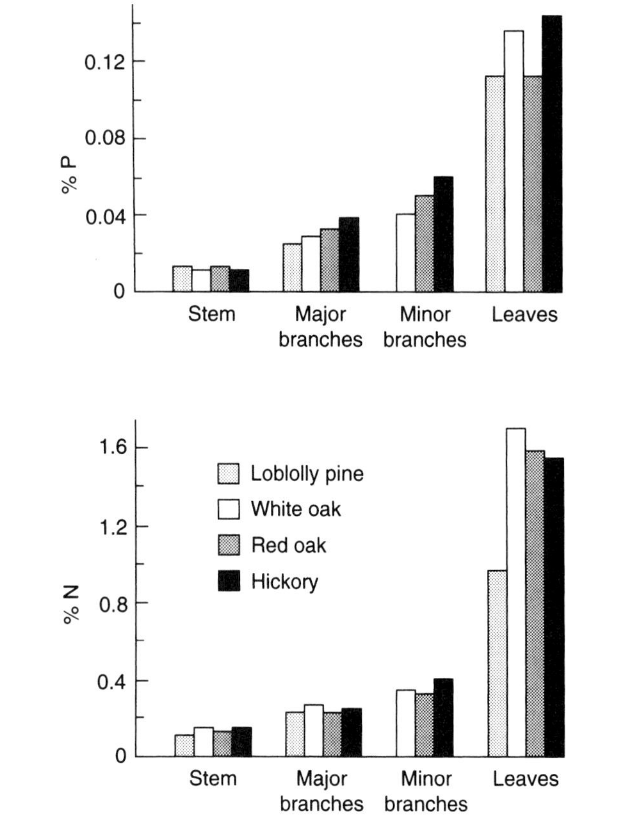

FIGURE 10.3. Average concentrations of N and P in various parts of trees. All data for loblolly pine branches are included under major branches. From Ralston and Prince (1965).

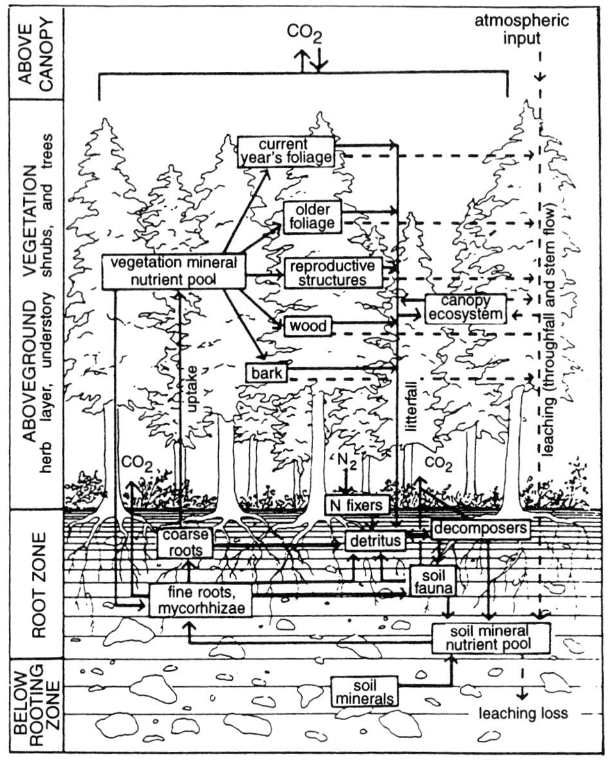

FIGURE 10.4. Model of nutrient cycling in conifer forests. From Johnson et al. (1982b). © Van Nostrand Rheinhold.

Trees store mineral nutrients in the leaves, stems, and roots. In stems the bark and ray parenchyma cells are major storage sites. As nutrient-storing tissues mature or senesce, mobile nutrients often are translocated to meristematically active tissues. Such transport is important for early growth in the following year and occurs well before rapid nutrient uptake from the soil occurs.

MINERAL CYCLING

The pool of soil minerals is maintained by continual recycling (Fig. 10.4) at three levels: (1) input and losses independent of vegetation, (2) exchange of mineral nutrients between the soil and plants, and (3) retranslocation within plants. The first two levels will be discussed here. Retranslocation of N is discussed in Chapter 9 of the present volume; retranslocation of mineral nutrients within plants is also discussed in Chapter 3 of Kozlowski and Pallardy (1997).

Variations in accumulation of nutrients in trees often are associated with differences in rates of nutrient cycling. In tropical forests uptake by roots is very efficient and a significant proportion of the mineral pool accumulates within the trees. In temperate forests, by comparison, more of the mineral nutrients cycled in litterfall accumulate in the slowly decomposing litter. The proportion of nutrients in trees of cold, boreal forests may be as low as 10% of the amount retained in tropical trees.

THE SOIL MINERAL POOL

Mineral nutrients in the soil are increased mainly by atmospheric deposition, weathering of rocks and minerals, decomposition of plant litter and roots, and exudation from roots. Soil nutrients are depleted by leaching away in drainage water, removal in harvested plants, and absorption by plants. Some N also is lost as gas by ammonification and denitrification (Chapter 9).

Atmospheric Deposition

Rain, snow, dew, and clouds contain appreciable amounts of mineral nutrients and add them, together with dry atmospheric deposits, to the soil (Fig. 10.5) (Jordan et al., 1980; Lovett et al., 1982; Lindberg et al.,

TABLE 10.2. **Distribution of N, P, K, and Ca as Percentage of Total Content, Total Nutrient Content, and Biomass of Five Species of Forest Trees**[a]

Species	Age (years)	Foliage (%)	Branches (%)	Bole Bark (%)	Bole Wood (%)	Roots (%)	Total (kg ha^{-1})
				N			
Pinus sylvestris	45	30	20	11	20	19	186
Picea glauca	40	34	28	10	13	15	449
Pseudotsuga menziesii	36	32	19	15	24	10	320
Betula verrucosa	55	14	31	—	27	28	543
Quercus alba	150	9	23	—	36	22	631
				P			
Pinus sylvestris	45	27	19	14	11	29	21
Picea glauca	40	42	27	13	8	10	64
Pseudotsuga menziesii	36	44	19	14	14	9	66
Betula verrucosa	55	12	35	—	32	21	34
Quercus alba	150	10	19	—	30	35	41
				K			
Pinus sylvestris	45	27	17	13	19	24	96
Picea glauca	40	34	31	12	13	10	254
Pseudotsuga menziesii	36	28	17	20	24	11	220
Betula verrucosa	55	22	23	—	32	23	200
Quercus alba	150	14	31	—	27	26	419
				Ca			
Pinus sylvestris	45	13	19	21	28	19	123
Picea glauca	40	32	27	20	10	11	809
Pseudotsuga menziesii	36	22	32	21	14	11	333
Betula verrucosa	55	6	28	—	42	24	651
Quercus alba	150	3	33	—	39	20	2,029
				Biomass (tons ha^{-1})			
Pinus sylvestris	45	4.4	10.6	5.3	55.6	19.3	95.2
Picea glauca	40	17.4	34.6	10.8	88.0	34.0	184.8
Pseudotsuga menziesii	36	9.1	22.0	18.7	121.7	33.0	204.5
Betula verrucosa	55	2.5	28.7	—	134.5	49.8	215.5
Quercus alba	150	5.4	52.8	—	129.0	95.6	282.8

[a]From Van den Driessche (1984).

TABLE 10.3. **N, P, K, Ca, and Mg (kg ha^{-1}) in Various Parts of Trees in a 16-Year-Old Loblolly Pine Plantation**[a]

Component	N	P	K	Ca	Mg
Needles, current	55	6.3	32	8	4.8
Needles, total	82	10.3	48	17	7.9
Branches, living	34	4.5	24	28	6.1
Branches, dead	26	1.5	4	30	3.0
Stem wood	79	10.7	65	74	22.7
Stem bark	36	4.2	24	38	6.5
Aboveground, total	257	30.9	165	187	46.2
Roots	64	16.9	61	52	21.9
Total	321	47.8	226	239	68.1

[a]From Wells et al. (1975).

1982, 1986; Braekke, 1990; Burkhardt and Eiden, 1990). In a mixed deciduous forest, dry deposition supplied Ca and N at rates approximating 40% of the annual requirement for wood production and more than 100% of the S requirement (Johnson et al., 1982a). During the growing season, interactions in the canopy influenced the amount of ion deposition on the forest floor. Deposition of nutrients by precipitation in a tropical watershed amounted to 139 kg ha^{-1} yr^{-1} of insoluble particulates. The rate of loading of soluble constituents was within the upper range of values reported for temperate-zone forests (Table 10.5). Loading of soluble cations varied in the following order: $Na^+ > Mg^{2+} > Ca^{2+} > H^+ > NH_4^+ > K^+$. Deposition of soluble anions varied as follows: $HCO_3^- > Cl^- > SO_4^{2-} > NO_3^- > PO_4^{3-}$.

TABLE 10.4. Mn, Zn, Fe, Ca, Al, Na, and Cu in Above-Ground Parts of Trees in a 6-Year-Old Loblolly Pine Plantation[a,b]

Component	Mn	Zn	Fe	Al	Na	Cu
Needles, current	1.222	0.666	0.334	2.178	0.258	21.5
Needles, total	2.544	0.327	0.650	4.116	0.356	31.6
Branches, living	1.716	0.345	0.915	2.519	1.384	63.7
Branches, dead	—	0.289	1.281	2.902	0.314	69.5
Stem wood	8.445	1.086	1.830	1.790	3.640	275.0
Stem bark	0.951	0.336	1.126	9.705	0.590	59.4
Total	13.656	2.383	5.802	21.032	6.284	499.2

[a]From Wells et al. (1975).
[b]Measurements of Mn, Zn, Fe, Al, and Na are in kg ha^{-1}; those of Cu are in g ha^{-1}.

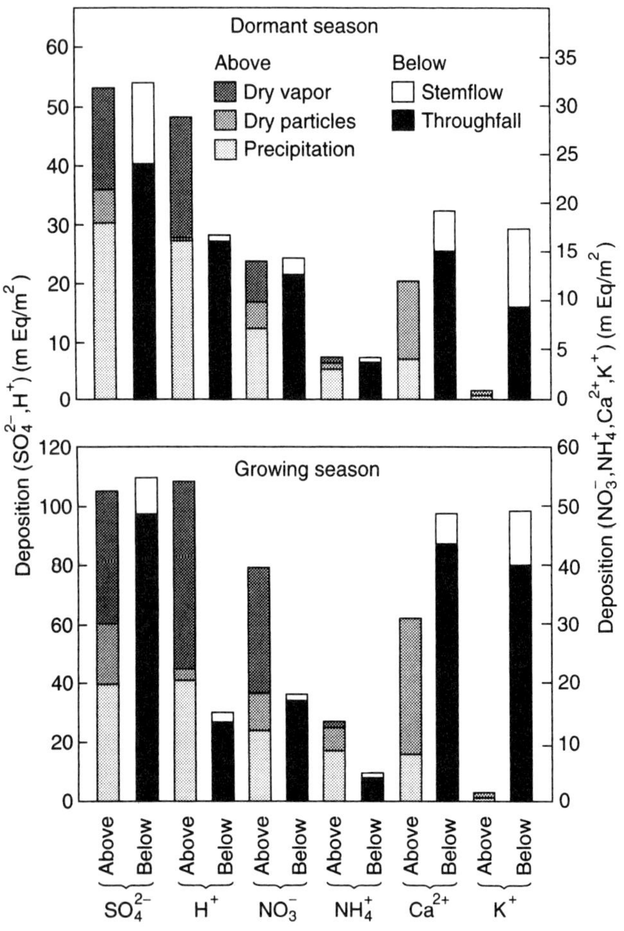

FIGURE 10.5. Contributions of atmospheric deposition and internal transfer processes to ion flux above and below the canopy of a mixed deciduous forest for the dormant and growing seasons. Reprinted with permission from Lindberg, S. E., Lovett, G. M., Richter, P. D., and Johnson, D. W. (1986). Atmospheric deposition and canopy interactions of major ions in a forest. *Science* **231**, 141–145; reprinted with permission of the American Association for the Advancement of Science.

Appreciable amounts of trace elements are deposited on the forest floor from the atmosphere. At various times during the year these may consist of comparable contributions from wet or dry deposits, or may be dominated by either.

The concentration of mineral nutrients usually is higher in cloud water than in rain or snow. Hence, at high elevations, rain combined with clouds often deposit large amounts of ions. In subalpine balsam fir forests in New Hampshire, which are often surrounded by clouds, about 40% of the time ion deposition from clouds greatly exceeded deposition from bulk precipitation (Table 10.6).

Leaching from Plants

As rainfall passes through a tree canopy its concentration of mineral nutrients increases. The increase is the result of leaching of mineral elements from plant tissues as well as washoff of atmospherically deposited materials and plant exudates on canopy surfaces.

Measurable amounts of mineral nutrients are lost from the free space or apoplast of leaves by the leaching action of rain, dew, mist, and fog. The capacity for nutrient losses by foliar leaching varies among species. For example, leaching of mineral nutrients from European ash crowns was substantially greater than from those of silver birch, littleleaf linden, and English oak (Table 10.7) (Hagen-Thorn et al., 2006). Losses also were higher from a deciduous forest in Tennessee than from loblolly and shortleaf pine stands (Luxmoore et al., 1981). Leaves with a thick waxy surface are wetted and leached with difficulty, and senescing leaves are more readily leached than young leaves. By eroding the leaf cuticles (Chapter 5, Kozlowski and Pallardy, 1997), acid rain may increase the rate at which

TABLE 10.5. Examples of Chemical Loading Rates (kg ha^{-1} year^{-1}) in Temperate and Tropical Ecosystems[a]

Site	Total insoluble particulates	Organic carbon			Phosphorus			Nitrogen				Soluble cations				Soluble anions		
		Total carbon	POC	DOC	Total P	DOP	PO_4^{3-}	Total N	DON	NH_4^+	NO_3^-	Ca^{2+}	Mg^{2+}	Na^+	K^+	HCO_3^-	Cl^-	SO_4^{2-}
								Temperate ecosystems										
North America																		
New Hampshire forest. Hubbard Brook				32.0	0.06		0.04			2.26	4.30	2.20	0.6	1.6	0.9	0	6.2	38.0
Colorado Mountains, Como Creek	115	32.8	23.9	8.9	0.26	0.03	0.04	4.80	0.99	1.04	1.62	3.35	0.40	1.59	1.27	13.8	1.69	7.8
								Tropical ecosystems										
South America																		
Lake Valencia	139	33.0	19.6	13.4	1.68	0.20	0.30	7.45	1.33	2.43	1.28	8.60	5.47	16.6	4.28	50.0	19.2	16.3
Amazon Basin, Brazil					0.27		0.04	9.95		3.15	2.52	3.59	3.00				11.2	16.8
Amazon Basin, Venezuela										21.4		27.7	3.4		24.2			136
Africa																		
Uganda and west coastal zone					1.20			19.1		6.6	4.9	2.44	2.91	21.7	21.6		11.4	22.8

[a]Modified from Lewis (1981). *Water Resour. Res.* 17, 169–181, copyright by the American Geophysical Union. POC, Particulate organic carbon; DOC, DOP, and DON, dissolved organic carbon, phosphorus, and nitrogen.

TABLE 10.6. Annual Deposition of Ions by Clouds and Bulk Precipitation at an Elevation of 1,220 m in a Balsam Fir Stand in New Hampshire[a]

Ion	Cloud deposition (kg ha^{-1} year^{-1})	Bulk precipitation (kg ha^{-1} year^{-1})	Percentage of sum contributed by clouds
H^+	2.4	1.5	62
NH_4^+	16.3	4.2	80
Na^+	5.8	1.7	77
K^+	3.3	2.1	61
SO_4^{2-}	137.9	64.8	68
NO_3^-	101.5	23.4	81

[a]Reprinted with permission from Lovett, G. M., Reiners, W. A., and Olson, R. K. (1982). Cloud droplet deposition in subalpine fir forests: Hydrologic and chemical inputs. *Science* 218, 1303–1304; reprinted with permission of the American Association for the Advancement of Science.

TABLE 10.7. Nutrient Amounts Leached from the Crowns of Different Species (mg/m² stand Leaf Area)[a]

	Birch	Ash	Oak	Linden
K (mg/m^2)	63.82	209.79	68.48	108.83
K (%)	*9.8*[b]	*29.1*	*8.3*	*11.3*
Ca (mg/m^2)	8.06	138.84	5.94	17.42
Ca (%)	*1.2*	*9.1*	*0.9*	*1.8*
Mg (mg/m^2)	9.42	52.72	2.62	4.06
Mg (%)	*4.1*	*14.1*	*2.0*	*3.9*
N (mg/m^2)	16.02	72.63	7.09	6.00
N (%)	*0.9*	*4.2*	*0.4*	*0.4*
S (mg/m^2)	2.89	66.82	4.09	7.55
S (%)	*2.6*	*18.8*	*3.3*	*5.3*
P (mg/m^2)	17.60	12.93	7.86	4.33
P (%)	*7.5*	*9.6*	*4.8*	*3.5*
Mn (mg/m^2)	5.16	0.47	0.84	1.48
Mn (%)	*3.3*	*12.8*	*1.4*	*2.5*
Fe (mg/m^2)	0.18	0.02	0.18	0.07
Fe (%)	*2.8*	*0.3*	*2.4*	*1.1*
Cu (mg/m^2)	0.00	−0.02	−0.01	0.02
Cu (%)	*−0.5*	*−2.9*	*−2.2*	*3.2*

[a]This table was published in *Forest Ecology and Management*, Vol. 228, Hagen-Thorn, A., Varnagiryte, I., Nihlgard, B., and Armolaitis, K. Autumn nutrient resorption and losses in four deciduous forest tree species, pp. 33–39. Copyright Elsevier 2006.

[b]Numbers in italics show the leached amounts as percentage of the total nutrient amounts in the crowns at the end of August.

cations are leached from foliage because leaching involves cation exchange reactions; with hydrogen ions in rainwater replacing cations on binding sites in the leaf cuticle and epidermis (Lovett et al., 1985). It also has been claimed that air pollutants such as O_3 can damage membranes of leaf cells, causing leakage of cellular contents, which then may be leached out of leaves by subsequent rains. In contrast to abundant loss of minerals by leaching from pollution-affected leaves, there are relatively small nutrient losses by leaching from healthy leaves.

Throughfall and Stemflow

The rain that falls on forests is partitioned into interception, throughfall, and stemflow. Rain that passes through tree crowns (throughfall) and water moving down the stem (stemflow) carry combined nutrients that originate outside the system (e.g., dry deposition between storms) together with nutrients from plants (e.g., leachates, exudates, and decomposition products). Hence, the amount of mineral nutrients that reaches the forest floor exceeds the amount received in incident precipitation. Amounts of minerals entering the soil in throughfall and stemflow compared with that in the incoming precipitation are usually elevated for most nutrients, and especially so for K (Parker, 1983; Duchesne et al., 2001; Tobón et al., 2004) (Fig. 10.5). There is some variation among forest ecosystems in specific patterns, however. For example, although ammonium and nitrate concentrations in throughfall and stemflow of Amazonian tropical rainforests were elevated (Tobón et al., 2004), throughfall collected under boreal Canadian black spruce forests was depleted in both forms of N (Morris et al., 2003). The latter authors suggested that microbial immobilization within the crown or capture of N by epiphytic lichens may have been responsible for N depletion in the black spruce ecosystem.

Of the water reaching the forest floor up to 85% is throughfall and 0 to 30% is stemflow (Parker, 1995). On an annual basis, throughfall often accounts for up to 90% of the nutrients released by leaching from plants. Nevertheless, the return of nutrients to the forest floor by stemflow is important because it deposits a relatively concentrated solution at the base of the stem (Tobón et al., 2004). Estimates of the width of the area around the stem affected by stemflow vary from 0.3 to 5 m.

Monovalent cations (e.g., Na^+, K^+) are readily leached from leaves and hence are transferred to the soil, primarily by throughfall. By comparison, divalent cations (e.g., Ca^{2+}, Mg^{2+}), which are more strongly bound, are transferred to the soil primarily through the shed leaves. Although decomposition of litter supplies very large amounts of mineral nutrients to the forest floor, such nutrients are released slowly by decomposition of organic matter, whereas nearly all throughfall and stemflow nutrients are in solution and immediately available for absorption by roots.

Deposition of mineral nutrients by combined throughfall and stemflow differs with season, amount, timing and type of precipitation, forest type, stand age, spacing of trees, soil fertility, soil type, and sources of deposits. Snowfall, especially dry cold powder, results in less foliar leaching (hence less nutrient flux) than does rainfall. Nutrient fluxes in throughfall and stemflow are much higher in tropical forests than in temperate forests. Annual nutrient depositions generally are higher in broadleaved stands than in pine stands, despite a shorter growing season (associated with deciduousness) of the former (Cole and Rapp, 1981). An example of the increase in mineral nutrients in throughfall and stemflow over that in incident precipitation is shown in Table 10.8.

TABLE 10.8. Mean Ratios of Solute Concentrations (μmol L^{-1}) in throughfall (Th) and stemflow (St) versus Those in Rainfall and Their Standard Deviation (n = 35) in Four Forest Ecosystems in the Colombian Amazonia, and Rainfall (Pg). Solute Concentrations in μmol L^{-1} (except for pH). DOC values are expressed in μmol L^{-1}. Values larger than 1.0 Indicate Element Enrichment, and Values below 1.0 μmol L^{-1} Indicate Depletion.[a]

		Sedimentary plain			High terrace		Low terrace		Flood plain	
		Pg	Th	St	Th	St	Th	St	Th	St
pH	Mean	5.03	1.03	0.87	1.09	0.93	1.05	0.93	1.1	0.94
	SD	0.3	0.06	0.06	0.11	0.24	0.1	0.1	0.08	0.12
H	Mean	13.6	0.9	5.7	0.8	6.0	1.2	4.2	0.5	3.8
	SD	12.0	0.7	4.3	1.4	8.6	2.3	5.7	0.4	4.6
K	Mean	9.0	3.6	5.5	5.0	7.6	4.9	6.4	4.6	8.7
	SD	7.5	2.1	2.6	4.8	7.2	4.0	5.5	2.9	5.7
Na	Mean	20.0	1.2	1.4	1.4	1.5	1.4	1.6	1.3	1.6
	SD	7.3	0.7	0.8	0.6	0.7	1.0	1.1	0.5	0.9
NH_4	Mean	11.5	2.4	5.8	3.1	5.4	2.8	4.5	2.8	8.1
	SD	8.4	2.4	8.3	5.3	7.3	4.3	7.3	2.8	19.4
Ca	Mean	6.8	1.3	2.2	1.4	2.9	1.5	2.0	2.2	4.9
	SD	2.3	0.5	2.6	0.9	3.9	1.1	1.0	1.4	3.3
Mg	Mean	2.8	2.9	3.7	2.7	6.1	3.2	6.7	5.1	11.2
	SD	1.7	1.5	1.9	1.5	4.6	1.6	3.9	3.0	6.7
Fe	Mean	0.7	1.5	4.8	1.9	6.7	2.1	3.7	1.7	7.1
	SD	0.4	0.7	5.5	2.0	15.4	1.9	6.2	1.4	7.4
Mn	Mean	0.3	3.0	3.4	0.5	2.2	2.1	6.7	1.7	7.3
	SD	0.4	3.0	4.4	0.5	1.9	1.8	12.1	2.7	6.6
Si	Mean	2.5	2.9	12.8	2.7	4.3	4.1	14.7	3.0	8.3
	SD	1.5	2.7	13.2	1.9	3.8	4.5	16.2	2.3	5.3
Cl	Mean	25.4	0.9	1.6	1.5	2.4	1.4	1.8	1.6	2.3
	SD	12.1	0.4	0.8	1.1	2.7	0.8	1.3	1.0	1.1
NO_3	Mean	6.4	3.7	5.2	2.6	8.1	5.2	7.4	6.1	11.4
	SD	4.0	4.7	4.7	3.9	21.7	8.4	13.7	15.4	22.1
OrthoP	Mean	0.3	3.3	5.1	2.4	6.0	3.7	3.7	5.1	8.0
	SD	0.3	6.2	8.9	2.4	7.2	5.3	5.7	12.9	12.7
SO_4	Mean	36.9	1.6	2.8	2.1	2.8	2.3	2.7	1.9	3.4
	SD	25.1	0.8	2.0	1.8	2.4	2.2	2.3	1.9	2.6
Cations	Mean	62.9	1.5	3.3	1.7	3.1	1.8	2.8	1.9	4.2
	SD	24.2	0.6	1.2	0.9	1.3	0.9	1.6	0.9	1.6
Anions	Mean	73.7	1.3	2.1	1.6	2.3	1.7	2.0	1.7	2.8
	SD	29.3	0.5	1.1	0.9	1.1	0.9	0.9	0.8	1.3
DOC	Mean	328.3	1.4	2.7	1.7	2.7	1.7	2.5	1.6	4.7
	SD	135.1	0.5	1.0	1.0	1.3	0.7	1.2	0.6	2.7

[a]From *Biogeochemistry*, Tobón, C., Sevink, J., and Verstraten, J. M. (2004). Solute fluxes in throughfall and stemflow in four forest ecosystems in northwest Amazonia. **70**, 1–25, Table 1. © 2004 with kind permission from Springer Science and Business Media.

Weathering of Rock and Minerals

Except for N, most of the soil mineral pool is derived from weathering of rocks and minerals. For example, 80 to 100% of the input of Ca, Mg, K, and P to forest ecosystems results from such weathering. Minerals may be released by weathering of parent bedrock or of transported materials such as glacial deposits, volcanic ash, windborne soils, or streamwater alluvium (Waring and Running, 1998). Weathering of rock may be a physical as well as a chemical process (Birkeland, 1984).

Several processes are involved in physical weathering, including unloading by erosion, expansion in cracks by freezing water or crystallizing salts, fire, thermal expansion and contraction of minerals, and rupture of rocks by growing roots. Processes of chemical weathering include simple dissolution of minerals, carbonation, oxidation, and hydrolysis (Birkeland, 1984). Important roles in chemical weathering have been ascribed to organic acids released by roots (Boyle et al., 1974), phenolic acids released by lichens (Tansey, 1977; Ascaso et al., 1982), and oxalic acids released by fungi (Fisher, 1972; Cromack et al., 1979). Birkeland (1984) and Waring and Running (1998) provide more detailed reviews of processes of weathering of rocks and release of mineral nutrients.

Decomposition of Organic Matter

Shed plant organs and tissues (consisting primarily of leaves and twigs as well as decaying roots and mycorrhizae) add large amounts of organic matter to the forest floor and soil. A small portion of the organic matter of the forest floor consists of epiphytic matter (e.g., live and dead vascular and nonvascular plants, microbes, invertebrates, and fungi) (Coxson and Nadkarni, 1995). Litter production typically is higher in most tropical forests than in temperate forests (Table 10.9). Broadleaved and coniferous forests of the temperate zone produce about the same amount of litter per unit of land area, but broadleaved litter generally contains a higher concentration of mineral nutrients. It has been estimated that deciduous trees lose about 85% of their annual uptake of mineral nutrients in litterfall, whereas conifers lose 10 to 25% depending on species (Monk, 1971). The less efficient internal cycling of nutrients in deciduous trees is associated with short internal turnover times for mineral nutrients and lower C gain per unit of nutrient turned over. Deciduous trees typically gain less than half as much C per gram of N turned over as evergreens do (Small, 1972; Schlesinger and Chabot, 1977).

Variable amounts of different mineral elements are returned to the soil by decaying organic matter. The relative abundance (mass basis) of mineral nutrients in the litter of a mature broadleaved forest in New Hampshire was N > Ca > K > Mn > Mg > S > P > Zn > Fe > Na > Cu. Nitrogen, Ca, and K accounted for 80.6% of the total; Zn, Fe, Na, and Cu for only 0.8%. The overstory, shrub, and herbaceous layers supplied 96.6, 1.7, and 1.6%, respectively, of the nutrient mass (Gosz et al., 1972).

TABLE 10.9. Litter Production by Various Forest Types[a]

Forest type	Forest floor mass ($kg\ ha^{-1}$)	Litter fall ($kg\ ha^{-1}\ year^{-1}$)
Tropical broadleaf		
Deciduous	8,789	9,438
Evergreen	22,547	9,369
Warm temperate broadleaf		
Deciduous	11,480	4,236
Evergreen	19,148	6,484
Cold temperate broadleaf		
Deciduous	32,207	3,854
Cold temperate needleleaf		
Deciduous	13,900	3,590
Evergreen	44,574	3,144
Boreal needleleaf		
Evergreen	44,693	2,428

[a]From Vogt et al. (1986).

The amount of mineral nutrients returned to the soil by death and decay of roots and mycorrhizae may exceed the return by leaf and twig litter. In a Pacific silver fir forest, most return of mineral nutrients to the soil was attributed to turnover of fine roots and mycorrhizae (Vogt et al., 1986). Return of mineral nutrients to the soil by mycorrhizae is particularly high. In a Douglas-fir forest the return of N, P, and K by mycorrhizae was 83 to 85% of the total tree return, and 25 to 51% of the return of Ca and Mg. The return of N, P, and K by mycorrhizae was four to five times greater than that by roots, nearly equal for Ca, and three times less for Mg (Fogel and Hunt, 1983). Mineral nutrient fluxes of Ca, Mg, and K from root turnover in a young sweetgum plantation exceeded those from aboveground litterfall, and litterfall contributed more N and P (Johnson et al., 2004). In contrast, fine root turnover was less important in releasing Ca and Mg than litterfall decomposition in a northern hardwood ecosystem (Burke and Raynal, 1994). In the latter system, root turnover contributions of N, P, K, and S were similar to those from litterfall. The return of minerals to the soil by fine roots may be expected to vary greatly because of differences in their rates of turnover in various tree stands. A greater proportion of total net primary productivity (TNPP) is allocated to fine roots in stands on poor sites than on good sites or as trees age (Keyes and Grier, 1981). Whereas fine roots contributed only 5% of TNPP in a fast-growing pine plantation on a fertile site (Santantonio and Santantonio, 1987), they accounted for 68% of the TNPP in a mature subalpine forest (Grier et al., 1981).

Because decomposition of litter is carried out by soil microorganisms, factors that control microbial activity regulate the rate of litter decomposition and release of mineral nutrients.

Temperature

Activity of microorganisms increases exponentially with increasing temperature. Because of slow litter decomposition in cool temperate forests, a much larger proportion of the minerals in the soil-plant system are present in the soil and litter than in tropical forests. In fact, in cool climates the accumulated litter contains so much of the total mineral pool that mineral deficiency in plants sometimes results (Chapter 9).

In Alaska, very low productivity of a black spruce forest was attributed to slow decomposition of litter and slow release of mineral nutrients (Van Cleve et al., 1983). Van Cleve et al. (1990) heated (to 8 to 10°C above ambient temperature) the forest floor of a black spruce stand in Alaska that had developed on permafrost. The elevated temperature increased the rate of litter decomposition and was followed by increased N and P concentrations in the forest floor, higher concentrations of N, P, and K in black spruce needles, and subsequently in increased rates of photosynthesis.

In the humid tropics litter decomposes rapidly, hence little organic matter accumulates on or in the soil. Zinke et al. (1984) showed that accumulation of organic matter on the forest floor was lower in the tropics, despite higher productivity, than in temperate regions under similar moisture regimes. The effect of temperature also is evident in slow decomposition of litter at high altitudes. In Malaysia, for example, accumulation of organic matter increased rapidly at altitudes between 5,000 and 5,600 ft (1525–1707 m), corresponding to a decrease in mean annual temperature from 65 to 63°F (18.3 to 16.6°C) (Young and Stephen, 1965).

Water Supply

In both temperate and tropical regions with seasonal rainfall, most litter decomposition occurs during the wet season. For example, in northwestern United States most conifer litter decomposes during the cool wet season and very little during the dry summer (Waring and Franklin, 1979). Decomposition of litter in a Nigerian rain forest was more than 10 times as high in the wet season as in the dry season (Swift et al., 1981).

Chemical Composition of Litter

The rate of litter decay and release of nutrients are regulated by the chemical composition of organic matter. Whereas most carbohydrates and proteins in litter degrade rapidly, cellulose and lignin decompose slowly. Phenolic compounds, which slow the rate of litter decomposition, are readily leached from litter by rain but bind with proteins to form a resistant complex (Schlesinger and Hasey, 1981). Hence, the overall effect of a high level of phenolic compounds in litter is to reduce the rate of turnover of organic matter. Plants on poor sites, in particular, produce large amounts of phenolic compounds, leading to low rates of turnover of soil organic matter (Chapin et al., 1986; Horner et al., 1988).

Litter that is rich in mineral nutrients decomposes faster than nutrient-poor litter. For example, the N in litter accelerates its decomposition (Fig. 10.6). As the ratio of C to N in litter decreases, the rate of decomposition increases. However, decomposition of Douglas-fir litter was influenced more by its lignin content than by its C:N ratio (Fogel and Cromack, 1977).

Wide variations have been reported in decomposition rates of various litter constituents. During the first year, the amount of sugars, steryl esters, and triglycerides in Scotch pine needle litter decreased greatly. Some isoprenoid alcohols, sterols, and acids were the most stable soluble components. Among the solid residue, the arabinans decomposed rapidly and lignin decomposed very slowly (Berg et al., 1982).

The specific effects of lignin and mineral nutrients on decomposition of litter apparently vary over time. For example, high nutrient levels accelerated decomposition of litter only initially. In later stages of decomposition, the lignin level had a retarding effect that apparently superseded the effect of nutrient level as decomposition continued (Berg et al., 1982).

The decomposition rate varies appreciably among litters of different species of plants in the same ecosystem. Generally the N-rich litter of broadleaved trees decomposes faster than the litter of conifers. However, there are differences in rates among species within each group. Macronutrients were released from decomposing litter of four species in Wisconsin in the following order (fastest to slowest): trembling aspen > northern pin oak = paper birch > jack pine (Bockheim et al., 1991). Leaf litter of red alder decomposed fastest followed by litters of Douglas-fir, western hemlock, and Pacific silver fir (Edmonds, 1980). In an oak-conifer

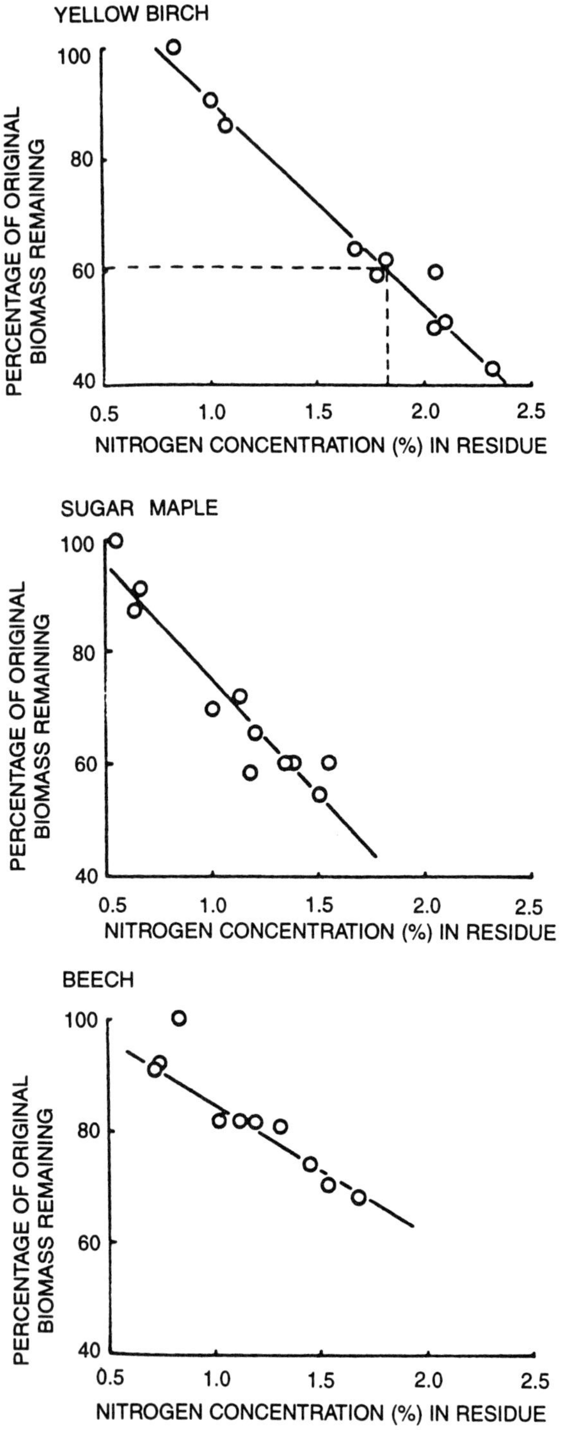

FIGURE 10.6. Percentage of the original leaf litter remaining as a function of its nitrogen concentration. Data are from Gosz et al. (1973), as replotted by Aber and Melillo (1980).

forest in Himalaya, the leaves of *Daphne cannalina* decomposed completely within six months whereas only 72% of *Cupressus torulosa* leaves decomposed in 18 months (Pandey and Singh, 1982). Mixed-species leaf litter may have different decomposition rates than mixed litter that has been sorted into single species components. In a synthesis of this literature, Gartner and Cardon (2003) concluded that leaf litter mixtures often show elevated mass loss rates and higher nutrient concentrations than separated litter of constituent species.

Exudation from Roots

Experiments with radioactive tracers show that labeled mineral nutrients supplied to the leaves often move in the phloem to the roots where small amounts leak out into the rhizosphere. Loss of nutrients by root exudation occurs largely in the region of meristematic activity behind the root tip and in the region of cell elongation. In addition to mineral nutrients, roots exude a variety of other compounds including carbohydrates (Chapter 7), amino acids, organic acids, nucleotides, flavonones, enzymes, and vitamins.

LOSSES OF MINERAL NUTRIENTS FROM ECOSYSTEMS

Ecosystem Disturbance

Forest ecosystems are subjected to frequent minor disturbances (e.g., surface fires, windthrow, lightning strikes, disease and insect attacks, and partial cuttings) as well as major disturbances (e.g., crown fires, hurricanes, soil erosion, and clear-cutting of trees) (Kozlowski et al., 1991). Changes associated with forest openings that follow disturbances include losses of nutrient capital. Following timber harvesting, reduced transpiration, increases in soil temperature and soil moisture (which accelerate decay of litter and release of nutrients), deposits of logging slash, soil compaction, erosion, and increased nitrification accelerate release of nutrients to drainage waters. However, the amounts of nutrients lost vary from negligible to catastrophic depending on the severity and duration of the disturbance and the forest type. Tropical forests are much more fragile than temperate forests and are more readily depleted of nutrient capital by disturbance.

Temperate Forests

With long rotations and conventional harvesting, involving removal of only some trees and leaving slash behind, only small amounts of mineral nutrients may be depleted from forest stands. Losses generally range from about 1.1 to 3.4 kg ha^{-1} for P, 11.2 kg for N and K, and more for Ca in many stands of broadleaved trees. Such losses generally often can be replaced by weathering of soil minerals, decomposition of organic matter, atmospheric deposits, and N fixation (Kozlowski et al., 1991). However, acidic deposition on certain sites in North America and Europe has resulted in depletion of soil Ca, Mg, and K through export from forest ecosystems in stream and ground water as these elements are exchanged for H^+ on the mineral and organic cation exchange complexes in the soil (e.g., Tomlinson, 2003; Watmough et al., 2005). Removal of all trees from a plantation or natural forest also may be expected to deplete more nutrients than partial cutting does, not only because increased amounts of nutrients are removed in the harvested trees but also because loss of soil minerals to drainage water is accelerated.

Clear-cutting (clear-felling) of temperate forests has variable effects on nutrient budgets depending on the harvested species and genotype, site, and harvest interval. Usually more nutrient capital is removed in harvested trees than is lost to drainage water. When a stand of trees is clear-cut and slash left on the site, smaller amounts of nutrients are lost from the ecosystem than when the slash is removed or burned. With some exceptions, clear-cutting of forests on long rotations often does not seriously deplete the nutrient capital of a site because replacement processes maintain the pool of available nutrients at adequate levels. For example, following clear-cutting of northern hardwood forests in New Hampshire, the combined losses of nutrients in the harvested trees and those leached to streamwater did not exceed 3% of the preharvest nutrient pool (Hornbeck et al., 1990).

The amount of nutrients lost by leaching to streamwater after a forest is clear-cut varies greatly with forest type, with small losses reported from conifer forests of the Pacific Northwest and larger losses from hardwood forests in the northeastern United States. Clear-cutting and burning of slash of old-growth western hemlock–western red cedar–Douglas-fir stands was followed by relatively small losses of nutrients to streamwaters. Nutrient exports were less than 10 kg ha^{-1} yr^{-1} for each of N, P, K, and Mg; less than 20 kg ha^{-1} yr^{-1} for Na and Cl; and less than 30 kg ha^{-1} yr^{-1} for Ca. These amounts were substantially lower than those removed in the harvested logs and lower than the amount lost to drainage water following clear-cutting of hardwood stands in the northeastern United States (Hornbeck et al., 1986), but similar to losses in other regions (Brown et al., 1973; Aubertin and Patric, 1974). Loss of dissolved N by drainage after clear-cutting a Douglas-fir forest was small (less than 2 kg ha^{-1} yr^{-1}) compared with nearly 400 kg ha^{-1} in the harvested tree stems (Sollins and McCorison, 1981). Another study showed that, after a mature Douglas-fir stand was clear-cut, nitrate N was the only constituent that increased substantially in the streamwater. However, the loss of N amounted to less than half the input of N in precipitation to the ecosystem (Martin and Harr, 1989).

The importance of rapid regeneration of clear-cut forest stands in preventing large losses of nutrients by leaching to streamwater is emphasized by studies conducted at the Hubbard Brook Forest in New Hampshire (Bormann and Likens, 1979). When this forest was clear-cut and regrowth prevented by herbicides for several years, both stream flow and loss of mineral nutrients in the drainage waters were dramatically increased. In the first growing season after clear-cutting, the concentration of almost all ions in streamwater rose appreciably. During a three-year period, the average concentrations of ions in streamwater from the devegetated system exceeded those of the forested ecosystem by the following multiples: NO_3^-, 40-fold; K^+, 11-fold; Ca^{2+}, 5.2-fold; Al^{3+}, 5.2-fold; H^+, 2.5-fold; Mg^{2+}, 3.9-fold; Na^+, 1.7-fold; Cl^-, 1.4-fold; and dissolved silica, 1.4-fold. The concentrations of most ions in the streamwater were highest during the second year after clear-cutting but declined during the third year. Over a three-year period, net losses of nutrients were approximately three times larger than those from the uncut forest. Differences in nutrient losses of the devegetated and uncut forests were regulated more by nutrient concentrations in the streamwater than by the amounts of water flowing through the soil.

Whole-Tree Harvesting

There has been much interest in harvesting most of the aboveground parts of trees including wood, bark, and leaves, and sometimes even root systems. Unfortunately, such whole-tree harvesting (WTH) may accentuate nutrient losses from a variety of forest types (Table 10.10). For example, in the pine and hardwood forests of the southeastern United States, nutrient removal by WTH was two to three times higher than by conventional harvesting (Phillips and Van Loon, 1984). In an upland mixed oak forest in Tennessee, WTH increased removal of biomass, N, P, K, and Ca by 2.6, 2.9, 3.1, 3.3, and 2.6 times, respectively, compared to conventional sawlog harvesting. However, WTH after the leaves were shed reduced the potential drain of N, P, K, and Ca by 7, 7, 23, and 5%, respectively, when compared to WTH during the growing season (Johnson et al., 1982a). In Quebec, Canada, WTH of jack pine on a poor site severely depleted N (Weetman and Algar, 1983). In some cases WTH causes increases in leaching of nutrients from cutover sites and into streams. However, the major deleterious effect of WTH is the removal of mineral nutrients in the

TABLE 10.10. Effects of Conventional and Whole-Tree Harvesting on Removal of Nutrients and Biomass from a Red Spruce-Balsam Fir Forest in Nova Scotia[a,b]

Plot	Compartment	Biomass (kg dry weight ha^{-1})	N (kg ha^{-1})	P (kg ha^{-1})	K (kg ha^{-1})	Ca (kg ha^{-1})	Mg (kg ha^{-1})
Conventional	Merchantable stem	105,200	98.2	16.3	91.7	180.9	17.0
Whole tree	Merchantable stem	117,700	120.1	18.2	76.2	218.9	20.4
	Tops, branches, foliage	34,800	119.0	17.0	56.4	117.6	16.5
	Total	152,500	239.1	35.2	132.6	336.5	36.9
	Increase[b]	29.6%	99.1%	93.4%	74.0%	53.7%	80.9%

[a]From Freedman et al. (1981).
[b]Percent increases in biomass or nutrient removals are calculated relative to the merchantable stem values for the whole-tree treatment.

TABLE 10.11. Percentage Increase in Depletion of Mineral Nutrients in Harvested Materials Accompanying a Change from Conventional to Whole-Tree Harvesting[a]

Forest type	% Increase				
	Aboveground biomass	N	P	K	Ca
Hemlock-cedar, <500 years old	43	165	117	77	95
Pine, 125 years old	15	53	54	14	15
Spruce-fir <350 years old	25	116	163	32	50
Hemlock-fir <550 years old	20	86	67	48	48
Spruce, 65 years old	99	288	367	236	179
Cottonwood, 9 years old	—	116	100	74	68

[a]From Kimmins (1977).

harvested trees rather than increased nutrient losses by leaching and runoff (Mann et al., 1988).

The extent to which WTH depletes minerals varies with tree species, tree age, and site (Tables 10.11 and 10.12). The biomass of temperate-zone deciduous trees contains more minerals than the biomass of conifers. Hence, more minerals usually are removed from deciduous than from conifer forests of similar biomass (Phillips and Van Loon, 1984). Whole-tree harvesting is more harmful on infertile than on fertile sites. Leaves usually have the highest concentrations of minerals followed by small roots and twigs, branches, and large roots and stems. Hence losses of mineral nutrients by WTH are smaller for species with small crowns (small leaf biomass) than for those with large crowns (Kozlowski, 1979).

Short Rotations

In recent years considerable interest has been shown in growing closely spaced forest trees on short rotations in order to increase biomass. Unfortunately, more mineral nutrients are depleted from forests on short than on long rotations. It has been estimated that conversion from one 30-year rotation for trembling aspen to three 10-year rotations (all with WTH) will increase depletions of N, P, K, and Ca by 345, 239, 234, and 173%, respectively (Boyle, 1975). Depletion of nutrient capital is faster in Monterey pine plantations than in native stands of *Eucalyptus* in Australia, to a large extent because the pines grow faster and their rotation length is shorter. Harvesting of Monterey pine (40-year rotations) removed approximately 4.5 times more P than harvesting of alpine ash (*Eucalyptus delegatensis*) (57-year rotation). When the rotation of Monterey pine was reduced to 18 years, 5.7 times more P was depleted than by a 57-year *Eucalyptus* rotation (Crane and Raison, 1981). Some effects of the age of trees at harvest, and hence rotation length, on losses of mineral nutrients with conversion from conventional harvesting to WTH are shown in Table 10.13.

The impact of short rotations on soil fertility will vary with the frequency of loss of mineral nutrients. A significant, sudden depletion of mineral nutrients will

TABLE 10.12. Removal of Mineral Nutrients and Biomass by Sawlog Removal with Clear-Cut and Whole-Tree Harvest[a]

Site[b]	SAW (kg ha^{-1})					WTH (kg ha^{-1})				
	SAW[c] biomass (mg/ha)	N	P	K	Ca	WTH[d] biomass (mg/ha)	N	P	K	Ca
Conifers										
Washington High Douglas-fir	281	478	56	225	23	318	728	96	326	411
Chesuncook	155	141	19	121	272	232	410	59	245	537
Washington Low Douglas-fir	134	161	27	81	NA[e]	165	325	56	140	NA
Clemson	85	63	6	35	71	110	123	10	56	111
Florida	58	59	5	20	80	106	110	10	35	138
Hardwoods										
Coweeta	43	58	7	48	130	178	277	41	216	544
Oak Ridge	64	110	7	36	410	175	323	23	128	1090
Cockaponset	121	162	5	108	442	158	273	19	162	530
Washington										
High alder	137	287	41	151	388	147	347	47	174	426
Low alder	111	311	22	122	NA	120	378	27	143	NA
Mt. Success	48	67	4	43	129	111	242	19	128	344

[a]Used with permission of the Society of American Foresters from Mann, L. K., Johnson, D. W., West, D. C., Cole, D. W., Hornbeck, J. W., Martin, C. W., Rieberk, H., Smith, C. T., Swank, W. T., Tritton, L. M., and Van Lear, D. H. (1988). Effects of whole-tree and stem only clearcutting on postharvest hydrologic losses, nutrient capital, and regrowth. *For. Sci.* **34**, 412–428; permission conveyed through Copyright Clearance Center, Inc.

[b]High, high-fertility site; low, low-fertility site.

[c]SAW, sawlog removal with clear-cut.

[d]WTH, above-stump whole-tree harvest. WTH removals were approximately equal to the total stand biomass.

[e]NA, not available.

TABLE 10.13. Effect of Age of Stand on Percentage Increase in Nutrient Losses on Conversion from Conventional to Whole-Tree Harvesting[a]

Species	Age (years)	% Increase N	P	K	Ca
Spruce	18	195	233	161	206
	50	114	115	26	40
	85	91	104	42	29
Pine	39	165	200	140	88
	44	124	133	108	84
	75	77	67	56	59
Pines, average	50	—	156	104	100
	100	—	87	59	52
Other conifers, average	50	—	170	127	138
	100	—	87	56	59
Hardwoods, average	50	—	122	92	67
	100	—	69	47	37

[a]From Kimmins (1977).

only temporarily inhibit tree growth if minerals are replaced by natural processes and the next harvest is delayed. However, nutrients can be progressively depleted by frequently repeated small losses. Hence, the shorter the rotation, the greater is the risk of mineral depletion from a site. Therefore, long rotations may be necessary for WTH on many infertile sites (Kozlowski, 1979).

Fertilizer applications sometimes are useful in plantations harvested on short rotations. Nitrogen-fixing species, such as alders and black locust, can be planted as species to be harvested or to supply N for other species (Hansen and Dawson, 1982; Dickmann and Stuart, 1983). Remedies for correction of mineral deficiencies are discussed in more detail in Chapter 7 of Kozlowski and Pallardy (1997).

Fire

Large amounts of mineral nutrients are lost from forests during fire. For example, as a result of volatilization, ash convection, and subsequent soil leaching and runoff, as much as 50 to 70% of the N and other nutrients may be lost from the ecosystem during hot fires (Wright and Bailey, 1982; Waring and Schlesinger, 1985).

The mineral nutrients in ash have several fates. They may be lost in surface runoff, leached into the soil and held, or leached through the soil profile. By removing vegetation and decreasing the infiltration capacity of soils, fires accelerate losses of nutrients by soil erosion. The amount of nutrients lost by water runoff

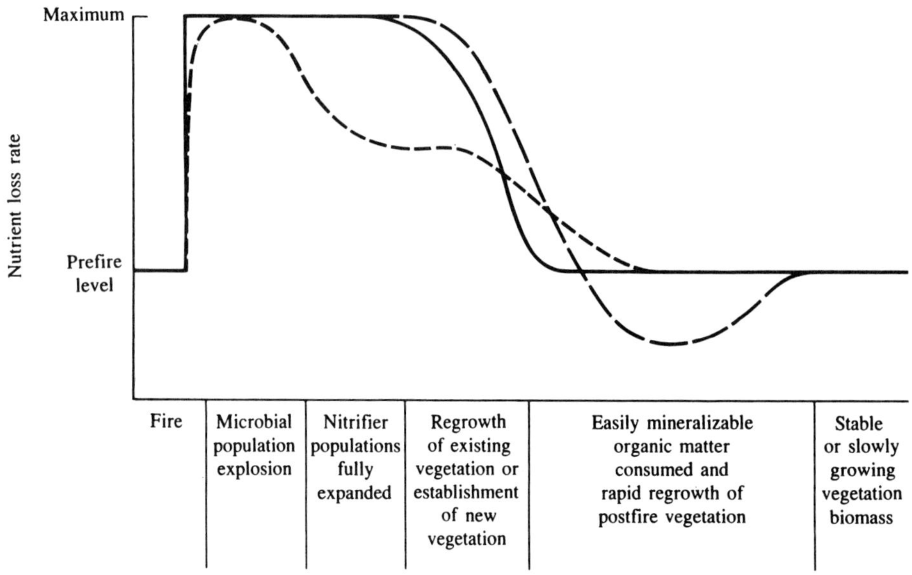

FIGURE 10.7. Hypothetical potential of losses of mineral nutrients from an ecosystem after fire due to wind erosion (—), leaching (— —), and water erosion (- - -). After Woodmansee and Wallach (1981).

depends on slope, amount of ash, soil infiltration capacity, and intensity and duration of rainfall after fire. When little ground cover is present shortly after a fire, losses from the exposed ash often are very high. As plants become reestablished, however, the rate of nutrient loss decreases (Fig. 10.7). Losses to groundwater usually occur in the following order: $K^+ > NH_4^+ > Mg^{2+} > Ca^{2+}$ (Wright and Bailey, 1982).

In many ecosystems the mineral nutrients lost during and shortly after fire by volatilization and runoff are rather rapidly replaced. In the southeastern United States, for example, pine forests have been burned for many years and have not undergone serious losses in fertility (Richter et al., 1982). Often the availability of mineral nutrients to plants is higher within a relatively short time after a fire than it was before the fire occurred (Kozlowski et al., 1991). Such increases may result from leaching of nutrients into the mineral soil, stimulation of microbial mineralization because of greater supplies of soil moisture, and decreased absorption of nutrients by plants as a result of root mortality. Nitrogen lost in combustion commonly is replaced by inputs in rain, increased microbial activity, and N fixation by free-living organisms (Fig. 10.7).

The fire-caused increase in availability of mineral nutrients may or may not improve site quality. If the soluble minerals that are concentrated in the ash are leached into the soil and absorbed by plant roots, site quality is temporarily increased. If, however, the minerals leach below the root zone or are removed by surface water flow, site quality may be lowered especially on sandy soils (Kozlowski et al., 1991).

Tropical Forests

Nutrient losses from undisturbed tropical forests are low (Jordan, 1985). Tropical forests produce a large root biomass that is concentrated near the soil surface. This permits efficient absorption from the soil volume in which the nutrients are concentrated after they are released by decomposing organic matter. It also provides a large surface area on which nutrients can be strongly adsorbed. Intermixing of surface roots with litter and litter-decomposing organisms near the soil surface facilitates nutrient recycling and prevents nutrient losses by leaching from the soil. In addition, mycorrhizal fungi, which are common in tropical forests, attach themselves to decomposing litter and wood, providing a direct pathway for nutrient transport to root systems.

Several other characteristics of tropical forests appear to be involved in nutrient conservation. These include (1) long-lived, tough, and resistant leaves (which prevent breaking of cuticular seals, hence decreasing leaching of minerals), (2) resorption of nutrients from leaves to twigs before the leaves are shed, (3) N fixation and scavenging of nutrients from rainwater by lichens and algae on leaves, (4) production of defensive chemicals against pathogens and herbivores, (5) thick bark that protects trees from invasion by bacteria, fungi, and insects, and (6) storage of a large proportion of the minerals in the biomass, from which they cannot be readily leached, rather than in the soil (Jordan, 1985).

The nutrient balances of tropical forests are much more fragile than those of temperate forests. Nutrient losses from tropical forests range from negligible amounts as a result of minor disturbances such as tree falls to almost total loss of nutrient capital in areas denuded by landslides. Although the effects of disturbances caused by humans are intermediate between these extremes, they often result in serious depletion of mineral nutrients from tropical ecosystems.

Formation of gaps in forests may be expected to increase leaching from the litter layer to the mineral soil because of the increased rates of litter decomposition associated with high temperatures. However, this effect may not be important when the gaps are small, possibly because of nutrient uptake by sprouts, saplings, microbes, and/or new seedlings (Uhl et al., 1988).

On small areas throughout the tropics, forest vegetation is felled, burned, and food crops planted in a system of shifting agriculture (also called "slash and burn" agriculture). The nutrient-rich ash increases the amounts of nutrients available for the first crop but N and S are lost by volatilization. Some of the ash may be blown from the soil surface or leached through the soil by rain. Nevertheless, because availability of soil nutrients is increased by burning, the yield of the first crop usually is high, but declines progressively during subsequent years. The number of crops that can be successfully grown after a tropical forest is cleared will vary with the specific crop, site, and management practices (Kleinman et al., 1995). Eventually fertilizer applications are needed as nutrients in the soil and vegetation become limited. If fertilizers are not added, the unproductive plot usually is abandoned and a new part of the forest is cleared and planted to crops. However, the nutrient capital of a plot can be maintained and soil properties improved by adding fertilizers. Changes in soil chemical properties after eight years of continuous cultivation of crops following clearing of a tropical rain forest were improved by additions of fertilizers and liming (Sanchez et al., 1983).

It is important to separate the effect of shifting cultivation on nutrients in the entire ecosystem from those in the soil compartment. Shifting cultivation results in loss of nutrients from the ecosystem. Nevertheless, nutrients in the soil may show only small changes for some time because nutrients that leach out of the soil are compensated by nutrients that leach into the soil from ash and decomposition of litter. For example, in a slash-and-burn site in an Amazonian rain forest in Venezuela, there was a net loss of nutrients from the ecosystem. Nevertheless, total amounts of nutrients in the soil increased after the burn (Jordan, 1985). Other studies showed that despite progressively decreasing crop yields under shifting cultivation, the nutrient capital in the soil of cultivated fields was as high as or higher than in the soil under undisturbed forest (Nye and Greenland, 1960; Brinkmann and Nascimento, 1973).

The progressive decline in crop yield under slash-and-burn agriculture has been attributed to a decrease in availability of nutrients to plants rather than to low total amounts of soil nutrients. This is emphasized by high productivity of successional species that absorb large amounts of nutrients that are much less available to crop plants. Low availability of N is important in decreasing crop yield on certain sites. Slow mineralization of N may cause N deficiency in plants despite high N levels in the soil.

Nutrients often are depleted early after trees are harvested and at various times thereafter. Ewel et al. (1981) quantified losses of mineral nutrients after cutting and burning of a Costa Rican wet forest. Harvesting of wood removed less than 10% of the total ecosystem nutrients to a soil depth of 3 cm. During drying and mulching (before burning), 33% of the K and 13% of the P were lost. Burning volatilized 31% of the initial amount of C, 22% of the N, and 49% of the S. Only small amounts of C and S were lost after the burn, probably because they had been volatilized during burning. Following the burn and with the onset of rain, losses of nonvolatile elements were high, amounting to 51% of the P, 33% of the K, 45% of the Ca, and 40% of the Mg. Losses of mobile elements including C, N, S, and K in harvested wood and by decomposition of organic matter, burning, and post-burn erosion are shown in Figure 10.8.

Jordan (1985) presented a useful model that summarizes the nutrient dynamics and productivity of tropical ecosystems during disturbances (Fig. 10.9). Mineral nutrients continually enter the ecosystem (a) primarily from precipitation, dry fall, N fixation, and weathering of minerals. Nutrients are concurrently lost (d) by leaching, erosion, and denitrification. In a closed forest (b) gain and loss of nutrients are balanced. Nevertheless, a steady-state is brief because of the impacts of tree fall gaps (c) or more severe disturbances (e.g., wind). Such disturbances release nutrients and make them available to plants although total nutrient stocks change little. When a forest is harvested and slash is burned (e), both N and S are lost by volatilization. Large amounts of macronutrients (Ca, K, Mg) enter the soil. The mineral nutrients in slash and organic matter decompose and are available for absorption by plants. If a cleared forest is used for annual cropping, fruit orchards, pulpwood plantations, or pasture (f, k), initial productivity is high. During cultivation, nutrient stocks are progressively depleted in the harvested crops, as well as by leaching, volatiliza-

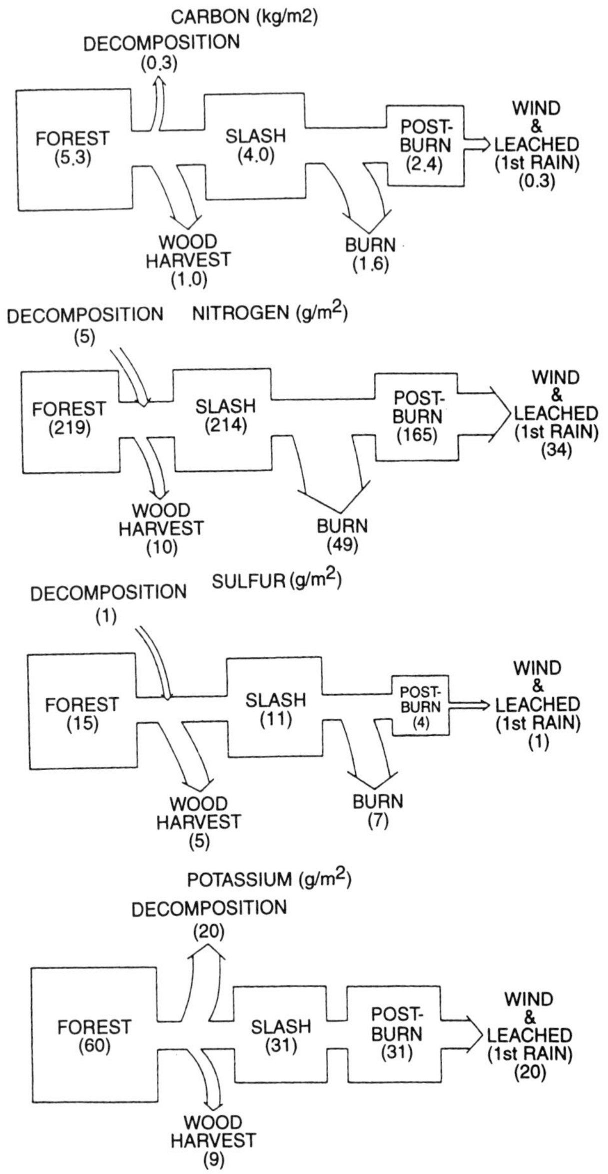

FIGURE 10.8. Storage and losses of carbon, nitrogen, sulfur, and potassium during wood harvesting, decomposition of organic matter, burning, and post-burn erosion in a Costa Rican forest. From Ewel et al. (1981).

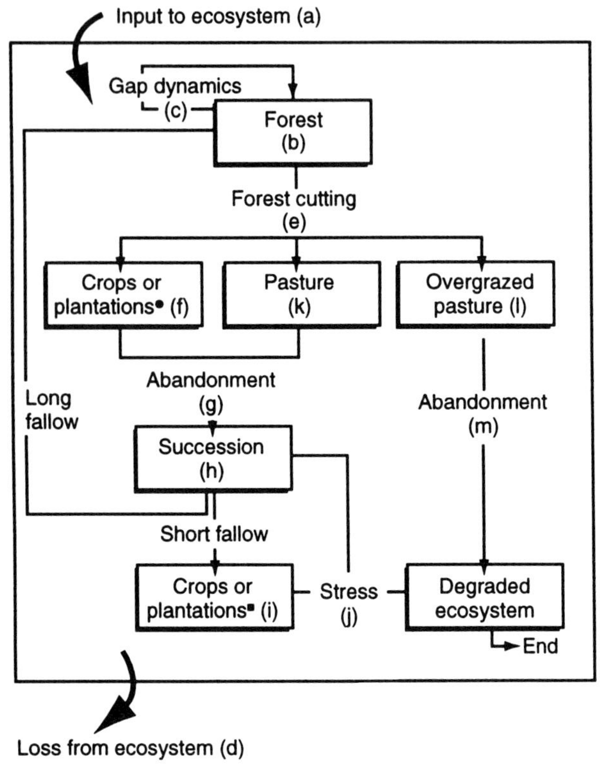

FIGURE 10.9. Model of ecosystem dynamics during disturbances of tropical forests: ●, soil high in organic matter; ■, soil low in organic matter. From Jordan (1985). *Nutrient Cycling in Tropical Forest Ecosystems*. Copyright 1985 John Wiley & Sons. Reprinted by permission of John Wiley & Sons, Ltd.

tion, and fixation. After a short period of cropping (g), it may be possible to restore nutrient stocks by fallowing (h). In parts of the Amazon Basin where replacement of nutrients occurs largely by atmospheric deposition, restoration of soil fertility by fallowing may require a long time. In contrast, in areas in which mineral substrates are only slightly weathered, the fallow vegetation restores nutrient stocks much faster. However, when fallow cycles are very short (i, j), replacement of soil nutrients may be inadequate.

With continual disturbances (l) (e.g., annually burned pasture lands), productivity may be expected to decrease over time because burning depletes soil organic matter and N and also converts Ca and K to soluble forms, thereby increasing their losses by leaching (Jordan, 1985).

Leaching from Soil

The large amounts of mineral nutrients leached from the soil are lost to groundwater and as surface runoff in streams. Nutrients are transported in streams as dissolved ions (largely reflecting chemical weathering) and particulates (primarily from mechanical weathering). The amount of nutrients in streamwater varies with plant species and the extent of ecosystem disturbance. It also varies with streamflow velocity and is high during years of greater discharge. Losses of nutrients in streamwaters are particularly high when snow melts in the spring and winter and little water is lost in evapotranspiration (Waring and Schlesinger, 1985).

Average annual losses of soluble NO_3^-, NH_4^+, PO_4^{3-}, Ca^{2+}, Mg^{2+}, and K^+ from five pine watersheds in Mississippi amounted to 0.32, 3.35, 0.04, 6.21, 3.05, and 3.31 kg ha^{-1}, respectively (Schreiber et al., 1974). In Minnesota more than 96% of the nutrients in surface runoff were transported by snowmelt. Organic N comprised 80% of the total N load in surface runoff; organic plus hydrolyzable P accounted for 45% of the total P load. Cations in surface runoff varied in the following order: $Ca^{2+} > K^+ > Mg^{2+} > Na^+$ (Timmons et al., 1977).

Maximum rates of leaching of nutrients from the forest floor occur from loblolly pine litter with low intensity rainfall and high temperatures. The concentrations of nutrients in the leachate peaked in the first 1.5 mm of rainfall after runoff began and stabilized after approximately 15 mm of rainfall fell (Duffy et al., 1985). Another study showed that the concentrations and amounts of most nutrients leached were greatest when subjected to low rainfall intensity. Except for NO_3^-, the leaching of nutrients was not related to rainfall intensity because the initially low levels of nutrients were rapidly removed with small amounts of simulated rain. The concentration of nutrients in the leachate increased rapidly to a maximum and then decreased to a constant value (Schreiber et al., 1990).

When the forest floor was exposed to progressively higher temperatures, leaching of nutrients increased with an increase in rainfall temperature. For example, with higher temperatures the percolate showed increased leaching losses of K^+, Mg^{2+}, and NH_4^+. However, losses of Al did not increase correspondingly, indicating a lag in mobilization and leaching of Al (Cronon, 1980). Increasing the temperature from 8 to 35°C nearly doubled the concentration of mineral nutrients in the soil leachate (Table 10.14). The accelerated leaching at the higher temperature was attributed to greater penetration of cell wall components by rainfall and solubilization of plant nutrients (Duffy and Schreiber, 1990).

TABLE 10.14. Effect of Temperature on Nutrient Losses ($mg\ m^{-2}$) in Leachate When 25.4 mm of Simulated Rain Was Applied to Soil Litter[a]

Treatment (°C)	PO^{3-}	NH_4^+	NO_3^-	Total organic carbon
8	18.10 ± 0.14	23.37 ± 0.79	2.26 ± 0.11	1,319 ± 7
23	22.88 ± 2.03	43.23 ± 1.54	2.36 ± 0.20	1,725 ± 101
35	31.31 ± 0.42	49.44 ± 2.89	2.51 ± 0.16	2,583 ± 17

[a]Used with permission of the Society of American Foresters from Duffy, P. D., and Schreiber, J. P. (1900). Nutrient leaching of a loblolly pine forest floor by simulated rainfall. II. Environmental factors. *For. Sci.* **36**, 777–789; permission conveyed through Copyright Clearance Center, Inc.

ABSORPTION OF MINERAL NUTRIENTS

Absorption of mineral nutrients is as important to the growth of plants as is absorption of water but is not as well understood to a large extent because it is more complex. Nutrients are absorbed into roots as ions dissolve in water. Absorption of nutrients by plants involves several steps including (1) movement of ions from the soil to root surfaces, (2) ion accumulation in root cells, (3) radial movement of ions from root surfaces into the xylem, and (4) translocation of ions from roots to shoots. These steps are discussed in more detail by Marshner (1995), Tinker and Nye (2000), and Epstein and Bloom (2005).

Terminology

As used here absorption and uptake are general terms applied to the entrance of substances into cells or tissues by any mechanism. Accumulation refers to the concentration of a specific substance within cells or tissues against a gradient in electrochemical potential or concentration, requiring expenditure of metabolic energy. Movement of materials against gradients of concentration or electrochemical potential brought about by the expenditure of metabolic energy is called active transport, in contrast to passive transport by diffusion along concentration gradients or mass flow caused by pressure gradients, such as the flow of water into roots and upward in the xylem of transpiring plants. However, the discovery of ion channel proteins that facilitate "passive" movement of minerals across membranes blurs the implicit lack of metabolic control over this mode of transport, as channels often open and close under metabolic influences (e.g., phosphorylation-dephosphorylation state of the channel protein, membrane potential changes on gated channels).

Accumulation of ions can be detected only behind relatively impermeable membranes, because substances leak out through permeable membranes by diffusion as rapidly as they are moved in by active transport. A membrane can be defined as a boundary layer that differs in permeability from the phases it separates. Membranes that permit some substances to pass more readily than others are termed differentially permeable, or less accurately, semipermeable. Cell membranes include those surrounding organelles such as nuclei and plastids, and the inner and outer boundaries of the cytoplasm, the vacuolar membrane or tonoplast, and the plasmalemma. In addition, multicellular structures such as the epidermis and endodermis and the bark play important roles in uptake and retention of various substances by plants. Some investigators treated the entire cortex of young roots as a multicellular membrane, but the endodermis usually is regarded as the critical membrane in roots with respect to the entrance of water and solutes because the Casparian strips on its radial walls render them relatively impermeable to water and solutes. The importance of the endodermis probably has been overemphasized, however, as shown later.

Absorption of ions by roots occurs in two steps. Step 1 is passive movement through all or part of the cortical free space or apoplast. This is the part of the root tissue that is penetrated by ions without crossing a living membrane. In the apoplastic pathway, ions can move through the free space in the cortex to the endodermis. Step 2 is active absorption through the plasmalemma of epidermal and cortical cells (Haynes, 1986; Barber, 1984). After passing through the endodermal cells in the symplast (protoplasts of adjacent cells connected by plasmodesmata), ions eventually are freed into the xylem sap and move upward in the transpiration stream to the leaves. There they move out of the xylem of the leaf veins into the leaf cell walls that comprise the free space. From this solution solutes are selectively accumulated by leaf cells. That most anions and most of the essential cations enter cells actively is shown by elimination of uptake by metabolic inhibitors or anaerobic conditions. The presence of appreciable amounts of ions in the free space of leaves explains the leaching of mineral nutrients from leaves and absorption of fertilizers applied to the foliage.

Movement of ions in free space is nonselective, reversible, and independent of metabolism. By comparison, uptake of ions by plant cells is relatively selective, nonreversible, and depends on metabolic activity. For example, some ions accumulate in cells to much higher concentrations relative to the external concentration than others, and it was shown many years ago that the ion content of cell sap is very different in composition from that of the medium in which the plants are growing. Thus absorption of ions by cells is largely controlled by a selective active transport mechanism. On the other hand, practically all ions present in the root medium are found in varying quantities in the shoots of plants, indicating that the ion barriers in roots are leaky.

The relationship between root nutrient ion uptake and concentration often indicates the presence of both high and low affinity uptake mechanisms, with the former dominating at low external nutrient concentrations and low affinity mechanisms becoming more active as concentrations increase (Wang et al., 2006). Nutrient uptake by both mechanisms often shows saturation kinetics typical of enzyme activities. Alternatively, the low affinity mechanism exhibits a complex "multiphasic" pattern that suggests the involvement of multiple transmembrane proteins (Epstein and Bloom, 2005). Active transport of ions across cell plasma membranes and accumulation in vacuoles involves a variety of membrane proteins (Sanders and Bethke, 2000). Pumps link the transport of ions (e.g., H^+, Ca^{2+}) across membranes directly to hydrolysis of ATP. Carrier proteins couple the energetically downhill flow of ions, such as H^+ created by proton pumping, to movement of ionic solutes across membranes. Carriers that conduct parallel flow of ions in the same direction (e.g., H^+ and K^+) are termed symporters; those that conduct in the opposite direction are called antiporters. An individual carrier protein transports about 10^3 solute molecules per second (Epstein and Bloom, 2005). Ion transport across membranes also involves the activity of channel proteins that provide a more-or-less conformationally specific pore through which ions can pass in response to concentration differences or electrochemical driving forces. Channels cannot be viewed as simple pipes, however, as chemical modifications and voltage changes in their immediate vicinity can induce conformational changes that open and close the pores.

In contrast to the transport capacity of carrier proteins, channels can pass up to 10^8 ions per second. Doyle et al. (1998) described the K^+ channel of prokaryotic organisms, which is essentially identical to that in plants (and all cellular organisms). The structure and function of this channel elegantly reveals how the charged K ion is stripped of its hydrating water molecules upon entering the channel protein, is sifted through a selectivity filter that places the carbonyl oxygen atoms of amino acid residues in strategic positions that both substitute for water and create an energetically unfavorable environment for competing ions (e.g., Na^+), and then is freed from the attraction of the filter by the repulsive forces of the next K^+ ion queuing up in the channel pore. Parenchyma cells in the stele likely release both anions and cations into the xylem through regulated channel proteins down concentration and electrochemical gradients (Véry and Sentenac, 2003; Roberts, 2006).

Wang et al. (2006) summarized the literature concerning root nutrient uptake mechanisms. Potassium, Mg^{2+}, and Ca^{2+} uptake was associated with both channel protein activity and proton driven antiporters. Chloride and NO_3^- uptake takes place by both the activities of channel proteins and H+/antiporter carrier proteins. Phosphate, SO_4^{2-}, and MoO_4^- uptake appears to be predominantly mediated by H+/anion symporter proteins. Uptake of NH_4^+ is associated with channel proteins. Micronutrient metals may enter the symplast via both metal/H^+ antiporters and a relatively nonselective cation channel. In B-adequate soils, the mechanism of uptake of B may be simple diffusion across membranes without channel protein facilitation. However, there is evidence of an active carrier-mediated process when B supply is low. It is worth noting that ion transport proteins also mediate efflux of ions from the symplast and so net uptake will be an integration of uptake and efflux activities of a suite of processes and proteins. Multiple gene families exist for many transport proteins and gene expression and protein function may vary among plant tissues and across species (Wang et al., 2006).

Ion Movement in Soil

Soils typically contain macronutrients in very dilute solutions (10^{-6} to 10^{-3} M). Such concentrations often are too low to supply the mineral requirements of plants. Mineral nutrients are made available to roots by interception of soil nutrients present at the soil-root interface as well as by ion movement toward roots by diffusion and by mass flow of water. Root hairs near the root tips can substantially extend the zone of root absorption radially in the soil (Tinker and Nye, 2000). The relative importance of diffusion and mass flow varies with the kind and concentration of ions in the soil solution, the rate at which ions are being accumulated by roots, and the rate of water flow to the roots.

Where nutrients immediately around the roots are depleted diffusion becomes the dominant mode of ion transport from the soil solution to the root. As plants continue to absorb nutrients and water, gradients are established in the soil water potential (Ψ) and nutrient concentration. Hence, ions subsequently diffuse along the gradient to the root surface. However, soil nutrients move by diffusion to the roots for distances of only 0.1 to 15 mm (Epstein, 1972). The amounts of individual nutrients available to roots by diffusion vary appreciably because of differences among ions in binding to the soil. For example, P is strongly adsorbed to the soil and nitrate is not.

Highly mobile elements (e.g., Mg) move readily to the root surface by both mass flow and diffusion and often accumulate near the roots when the supply exceeds the amount absorbed. The concentrations of N, P, and K often are so low in the soil solution that mass flow of water provides only a small amount of a plant's needs. Hence, most of these elements move to the root surface by diffusion (Chapin, 1980).

The Absorbing Zone

Most measurements of absorption of mineral nutrients and water have been made with young unsuberized roots. During secondary growth of roots, the epidermis, cortical parenchyma, and endodermis are lost (Chapter 3) and the outer phloem becomes suberized. Such roots often have been considered to be impermeable. However, significant amounts of minerals and water are absorbed at some distance from the root tip through the suberized portions of roots. Entry, probably by both mass flow and diffusion, occurs through lenticels, tissues between plates of bark, and openings created by the death of small roots.

When the soil is cold or dry, few or no unsuberized roots can be found (Kramer, 1983). In midsummer less than 1% of the surfaces of loblolly pine and yellow-poplar roots in the upper 10 cm of soil was unsuberized (Kramer and Bullock, 1966). Unsuberized roots are more permeable than suberized roots to solutes and water. For example, the resistance of unsuberized roots of loblolly pine to water movement was about half that of suberized roots (Sands et al., 1982). However, unsuberized roots comprise only a small portion of the root surface and do not absorb all the ions and water needed by trees. Hence appreciable absorption often occurs through suberized roots (Table 10.15). During active root growth, absorption

TABLE 10.15. Effects of Removal of Unsuberized Roots on Uptake of Water and ^{32}P through 1-Year-Old Loblolly Pine Seedlings under a Pressure of 31 cm Hg[a,b]

Description	Total surface area (cm^2)	Rate of H_2O uptake ($cm^3\ cm^{-2}\ s^{-1}$)	Rate of ^{32}P uptake ($cpm\ cm^{-2}\ hr^{-1}$)	Concentration factor
Unpruned root systems	147.3	4.69	333	0.478
Part of unsuberized root surfaces removed	112.0 (24%)	4.28 (9%)	239 (28%)	0.365 (24%)
All unsuberized roots removed	86.0 (42%)	3.61 (23%)	178 (47%)	0.324 (32%)

[a]From Chung and Kramer (1975); after Kramer (1983).

[b]Numbers in parentheses are percentage reductions caused by pruning. Removal of all unsuberized roots reduced root surface by 42%, rate of water uptake by 23%, and ^{32}P uptake by 47%, indicating a high rate of uptake of water and salt through the suberized roots. The low concentration factor indicated existence of an effective ion barrier in suberized roots.

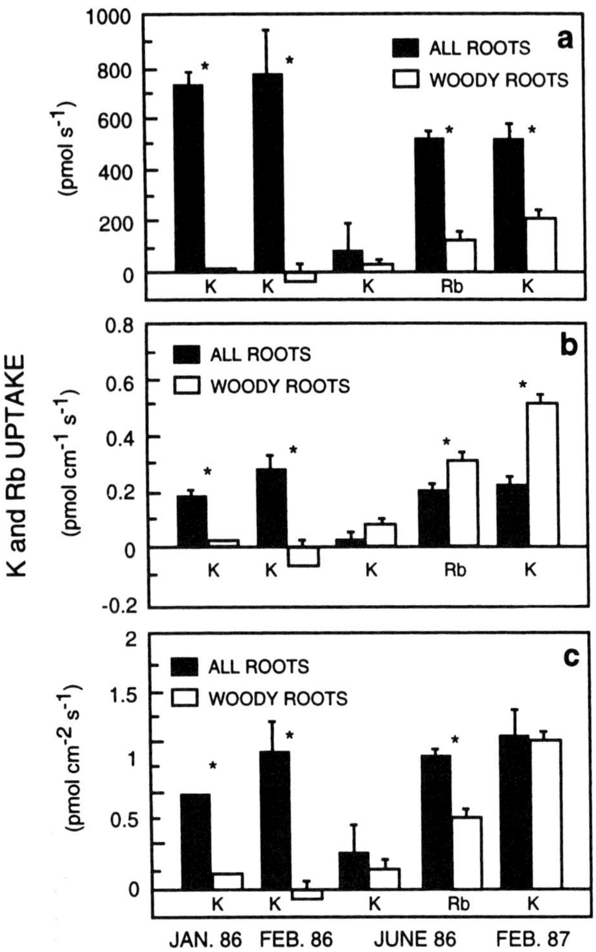

FIGURE 10.10. Uptake of K^+ and Rb^+ by woody roots and entire root systems (woody plus newly grown roots) of slash pine seedlings in solution culture. Uptake expressed: (a) per pot, (b) per unit root length, and (c) per unit of root surface area. There was no new shoot growth in the January or in both February experiments. Error bars denote standard errors of the means; asterisks indicate significant differences in K^+ or Rb^+ absorption between treatments in each experiment. From Van Rees and Comerford (1990).

of water, K, and Rb per unit of root length of slash pine seedlings with only woody roots (nonwoody roots removed) was comparable to or greater than in seedlings with both woody and nonwoody roots (Fig. 10.10).

Factors Affecting Absorption

The amounts and kinds of ions absorbed by plants are influenced by the presence of mycorrhizae and vary with plant species and genotype as well as with environmental conditions such as soil fertility, soil moisture supply, and root metabolism.

Species and Genotype

Large variations occur among plant species and genotypes in capacity for absorption and utilization of mineral nutrients. More nutrients are required and absorbed by many broadleaved trees than by conifers (Ralston and Prince, 1965) as shown by dominance by evergreens of infertile soils and ridgetops and growth of deciduous broadleaved trees on adjacent, more fertile soils (Monk, 1966). Sixteen-year-old Aigeiros poplars contained more than twice as much N, P, K, Ca, and Mg combined as southern pines of the same age (Switzer et al., 1976). The ash content of flowering dogwood, white oak, and sweetgum was about twice as high (7.0 to 7.2%) as that for loblolly and shortleaf pines (3.0 to 3.5%) on the same site. Leaves of dogwood, tulip poplar, white oak, and hickory contained approximately 2% Ca; those of scarlet oak, post oak, and loblolly pine contained less than 1% (Coile, 1937). Annual accumulation of mineral nutrients was greater in the evergreen chaparral shrub, *Ceanothus megacarpus*, than in the drought-deciduous coastal sage shrubs, *Salvia leucophylla* and *Artemisia californica* (Gray, 1983).

Clonal and provenance variations in absorption of mineral nutrients are well documented and correlated with differences in growth rates. Accumulation of N, P, Na, Mg, and B varied among 45 Scotch pine provenances (Steinbeck, 1966). Most studies of ion uptake by different genotypes have been conducted with plants growing in nutrient solutions. However, as Bowen (1985) emphasized, the limiting factor for ion uptake from soil is not the absorbing capacity of plants but rather ion transfer through the soil. Hence, selection of genotypes for high rates of ion uptake from nutrient solutions may not be useful for selecting plants for planting in mineral deficient soils. On the other hand, genotypic differences in rates of ion absorption may be important where sudden flushes of nutrients occur following application of fertilizers. The genetic basis of mineral nutrition of forest trees was reviewed by Goddard and Hollis (1984).

Mycorrhizae

Mycorrhizal fungi play a very important role in increasing mineral uptake from the soil (see Harley and Smith, 1983, for review). Kramer and Wilbur (1949) showed that larger amounts of radioactive P were accumulated by mycorrhizal pine roots than by nonmycorrhizal roots. Melin and his coworkers demonstrated that mycorrhizal fungi transferred P, N, Ca, and Na from the substrate to tree roots (Melin and Nilsson, 1950a,b; Melin et al., 1958).

The rate of absorption of mineral nutrients is determined by nutrient transfer in the soil, the extent of the root system, and the absorbing capacity of roots. Contact between root surfaces and soil nutrients is necessary for absorption. The contact can be the result of root growth to where the nutrients are located or to transport of nutrients to the root surface. The absorption of nutrients by roots varies with plant species and genotype as well as with environmental conditions.

Both inoculation of red pine seedlings with *Hebeloma arenosa* and amendment of soil with P influenced seedling growth. In P-unamended soil the inoculated plants formed abundant mycorrhizae and, after 19 weeks, had 12 times the root dry weight and 8 times the shoot dry weight of nonmycorrhizal seedlings (MacFall et al., 1991). In another study, mycorrhizal red pine seedlings grown in P-unamended soil had higher root and shoot P concentrations than did nonmycorrhizal seedlings growing in similarly unfertilized soil, but the concentrations were lower than for either mycorrhizal or nonmycorrhizal seedlings grown in P-amended soil (Fig. 10.11). Hence, even though the mycorrhizae increased both the P concentration of seedlings and seedling dry weights when grown in P-unamended soil, the amount of available P in the soil was too low for the seedlings to achieve their full growth potential (MacFall et al., 1992).

The increased mineral uptake by roots of plants with mycorrhizae is traceable largely to their extensive absorbing surface. The fungal hyphae often extend into the portion of the soil that is not penetrated by roots or root hairs. Often the hyphae enter spaces between soil particles that are too small to be invaded by roots. Bowen and Theodorou (1967) estimated that the volume of soil exploited by a mycorrhizal root may be as much as 10 times greater than that exploited by a nonmycorrhizal root. Increased efficiency of mineral uptake by mycorrhizal roots also may be associated with reduction in air gaps between soil particles and plant roots because of decreased shrinkage of mycorrhizal roots, a low-resistance pathway of the fungus for movement of ions throughout the root cortex, and increased root growth. The greater rooting intensity associated with mycorrhizal infection increases absorption of immobile nutrients such as P much more than that of highly mobile nutrients (Bowen, 1984). Mycorrhizal fungi also may increase nutrient availability by hydrolyzing certain nutrients in the soil. For example, surface acid phosphatases in mycorrhizae may hydrolyze organic and inorganic forms of phosphate (Reid, 1984; Marschner, 1995).

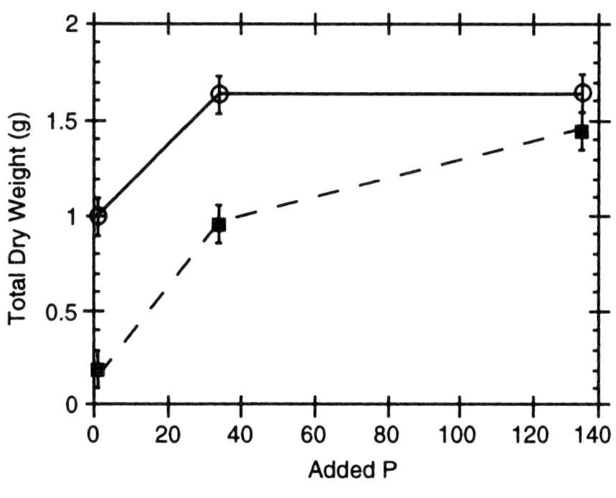

FIGURE 10.11. Dry weights of (○) red pine seedlings inoculated with *Hebeloma arenosa* and (■) uninoculated seedlings grown over a range of P amendments. The uninoculated seedlings did not form mycorrhizae at the highest level of applied P. From MacFall et al. (1992).

In addition to affecting establishment and growth of individual trees, mycorrhizae often influence the structure of entire ecosystems by at least three mechanisms (Perry et al., 1987):

- Enabling trees to compete with grasses and herbs for resources
- Decreasing competition among plants and increasing productivity of species mixtures, especially those on infertile sites
- Increasing interplant transfers of compounds essential for growth of higher plants

The hyphae of the external mycelium can initiate mycorrhizal infections within and between species. In this way a persistent network of hyphal interconnections is established among plants within an ecosystem. The mycelial strands comprise a network of direct pathways through which some minerals, water, and carbohydrates can move in channels that are functionally analogous to xylem and phloem (see also Chapter 12) (Taber and Taber, 1984; Francis and Read, 1984; Read et al., 1985). Griffiths et al. (1991) reported that the ectomycorrhizal fungi *Gautieria monticola* and *Hysterangium setchellii* formed dense hyphal mats in Douglas-fir stands. All seedlings under the canopy of a 60- to 75-year-old stand were associated with mats formed by ectomycorrhizal fungi. The mats apparently acted as nurseries for seedlings by providing them with carbohydrates and suppressing infection by pathogens. Because Douglas-fir is a relatively shade-intolerant species, it appeared unlikely that a seedling could support the mass of hyphae with which it was associated. Hence the mycorrhizal fungi probably were a conduit for carbohydrate transport from the overstory trees to the shaded seedlings.

Simard et al. (1997) supplied $^{13}CO_2$ and $^{14}CO_2$ to paper birch and Douglas-fir seedlings planted in close proximity in the field. These two species shared several types of ectomycorrhizal associations on over 90% of their root tips. Planted equidistant from this pair was a seedling of western red cedar, which does not form ectomycorrhizae and served as a control for soil-based movement of C. Isotope transfer between birch and Douglas-fir was detected in the second year after planting but there was no net transfer. However, in the third year, net positive transfer was observed in Douglas-fir, with the greatest amount occurring in heavily shaded seedlings indicating this species behavior as a sink for the birch source plant. There was some transfer of labeled C to western red cedar seedlings, but it was small enough to allow the authors to conclude that the birch-Douglas-fir transfers were primarily across ectomycorrhizal bridges. Transfer of ^{15}N from *Frankia*-infected *Alnus glutinosa* seedlings provided ^{15}N to *Pinus contorta* seedlings, presumably across *Paxillus involutus* fungal connections, also has been demonstrated in laboratory studies (Arnebrant et al., 1993).

Soil Fertility

Plants absorb more nutrients from fertile than from infertile soils. Nutrient mobility in the soil, which affects nutrient uptake, depends on the nutrient concentration in the soil solution. The rate of diffusion of ions toward roots usually is faster the higher the concentration of nutrients in the soil solution. The higher nutrient accumulation in leaves of plants growing in fertile soils forms the basis of the foliar analysis method of evaluating the supply of soil nutrients. However, there are important differences among species and genotypes in rates of mineral uptake and these often are maintained when plants are grown in soils of widely different mineral content.

The supply of soil nutrients influences not only the total increase in plant dry weight but also the partitioning of dry matter in plants, with high levels of available nutrients associated with greater distribution to shoots than to roots. For example, root production accounted for 23% of the total annual production of biomass of 40-year-old Douglas-fir trees on a mineral-rich site and 53% of the total on an infertile site (Keyes and Grier, 1981). Approximately 65% of the photosynthate of Monterey pine stands on infertile soil was used belowground, and much less in trees on fertile soils (Linder and Axelsson, 1982). Similar differences have been reported in other studies (McMurtrie, 1985). Cromer and Jarvis (1990) found that the ratio of leaf dry weight to root dry weight in *Eucalyptus grandis* seedlings fell from over 3 to less than 1 as nitrogen addition rate fell from 35 mg g^{-1} to 10 mg g^{-1}. High nutrient addition rates favored leaf development in three species of *Salix*, and low nutrient additions stimulated root growth (Ericsson, 1981).

Soil Moisture

Very low or very high soil moisture contents affect root growth, making it difficult to separate the direct effects of water supply to roots on ion uptake from the indirect effects associated with changes in rates of root growth and differentiation.

Water Deficits

The movement of ions in the soil and plant is correlated with water movement. For many ions mass transport is inadequate and diffusion becomes necessary. As the soil dries, the root surface in contact with soil water decreases. Shrinkage of both soil and roots may create vapor gaps in the ion translocation pathway. Soil water content at or near field capacity in a medium-textured soil permits the most ideal conditions of enough air space for O_2 diffusion, most nutrients in soluble form, greatest cross-sectional area for diffusion of ions and mass flow of water, and favorable conditions for root extension. As soil dries from field capacity, conditions become less favorable for ion availability and absorption (Viets, 1972), and uptake of nutrients essentially stops in dry soils.

In soil near field capacity, movement of water into roots is rapid and the rate of transpiration is controlled largely by atmospheric factors. However, as the soil dries the supply of water to the roots becomes a limiting factor and the rate of transpiration decreases (Chapter 12). A high transpiration rate may be expected to increase both active and passive absorption of ions. Rapid flow of water through the root xylem when the rate of transpiration is high tends to sweep ions upward from the roots, and the decreased concentration in the roots should increase active transport. In older roots, in which lenticels and openings caused by death of branch roots permit some flow of water, more ions enter the stele in the moving water stream when the rate of transpiration is high than when it is low.

A deficient soil water supply leads to leaf dehydration, stomatal closure, and reduced transpiration. The stomata begin to close when the turgor of guard cells decreases, often long before leaves wilt (Kozlowski, 1982a). Some investigators emphasized that, as the soil dries, there is increased movement from the roots to the shoots of a signal (possibly ABA) that induces stomatal closure (Chapter 12). Absorption slows and finally stops because of lack of a sufficient gradient in Ψ from the soil to the roots. Increase in resistance to water flow in the soil and in the roots, and possibly decreasing contact between the roots and soil, also reduces the rate of water movement in the soil-root interface, all contributing to decreased absorption of ions (Kramer, 1983).

Flooding

Inundation of soil affects ion absorption through its effects on soil conditions and uptake responses of plants. The specific effect of flooding varies greatly among plant species and specific ions absorbed.

In flood-intolerant species, absorption of N, P, and K decreases as the amount of energy released in root respiration becomes insufficient to sustain uptake in amounts needed for growth. Under anaerobic conditions the permeability of root cell membranes also is affected, leading to increased loss of ions by leaching. Reduction in ion uptake also is associated with suppression of mycorrhizal fungi in flooded soils (Kozlowski and Pallardy, 1984).

Flooding of soil reduces both the N concentration and total N content of plant tissues. This is partly the result of rapid depletion of nitrate, which is unstable in anaerobic soil, and lost after conversion to N_2O or N_2 by denitrification. The low N concentration of flooded plants also is associated with inhibitory effects of anaerobiosis on root respiration. Uptake of P and K also is reduced in flooded soil. Flooding has less effect on absorption of Ca and Mg than on N, P, and K (Kozlowski and Pallardy, 1984). In contrast to reduced uptake of N, P, and K by flood-intolerant plants, absorption of Fe and Mn is increased as ferric and manganic forms are converted to the more reduced soluble ferrous and manganous forms (Ponnamperuma, 1972). However, despite the increase in concentration of Fe and Mn, total uptake of these elements is reduced in accordance with slower growth of the flooded plants (Kozlowski and Pallardy, 1984).

Mineral uptake is affected much less by flooding of flood tolerant species than of flood intolerant species (Dickson et al., 1972). An important factor in this difference is the formation of adventitious roots in many flood tolerant species, thereby compensating for the loss of absorbing capacity as a result of decay of part of the original root system (Sena Gomes and Kozlowski, 1980; Kozlowski, 1984a,b, 1986b; see also Chapter 5 of Kozlowski and Pallardy, 1997).

Root Metabolism

Mineral uptake involves active transport of ions, which depends on expenditure of metabolic energy (Chapter 6). Hence, absorption of mineral nutrients is influenced by environmental factors such as aeration and temperature that affect metabolism, and lowering the O_2 level in solutions from near 90 to 50% equilibrium saturation with air decreased uptake of P, K, Ca, and Mg by roots of slash pine (Shoulders and Ralston, 1975). In another study, active uptake of K by slash pine roots depended on transport of O_2 from upper parts of roots and/or stems exposed to air. Roots absorbed K when they were exposed to an aerobic environment, and absorption stopped when N_2 replaced air in enclosures surrounding the lower stem and basal roots (Fig. 10.12). Active K uptake responded directly to levels of available O_2 (Fisher and Stone, 1990a,b). At solution O_2 concentrations of less than 1%, K leaked from the roots of plum trees, but when the soil was reaerated K uptake resumed (Rosen and Carlson, 1984).

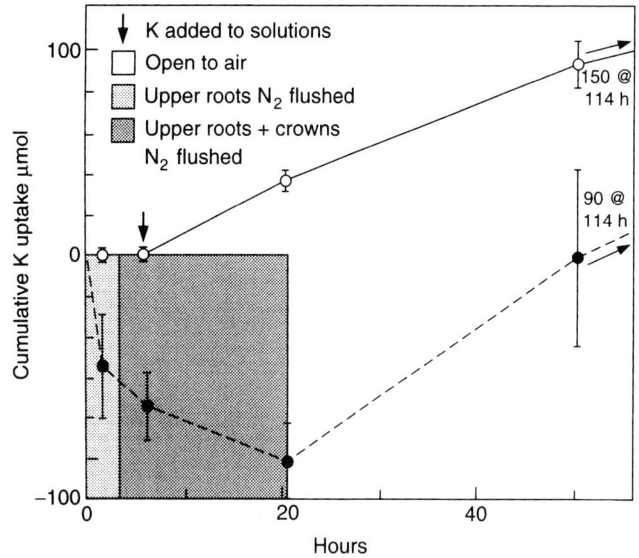

FIGURE 10.12. Effect of aeration of lower stems and roots of slash pine seedlings on absorption of potassium. ○, seedlings with upper roots and tops open to air; ●, upper roots and crown deprived of O_2 by exposure to N_2. Shading shows the duration of exposure to N_2. From Fisher and Stone (1990b).

Root respiration and mineral uptake vary with soil temperature. Both total and maintenance root respiration increased as an exponential function of soil temperature (Lawrence and Oechel, 1983). A rise in temperature from 5 to 25°C increased root respiration of green alder and balsam poplar by 4.6 and 5.0 times, respectively, and increased that of trembling aspen and paper birch by 2.9 and 3.9 times, respectively.

Absorption by Leaves and Twigs

Some mineral nutrients are absorbed by leaves and twigs. The mineral nutrients that are deposited on leaves from the atmosphere or applied as foliar sprays may sequentially (1) be transported through the cuticle and epidermal cells by diffusion, (2) be adsorbed on the surface of the plasmalemma, and (3) be moved through the plasmalemma to the cytoplasm. In addition, some nutrients enter the intercellular spaces of leaves through stomatal pores. To enter mesophyll cells from the intercellular spaces of leaves, ions must penetrate the cuticles that cover epidermal cells. Unlike epidermal cells, however, the mesophyll cells are easily entered. Some of the nutrients on leaf surfaces also are absorbed through trichomes (Swietlik and Faust, 1984). When present in trees, epiphytes can obtain atmospheric nutrients, which subsequently are recycled to other ecosystem members (Coxson and Nadkarni, 1995).

The capacity for foliar absorption of nutrients varies among species and is less efficient in peach, plum, and sour cherry than in apple or citrus. Apple leaves may absorb more than twice as much N per unit of leaf dry weight as sour cherry leaves. Rapid leaf absorption of urea N has been reported for banana, coffee, and cacao (Swietlik and Faust, 1984).

Absorption of foliar-applied nutrients is influenced by several factors that affect development of the cuticle and the process of active ion uptake. Among the most important of these are light, temperature, relative humidity, age of leaves, nutritional status of plants, formulation and concentration of foliar sprays, and surfactants (Swietlik and Slowik, 1984).

Diffusion of mineral elements through the cuticle is influenced by the amount, distribution, and composition of epicuticular waxes. Waxes are much more impermeable than other cuticular components such as pectinaceous compounds and proteins. The importance of leaf waxes as a barrier to diffusion is emphasized by accelerated penetration of ions into dewaxed leaves. The chemical composition of leaf waxes also affects ion uptake. For example, permeability varied in the following order: esters > fatty acids > alcohols > triterpenoids > hydrocarbons (Baker and Bukovac, 1971). Diffusion of specific ions through the cuticle also differs. For example, penetration of cations varied as follows: $Cs^+ > Rb^+ > Na^+ > Li^+ > Mg^{2+} > Sr^{2+} > Ca^{2+}$ (Halevy and Wittwer, 1965).

Cations that penetrate the leaf cuticle may move in the free space of cell walls to vascular tissues and may then be actively loaded into the phloem (see Chapter 7 of Kozlowski and Pallardy, 1997). Katz et al. (1989) identified the pathway for transport of mineral nutrients through the bark and along the rays of Norway spruce twigs. In addition to uptake by mass flow, Mg diffused along a concentration gradient from the twig surface into the xylem. The rate of transport through twigs may be expected to vary with the concentration of elements deposited by atmospheric precipitation, the concentration gradient between the plant surface and the xylem sap, the xylem water potential, and the intensity and duration of rainfall.

SUMMARY

Mineral nutrient deficiencies are common and often limit the growth of woody plants. Elements essential for growth include macronutrients (N, P, Ca, Mg, and S) and micronutrients (Fe, Mn, Zn, Cu, B, Mo, Ni, and Cl). Mineral nutrients are important as constituents of plant tissues, catalysts, osmotic regulators, constituents of buffer systems, and regulators of membrane permeability. Mineral deficiencies inhibit vegetative and reproductive growth by causing changes in physiological processes. Visible symptoms of mineral deficiency include chlorosis, leaf necrosis, rosetting, bark lesions, and excessive gum formation. The amounts of mineral nutrients in woody plants vary with species and genotype, site, season, and in different parts of the same plant. Total nutrient contents of forests vary in the following order: tropical > temperate broadleaf > temperate conifer > boreal forests. Deciduous trees generally accumulate more minerals than evergreen trees. The concentration of minerals in trees usually varies as follows: leaves > small branches > large branches > stems.

The pool of soil minerals in forests is maintained by cycling through the trees, understory vegetation, forest floor, and mineral soil. In tropical forests nutrient cycling is rapid and a large proportion of the mineral pool is in the trees. In temperate and boreal forests, a higher proportion of the nutrient pool is in the soil and litter. Mineral nutrients in the soil are increased by atmospheric deposition, weathering of rocks and minerals, decomposition of litter and roots, leaching from plants, and exudation from roots. Nutrients are depleted from ecosystems by leaching in drainage water, removal in harvested plants, absorption by plants, and volatilization of N and S during fire.

Forests are periodically subjected to minor disturbances (e.g., windthrow, lightning strikes, partial cuttings) and major disturbances (e.g., crown fires, soil erosion, clear-cutting) that accelerate loss of mineral nutrients from the ecosystem. The amounts lost vary from negligible to catastrophic and depend on the severity and duration of the disturbance as well as the forest type. Clear-cutting (clear-felling) removes mineral nutrients in the harvested wood and results in increased leaching losses by drainage to streamwaters. Whole-tree harvesting removes more nutrients from a site than harvesting logs only while leaving the slash. More mineral nutrients are depleted by short rotations than by long ones. Tropical forests are more fragile and more easily depleted of nutrients than temperate forests. Shifting cultivation in the tropics, involving cutting and burning of forests and planting of food crops, may be followed by loss of nutrients from the site and progressively reduced growth of planted crops.

Absorption of nutrients by plants involves movement of ions from the soil to root surfaces by diffusion and mass flow, ion accumulation in root cells, radial movement of ions from root surfaces into the xylem, and translocation of ions from roots to shoots. Absorption of ions by roots occurs by passive movement through the apoplast followed by active absorption through the plasmalemmas of epidermal and cortical cells. Active uptake on nutrient ions is complex and involves numerous membrane proteins that variously function as pumps, carriers, and channels. The amounts and kinds of ions absorbed by roots are influenced by mycorrhizae and also vary with plant species and genotype, soil fertility, soil moisture supply (including drought and flooding), and root metabolism.

Some mineral nutrients are absorbed by leaves and twigs. Mineral nutrients that are deposited on leaves from the atmosphere or applied as foliar sprays are transported through the cuticle and epidermal cells by diffusion, adsorbed on the surface of the plasmalemma, and moved through the plasmalemma to the cytoplasm. The capacity for foliar absorption of nutrients varies with species and several factors that influence development of the cuticle and the process of active ion uptake. Absorption of some mineral nutrients by twigs and transport into the xylem has been demonstrated.

General References

Allen, M. F. (1991). *The Ecology of Mycorrhizae*. Cambridge Univ. Press, Cambridge.

Atkinson, D., Jackson, J. E., Sharples, R. O., and Waller, W. M., eds. (1980). *Mineral Nutrition of Fruit Trees*. Butterworth, London.

Attiwill, P. M. and Adams, M. A. (1993). Nutrient cycling in forests. *New Phytol* **124**, 561–582.

Barber, S. A. (1995). *Soil Nutrient Bioavailability: A Mechanistic Approach, 2nd edition*. Wiley Interscience, New York.

Bergmann, W., ed. (1992). *Nutritional Disorders of Plants*. Gustav Fisher, Jena.

Bowen, G. D. and Nambiar, E. K. S., eds. (1984). *Nutrition of Plantation Forests*. Academic Press, London.

Chen, Y. and Hadar, Y., eds. (1991). *Iron Nutrition and Interactions in Plants*. Kluwer, Dordrecht, Boston, and London.

Epstein, E. and Bloom, A. J. (2005). *Mineral Nutrition of Plants: Principles and Perspectives, 2nd edition*. Sinauer Associates, Inc., Sunderland, Massachusetts.

Faust, M. (1989). *Physiology of Temperate Zone Fruit Trees*. Wiley, New York.

Harrison, A. F., Ineson, P., and Heal, O. W., eds. (1990). *Nutrient Cycling in Terrestrial Ecosystems*. Elsevier, London and New York.

Johnson, D. W. and Lindberg, S. E., eds. (1992). *Atmospheric Deposition and Forest Nutrient Cycling*. Springer-Verlag, New York and Berlin.

Jordan, C. F. (1985). *Nutrient Cycling in Tropical Forest Ecosystems*. Wiley, Chichester and New York.

Marschner, H. (1995). *The Mineral Nutrition of Higher Plants, 2nd edition*. Academic Press, London.

Mengel, K. and Kirkby, E. A. (1978). *Principles of Plant Nutrition*. International Potash Institute, Berne, Switzerland.

Proctor, J., ed. (1989). *Mineral Nutrients in Tropical Forest and Savanna Ecosystems*. Blackwell, Oxford.

Rendig, V. V. and Taylor, H. M. (1989). *Principles of Soil-Plant Interrelationships*. McGraw-Hill, New York.

Smith, S. E. and Gianinazzi-Pearson, V. (1988). Physiological interactions between symbionts in vesicular arbuscular mycorrhizal plants. *Annu Rev Plant Physiol Plant Mol Biol* **39**, 221–246.

Smith, S. E. and Read, D. J. (1997). *Mycorrhizal Symbiosis, 2nd edition*. Academic Press, London.

Tinker, B. and Läuchli, A., eds. (1984–1988). *Advances in Plant Nutrition*. Vols. 1–3. Praeger, New York.

Tinker, P. B. and Nye, P. H. (2000). *Solute Movement in the Rhizosphere*. Oxford University Press, New York.

Wild, A. (1989). *Russell's Soil Conditions and Plant Growth*. Wiley, New York.

CHAPTER

11

Absorption of Water and Ascent of Sap

INTRODUCTION

Over that part of the earth's surface where temperatures permit plant growth the occurrence of trees is controlled chiefly by the water supply. Most grasslands and deserts could support forests if the quantity and seasonal distribution of precipitation were favorable. Other large areas support only sparse stands of trees because of limited water supplies. The ecological significance of water arises from its physiological importance. An adequate supply of water is just as essential to the successful growth of plants as photosynthesis and the other biochemical processes involved in the synthesis of food and its transformation into new tissues. An essential factor in plant water relations is maintenance of an amount of water sufficient to sustain cell turgor and permit normal functioning of the physiological and biochemical processes involved in growth. Plant water status is controlled by the relative rates of water absorption, water loss, and internal storage, as discussed later.

Importance of Water

The importance of an adequate water supply for growth of woody plants has been well documented by Zahner (1968). He reported that up to 80% of the variation in diameter growth of trees in humid areas (and up to 90% in arid areas) can be attributed to variations in rainfall and plant water stress. Bassett (1964) found a very high correlation between wood production and available soil water in a pine stand in Arkansas over a 21-year period. In fact, prediction of past climatic conditions from tree ring widths for many arid and some humid regions is well established as the discipline of dendrochronology (Fritts, 1976; Cook and Jacoby, 1977). These relationships are discussed in more detail in Chapters 3 and 5 of Kozlowski and Pallardy (1997). Readers are referred to the book by Schweingruber (1988) and the two volumes edited by Cook and Kairiukstis (1990) and Lewis (1995) for good discussions of dendrochronology.

The importance of water in the life of woody plants can be shown by listing its more important functions. These can be grouped in four categories:

- Water is an essential constituent of protoplasm and forms 80 to 90% of the fresh weight of actively growing tissues.
- Water is the solvent in which gases, salts, and other solutes move within and between cells and from organ to organ.
- Water is a reagent in photosynthesis and a substrate or product in many other metabolic reactions.
- Water is essential for maintenance of turgidity of cells and tissues, assuring the presence of a driving force for cell enlargement, stomatal opening, and maintenance of the form of young leaves and other slightly lignified structures.

An understanding of plant water relations requires consideration of both soil and atmospheric water. However, I will first concentrate on two interrelated aspects of plant water relations. One deals with the water relations of cells and tissues within the plant; the other deals with the water relations of the plant as a whole. Plant water relations involve the absorption of water, ascent of sap, loss of water by transpiration, and the internal water balance of the tree.

Cell Water Relations

The water relations of plants are controlled primarily by cell water relations; hence, consideration will be given to cell structure and functioning in relation to water movement.

Cell Structure

Living cells of plants consist of protoplasts surrounded by constraining walls that severely limit changes in volume, particularly in older tissues in which the cell walls are lignified. In mature living cells, the protoplasts consist of large central vacuoles enclosed in thin layers of cytoplasm next to cell walls. A nucleus and various other organelles such as plastids and mitochondria are embedded in the cytoplasm (Fig. 11.1). Electron microscopy reveals various other structures in the cytoplasm such as ribosomes, peroxisomes, dictyosomes, microtubules, and the complex system of membranes known as the endoplasmic reticulum. Large amounts of water are bound to the protein framework of cells, and the surface membranes of protoplasts (plasmalemma and tonoplast) are permeable to water but relatively impermeable to solutes. As a result, mineral ions and organic solutes can accumulate in vacuoles, producing osmotic potentials between −0.5 and −5 MPa. Often the protoplasts of adjacent cells are connected by strands of cytoplasm called plasmodesmata, forming a continuous system called the *symplast*. Vacuoles vary in size from tiny rod-shaped or spherical structures in meristematic tissues to large central vacuoles of mature parenchyma cells that can occupy up to 90% of the cell volume (Nobel, 1991).

Water Status Quantification and Terminology

Water Content

It has not proven easy to identify a single measure of water status in plants that is applicable and useful in every situation. The use of concentration of substances that is common in the chemical sciences does not translate well in water relations because of its insensitivity over the range in water status that is relevant in plant water relations. For example, pure water is approximately 55.5 M, but severely desiccated tissues maintain water concentrations of 54 M, a reduction of only 3%.

For many years water contents commonly were measured in water relations research:

$$\text{Water Content} = \frac{\text{Fresh Weight}}{\text{Dry Weight}} \times 100 \qquad (11.1)$$

This measure is easily obtained by gravimetric measurements and is more sensitive to variations in how much water is present than is concentration of water. However, it has several deficiencies (Kramer, 1983). First, water content may vary independently of the mass of water present because of changes in dry weight of tissues (Kozlowski and Clausen, 1965). Such variations are common occurrences during plant ontogeny (Fig. 11.2). It also is quite difficult or impossible to compare water contents among plant tissues, such as roots and leaves and between the soil and plant. This is because water content is independently influenced by variations in dry weight between organs in the former case, and the grossly different density and composition of plant dry matter and soil minerals in the latter case.

Another measure of plant water status, Relative Water Content (RWC), is uninfluenced by dry weight changes:

$$\text{Relative Water Content} = \frac{(\text{Fresh Weight} - \text{Dry Weight})}{(\text{Saturated Weight} - \text{Dry Weight})} \times 100 \qquad (11.2)$$

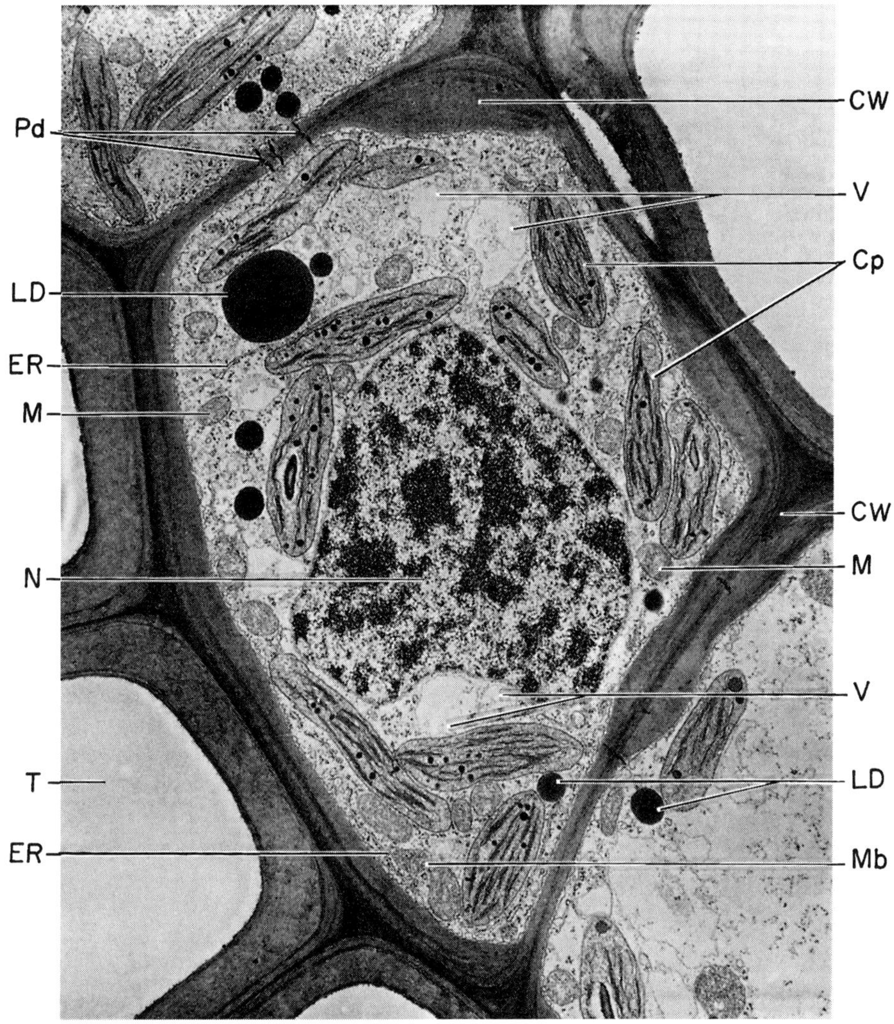

FIGURE 11.1. An electron micrograph of a xylem ray parenchyma cell of red pine showing its principal structures (×8,000). CW, cell wall; Cp, chloroplast; LD, lipid droplet; ER, endoplasmic reticulum; Mb, microbody; N, nucleus; V, vacuole; T, tracheid; Pd plasmodesmata; M, mitochondrion. Photo courtesy of D. Neuberger.

Relative water content is nearly synonymous with relative water volume of the symplast, and can serve in some instances as a useful physiological indicator of dehydration levels of cells and organelles. Saturated weights most often are obtained by allowing tissue samples to equilibrate with pure water. Pallardy et al. (1991) reviewed and evaluated procedures for measuring RWC in woody plants. Although RWC generally has more utility in characterizing plant water status than does water content, it is not directly related to some important physiological states of plants, such as the degree of turgor. Additionally, it is impossible meaningfully to compare measurements of soil water content and RWC values of plants.

Water Potential

Development of water potential concepts and methods has been quite useful in quantification of the water status of plants. This approach to plant water relations focuses on the chemical potential of water in the plant and soil as a measure of water status. Chemical potential (μ) is related to free energy of a system or a component of a system and refers to its capacity to do work. The chemical potential of a substance is an intrinsic property of a substance, like its temperature. Measuring the absolute chemical potential of water is difficult, but it is easy to measure the difference in potential between a standard (pure water, μ_w*) and water in a solution such as the cell sap (μ_w).

In plant water relations, the chemical potential of water is converted to water potential (Ψ_w) by dividing chemical potential by the partial molal volume of water ($\bar{V}_w$, m^3 mole^{-1}):

$$\Psi_w = \frac{\mu_w - \mu_w^*}{\bar{V}_w} \tag{11.3}$$

The units for chemical potential are in energy terms (J mole^{-1}) and, as J mole^{-1} = N m mole^{-1}, Ψ_w may be

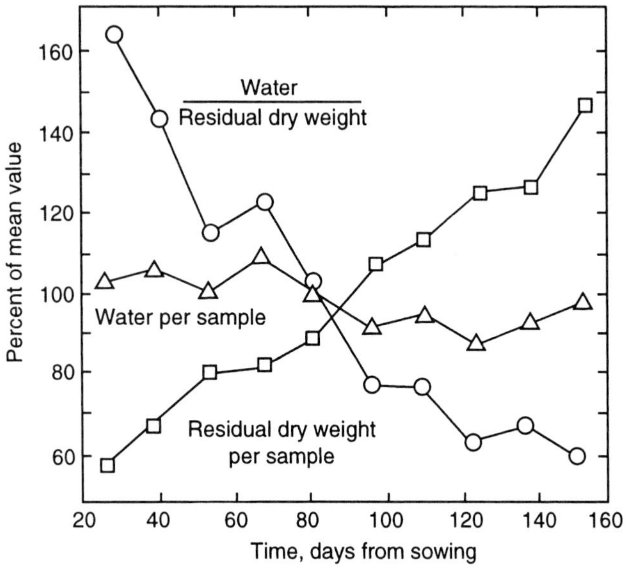

FIGURE 11.2. Change in water content of cotton leaves based on dry weight. Increased dry weight creates the appearance that water mass of the leaf is declining when it actually is almost stable. From Weatherly (1950); adapted from Kramer (1983).

expressed as force/area (N m^{-2}), which is pressure. Currently preferred as a System Internationale (SI) unit of pressure is the Pascal (Pa) (1 Pa = 1 N m^{-2}; 1 MPa = 1×10^6 Pa = 10 bar = 9.87 atmosphere; 1 kPa = 10 mbar).

Several separable component potentials can contribute to Ψ_w:

$$\Psi_w = \Psi_\pi + \Psi_p + \Psi_g \tag{11.4}$$

where Ψ_π = osmotic or solute potential, Ψ_p = turgor or pressure potential, and Ψ_g = gravitational potential. The osmotic contribution arises from dissolved solutes and lowered activity of water near charged surfaces. Separating these two effects is sometimes useful, particularly in the soil and in cell walls, and some researchers consequently distinguish a matric potential component (Ψ_m) associated with surface effects. The turgor potential derives from xylem tension and positive pressure inside cells as water presses against the walls. The gravitational component potential varies with plant height at a rate of 0.1 MPa per 10 m vertical displacement.

Values of Ψ_π and Ψ_m are negative, but Ψ_p may be positive, as in turgid cells, or negative as it frequently is in xylem sap under tension. The sum of the component potentials on the right-hand side of Eq. 11.4 is negative, except in fully turgid cells when it becomes zero.

Water Movement

Several advantages have led to wide acceptance of water potential measurements. Most important was the fact that water movement occurs along gradients of decreasing free energy, often expressed as differences in Ψ_w. Hence if Ψ_w is measured at two points of a system (e.g., between soil and plant, or roots and leaves), the direction of water flow and gradient driving flows are easily inferred. Component potentials, particularly Ψ_π and Ψ_p, also have inherent physiological meaning, indicating, respectively, the level of solute accumulation and turgor in plant tissues.

If the difference in potential is produced by some external agent, such as pressure or gravity, the movement is termed mass flow. Examples are the flow of water in pipes under pressure gradients, flow of water in streams caused by gravity, and the ascent of sap in plants, caused by evaporation from the shoots (Chapter 12). If movement results from random motion of molecules caused by their own kinetic energy, as in evaporation, the process is called diffusion. Osmosis is an example of diffusion induced by a difference in potential of water on two sides of a membrane, usually caused by differences in concentration of solutes.

Diffusion rates of molecules in liquid water are adequate to support rapid transport across the distances (μm) involved at the cellular level, although cytoplasmic streaming operates at the cellular level to further elevate the internal velocity of solutes. However, it is worth noting the much greater importance of mass flow in long distance transport as compared with diffusive movement. For example, Nobel (1991) estimated that small solute molecules in aqueous solutions would require eight *years* to diffuse a distance of 1 m. In contrast, solutes, and the water carried with them, may move many m hr^{-1} by mass flow in the xylem.

There has been a lively discussion concerning the utility of water potential in plant water relations (Sinclair and Ludlow, 1985; Kramer, 1988; Passioura, 1988a; Schulze et al., 1988; Boyer, 1989). Several potential deficiencies of water potential have been noted. Most significantly, water flow within the soil-plant system often is governed by component potentials, rather than Ψ_w. Additionally, reductions in the chemical activity of water associated with severe water deficits in plants usually are insufficient to account for inhibition of enzyme-catalyzed reactions and biochemical processes such as photosynthesis and respiration (Boyer, 1989). Kramer and Boyer (1995) hypothesized that changes in the concentration of regulatory molecules in dehydrating plants were responsible for drought-related reductions in enzyme activity. The central role of Ψ_w of the leaf (often denoted as Ψ_l) as an indicator of physiological responses during water stress also has been questioned (Blackman and Davies, 1985; Gollan et al., 1986; Schulze, 1986). Most important have been reports of stomatal closure during early stages of soil drying without changes in leaf Ψ_w or in component potentials (Blackman and Davies, 1985).

These criticisms have been considered by Schulze et al. (1988), Kramer (1988), and Boyer (1989). Although the component potentials providing the driving force for water flow in the soil-plant system do vary, only in unusual circumstances do these gradients diverge significantly from measured gradients in Ψ_w (Fig. 11.3). For example, Ψ_p and Ψ_g gradients within the xylem drive water flow, but because xylem Ψ_π values usually are near 0, there is rarely a large deviation of the actual driving gradient for water flow from the gradient in Ψ_w. Additionally, because water must pass through the plasmalemmas of root cells as it travels between the soil and root xylem, flow between soil and the root xylem depends on both Ψ_π and Ψ_p (Ψ_g gradients are negligible), and hence this flow also is described by differences in Ψ_w.

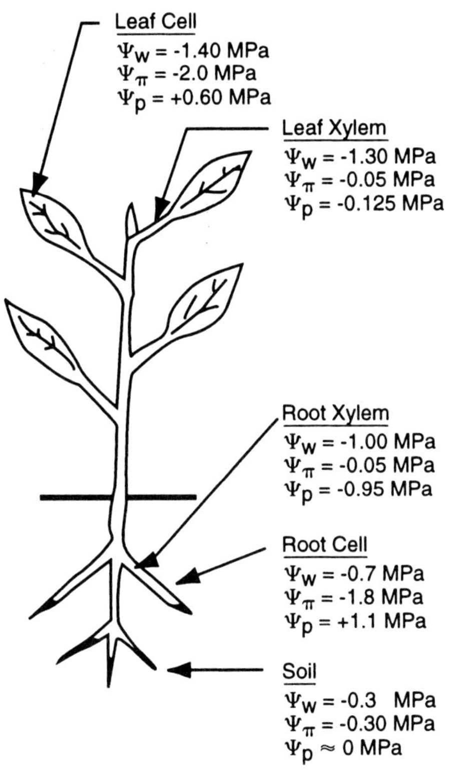

FIGURE 11.3. Illustration of the variation in total Ψ_w and its components in a transpiring plant that might be expected at the indicated points along the Soil–Plant–Atmosphere Continuum. Modified from Jones (1983). © Cambridge University Press. Reprinted with permission of Cambridge University Press.

Measurement of Water Potential and Its Components

Water potential and its components can be measured by a variety of techniques, including: (1) vapor pressure methods, primarily with thermocouple psychrometers and dew point hygrometers, (2) tensiometry and the pressure plate apparatus, and (3) the pressure chamber. Most estimates of total water potential are made using thermocouple psychrometers or hygrometers, or the pressure chamber. Psychrometers and hygrometers measure wet bulb or dew point temperature depression, respectively, in thermal and vapor equilibrium with a plant (or soil) sample. Osmotic potential and bulk Ψ_p of plant tissues can be estimated by freezing a sample after an initial measurement of total water potential. Freezing disrupts cell membranes, eliminating the contribution of Ψ_p, and makes subsequent Ψ_w determinations equivalent to Ψ_π. Turgor potential can be determined by difference (see Eq. 11.4).

Tensiometers frequently are used to monitor soil Ψ_w in the field. Most commonly, these instruments consist of a reservoir placed in hydraulic contact with the soil across a porous ceramic cup. Tension develops within the water reservoir as it equilibrates with an unsaturated soil. A gauge or pressure transducer measures the tension within the instrument. Tensiometers are useful and inexpensive instruments, but they are limited to the maximum tension that can be developed without breakage of water columns (usually about 0.08 MPa).

The pressure plate apparatus is used to develop soil water retention curves. In a pressure-tight vessel, saturated soil is placed in close contact with a porous plate. Pressure is gradually increased in the vessel and the amount of water forced through the plate is recorded at various applied pressures. The measurement provides the relationship between Ψ_w and water content of the soil.

The pressure chamber technique of Ψ_w measurement is based on the fact that the effect of pressure on water potential is equivalent to that of solutes and other components. The instrument consists of a pressure-safe vessel attached to a pressure gauge. Pressure is slowly increased within the chamber and the protruding part of the enclosed, previously excised plant sample (usually the petiole or rachis) is observed for the appearance of xylem sap. The gauge pressure associated with the initial appearance of sap is termed the balance pressure or "end point."

The measurement made with a pressure chamber can be related to water potential in the cell walls of a leaf:

$$\Psi_w = \Psi_m^{apo} + \Psi_\pi^{apo} = -P + \Psi_\pi^{apo} \quad (11.5)$$

where Ψ_m^{apo} and Ψ_π^{apo} represent matric and osmotic components of water potential in the apoplastic spaces, respectively, and P is the chamber balance pressure (Boyer, 1969). Thus, the system described by the cellulose matrix of cell walls and the pressure chamber is analogous to the pressure plate-soil system used to measure Ψ_w of soils. Matric potential of the apoplast is similar to Ψ_w if there is equilibrium between cellular and apoplastic water and if the osmotic potential of apoplastic water is high (i.e., close to 0). The extracellular solution is normally so dilute that Ψ_m^{apo} nearly always exceeds –0.3 MPa, except in plants growing in saline media (Barrs and Kramer, 1969; Duniway, 1971). In the latter case, a psychrometer or hygrometer measurement of the Ψ_π of expressed xylem sap can be used with the pressure chamber to obtain Ψ_w. With some practice the pressure chamber can become one of the most reliable instruments available for plant water relations research. Simple construction and durability have made it a preferred instrument for measuring Ψ_w in the field (Pallardy et al., 1991).

Balling and Zimmermann (1990) questioned the validity of pressure chamber measurements of Ψ_w because values obtained did not correspond to the values of xylem tension derived from a miniature pressure probe inserted into xylem elements. However, other experiments have shown close correspondence between pressure chamber measurements of Ψ_w and other methods (e.g., dew point hygrometry; Baughn and Tanner, 1976). Passioura (1991) presented an analysis of Balling and Zimmermann's experiment reconciling most of their results with the theory and procedures of pressure chamber measurements.

The pressure chamber also is used to derive the relationship between Ψ_w and expressed water volume (or RWC) of plant tissues. The procedure has been termed pressure-volume analysis. Repeated measurements of P are made during leaf dehydration induced by sap expression (Scholander et al., 1964) or through transpirational water loss (Richter, 1978). If the inverse of P (i.e., $1/\Psi_w$) is plotted versus expressed volume (V_e) or RWC, a pressure-volume curve of the plant tissue is obtained (Fig. 11.4). The relationship between $1/\Psi_w$ and V_e or RWC initially is curvilinear because dehydration reduces both Ψ_p and Ψ_π, but becomes linear at the point of turgor loss, when cells in the tissue essentially behave as osmometers. In the linear region, increases in pressure force water from cells, increasing solute concentrations and, therefore, Ψ_π by an amount equivalent to the change in pressure. In the region of positive turgor, Ψ_p can be calculated by subtracting the estimated value of Ψ_π, providing the means for assessing the response of Ψ_p as dehydration proceeds. The bulk elastic modulus of a tissue also can be calculated over any desired interval (although it most often is reported near full turgor):

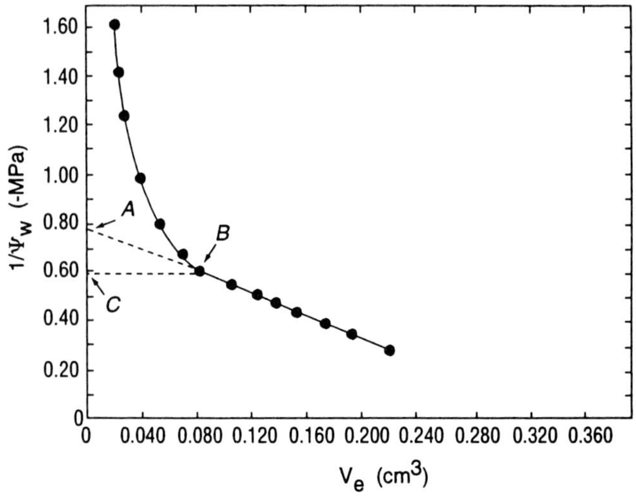

FIGURE 11.4. Example of a pressure-volume curve illustrating the relationship between $1/\Psi_w$ and volume of expressed sap (V_e). The y-axis intercept at point A estimates the reciprocal of osmotic potential at full turgor; B is the point of turgor loss and C is the estimate of the reciprocal of the osmotic potential at the turgor loss point. The elastic modulus may be calculated from changes in the relationship between $1/\Psi_w$ and V_e and the slope of the line AB (see text). Reprinted with permission from Pallardy, S. G., Pereira, J. S, and Parker, W. C. (1991). Measuring the state of water in tree systems. In *Techniques and Approaches in Forest Tree Ecophysiology*, J. P. Lassoie and T. M. Hinckley, eds., pp. 28–76. Copyright CRC Press, Boca Raton, Florida.

$$\varepsilon = \frac{\Delta\Psi_p}{\Delta V} \quad \text{or} \quad \varepsilon = \frac{\Delta\Psi_p}{\Delta RWC} \quad (11.6)$$

The curves obtained by pressure-volume analysis have proven very useful in documentation of several aspects of tissue water relations, including measurement of osmotic potentials, osmotic adjustment, and elastic properties of plant tissues. Reviews of methods of measuring water status and pressure-volume analysis can be found in Slavik (1974), Ritchie and Hinckley (1975), Turner (1981, 1988), Tyree and Jarvis (1982), Pallardy et al. (1991), and Boyer (1995).

The Soil–Plant–Atmosphere Continuum

One important contribution to plant water relations is the treatment of water movement through the soil, into roots, through the plant, and out into the air as a series of closely interrelated processes. This concept is sometimes called the Soil–Plant–Atmosphere Continuum (SPAC) (Philip, 1966), and it is useful in emphasizing the necessity of considering all aspects of water relations in studying the water balance of plants. This concept leads to treatment of water movement in the SPAC system as analogous to the flow of electricity in a conducting system, and it therefore can be described by an analog of Ohm's Law where:

$$\text{Flow} = \frac{\text{difference in } \Psi_w}{\text{resistance}} \qquad (11.7)$$

This concept can be applied to steady-state flow through a plant as follows:

$$\text{Flow} = \frac{\Psi_{\text{soil}} - \Psi_{\text{root surface}}}{r_{\text{soil}\rightarrow\text{root}}} = \frac{\Psi_{\text{root surface}} - \Psi_{\text{xylem}}}{r_{\text{root}}} = \frac{\Psi_{\text{xylem}} - \Psi_{\text{leaf cells}}}{r_{\text{xylem}\rightarrow\text{leaf cells}}} = \frac{C_{\text{leaf}} - C_{\text{air}}}{r_{\text{leaf}} + r_{\text{air}}} \qquad (11.8)$$

where C corresponds to concentration of water vapor.

The continuum concept provides a useful, unifying theory in which water movement through soil, roots, stems, and leaves, and its evaporation into the air, can be studied in terms of the driving forces and resistances operating in each segment. The concept also is useful in analyzing the manner in which various plant and environmental factors affect water movement by influencing either the driving forces or the resistances, or sometimes both. For example, drying of soil causes both an increase in resistance to water flow toward roots and a decrease in driving force or water potential; deficient aeration and reduced soil temperature increase the resistance to water flow through roots; an increase in leaf and air temperature increases transpiration because it increases the vapor concentration gradient or driving force from leaf to air (see Tables 11.1 and 11.2). Closure of stomata increases the resistance to diffusion of water vapor out of leaves. The continuum concept and its application were discussed by Richter (1973a), Jarvis (1975), Weatherley (1976), Hinckley et al. (1978a), Boyer (1985), Kaufmann and Fiscus (1985), Pallardy (1989), Pallardy et al. (1991), Kirkham (2002), and in several symposium papers in *Agronomy Journal* (Volume 29, Issue 6, 2003).

The continuum concept also facilitates modeling of water movement, as in the example shown in Figure 11.5. Models range from those for individual stomata

TABLE 11.1. Effect of Increasing Temperature on Vapor Concentration of Water in Leaves and Vapor Concentration Gradient from Leaf to Air at an Assumed Constant Relative Humidity of 60%[a]

Parameter	Temperature (°C)		
	10	20	30
Vapor concentration of tissue (g m^{-3})	9.41	17.31	30.40
Vapor concentration of air at 60% relative humidity (g cm^{-3})	5.65	10.39	18.24
Vapor concentration gradient (g cm^{-3})	3.76	6.92	12.16

[a]The vapor concentration of the leaf tissue is assumed to be the saturation vapor concentration of water, because the lowering caused by cell solutes is only about 3%.

TABLE 11.2. Effect of Increasing Temperature of Leaf and Air with No Change in Absolute Humidity of Vapor Concentration Gradient from Leaf to Air

Parameter	Leaf and air temperature (°C)		
	10	20	30
Relative humidity of air assuming no change in absolute humidity (%)	80	44	25
Vapor concentration at evaporating surface of leaf (g m^{-3})	9.41	17.31	30.40
Vapor concentration in air at indicated temperatures (g m^{-3})	7.53	7.62	7.60
Vapor concentration gradient from leaf to air (g m^{-3})	1.88	9.69	22.80

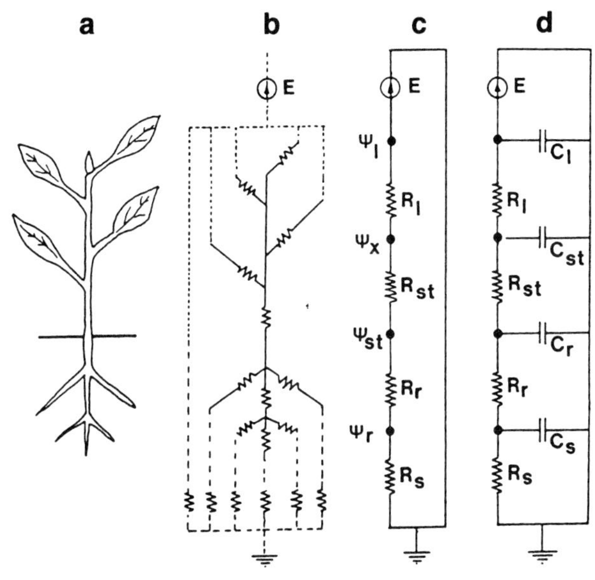

FIGURE 11.5. (a) Simplified representation of a plant. (b) Corresponding network of flow resistances, including resistances in the soil, roots, stem, and leaves. (c) Simplified catenary model with complex branched pathway of (b) represented as a linear series with hydraulic resistances in the soil (R_s), roots (R_r), stem (R_{st}), and leaves (R_l) each represented by a single resistor. (d) Same as in (c), but including capacitances (C) of corresponding tissues. E represents transpiration demand and direction. After Jones (1983); adapted from Pallardy (1989).

(DeMichele and Sharpe, 1973) to those for whole stands of trees (Waggoner and Reifsnyder, 1968). Modelers hope eventually to be able to predict plant behavior over a wide range of environmental conditions, but much more information will be needed before this is possible.

Readers are cautioned that this elementary discussion of the continuum concept is, for a number of reasons, an oversimplification. First, it assumes steady-state conditions that seldom exist in plants. Even within a plant, flow may vary among different segments of the continuum as different parts of a tree crown are subjected to varying regimes of radiation and evaporative demand (Richter, 1973a). Also, complications occur because water movement in the liquid phase is proportional to the difference in water potential, whereas movement in the vapor phase is proportional to the gradient in water vapor concentration. The woody plant body also serves as a complex reservoir of water that is depleted and replenished diurnally and seasonally. This water "storage" can be incorporated into the electrical analogy if capacitors are considered as part of the system (Fig. 11.5).

Progress has been made in constructing models that incorporate segmented flow pathways and capacitance of plant tissues that can simulate Ψ_w and flow dynamics well. For example, Tyree (1988) developed a model incorporating both comprehensive data of hydraulic pathways and water storage properties in northern white cedar trees that provided simulated diurnal Ψ_w values that corresponded well to those measured (Fig. 11.6). Similar models have been developed by Calkin et al. (1986), Edwards et al. (1986), Hunt and Nobel (1987), and Milne (1989). Whitehead and Hinckley (1991) cautioned that most of these attempts should be considered exploratory because of inadequate knowledge of the complexity of flow pathways in woody plants and because of technical difficulties in obtaining critical Ψ_w values of plant tissues (e.g., *in situ* measurements on roots).

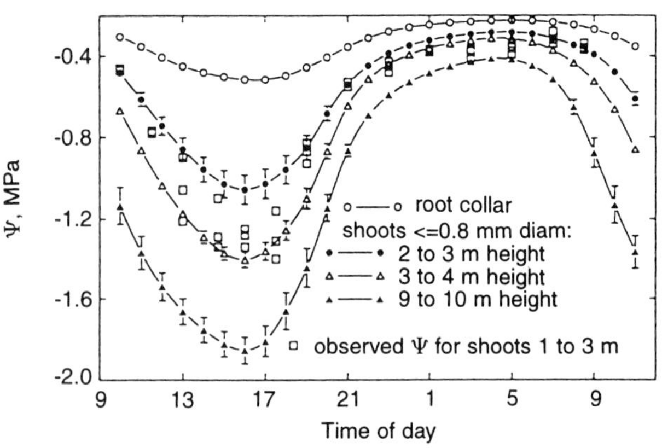

FIGURE 11.6. Predicted and measured diurnal changes in Ψ_w for the root collar and minor shoots of northern white cedar at indicated heights calculated by a model. From Tyree (1988).

It should be noted that application of the Ohm's Law equations to water flow in plants often suggests that resistance to flow may change with the rate of flow, a puzzling result that remains incompletely understood (Fiscus and Kramer, 1975). This problem is discussed further in a later section concerning water absorption.

Given the coherence the SPAC concept brings to plant water relations, I will discuss in this and the next chapter the absorption, transport, and loss of water from woody plants within this framework.

ABSORPTION OF WATER

Soil Water

In moist soil the rate of water absorption is controlled primarily by two factors—the rate of transpiration, because it largely controls Ψ_w in the root xylem, and the efficiency of root systems as absorbing surfaces. As soil dries the availability of water begins to be limited by decreasing water potential and soil hydraulic conductivity. Soil aeration, soil temperature, and the concentration and composition of the soil solution also may sometimes limit absorption of water.

Overall, the rate of water absorption depends on the steepness of the gradient in Ψ_w from soil to root and the resistance in the soil–root portion of the SPAC. As the soil dries, water becomes progressively less available because its potential decreases and resistance to movement toward roots increases. The relationship between soil Ψ_w and moisture content is shown in Figure 11.7. The readily available water is traditionally defined as that between field capacity and the permanent wilting percentage. Field capacity is the soil water content a few days after the soil has been thoroughly wetted, and downward movement of gravitational water has become very slow. The permanent wilting percentage is the soil water content at which plants remain wilted unless the soil is rewetted. It is obvious from Figure 11.7 that there is much more readily available water in fine-textured than in coarse-textured soils. Neither field capacity nor permanent wilting percentage are physical constants, but merely convenient regions on the water potential-water content curve. The permanent wilting percentage is usually said to occur at about −1.5 MPa, but this is because sunflowers or similar mesophytes were used to determine it (Slatyer, 1957). In reality, plants will continue to absorb some water from a soil until the bulk soil Ψ_w reaches

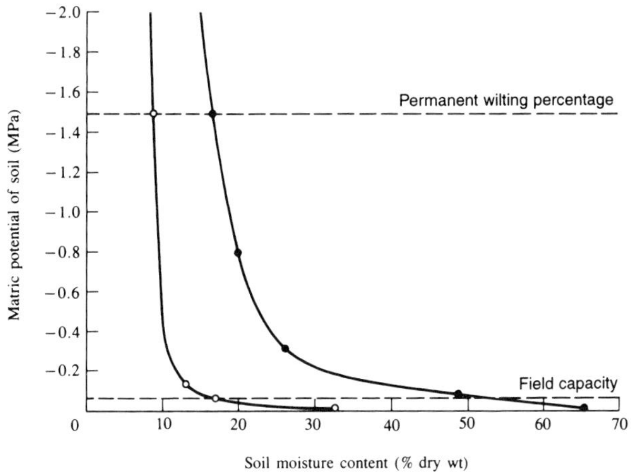

FIGURE 11.7. Matric potentials of (○) sandy loam and (●) silty clay loam soils as a function of soil water content. The curve for Panoche sandy loam is from Wadleigh et al. (1946) and that for Chino silty clay loam is from Richards and Weaver (1944).

the Ψ_π of the plant. Slatyer found the water potential of severely wilted privet plants to be as low as −7 MPa, and values much lower than −1.5 MPa have been reported for other plants (Williams et al., 1997; Sperry and Hacke, 2002).

It is well known that the resistance to water flow through soil increases rapidly as soil dries because of decreased cross section for flow and increased pathway tortuosity as the films of water decrease in thickness and discontinuities develop. Soil hydraulic conductivity may decrease by several orders of magnitude as soil dries from field capacity to −1.5 MPa (Kramer and Boyer, 1995). Additionally, there is evidence that soil-root air gaps may form in drying soil as both roots and soil shrink (Huck et al., 1970; Faiz and Weatherley, 1982; Taylor and Willat, 1983). North and Nobel (1997) noted up to 34% shrinkage roots of the desert succulent *Agave deserti* subjected to drought, reducing predicted hydraulic conductivity at the soil-root surface by a factor of five. However, by vibrating plant containers predicted hydraulic conductivity could be restored to that of moist soil. The mode of movement of water across this gap is restricted to diffusion, which is far slower that normal mass flow-based transport of liquid water.

One area of uncertainty concerns the relative size of this external resistance ($r_{soil \rightarrow root\ surface}$) to that associated with the plant in drying soil. Blizzard and Boyer (1980) found that although resistance to water flow in both soil and plant segments of the SPAC increased as the soil dried, resistance in the plant was always greater than that in the soil regardless of soil water content. Boyer (1985) suggested that reduced vascular system transport capacity, arising from xylem vessel cavitation, was responsible for such responses. In contrast, Nobel and Cui (1992) reported far greater sensitivity of the soil-root surface segment of the SPAC to drying soil compared with that of roots of several desert succulents growing in a sandy loam soil. However, it must be noted that soil texture has a great influence on hydraulic properties of the soil and the geometry of root-soil particle contact. Sandy soils, which often are used in culture of experimental plants, exaggerate soil resistance responses to water depletion (e.g., Sands and Theodorou, 1978). Finally, if plant Ψ declines sufficiently under severe water stress, hydraulic disconnection of roots from shoots (and essentially infinite transport resistance) will result from runaway cavitation of xylem. This event will be attended by shoot desiccation and death, and there is some evidence that this phenomenon occasionally occurs in nature (Davis et al., 2002).

Water may be translocated from regions of moist soil to those that are dry by movement within and release from root systems. Early experiments by Magistad and Brezeale (1929) showed that roots of *Opuntia discata* lost more water to dry soil than shoots did to the atmosphere. These authors postulated that some water absorbed by roots of plants growing in moist, deep soil could be lost to dry, upper soil layers. This early work was supported by research showing that water could be released to dry soil, but often at rates that were lower than those associated with water uptake (Molz and Peterson, 1976; Nobel and Sanderson, 1984). Water transfer between root systems also has been documented (Bormann, 1957; Hansen and Dickson, 1979). Corak et al. (1987) showed that tritiated water absorbed by deep roots of alfalfa could be detected in the leaves of corn planted in the same shallow soil volume with alfalfa.

Research that capitalizes on variation in isotopic composition of water with soil depth has shown that water absorbed by deep tree roots can be released into the upper soil. This phenomenon, called hydraulic lift or hydraulic redistribution, can provide ecologically significant quantities of water to shallowly rooted understory plants. Hydraulic lift can be evaluated for samples of water extracted from various soil layers and from xylem sap by measuring the ratio of stable isotopes of hydrogen (Deuterium, D, (^{2}H) vs. ^{1}H) usually reported as δD (‰)(Dawson and Ehleringer, 1991):

$$\delta D = \left[(D/H)_{sample} / (D/H)_{SMOW} - 1 \right] \times 1{,}000 \quad (11.9)$$

where D/H_{SMOW} represents the ratio of hydrogen isotopes in standard mean ocean water (White et al., 1985; Ehleringer and Dawson, 1992). Larger negative values of δD indicate depletion of D compared with

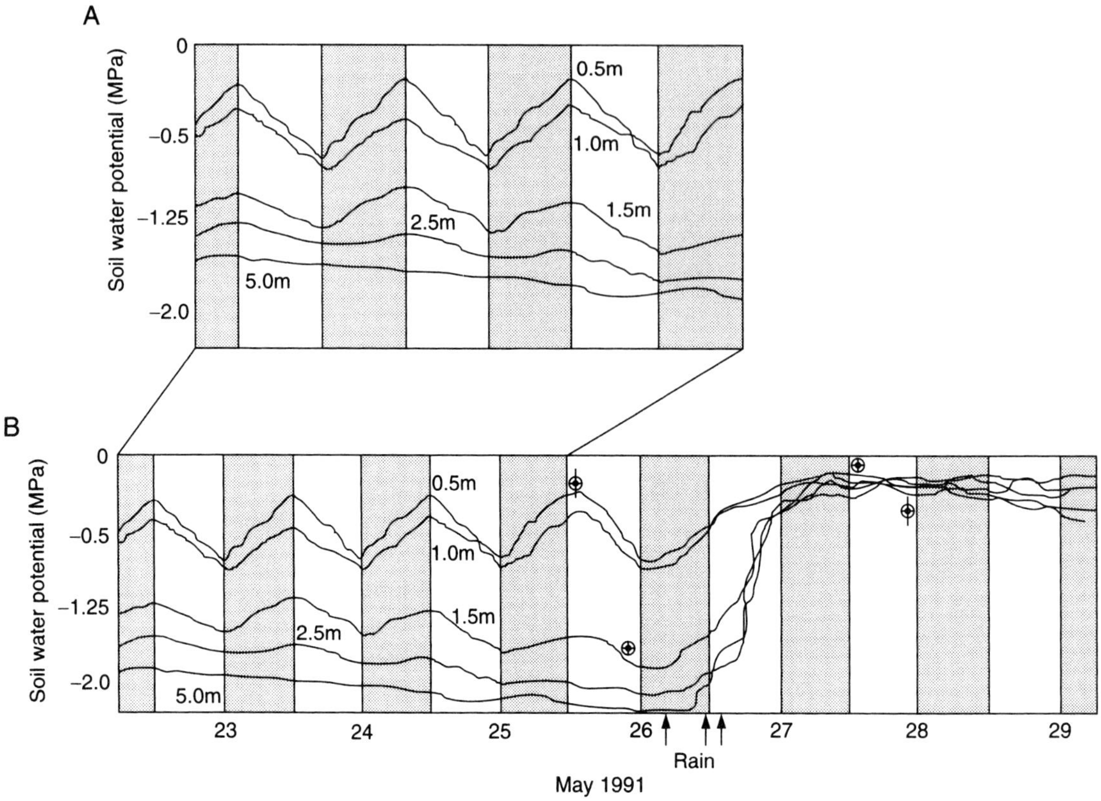

FIGURE 11.8. Diurnal patterns of mean soil Ψ_w (measured for 20 and 30 cm depths in the four cardinal compass directions) at five distances from stems of mature sugar maple trees. (A) Patterns over a 30-hr period at the end of a 16-day drought; (B) patterns over a 6.5 day period including the period displayed in (A) and periods of rain. Dark areas indicate night; light areas indicate day. Plotted symbols indicate mean values of Ψ_w at dawn and dusk on May 25 and May 27, 1991. From *Oecologia*, Hydraulic lift and water use by plants—Implications for water balance, performance and plant-plant interactions. Dawson, T. E. **95**, 565–574. Figure 1. Copyright 1993 Springer-Verlag.

^{1}H. In applications, the procedure takes advantage of differences between H isotope composition of deep soil water, which represents a rough average of annual precipitation, and that of growing season precipitation, which is enriched in D when compared with annual averages (Dansgaard, 1964). Xylem sap δD should be a weighted average of hydrogen isotopic composition of the soil solution in the rhizosphere of a plant, as there is apparently no significant discrimination in uptake by roots and translocation in the xylem (see reviews by White, 1989; Dawson and Ehleringer, 1991; and Ehleringer and Dawson, 1992). Hence if δD of xylem sap of plants sampled during the growing season is closer to the δD value of deep soil water than it is to that of summer precipitation, it constitutes evidence that the plant is obtaining water, directly or via deep-rooted plants, from deep soil sources.

Dawson (1993) employed stable hydrogen isotope analysis to show that water absorbed by deep roots of sugar maple trees was released into shallow soil layers where it was absorbed by understory plants. Soil Ψ_w measurements next to trees showed a cyclic diurnal pattern of water depletion during the day followed by replenishment at night (Fig. 11.8). The cycle damped at increasing distance from the stem. Soil Ψ_w and δD values for shallow soil water (30 cm depth) also clearly showed a gradient of decreasing availability of water away from the stem and a decreased proportion of translocated deep soil water in shallow soil (Fig. 11.9). Hydrogen isotope composition, Ψ_l, and leaf diffusion conductance (g_l) measurements of several species of herbaceous and woody plants growing at various distances from sugar maple tree stems further showed that released water was absorbed by plants and consequently increased Ψ_w and g_l values and shoot growth in most species.

Other work employing both isotopic methods and careful study of soil water potential and water content dynamics has shown that transfer of water within a

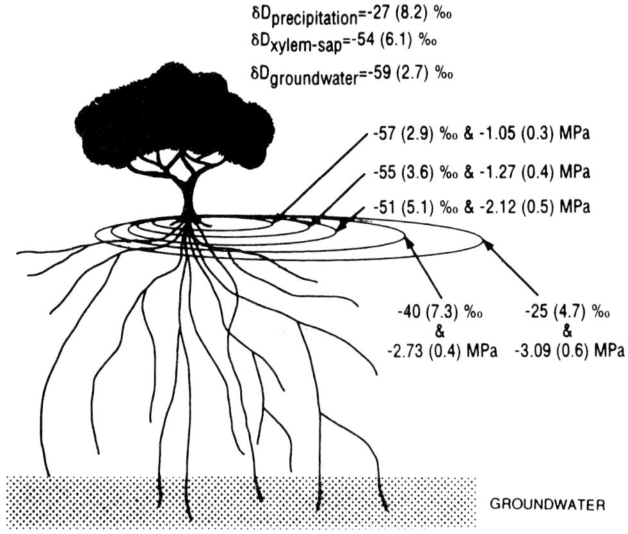

FIGURE 11.9. Mean (±SD) stable hydrogen isotopic composition (δD, ‰) and soil Ψ_w at 30 cm depth at 5 distances (0.5, 1.0, 1.5, 2.5, 5.0 m) from stems of mature sugar maple trees. Also shown are δD values of precipitation, xylem sap of sugar maple trees, and groundwater. From *Oecologia*, Hydraulic lift and water use by plants—Implications for water balance, performance and plant-plant interactions. Dawson, T. E. **95**, 565–574. Figure 3. Copyright 1993 Springer-Verlag.

soil profile via plant roots systems is widespread and need not be limited to upward movement. For example, Burgess et al. (2001) identified downward sap flow in tap roots of *Eucalyptus camaldulensis* trees after wetting rains broke a seasonal drought in southwest Australia. This movement was associated with increased soil water content between 180 and 260 cm depths in the soil and no change in soil water content between 90 and 140 cm. The latter observation indicated that the deep soil changes did not arise from a wetting front moving downward in the soil after rain. Lateral movement has also been reported in ponderosa pine and Douglas-fir-dominated ecosystems of the Pacific Northwest United States (Brooks et al., 2002). Hence "hydraulic redistribution, HR" has superseded hydraulic lift to describe this phenomenon (Burgess et al., 1998). More than 50 cases of HR have been documented in both woody and herbaceous species (Caldwell et al., 1998; Jackson et al., 2000).

The amount of redistribution and quantity of redistributed water contributing to subsequent transpiration of trees varies with total soil water content. Not surprisingly, redistribution is minimal early in the growing season in a moist soil, as strong water potential gradients to drive flow are lacking (Fig. 11.10). Warren et al. (2007) estimated that HR amounted to between 3 and 9% of total seasonal water use in ponderosa pine and Douglas-fir ecosystems in the Pacific Northwest. However, the fractional contribu-

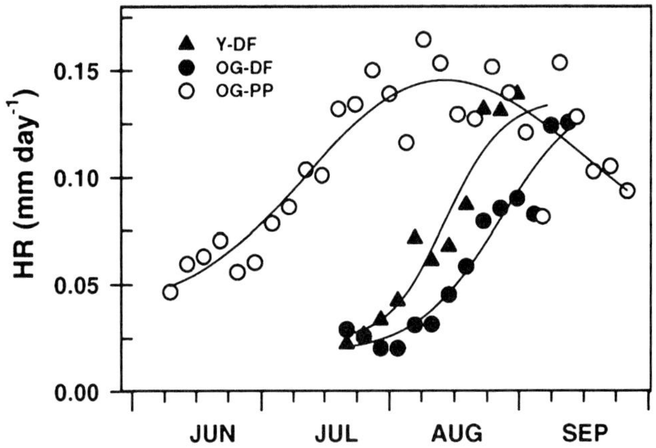

FIGURE 11.10. Hydraulic redistribution (HR) of soil water by roots for young- and old-growth Douglas-fir stands (Y-DF, OG-DF) and for old growth ponderosa pine (OG-PP) based on diurnal fluctuations in soil water content in the upper 15–60 cm during a seasonal drought. This figure was published in *Agric. For. Meteorol.*, Vol. 130, Warren, J. M., Meinzer, F. C., Brooks, J. R., and Domec, J. C., Vertical stratification of soil water storage and release dynamics in Pacific Northwest coniferous forests, 39–58. Copyright Elsevier 2005.

tion of redistributed water to whole-stand transpiration during seasonal drought is greater. For example, during seasonal droughts in temperate coniferous and tropical savanna ecosystems up to 90% of water withdrawn from shallow soil was replaced by redistributed water (Meinzer et al., 2004). Maximum HR of water in temperate coniferous forests approached 0.15 mm day^{-1} (Fig. 11.10) when total stand transpiration ranged from 0.5 to 1.0 mm day^{-1}, suggesting that if all this water were absorbed following transfer, 15 to 30% of daily stand transpiration during late-season droughts could potentially arise from HR water. Brooks et al. (2002) similarly estimated that 28 and 35% of daily transpiration of Douglas-fir and ponderosa pine forests, respectively, could be replaced by HR water during late season droughts. Lee et al. (2005) employed an atmospheric general circulation model to estimate the impact on transpiration of HR where it likely is a significant factor in soil water relations. Over Amazon forests, for example, the HR-inclusive model resulted in an increase in transpiration of 40% during the dry season, reconciling much of the underestimation by models of transpiration in these regions when compared to eddy covariance measurements.

There may be adaptive costs to hydraulic redistribution, as release of HR to the soil makes it available to competitors. However, in addition to increasing the subsequent water available in the rhizosphere the next day, the release of water in the vicinity of roots may have benefits of increased nutrient acquisition in otherwise dry soil that outweigh these costs (Caldwell et al., 1998).

Concentration and Composition of Soil Solution

The soil water potential is controlled by the surface forces that bind water in capillaries and on surfaces and the reduction in water activity produced by dissolved solutes. If the osmotic potential is lower than −0.2 or −0.3 MPa, plant growth is likely to be retarded even in soils with a water content near field capacity. However, excessive salinity is common only in arid regions where evapotranspiration greatly exceeds rainfall, and it is seldom a problem in forested areas. It sometimes is a problem for fruit trees in dry areas where the irrigation water often contains appreciable amounts of salts. Also, there is some interest in identifying woody plants suitable for use in coastal areas where salt spray causes injury. Effects of salinity on plants are discussed in more detail in Chapter 5 of Kozlowski and Pallardy (1997).

Soil Aeration

The growth and physiological activity of roots often are reduced by a deficiency of oxygen. Though this is most severe in flooded soil, a chronic but moderate deficiency often exists in heavy clay soil that limits root penetration and possibly the uptake of mineral nutrients. Flooding soil with water usually drastically reduces water absorption (Fig. 11.11) because it increases the resistance to water flow into roots and because flooding induces stomatal closure (Kozlowski and Pallardy, 1979; Sena Gomes, 1986). Hanson et al. (1985) showed that such reductions in absorption by seedlings could largely be attributed to anaerobiosis in the root zone, as water flux into red pine seedlings grown in solution culture was drastically and reversibly curtailed when aerating gas was replaced with N_2 (Fig. 11.12). Oxygen deficiency may reduce the capacity of water channel proteins to pass water through root membranes in response to water potential gradients (see later). Inhibition of water channel activity in the model plant *Arabidopsis thaliana* was mediated by acidification of the cytoplasm that accompanied anoxic conditions (Tournaire-Roux et al., 2003).

Hook (1984) emphasized the capacity of highly flood tolerant species to develop new roots in flooded soil from primary and major secondary roots and to

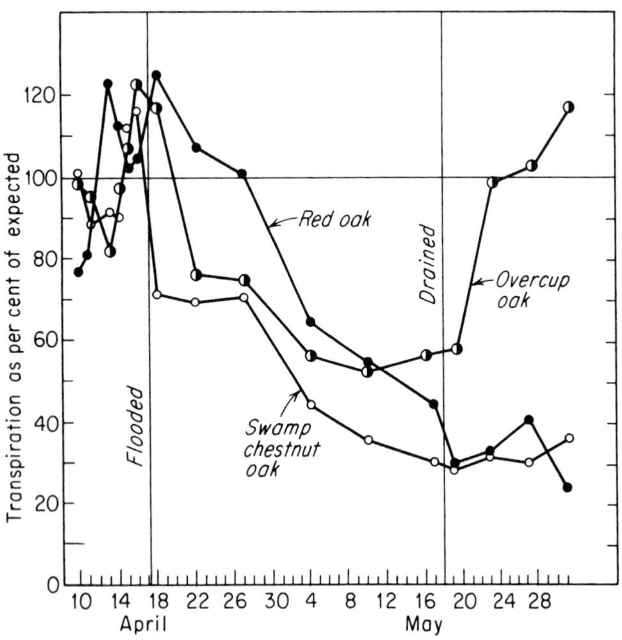

FIGURE 11.11. Effects of flooding on soil water absorption as indicated by changes in the rate of transpiration. Seedlings lost most leaves during flooding, but overcup oak leafed out again when the soil was drained. From Parker (1950). © American Society of Plant Physiologists.

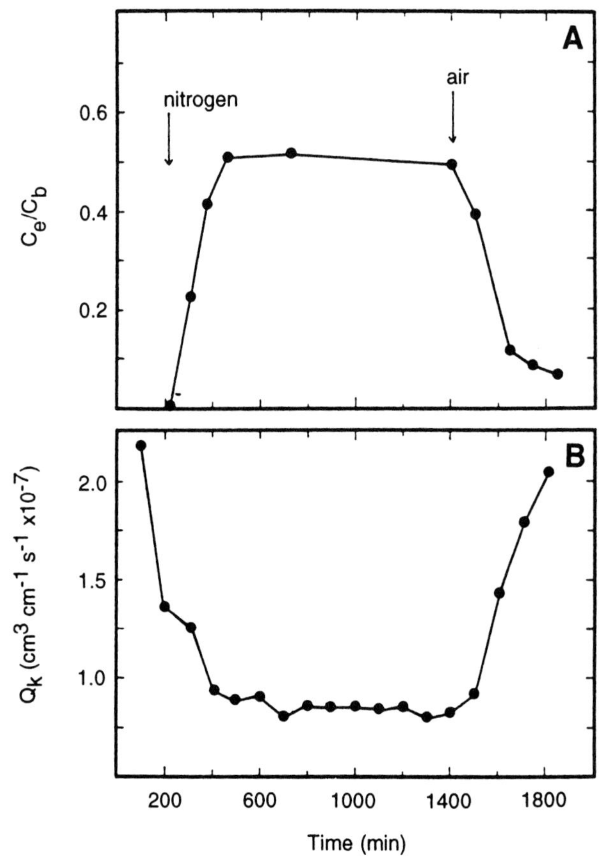

FIGURE 11.12. (A) Relative apoplastic flow of water into root xylem (C_e/C_b) under pressure in red pine seedlings suspended in a pressurized, solution-filled chamber. (B) Water flux per unit root length (Q_k) through the red pine root system. Arrows indicate when the solution was rendered hypoxic (by bubbling of N_2) and when air was reintroduced. Results are shown for a typical seedling. From Hanson et al. (1985).

transport O_2 by diffusion to the rhizosphere through intercellular spaces (see also Coutts and Phillipson, 1978a,b; Fisher and Stone, 1990). Sena Gomes and Kozlowski (1980) observed that water uptake by flooded plants of green ash with adventitious roots growing into flood water above the soil was 90% greater than by those from which these roots had been removed. Root system adaptations thus may be the most important factors in determining flood tolerance and water uptake properties in flooded plants. In response to low oxygen concentration in the root zone, many species form aerenchyma tissues in roots and stems, through which oxygen moves readily. Aerenchyma tissue, with large intercellular spaces, forms by both cell separation (schizogeny) and disintegration (lysigeny) (Angeles, 1992). Species showing such responses include *Pinus contorta* (Coutts and Philipson, 1978b) and *P. serotina*. Some species also show capacity for internal stem aeration by high cambial permeability (Topa and McLeod, 1986). Lysigenous air-space formation arises from programmed cell death of certain cells in a process that is mediated by ethylene accumulation under low CO_2 concentration (Jackson and Armstrong, 1999). Ethylene production is stimulated because of increases in both its immediate precursor (ACC, 1-aminocyclopropane-1-carboxylic acid) and the enzymes involved in ACC synthesis and subsequent conversion to ethylene. Kozlowski (1982b) and Kozlowski and Pallardy (1984, 2002) reviewed the water relations of flooded plants (see also Chapter 5, Kozlowski and Pallardy, 1997).

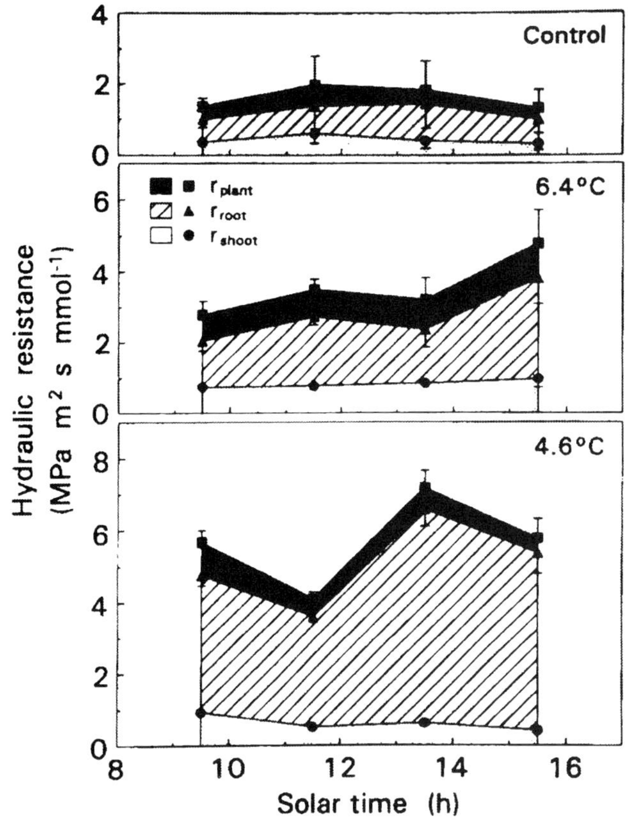

FIGURE 11.13. Diurnal pattern of apparent plant (r_{plant}), root (r_{root}), and shoot (r_{shoot}) hydraulic resistance at two low soil temperatures (6.4°C, 4.6°C) compared to values at a control soil temperature of 11.5°C. Values for all resistances are plotted from the abscissa; bars indicate 1 standard deviation. From Pavel and Fereres (1998).

Soil Temperature

Many writers regard cold soil as an important ecological factor, and the decreased availability of water in the cold soil at high altitudes may affect vegetation (Whitfield, 1932; Clements and Martin, 1934) and location of timberline (Michaelis, 1934). Poorly drained soils are slow to warm in the spring, and Firbas (1931) and Döring (1935) stated that the cold soils of European high moors limit plant growth. Cameron (1941) reported that orange trees often wilt during the winter in California because of slow absorption of water from cold soil. Pavel and Fereres (1998) reported that root hydraulic resistance of olive seedlings was far higher at 6.4°C soil temperature compared to that at 11.5°C, and root resistance dominated whole-plant resistance to liquid water flow at the low soil temperature (Fig. 11.13). As the air temperature rises above freezing during sunny or spring days, the vapor pressure gradient between the leaves of evergreens and air is increased, leading to increase in transpirational water loss and dehydration of leaves. Winter desiccation can be a serious problem in conifers (Chapter 5, Kozlowski and Pallardy, 1997).

Considerable differences exist among species in effects of low temperature on water absorption. Usually, species from warm climates show greater reduction than species from cold climates. Kozlowski (1943) found that water absorption was reduced more in loblolly pine than in eastern white pine as the soil temperature was reduced from 15 to 5°C. Similar results were obtained by Day et al. (1991), who noted that the effects of low root system temperatures had a larger impact on shoot Ψ_w of loblolly pine compared with lodgepole pine, which is native to cooler montane regions (DeLucia et al., 1991). Kaufmann (1975) showed that subalpine Engelmann spruce showed much less increase in liquid flow resistance at 5°C than did subtropical citrus.

Cold soil reduces water uptake in two ways—directly, by decreasing the permeability of roots to water, and indirectly, by increasing the viscosity of water, which slows its movement through both soil and roots. Lee and Chung (2005) reported similarities between low temperature effects on root water uptake on two herbaceous cucurbit species and inhibitory effects of mercuric chloride (a known inhibitor of water channel activity), suggesting that water channel activity (open or closed state) may respond to root temperature. The capacity for at least some water absorption in cold soil would appear to provide the necessary capacity for winter transpiration needs of many conifers (Lassoie et al., 1985) and to prevent severe water stress from developing during the early part of the growing season in cold regions, when soil temperatures lag behind air temperatures because of high soil heat capacity and the presence of residual snow packs. The presence of a thick snowpack in winter often keeps soil water from freezing, thereby rendering it available for uptake.

Comparatively few data are available concerning the effects of high soil temperatures on water absorption and plant water relations. Bialoglowski (1936) and Haas (1936) reported that temperatures above 30°C reduced water absorption of lemons, grapefruit, and Valencia orange trees. More recently, McLeod et al. (1986) observed that, whereas floodwater temperatures between 30 and 35°C had little influence on stomatal conductance and water use efficiency in several flood tolerant species (water tupelo, baldcypress, button bush, and black willow), higher temperature (40°C) floodwater induced stomatal closure and reduced water use efficiency. In contrast, Nobel and Lee (1991) found that elevating the soil temperature from 5 to 45°C resulted in an increase in root Ψ_w and root hydraulic conductivity in two succulent species (*Agave deserti* and *Opuntia ficus-indica*). The apparent beneficial influence of high temperature in water relations of succulent species is consistent with the environment in which they exist.

Secondary effects of soil temperature, such as decreased root extension, have an impact on whole-plant water relations and these restrictions reach their extreme limits in the shallow permafrost depths in taiga forests (Oechel and Lawrence, 1985). Low soil temperatures also may alter root metabolism; however, in most situations these indirect influences are believed to be much less important than the direct effects on resistance to water flow. Effects of temperature on root growth are discussed further in Chapter 5 of Kozlowski and Pallardy (1997).

Absorption through Leaves and Stems

Although the quantity of water absorbed through leaves and stems is very small, it has received considerable attention. The early literature, reviewed by Miller (1938, pp. 188–190), showed that significant amounts of water can enter the leaves of many kinds of plants. The fact that the cuticle, when wetted, is moderately permeable permits foliar fertilization, which is discussed in Chapter 7 of Kozlowski and Pallardy (1997). There also is some absorption of mineral nutrients and presumably of water through lenticels and other gaps in the bark, and even through leaf scars.

Atmospheric Moisture

Woody plants absorb some liquid water, water vapor, and dew from the atmosphere. These sources are sometimes ecologically important because they influence leaf hydration. Fog consists of small water drops suspended in air supersaturated with water vapor. Dew consists of condensed water deposited when the temperature of a surface decreases below the temperature of the dew point of the surrounding air.

Water on leaves influences plant hydration by entering the plant and by decreasing transpiration. Hence leaf water deficits are reduced and turgor is increased, thereby stimulating plant growth. Interestingly, constant leaf wetness induced by misting may reduce photosynthesis and growth compared with well-watered plants with dry leaves (Ishibashi and Terashima, 1995; see also Chapter 5). Some such effects may actually be indirect as a result of pollutants in the dew (e.g., OH radical) (Yoon et al., 2006), but in water-stressed plants the effect of leaf wetting with unpolluted water will nearly always be beneficial. Small amounts of water are absorbed through the leaf cuticle, which is moderately permeable when wetted, by twigs through lenticels, other gaps in the bark, and through leaf scars. Uptake of water by leaves depends on a gradient of decreasing Ψ_w from the atmosphere to the leaves. Such gradients often occur in plants of arid regions and those in humid regions experiencing drought.

Wettability of leaves depends on the distribution of surface waxes and on the contact angle between liquid droplets and the leaf surface. The greater the contact angle the more difficult it is to wet the leaf surface. Needles of Monterey pine are better adapted than those of Scotch pine for foliar absorption because waxy outgrowths of the former species cover less of the needle surface (Leyton and Armitage, 1968). Foliar uptake of water is increased by such leaf structures as trichomes, hydathodes, and specialized cuticle structures (Rundel, 1982). Although Grammatikopoulos and Manetas (1994) could not detect direct uptake of water through hairy leaves of *Phlomis fruticosa* and two species of mullein, there was greater uptake of water sprayed on the upper leaf surface than in species with nonhairy leaves. There also was greater retention of water on sprayed leaves and the authors also suggested that hairy surfaces served to retard evaporation through creation of a thicker boundary layer.

Some foggy coastal areas support luxuriant vegetation. For example, dense cloud forests are found at high elevations where drizzles and fogs are frequent throughout the year. On the upper windward slopes of the Sierra Madre of eastern Mexico, for example, fog droplets collect on leaves and branches of trees, coalesce into larger drops, and fall to the ground, thus increasing the soil water content (Vogelmann, 1982). The soil under the crowns of pine trees was saturated to a depth of 8 to 10 cm, whereas beyond the crowns the soil was powder dry. Azevedo and Morgan (1974) collected more than 40 cm of fog water under the crowns of single Douglas-fir trees in the coastal fog belt of northern California. Ellis (1971) reported that fog drip supplemented annual rainfall by as much as 40% in a high-altitude eucalyptus forest in Australia.

Several investigators found evidence of water uptake by leaves in saturated water vapor or from liquid water. Stone and Fowells (1955) showed that in greenhouse experiments dew increased the survival of seedlings of ponderosa pine. Absorption of water by pine seedlings from a saturated atmosphere also was observed by Stone et al. (1950). Went and Babu (1978) showed water uptake and elevation of Ψ_l by up to 0.3 MPa in *Cucumis* and *Citrullus* plants on which formation of dew was induced by energy exchange between the leaves of the plants and cold surfaces. Sharma (1976) found that dew could be present from early evening (7 P.M.) to midmorning (9 A.M.) in *Paspalum* pastures. However, usually the amounts absorbed are very small and translocation of water within plants is very slow.

Some investigators claimed that plants could absorb water from a saturated atmosphere and transport it downward through the plant and into the soil (Slatyer, 1956). For example in the Atacama Desert of southern Chile, an essentially rainless area, *Prosopis tamarugo* plants can absorb water vapor from the air. Because of the very low Ψ_w of the salty surface soil, the foliar-absorbed water may move downward through the plant and into the soil (Went, 1975).

Given the effective barrier to water loss provided by the cuticle of most species, it is unlikely that water uptake via this pathway is substantial, although it may be important in certain circumstances. The ecological significance of dew depends on: (1) its amount and duration, (2) the uptake by plants, which depends on leaf display patterns and cuticle transport properties, (3) the physiological responses of plants to elevated nighttime Ψ_l, and (4) alternative sources of plant water (i.e., soil water). These criteria suggest that the greatest potential direct benefit of dew might accrue to actively growing seedlings in openings during periods of low soil water availability. This situation would present the conditions for maximum leaf area exposed to the night sky, and dew accumulations under these conditions might permit average diurnal, and especially nighttime Ψ_l, to be significantly greater than otherwise would be possible. The influence of dew in such situations deserves further study.

Dew also may exert some indirect ecological influence by moving the evaporating water surface *outside* the plant body for a time, substituting evaporation of external water for internal water loss by transpiration. Chaney (1981) provided a good review of atmospheric sources of water and the potential impact on plant water relations.

The converse of absorption through leaves is the leaching of minerals out of leaves by rain or sprinkler irrigation (Tukey et al., 1965). Dew and fog drip also cause leaching (Chapter 10). According to Madgwick and Ovington (1959), deciduous species lose more nutrients than conifers by leaching during the summer, but leaching from conifer leaves continues during the winter. Obviously if solutes can be leached out, uptake of water and solutes can also occur when liquid is present on leaf surfaces and the gradient in water potential is favorable.

Absorption through Roots

Root systems of woody perennial plants consist of roots in all stages of development from delicate, newly formed, unsuberized tips less than 1 mm in diameter to old woody roots covered with a thick layer of bark and having a diameter of many centimeters (Chapter 2). Furthermore, roots often are modified by the presence of mycorrhizal fungi. As a result, there are wide variations in permeability of roots to water and ions, as is shown in Table 11.3. Figure 11.14 shows the tissues through which water must pass to enter young roots of yellow-poplar seedlings. The walls of the epidermal and cortical cells are composed largely of cellulose at this stage, but the walls of the endodermal cells are already beginning to thicken. Strips of suberized tissue, the Casparian strips, develop on the radial walls and generally are assumed to decrease their permeability to water and solutes. However, research suggests that the endodermis does not always form a permanent impermeable barrier to water and solutes. For example, as young roots develop in the pericycle and push out through the endodermis, gaps are produced through which water and solutes can enter freely until the endodermis of the branch and parent roots is con-

TABLE 11.3. Relative Permeabilities of Grape Roots of Various Ages to Water and ^{32}P[a,b]

Zone and condition of roots	Relative Permeabilities	
	Water	^{32}P
Roots of current season		
Growing		
Terminal 8 cm, elongating, unbranched, unsuberized	1	1
Unsuberized, bearing elongating branches	155	75
Dormant		
Main axis and branches dormant and partially suberized before elongation completed	545	320
Main axis and branches dormant and partially suberized	65	35
Roots of preceding seasons		
Segments bearing branches		
Heavily suberized main axis with many short short suberized branches	0.2	0.04
Segments unbranched		
Heavily suberized, thick bark, and relatively small xylem cylinder		
Intac	0.2	0.02
Decorticated	290	140

[a]From Queen (1967).
[b]Measurements were taken under a pressure gradient of 660 mbar.

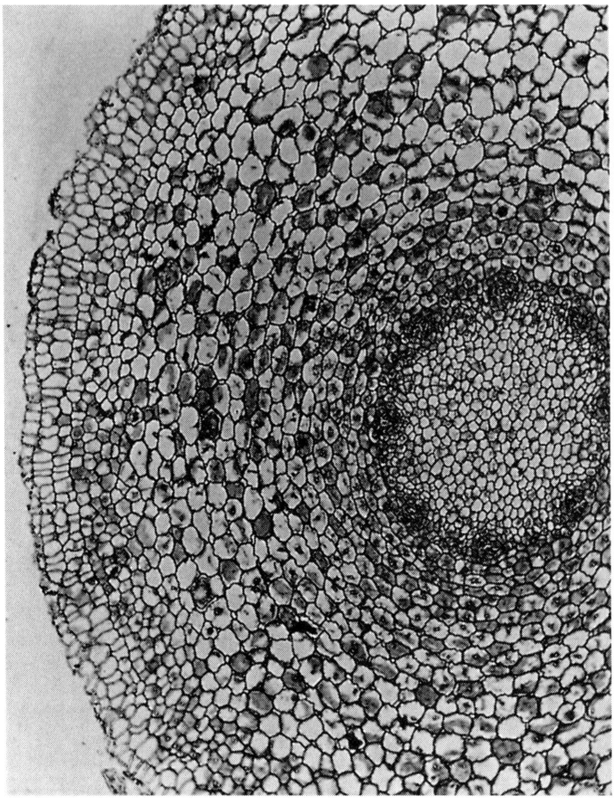

FIGURE 11.14. Cross section of a young root of yellow-poplar (×80) about 0.6 mm behind the apex. Note the thick layer of cortical parenchyma surrounding the stele. From Popham (1952), by permission of the author.

nected (Queen, 1967; Dumbroff and Peirson, 1971; Skinner and Radin, 1994). Clarkson et al. (1971) suggest that water and solute movement may occur through plasmodesmata in the endodermal cell walls. Whatever the details, it seems certain that significant quantities of water and ions cross the endodermis many centimeters behind the root tip in herbaceous plants and probably also in woody plants.

As roots grow older, the epidermis, root hairs, and part of the cortex are destroyed by a cork cambium that develops in the outer part of the cortex (Fig. 11.15, see also Chapter 3). Eventually, even the endodermis is lost because of cambial activity and the root consists of xylem, cambium, phloem, and a suberized layer in the outer surface of the phloem.

The pathway of radial water movement in roots has long been a subject of debate and remains so to this day. Early workers assumed that water and solutes move from cell to cell across the vacuoles of the cells lying between the root surface and the xylem, but the experiments of Strugger (1949) suggested that considerable movement of water may occur in the cell walls. This view was questioned by Newman (1976) who argued that most water movement probably occurs in the symplast. The relative importance of these pathways depends upon both the comparative hydraulic conductivities and cross-sectional areas (presented to inward-moving water) of cell membranes and cell walls. Some evidence from root pressure probe experiments (Steudle and Jeschke, 1983; Steudle et al., 1987), and studies of rehydration kinetics of tissues suggest that water flow through roots may be primarily symplastic when transpiration is low but apoplastic at high flow rates (Boyer, 1985; Steudle, 2000). However, the issue remains unsettled.

Symplastic water flow requires movement across at least the plasma membrane. It was long known that hydraulic conductivity of membranes was substantially higher than could be accommodated by a simple lipid bilayer. In the last 15 years, the existence of water channel or aquaporin proteins in membranes has been elucidated. In the plasma membrane these proteins have six cylindrical membrane-spanning α-helix regions that define a central pore (Fig. 11.16).

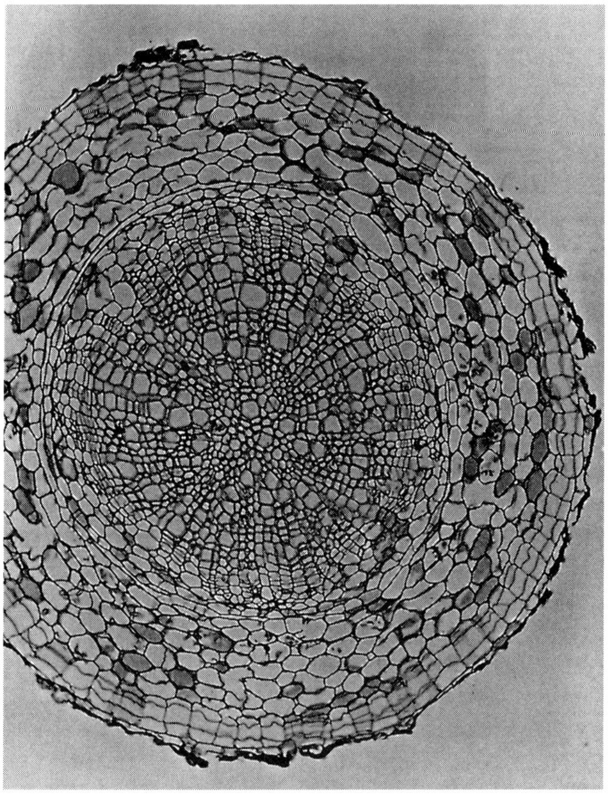

FIGURE 11.15. Cross section of an older root of yellow-poplar (×60) from which most of the outer parenchyma has sloughed off. A layer of suberized tissue is developing at the outer surface. From Popham (1952), by permission of the author.

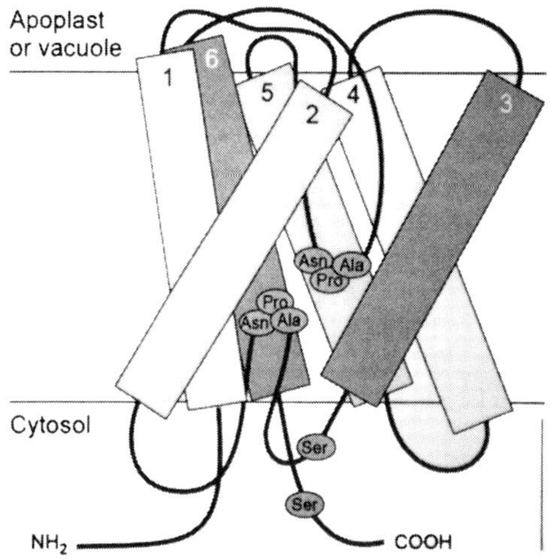

FIGURE 11.16. Basic structure of water channel proteins. The side with the N- and C-termini faces the cytosol, whereas the other side faces the apoplast. There are six transmembrane helices. The first cytosolic loop and the third extracytosolic loop, each of which contain a conserved Asn-Pro-Ala (NPA) loop, are relatively hydrophobic and probably dip into the membrane from opposite sides creating a seventh transmembrane structure. The shading of the helices reflects the internal homology, where the N-terminal half of the protein is homologous to the C-terminal half, although the two halves are inserted inversely in the membrane. Thus, helix 1 corresponds to helix 4, helix 2 to helix 5, and helix 3 to helix 6. Conserved serine residues that are known to be phosphorylated and dephosphorylated are also depicted. This schematic model is based on the structure for human aquaporin 1. This figure was published in *Trends Plant Sci.*, Vol. 4, Kjellbom, P., Larsson, C., Johansson, I., Karlsson, M., and Johanson, U., 308–314. Copyright Elsevier 1999.

Their amino and carboxyl ends are in the cytoplasm. Within the narrow pore there are two additional half-spanning helices, each with a key sequence of the three amino acids asparagine-proline-alanine. The asparagines in these loops, combined with facing hydrophobic amino acids, orient the water molecule as it passes in a single file such that it is stripped of protons (a key requirement that sustains important proton gradients across membranes) (Murata et al., 2000). Within the membrane, four proteins associate as a tetramer. The movement of water through the channels is a passive process in response to water potential gradients.

Genome searches have identified more than 30 aquaporin genes in maize and *Arabidopsis*, which number suggests that slightly different forms of the protein might be expressed in different tissues and circumstances (Luu and Maurel, 2005). Certain aquaporin proteins pass only water; others are know to pass other small molecules such glycerol, urea, NH_3/NH_4^+ and CO_2, either specifically or along with water (Kaldenhoff and Fischer, 2006). Research also has indicated that water channels can be configured in an open or closed conformation by at least two cellular conditions (Tornroth-Horsefield et al., 2006). Phosphorylation and dephosphorylation of serines in the cytoplasmic domain (Fig. 11.16) cause the pores to open and close, respectively. There also is evidence that protonation of a cytoplasmic histidine residue also causes closure of the pore. Study of cellular and environmental regulation of pore activity is just beginning (Luu and Maurel, 2005; Tornroth-Horsefield et al., 2006). Suppression and enhancement of expression of aquaporin genes confirm the importance of these proteins in water relations at the whole-plant level (Siefritz et al., 2002; Lian et al., 2004).

Presumably the Casparian strips block much of the water movement through the walls of endodermal cells and thus significant amounts of water flow through the symplast at *some* point. There also is some evidence that the exodermis sometimes found at the periphery of the root cylinder provides a barrier to apoplastic flow of water (Steudle and Peterson, 1998). Hanson et al. (1985) demonstrated that apoplastic flow into the root xylem of young red pine seedlings remained very low unless the roots were deprived of adequate oxygen (Fig. 11.12). However, as stated earlier, there probably is considerable mass flow through gaps in the endodermis, particularly in older roots. In support of this idea, Steudle et al. (1993) found little increase in hydraulic conductivity in maize roots in which the endodermis was mechanically punctured. Boyer (1985), Passioura (1988b), Steudle and Peterson (1998), and Steudle (2000, 2001) reviewed water flow into and through roots.

Considerable absorption of water occurs through the older suberized roots of woody plants. Newly planted tree seedlings bear few or no unsuberized roots and some survive for many months without producing new roots (Lopushinsky and Beebe, 1976). Kramer and Bullock (1966) found that during the summer less than 1% of the root surface under stands of yellow-poplar and loblolly pine was unsuberized, and Head (1967) found a marked reduction in production of new roots on apple and plum trees during the summer. All these investigators concluded that considerable amounts of water and mineral nutrients must be absorbed through suberized roots. Direct measurements also show absorption of water and solutes through older, suberized roots (Table 11.3; Chung and Kramer, 1975; van Rees and Comerford, 1990; Nobel et al. 1990). MacFall et al. (1990), using magnetic resonance imaging, provided some evidence showing depletion of soil water in proximity to woody taproots, lateral roots, and mycorrhizas of loblolly pine. Depletion of water occurred first around the taproot.

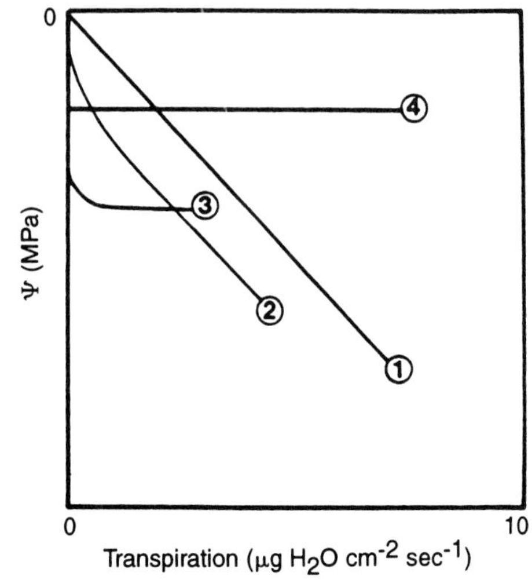

FIGURE 11.17. Four frequently observed patterns of response between leaf water potential (Ψ_w) and transpiration rate. Patterns of response are as follows: 1, linear decreasing, 2 and 3, curvilinear decreasing, and 4, horizontal (completely compensating). From Wenkert (1983).

Root Resistance

Besides the changes in resistance to water flow caused by maturation of roots, there are many examples of changes in apparent plant resistance related to rates of transpiration and water flow through the plant. Evidence of these changes often is derived from plots of the relationship between Ψ_w and transpiration rate based on a rearrangement of Eq. 11.7:

$$R_{(soil \rightarrow leaf)} = -\frac{(\Psi_{w(leaf)} - \Psi_{w(soil)})}{T} \qquad (11.10)$$

There is diversity in apparent flow resistance patterns, with the relationship between Ψ_l and transpiration ranging from horizontal, to linearly declining, to curvilinear (Wenkert, 1983; Pallardy, 1989) (Fig. 11.17). Curves for certain species, particularly herbaceous annuals (Kaufmann, 1976), exhibit essentially flat slopes, suggesting a completely compensating decline in apparent flow resistance as transpiration increases such that Ψ_l is maintained constant across a broad range of transpiration rates. For other species, including woody species such as poplars (Pallardy and Kozlowski, 1981) and spruce (Kaufmann, 1975), the relationship between Ψ_l and transpiration is curvilinear, indicating decreasing, but not compensating, resistance with increasing flux. Although the type of response observed for a species may not always be consistent (for example two types of response were shown in different studies of *Helianthus*, *Zea*, and *Gossypium*) (Kaufmann 1976), these observations of varying resistance are too frequent to dismiss.

The reasons for this response and even the location in the plant associated with it are not known. Boyer (1974) concluded that apparent variation in flow resistance was associated with sequential operation of a high-resistance symplastic pathway in growing leaves under low transpiration and a low resistance apoplastic pathway when transpiration quickened. However, most evidence indicates roots as most likely responsible for variable flow resistance. Several possible explanations for these apparent changes have been advanced, including failure of the simple Ohm's Law Model to account for coupled water and solute uptake by roots (Fiscus and Kramer, 1975; Dalton et al., 1975; Markhart and Smit, 1990), turgor-linked changes in root hydraulic conductivity and in the pattern of water uptake by whole root systems (Brouwer, 1953; Pallardy and Kozlowski, 1981; Reid and Huck, 1990), and shifts in root water transport from a primarily symplastic to an apoplastic pathway (Koide, 1985; Hallgren et al., 1994; Steudle, 1994, 2000). The mechanism(s) responsible for variable flow resistance certainly deserve more attention. Whatever the mechanism, the observed changes in apparent resistance have implications for whole-plant water relations because a certain level of homeostasis (in some species complete homeostasis) is conferred, by which water potential depression is moderated at high transpiration rates. Diurnal changes in apparent root resistance also have been reported, with a minimum at midday and a maximum near midnight (Parsons and Kramer, 1974). These cycles may be related to diurnal rhythms in water channel gene expression and regulation (Lopez et al., 2003; Uehlein and Kaldenhoff, 2006).

Extent and Efficiency of Root Systems

Success of all kinds of plants with respect to water and mineral absorption depends on the extent and permeability of roots. These problems are dealt with in detail by Kramer and Boyer (1995) and by Caldwell (1976). As was mentioned in Chapter 2, most trees have root systems that extend beyond the spread of the crown and as deeply into the soil as aeration and soil physical structure permit. Root systems often are concentrated at shallow depths; often root density is greatest in the 30 cm below the soil surface (Table 11.4).

The limits on roots system depth can range from a few centimeters, as in certain *Nothofagus* forests of Tierra del Fuego (Fig. 11.18) to more than 50 m as in certain juniper, mesquite, and eucalyptus species (Stone and Kalisz, 1991). The latter authors surveyed the literature for maximum extent of root systems of a wide variety of gymnosperms and angiosperms (Table 11.5). General correlations of maximum depth of rooting with climate are apparent, as one finds that species native to semi-arid regions (e.g., acacia, mesquite, pine, and juniper) send down roots far below the soil surface, whereas other species may possess very superficial root systems. Similarly, taxa with wide eco-

TABLE 11.4. Total Root Lengths of Scotch Pine Trees per Unit Soil Area (L_a) and Soil Volume (L_v) with Average Root Diameters[a,b]

	Horizon (cm)	L_a ($cm\ cm^{-2}$)	L_v ($cm\ cm^{-3}$)	Root diameter (mm)
1	(0–15)	80.17 ± 6.08	5.26	0.278 ± 0.04
2	(15–30)	19.74 ± 1.38	1.25	0.325 ± 0.04
3	(30–45)	9.30 ± 0.79	0.61	0.475 ± 0.07
4	(45–61)	5.17 ± 0.79	0.34	0.508 ± 0.08
5	(61–76)	3.00 ± 0.40	0.19	0.553 ± 0.08
6	(76–91)	2.85 ± 0.10	0.18	0.531 ± 0.10
7	(91–106)	1.29 ± 0.35 [2.51]	0.08 [0.16]	0.618 ± 0.21
8	(106–122)	1.38 ± 0.40 [3.11]	0.09 [0.20]	0.640 ± 0.18
9	(122–137)	0.94 ± 0.11 [2.05]	0.06 [0.13]	0.752 ± 0.24
10	(138–153)	1.06 ± 0.13 [2.75]	0.06 [0.18]	0.988 ± 0.21
11	(153–168)	0.56 ± 0.20 [1.52]	0.03 [0.09]	0.775 ± 0.32
12	(168–183)	0.69 ± 0.23 [2.84]	0.04 [0.18]	0.775 ± 0.32
Core sum		126.15		

[a]From Roberts (1976a).

[b]Figures in brackets represent the means of only those cores containing roots.

FIGURE 11.18. The very shallow rooting zone of *Nothofagus* forests in Tierra del Fuego, Argentina. Roots may be restricted to 10 cm depth penetration, rendering whole stands of trees exceedingly susceptible to windthrow. Photograph provided with permission by A. Rebertus, University of Missouri.

TABLE 11.5. **Summary of Maximum Vertical and Horizontal Root System Extent for Several Conifer and Angiosperm Taxa**[a]

Taxon	Root system extent: Depth (m)	Root system extent: Radius (m)	Number of studies
Conifers			
Cupressaceae (s.1)			
Cupressus	4.6–4.9	—	2
Juniperus	2.2–>61.0	6.1–12.0	8
Thuja	—	6.4–10.0	3
Sequoia	5.0	—	1
Sequoiadendron	—	38.1	1
Pinacceae			
Abies	1.5–4.0	14.0	4
Larix	1.2–4.5	7.1–>9.1	6
Picea	1.4–6.0	7.5–20.0	14
Pinus	1.0–24.0	5.5–25.6	81
Pseudotsuga	1.5–10.0	6.4–13.0	7
Tsuga	>1.9	10.0	3
Podocarpaceae			
Podocarpus	—	>19.5	1
Angiosperms			
Aceraceae			
Acer	>1.2–4.0	3.6–20	14
Betulaceae			
Alnus	1.7–3.8	—	2
Betula	1.1–4.0	>8.0–23.8	9
Fabaceae (s.1.)			
Acacia	1.2–35.0	8.0–30.5	13
Gleditsia	1.6–3.3	5.0–15.0	4
Prosopis	3.0–>53.0	15.0–27.0	9
Robinia	2.1–>7.9	2.7–14.0	6
Fagaceae			
Fagus	>1.5–2.3	4.3–15.0	4
Nothofagus	>2.0	—	1
Quercus	1.1–24.2	3.0–30.5	39
Juglandaceae			
Carya	>1.8–>3.0	1.8–16.6	6
Juglans	1.6–>3.6	4.5–34.1	9
Magnoliaceae			
Liriodendron	>2.2–>2.9	—	2
Magnolia	>2.0	—	1
Moraceae			
Ficus	4.8–5.9	—	2
Maclura	2.4–8.2	4.3–>19.3	3
Morus	—	12.7	1
Myrtaceae			
Eucalyptus	1.5–6.0	>5.8–20.0	23
Melaleuca	>2.5	—	1
Meterosideros	5.0	>30.0	1
Oleaceae			
Fraxinus	1.8–2.2	7.3–21.0	5
Platanaceae			
Platanus	2.1	2.7–15.0	2
Rosaceae			
Five genera	1.8–10.7	2.0–11.0	27
Salicaceae			
Populus	1.3–3.6	1.5–30.5	13
Salix	>3.6–4.2	6.7–40.0	3
Ulmaceae			
Celtis	1.3–2.5	6.1–12.6	2
Ulmus	1.2–8.2	>3.7–34.1	8

[a]Adapted from Stone and Kalisz (1991).

logical distribution across climatic regions (e.g., oak, pine, and eucalyptus) show a far greater range in maximum root depth than taxa that are for the most part restricted to humid regions and habitats (e.g., maple, birch, ash, poplars, fir, spruce, larch, hemlock) (Burns and Honkala, 1990a,b; Stone and Kalisz, 1991). The importance of deep roots, even if sparse, to water relations will be discussed later.

Strong and La Roi (1983) noted that vertical distribution of roots of jack pine, black and white spruce, tamarack, and balsam fir responded more to soil conditions, particularly soil texture, than did horizontal root system development. Lateral root system extent thus appears less closely related to climatic and habitat factors, but the data emphasize the potential for woody plants to exploit soil resources at great distance, substantially beyond the spread of the crown. Stone and Kalisz (1991) found numerous reports of maximum lateral root extent of more than 30 m from the trunk (e.g., in oaks, giant sequoia, acacia, willow, poplar, and elm) and a few species apparently can extend lateral roots to 50 m and beyond (e.g., *Nuytsia floribunda*).

The extent of root development will depend both upon genetic and stand factors, including the influence of competing species and stand density. Shainsky et al. (1992) showed that root biomass of five-year-old Douglas-fir trees was greatly reduced by increasing red alder density in mixed plantings of the two species (Fig. 11.19). In contrast, root biomass of red alder was far less sensitive to increasing density of Douglas-fir. Similar impacts of *Eucalyptus obliqua* invasion on rooting density of fine roots of Monterey pine also have been reported (Bi et al., 1992).

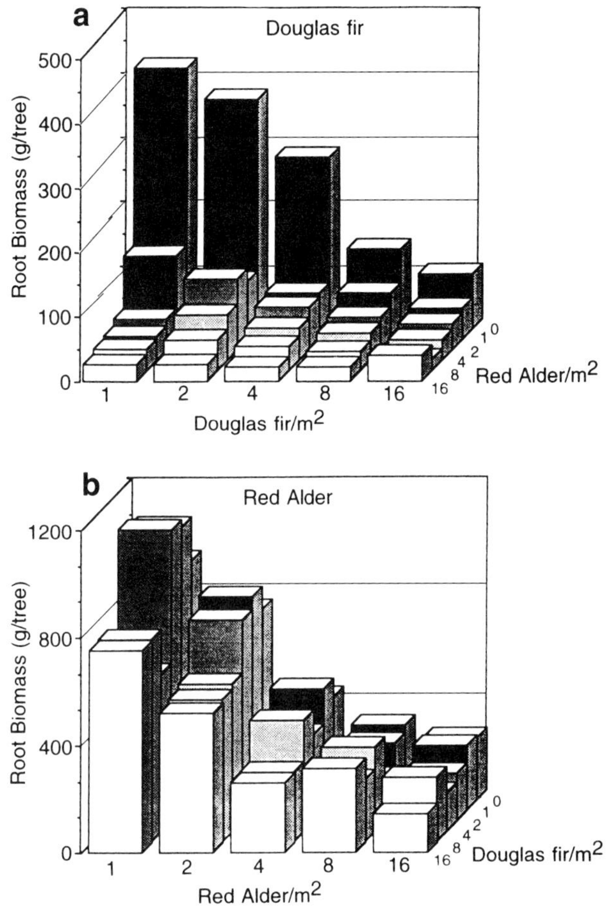

FIGURE 11.19. Root biomass of five-year-old Douglas-fir (a) and red alder (b) trees growing in mixed plantations at several spacings. Note the impact of the density of one species on the root biomass of the other. The inhibitory effect of red alder competition on belowground growth of Douglas-fir is particularly striking. From Shainsky et al. (1992).

The efficiency of roots in absorption depends on the amount of surface in contact with the soil and on the permeability of the surface. Obviously, root systems bearing numerous small branches should be more efficient than systems consisting of fewer large, sparsely branched roots.

Mycorrhizas and Water Relations

Mycorrhizas presumably increase the efficiency of mineral absorption, chiefly because the hyphae extend out into the soil and thereby increase the absorbing surface (Chapter 10). They also maintain an active absorption system on the older roots long after they have become suberized (Bowen, 1973). The effects of mycorrhizal roots on water absorption are more difficult to evaluate. Reports of mycorrhizal influences on plant water status have been diverse, ranging from maintenance of higher Ψ_w (Allen and Allen, 1986; Walker et al., 1989) to an apparent increase in water stress during drought where Ψ_w of mycorrhizal plants was reduced relative to those of nonmycorrhizal controls (Dixon et al., 1981; Sweatt and Davies, 1984). Additionally, an absence of any influence of mycorrhizal colonization on Ψ_w sometimes has been reported (Augé et al., 1986a; Dosskey et al., 1991). The extensive review of vesicular-arbuscular mycorrhizal effects on plant water relations by Augé (2001) reported a similar diversity of responses.

It must be emphasized that the benefits of mycorrhizas cannot be evaluated solely by interpretation of trends in Ψ_w of plants. This is because many potentially beneficial effects of mycorrhizas may cause apparent increased water stress. For example, stimulation of transpiration (and photosynthesis) may cause lower leaf water potentials because of the greater rates of water flux through the plant. Additionally, in pot studies mycorrhizal plants, which often are larger than nonmycorrhizal controls, usually deplete soil water before nonmycorrhizal plants do because of greater leaf areas and transpiration rates. Lower water potentials are significant and unambiguous indicators of water stress only under conditions where the influences of plant size, container, and other effects have been eliminated or taken into account. Further, even in these compensated cases, lower Ψ_w in mycorrhizal plants may not indicate disadvantage per se. The physiological responses to mild water stress may be far outweighed by improvements in other aspects of plant metabolism (net photosynthesis or mineral nutrition).

Several investigators have reported influences of mycorrhizas on intrinsic hydraulic properties of the root (apart from transport through hyphae, see later). For example, Hardie and Leyton (1981) observed that hydraulic conductivity of mycorrhizal red clover plants was three times that of nonmycorrhizal plants. Dixon et al. (1983) and Bildusas et al. (1986) found that whole-plant liquid flow resistance was reduced in mycorrhizal black oak and *Bromus* plants. According to Huang et al. (1985), soil-to-leaf Ψ_w differences were reduced in mycorrhizal *Leucaena leucocephala* plants despite higher transpiration rates. Marjanović et al. (2005) linked increased hydraulic conductivity in ectomycorrhizal roots of *Populus tremuloides* × *P. tremula* with increased expression of water channel proteins. Expression of a gene particularly effective in increasing water permeability in a *Xenopus* oocyte assay was increased five-fold in mycorrhizal fine roots. Others, however, have observed that when other confounding factors are carefully excluded (e.g., plant size effects, root length differences, P nutrition), no fundamental changes in root water uptake and transport properties could be detected (Safir and Nelsen, 1985; Andersen et al., 1988; Augé, 2001). Mycorrhizal root systems also may have lower hydraulic conductivity than nonmycorrhizal roots (Sands and Theodorou, 1978; Sands et al., 1982; Newman and Davies, 1988; Coleman et al., 1990).

Safir et al. (1972) originally proposed that the sole effect of vesicular-arbuscular mycorrhizas (VAM) on improvement of water relations (as indicated by lower whole-plant liquid flow resistance) was mediated through improved phosphorus nutrition of the host. This conclusion was based on the fact that P fertilization greatly reduced or eliminated differences associated with mycorrhizal colonization in low-P plants. Supporting evidence for a relationship between P nutrition and whole-plant flow resistance has been obtained for apple and Douglas-fir trees (Runjin, 1989; Coleman et al., 1990). The mechanism of this influence was not explored, but could be associated with changes in root system morphology and hydraulic conductivity.

Several experiments indicate that external hyphae also may facilitate water absorption by roots. There is strong evidence that at least some substances pass from hyphae to plant (Duddridge et al., 1980). For example, many studies have shown plant-to-plant transfer of mineral nutrients in mycorrhizal associations. However, whether the additional amount of water absorbed by mycorrhizal roots could be considered physiologically and ecologically significant is not well established. A few studies do appear to show detectable influences of hyphae on water relations of the host plant. For example, Faber et al. (1991) grew cowpea seedlings in culture conditions that permitted the extra-matrical hyphal connections with moist soil to be severed without disturbing the rest of the system. When these spanning hyphae were severed there was a 35% reduction in transpiration compared to that of controls, and it is difficult to interpret these data other than indicating a direct contribution of water flow along the hyphal threads. Similar results had been reported earlier by Brownlee et al. (1983) and by Hardie (1985, 1986). Finally, Boyd et al. (1986) observed that transpiration declined substantially within 10 minutes after hyphal contact to moist soil was severed in silver birch-*Paxillus involutus* and Scotch pine-*Suillus bovinus* associations. Transfer of water along hyphae in ectomycorrhizal species would appear to be most feasible because they frequently possess internal channels that could function as low-resistance "vessels" for water flow. In VAM species, the fungal hyphae do not have such channels and a more high-resistance protoplasmic route would apparently have to be traversed by water.

WATER ABSORPTION PROCESSES

Water absorption occurs along gradients in water potential or its components, from the medium in which the roots are growing to the root xylem. Two absorption processes can be identified, osmotically driven absorption, which is common in slowly transpiring plants, and passive absorption, which predominates in actively transpiring plants and is responsible for most of the water absorption by woody plants. The difference between osmotically driven and passive absorption is in the manner in which these gradients are produced.

Osmotically Driven Absorption

The roots of plants growing in warm, well-aerated, moist soil function as osmometers when the plants are transpiring slowly, because accumulation of solutes in the xylem sap lowers the Ψ_π and consequently the Ψ_w below Ψ_w of the soil. The resulting inward movement of water produces the root pressure that is responsible for guttation and the exudation from wounds observed in some plants such as birch trees and grape vines. In other species, such as conifers, root pressures have not been measured or are only rarely reported (e.g., Lopushinsky, 1980). Attempts have been made to explain root pressure as caused by direct, active secretion of water or by electroosmosis, but a simple osmotic theory seems to provide an adequate explanation (Kramer and Boyer, 1995). There is no evidence that active transport of water even occurs in plants and electroosmosis probably could not move the volume of water that exudes from detopped root systems. Whether osmotically driven water movement is by diffusion or mass flow is not certain, but Boyer (1985) argued that sufficient tension could be developed within membrane channels to support mass flow. Kramer and Kozlowski (1979, pp. 451–452) and Kramer and Boyer (1995, pp. 170–178) discuss the process of osmotically driven water absorption further.

Passive Absorption

As the rate of transpiration increases and tension develops in the xylem sap, the gradient for water uptake switches from dominance of Ψ_π to Ψ_p gradients. The greater water volume flux under these conditions sweeps out accumulated solutes in the root xylem sap (Lopushinsky, 1964) and decreases the amount of osmotically driven absorption. The roots become passive absorbing organs through which water is pulled in by mass flow generated in the transpiring shoots. It seems likely that practically all water absorption by transpiring plants, both woody and herbaceous, occurs passively.

ROOT AND STEM PRESSURES

From very early days the exudation of sap from injured plants has been observed. In the Far East sap has been obtained from palms to make sugar and wine from before the beginning of recorded history. According to Evelyn (1670), birches had long been tapped in England and on the Continent and the sap was used for various purposes, including fermenting beer. The first Europeans to visit Canada and New England found the Indians tapping maple trees and boiling down the sap to make sugar, and in Mexico the Spanish conquistadors found the natives collecting the sugar sap from agave and fermenting it into an alcoholic beverage known as pulque. Unfortunately, early writers indiscriminately grouped together all examples of "bleeding" or "weeping" without regard to their origin. Wieler (1893) listed nearly 200 species belonging to many genera, but his list included examples of plants showing true root pressure, sap flow from wounds, guttation, and even secretion from glandular hairs. It is necessary to distinguish between sap flow caused by root pressure, as in grape and birch, and that caused by stem pressure, as in maple, or by wounding, as in palms. Milburn and Kallarackal (1991) summarized the literature concerning sap exudation.

Root Pressure

Root pressure is not common among trees of the Temperate Zone and occurs chiefly in the spring before leaves develop and transpiration is rapid. However, Parker (1964) reported copious exudation from black birch in New England in October and November, after leaf fall. There was no exudation following a dry summer. Hales (1727) made the first published measurements of root pressure and reported a pressure of 0.1 MPa in grape. Clark (1874) tested over 60 species of woody plants in Massachusetts and found exudation from only a few species, including maple, birch, walnut, hop hornbeam, and grape. Sap flow ceases as leaves develop and increasing transpiration produces negative pressure or tension in the xylem sap. The sugar content of birch sap often is about 1.5%, lower than that of maple sap (Chapter 7), and consists chiefly of reducing sugars. Detopped conifer seedlings can be induced to exude sap if intact seedlings are kept well moistened while being subjected to a preconditioning period of cold storage (Lopushinsky, 1980). However, reports of sap exudation in conifers under natural conditions are rare (Milburn and Kallackaral, 1991). Oleoresin flow is discussed in Chapter 8.

Guttation

In herbaceous plants the most common evidence of root pressure is the exudation of droplets of liquid from the margins and tips of leaves. The quantity of liquid exuded varies from a few drops to many milliliters, and the composition varies from almost pure water to a dilute solution of organic and inorganic substances. Guttation usually occurs through stomalike openings in the epidermis called hydathodes, which are located near the ends of veins. In tropical rain forests, guttation is common at night, but it is uncommon in woody plants of the Temperate Zone because the necessary combination of warm, moist soil and very humid air is less common than in the tropics. A few instances of guttation from the twigs of trees have been reported (Büsgen and Münch, 1931). Raber (1937) observed sap flow from leaf scars of deciduous trees in Louisiana after leaf fall, and Friesner (1940) reported exudation from stump sprouts of red maple in February in Indiana. Exudation of liquid from roots and root hairs of woody plants also has been reported (Head, 1964), and, since this probably is caused by root pressure, it may be termed root guttation. No guttation has ever been reported in conifers, as would be expected because of the absence of root pressure, but artificial guttation can be caused by subjecting the root system to pressure (Klepper and Kaufmann, 1966).

Guttation is of negligible importance to plants. Occasionally, injury to leaf margins is caused by deposits of minerals left by evaporation of guttated water and it is claimed that the guttated liquid provides a pathway for the entrance of pathogenic organisms. In general, however, guttation can be regarded as simply an incidental result of the development of hydrostatic pressure in the xylem of slowly transpiring plants.

Maple Sap Flow

Maple sap flow deserves special attention both because it forms the basis of an important industry in the northeastern United States and because it is interesting physiologically. There are several reasons for believing that it occurs quite independently of root pressure, including the fact that pressure gauges attached to roots of maple often show negative pressure when stems show positive pressure (see Fig. 11.20). More convincing is the fact that segments of stems and branches removed from maple trees show sap flow if supplied with water and subjected to temperatures that rise above and fall below freezing

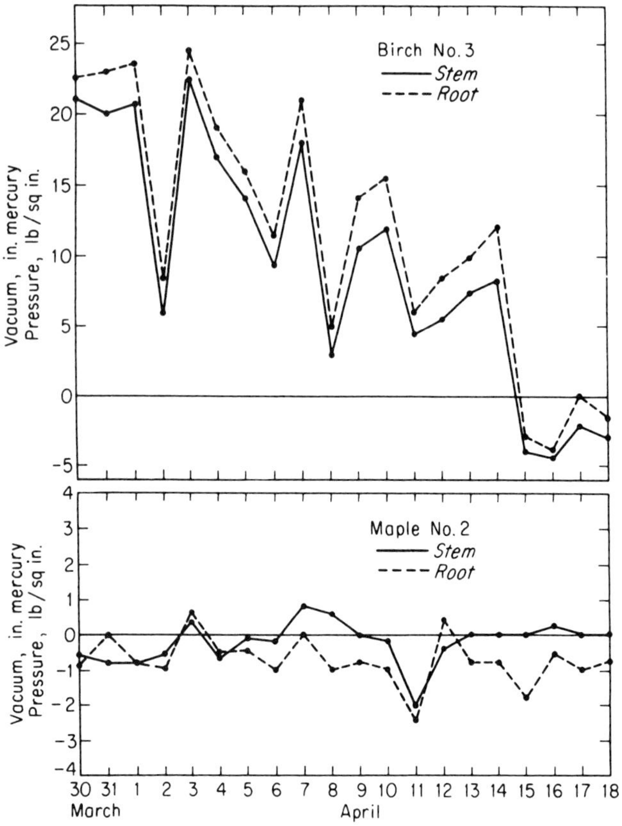

FIGURE 11.20. Simultaneous measurements of root and stem pressure in river birch and red maple. Root pressures in birch exceed stem pressures, and the two change almost simultaneously. Root pressure is usually absent in maple, even when positive pressure exists in the stems.

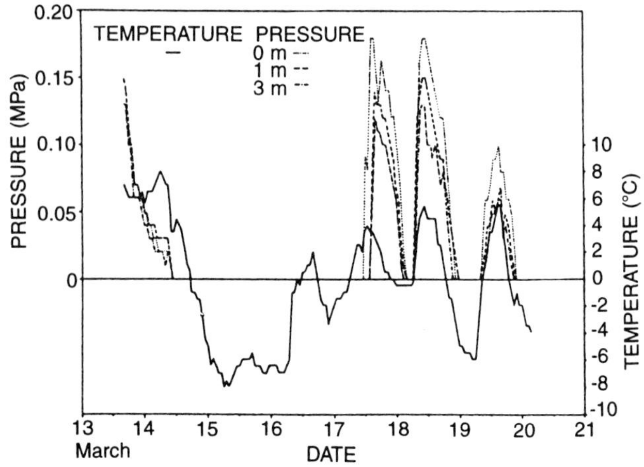

FIGURE. 11.21. Sap hydrostatic pressure at three heights in a tree trunk of sugar maple and air temperature at 1 m between March 13 and 21, 1979. Note that positive hydrostatic pressure requires both below-freezing temperature the previous night and above-freezing temperature the following day. From Cortes and Sinclair (1985).

(Stevens and Eggert, 1945; Marvin and Greene, 1951; O'Malley and Milburn, 1983; Tyree, 1983).

The extensive observations on maple sap flow made by Clark (1874, 1875) over a century ago are still applicable. In Massachusetts, maple sap flow can occur any time from October to April if freezing nights are followed by warm days. Sap flow ceases if temperatures are continuously above or below freezing; it stops in the spring when night temperatures no longer fall below freezing and it usually ceases in the afternoon and does not start again until the temperature rises above freezing the next morning. Failure to understand that sap pressure in the stems of trees often undergoes daily variations from positive to negative has led to unfortunate errors in the interpretation of experimental data (Kramer, 1940). In contrast to the situation in maple, the root-pressure-generated flow of birch and grape sap increases as the soil warms until increased transpiration caused by opening of leaves brings an end to root pressure.

Because of its dependence on weather, maple sap flow usually is intermittent and two or three to 10 or 12 runs may occur in a single spring. Many producers of maple syrup have used vacuum to increase the sap flow up to three times the normal amount (Koelling et al., 1968). The sugar content of maple sap varies from 0.5 to 7 or even 10%, but it usually is 2 or 3%, much lower than the sugar content of palm sap (Chapter 7).

Trees on infertile or dry soil will yield less than those growing on fertile, moist soil. The sugar yield obviously is related to photosynthesis and large, well-exposed crowns are advantageous. Fertilization is also said to increase the yield. Trees grown for sap production should be more widely spaced than those grown for timber, and roadside trees are said to produce large quantities of sap. Jones et al. (1903) reported that defoliation during the summer greatly reduced maple syrup yield the next spring.

There has been significant progress in our understanding of sap flow. It is caused by stem pressure produced during alternating diurnal cycles of below- and above-freezing temperatures (Fig. 11.21). Stem pressure does not develop during the day unless the temperature regime permits both freezing and thawing. During freezing, maple twigs take up solutions and upon thawing they exude a slightly smaller amount than was absorbed. This cycle can be repeated for a few cycles as twigs hydrate, but at very high twig water content freezing-induced absorption disappears (Milburn and O'Malley, 1984; Johnson and Tyree, 1992).

Milburn and O'Malley (1984) proposed a cellular mechanism to explain these observations (Fig. 11.22). Xylem of sugar maple stems has abundant gas-filled

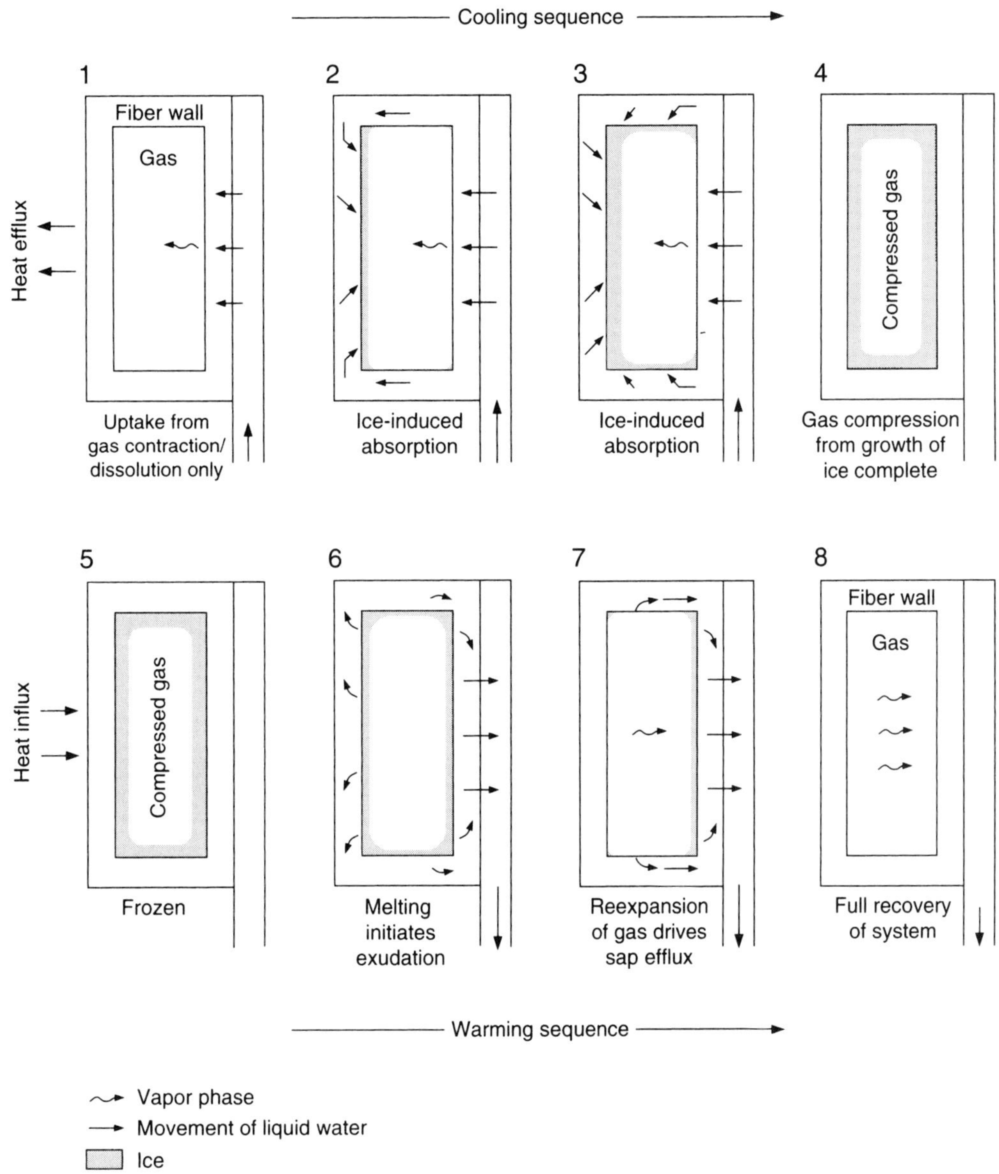

FIGURE 11.22. Schematic diagram of a proposed mechanism for stem pressure in maples. Declining temperature causes water uptake into fiber-tracheids as gas contracts and some dissolves into the liquid phase (top left); on freezing, ice vapor distills onto fiber-tracheid walls, compressing gas (top middle and right); on thawing gas pressure forces sap out under positive pressure (bottom middle). From Milburn and O'Malley (1984).

fibers with liquid-filled vessels and few intercellular spaces. As stem xylem cools, internal gas space declines in conformance with gas laws and with increased dissolution of gases into the liquid phase. When ice-nucleating temperatures are reached during a freeze–thaw cycle, liquid water moving to the interior surface of fiber cells begins to freeze on the walls, accumulating by vapor distillation and decreasing the air space within the fiber lumen and substantially elevating internal pressure. Upon thawing, the high gas pressure within the fibers forces liquid water from the lumen and, in bulk, to any region of lower pressure. If the moisture content is elevated to the point at which fibers fill with water, the absorption–exudation cycle disappears. Most experimental results are consistent with this hypothesis, but a reported requirement for sucrose (or other di- or oligosaccharides) in the sap (Johnson et al., 1987; Johnson and Tyree, 1992) has not been adequately reconciled to this simple physical model.

Other Examples of Stem Pressure

Two other plants that yield commercial quantities of sap are palms and agaves. In the tropical regions of India and Asia palm sap probably was used as a source of sugar before sugar cane was cultivated. Palm sap also is fermented to make palm wine. According to Molisch (1902), who studied the process in Java, sap flow usually is caused by cutting out the inflorescence and it can be maintained for weeks or even months by repeatedly cutting and pounding the stem. Sap also is obtained from the Palmyra palm by making incisions into the bark, and this process can be repeated year after year. When the central bud is cut out of date palms, sap flow ceases after several weeks and the palm dies (Corner, 1966). Davis (1961) thought root pressure was important in palms, but he later reported that it is rare and probably plays no part in palm sap flow (Milburn and Davis, 1973). The sap apparently originates from the phloem and the sugar probably was mobilized for use in the developing inflorescence or stem tips. Sap flow from agaves and palms is discussed in more detail by Van Die and Tammes in Zimmermann and Milburn (1975) and by Milburn and Kallarackal (1991).

ASCENT OF SAP

The existence of tall land plants became possible only after plants evolved a vascular system that permitted rapid movement of water to the transpiring shoots. It is difficult for terrestrial plants more than 20 or 30 cm in height to exist in any except the most humid habitats without a vascular system, because water movement from cell to cell by diffusion is much too slow to keep the tops of transpiring plants from being dehydrated. The magnitude of the problem in trees is indicated by the fact that on a hot summer day 200 liters or more of water may move from the roots to the evaporating surfaces in the leaves 20, 30, or even 100 m above.

Hales (1727) made careful observations on the absorption and loss of water and wrote, "The last three experiments all show that the capillary sap vessels imbibe moisture plentifully; but they have little power to protrude it farther without the assistance of the perspiring leaves, which do promote its progress." Hales' explanation foreshadowed our current explanation, although he did not specify how transpiration could "promote its progress." Toward the end of the nineteenth century, Boehm, Sachs, and Strasburger concluded that loss of water produces the pull causing the ascent of sap, but they also lacked an essential fact for a complete explanation. The final step was supplied by Askenasy (1895) and by Dixon and Joly (1895), who pointed out that water confined in small tubes such as the xylem elements has a very high cohesive force and can be subjected to tension. The history of study of the ascent of sap can be found in Miller (1938, pp. 855–872) and the ascent of sap is discussed in detail in Zimmermann and Brown (1971), Pickard (1981), and Tyree and Zimmermann (2002).

Although the cohesion-tension theory of the ascent of sap has existed since the end of the nineteenth century, it has been rather reluctantly accepted and occasionally still is questioned (e.g., Zimmermann et al., 1993; see later). Reluctance to accept the theory probably springs partly from an almost instinctive difficulty in believing that water can be subjected to tension, but it also arises from doubts about the possibility of maintaining a fragile, stressed system in a swaying tree trunk.

The cohesion-tension theory is based on the following premises:

1. Water has high internal cohesive forces, and, when confined in small tubes with wettable walls such as the xylem elements, it can sustain a tension of 3 to possibly 30 MPa.
2. The water in a plant forms a continuous system in the water-saturated cell walls from the evaporating surfaces of the leaves to the absorbing surfaces of the roots.
3. When water evaporates from any part of the system, but chiefly from the leaves, the reduction in water potential at the evaporating surfaces causes movement of water out of the xylem to the evaporating surfaces.
4. Because of the cohesive attraction among water molecules, the loss of water produces tension in the xylem sap that is transmitted through the continuous water columns to the roots, where it reduces the water potential and causes inflow of water from the soil.

Thus, as mentioned earlier in the section on passive absorption, in transpiring plants water absorption is controlled directly by the rate of transpiration.

The theoretical intermolecular attractive forces in water are extremely strong, and Ursprung (1915) measured a tension of more than 31 MPa in annulus cells of fern sporangia, whereas Briggs (1949) demonstrated a tension of 22.3 MPa in water subjected to centrifugal force. Greenidge (1954) suggested that the highest tensions in trees would average only about 3 MPa. However, measurements of xylem water potential made with the pressure chamber indicate the existence of tensions up to 8 MPa, and a tension of only about 2 MPa should suffice to overcome both gravity and the resistance to flow required to move water to the top of a tree 100 m in height (Dixon, 1914).

Hence, the tensile strength of water appears adequate to sustain liquid continuity within the xylem in the absence of events leading to cavitation. However, criticism of the cohesion–tension model of sap ascent often is based on arguments relating to other factors, such as the purity of water and the weakness of adhesive forces between water and xylem element walls, as these also influence the maximum sustainable tension in xylem elements. Smith (1994), presenting data from experiments that tested the tensile strength of artificial seawater and distilled water in capillary tubes with different wall wettabilities, argued that water in xylem elements might not be sufficient to sustain tensions of more than –0.6 MPa because of impurities and areas of low wettability in cell walls associated with high lignin content.

Balling and Zimmermann (1990) used a miniature pressure probe to make measurements of xylem pressure in tobacco and willow. Only subatmospheric positive pressures and slight tensions (>–0.3 MPa) were measured, leading the authors to question the validity of the cohesion–tension model of sap ascent (Zimmermann et al., 1993). Wei et al. (1999a,b, 2001) pointed out several instrument and methodological problems associated with use of the traditional pressure probe to measure tensions in the xylem (it was designed to measure positive pressure in living cells). With modification to the probe, Wei et al. (1999b) observed good agreement between xylem tensions and pressure chamber balance pressures in maize leaves to nearly –0.8 MPa. Because of the apparently inherent vulnerability of the probe itself to air seeding and consequent induction of cavitation, measurements are only possible to moderate xylem tensions (above –1 to –1.6 MPa).

To move water upward against gravity, a pull of –0.01 MPa per meter is required, plus whatever pull is required to overcome frictional resistance to upward flow in the xylem. Field measurements of vertical Ψ_w gradients sometimes are lower than the required minimum. For example, Tobiessen et al. (1971), Connor et al. (1977), and Ginter-Whitehouse et al. (1983) mentioned situations in which the Ψ_w gradient in the stem appeared to be less than the required minimum. Richter (1973a,b) pointed out that these results could largely be explained by the common practice of sampling Ψ_w at the periphery of the crown and to vertical changes in the environmental conditions that control transpiration (Fig. 11.23). Differential flow resistances to and transpiration rates at the sampling points in the crown

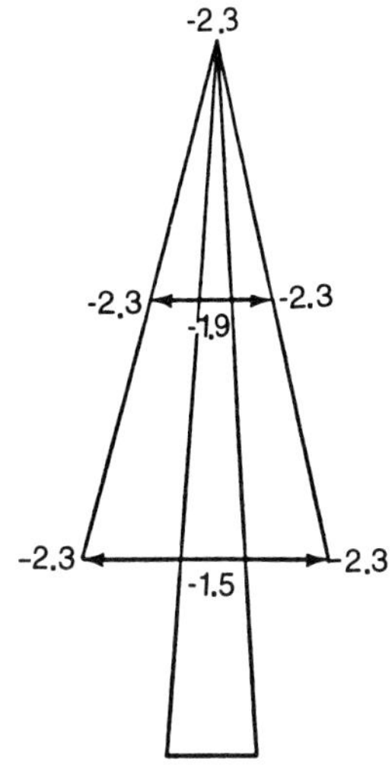

FIGURE 11.23. Diagram of leaf and stem water potentials in a tree, illustrating an apparent absence of vertical gradients because of sampling location. From Richter (1973b).

can result in an apparent absence of a Ψ_w gradient when one is present in the stem of the tree.

More research will be required to resolve the issues surrounding the cohesion–tension theory of sap ascent (for a recent exchange between those holding opposing views see Zimmermann et al. (2004) and Brooks and cosigners (2004)), but at present it seems unwise to abandon current concepts of sap flow in the Soil–Plant–Atmosphere Continuum. It could very well be that the apparent problems of the cohesion–tension theory have been overemphasized and based on faulty experiments. In any event, as Renner (1912) pointed out long ago, the cohesion–tension theory is the only one that explains how absorption and transpiration are effectively coupled together. The continuous water columns extending from leaves to roots provide the feedback mechanism by which changes in rates of water loss and absorption control one another. This mechanism is essential for the survival of transpiring plants, although its importance has been neglected by some critics of the cohesion–tension theory.

THE WATER CONDUCTING SYSTEM

Essentially all the water that is absorbed by roots moves upward through the stem and branches to the leaves in the xylem. However, most water must cross from one to several layers of living cells to enter the root xylem, and in the leaves it may pass through several cells before reaching the evaporating surfaces. As is the case in roots, there is uncertainty concerning the pathway between xylem and the evaporating surfaces. In principle, the pathway could be entirely apoplastic, as there is no structure corresponding to the root endodermis in leaves. However, there is evidence that water movement out of leaf xylem is primarily symplastic, as apoplastic tracers in the transpiration stream accumulate at specific "sumps" located near vascular bundles (Canny, 1990).

In the xylem, water moves through dead elements. The xylem consists of wood and ray parenchyma cells, fibers, tracheids, and in angiosperms, vessels (Chapter 2), but upward water movement occurs principally in the tracheids of conifers and in the long vessels of angiosperms because they offer the least resistance to flow. In conifers, the tracheids are single cells up to 5 mm in length and 30 μm in diameter, and water must pass through thousands of cell walls as it moves up the stems. However, as will be discussed in detail later, tracheid-bearing xylem with elements that are similar in diameter to vessel-bearing xylem may have equivalent hydraulic conductivities because the bordered pits of tracheids offer much less flow resistance between vascular elements than intervessel pits (Pittermann et al., 2005).

Within a species there usually is a close relationship between the xylem conducting area and the amount of leaf area that must be supported by that xylem (Kaufmann and Troendle, 1981). However, the proportion of the total cross section of the tree stem that is involved in upward water transport varies widely among species. The weight of evidence indicates that the heartwood does not conduct water and, in mature trees, water moves upward in only a portion of the sapwood. In stems of ring-porous species such as oak, ash, and elm, most of the water moves in the vessels of the outer annual ring only. In American elm, for example, more than 90% of upward water movement took place in the outermost annual ring (Ellmore and Ewers, 1986). In diffuse-porous species, such as birch, poplar, and maple, the vessels in more than one annual ring of sapwood conduct water. Anfodillo et al. (1993) used infrared thermography to study sap flow in stems of a variety of trees, including hybrid poplar. In poplar, the three most recent annual rings conducted most water in five-year-old trees, and the two oldest annual rings conducted little or no water. In gymnosperms, many tracheids in several annual rings conduct water. The large diameter earlywood tracheids comprise the major path for water movement, with the small latewood tracheids often conducting little or no water. The high resistance to water movement in the latewood tracheids may be associated with their small diameter and few and small bordered pits (Kozlowski et al., 1966, 1967).

The path of water transport in seedlings may be somewhat different than in large trees. In white ash seedlings, water moved upward primarily in the large earlywood vessels of the current annual ring, except in the current shoot in which the ring-porous character was not developed and water moved through large vessels scattered throughout the xylem (Chaney and Kozlowski, 1977). In maple seedlings, water moved in the large vessels of the current annual ring and in the outer two-thirds of the annual ring of the prior year. In large trees, the central core of heartwood does not conduct water. However, in four-year-old gymnosperm seedlings in which heartwood had not formed, some of the tracheids in each annual ring conducted water (Kozlowski et al., 1966).

The path of upward transport of water may be essentially vertical, or there may be considerable deviation from a vertical path depending on xylem structure and orientation of pits. Rudinsky and Vité (1959) identified five general types of water uptake in gymnosperms (Fig. 11.24). In many trees, ascent of water

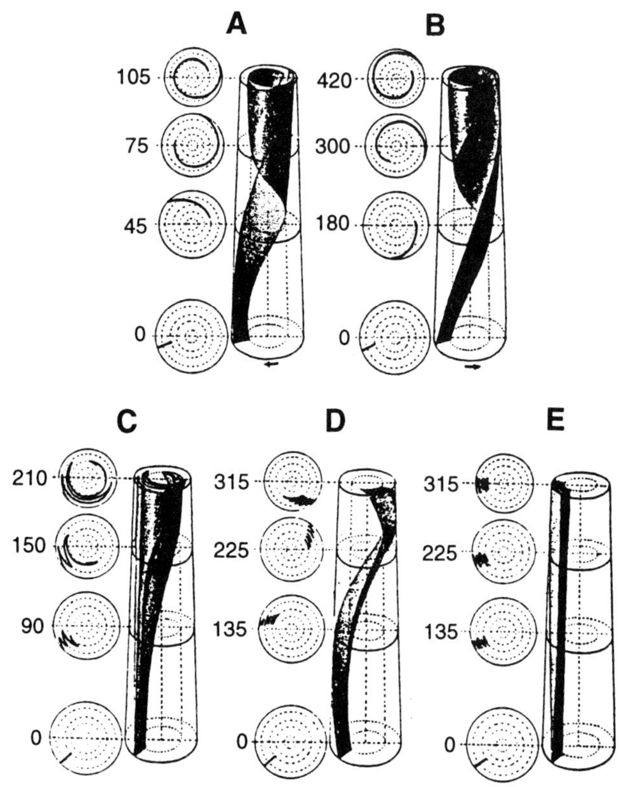

FIGURE 11.24. Types of water conducting systems as shown by dye stains in tracheids. The numbers give the stem height in centimeters. (A) Spiral ascent turning right; (B) spiral ascent turning left; (C) interlocked ascent; (D) sectorial winding ascent; (E) sectorial straight ascent. Used with permission of the Society of American Foresters from Rudinsky, J. A., and Vité, J. P. (1959). Certain ecological and phylogenetic aspects of the pattern of water conduction in conifers. *For. Sci.* 5, 259–266; permission conveyed through Copyright Clearance Center, Inc.

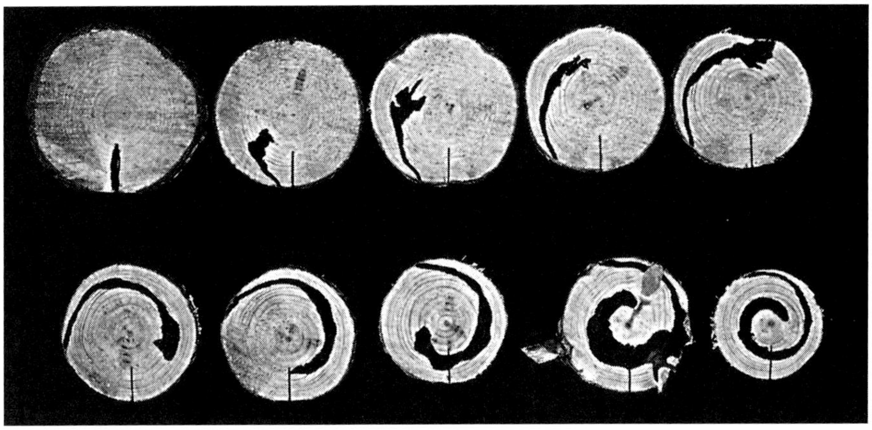

FIGURE 11.25. Spiral path of sap ascent in a red pine tree. Acid fuchsin dye was injected into the stem base and rose in a spiral pattern. The vertical line is above the point of injection. The sections were cut at intervals of 60 cm, the lowest section being at the top left. From Kozlowski and Winget (1963), © 1963 University of Chicago Press.

follows a spiral pathway more often than a strictly vertical one (Fig. 11.25). This pathway has been shown in angiosperm species, as in eucalyptus (Anfodillo et al., 1993), and is especially common in conifers. Often there is more spiraling of the ascending sap stream than can be accounted for by the structural spiral of the xylem. This probably is associated with the way in which the bordered pits are arranged in the tracheids (Kozlowski et al., 1966, 1967).

Wide interspecific and intraspecific variations have been shown in water uptake patterns associated with differences in spiral grain. The inclination of xylem elements in some species may be predominantly to the left, in others to the right. Differences in direction of spiral grain even occur between trees of the same species on the same site as well as on different sites. The direction of spiraling of xylem elements is correlated with the direction of pseudotransverse divisions in cambial cells. The degree of spiraling varies with species, heredity, growth rate, stem height, and age of trees (Noskowiak, 1963; Kubler, 1991).

Variations in paths of water movement have very important practical implications. Dye solutions spread tangentially within growth rings as sap moves upward from the point of injection (Zimmermann and Brown, 1971). Kubler (1991) claimed that spiral grain allowed a distribution of water from individual roots to many branches, thus eliminating the possibility of a branch being cut off from water by partial destruction of the root system. Many systemic chemicals move upward in tree stems along the path of ascent. Hence, distribution of water and chemotherapeutants in tree crowns varies greatly with the specific pattern of water uptake. For example, the most complete distribution of water into tree crowns was shown by a system of spiral ascent and the least effective distribution by vertical ascent (Rudinski and Vité, 1959). Paths of water conduction also influence host-parasite relations of vascular wilt diseases. White oak, in which the sap moved vertically upward, showed less injury from oak wilt than northern pin oak, in which the transpiration stream spiraled and spread out in the top (Kozlowski et al., 1962; Kozlowski and Winget, 1963). Kubler (1991) provided a good discussion of the functional significance of spiral grain in trees.

Efficiency of Water Conduction

To characterize patterns of water conduction, investigators often compare the leaf specific conductivity (LSC or K_L) among species and in different parts of the same tree. Leaf specific conductivity is defined as the rate of water flow (kg s^{-1}) through a stem or branch caused by a unit of pressure potential gradient (MPa m^{-1}) per unit of leaf surface area supplied by the stem (m^2) (Zimmermann, 1983). Leaf specific conductivity, which varies widely among species and growth forms as well as within plants, is greatly influenced by xylem anatomy (Zimmermann, 1983; Ewers, 1985; Tyree and Ewers, 1991; Chiu et al., 1992; Zotz et al., 1994; Patiño et al., 1995) (Fig. 11.26). Specific conductivity of the sapwood, K_s, also may be estimated by dividing the rate of water flow (kg s^{-1}) through a stem or branch caused by a unit of pressure potential gradient (MPa m^{-1}) by the sapwood transverse area, allowing comparisons of the water transport capacity of unit areas of various xylem types (Ewers and Cruziat, 1992).

In angiosperms, wide vessels, long vessels, and numerous vessels are associated with high specific conductivity. In gymnosperms, conductivity varies appreciably with differences in tracheid diameter. Such differences can be predicted by the Hagen-Poiseuille law for volume flow through ideal capillary tubes. In circular tubes with rigid walls and laminar flow, the volume flow rate, q_v (m^3 s^{-1}) can be described by:

$$q_v = \frac{\pi r^4}{8\eta l}\Delta p \qquad (11.11)$$

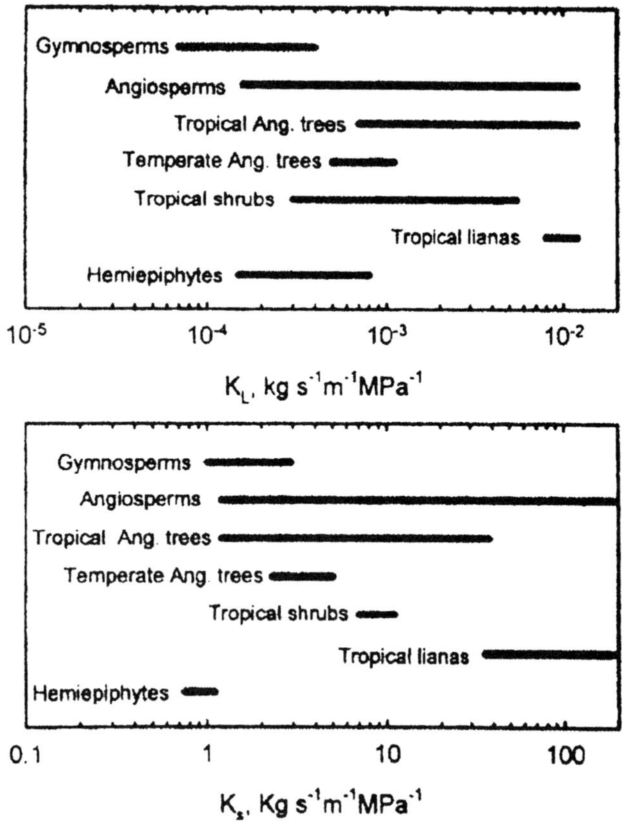

FIGURE 11.26. Comparative leaf-specific (K_L, top) and wood-specific (K_S, bottom) hydraulic conductivities in various wood types. From "Xylem Structure and the Ascent of Sap." (2002), Chapter 5—Hydraulic Architecture of Woody Shoots, M. T. Tyree and M. H. Zimmermann, p. 156, Figure 5.11. © 2002 with kind permission of Springer Science and Business Media.

Hence, for a given pressure gradient, $\Delta p/l$, the flow rate increases as the fourth power of the radius (r) of the tube and inversely as the dynamic viscosity, η, of the liquid. Thus, under the same pressure gradient, the flow rate will be 10,000 times faster in a tube with a 1 mm radius than in one with a 0.1 mm radius (Leyton, 1975). Strictly from a hydraulic perspective it thus would appear that xylem should contain just a few very large vessels. However, few vessels wider than 0.5 mm are found in woody plants, including vines, and even xylem of ring-porous trees possesses many small vessels in addition to those that are large (Tyree et al., 1994). Tyree et al. (1994) suggested that the observed mixture of element sizes in xylem might be an evolutionary response to the need for both hydraulic efficiency and support if large vessels were mechanically weaker than small ones. It also must be remembered that xylem conduits are leaky tubes and that in the small veins of leaves the velocity of sap movement will decline to zero at positive values of volume flow (Canny, 1993b).

Conductivity also is influenced by between-element pit resistance and the structure of the perforation plates of vessels (Bolton and Robson, 1988). Partitioning of resistance between the lumen and resistance encountered by water moving to adjacent conduits indicated that mean end-wall resistance constituted 64% of total resistivity of tracheids of gymnosperms and 52% of total resistivity of vessels (Pittermann et al., 2005; Hacke et al., 2006). Species differences in xylem anatomy characteristics often are compensatory. For example, conductivity in stems of white ash and sugar maple seedlings was similar as a result of conduction in white ash in a small number of large diameter vessels and in maple in a large number of small-diameter vessels (Chaney and Kozlowski, 1977).

The specific conductivity of stems, K_S, generally is higher in angiosperms than in gymnosperms and is higher for vines than for trees (Fig. 11.26). However, for a given conduit diameter, tracheids of conifers may contribute the same conductivity despite being much shorter that comparable-diameter vessels. Pittermann et al. (2005) showed that pit area resistance of bordered pits of gymnosperms was much lower than that of angiosperms and could largely compensate hydraulically for the shorter lengths of tracheids in conifer wood. The lower pit area resistance of conifers was associated with the more porous margo of their bordered pits. Plants of dry seasons or dry sites tend to have lower conductivities than plants of the same growth form in habitats with wetter seasons or wetter sites (Gartner et al., 1990). In seasonally dry tropical forests, species with drought-deciduous leaves exhibit much higher maximum specific conductivities than do evergreen species. However, drought-deciduous species also showed great loss of specific conductivity of the xylem by embolism during the dry season, resulting in lower minimum specific conductivity in this group than in evergreen species (Sobrado, 1993).

Conductivity varies greatly within trees. It is higher in the stem than in the branches and is particularly low in second-order branches and at branch insertions (Fig. 11.27) (Ewers and Zimmermann, 1984). In northern white cedar the LSC was 30 times higher in stems than in small twigs (Tyree et al., 1983).

Although resistance to radial water movement from soil to the xylem is high in roots, the resistance to longitudinal flow in the xylem is lower in woody roots than in stems (Jones, 1989). Stone and Stone (1975a) found that the conductivity of red pine roots was up to 50 times higher than that of stems, and that conductivity increased with distance from the base of the stem. No spiral movement was observed in roots, although it is common in stems. It is reported that there is a constricted region in the xylem supplying the

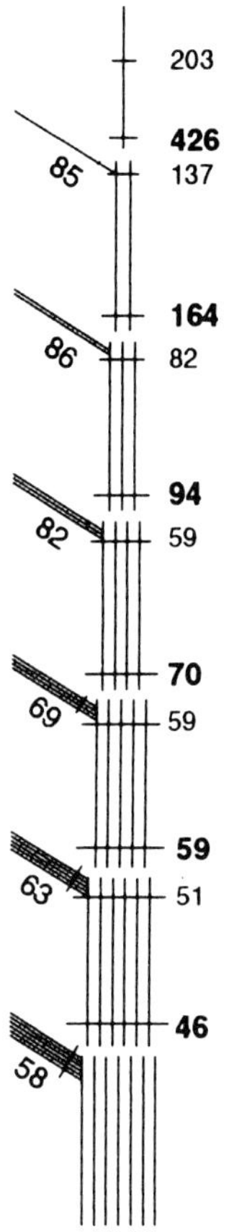

FIGURE 11.27. Differences in relative amounts of water-conducting surface compared to leaf surface along the main stem and in the branches of a 6-year-old white fir tree, expressed as hundredths of square millimeters of xylem cross section per gram of fresh needle weight. The relative conductivity increases from base to apex, but it is lower at the point where branch whorls are attached (numbers in light-face type) than between nodes (numbers in boldface type). After Huber (1928).

leaves of some trees, caused by reduction in the number and diameter of vessels (Larson and Isebrands, 1978), that increases resistance to water flow into leaves. Yang and Tyree (1994) estimated that resistance to water flow of whole shoots of maple trees was partitioned approximately 50% to leaves and petioles, 35% to branches in the crown, and 15% to the trunk. Sack and Holbrook (2006) reported that leaf hydraulic conductance varied 65-fold across numerous plant species and accounted for more than 30% of the total resistance to water flow.

Substantial diurnal variation in hydraulic conductivity is sometimes reported, implying substantial reversal of embolism under conditions of xylem tension that would appear to be unfavorable for refilling (Yang and Tyree, 1992). For example, Zwienecki and Holbrook (1998) reported percent loss of conductivity (PLC) in the afternoon of 45 and 70%, respectively, in current year shoots of *Fraxinus americana* and *Acer rubrum*. Morning PLC values were substantially higher and suggested about 50% recovery of conductivity overnight. Shoots of *Picea rubens* showed less loss of conductivity and less diurnal variation in the same study. In contrast, Clearwater and Clark (2003), using *in vivo* magnetic resonance imaging to visualize water in vessels, could detect no refilling in three woody vines that had embolized vessels. Various mechanisms have been advanced to explain the refilling phenomenon (Salleo et al., 2004; Höltta et al., 2006), but its confirmed existence, generality, and mechanism of refilling at substantial tensions await further exploration.

Air Embolism and Xylem Blockage

Tyree and Sperry (1988, 1989) proposed that rupture of stressed water columns followed by air embolism in the conducting xylem elements often decreases water conductivity. Embolism commonly is induced by drought, excessive transpiration, and winter freezing. For a long time rupture of water columns under tensions was attributed to the inherent instability of water in the metastable state. It followed that vulnerability to embolus formation was thus a function of element volume, given the greater likelihood of random bubble formation in a large volume than a small one. Hence it was thought that ring-porous trees with wide xylem elements, such as oaks, elms, and ashes would be more vulnerable to cavitation than diffuse-porous angiosperms and gymnosperms, which have narrow xylem elements (Carlquist, 1983). However, Pickard (1981) pointed out that the tensions developed in xylem sap were low enough that spontaneous rupture would be very improbable statistically, even in wide vessels.

Evidence now suggests that most often embolism appears to be caused by air aspirated into a vessel or tracheid by way of the pores in pit membranes that adjoin an adjacent air-filled xylem element or air space (Crombie et al., 1985). This hypothetical mechanism was called "air-seeding" by Zimmermann (1983). Once air enters a vessel, it disrupts the cohesion of the water molecules and the water column breaks and retracts, filling the element first with water vapor. Eventually, as air comes out of solution from the surrounding water, the vessel completely fills with air. The adjacent elements do not embolize as long as the pressure difference does not exceed the surface tension of the air-water interface in pores connecting the embolized and filled elements. Because larger pores can only sustain smaller pressure differences, they are suggested as those most likely to become seeded with air (Crombie et al., 1985; Sperry and Tyree, 1988; Jarbeau et al., 1995). Sperry et al. (1991) found that older vessels of trembling aspen, which were more likely to be embolized, had developed large holes in pit membranes. It was also possible to reduce hydraulic conductivity by increasing external gas pressure on aspen stem segments in which the xylem was under tension, thereby presumably increasing the pressure difference across pores near the critical size for seeding. Similar decreases in hydraulic conductivity under gas pressure also have been shown for willow (Cochard et al., 1992) and two chaparral shrub species (Jarbeau et al., 1995).

In tracheid-bearing species air movement between embolized and adjacent elements is prevented by movement (or aspiration) of the pit membrane and torus against the pit border that occurs in response to the pressure differences initiated with cavitation. Between-tracheid movement of air bubbles does not occur unless the pressure difference becomes great enough to force the torus through the pit border (Tyree et al., 1994).

Work by Tyree et al. (1994) and Hacke et al. (2006) has indicated moderate to strong correlations between conduit diameter and vulnerability to cavitation in both conifer and angiosperm species, but the reason for this relationship remains incompletely understood. Because of limitations of carbohydrate delivery capacity to the site of cell wall synthesis, rapidly growing elements destined to a large mature size may lack adequate materials to build fine-mesh pit membranes and so would be more likely to contain a large air-seeding pore. Hargrave et al. (1994) observed that the diameters of embolized vessels of coastal sage were significantly larger than unembolized vessels in dehydrated stems (29 μm versus 20 μm). However, in a study incorporating many species, Hacke et al. (2006) found that pit area resistance was weakly, but significantly, positively correlated with vulnerability to cavitation, a result opposite to the expected relationship. Alternatively, Wheeler et al. (2005) and Hacke et al. (2006) presented evidence that supports the hypothesis that vulnerability to cavitation might be proportional to pit surface area. This would follow because a large, air seeding pore is more likely (independent of mean pore area) where more pit area is present in an element. Larger total pit areas would tend to be associated with large vessel volumes. Hacke et al. (2006) presented data that showed a strong correlation between pit surface area and cavitation vulnerability for 29 angiosperm species.

The extent of embolism of the conducting conduits varies with species, season, and in different parts of the same tree. Tyree and Ewers (1991), Sperry and Sullivan (1992), and Cochard (1992) argued that vulnerability to xylem embolism under tension was correlated with species distribution across habitats of varying water availability (Fig. 11.28). For several gymnosperm and angiosperm species, loss of hydraulic conductivity was seen at higher Ψ_w in mesic or wet-site species such as eastern cottonwood, *Schefflera morototoni*, northern white-cedar, eastern hemlock, and numerous species of true firs than in xeric species such as eastern red cedar, Rocky Mountain juniper, and Gambel oak. American mangrove develops high xylem tensions in its natural saltwater habitat and also shows high resistance to embolus formation (Sperry et al., 1988a).

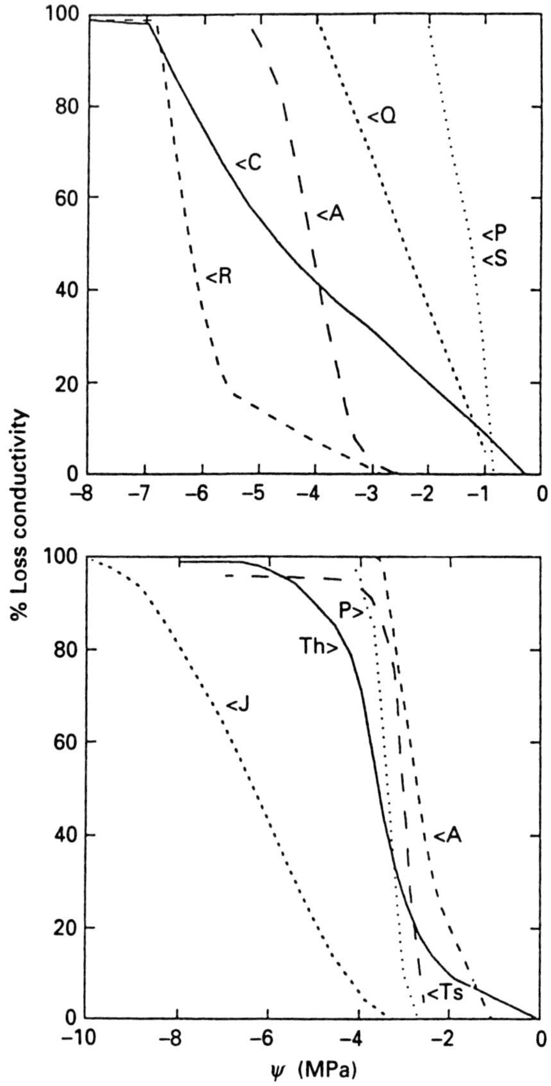

FIGURE 11.28. Vulnerability to cavitation of stem sections of various species as indicated by the percent loss of hydraulic conductivity plotted versus Ψ_w. (Top) Angiosperms: R, *Rhizophora mangle*; A, sugar maple; C, *Cassipourea elliptica*; Q, northern red oak; P, eastern cottonwood; S, *Schefflera morototoni*. (Bottom) Gymnosperms: J, eastern red cedar; Th, northern white-cedar; Ts, eastern hemlock; A, balsam fir; P, red spruce. From Tyree and Ewers (1991).

Embolism should be expected from freezing of xylem sap because air that is dissolved in water is not soluble in ice (Tyree et al., 1994). Loss of hydraulic conductivity in xylem by embolism that results from the freezing of water in stems appears to be a significant annual occurrence in most temperate angiosperm trees, but not in many gymnosperms. For example, embolism of vessels in sugar maple increased as winter approached (Sperry et al., 1988). By February, there was an average 84% reduction in twig hydraulic conductivity. Beginning in late March, loss of hydraulic conductivity decreased to approximately 20% into June. Recovery of hydraulic conductivity after winter may be related to dissolution of air bubbles in stems of sugar maple under slight positive pressure (Tyree and Yang, 1992). Air bubbles also may dissolve at low tensions, as was shown for Scotch pine (Borghetti et al., 1991; Sobrado et al., 1992; Edwards et al., 1994). Gas in bubbles may redissolve because of pressure imposed on entrapped gases by surface tension of water surrounding the bubble or because concentration gradients exist for diffusion of gas molecules between bubble and adjacent liquid water (Edwards et al., 1994). The xylem vessels of wild grapevine commonly are filled with gas during the winter. Before the leaves expand in the spring, the vessels become filled with water by root pressure. Some air in the vessels apparently is dissolved in the ascending xylem sap and some is pushed out of the vessels and out of the vine (Sperry et al., 1987). Tyree and Yang (1992), Grace (1993), and Edwards et al. (1994) discussed these and other possible mechanisms of xylem element refilling after embolism.

In conifers, there is less tendency for persistent embolism of tracheids with freezing than is the case for the vessels of angiosperms (Taneda and Tateno, 2005). Sperry and Sullivan (1992) demonstrated that incremental loss of hydraulic conductivity at a given xylem tension of subalpine fir and Rocky Mountain juniper during freeze-thaw cycles was much lower than it was in diffuse-porous angiosperms. Additionally, the ring-porous Gambel oak was much more susceptible to freeze-induced embolism than conifers or diffuse-porous species. In Gambel oak, nearly total embolism occurred at any xylem tension under which stems were frozen, although this species was very resistant to embolism under tension alone. These results suggest that loss of xylem function in oaks is largely attributable to freeze-induced embolism rather than to growing season water stress. On the other hand, resistance of conifers to freeze-induced embolism may assure capacity for water transport during the winter and promote retention of functional xylem in the sapwood from year to year. Tyree et al. (1994) suggested that the tendency for greater freeze-induced dysfunction in plants with large-diameter xylem elements might be related to the greater time required for the large gas bubbles they contain to dissolve. If sufficient tensions in the thawed xylem developed before bubbles dissolve, bubble expansion would result.

Some research has shown that significant reductions in hydraulic conductivity of xylem occur at Ψ_w values that are characteristic of those encountered in the field (Tyree and Sperry, 1988). Consequently, some investigators believe that plants may function near the point of "runaway cavitation," maximizing photosynthesis by maintaining stomatal opening (and hence transpiration) near the point at which each incremental xylem conduit loss leads to a cascading spiral of increased tension and additional conduit embolism (Tyree and Sperry, 1988; Tyree and Ewers, 1991; Davis et al., 2002). If this is a general feature of plant function, xylem transport characteristics likely play a primary role in evolutionary adaptation to water stress in woody plants.

Although this response pattern is plausible, it is difficult to reconcile with other water relations phenomena such as drought-induced osmotic adjustment of leaves, which lowers Ψ_π and Ψ_w of leaves (and hence increases maximum xylem tension) (Parker et al., 1982). There also is an intuitive recognition of the evolutionary requirement for some level of excess xylem transport capacity to allow large woody plants to persist until they reach reproductive age in an environment of periodic severe droughts. There is observational evidence that the vascular system of intact woody plants provides greater capacity for water transport than is necessary for their survival. As an extreme example, Kubler (1991) described a felled pear tree that was nearly completely severed from the stump. The tree remained alive for several years, apparently receiving sufficient water and minerals through only a small segment of the original stem.

It also must be noted that embolism of some water-conducting elements does not necessarily cause immediate collapse of xylem water transport, as considerable redundancy is built into the system (Tyree et al., 1994). Radioactive iodine and phosphorous were used to trace the transpiration stream in hickory, red pine, and blue beech trees that had overlapping horizontal saw cuts at different stem heights. These cuts embolize relatively few vessels or tracheids (Zimmermann, 1983). The isotopes moved vertically up to a cut and then readily passed around it. The isotopes then moved vertically to another cut and around it. Opposing cuts only six inches apart did not block the transpiration stream (Fig. 11.29). Scholander et al. (1957) made double horizontal saw cuts in *Tetracera* vines and observed that only the vessels that were severed by the cuts failed to conduct water. Other vessels above and below these contained water and their conduction was not impaired. Hence, upward water movement occurs in vessels and tracheids along a path of least resistance. If these are blocked, water may move laterally through shorter elements where it encounters greater resistance. This, however, often does not incapacitate the entire conducting system (Scholander, 1958; Richter, 1974).

The ecological significance of xylem embolism probably is greatest in mortality of shallow-rooted seedlings under drought and in protective "trip-wire" embolisms in conduits connecting peripheral branches with the main stem xylem. Young seedlings, being very shallowly rooted in the soil, are far more likely to develop the xylem tensions necessary to cause cavitation. Kavanaugh (1992) reported that cavitation events measured in western hemlock seedlings occurred at Ψ_w between −1.9 to −3.4 MPa, values that also could be observed in newly planted seedlings in the field. The rapid desiccation and browning of seedlings commonly observed during summer droughts also are consistent with an abrupt hydraulic disconnection of shoots from roots. In older plants, embolism in high-resistance elements of branch-stem junctions may protect the main stem from xylem tensions sufficient to cause cavitation. This phenomenon may be especially important for woody monocotyledons that have a fixed vascular transport capacity. For example, in the palm *Rhapis excelsa* cavitation and xylem emboli were nearly completely restricted to leaf petioles (Fig. 11.30). Salleo et al. (1984) noted that a higher percentage of vessels terminated in nodal regions of diffuse-porous angiosperm trees than in internodes, thereby lessening the chance of embolism introduction from adjacent leaves or branches. These results support the "segmentation" hypothesis of Zimmermann (1983), who proposed that hydraulic systems in plants have important "failure points" that isolate essential transport tissues from catastrophic cavitation.

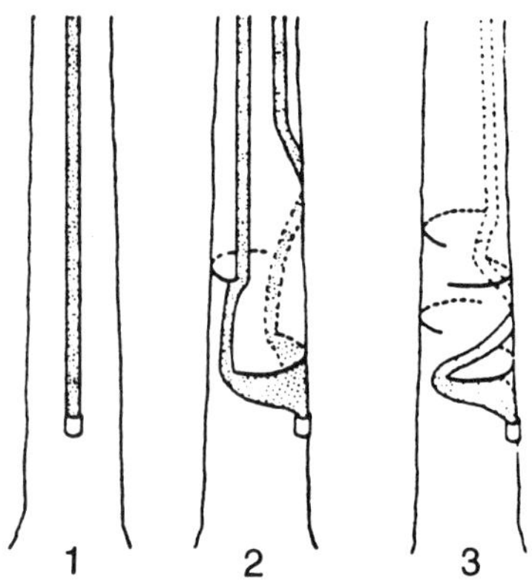

FIGURE 11.29. Movement of ^{32}P around horizontal saw cuts in red pine stems. With no cuts, the isotope (stippled area) moved vertically upward (tree 1). With two opposite cuts, the isotope moved around the cuts and then vertically (tree 2). With four differently oriented cuts only six inches apart (tree 3), the isotope moved around the cuts in its ascent. From Postlethwait and Rogers (1958).

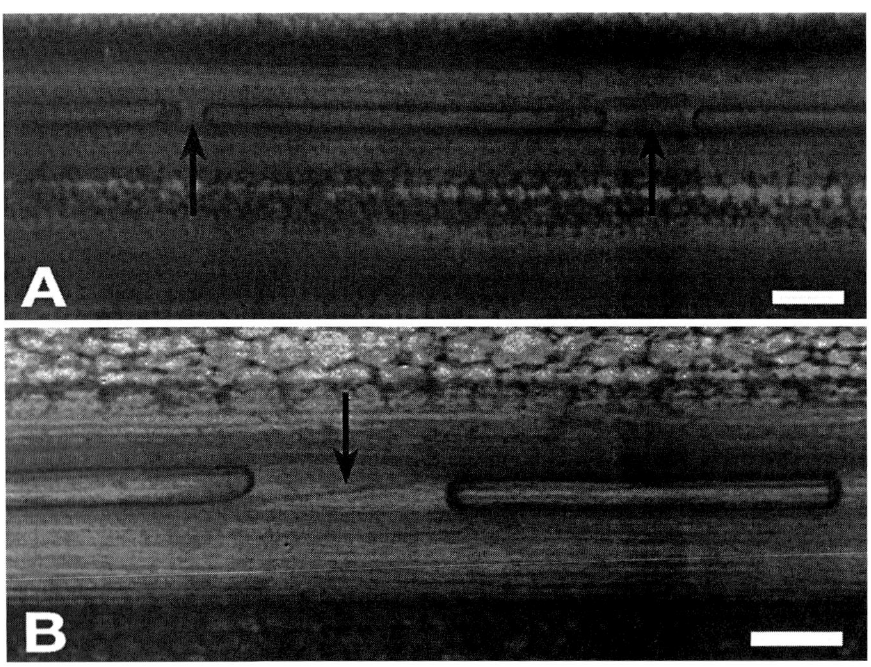

FIGURE 11.30. Longitudinal sections of petioles of the palm *Rhapis excelsa* showing resistances indicative of cavitation. (A) Intact vessel with bubbles arranged in series in each vessel member with scalariform liquid-filled perforation plates marked by arrows. Bar: 200 μm. (B) Close-up of the scalariform perforation plate region. Bar: 100 μm. From Sperry (1986).

Disease

Activities of bacteria, fungal pathogens, nematodes, and insects often lead to blocking of xylem conduits. Obstructions to water transport in plants with vascular wilt diseases may include the mycelia of fungal pathogens and cells of bacteria, accumulation of substances resulting from partial breakdown of host tissues, compounds secreted by the pathogen or host, and structures formed by renewed growth of living cells of the xylem (Talboys, 1978).

Vascular plugging has been reported in several wilt diseases of trees including Dutch elm disease, oak wilt, verticillium wilt of elm and maple, and mimosa wilt (Kozlowski, 1979). In some diseased plants, gums are extruded through pits connecting parenchyma cells with vessels, resulting in masses that line the vessels and reduce hydraulic conductivity or completely plug the vessels. In peach, for example, xylem dysfunction caused by *Cytospora leucostoma* is associated with gum formation (Hampson and Sinclair, 1973).

The vessels of trees infected with vascular wilt disease often become occluded by tyloses, which develop from parenchyma cells via the pits, with the pit membrane distended to form a balloon-like intrusion into a vessel, thereby impeding water transport (Fig. 11.31). In trees with oak wilt disease, the vascular system becomes plugged with gums and tyloses that impede water transport (Ayres, 1978). Resistance to water flow in stems of northern red oak seedlings was substantially increased after inoculation with *Ceratocystis fagacearum*, the incitant of oak wilt (Gregory, 1971). Formation of tyloses and gums also was associated with wilt in northern pin oaks. Tyloses formed abundantly in large vessels and less commonly in small vessels. Whereas tyloses were not formed in tracheids, gummosis was observed in both tracheids and small vessels (Struckmeyer et al., 1954). In northern pin oak trees inoculated with the oak wilt fungus, the vessels were occluded with tyloses and gum three to five days prior to wilting of leaves (Kozlowski et al., 1962). The overall effects of gummosis and tylosis on water transport depend on the frequency and extent of the obstructions. Limited distribution of these materials may have little effect on total water transport, with the occluded parts of the hydrostatic system bypassed and compensated by more rapid water flow in unoccluded vessels. In contrast, extensive occlusion often causes injury and/or mortality of trees by dehydration (Talboys, 1978).

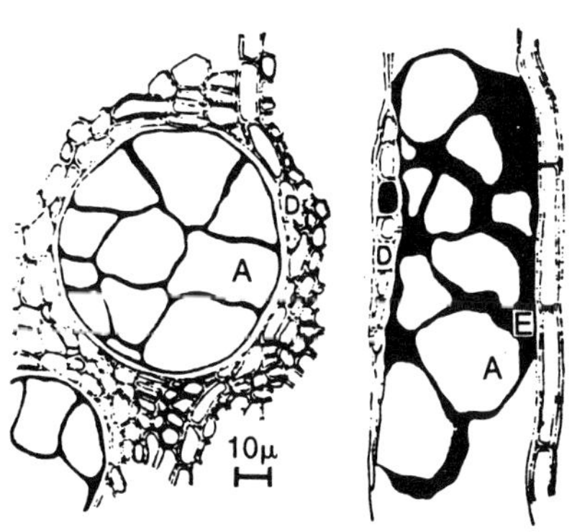

FIGURE 11.31. Effect of oak wilt on occlusion of vessels. (Left) Vessel of diseased tree blocked by tyloses (A); D, ray cell. (Right) Vessel of diseased tree occluded with tyloses (A) and gum deposits (E). From Struckmeyer et al. (1954).

Infection of elms of a clone susceptible to Dutch elm disease reduced the hydraulic conductance of three- and four-year-old branches by 66% within 11 days (Melching and Sinclair, 1975). In infected elms, fungal growth, fungal metabolites, and formation of gums and tyloses were implicated in disrupting water flow (Newbanks et al., 1983). These observations are consistent with the view that injuries from vascular wilt diseases are complex and not due to a single cause (Talboys, 1978).

Similarly, responses of Japanese black pine to infection by the nematode that causes pine wilt disease, the pine wood nematode (*Bursaphelenchus xylophilus*) (Steiner and Buhrer) Nickle, involved both hydraulic conductivity and tissue necrosis (Ikeda and Kiyohara, 1995). Pine seedlings inoculated with both avirulent and virulent strains of the nematode showed declines in hydraulic conductance of the xylem, but loss of hydraulic conductance was much greater if the virulent strain was present. Loss of xylem conductance was linked with aspiration of pits in tracheids. Conductivity losses in virulent strain-infected plants were associated with reductions in transpiration and steep depressions in predawn Ψ_w, indicating shoot water stress. In addition to effects on water relations, nematode infection caused death of parenchyma cells of the xylem. Cell death was total for seedlings inoculated with virulent strains of nematode, and the vascular cambium of these plants also was killed. The authors concluded that tree death caused by pine wilt disease likely was related both to hydraulic dysfunction and cellular injury. Desprez-Loustau et al. (2006) reviewed drought-pathogen interactions in forest trees.

SUMMARY

Water is essential for survival, growth, and proper metabolic function in plants. Appropriate quantification of water status of plants depends on research objectives, but relative water content and water potential concepts have proven most useful to investigators. Whereas relative water content is derived from the amount of water in a tissue compared with that contained at full hydration, water potential concepts are based on the free energy of water and consequent ability to predict flow directions down gradients of free energy. Total water potential includes the influence of several component potentials, including those attributable to solute and surface effects (osmotic potential), pressure effects (pressure or turgor potential), and those attributable to gravity (gravitational potential). There are numerous ways of quantifying water potential and its components, including vapor pressure, tensiometry, and pressure chamber methods.

Water movement within plants is governed by gradients in water potential or certain component potentials. Integration of water potential concepts and relevant flow pathways has produced the concept of the Soil–Plant–Atmosphere Continuum, which has provided a useful, unified model of water flow from soil to the atmosphere. Water absorption in slowly transpiring plants may be osmotically driven, but in rapidly transpiring plants water uptake is largely passive. Osmotically driven water uptake is responsible for root pressure, but stem pressure also is thought to be responsible for many episodes of sap exudation from stems.

Water uptake by roots depends on root system architecture and hydraulic properties, on soil properties and water status, and on environmental conditions. Water absorption will occur in woody, suberized, and unsuberized portions of a root system. Poor soil aeration and low temperatures reduce the efficiency of uptake by roots and can inhibit the normal root development in soil. Low soil water content reduces water absorption because it results in greatly increased resistance to water flow to roots and in a reduction in the gradient in water potential between roots and soil. Mycorrhizas may promote water absorption by increasing the longevity of unsuberized roots, increasing root hydraulic conductivity, and by transporting soil water through hyphae. Although some absorption of water through leaves and stems has been demonstrated, uptake of water from dew and fog is not physiologically and ecologically significant except under unusual circumstances.

Water transport within the plant is largely governed by gradients in pressure and gravitational potential and on the anatomical features of the xylem that control hydraulic properties. Flow capacity in capillaries that resemble xylem conduits increases with the fourth power of the radius. Hence wide xylem elements have vastly greater flow capacity. The relatively narrow, short tracheids of gymnosperms offer inherently greater resistance than angiosperms with large-diameter vessels. However, tracheid-bearing wood offers about the same resistance to flow as vessel-bearing wood of equivalent element diameter because the shorter tracheids have lower pit resistances for element-to-element transfer than do vessels. The fraction of xylem involved in water transport also varies widely, with ring-porous angiosperms often nearly totally dependent on the current year's xylem for water transport, whereas diffuse-porous angiosperms and conifers may have several annual rings active in water transport.

Tensions or freezing of water in the xylem can induce rupture of water columns and embolisms in xylem conduits. Susceptibility to formation of embolisms under tension is thought to be related to the size of pores in the pits located in cell walls. Substantial loss of xylem function through cavitation can occur during the growing season, but may be regained by bubble dissolution and/or dormant season refilling of xylem conduits by root pressure. Species from arid habitats appear to possess greater resistance to cavitation than those of mesic habitats. Conifer anatomy appears to confer higher resistance to cavitation at freezing temperatures than does angiosperm anatomy; ring-porous species undergo especially substantial cavitation when xylem is frozen.

Some diseases may induce dysfunction in xylem transport capacity. The injury caused by many vascular wilt diseases, including Dutch elm disease, oak wilt, and *Verticillium* wilt, has been associated with effects on host plant xylem. Such effects include vascular plugging by fungal hyphae or through induction of tylosis formation and gum production by the plant.

General References

Boyer, J. S. (1995). *Measuring the Water Status of Plants and Soils*. Academic Press, New York.

Ehleringer, J. R., Hall, A. E., and Farquhar, G. D., eds. (1993). *Stable Isotopes and Plant Carbon-Water Relations*. Academic Press, San Diego.

Griffiths, J. A. C. and Griffiths, H., eds. (1993). *Water Deficits: Plant Responses from Cell to Community*. BIOS, Oxford.

Hinckley, T. M., Lassoie, J. P., and Running, S. W. (1978).Temporal and spatial variations in the water status of forest trees. *Forest Science Monograph* **20**, 72 pp.

Kozlowski, T. T., ed. (1968–1983). *Water Deficits and Plant Growth*. Vols. 1–7. Academic Press, New York.

Kozlowski, T. T., Kramer, P. J., and Pallardy, S. G. (1991). *The Physiological Ecology of Woody Plants*. Academic Press, New York.

Kramer, P. J. and Boyer, J. S. (1995). *Water Relations of Plants and Soils*. Academic Press, San Diego.

Kreeb, K. H., Richter, H., and Hinckley, T. M., eds. (1989). *Structural and Functional Responses to Environmental Stresses: Water Shortage*. SPB Academic Publishing, The Hague, The Netherlands.

Lassoie, J. P. and Hinckley, T. M., eds. (1991). *Techniques and Approaches in Forest Tree Ecophysiology*. CRC Press, Boca Raton, Florida.

Levitt, J. (1980). *Responses of Plants to Environmental Stresses. Volume 2: Water, Radiation, Salt and Other Stresses*. Academic Press, New York.

Lewis, T. E., ed. (1995). *Tree Rings as Indicators of Ecosystem Health*. CRC Press, Boca Raton, Florida.

Mooney, H. A., Winner, W. E., and Pell, E. J., eds. (1991). *Response of Plants to Multiple Stresses*. Academic Press, New York.

Morison, J. I. and Morecroft, M., eds. (2007). *Plant Growth and Climate Change*. Blackwell Publishing, Oxford.

Nobel, P. S. (2005). *Physicochemical and Environmental Plant Physiology, 3rd edition*. Elsevier/Academic Press, San Diego.

Pearcy, R. W., Ehleringer, J., Mooney, H. A., and Rundel, P. W., eds. (1989). *Plant Physiological Ecology: Field Methods and Instrumentation*. Chapman and Hall, New York.

Rundel, P. W., Ehleringer, J. R., and Nagy, K. A., eds. (1989). *Stable Isotopes in Ecological Research*. Springer-Verlag, New York.

Schweingruber, F. H. (1988). *Tree Rings: Basics and Applications of Dendrochronology*. Kluwer Academic Publishers, Dordrecht, the Netherlands.

Tyree, M. T. and Zimmermann, M. H. (2002), *Xylem Structure and the Ascent of Sap*, 2nd Edition. Springer, Berlin and New York.

CHAPTER

12

Transpiration and Plant Water Balance

INTRODUCTION

Transpiration is the loss of water vapor from plants. The first quantitative measurements of water loss from plants appear to have been made by Stephen Hales, prior to 1727. He measured water loss by weighing potted grapevines, apple and lemon trees, and various herbaceous plants. Several studies were made during the second half of the nineteenth century, of which the best known are those by von Höhnel, published in 1881 and 1884. The early work on transpiration of trees was summarized by Raber (1937). Slatyer (1967), Kramer (1983), Pearcy et al. (1989), and Kaufmann and Kelliher (1991) present discussions of various methods of measuring transpiration and their advantages and disadvantages.

Transpiration is a dominant factor in plant water relations because evaporation of water produces the energy gradient that causes movement of water through plants. It therefore controls the rate of absorption and the ascent of sap and causes almost daily leaf water deficits. A single isolated tree may lose 200 to 400 liters of water per day, and a hardwood forest in the humid Appalachian Mountains of the southeast United States loses 42 to 55 cm of water per year (Hoover, 1944). Several hundred kilograms of water are used by plants for every kilogram of dry matter produced; about 95% of the absorbed water simply passes through the plant and is lost by transpiration.

Rapidly transpiring plants lose so much water on sunny days that the cells of young twigs and leaves may lose turgor and wilt, stomata close reducing photosynthesis, and growth declines or ceases. The harmful effects of water stress will be presented later in this chapter. Although transpiration sometimes may provide beneficial cooling of leaves, it must be regarded largely as an unavoidable drawback. Transpiration is unavoidable because of the structure of leaves and a drawback because it often induces water deficits, inhibits growth, and also causes injury and death of plants by dehydration.

The Process of Transpiration

Because of the great importance of transpiration in the overall water economy of plants, the nature of the process and the factors affecting it deserve careful attention. Transpiration is basically a process of evaporation that is controlled by physical factors. However, transpiration also is a physiological process, and as such it is affected by plant factors such as leaf structure and exposure and the responses of stomata. It usually occurs in two stages: evaporation of water from cell walls into intercellular spaces and diffusion of water vapor into the outside air. Although it might seem reasonable to expect that most water evaporates from the walls of mesophyll cells, there is no consensus regarding the pattern of evaporation within the leaf (Kramer and Boyer, 1995).

Transpiration as a Physical Process

The rate of evaporation of water from any surface depends on (1) the energy supply available to vaporize water, (2) the vapor concentration gradient that constitutes the driving force for movement of water vapor, and (3) the resistances in the diffusion pathway. Solar radiation serves as the primary source of energy for evaporation of water. Most water vapor escapes through the stomata, some passes out through the epidermis of leaves and its cuticular covering, and some escapes from the bark of stems, branches, and twigs of woody species.

Evaporation can be described by a simple equation:

$$E = \frac{C_{water} - C_{air}}{r_{air}} \tag{12.1}$$

where E is the evaporation in kg m^{-2} s^{-1}; C_{water} and C_{air} are the concentrations of water vapor at the water surface and in the bulk air, respectively, in kg m^{-3}, and r_{air} is the boundary layer resistance encountered by diffusing molecules in s m^{-1}. Because transpiration is controlled to a considerable degree by leaf resistance, additional terms must be added to describe it:

$$T = \frac{C_{leaf} - C_{air}}{r_{leaf} + r_{air}} \tag{12.2}$$

In this equation T is transpiration, C_{leaf} is the water vapor concentration at the evaporating surfaces within

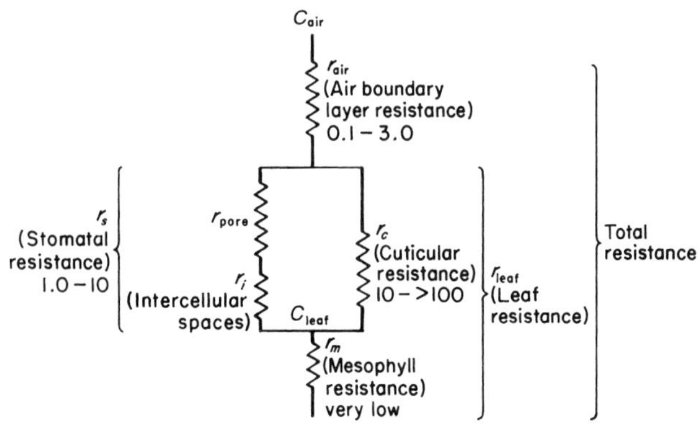

FIGURE 12.1. Diagram showing resistances to the diffusion of water vapor from a leaf. The rate of transpiration is proportional to the steepness of the gradient in water vapor concentration, C_{leaf} to C_{air}, and is inversely proportional to the resistances. Resistances are given in s cm^{-1}. These vary widely among species and with environment, and can differ substantially from those shown here.

the leaf, and r_{leaf} is the additional resistance to diffusion in the leaf. This equation states that the rate of transpiration in kg water m^{-2} s^{-1} is proportional to $C_{leaf} - C_{air}$ (ΔC), the difference in concentration of water vapor between the evaporating surfaces in the leaf and the bulk air outside the leaf, divided by the sum of the resistances to diffusion ($r_{leaf} + r_{air}$) in s m^{-1}. The situation with respect to the conceptual resistance network is indicated in Figure 12.1.

Although mass and resistance terms are convenient to use in discussions of water flow through the SPAC, other measures also are used. For example, transpiration rates may be reported in molar flux terms (mmol H_2O m^{-2} s^{-1}). Gas phase limitation to vapor flow may be reported as conductance, g, where $g = 1/r$. Units for conductance also may vary. For example, g_{air} as reported earlier would be expressed in m s^{-1}. However, r and g in this system are not constant with pressure and temperature, as $r = \ell/D_{H_2O}$, where ℓ is the path length (m) over which water vapor must move and D is the diffusion coefficient of water vapor (m^2 s), which varies with temperature and pressure. To avoid this problem, units of g may be derived from the following equation:

$$E = g_{air} \times (c_{leaf} - c_{air}) \tag{12.3}$$

where c is the mole fraction of water vapor (a unitless parameter). In this case g has the same units as evaporation (e.g., mmol m^{-2} s^{-1}) and is independent of pressure and much less dependent on temperature (Nobel, 1991). At one atmosphere and 20°C, g of 1 mm s^{-1} is equivalent to 41.6 mmol m^{-2} s^{-1}.

Energy Use in Transpiration

The energy input to a stand of plants or individual leaves comes from direct solar radiation, radiation reflected and reradiated from the soil and surrounding vegetation, and advective flow of sensible heat from the surroundings. The energy load is dissipated by three mechanisms: reradiation, convection of sensible heat, and dissipation of latent heat by evaporation of water (transpiration). The energy load on a leaf is partitioned as follows:

$$\underbrace{S + G - rS - R}_{\text{Net Radiation}} \pm H \pm 1E + A = 0 \tag{12.4}$$

Net radiation, the actual radiation available to leaves, consists of S, the total solar radiation, plus G, the long-wave radiation from the environment, minus rS, the radiation reflected from leaves, and R, that reradiated from leaves. H is the sensible heat exchange with the environment by convection and advection, lE the latent heat lost in transpiration or gained in dew or frost formation, and A the energy used in metabolic processes, especially photosynthesis. The latter is only 2 or 3% of the total.

Since the value of all the terms in Eq. (12.4) can vary considerably, the energy relations of individual leaves are rather complex. Nobel (1991) estimated that more than two-thirds of the incident radiation was balanced by reradiation of leaves at longer wavelengths. When leaf and air temperature are equal, reradiation and transpiration dissipate the entire energy load. If stomatal closure reduces transpiration, the energy load must be dissipated by reradiation and transfer of sensible heat, but usually there is a dynamic equilibrium in which all three mechanisms operate. Occasionally, advective energy transfer from the surroundings to a small, isolated mass of vegetation results in a rate of transpiration exceeding that explainable in terms of incident radiation; this is the so-called oasis effect. At night leaves often are cooled to below air temperature by radiation to the sky. This process results in flow of sensible and latent heat toward them and, if leaf temperature falls low enough, condensation of dew or frost. This complex topic is discussed in more detail by Gates (1980), Kozlowski et al. (1991, Chapter 4), and Nobel (1991).

Vapor Concentration Gradient from Leaf to Air

As mentioned earlier, the driving force for movement of water vapor out of plants is the difference in vapor concentration between plant tissue and air, $C_{leaf} - C_{air}$. This difference depends on two variables, the vapor concentration at the evaporating surfaces and that of the surrounding bulk air.

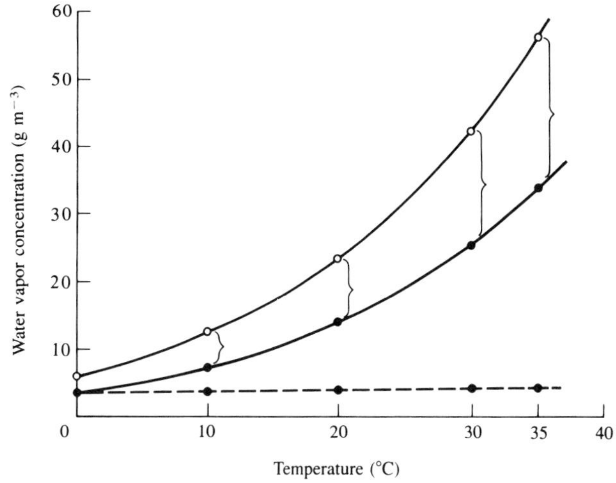

FIGURE 12.2. Effect of increasing temperature on water vapor concentration difference (braces) between leaf (–○–)and air (–●–) if the air in the leaf is assumed to be saturated and external air is maintained at 60% relative humidity. (--●--), water vapor concentration at 60% relative humidity, and air temperature of 0°C.

The vapor concentration at the evaporating surfaces of cells is influenced chiefly by the temperature and Ψ_w at the surfaces. If the water potential at the cell surfaces is taken as zero (i.e., the cells are turgid), the vapor concentration can be taken as the saturation vapor concentration at that temperature. The effect of temperature on the vapor concentration of water is shown in Figure 12.2, where it can be seen that increasing the temperature from 10 to 30°C more than triples the saturation vapor concentration and the difference between C_{leaf} and C_{air} at constant relative humidity. Thus even small changes in leaf temperature can produce considerable changes in rates of transpiration, even when r_{air}, r_{leaf}, and relative humidity remain constant.

The effect of a reduction in Ψ_l on vapor concentration is quite small, a decrease in leaf Ψ_w to –4 MPa reducing the vapor concentration gradient only about 5% at 50% relative humidity and 30°C. Thus, moderate changes in Ψ_l are unimportant compared to other factors affecting ΔC and have little effect on transpiration. Although it generally is assumed that the water potential at the mesophyll cell surfaces is similar to that of the bulk tissue, this apparently is not always true. Several investigators have reported humidity in the intercellular spaces of rapidly transpiring plants equivalent to a Ψ_w of –10 MPa or lower (Shimshi, 1963; Ward and Bunce, 1986; Egorov and Karpushkin, 1988).

The vapor concentration of the bulk air surrounding plants depends on temperature and humidity, as shown in Tables 11.1 and 11.2. The actual amount of water present in the atmosphere is known as the "absolute humidity." More often the moisture content of the air is expressed in terms of "relative humidity" which is the percentage of saturation at a given temperature. The reason evaporation and transpiration increase with increasing temperature is because the vapor concentration of the water in leaves increases more rapidly than that in the unsaturated air (Fig. 12.2), not as is often erroneously supposed because the relative humidity decreases.

Resistances in the Water Vapor Pathway

Small amounts of water vapor escape through the bark, chiefly through lenticels. Some escapes through the epidermis, but about 90% or more escapes through the stomata because the resistance to diffusion through stomata is much lower when they are open than resistance to diffusion through other parts of the epidermis. There are two kinds of resistances to the diffusion of water vapor, the resistances associated with the leaf and the external boundary layer resistance in the air adjacent to the leaf surface. Nobel (1991) discussed resistances to gas exchange in detail (see also Chapter 5).

Leaf Resistances

The outer epidermal surfaces usually are covered by a layer of cuticle that is relatively impermeable to water. It differs greatly in thickness and permeability among leaves of different species and those developed in different environmental conditions. The cuticle often is covered by a deposit of wax that probably diffuses through the wall and the cuticle (Shepherd and Griffiths, 2006) and accumulates on the outer surface, as was discussed in Chapter 2. The role of wax in the cuticle was discussed in Chapter 8. Leaf waxes may form heavy deposits on stomata, sometimes even occluding the stomatal pores. The waxes in the stomatal antechambers of conifers with sunken stomata may be more important in controlling water loss than the waxes on the epidermal surface. Wax that occluded the stomatal pores of Sitka spruce was estimated to decrease transpiration by nearly two-thirds when the stomata were fully open (Jeffree et al., 1971).

Cuticular resistance varies widely and usually increases materially as leaves become dehydrated or are exposed to low humidity (Kerstiens, 1996, 2006). The resistances to diffusion of water vapor through the stomata and the cuticle of several species are shown in Table 12.1 and differences in cuticular and stomatal transpiration among several species are shown in Table 12.2. Comprehensive treatment of plant cuticles can be

TABLE 12.1. Resistances to Movement of Water Vapor and Carbon Dioxide through the Boundary Layer (r_{air}), Cuticle (r_c), and Open Stomata (r_s), and the Mesophyll Resistance for CO_2 (r_m)[a]

	Resistances (s cm^{-1})				
	Water vapor			CO_2	
Species	r_{air}	r_s	r_c	r_s	r_m
European silver birch	0.80	0.92	83	1.56	5.8
English oak	0.69	6.70	380	11.30	9.6
Norway maple	0.69	4.70	85	8.00	7.3
Sunflower	0.55	0.38	—	0.65	2.4

[a]From Holmgren et al. (1965).

TABLE 12.2. Total and Cuticular Transpiration of Leaves of Various Kinds of Plants under Standard Evaporating Conditions[a,b]

Species	Transpiration with open stomata	Cuticular transpiration with closed stomata	Cuticular transpiration (% of total)
Woody plants			
European silver birch	1.20	0.15	12.0
European beech	0.65	0.14	21.0
Norway spruce	0.74	0.02	3.0
Scotch pine	0.83	0.02	2.5
Alpine-rose	0.93	0.09	10.0
Herbaceous plants			
Crown vetch	3.09	0.29	9.5
Woundwort	2.78	0.28	10.0
Locoweed	2.62	0.15	6.0

[a]From Larcher (1975), with permission of Springer-Verlag.
[b]Rates are given in mmol m^{-2} s^{-1}; the surface area includes both sides of the leaves.

found in the book by Juniper and Jeffree (1983) and the volume edited by Riederer and Müller (2006).

When the stomata are open the stomatal resistance, r_s, is so much lower than the cuticular resistance that most of the water escapes through the stomata. The principal factor affecting stomatal resistance is stomatal aperture, which responds to many endogenous and environmental influences including light intensity, CO_2 concentration within leaf air spaces, vapor concentration difference between leaf and air, temperature, plant water status, plant hormones and age (see following section on stomatal control of transpiration).

It also is possible to identify resistances to water vapor diffusion in the intercellular spaces and the mesophyll cells. The resistance in the intercellular spaces (r_{ias}), as measured with a diffusion porometer, appears to be significant only when the stomata are wide open (Jarvis and Slatyer, 1970). Farquhar and Raschke (1978) did not find any mesophyll cell wall resistance in leaves of several herbaceous species and concluded that the water vapor concentration at the evaporating surfaces is equal to its saturation vapor concentration.

Stomatal resistance and r_{ias} are connected in series, whereas r_s and r_c are in parallel (Fig. 12.1). The total resistance of a leaf surface can be described as follows:

$$\frac{1}{r_{leaf}} = \frac{1}{r_c} + \frac{1}{r_s + r_{ias}} \quad \text{or} \quad r_{leaf} = \frac{(r_c)(r_s + r_{ias})}{r_c + r_s + r_{ias}} \quad (12.5)$$

External Resistances

The external resistance depends largely on leaf size and shape and on wind speed. Increasing air movement acts directly to increase transpiration by thinning the boundary layer of water vapor surrounding leaves in quiet air and reducing r_{air} in Eq. (12.2). Wind also acts indirectly to decrease transpiration by cooling leaves and decreasing C_{leaf}. Most of the effect occurs at low velocities. Knoerr (1967) pointed out that, although a breeze should increase transpiration of leaves exposed to low levels of radiation, at higher levels when they tend to be warmer than the air a breeze might decrease transpiration, by cooling the leaves as shown in Figure 12.3. However, the actual responses of plants seem rather variable.

Davies et al. (1974b) reported significant differences in the reaction of seedlings of three species in a wind tunnel under artificial illumination of 28,700 to 35,500 lux and air speeds of 5.8 to 26 m s^{-1}. The overall transpiration rates in both wind and quiet air for three species of trees varied in the following order: white ash

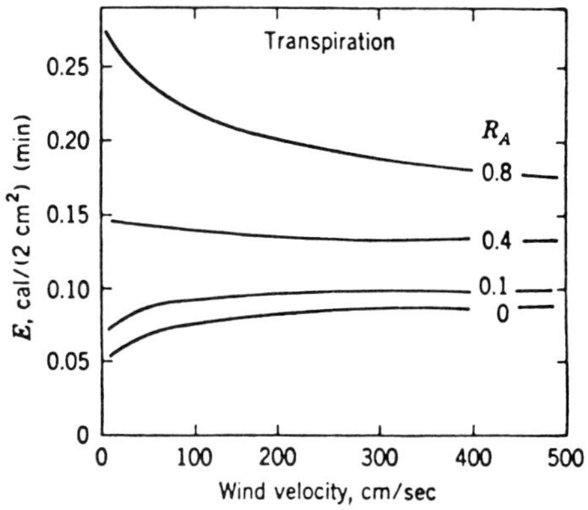

FIGURE 12.3. Curves showing theoretical latent heat exchange (E) of a leaf by transpiration at various wind speeds and net radiation (R_A). Under intense insolation wind may decrease transpiration by cooling leaves; at low insolation it might increase transpiration by increasing energy supply available to the leaf. From Knoerr (1967).

> sugar maple > red pine. Wind increased transpiration of ash seedlings at all speeds, decreased transpiration of maple, and had no significant effect on pine. The stomata of maple leaves closed promptly when subjected to wind, but those of ash did not close until considerable dehydration had occurred. Grace (1977) and Coutts and Grace (1995) reviewed plant responses to wind.

Resistance to water vapor movement across the boundary layer is in series with r_{leaf} and hence we can describe the entire resistance from internal air spaces to the bulk air surface for a single side of a leaf as:

$$r_{total,\ one\ side} = r_{leaf} + r_{air} \quad (12.6)$$

Finally, it is useful to note that water vapor may escape from either side of a leaf, and because these two pathways are in parallel an appropriate equation can be developed to totally describe the resistance of a leaf to water vapor loss:

$$r_{total,\ both\ sides} = \frac{r_{total,\ upper\ side} \times r_{total,\ lower\ side}}{r_{total,\ upper\ side} + r_{total,\ lower\ side}} \quad (12.7)$$

Total resistance to water vapor loss through two leaf surfaces will always be lower than that through either surface alone. Table 12.3 presents representative values of component and total resistances for various leaf types. In the case of leaves with stomata on a single leaf surface, most water vapor escapes through the surface that has stomata. In amphistomatous species with equal stomatal density and aperture on both surfaces, $r_{total,\ both\ sides}$ will be one-half that of either surface alone. The strong influence of stomatal aperture on water loss from leaves is illustrated by the large impact of stomatal resistance changes on r_{total}. As noted earlier,

TABLE 12.3. Representative Conductances and Resistances for Water Vapor Diffusing Out of Leaves[a]

Component condition	Conductance		Resistance	
	mm s^{-1}	mmol m^{-2} s^{-1}	s m^{-1}	m^2 s mol^{-1}
Boundary layer				
Thin	80	3,200	13	0.3
Thick	8	320	130	3
Stomata				
Large area, open	19	760	53	1.3
Small area, open	1.7	7	600	14
Closed	0	0	∞	∞
Mesophytes, open	4–20	160–800	50–250	1.3–6
Xerophytes and trees, open	1–4	40–160	250–1,000	6–25
Cuticle				
Crops	0.1–0.4	4–16	2,500–10,000	60–250
Many trees	0.05–0.2	2–8	5,000–20,000	125–500
Many xerophytes	0.01–0.1	0.4–4	10,000–100,000	250–2,500
Intercellular air spaces				
Calculation	24–240	1,000–10,000	4.2–42	0.1–1
Waxy layer				
Typical	50–200	2,000–8,000	5–20	0.1–0.5
Certain xerophytes	10	400	100	2.5
Typical	40–100	1,600–4,000	10–25	0.2–0.6
Leaf (lower surface)				
Crops, open stomata	2–10	80–400	100–500	2.5–13
Trees, open stomata	0.5–3	20–120	300–2,000	8–50

[a]From Nobel (1991).

wind may influence r_{total} occasionally, especially in still air or very strong winds that may affect both the boundary layer resistance and leaf temperature (and hence $C_{leaf} - C_{air}$). Cuticular resistance has its largest impact on the potential survival of plants under long-term drought, as it represents the only remaining barrier to lethal desiccation of leaves after the stomata close.

TABLE 12.4. Transpiration Rates per Unit of Leaf Surface and per Seedling of Loblolly Pine and Harwood Seedlings for the Period August 22 to September 2[a,b]

Rate	Loblolly pine	Yellow-poplar	Northern red oak
Transpiration (g day^{-1} m^{-2})	508	976	1245
Transpiration (g day^{-1} $seedling^{-1}$)	106.70	59.10	77.00
Average leaf area per tree (m^2)	0.21	0.061	0.062
Average height of trees (m)	0.34	0.34	0.20

[a]From Kramer and Kozlowski (1960).
[b]Average of six seedlings of each species.

FACTORS AFFECTING TRANSPIRATION

Although the rate of transpiration basically is controlled by physical factors, it is influenced by several plant factors that affect both driving forces and resistances.

Leaf Area

The total leaf area has significant effects on water loss of individual plants, as plants with large leaf areas usually transpire more than those with small leaf areas. Some woody species such as creosote bush shed most of their leaves when subjected to water stress, greatly reducing the transpiring surface. This also occurs in some mesophytic woody species such as buckeye, black walnut, willow, true poplars, and yellow-poplar. Curling and rolling of wilting leaves also reduces the exposed surface and increases resistance to diffusion of water vapor, especially if most of the stomata are on the inner surface of the curved leaf. Seasonal and developmental changes in the transpiring surfaces of a number of species are discussed by Addicott (1991).

As shown in Table 12.4, there are wide variations in transpiration per unit of leaf area among species. However, such variations can be quite misleading because the differences in total leaf area may compensate for differences in rate per unit of leaf area. Transpiration per unit of land area generally increases with greater Leaf Area Index unless canopy boundary layer resistance is so high that energy input controls evaporation (Landsberg, 1986).

Root–Shoot Ratio

The ratio of roots to shoots or, more accurately, the ratio of absorbing surface to transpiring surface is of greater importance than leaf surface alone because if absorption lags behind transpiration a leaf water deficit develops, the stomata may close and transpiration is reduced. Parker (1949) found that the rate of transpiration per unit of leaf area of northern red oak and loblolly pine seedlings growing in moist soil increased as the ratio of root to leaf surface increased. Pereira and Kozlowski (1977b) found that sugar maple seedlings with a large leaf area developed more severe water stress than partly defoliated seedlings. Trees with extensive, much-branched root systems survive droughts much better than those with shallow or sparsely branched root systems (Chapter 5, Kozlowski and Pallardy, 1997).

Loss of roots during lifting of seedlings is a serious problem because the most common cause of death of transplanted seedlings is desiccation caused by lack of an effective absorbing surface. Undercutting and "wrenching" of seedlings in seed beds is intended to produce compact, profusely branched root systems that can be lifted with minimum injury (see Chapter 7, Kozlowski and Pallardy, 1997). Lopushinsky and Beebe (1976) reported that survival of outplanted Douglas-fir and ponderosa pine seedlings in a region with dry summers was improved by large root systems and high root–shoot ratios. Allen (1955) found that clipping the needles of longleaf pine planting stock back to 12.5 cm in length reduced mortality of outplanted seedlings. It is common practice to prune back the tops of transplanted trees and shrubs to compensate for loss of roots during the transplanting process. However, reduction in transpiring surface also reduces the photosynthetic surface, which is undesirable except when conservation of water is more important than loss of photosynthetic capacity. Other treatments are discussed in the section on control of transpiration.

Leaf Size and Shape

The size and shape of leaves affect the rate of transpiration per unit of surface. Slatyer (1967) reported that the boundary layer resistance, r_{air}, is about three times greater for a leaf 10 cm wide than for one only 1 cm wide. Also small, deeply dissected leaves and compound leaves with small leaflets tend to be cooler than large leaves because their thinner boundary layers permit more rapid transfer of sensible heat. Tibbals et al. (1964) reported that in quiet air broad leaves are considerably warmer than pine needles exposed to the same incident radiation. Nobel (1976) found that the large shade leaves of *Hyptis emoryi* become much warmer than the small sun leaves. Within tree crowns there can be substantial morphological variation in leaf shape that is correlated with capacity for sensible heat exchange. Baranski (1975) studied variation in leaf size and shape of white oak from a number of trees growing in the eastern United States. Even within the crown of a single tree there were large gradients in leaf shape, with leaves from more sun-exposed upper and outer crown positions more dissected, as reflected in a larger lobe length/width ratio and greater indentation (Fig. 12.4; see also Chapter 2). More dissected leaves would have narrower effective leaf widths, and consequently thinner boundary layers and lower r_{air}. This leaf type presumably would have the capacity to avoid heat injury from high solar irradiance in exposed canopy positions.

The thin boundary layer and lower r_{air} of small, dissected leaves also is more favorable to water vapor loss so the two effects may tend to compensate one another with respect to sensible heat transfer and transpiration if stomatal resistance is similar (Raschke, 1976). The adaptive value of dissected leaves may be most important after stomata have closed and leaf temperatures rise as sensible heat transfer dominates leaf energy exchange. The energy relations of leaves are discussed further by Gates (1965, 1980), Slatyer (1967, pp. 237–247), Knoerr (1967), and Nobel (2005).

Leaf Orientation

Most leaves grow in such a manner as to be more or less perpendicular to the brightest light that strikes them. This is noticeable on vines covering walls and on isolated trees where there is a complete mosaic of leaves on the outer surface neatly arranged so they intercept as much light as possible. On the other hand, the leaves of a few species such as *Silphium* and turkey oak are oriented vertically, and the needles of longleaf pine seedlings in the grass stage grow upright. Such an orientation obviously decreases energy absorption and tends to decrease the midday leaf temperature, which may in turn decrease water loss, but it has never been demonstrated that this unusual orientation really has significant survival value. Needles of most pines occur in fascicles and shade one another. This decreases the rate of photosynthesis (Kramer and Clark, 1947)

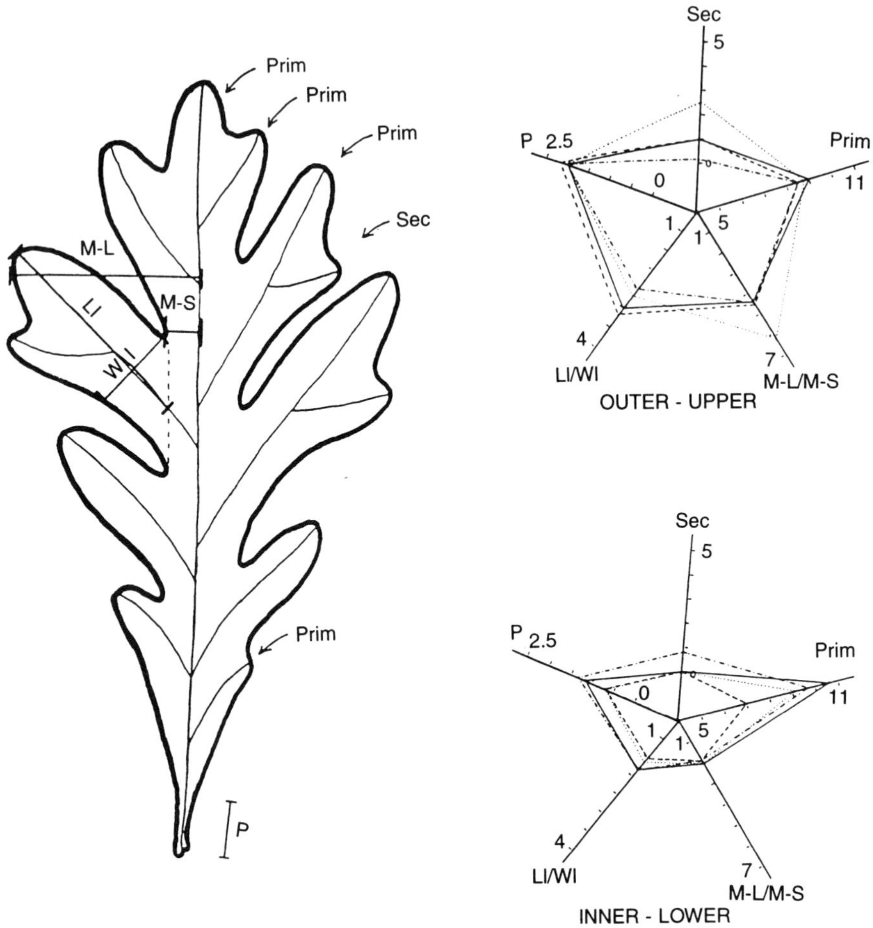

FIGURE 12.4. Polygonal graphs for five leaf characters taken within the outer portion of the upper crown and inner portion of the lower crown of four white oak trees. Degree of dissection is best indicated by the L1/W1 axis (lobe length/width ratio), and the indentation index (M-L/M-S, distance between midvein and lobe apex/distance between midvein and sinus base). Leaves from shaded positions exhibit much reduced dissection. Prim, primary lobe; Sec, secondary lobe; P, petiole length. Adapted from Baranski (1975).

and presumably also decreases transpiration. The drooping and rolling characteristic of wilted leaves also decreases the amount of radiation received. These changes in leaf orientation are an indication of water stress, but probably cause sufficient decrease in further water loss to prolong life. Caldwell (1970) reported that wind changes the leaf orientation of Swiss stone pine enough to reduce photosynthesis.

Leaf Surfaces

I already have mentioned the importance of the cuticle and epicuticular waxes in increasing resistance to water loss through the epidermis. The net effect of the thick coat of hairs found on some kinds of leaves is less certain. Pubescence would be expected to increase the boundary layer resistance, r_{air}, thereby decreasing both heat loss and escape of water vapor in moving air. However, conductance of CO_2 into the pubescent leaves of *Encelia farinosa* is not reduced (Ehleringer et al., 1976) as compared with the nonpubescent leaves of *E. californica*. If r_{air} is not increased for CO_2 it presumably would not be increased for water vapor. Living hairs might increase transpiration by increasing the evaporating surface.

The albedo or reflective characteristics of leaf surfaces materially affect leaf temperature. Vegetation usually reflects 15 to 25% of the incident radiation. Holmes and Keiller (2002) observed such a range in reflectance at 680 nm for waxy leaves of *Eucalyptus gunnii*, *E. cinerea*, and *Kalanchoe pumila*. Removal with chloroform of surface wax from leaves of these species decreased reflectance by about 50%. Three species with wooly hair on leaves (*Campanula elationoides*, *K. tomentosa*, and *Verbascum dumulosum*) reflected about 15% of incident 680 nm radiation, with similar reductions if hairs were removed by gentle abrasion with emery cloth. Billings and Morris (1951) reported that leaves of desert plants usually reflect more light than leaves of plants from less exposed habitats. The white, densely pubescent leaves of the desert shrub, *Encelia farinosa*, reflect about twice as much radiation as the green, nonpubescent leaves of *E. californica* native to the moist coastal region. In fact, so much light is reflected that net photosynthesis of *E. farinosa* is reduced (Ehleringer et al., 1976). However, Ehleringer and Mooney (1978) concluded that the net impact of increased pubescence of *E. farinosa* leaves was to increase the potential capacity for carbon gain, as the reduction in photosynthesis associated with decreased absorption of radiation was more than offset by the potential loss of photosynthate of leaves subjected to higher temperatures and greater transpiration rates. The reduction in transpiration reported to occur after the application of Bordeaux mixture (Miller, 1938) probably results from the lower leaf temperature caused by the white coating it produces on leaves.

Stomata

Most of the water lost from plants escapes through the stomata of the leaves (see Figs. 2.5, 8.4) and most of the carbon dioxide used in photosynthesis enters by the same pathway. Although the stomatal pores usually occupy no more than 1% of the leaf surface, diffusion of water vapor through stomata may amount to 50% of the rate of evaporation from a free water surface. This is because their size and spacing result in their functioning as very efficient pathways for diffusion of gases. Further information on stomata can be found in Chapters 2 and 5.

The size of stomatal pores is controlled by turgor of the guard cells and adjacent epidermal cells, increasing with increased turgor and decreasing with decreased turgor of guard cells. Turgor changes in adjacent epidermal cells tend to have the opposite influence on stomatal aperture. Analysis of mechanical models indicates that both asymmetry in cell wall thickness and radial orientation of cellulose microfibrils in the walls cause guard cells to change shape in their peculiar fashion (Aylor et al., 1973). Stomatal aperture usually is constrained to some extent by adjacent epidermal cells, as the guard cells encounter increasing resistance to opening the more they deform. In some species with modified subsidiary cells, transport of solutes into the guard cells (see later) may reduce turgor in adjacent cells and promote wider apertures (Franks and Farquhar, 2007).

Turgor changes within the guard cells are controlled by changes in content of osmotic constituents. Potassium ions move in large quantities into and out of guard cells during stomatal movements (MacCallum, 1905; Imamura, 1943; Fujino, 1967; Fischer, 1968; Humble and Hsiao, 1970; Humble and Raschke, 1971; Raschke and Humble, 1973). The balancing counter-ions for K^+ may be organic acids, particularly malic and citric (Outlaw, 1987) or external anions such as Cl^-. Organic acids are produced within the guard cells through degradation of starch or importation of sugars and from glycolytic production of pyruvate (phosphoenolpyruvate). The latter then combines with CO_2 in a reaction catalyzed by PEP carboxylase to form oxaloacetic acid and ultimately, through the Krebs cycle, malic and citric acids. Most fixation of CO_2 by guard cells probably occurs by this reaction.

Although there is some uncertainty concerning the capacity of guard cells to fix CO_2 via Rubisco (Outlaw, 1987; Vaughan, 1988), most evidence supports only low rates of photosynthetic carbon fixation by this enzyme (Reckmann et al., 1990). If Rubisco is absent or minimally present, guard cell starch must be synthesized from sugars translocated from the mesophyll. The association between K^+ uptake by guard cells and stomatal opening in the morning may not persist throughout the day. Talbott and Zeiger (1996) demonstrated that although stomatal movements were closely linked with guard cell K^+ contents in the morning, later in the day this association was lost. Instead, guard cell sucrose content was linked with stomatal movements induced by adjusting the CO_2 concentrations around the leaf. The role of sugars in stomatal movements is supported by documented presence and diurnal modulation of sugar transporters in the guard cell plasma membrane (Stadler et al., 2003).

Uptake of K^+ has been linked with K^+-specific channels in the guard cell membranes (Schroeder et al., 1984; Schroeder et al., 1987; Hedrich and Schroeder, 1989; Maathius et al., 1997). For example, the K^+ channel protein KPT1 form *Populus tremula* × *P. tremuloides* promoted uptake of K^+ when the gene for this protein was transferred to a K^+ uptake-deficient mutant of the bacterium *Escherichia coli* (Langer et al., 2004). These channels are opened, or gated, by changes in electrical potential across the plasmalemma (Maathius et al., 1997). The voltages arise from the pumping of H^+ out of guard cells, which may be supported by several processes including photosynthetic electron transport, respiration, and by a blue light-induced activation of H^+-ATPases in the plasmalemma of guard cells (Assmann and Zeiger, 1987; Roelfsema and Hedrich, 2005). A Cl^-/H^+ symporter in the guard cell plasma membrane allows movement of chloride as a counter-ion. Stomatal closure occurs by loss of anions through an anion channel that promotes depolarization of the membrane, a response that both closes the inward-directed K^+ channel and allows efflux of K^+ through outward-directed K^+ channels. Guard cell regulation of internal osmotic relations is quite complex. Cytosolic calcium ions appear to play a major role in regulation of ion transport processes both in opening and closing movements (McAinsh et al., 2000) (Table 12.5).

Stomata respond to many environmental and endogenous stimuli (Fig. 12.5). In general, stomata open in light or in response to low carbon dioxide concentration in the intercellular spaces and they close in darkness. Response to high CO_2 concentration may be mediated through the activation of anion efflux channels in the guard cell plasma membrane (Roelfsema et al., 2002). Hence the light-induced stomatal opening is likely a concerted response to both the blue light stimulation of plasmalemma H+ ATPase noted earlier and inhibition of anion efflux by low CO_2 that attends photosynthesis in the light (Roelfsema and Hedrich, 2005).

TABLE 12.5. Agents Causing Increased Cytoplasmic Concentrations of Ca^{+2} Associated with Stomatal Movements[a]

Stomatal closure	Stomatal opening
ABA	Auxin
Elevated CO_2 concentration	Fusicoccin
Oxidative stress	
External Ca^{2+}	
Low external K^+ concentration	

[a]Reproduced with permission, from McAinsh, M. R., Gray, J. E., Hetherington, A. M., Leckie, C. P., and Ng, C. (2000). *Biochem. Soc. Trans.* **28**, 476–481. © The Biochemical Society.

Exposure of the epidermis to dry air causes closure of stomata in turgid leaves of many species (Lange et al., 1971; Schulze et al., 1972; Sheriff, 1977; Sena Gomes et al., 1987). The response can be very rapid. Fanjul and Jones (1982) found that stomata of apple leaves exhibited nearly complete response to a change in atmospheric water vapor within 15 seconds, the time constant of the porometer used in measurements. The initial movements of stomata in responding to changes in water vapor content of the air apparently are independent of K^+ transport, as Lösch and Schenk (1978) demonstrated that K^+ changes lagged behind changes in stomatal aperture in both closing and opening movements. One often can observe responses to water vapor content in the air with no change in bulk leaf Ψ_l (Pallardy and Kozlowski, 1979a).

Recent work employing mutants in ABA synthesis and ABA-signaling networks in *Arabidopsis thaliana* has provided contradictory findings of the potential role of ABA in stomatal humidity responses. Assmann et al. (2000) observed no attenuation of the humidity response of stomata in such mutants, whereas Xie et al. (2006) associated humidity-insensitive mutants with genes that encode an enzyme involved in ABA synthesis and a protein kinase involved in the ABA signaling pathway that leads to stomatal closure. However, the apparent independence from ion transport and rapidity of the response to water vapor content of the air suggest a mechanism that involves direct water loss from guard or epidermal cells, perhaps in the stomatal area (i.e., peristomatal trans-

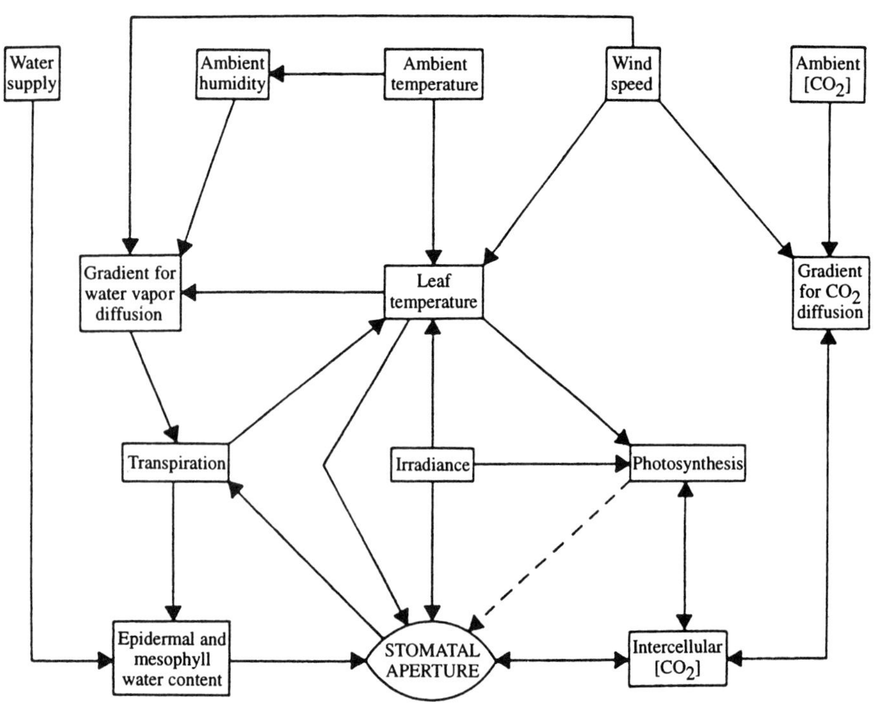

FIGURE 12.5. Important interrelationships among major environmental variables and physiological processes influencing stomatal aperture. Arrows indicate influences between factors; dashed line indicates a relationship based on weak evidence. From *Methods in Stomatal Research* by Weyers, J. and Meidner, H., Pearson Education Limited. © Longman Group UK Limited 1990.

piration). Stomatal conductance appears closely linked with changes in transpiration rate induced by vapor concentration changes (Mott and Parkhurst, 1991; Monteith, 1995). Appleby and Davies (1983) claimed that a specific portion of the guard cell wall could serve as a sensor of water vapor status of the air. Other work suggests that wax layers covering the epidermis extend to the internal surfaces of substomatal chambers. (Nonami et al., 1990, cited in Schulze, 1993). This structural feature would make the mesophyll cells adjacent to the chamber the source of most of the water vapor that diffuses through the stomata. Given this situation, evaporation from mesophyll cells would increase in dry air and local reductions in Ψ_w in the mesophyll could rapidly be propagated to neighboring epidermal cells, thus causing stomatal closure (Schulze, 1993). More work is needed to adequately resolve conflicting evidence concerning the mechanism involved. Schulze (1986, 1993) and Grantz (1990) discussed stomatal responses to water vapor content of the air.

Stomata also close if leaves become dehydrated. This has been attributed to simple loss of turgor in leaves and to hormonal influences. It is generally accepted that stomata close when exposed to increased levels of abscisic acid (ABA) and bulk leaf ABA concentrations increase in detached dehydrated leaves (Hetherington and Quatrano, 1991; Liu et al., 2001a; Dodd, 2003; Davies et al., 2005; Israelsson et al., 2006). Abscisic acid synthesis in the cytoplasm of mesophyll cells increases as bulk turgor of the leaf approaches zero (Pierce and Raschke, 1980; Sauter et al., 2001). However, most evidence points to ABA moving from roots in the xylem sap as a more important source of stomatal regulation in leaves, at least in woody plants (see later). A membrane receptor for ABA has not been identified and the mechanisms by which ABA acts to effect stomatal aperture change are not completely known (Roelfsema and Hedrich, 2005). ABA induces an increase in cytoplasmic free Ca^{2+}, which response may be linked with effects on ion channels in the plasma membrane possibly through activation of anion efflux channels and inhibition of plasma membrane H^+ ATPases that promote K^+ uptake via membrane hyperpolarization.

Although leaf water status would appear to have an overriding influence on stomatal aperture, as noted earlier, soil water status has emerged as an important influence on stomatal aperture, particularly in the initial stages of soil drying (Gollan et al., 1986; Schulze, 1986; Zhang and Davies, 1989; Wartinger et al., 1990; Davies et al., 2005). This response need not be considered predominant under all conditions, particularly during a prolonged drought when soil Ψ_w drops very low, consequently forcing down Ψ_l. Hypotheses have been developed involving production in the roots of either a negative signal (i.e., reduced translocation of a substance promoting opening such as cytokinins, Blackman and Davies, 1985) or a positive signal (i.e., increased translocation of a substance promoting closure). Most evidence to date supports the latter type of signal (Blackman and Davies, 1985; Zhang et al., 1987; Zhang and Davies, 1989, 1990a, 1991; Gowing et al., 1990; Tardieu et al., 1991; Davies and Zhang, 1991; Gowing et al., 1990, 1993a,b; Gallardo et al., 1994).

Abscisic acid (ABA) has received the most attention as a positive signal. Abscisic acid (1) is produced in dehydrating roots (Cornish and Zeevart, 1985; Robertson et al., 1985; Zhang et al., 1987, Zhang and Davies, 1989), (2) increases in the xylem sap after root system dehydration (Zhang and Davies, 1990a,b; Jackson et al., 1995: Correia et al., 1995; Liu et al., 2001a), and (3) causes stomatal closure when introduced into the transpiration stream (Kriedemann et al., 1972; Atkinson et al., 1989; Correia and Pereira, 1994; Correia et al., 1995; Liu et al., 2001b). For example, Loewenstein and Pallardy (1998a,b) observed strong correlations between xylem sap ABA concentrations and leaf diffusive conductance in both mature canopy trees and seedlings of temperate deciduous species during natural and imposed drought, respectively (Fig. 12.6). Stomatal closure could also be induced in seedlings of these species by feeding detached leaves artificial xylem sap containing ABA (Loewenstein and Pallardy 1998b).

Factors in xylem sap other than ABA also may influence stomatal conductance. For example, when Munns and King (1988) supplied to wheat plants the expressed xylem sap of plants that had been exposed to drying soil and from which any ABA had been removed, transpiration was still somewhat suppressed. Xylem sap pH also may play a role in regulation of stomatal aperture by stimulating release of ABA from chloroplast anion traps within the leaf mesophyll (Hartung and Slovick, 1991), or by promoting movement of ABA into mesophyll chloroplasts where it may be metabolized (Schulze, 1993). Normally ABA migrates to the chloroplasts in turgid leaves because of their tendency to serve as anion traps. This happens because photosynthesis raises the pH of the stoma relative to that in the cytoplasm, and thus a greater fraction of ABA in the cytoplasm is protonated and uncharged. Uncharged molecules move relatively easily through the membranes of the chloroplast envelope, but tend to become trapped once inside because of the tendency to assume anionic form as they dissociate at higher pH. If xylem sap becomes more alkaline, cellular anion traps release some ABA to the cytoplasm and apoplast.

Sobeih et al. (2004) observed that early stomatal closure of tomato plants subjected to partial root zone

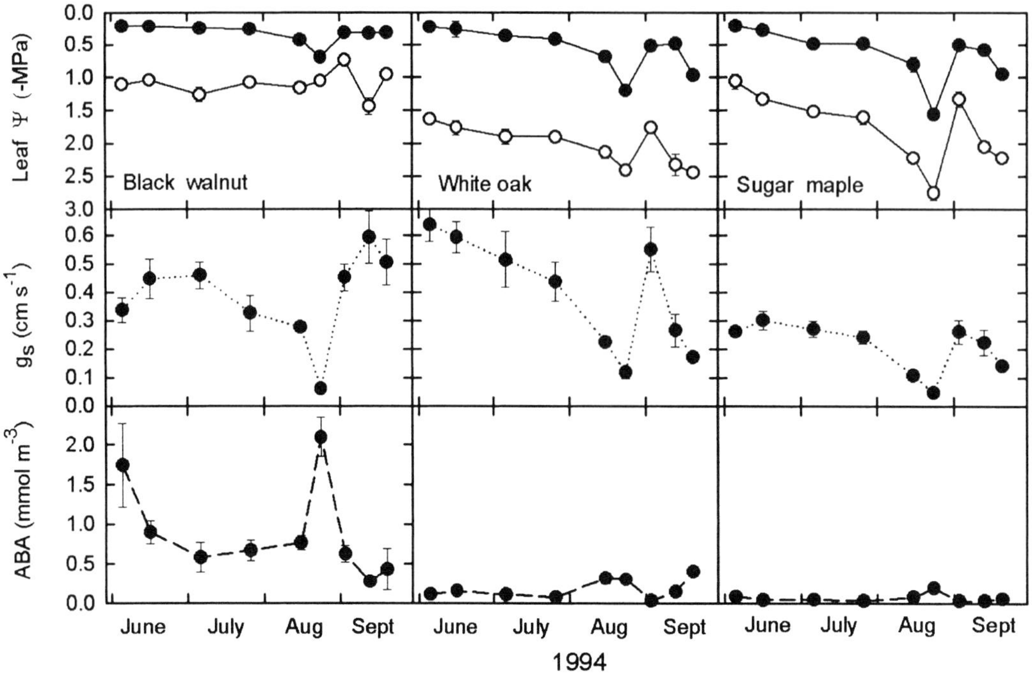

FIGURE 12.6. Seasonal trends in predawn (●) and midday (○) leaf water potential (Ψ), stomatal conductance (g_s), and xylem sap ABA concentration of black walnut, white oak, and sugar maple canopy trees in a central Missouri. forest during a growing season with an August drought. From Loewenstein and Pallardy (1998a).

drying was better associated with increased xylem sap pH, compared to variable xylem sap ABA concentrations in different experiments. At more severe levels of water stress, xylem sap ABA was consistently elevated and better correlated with persistent stomatal closure. By this time xylem sap pH was similar in well-watered plants and those subjected to root drying. Alteration of xylem sap pH might even allow stomatal regulation by inducing changes in ABA available to guard cells with no change in xylem sap levels of this hormone (Wilkinson and Davies, 2002). Xylem sap pH may have an additional direct effect on the sensitivity of stomata to ABA (Gollan et al., 1992; Schurr et al., 1992). Calcium and potassium levels in xylem sap also may directly influence stomatal aperture or stomatal sensitivity to ABA (e.g., DeSilva et al., 1985; Atkinson et al., 1990; Davies et al., 1990; Snaith and Mansfield, 1982; Atkinson, 1991; Schurr et al., 1992; Ruiz et al., 1993).

With respect to woody plants, it is worth noting that water stress apparently has less influence on xylem sap pH than in herbaceous plants (Loewenstein and Pallardy, 1998b, Augé et al., 2000; Johnson et al., 2001; Wilkinson and Davies, 2002; Davies et al., 2005). Hence in woody plants, root-sourced ABA may play a dominant long-distance signaling role in stomatal closure during developing drought. Whereas in some species root-sourced ABA may be synthesized in the roots (e.g., Juglans nigra, Loewenstein and Pallardy, 1998b), in others some xylem sap ABA may be recycled ABA from leaves provided to roots via the phloem.

There has been some discussion as to whether xylem sap concentration, total flux of ABA to leaves, or the amount of ABA moving to the stomatal region itself correlates best with stomatal aperture (Zhang et al., 1987; Schurr et al., 1992; Tardieu and Davies, 1993; Gowing et al., 1993a,b). Experiments generally indicate that in woody plants xylem sap ABA concentration has a closer relationship to stomatal aperture than does total ABA flux to the leaf (Gowing et al., 1993a; Jackson et al., 1995; Correia et al., 1997; Loewenstein and Pallardy, 1998b). As ABA moving to the leaf may quickly segregate among compartments, as noted earlier, and also be rapidly metabolized (e.g., Gowing et al., 1993a), recently delivered ABA in the xylem sap may best represent the quantity of this hormone in the apoplast adjacent to the guard cell. When plants are exposed to natural diurnal conditions during which Ψ_l declines, stomata may exhibit a complex relationship with xylem sap ABA, Ψ_l, and vapor pressure deficit (Tardieu and Davies, 1992, Tardieu et al. 1993) (Fig. 12.7). Hence it is very probable that both leaf and root water status influence stomatal aperture, with the dominant mechanism varying with environmental conditions and species. Davies and Zhang (1991) provided a general hypothetical scheme for root-origin control of shoot responses (Fig. 12.8).

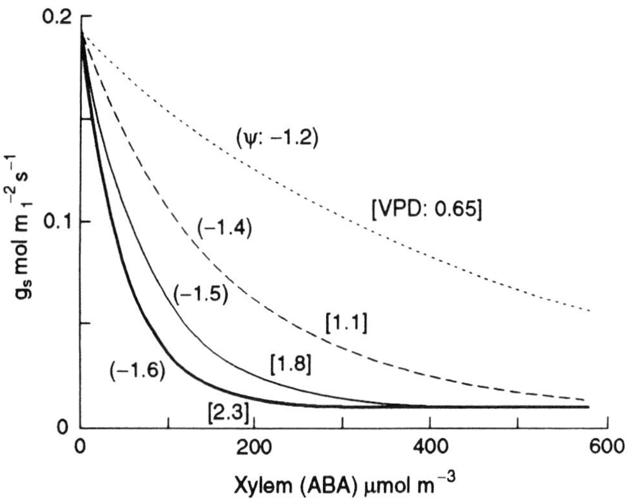

FIGURE 12.7. Relationship between stomatal conductance (g_s) and xylem abscisic acid concentration (ABA) at several times of the day for field-grown maize plants. Numbers indicate mean leaf Ψ_w (MPa, in parentheses) and mean vapor pressure deficit (VPD, in brackets). . . . , 7:30 to 9:00 A.M.; ---, 9:00 to 11:30 A.M.; —, 11:30 A.M. to 1:00 P.M.; ——, 1:00 to 4:30 P.M. From Tardieu and Davies (1992).

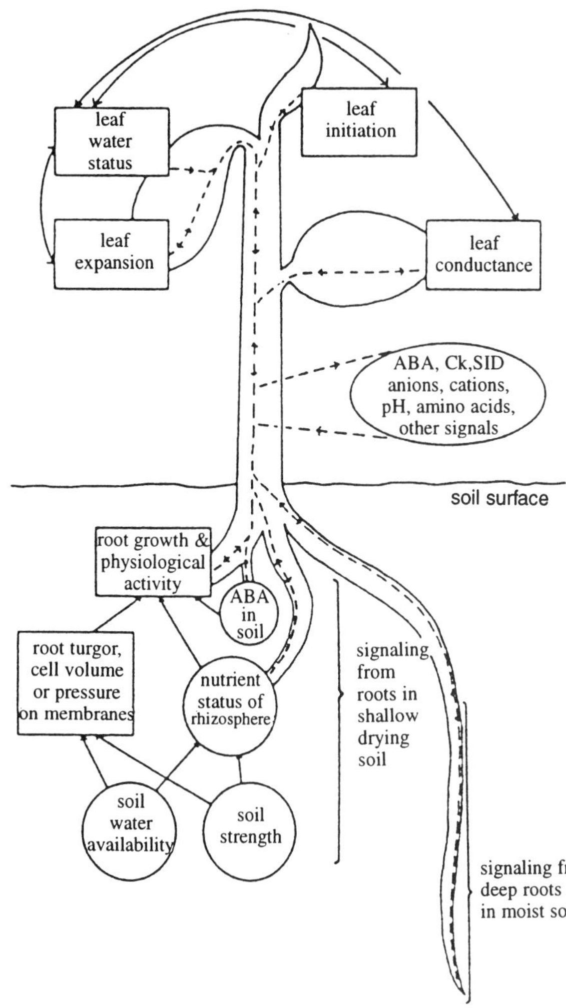

FIGURE 12.8. Proposed relationships among factors influencing the production of chemical signals (dashed lines) in roots in drying soil. Soil effects are shown as circles and physiological and developmental processes as rectangles. Hydraulic impacts of imposed leaf water deficits include direct effects (indicated by solid lines) on shoot processes and chemical signals moving to the roots (indicated by bidirectional arrows on dashed lines). ABA, abscisic acid; Ck, cytokinin; SID, strong ion difference. From Davies and Zhang (1991). Reproduced, with permission, from the *Annual Review of Plant Physiology and Molecular Biology*, Volume 42, ©1991, by Annual Reviews, Inc.

Numerous other guard cell metabolic processes exist that influence stomatal aperture either directly or indirectly (e.g., in response to ABA), including the activity of the cell cytoskeleton (Eun and Lee, 1997; Hwang et al., 1997), nitric oxide metabolism (Desikan et al., 2002; Garcia-Mata and Lamattina, 2002), and reactive oxygen species (Kwak et al., 2006). The nexus of stomatal control by water status via guard cell metabolism is exceedingly complex, involving intricate signaling networks that overlap with response networks to other environmental factors such as salinity and cold (McAinsh et al., 2000; Schroeder et al., 2001; Hetherington, 2001; Hetherington and Woodward, 2003; Israelsson et al., 2006). Much more research will be required before these networks are completely characterized.

Wind often affects transpiration by altering stomatal aperture in a pattern that varies among species (Chapter 5, Kozlowski and Pallardy, 1997). Stomatal closure by wind may result from leaf dehydration and also from lowered air humidity as shown for sugar maple (Davies et al., 1974b) and cacao (Sena Gomes et al., 1987). Changes in stomatal aperture also may involve the CO_2-sensing mechanism of the guard cells. The CO_2 concentration near the leaf surface is higher when the wind is strong than when it is weak (Mansfield and Davies, 1985). Stomata may open or close through a feedback response that depends on the partial CO_2 concentration in the intercellular spaces of leaves.

The capacity of guard cells to control stomatal aperture changes during leaf development. Senescent leaves often show drastic losses in capacity of stomata to open under normal stimuli. Gee and Federer (1972) noted such increases in average diffusion resistance of leaves and in the variability of measured resistances in canopy yellow birch and American beech trees growing in New Hampshire. Similar declines have been observed in other woody and herbaceous species (e.g., tobacco and barley (MacDowall, 1963; Friedrich and Huffaker, 1980), northern red oak and red maple, (Turner and Heichel, 1977)).

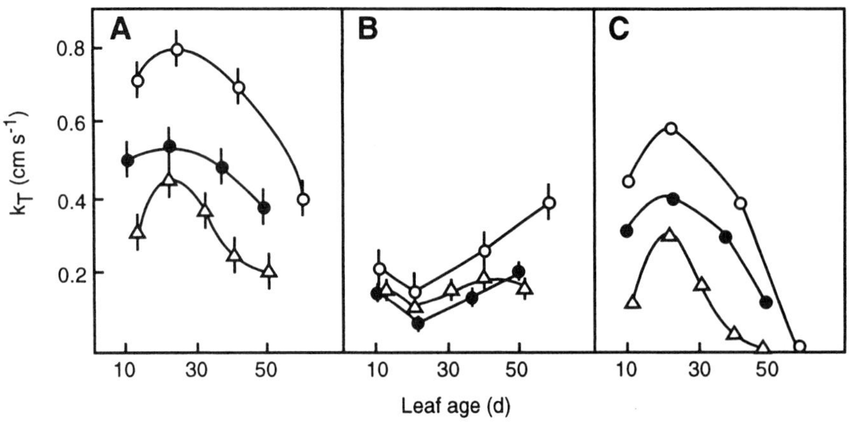

FIGURE 12.9. Mean leaf conductance (both sides, $k_T \pm$ S.E.) of leaves of different ages for three poplar clones (○, ●, △) in a growth chamber environment. (A) In constant light; (B) In constant dark; (C) Difference between k_T in light and dark. Symbols represent between 36 and 54 observations. From Reich (1984a).

Less apparent is a similar, but less extreme limitation in capacity for stomatal control early in leaf development and a gradual decline in control during the post-maturation phase of leaf development during which most photosynthesis occurs (Burrows and Milthorpe, 1976). Reich (1984a) observed such changes in the range of stomatal conductance in leaves of three poplar clones (Fig. 12.9). Stomatal conductance under light and dark conditions showed maximum difference soon after full leaf expansion. Earlier and later, stomata neither opened as widely nor closed as tightly. Similarly, Martin et al. (1994) observed reduced capacity for stomatal closure in the dark in old leaves of three Costa Rican tree species.

Stomatal sensitivity to leaf Ψ_w may shift appreciably over a growing season if developmental changes in Ψ_π and drought-related osmotic adjustment occur. In these cases, leaf turgor would be sustained to lower total Ψ_w, which presumably would prevent direct loss of guard cell turgor or ABA action. For example, Richter et al. (1981) and Parker et al. (1982) noted shifts in stomatal closure to lower leaf Ψ_w for several woody angiosperm species that were correlated with reductions in Ψ_π. Progressive declines in Ψ_π during the growing season also may occur even in the absence of drought and would be expected to have similar impacts on stomata (e.g., Parker et al., 1982; Kwon and Pallardy, 1989; Abrams, 1990). Stomatal aperture and its control also are discussed in Chapter 5.

Stomatal Control of Transpiration

Earlier in the twentieth century much effort was expended attempting to determine the effect of partial closure of stomata on the rate of transpiration. Investigators were influenced by the experiments of Brown and Escombe (1900), which were conducted in quiet air where the boundary layer resistance (r_{air}) was as high as the stomatal resistance (r_s). Their results indicated that large changes in stomatal aperture should have little effect on the rate of transpiration. However, it has been shown repeatedly that in moving air where r_{air} is low there is a strong relationship between stomatal aperture and transpiration (Stålfelt, 1932; Bange, 1953). It is now generally agreed that although partial closure has little effect on transpiration in quiet air, it greatly reduces the rate in moving air, as shown in Figure 12.10. This subject was discussed in detail by Slatyer (1967, pp. 260–269), who emphasized that the effect of stomatal closure varies with the relative values of r_s and r_{air}.

Scaling to the level of a plant canopy, there are similar relationships between stomatal control of transpiration and the physical characteristics of the canopy. The canopies of many agricultural crops are short and of uniform height and structure. This results in a large aerodynamic resistance above the canopy and reduced linkage of canopy transpiration rate with the integrated stomatal aperture of the crop. In contrast, forest canopies are tall, open, aerodynamically rough, and thus generally well coupled to the bulk atmosphere above the canopy. This canopy structure results in a stronger correlation between stomatal aperture and transpiration rates in forests than in field crops. Multilayered forest canopies, especially those of tropical forests, possess intermediate physical characteristics and responses (Meinzer et al., 1993). McNaughton and Jarvis (1983), Jarvis (1985, 1993), and Jarvis and McNaughton (1986) discussed transpiration in forest stands.

INTERACTION OF FACTORS AFFECTING TRANSPIRATION

The important environmental factors affecting transpiration are light intensity, vapor concentration gradient between leaf and air, wind, and soil water supply.

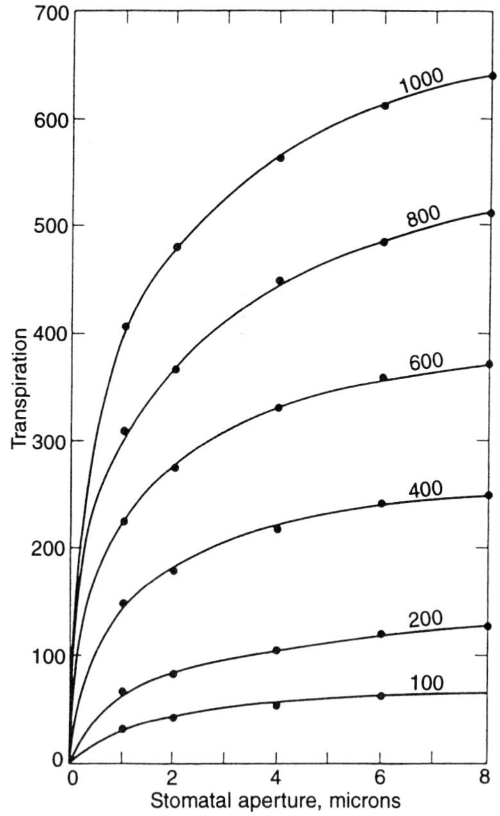

FIGURE 12.10. Relationship between stomatal aperture and rate of transpiration at various potential rates of transpiration. Numbers above each curve indicate appropriate rate of evaporation from blotter paper atmometers (mg H_2O per 25 cm^2 evaporating surface per hr). At low rates of evaporation and potential transpiration, stomatal aperture has much less control over the rate of transpiration than at potentially high rates. From Stålfelt (1932), with permission of Springer-Verlag.

Plant factors include leaf area, leaf exposure, canopy structure, stomatal aperture, and the effectiveness of the roots as absorbing surfaces.

There are complex interactions among the various controlling factors that can be summarized in terms of their effects on various terms of Eq. (12.2). For example, changes in light intensity affect r_{leaf}, by influencing stomatal aperture and C_{leaf} by effects on leaf temperature. Atmospheric temperature affects C_{leaf} (and hence ΔC). As noted earlier, although increasing the temperature of a given air mass from 20 to 30°C will decrease the relative humidity, it will not significantly increase the water vapor concentration of the air (see Tables 11.1 and 11.2). An increase in temperature therefore causes an increase in transpiration because it increases the vapor concentration gradient, ΔC, from leaf to air, not because it is accompanied by a decrease in relative humidity of the air. Likewise, at a constant temperature, a change in atmospheric humidity affects transpiration by changing C_{air} and ΔC from leaf to air. Martin (1943) found a very close relationship between transpiration of plants in darkness at a constant temperature and the vapor concentration of the atmosphere. Cole and Decker (1973) also found that transpiration is a linear function of ΔC. This will generally be the case unless there is pronounced stomatal closure as ΔC steepens (Farquhar, 1978).

It was mentioned earlier that wind increases transpiration by reducing boundary layer thickness and consequently r_{air}, and that it acts indirectly to decrease transpiration by cooling the leaves, thereby reducing C_{leaf}. Other effects of wind such as increased ventilation of the intercellular spaces by flexing of leaves and increasing the passage of air through amphistomatous leaves (leaves with stomata on both surfaces) probably are of minor importance (Woolley, 1961; Roden and Pearcy, 1993a). However, Shive and Brown (1978) found that flexing of cottonwood leaves caused bulk flow of gas through them and decreased total resistance by about 25%.

Leaf arrangement affects exposure to the sun and leaf temperature. In turn, changes in leaf temperature alter C_{leaf}. Upright leaves receive less energy at the hottest time of day than horizontal leaves, and clustered leaves such as fascicles of pine needles receive less energy than those arranged separately. Variations in internal geometry and volume of intercellular spaces may affect the resistance in the intercellular spaces r_{ias}, cuticular wax content affects r_c, and extent of stomatal opening affects r_s, as was indicated in Figure 12.1.

It should be emphasized that a change in one of the factors affecting transpiration does not necessarily produce a proportional change in transpiration, because the rate is not controlled by any single factor. For example, a breeze tends to increase transpiration by lowering r_{air} but decreases it if the leaf is cooled enough to substantially lower C_{leaf}. In general, the effects of various factors on the rate of transpiration can be explained in terms of their influences on the differences in vapor concentration between leaf and air ($\Delta C = C_{leaf} - C_{air}$) and the resistances in leaf and air pathways ($r_{leaf} + r_{air}$).

The supply of water to the roots also affects transpiration because, as discussed previously, a deficient water supply causes stomatal closure. In soil near field capacity, movement of water into roots is rapid and the rate of transpiration is controlled largely by atmospheric factors, except for the occasional midday wilting of rapidly transpiring plants caused by high root resistance.

However, as the soil water content decreases, the supply of water to the roots becomes a limiting factor and the rate of transpiration decreases. It has been shown by various investigators, from Hartig and von Höhnel in the nineteenth century to Kozlowski (1949), Bourdeau (1954), Slatyer (1956), Jackson et al. (1973), Running (1976), Gowing et al. (1990), and Wartinger et al. (1990) that decreasing soil moisture was correlated with decreases in the transpiration of trees. Lopushinsky and Klock (1974) found that the transpiration rate of several conifers began to decrease at a soil moisture potential of –0.1 or –0.21 MPa, but the decrease at –1 MPa was much greater in ponderosa and lodgepole pine than in Douglas-fir or grand fir. According to Ringoet (1952), low soil moisture reduced the transpiration of oil palms so much that the trees transpired more during the rainy season than during the dry season, although atmospheric factors were more favorable for transpiration during the dry season.

As noted previously, stomatal closure associated with drying soil may either be associated with root-sourced chemical signals that arise from perception of soil drying directly by roots (Davies and Zhang, 1991) or with leaf water deficits (Pierce and Raschke, 1980), or both (Tardieu and Davies, 1992).

TRANSPIRATION RATES

Numerous measurements have been made of transpiration rates of trees and shrubs of various species and ages under a wide range of conditions. Many earlier results were summarized by Kramer and Kozlowski (1979). Transpiration data for seedlings of two deciduous hardwoods and loblolly pine in Table 12.4 show that although the hardwoods transpired about twice as rapidly as pine per unit of leaf surface, the transpiration per seedling of similar size was greater for the pine because of its greater leaf surface.

Seasonal cycles of transpiration of an open evergreen *Callitris/Eucalyptus* woodland in Southeast Australia are shown in Figure 12.11. Winter transpiration rates were only a small fraction of those observed during the summer. Weaver and Mogensen (1919) in Nebraska and Ivanov (1924) at Leningrad reported that the winter transpiration rate of conifers was less than 1% of the summer rate. Exposure to low temperature is said to greatly reduce transpiration of conifers (Christersson, 1972). Even in some parts of the tropics, seasonal cycles in transpiration occur because of variations in rainfall, humidity, and soil moisture. Ringoet (1952), for example, found large seasonal differences in transpiration of oil palms growing in the Belgian Congo.

As long as a gradient in water vapor pressure exists at night, plants will continue to transpire. Unless the surrounding air is saturated, stomatal closure and an effective cuticle can greatly reduce but not completely eliminate nighttime leaf water loss. Stomata may also not close completely at night. Persistent stomatal opening into the night sometimes has been reported for well-watered plants growing in moist air, particularly

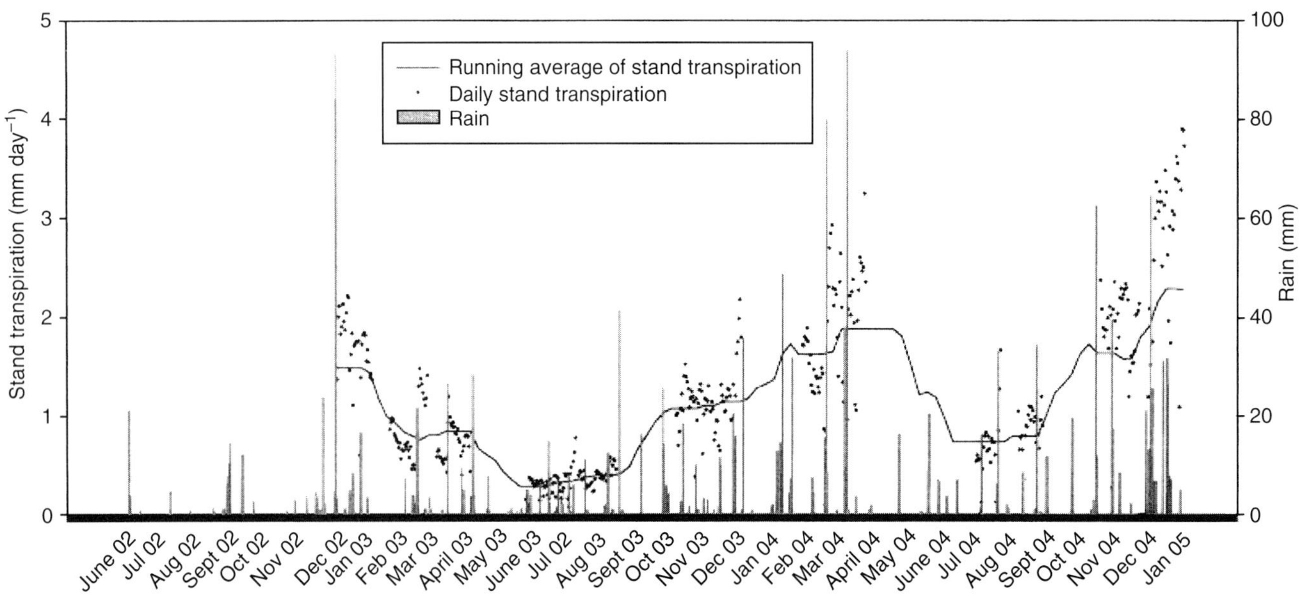

FIGURE 12.11. Daily stand transpiration over two years in a mixed *Eucalyptus-Callitris* woodland in southeast Australia. Data points represent the day running average of daily transpiration. Grey columns represent daily rainfall (mm). From Zeppel et al. (2006).

for conifers such as Douglas-fir (Blake and Ferrell, 1977) and Pacific silver fir (Hinckley and Ritchie, 1973). Benyon (1999) also observed high rates (maximum 0.3 mm hr^{-1}) of nighttime water loss on one night in an irrigated young *Eucalyptus grandis* plantation and estimated that nighttime leaf conductance was 20 times cuticular conductance. On 24 other nights, lower rates were observed and transpiration was tightly correlated with vapor pressure deficit and nighttime wind speed. In tropical savanna trees, Bucci et al. (2004, 2005) reported that between 13 and 28% of total daily water loss occurred at night during the dry season when vapor pressure deficit remained high. Stomatal conductance never declined below 40 mmol m^{-2} s^{-1} from typical daytime values of 100 to 200 mmol m^{-2} s^{-1}. In some cases stomatal conductance rose at the end of the night to near daytime levels (Bucci et al., 2004).

Caird et al. (2007) summarized findings of numerous studies of nighttime stomatal conductance and transpiration in diverse C_3 and C_4 plants, finding that nighttime transpiration rates were typically 5 to 15% of daytime rates but could be as high as 30%. Although nighttime transpiration would seem wasteful, it may be an unavoidable consequence of exposing an imperfectly sealed organ in a dry environment. Alternatively, Caird et al. (2007) suggested that nighttime transpiration, by its promotion of root water absorption, may enhance nutrient availability to the plant. They also suggested that predawn stomatal opening might increase early-morning photosynthesis. Neither of these potential benefits has been thoroughly studied. Nighttime transpiration may partially account for occasional reports of differences between predawn leaf water potentials and soil water potential (e.g., Donovan et al., 1999, 2001, 2003), a situation that prevents the requisite equilibration between soil and plant.

In many studies transpiration is indirectly monitored by measurement of sap flow velocity using various techniques (Kaufmann and Kelliher, 1991). Sap flow velocity at the base of large trees frequently lags behind crown transpiration in the morning and exceeds it in the evening, reflecting the capacitance of stem and crown portions of the Soil–Plant–Atmosphere Continuum (Cohen et al., 1985) (Fig. 12.12). Maximum rates of sap flow in trees are reported to vary between 1 and 2 m hr^{-1} in conifers, 1 to 6 m hr^{-1} in diffuse-porous trees, and 4 to 40 m hr^{-1} in ring-porous trees (Zimmermann and Brown, 1971) The velocity of flow in conifers and diffuse-porous trees is low, as water moves through conducting elements in a number of annual rings of sapwood, whereas in ring-porous broadleaved trees it moves rapidly through relatively few vessels located in only one or two annual rings (Chapter 11).

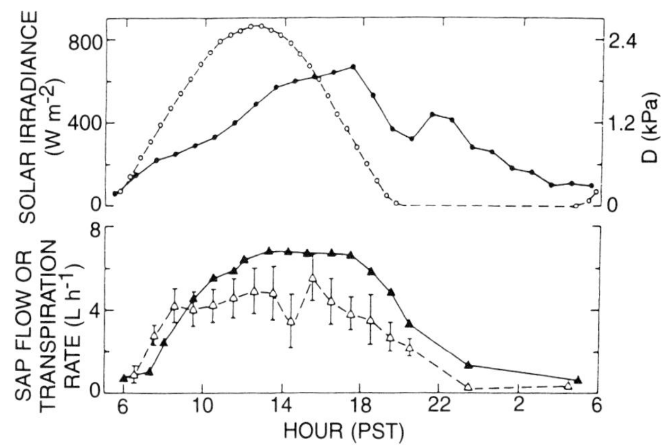

FIGURE 12.12. Diurnal pattern of stem sap flow (▲) estimated from heat pulse velocity measurements and foliage transpiration rate (△) of a Douglas-fir tree in August. Also shown are within-canopy water vapor pressure deficit, D (●) and solar irradiance (○) above the canopy. From Cohen et al. (1985).

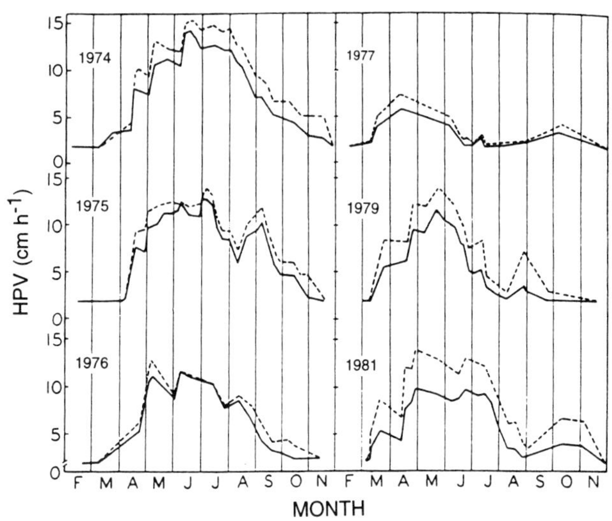

FIGURE 12.13. Seasonal variation in heat pulse velocity (HPV) in Douglas-fir (---) and ponderosa pine (—) for six years. From Lopushinsky (1986).

The unusually low rate of sap movement in conifers in 1977 (Fig. 12.13) was attributed to a low winter snowpack followed by a dry spring (Lopushinsky, 1986). When soil moisture was not limiting, seasonal patterns of sap movement in Douglas-fir and ponderosa pine stems were regulated by air temperature, solar radiation, and vapor pressure deficit. As soil moisture decreased during the summer, the rate of sap movement no longer followed evaporative

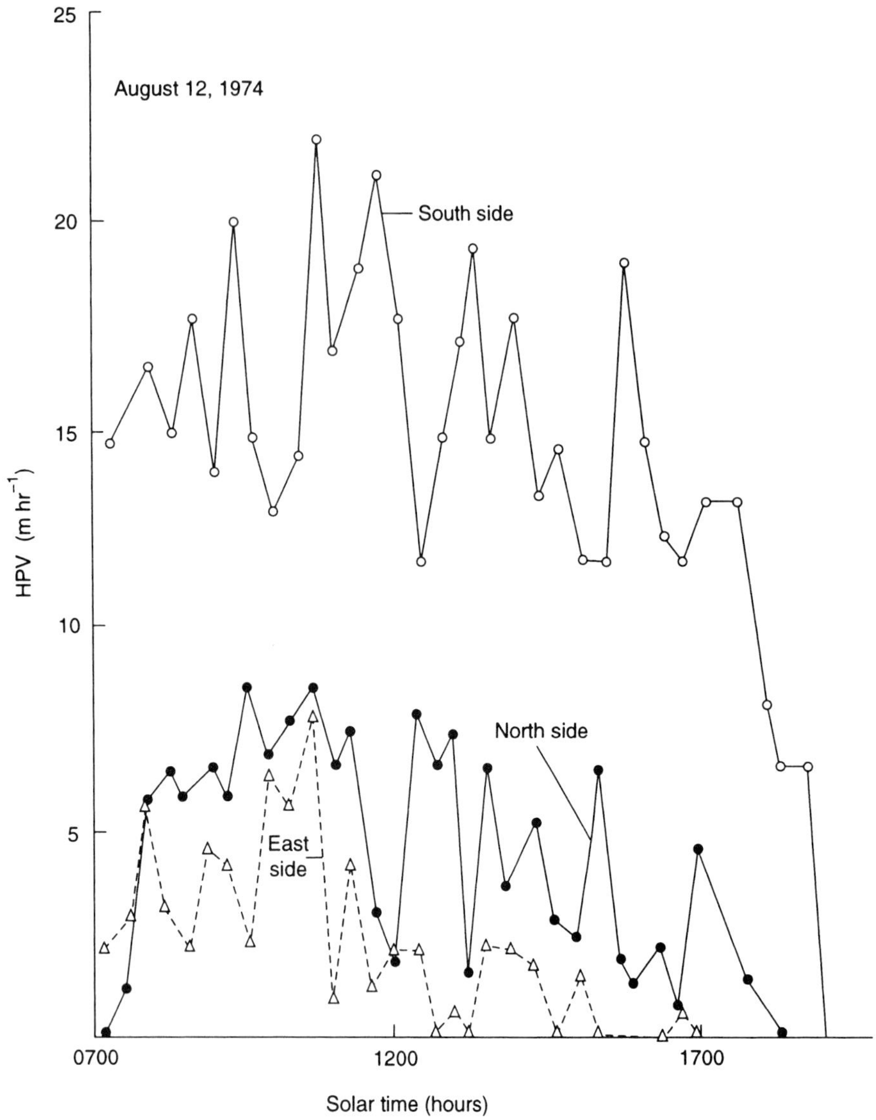

FIGURE 12.14. Heat pulse velocities (HPV) on north-, south-, and east-facing sides of a black oak tree in August. The sample tree had a diameter at breast height of 13.3 cm and was located on a south-southwest-facing slope. Used with permission of the Society of American Foresters from Miller, D. R., Vavrina, C. A., and Christensen, T. W. (1980). Measurement of sap flow and transpiration in ring-porous oaks using a heat pulse velocity technique. *For. Sci.* **26**, 485–494; permission conveyed through Copyright Clearance Center, Inc.

demand. Early in the summer, the rate of daily sap movement was highest near midday; in the autumn maximum rates occurred later in the day (Lopushinsky, 1986).

Changes in xylem conductivity may also affect the velocity of sap movement. As the water content of the sapwood of Douglas-fir decreased, stem conductivity also decreased (Waring and Running, 1978). Miller et al. (1980) found that sap flow in black and white oaks was most responsive to solar radiation up to 0.6 cal cm^{-2} min^{-1} flux density; thereafter it was more responsive to changes in vapor pressure deficit of the air. Sap flow in oaks also was quite variable around the trunk, with sections below well-lit portions of the crown having far higher flow rates than shaded portions (Fig. 12.14).

WATER LOSS FROM PLANT STANDS

Foresters, horticulturists, and agronomists often are more interested in the amount of water lost from stands of plants than in the loss from individual plants. The loss from a forest, orchard, field, or grassland includes both transpiration from the vegetation and evaporation from the soil surface, and the combined losses usually are termed *evapotranspiration*.

Factors Controlling Evapotranspiration

Evaporation from soil and transpiration from plants may be regarded as alternative pathways for water movement into the bulk air. As mentioned earlier, evaporation depends on the supply of energy required to vaporize water, the difference in vapor concentration between the evaporating surface and the bulk air, and resistances in the particular pathway. Escape of water vapor from vegetation (transpiration) is complicated by plant control over internal resistance and the variable nature of the evaporating surfaces. However, evapotranspiration can be described by a slight modification of Eq. 12.2:

$$E = \frac{C_{int} - C_{air}}{r_{int} + r_{air}} \tag{12.8}$$

The subscript int refers to the internal part of the pathway between the evaporating surface and the soil or plant surface. In wet soils r_{int} is negligible and C_{int} is at the soil surface, but as the soil dries and the evaporating surface retreats into the soil mass, r_{int} becomes important and the rate of evaporation decreases. When leaf surfaces are dry the evaporating surfaces are within them, but in stands of plants the surface of the stand sometimes is treated as the evaporating surface. However, this is an oversimplification because both air and leaf conditions vary from bottom to top of a crop canopy. The complex problems involved in the water relations of forest stands were discussed by Kaufmann and Fiscus (1985), Landsberg (1986), and McNaughton and Jarvis (1983) and will be dealt with only briefly.

The amount of water evaporated from a unit area of the earth's surface depends first of all on the energy available. Since on a bright summer day 320 to 380 cal cm^{-2} are available to evaporate water and 570 calories are required to evaporate 1 g of water, the maximum possible amount of evaporation is approximately 6 mm day^{-1}. However, incident energy is sometimes supplemented by advection (horizontal flow) of energy from the surroundings, as in the case of isolated, exposed trees or cultivated fields surrounded by desert. Although the leaf area of stands of plants commonly is four to six times that of the soil on which they are growing, the rate of water loss cannot exceed that from moist soil or a water surface receiving the same amount of energy.

Effects of Changes in Plant Cover

Foresters, hydrologists, and agronomists are concerned with the effects of changes in composition, height, and density of vegetation in stands on the rate of water loss. It often is stated that the rate of water loss from closed stands of all sorts, including grass, cultivated crops, and forests, is similar as long as the soil is moist. However, as Rider (1957) warned, this statement should be viewed with caution because

TABLE 12.6. Comparison of Annual Evaporation from Forest and Adjacent Grasslands Growing in Moist Soil[a]

	Evaporation (mm)		
Forest type	**Forest**	**Adjacent grass**	**Grass/forest**
Sitka spruce	800	416	0.52
Norway spruce	579	521	0.90
Mixed conifer-deciduous	861	696	0.81
Mixed deciduous	—	—	0.8–1.0
Snow gum	—	—	1.0

[a]From Rutter (1968).

the albedo and the internal resistance vary among different types of vegetation. A light-colored canopy presumably would lose less water than one that is dark. According to Rutter (1968, p. 66) the stomatal resistance in forest stands probably is somewhat higher than in stands of herbaceous plants. However, the surfaces of forest canopies often are more irregular than those of low-growing crops and this should result in greater turbulence, reducing r_{int} as compared with crops having a very uniform surface (Jarvis, 1993). Nevertheless, most studies show small differences among tree species and between forests and grasslands so long as water supply is not limiting, water loss from grassland being about 80% or more of that from forests. When soil water deficits develop, evapotranspiration from grassland decreases more than that from forests, perhaps partly because of shallower roots characteristic of the former type of vegetation. The data from a number of studies are presented by Rutter (1968, pp. 45–57) and are summarized in Table 12.6.

Thinning

The thinning of forest stands may decrease evapotranspiration. Knoerr (1965) studied soil water depletion under stands of California red fir varying from 50 to 100% cover and found maximum water uptake by stands with 70 to 80% cover. In 31-year-old stands of Hinoki cypress thinned from 1,750 to 1,325 trees ha^{-1}, daily transpiration of individual trees increased at a given level of solar radiation, but stand transpiration decreased by 21% (Morikawa et al., 1986). Similarly, heavily thinned (100 trees ha^{-1}) *Eucalyptus nitens* plantations had stand transpiration rates that were less than 25% those of unthinned stands (1,256 trees ha^{-1}), and transpiration rate per tree tripled in the thinned stand (Medhurst et al., 2002). Individual trees lose more water because they have greater canopy exposure to solar radiation, increased crown boundary layer conductance (Jarvis, 1993), and subsequently, greater amounts of foliage per tree. Hewlett and Hibbert (1961) found that the increase in water yield from a watershed at Coweeta, North Carolina, was approximately proportional to the reduction in basal area after thinning. Baker (1986) reported that annual water yields from ponderosa pine stands in Arizona were larger and that the increase persisted longer as the degree of overstory removal increased. However, persistent elevated water yields on southern aspect pine stands that had been strip cut and thinned were not observed.

A portion of the increase in water yield associated with thinning of stands may be caused by decreased interception. For example, part of the increase in water yield observed by Hewlett and Hibbert (1961) may have been attributable to reduced evaporative loss from intercepted precipitation. Similarly, in dense lodgepole pine forests in the Rocky Mountains, greater annual snowpack equivalents were found in thinned stands because there was less interception and sublimation loss (Gary and Watkins, 1985). A thicker snowpack provided greater water yield the following summer. Troendle (1988), summarizing the results of several studies, indicated that peak water equivalent increased up to 35% as the % basal area removed from lodgepole pine stands increased. Removal of understory vegetation also appears to reduce water loss (Rutter, 1968, p. 43) (Chapter 7, Kozlowski and Pallardy, 1997).

Relative Losses by Evaporation and Transpiration

Where water yield is important, there is much interest in the relative amounts of water lost by evaporation and transpiration. Over the whole United States about one-fourth of the total precipitation escapes as stream flow and three-fourths is returned to the atmosphere by evapotranspiration (Ackerman and Loff, 1959). It was stated in the preceding sections that there appear to be no large differences in water loss from different types of vegetation growing in moist soil. However, as the soil dries, differences in depth of rooting and plant control of transpiration may become important. It is claimed that both in the Central Valley of California (Biswell and Schultz, 1957) and in southern California (Pillsbury et al., 1961) conversion of chaparral to grass greatly increases water runoff.

Clear-cutting of lodgepole pine in Colorado increased stream flow by about 30% (Wilm and Dunford, 1948) and removal of all woody vegetation from a watershed at Coweeta, North Carolina, increased stream flow more than 70% the first year. The increase was greater on north- than on south-facing slopes. The effects of removing forest cover on stream flow varied with the amount and seasonal distribution of precipitation and the amount per storm (Hewlett and Hibbert, 1967), and hence it was not an accurate indicator of the rate of transpiration. The actual loss of water by transpiration from the Coweeta forests probably was considerably greater than the amount indicated by the increase in runoff after clear-cutting. Patric et al. (1965) measured soil water depletion under stands of trees growing in soil covered with plastic to prevent loss by evaporation and estimated transpiration from April through October to be 41 cm for a 21-year-old loblolly pine stand and 37.2 cm for an oak-hickory stand. Stålfelt (1963) reported that evaporation of water from the surface soil under an open stand of Norway spruce made up 20% of the total water loss, but Rutter (1968) reported only 8% lost by evaporation under a stand of Scotch pine. Baumgartner (1967), using an energy balance method, calculated evaporation from the soil under forests, meadows, and cultivated crops to be 10, 25, and 45%, respectively, of the total water loss. Experiments with corn at Urbana, Illinois, suggest that about 50% of the total evapotranspiration was by evaporation from the soil surface. Higher evaporation losses are to be expected from crops where the soil is much more exposed early in the growing season than in forests.

It is fairly common for evapotranspiration to exceed precipitation during the growing season, as in the Illinois corn field where total evapotranspiration from the control plants exceeded rainfall during the growing season by 9.7 to 13.7 cm in a three-year experiment. Trees sometimes can be established in areas of limited rainfall but they die when moisture in the soil is depleted. Bunger and Thomson (1938) observed this situation in the panhandle of Oklahoma. Wiggans (1936, 1937) reported that in a 20-year-old apple orchard in eastern Nebraska, evapotranspiration was removing 28 to 38 cm more water per year than was replaced by precipitation. He predicted the imminent death of these trees, which already were removing water to a depth of about 10 m. A severe autumn freeze killed them before the soil water reserve was exhausted. Many plantations of forest trees were established in the prairie and plains states in the 1880s and 1890s. Most of these trees grew well for a number of years, but eventually began to die because they had exhausted all the reserve soil water and could not survive on the current rainfall during dry cycles.

Changes in Species Composition

The effects of changing the species composition of a forest stand on water yield are complicated by the influences on interception of precipitation, and by differences in depth of rooting and length of the transpiring season. Most of the experiments cited by Rutter (1968, pp. 45–50) show little difference among species, at least in the summer. Loss of intercepted water from conifer shoots over a year exceeds that from deciduous species (Jarvis, 1993; Putuhena and Cordery, 2000). In an experiment at Coweeta, North Carolina, 15 years after two experimental watersheds had been converted from mature, deciduous hardwood forest to eastern white pine, annual stream flow was reduced by 20% (Swank and Douglass, 1974). Reduction in stream flow occurred during every month, but the largest reductions were in the dormant- and early growing season when the leaf area index was 9.9 for pine, but less than 1 for hardwoods, resulting in greater interception and subsequent evaporation from the pine than from the bare hardwoods. The stream flow data indicate that combined interception and transpiration losses also are greater for pine during the growing season.

Methods for Reducing Transpiration

In view of the damage caused by excessive transpiration there has been much interest in finding ways to reduce the rate of water loss from plants. Efforts have been centered on three problems. Reduction of transpiration following transplanting would enable plants to maintain turgor until their root systems are reestablished. Reduction during droughts would enable plants to survive with minimal injury, and reduction of transpiration of the plant cover on watersheds would increase the water yield usable for other purposes. The methods used consist basically of application of waterproof coatings or of materials that cause closure of stomata. Substances intended to reduce transpiration commonly are termed antitranspirants or antidesiccants.

Some early studies involved dipping or spraying the tops of seedlings with various substances such as latex emulsions, polyvinyl, polyethylene, and vinyl acrylate compounds. Some reduction in transpiration and some increase in survival were observed, but Allen (1955) suggested that further study of such coatings should be made before recommending their use. Lee and Kozlowski (1974) and Davies and Kozlowski (1975a) discussed problems encountered in the use of antitranspirant coatings on tree seedlings. Their effectiveness seems to depend on the plant species, stage of development, and atmospheric conditions during the test period (Gale and Hagan, 1966). According to Turner and DeRoo (1974), antitranspirants do not reduce injury to evergreen trees when the air temperature is below freezing, but may be effective in the spring when air temperatures rise but the soil remains cold or frozen.

Films have limited usefulness on growing plants because repeated applications are necessary on plants with increasing amounts of foliage and because substances impermeable to water vapor also are usually quite impermeable to carbon dioxide and reduce the rate of photosynthesis. Films seem most useful where reduction of transpiration is more important than a high rate of photosynthesis. Examples might be their use on leaves while recently transplanted trees initiate root growth into adjacent soil and on fruit trees to increase the size of ripening fruit and improve its keeping qualities after picking (Uriu et al., 1975).

Application of substances that bring about closure of stomata is attractive because partial closure of stomata should reduce transpiration more than it reduces photosynthesis. This claim is based on the observation of Gaastra (1959) that the mesophyll resistance for entrance of carbon dioxide is considerably higher than the resistances affecting the exit of water vapor. Thus, a large increase in r_s should have a smaller effect on carbon dioxide uptake than on water loss. Slatyer and Bierhuizen (1964) reported that phenylmercuric acetate reduced transpiration more than photosynthesis by closing stomata. On the other hand, it is reported that photosynthesis and transpiration were decreased to the same extent by stomatal closure in cotton, tomato, and loblolly pine (Barrs, 1968; Brix, 1962).

Phenylmercuric acetate was reported by Waggoner and Bravdo (1967) to significantly reduce transpiration of an entire pine stand by closing stomata. Unfortunately, it would be undesirable to add more mercury compounds to our already polluted environment. Furthermore, Waisel et al. (1969) reported that phenylmercuric acetate injured leaves of paper birch. Keller (1966) noted that both phenylmercuric acetate and a film-forming polyvinyl compound reduced photosynthesis and root growth of spruce seedlings and the phenylmercuric acetate injured the needles. Phenylmercuric acetate also is said to reduce vegetative growth of tea (Nagarajah and Ratnasooriya, 1977). After the discovery that abscisic acid inhibits stomatal opening it was tested as an antitranspirant (Jones and Mansfield, 1972). Davies and Kozlowski (1975a) found that ABA reduced transpiration of sugar maple, white ash, and Calamondin orange seedlings as much as 60%, and some reduction persisted up to 21 days after treatment. No toxic effects were observed.

In another series of experiments a silicone coating was more effective than abscisic acid in reducing transpiration of white ash and red pine seedlings (Davies and Kozlowski, 1975b). However, it was concluded that film-forming antitranspirants may be unsatisfactory for use on some gymnosperms because they reduced photosynthesis by 95% for at least 12 days. Chitosan, a glucosamine polymer derivative of crustacean chitin, has been shown to reduce stomatal conductance of pepper plants when applied as a spray (Bittelli et al., 2001). Chitosan application reduced water use from 26 to 43% while maintaining biomass production and yield. Kramer and Boyer (1995) discussed the potential utility and disadvantages of both film-type and metabolic antitranspirants.

Transpiration Ratio and Water Use Efficiency

There is considerable interest in the relationship between plant production and various measures of evaporation from the land area on which plants grow. In a water-scarce environment, there is an obvious need in managed systems to maximize growth with the amount of water available (e.g., Davies et al., 2002). In natural systems, there also is interest in relationships between biomass and productivity and water loss. Traditionally, the amount of water required to produce a unit of plant dry matter often has been termed the transpiration ratio. In contrast, water use efficiency often was used in agriculture and ecosystem ecology to indicate the amount of dry matter production per unit of combined evaporation and transpiration (Begg and Turner, 1976; Kramer and Kozlowski, 1979; Kramer, 1983):

$$\mathrm{WUE} = \frac{\text{Dry matter or crop yield } (\text{kg ha}^{-1})}{\text{Water consumed in evapotranspiration } (\text{kg ha}^{-1})} \tag{12.9}$$

Plant physiologists have defined water use efficiency in a fashion that is somewhat similar to transpiration ratio (Fischer and Turner, 1978; Kramer, 1983):

$$\mathrm{WUE} = \frac{\text{Net } CO_2 \text{ uptake } (\mu\text{mol m}^{-2}\text{ sec}^{-1})}{\text{Transpiration rate } (\text{mmol m}^{-2}\text{ sec}^{-1})} \tag{12.10}$$

It important to note that the two forms of WUE are not equivalent, and that the definition of WUE may even differ among plant physiologists depending upon the time scale involved in measurements. In many studies of gas exchange of woody plants (e.g., Seiler and Johnson, 1985; DeLucia and Heckathorn, 1989; Ni and Pallardy, 1991) WUE is expressed as an "instantaneous" measure derived from photosynthesis and transpiration rate (or leaf diffusion conductance) data obtained from concurrent measurements of these parameters. The *daily* value of WUE from any such study would be different because of nightly respiratory consumption of photosynthate produced during the day and because gas exchange may be measured under conditions that are not representative of the growth environment of the whole plant. Hence it is important that WUE be carefully defined and cautiously interpreted.

There appear to be strong relationships between WUE and mode of photosynthesis for Crassulacean Acid Metabolism (CAM), C_3 and C_4 species. Species with the C_4 photosynthetic mechanism tend to exhibit higher WUE than C_3 species, especially under hot, high-light, water-stressed conditions. Such conditions maximize differences in light saturation characteristics of photosynthesis, stimulate photorespiration in C_3 plants, and increase the influence on photosynthesis of differences between C_3 and C_4 species in CO_2 transfer resistances (Nobel, 1991). Robichaux and Pearcy (1984) documented consistently higher instantaneous WUE values for C_4 *Euphorbia* shrubs than for C_3 *Scaveola* shrubs across a range of habitats in Hawaii (Fig. 12.15). Species possessing CAM photosynthesis show much greater water use efficiency when operating in CAM mode (i.e., absorbing CO_2 from the air at night and refixing it internally during the day) because stomata of CAM plants are open only at night when vapor

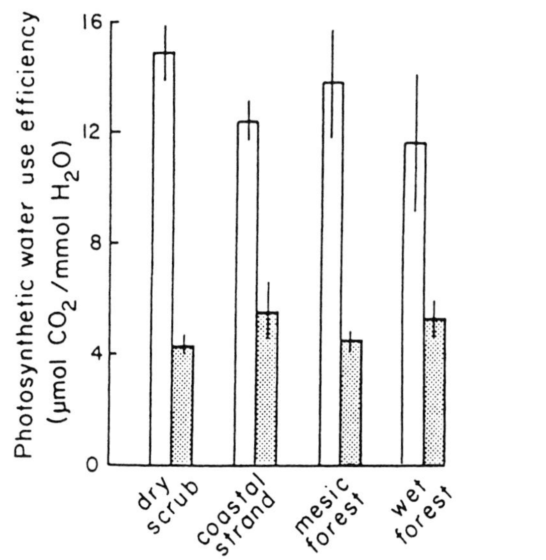

FIGURE 12.15. Instantaneous water use efficiency values of C_4 *Euphorbia* spp. (open bars) and C_3 *Scaevola* spp. (stippled bars) across a range of habitats. From Robichaux and Pearcy (1984).

concentration gradients that drive transpiration are much lower. When the tropical C_3-CAM tree *Clusia minor* was induced to CAM photosynthetic mode by water stress, daily WUE increased by a factor of 2.26 to 4.57, depending upon the light regime and N fertilization (Franco et al., 1991). The greatest increase in WUE was associated with plants subjected to high light intensity or high rates of N addition. Nobel (1985) noted that whereas annual WUE of a plant of *Ferocactus acanthoides* was 14 g CO_2/kg H_2O, it was much lower in C_3 and C_4 species (1–3 g CO_2/kg H_2O).

There also appear to be substantial differences in WUE among species that are independent of the mode of photosynthesis. For example, DeLucia and Heckathorn (1989), Ni and Pallardy (1991), and Guehl et al. (1991) observed differences in instantaneous WUE among several C_3 conifer and hardwood species. It sometimes has been asserted that high WUE is a general adaptation of drought-tolerant plants, but this claim often is not substantiated by experimental data (DeLucia and Heckathorn, 1989; Ni and Pallardy, 1991). It also has been claimed that plants are adapted such that stomatal aperture is modulated on some time scale (often diurnally) so as to maximize the gain of carbon relative to water loss (Cowan, 1977). Optimization theory predicts that WUE is maximized over a time period if the ratio of the partial derivatives of transpiration (T) and photosynthesis (A) with respect to stomatal conductance is a constant:

$$\frac{\partial T}{\partial A} = \lambda = \text{constant} \qquad (12.11)$$

This is an interesting hypothesis, but it remains to be adequately tested as a general plant response. Additionally, implicit in this analysis is an assumption that conserved water remains available to the "prudent" plant. This is almost certainly not the case in mixed stands where competing plants that do not show identical stomatal characteristics may absorb and transpire the water savings of a neighboring plant (Davies and Pereira, 1992; Jones, 1993).

Annual WUE also is influenced by the type of vegetative cover. Webb et al. (1978) noted that the ratio of aboveground net primary production to actual evapotranspiration in temperate North America was highest in conifer and hardwood forests, lower in shortgrass prairie and cold desert regions, and lowest in hot deserts. Much of the reduction in this measure of WUE could be attributed to reduced vegetative cover and consequently greater evaporative loss of water.

There has been considerable interest in using carbon isotope discrimination ($\delta^{13}C$) values (see Chapter 5) to estimate integrated WUE in both agronomic plants and trees. This application is based on discrimination against fixation of $^{13}CO_2$ by the photosynthetic apparatus, primarily in the initial carboxylation reaction. Because total transfer resistances to CO_2 are greater than those for water vapor loss, if other factors are equal a reduction in stomatal aperture should reduce transpiration relatively more than it does photosynthesis. Stomatal closure thus should be reflected in higher $\delta^{13}C$ of photosynthates because intercellular spaces beneath closed stomata will become increasingly enriched in $^{13}CO_2$, resulting in greater levels of its fixation despite enzyme discrimination. Hence WUE would be expected to be higher in plants exhibiting higher $\delta^{13}C$.

Experimental results based on applications of these ideas indicate that $\delta^{13}C$ values are often linked with WUE, but this is not always the case. For example, in agreement with the theoretically expected pattern, $\delta^{13}C$ in plant materials was positively associated with the imposition of water stress (suggesting higher WUE) in *Pinus pinaster* (Picon et al., 1996), *Juniperus* spp. (Moore et al., 1999), *Prunus persica* (Arndt et al., 2000), *Acer rubrum* (Bauerle et al., 2003) and *Salix arctica* (Sullivan and Welker, 2007). Meinzer et al. (1992) also found that $\delta^{13}C$ of coffee leaves was highest in potted plants that were watered weekly, intermediate in plants watered twice weekly, and lowest in those watered twice daily, suggesting an increase in WUE as plant water stress intensified. However, instantaneous whole-plant WUE measurements based on photosynthesis and transpiration values were the opposite of those predicted by $\delta^{13}C$ data. Leaf area and self-shading were much greater in plants that were watered more frequently. In explanation of these contradictory findings, the authors suggested that, for a shade plant such as coffee, the increased insolation of the canopy in sparingly watered plants increased transpiration significantly more than it increased photosynthesis. This influence would not substantially affect $\delta^{13}C$ (which depends largely on c_i/c_a, Chapter 5), but it presumably would have a large effect on actual WUE.

Garten and Taylor (1992) sampled $\delta^{13}C$ of a wide variety of deciduous tree species and loblolly and shortleaf pines in Tennessee at various topographic positions. Pines tended to have higher $\delta^{13}C$ than did deciduous trees during a dry year, suggesting higher WUE for coniferous species (Fig. 12.16). $\delta^{13}C$ was higher in leaf tissues taken from trees of deciduous species on ridge sites than those growing on floodplains, and higher in dry than wet years on both sites. These patterns would be expected if stomatal closure (and hence WUE) was greater on xeric sites and in dry years. However, when WUE was estimated from calculated c_i/c_a values and relative humidity data, WUE of trees on ridge sites was lower in a dry year than a

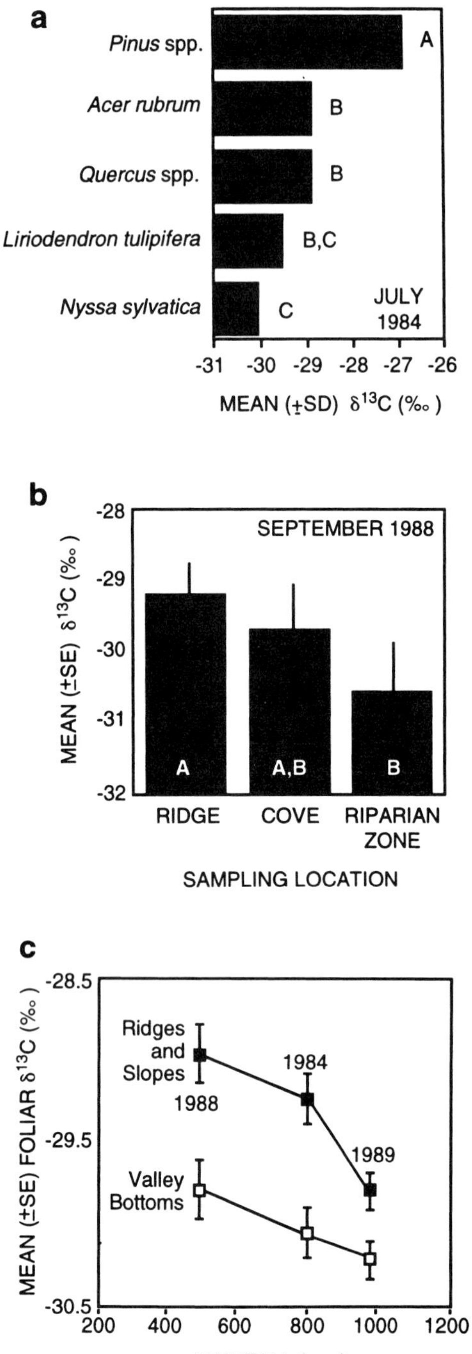

FIGURE 12.16. (a) Mean carbon isotope discrimination ($\delta^{13}C$) values from pine, red maple, oak, yellow-poplar, and black tupelo trees growing on a ridge site at Walker Branch Watershed in Tennessee in July, 1984. Bars having no adjacent letters in common are significantly different ($p < 0.05$). (b) Mean $\delta^{13}C$ values in September, 1988, from five deciduous species growing on ridge, cove, or riparian sites on the Walker Branch Watershed. Columns lacking common letters are significantly different ($p < 0.05$). (c) Mean $\delta^{13}C$ values of deciduous trees on xeric (■) or mesic (□) habitats on Walker Branch Watershed versus annual precipitation during the first half of 1984, 1988, and 1989. Bars indicate ±1 S.E. From *Oecologia*, Foliar delta C-13 within a temperate deciduous forest—Spatial, temporal, and species sources of variation. Garten, C. T. and Taylor, G. E. (1992). **90**, 1–7, © 1992 Springer-Verlag.

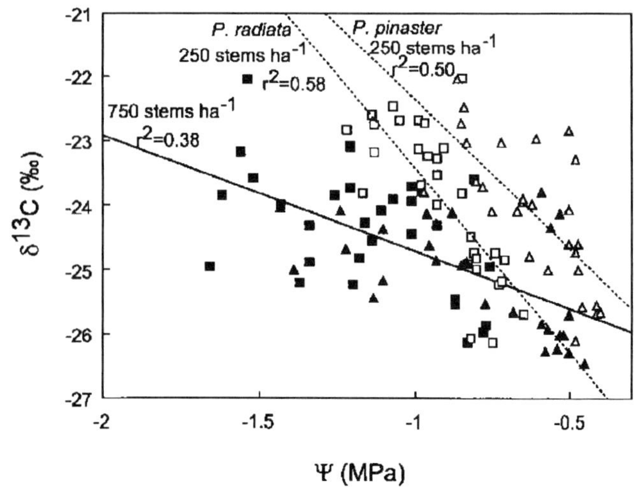

FIGURE 12.17. The relationship between Ψ (MPa) and stemwood $\delta^{13}C$ (‰) in *P. radiata* (■, □) and *P. pinaster* (▲, △) thinned to 750 stems ha^{-1} (■, ▲) or 250 stems ha^{-1} (□, △). Data are derived from monthly Ψ measurements and seasonal $\delta^{13}C$ measurements. Each point is an average of several monthly Ψ measurements, which correspond to the time when the wood for the $\delta^{13}C$ measurement was laid down. From *Oecologia*, Water availability and carbon isotope discrimination in conifers. Warren, C. R., McGrath, J. F., and Adams, M. A. (2001). **127**, 476–486, Figure 6 © 2001 with kind permission from Springer Science and Business Media.

wet year, although $\delta^{13}C$ values predicted the opposite. The authors attributed this difference to lower mean relative humidity and higher temperature in the wet year, conditions that influenced actual WUE but were not reflected in the $\delta^{13}C$ data.

Results of Warren et al. (2001) illustrate well the complexity of the relationship among $\delta^{13}C$, biotic, and environmental influences (Fig. 12.17). In this study the influences of water stress and thinning on $\delta^{13}C$ of two plantation-grown pine species was examined. There was a positive relationship between $\delta^{13}C$ and dry season water stress level ($-\Psi_l$), implying greater WUE, in both species in lightly thinned stands. However, in heavily thinned stands, $\delta^{13}C$ tended to be higher at a given dry season average Ψ_l, and species showed different responses. The authors suggested that heavy thinning might increase $\delta^{13}C$ by providing more nutrients to foliage that could sensitize stomata to water deficits, or by reductions in stomatal conductance caused by increased evaporative demand in the more open stands.

These results indicate a potential for stable carbon isotope analysis, particularly in providing long-term, integrated measures of WUE that are difficult to obtain with gas exchange techniques, but also emphasize that $\delta^{13}C$ values are determined only by properties of the movement and fixation of CO_2 in leaves. If other environmental and plant conditions that influence WUE (e.g., humidity, leaf and air temperature, photosynthetic capacity, conductance to CO_2 in the mesophyll) are not constant between samples, $\delta^{13}C$ data may provide misleading estimates. Other applications and precautions of $\delta^{13}C$ analysis were discussed in Chapter 5 and in Peterson and Fry (1987), Farquhar et al. (1989), Griffiths (1991), Dawson et al. (2002), and Fry (2006).

Recently, researchers have employed tandem measurements of $\delta^{13}C$ and $\delta^{18}O$ in leaves and wood to account for the effects of humidity and photosynthetic capacity on isotopic estimates of WUE (Sauer et al., 1997) (see also Chapter 5). This strategy capitalizes on the fact that water (of which some naturally contains ^{18}O) is enriched during evaporation from the leaf (being more massive that ^{16}O it evaporates more slowly). Carbonic anhydrase in the chloroplast catalyzes exchange of oxygen atoms from ^{18}O-enriched water to CO_2 resulting in higher ^{18}O contents in substrate CO_2 content. Subsequent fixation of ^{18}O-enriched CO_2 into carbohydrates provides a signal of the level of enrichment of this atom in leaf water. As ^{18}O enrichment of leaf water is a function of integrated evaporation rate, its ^{18}O content is a proxy indicator for the integrated vapor pressure deficit to which the leaf is exposed. In this way the effects of humidity on $\delta^{13}C$ can be evaluated. If $\delta^{13}C$ and $\delta^{18}O$ change in the same direction across samples of interest, a humidity effect on $\delta^{13}C$-based estimates of WUE may be present. If such a relationship is absent, changes in WUE are indicated by changes in $\delta^{13}C$ as long as the light environment and ^{18}O isotopic composition of source water do not vary (Sauer et al., 1997).

THE WATER BALANCE

The growth of both woody and herbaceous plants is reduced more often by water deficits than by any other environmental factor. The extensive evidence summarized by Zahner (1968), showing correlations between both height and diameter growth and available water, indicates that 70 to 80% of the variation in the width of annual rings in humid regions and 90% in arid regions can be attributed to differences in water stress (see also Chapter 5, Kozlowski and Pallardy, 1997). The degree of water stress in plants is controlled not only by the Ψ_w of soil water with which the root system is in contact but also by the relative rates of water absorption and water loss. Hence, water deficits can be caused by low soil Ψ_w, slow absorption, rapid water loss, or most often by a combination of the three. Thus the study of factors affecting water absorption and transpiration is important because it contributes to an understanding of the internal water balance of plants, which in turn affects the physiological processes and conditions controlling the quantity and quality of growth.

By plant water stress I mean a condition in which the cells are less than fully turgid and the water potential is substantially less than zero. The first visible effects of water stress are cessation of growth, closure of stomata, wilting of leaves and of young stems (Hsiao, 1973), but there are many important invisible effects that will be discussed later.

The Dynamics of Plant Water Status

Use of such terms as water balance and water economy emphasizes that the internal water relations of plants may be regarded as resembling a budget in which water status is controlled by the relative rates of water absorption (income) and water loss (expenditure).

The water status of plants growing under natural conditions can be exceedingly variable over temporal scales ranging from minutes to months. Diurnally, stomata of well-watered plants open soon after sunrise and increasing evaporative demands create the water vapor concentration gradients necessary for transpiration to proceed (Fig. 12.18). As water is removed from mesophyll cells during transpiration, Ψ_l declines. This creates a gradient in Ψ_w between the leaf and all other parts hydraulically connected to the leaves, and this gradient is responsible for water movement to transpiring organs. However, resistance in the SPAC prevents full recovery in Ψ_l as long as transpiration continues. On perfectly clear days, Ψ_l shows a temporal pattern similar to that of solar radiation. However, the balance between transpirational water loss and its replacement is exceedingly dynamic and small changes in environmental conditions can result in wide swings in Ψ_l in just minutes. This most often is observed on partly cloudy days in plants rooted in moist soil (Klepper, 1968; Stansell et al., 1973). These rapid changes result as transpiration varies with radiation effects on the leaf-to-air vapor concentration gradient. In passing from full sunlight to shade the temperature of a leaf declines rapidly because of convectional and transpirational cooling, thereby reducing ΔC. Late afternoon recovery of Ψ_l in well-watered plants follows reductions in solar radiation, but may show a slightly different pattern than that in the early morning because of higher vapor pressure deficits that are characteristic of the late afternoon (Fig 12.18).

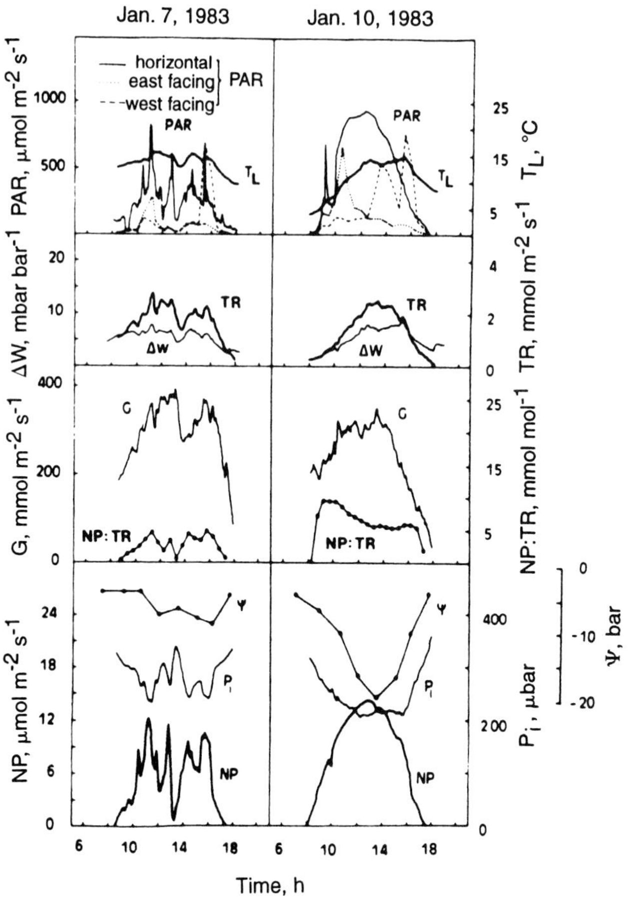

FIGURE 12.18. Daily time courses of photosynthetically active photon flux density (PAR), leaf temperature (T_L), leaf to air water vapor mole fraction difference (ΔW), transpiration (TR) and net photosynthesis (NP) rates, leaf conductance to water vapor (G), water use efficiency as NP:TR ratio (half hour-averages), intercellular CO_2 partial pressure (P_i), and leaf water potential (Ψ) of Tasmanian blue gum trees on January 7 and 10, 1983. From Pereira et al. (1986).

As soil moisture is depleted, the recovery of Ψ_w to that of the bulk soil overnight may be inhibited by progressive increases in liquid flow resistance in the soil (Slatyer, 1967) or plant (Tyree and Ewers, 1991). In any event, the upper limit of nocturnal recovery is set by soil Ψ_w and during drought a gradual depression commonly is observed in Ψ_l measured at dawn. This pattern may be repeated through the growing season as the soil is periodically wetted by summer rains (Fig. 12.19). Diurnal depression of Ψ_l during drought is reduced as stomata close; in cases of severe drought there may be very little change in Ψ_l from morning to evening (Fig. 12.20).

The Absorption Lag

There often are marked decreases in the water content of stems of plants near midday in sunny weather. This is demonstrated by the decrease in stem diameter reported by MacDougal (1938), Kozlowski and Winget (1964), Braekke and Kozlowski (1975), Neher (1993), and others, and shown in Figure 5.8 of Kozlowski and Pallardy (1997). Gibbs (1935) made a careful study of diurnal changes in water content of the wood in birch tree trunks, and some of his data are summarized in Table 12.7. He found that the maximum water content occurred near sunrise, decreased during the morning and midday, and rose in the afternoon and evening. This pattern seems to be characteristic of many kinds of plants in warm, sunny weather and indicates that tree trunks act as a storage place for water that is withdrawn when transpiration exceeds absorption and is replaced when the reverse situation occurs.

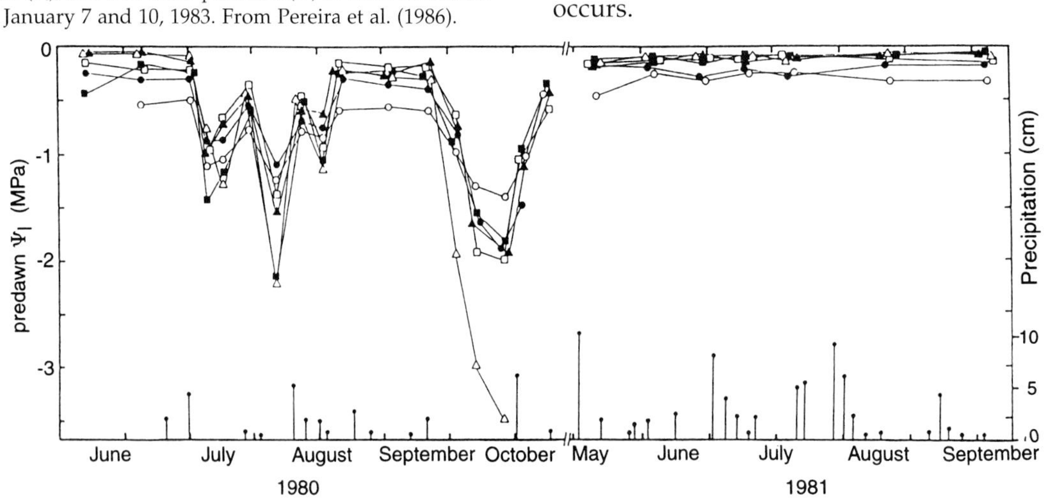

FIGURE 12.19. Seasonal patterns of precipitation (histogram) and predawn leaf water potential (Ψ_l) for the 1980 and 1981 growing seasons for plants of white (■), northern red (□) and black oak (▲), flowering dogwood (△), sugar maple (●), and eastern red cedar (○). The growing season of 1980 was unusually hot and dry, and that of 1981 was mild and wet. Used with permission of the Society of American Foresters from Bahari, Z. A., Pallardy, S. G., and Parker, W. C. (1985). Photosynthesis, water relations and drought adaptation in six woody species of oak-hickory forests in central Missouri. *For. Sci.* **31**, 557–569; permission conveyed through Copyright Clearance Center, Inc.

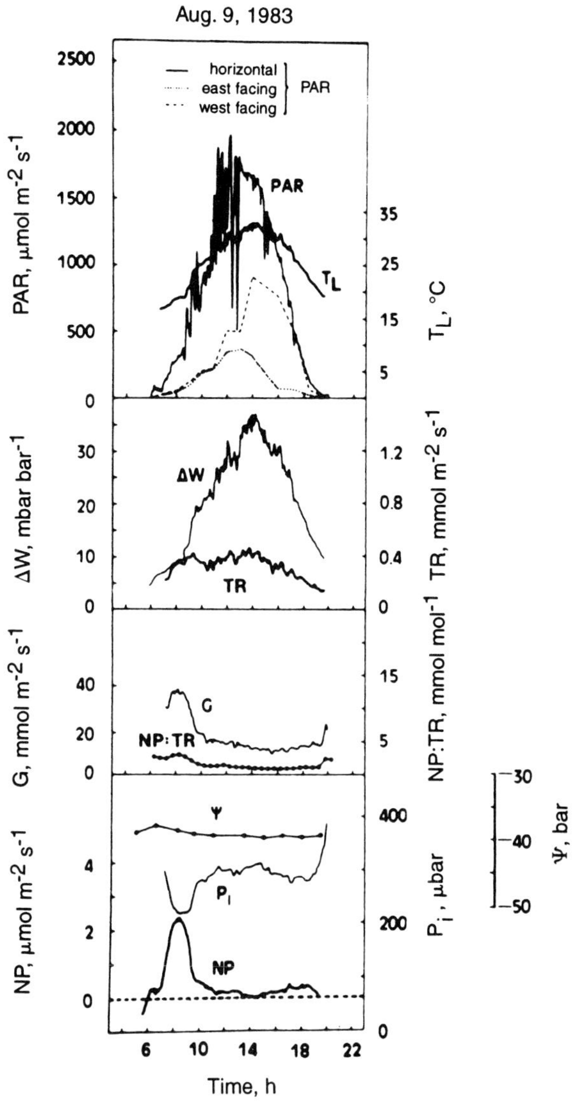

FIGURE 12.20. Daily time courses of photosynthetically active photon flux density (PAR), leaf temperature (T_L), leaf to air water vapor mole fraction difference (ΔW), transpiration (TR) and net photosynthesis (NP) rates, leaf conductance to water vapor (G), water use efficiency as NP:TR ratio (half hour-averages), intercellular CO_2 partial pressure (P_i), and leaf water potential (Ψ) of Tasmanian blue gum on August 9, 1983. From Pereira et al. (1986).

TABLE 12.7. Diurnal Variations in Water Content of Wood in Birch Trunks[a]

Date	Time	Weather	Water content (% dry weight)
August 24	5:00 A.M.	Clear	65
	1:00 P.M.	Clear, hot	54
	7:00 P.M.	Clear	58
August 25	5:00 A.M.	Clear	59
	1:00 P.M.	Slightly overcast	50
	7:00 P.M.	Clear	53

[a]From Gibbs (1935).

The cause of this fluctuation in water content is the fact that the resistance to movement of water from turgid plant tissue to the transpiring leaves is lower than the resistance to intake through the roots (Fig. 12.21). Thus, as transpiration increases in the morning, absorption does not begin to increase until the decreasing Ψ_l produces sufficient tension in the xylem sap to overcome the resistance to water flow through the xylem and the even larger resistance to radial movement from the soil into the root xylem. In the meantime, water is removed from tissue such as the sapwood of stems that offers lower resistance to flow. Hellkvist et al. (1974) reported leaf water potentials of –1.2 to –1.5 MPa and daily reduction in turgor to 40% of the maximum in Sitka spruce growing in moist soil in Scotland.

That the removal of water from a tree trunk lags behind loss from the leaves is shown by the observation of Waggoner and Turner (1971) that stem shrinkage at breast height in pine trees lags about two hours behind decrease in leaf water potential (see Chapter 5, Kozlowski and Pallardy, 1997). Zaerr (1971) observed a similar lag in Douglas-fir. This lag indicates a significant resistance to water flow through the stem and branches and into the evaporating surfaces of the leaves. Late in the day as the temperature decreases and stomata close transpiration is rapidly reduced but absorption continues until Ψ_w in the plant increases to approximately that in the soil as noted earlier.

Internal Competition for Water

During the growing season the various parts of trees and large herbaceous plants are often in competition for water. Because of differences in shading and in concentration of solutes, various parts of the shoots lose water at dissimilar rates and different levels of water deficits and water potential develop. This is especially important in drying soil when those regions in plants that develop the lowest Ψ_w obtain water at the expense of older tissues. Although young leaves may wilt first, they usually are the last to die on plants subjected to water stress. Water stress hastens senescence, possibly in part because it reduces the supply of cytokinins and changes the balance of growth regulators in the leaves (Nooden and Leopold, 1988; Srivistava, 2001). Lower, shaded leaves also are affected because they produce less carbohydrate than upper, better exposed leaves do, and they may be less able to compete osmotically for water, thereby enduring longer periods of low or zero turgor. Thus, dehydration may be a factor in the death of the lower, shaded branches of trees. According to Chalmers and Wilson (1978), the demand of developing peach fruits for car-

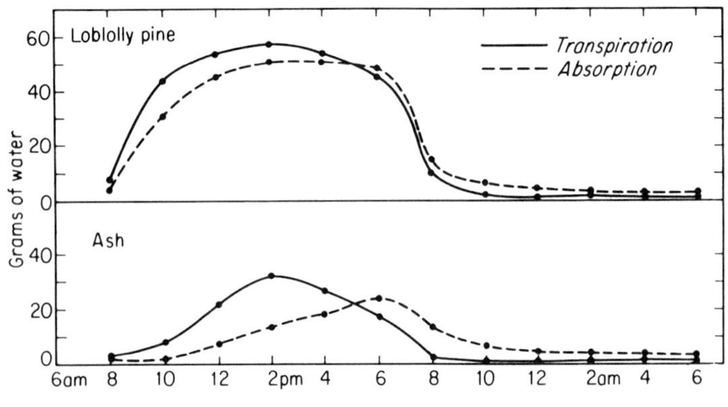

FIGURE 12.21. The relationship between water absorption and transpiration of white ash and loblolly pine. Note the lag between absorption and transpiration in both the morning and evening. From Kramer (1937).

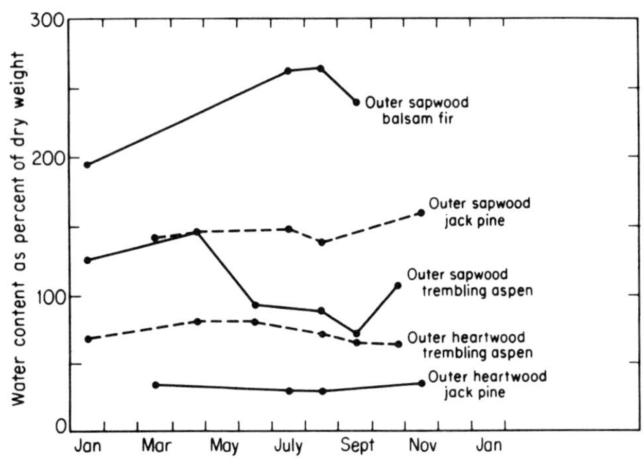

FIGURE 12.22. Seasonal changes in water content of the wood of conifer and deciduous species in eastern Canada. In general, the water content of conifer wood undergoes smaller variations compared to wood of deciduous species. From Clark and Gibbs (1957).

bohydrates and the increased water stress reduced branch growth of fruiting peach trees.

Long-Term Variations in Water Content

Over 50% of the total fresh weight of a tree consists of water, but the water concentration varies widely in different parts of a tree and with species, age, site, and season. The water content of well-developed heartwood usually is much lower than that of the sapwood (Chapter 3). Some data are shown in Figure 12.22. According to Ovington (1956) the water content usually increases from the base to the top of trees, but Ito (1955) reported that in Japanese chestnut the water content decreased from the base upward. Luxford (1930) noted that the water content of the heartwood of redwood is greatest at the base and lower toward the top, but the situation in the sapwood is reversed, being lowest at the base and highest toward the top.

Seasonal Variations in Water Content

Large seasonal variations occur in the water content of the trunks of trees of some, but not all, species. Although the largest seasonal variations in water content usually occur in hardwoods, Ito (1955) found larger seasonal variations in the water content of the wood of Japanese red pine than in Japanese chestnut, the minimum occurring in August in both species. According to Gibbs (1935), R. Hartig and E. Münch reported that in Europe conifers show significant seasonal variations in water content, but Gibbs found rather small variations in conifers in eastern Canada.

The typical seasonal pattern of change in water content for a diffuse-porous species, birch, is shown in Figure 12.23. Generally, in eastern Canada, tree trunks of birches, cottonwoods, and some aspens and willows attain their highest water content in the spring just before the leaves open. The water content decreases during the summer to a minimum just before leaf fall, then increases during the autumn after leaf fall reduces transpiration but before the soil becomes cold enough to hinder water absorption. Some species show another decrease during the late winter, presumably because cold soil hinders water absorption, followed by an increase in water content to the maximum after the soil thaws, but before the buds open and leaves expand. Among the variants from this pattern, white ash and American elm show no autumn increase, silver maple and American beech attain maximum water content in the autumn, and beech is unique among the species studied by showing no spring increase in water content. Those interested in more details should consult the papers by Gibbs (1935, 1939, 1953) and Clark and Gibbs (1957). Few data are available for milder climates, but the winter decrease in water content is less likely to occur where the soil does not freeze.

The pattern of release of stored water in tree stems follows a curvilinear trend (Fig. 12.24). Release of water near full stem hydration is at first rapid and

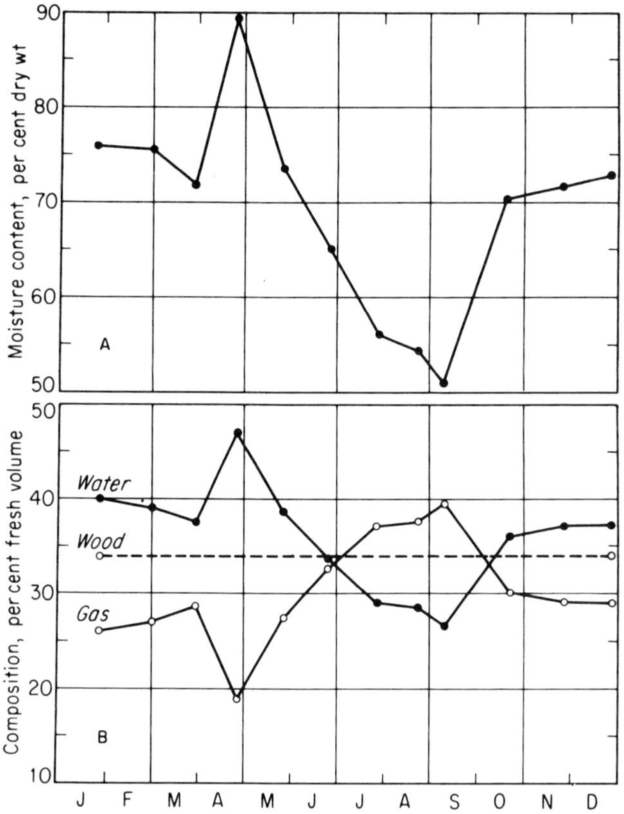

FIGURE 12.23. (A) Seasonal changes in water content of yellow birch tree trunks determined from disks cut from the base, middle, and top of the trunks. (B) Seasonal changes in water and gas content of yellow birch tree trunks calculated as percentages of fresh volume. Note the midsummer decrease in water content and increase in gas content during the period of rapid transpiration, also the autumn increase after leaf fall. From Clark and Gibbs (1957).

presumably reflects loss of capillary water (i.e., that portion of stem water held in the xylem matrix by capillarity), followed by relatively slow water release associated with elastic changes in xylem tissue. A final rapid water release is associated with cavitation in xylem elements and replacement of liquid water by gas bubbles (Tyree and Yang, 1990).

There has been much speculation and some study of the potential contribution of stored water to the transpiration stream of a tree. The ecological significance of stored water will be determined by the amount of water stored in the plant body, transpiration demands, and the rate of release of water from storage.

The relative amount of water stored in leaves is small except in seedlings. For example, 33% of the water in two-year-old lodgepole pine seedlings was in the foliage compared to only 4% in the foliage of 10- to 60-year-old trees (Running, 1979). The water in foliage can supply transpiration needs for only a relatively short time (Table 12.8). In an old Douglas-fir forest less than 0.1% of the amount of stored water was in the

TABLE 12.8. Potential Hours of Water Used in Transpiration Which Could Come from Various Plant Tissues in Woody Angiosperm, Conifer, and Herbaceous Species[a]

	Potential transpiration supply (hr)		
Storage zone	**Conifers**	**Hardwoods**	**Herbs**
Roots	14.0	4.9	2.8
Stem			
Sapwood	50–180	12.2	1.5
Extensible	1.0	6.7	1.3
Foliage	1.1	2.4	0.3

[a]From Hinckley et al. (1978a). Copyright 1978 by the Society of American Foresters, reprinted with permission.

needles, enough to supply transpiration for only a few minutes (Waring and Running, 1978). In Scotch pine, however, the foliage contained enough water to supply transpiration for several hours (Waring et al., 1979).

Considerable water is stored in tree stems. The availability of substantial amounts of water in stems is emphasized by the long time lag in propagation of changes in Ψ_w from transpiring leaves to the roots. Most of the stored water is localized in the sapwood (Table 12.8). In England, more than 70% of the water stored in aboveground tissues by conifers was in stems (Whitehead and Jarvis, 1981). Only a small amount of water is stored in the cambium and phloem. In a 20-year-old conifer stand the water stored in the cambium and phloem could supply transpiration for only about an hour (Jarvis, 1975).

More water is stored in sapwood of gymnosperms than angiosperms, and more in diffuse-porous than ring-porous angiosperms. Stewart (1967) estimated average sapwood moisture contents of ring-porous trees, diffuse-porous species, and conifers as less than 75, 100, and over 130%, respectively. Seasonal depletion of stem moisture was estimated to reduce sapwood moisture contents of ring-porous species to 35 to 60%, diffuse-porous species to 60 to 75%, and conifers to 75 to 100%.

There is evidence that conifers can recharge their sapwood and use stored water better than hardwoods can (Woodward, 1987). According to Siau (1972), the entire water-conducting column in hardwoods is disrupted by cavitation following withdrawal of water. By comparison, in conifers only individual tracheids are embolized because the bordered pits between tracheid walls close as a pressure gradient is created, hence confining gas bubbles to individual tracheids (Gregory and Petty, 1973) (Chapter 11). Hence conifers may better buffer leaf water deficits with stored water in the sapwood.

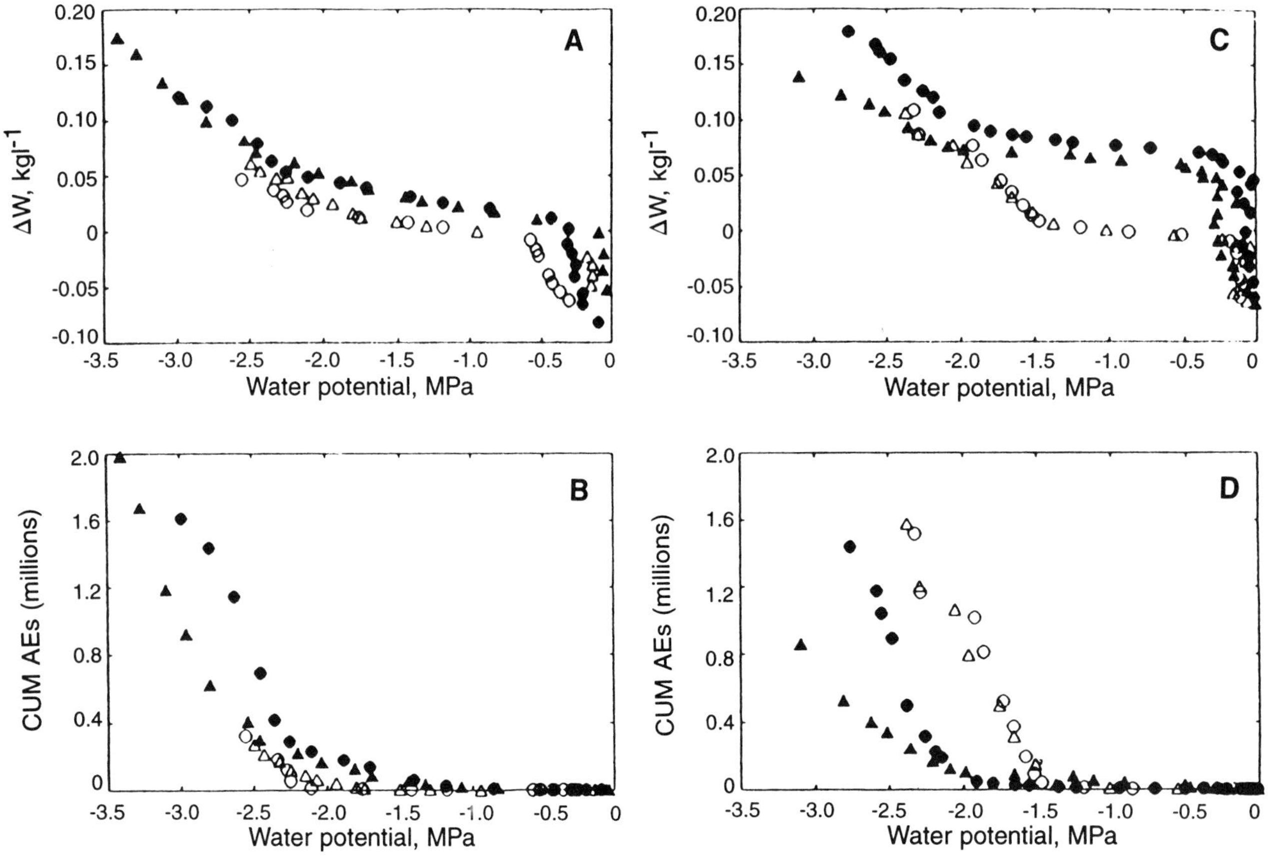

FIGURE 12.24. Dehydration isotherms of northern white cedar stems. (A, C) Water loss plotted against stem water potential. (B, D) Occurrence of acoustic emission events as a measure of the degree of xylem cavitation. Open symbols are for the first dehydration of stems, and closed symbols are for a second dehydration. Rehydration proceeds either for 1.5 days before the second dehydration (A, B) and for 2.5 days in (C, D). From *Planta*, Water-storage capacity of *Thuja*, *Tsuga* and *Acer* stems measured by dehydration isotherms—The contribution of capillary water and cavitation, Tyree, M. T. and Yang, S. D., **182**, 420–426, Figure 1 © 1990 with kind permission from Springer Science and Business Media.

Several investigators calculated that the water stored in the sapwood of conifers could supply transpiration needs for several days. Examples are Douglas-fir (Waring and Running, 1978) and Scotch pine (Waring et al., 1979). Approximately two-thirds of the water available for transpiration was located in the stem sapwood and less than 5% in the foliage, cambium, and phloem (Waring et al., 1979). Tyree (1988) estimated that trunks of northern white cedar trees contributed 5 to 6% of daily water use. Tyree and Yang (1990) suggested that the large release of capillary water at high Ψ_w was of little ecological significance, as it occurred under conditions when soil moisture would be readily absorbed to replace daily transpirational needs. Under severe drought, however, cavitation-induced water release to the foliage might delay lethal desiccation, but at the cost of reduced xylem hydraulic conductance. This might permit survival of a tree, especially if the hydraulic conductance could be recovered by the next growing season.

Both the large perennial roots and small fine roots also store some water. In seedlings, water resources in roots are important in preventing severe water deficits in leaves (Pallardy et al., 1982). Jarvis (1975) calculated that in conifers growing in dense stands enough water might be stored in the root system to supply transpiration for up to 14 hours.

Fruits also function as water reservoirs as shown by their shrinkage when the rate of transpiration is high (Chapters 4 and 6, Kozlowski and Pallardy, 1997). Young fruits commonly shrink less than old fruits either because young fruits store little water, or are less elastic than larger, more mature fruits.

EFFECTS OF WATER STRESS

Water deficits affect every aspect of plant growth, modifying anatomy, morphology, physiology, and biochemistry (Kozlowski, 1985b; Kozlowski and Pallardy,

1997). Trees are smaller on dry sites, their leaves usually are smaller, thicker, and have heavier wax deposition in the cuticle; moreover, vessel diameter of earlywood often is smaller and the cell walls usually are thicker and more lignified. An extreme example of reduced growth on a dry site is that of bristlecone pines on the White Mountains of California (Schulman, 1958). As mentioned at the beginning of this section, the amount of growth made by trees is closely correlated with the availability of water. In general, cell division is reduced less by water stress than cell enlargement. Both turgor and extensible cell walls are required for growth, but recent research, primarily with herbaceous crop plants, indicates that turgor often is maintained under water stress and that growth inhibition arises because of reduced hydraulic conductivity and cell wall extensibility (Michelana and Boyer, 1982; Nonami and Boyer, 1989, 1990a,b). In woody plants, Roden et al. (1990) similarly showed that although leaf growth was substantially reduced in leaves of unirrigated poplars, turgor was maintained by downward shifts in osmotic potential. However, extensibility of leaf cell walls was greatly diminished in unirrigated plants and was therefore primarily responsible for leaf growth inhibition (Fig. 3.3). Environmental and physiological aspects of growth in water-stressed plants are discussed further in Chapters 3 to 6 of Kozlowski and Pallardy (1997).

Research has shown that water stress affects many enzyme-mediated processes. This effect may be direct or an indirect response to dehydration-linked increases in concentrations of enzyme inhibitors (Kramer and Boyer, 1995). Respiration usually is reduced under severe water stress, although there may be a transient increase under mild stress (Brix, 1962; Mooney, 1969). Water stress affects carbohydrate metabolism and utilization in a variety of ways. During drought, sugar concentrations may increase in leaves (Kuhns and Gjerstad, 1988; Grieu et al., 1988) and roots (Parker and Patton, 1975). Increased sugar levels may be associated with osmotic adjustment of plant tissues (Morgan, 1984; Kuhns and Gjerstad, 1988). Partitioning of dry matter between roots and shoots often is significantly altered under moderate water stress with greater root growth relative to that of shoots (Sharp and Davies, 1979; Keyes and Grier, 1981; Axelsson and Axelsson, 1986; Gower et al., 1992; also Chapter 4, Kozlowski and Pallardy, 1997). Water deficits also have a profound influence on photosynthesis and nearly all of its constituent processes (Chapter 5).

In general, protein synthesis is inhibited by water stress. For example, Hulbert et al. (1988) showed that protein synthesis, indicated by incorporation of ^{35}S, in loblolly pine hypocotyls was reduced when tissues were exposed to osmotic stress. However, work with some plants has shown that some proteins may increase or be newly synthesized under water stress (Heikkela et al., 1984; Piatkowski et al., 1990). In the latter study a protein was isolated from the resurrection plant (*Craterostigma plantagineum*) that was quite similar to one temporally associated with dehydration that occurs late in seed development of several crop plants. Interestingly, the secondary structure of this protein predicted by computer analysis revealed distinct hydrophilic and hydrophobic regions. The authors suggested that the protein might thus function in a protective role, maintaining spatial separation of charged molecules, as does water, by its dipolar structure. These proteins are discussed in more detail later. There is evidence in many, but not all, cases of water-stress induced protein synthesis that abscisic acid plays a role in gene expression leading to the synthesis of these proteins (Chandler and Robertson, 1994; Giraudat et al., 1994; Bray, 1997).

As noted previously in this chapter, the balance of growth regulators also is affected by water stress. Abscisic acid levels in both roots and shoots increase with tissue dehydration (Blake and Ferrell, 1977; Cornish and Zeevart, 1985; Lin et al., 1986). Cytokinin and gibberellin contents of shoots appear to decline under water stress in some cases, but not always (Morgan, 1990). Much of the relevant research was done with herbaceous plants but the results probably are equally applicable to woody plants.

ADAPTATION TO DROUGHT

It should be emphasized that drought is a meteorological phenomenon, usually described as a period without rainfall of sufficient duration to cause depletion of soil moisture and reduction in plant growth. Drought may be essentially permanent, as in arid regions; seasonal, as in areas with well-defined wet and dry seasons; or random, as in most humid areas. The length of the period without rainfall required to produce drought conditions depends chiefly on the water storage capacity of the soil and the rate of evapotranspiration, and to a lesser degree on the kind of vegetation present. Even in such humid regions as Western Europe and the Southeastern United States, injurious droughts are common (Decker, 1983).

Drought is an environmental factor that produces water deficit in plants. Water deficit is initiated when low Ψ_w develops and cell turgor begins to fall appreciably below its maximum value. Plant water deficits nearly always accompany droughts, but also occur at other times either because of excessive transpiration or when absorption is hindered by cold soil, soil salinity,

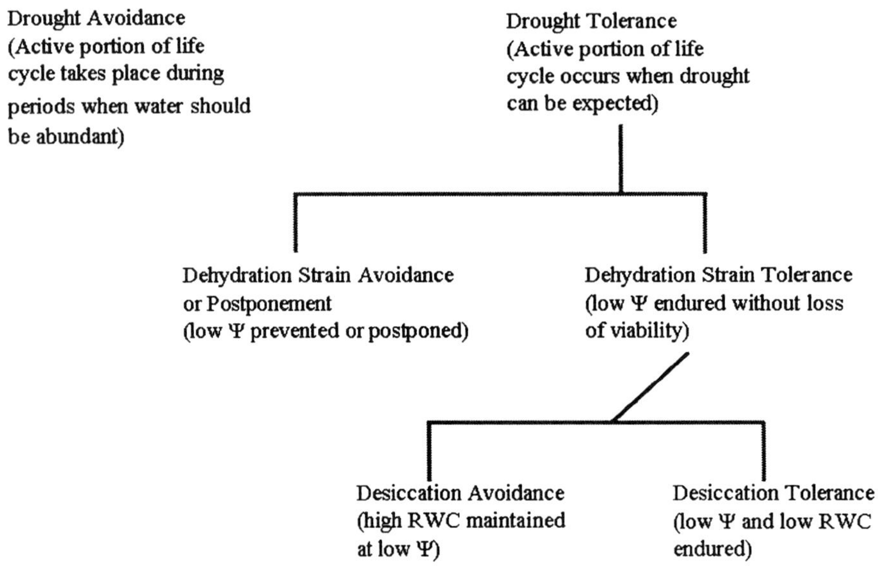

FIGURE 12.25. General scheme of mechanisms of adaptation to drought.

or damage to root systems. The effects of water deficits produced by drought or other causes are just as important to the growth of forest, fruit, and ornamental trees as for annual herbaceous crop plants. The capacity of plants to survive drought depends on a variety of phenological, morphological, and physiological factors. Farmers and foresters, as well as ecologists and physiologists, know that trees of some species survive drought with less injury than those of other species.

The adaptations responsible for these differences are discussed in some detail later. A convenient classification system can be developed that illustrates how adaptations are related to maintenance of plant water status (either Ψ_w or RWC) during drought (Fig. 12.25). Although these categories appear mutually exclusive and species tend to depend on one type of adaptation more than others, an individual plant can exhibit several adaptations simultaneously or at different times during a drought. Because of their perennial habit, drought avoidance adaptations in woody plants are consequently rare. In cases where plants may be exposed to drought (which can be considered analogous to a physical stress; Levitt, 1980a), there is a potential for a corresponding dehydration "strain" associated with reduced Ψ_w. Many tree species have well-developed adaptations for dehydration strain avoidance or postponement, and others may exhibit substantial dehydration strain tolerance capacity if they lack avoidance features or if avoidance adaptations are unable to prevent development of low Ψ_w. Kozlowski (1976b), Hinckley et al. (1978a), Kramer (1980), Levitt (1980b), Pallardy (1981), Turner (1986), Ludlow (1989), and Kramer and Boyer (1995) discuss drought adaptation mechanisms in both woody and herbaceous plants.

Drought Avoidance

Drought-avoiding plants occur in regions with well-defined dry seasons. They include the desert ephemerals with such short life cycles that they are completed in a few weeks after winter rains, and plants that mature early in the summer before the soil dries. The ephemerals include chiefly annual plants, although drought avoidance may be important to some perennial plants in Mediterranean climates.

Drought Tolerance

The term drought tolerance accurately describes the capacity of plants to pass through the active portion of their life cycle during periods when drought is expected.

Dehydration Strain Avoidance or Postponement

When the supply of water to the roots is reduced, the occurrence of injurious Ψ_w can be postponed in several ways. The most obvious method is by the storage of a large volume of water in fleshy roots or in stems, but the usefulness of this is limited to a few species such as cacti, which have a large storage capacity and efficient control of the transpiration rate. Although considerable water is stored in tree trunks the volume is small in comparison to the seasonal loss by transpiration from woody plants (Roberts, 1976b). It was noted earlier that some water released from

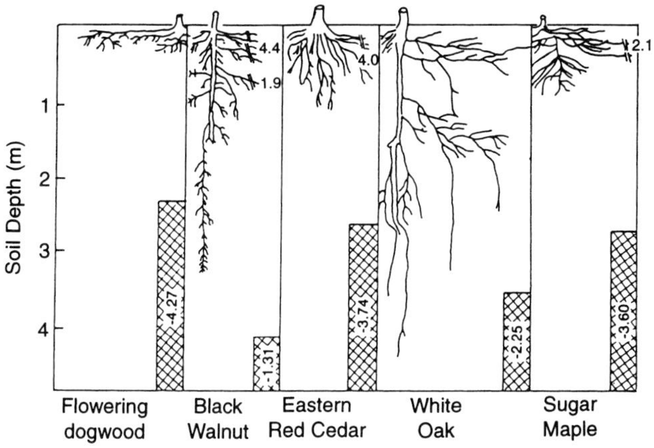

FIGURE 12.26. Distribution of roots of flowering dogwood, black walnut, eastern red cedar, white oak and sugar maple, and minimum predawn leaf water potential (in MPa, histogram) during severe drought. Numbers adjacent to root systems indicate horizontal extension of roots (m). From Hinckley et al. (1981) and based on data of Biswell (1935), Hinckley et al. (1979), and Sprackling and Read (1979).

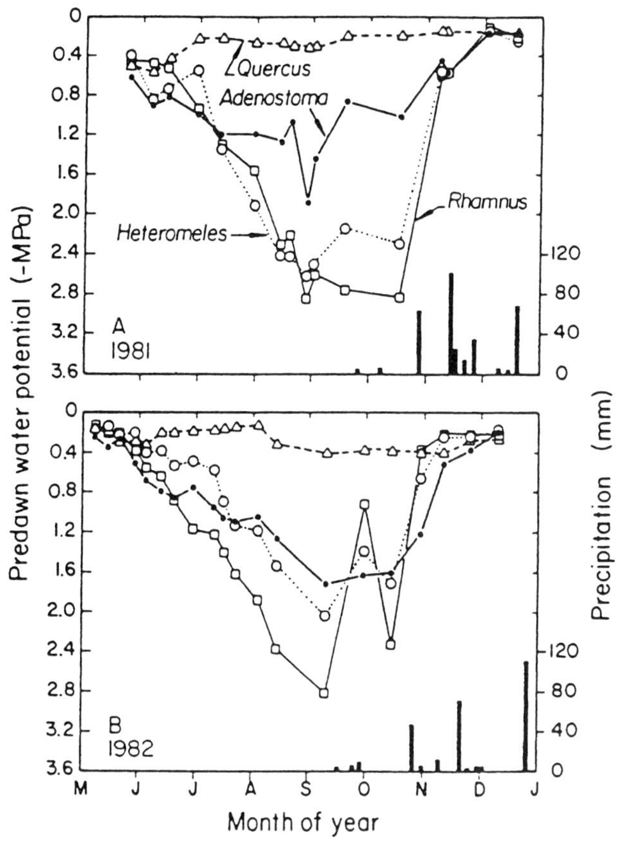

FIGURE 12.27. Predawn water potential of chaparral shrubs during seasonal drying and rewetting cycles of 1981 (A) and 1982 (B). From *Oecologia*, Tissue water relations of four co-occurring chaparral shrubs. Davis, S. D. and Mooney, H. A., **70**, 527–535, © 1986 Springer-Verlag.

stems during cavitation may serve to rehydrate leaves, but at the cost of reduced hydraulic conductance (Tyree and Yang, 1990).

Other means by which low Ψ_w can be avoided involve morphological adaptations to better exploit soil water resources by root systems; low xylem flow resistance (for a given rate of transpiration), which reduces the Ψ_w gradient between roots and leaves; and leaf adaptations, which reduce water loss to the atmosphere under drought.

Root Systems

Most observers agree that deep, wide-spreading root systems are important in postponing dehydration. It was noted earlier that root system depth and extent frequently are correlated with species distribution (Stone and Kalisz, 1991). Oppenheimer (1951) found this important in distribution of plant species in Israel. Hinckley et al. (1981) showed that maximum rooting depth was fairly well correlated with predawn Ψ_l developed during severe drought in a number of temperate North American tree species (Fig. 12.26). Drought-sensitive species such as flowering dogwood and sugar maple exhibited both restricted rooting capacity and very low predawn Ψ_l, whereas drought-tolerant oak species had much deeper roots and higher predawn Ψ_l. Davis and Mooney (1986) demonstrated how chaparral species partitioned soil moisture throughout the year by differential root growth at various depths (Fig. 12.27).

Genetic variation in root system development within species also may influence dehydration strain avoidance. For example, extensive root proliferation is said to be the chief cause of the greater dehydration avoidance of some strains of loblolly pine in Texas, with regulation of transpiration a secondary factor (van Buijtenen et al., 1976). Pallardy (1981) reviewed the literature on intraspecific variations in drought adaptation in woody plants and noted that similar genotypic variations in rooting capacity were evident among seed sources of Douglas-fir (Heiner and Lavender, 1972), sugar maple (Kriebel, 1963), Caribbean pine (Venator, 1976), and clones of tea (Carr, 1977; Othieno, 1978a,b).

Resistance to Liquid Water Flow

By minimizing the flow resistance between the roots and leaves, the Ψ_w gradient within a plant is reduced for a given level of transpiration. Differences in xylem structure and function, discussed earlier in this chapter, suggest that liquid flow resistance (and hence xylem transport capacity) varies widely among taxonomic groups and can have a major impact on maximum

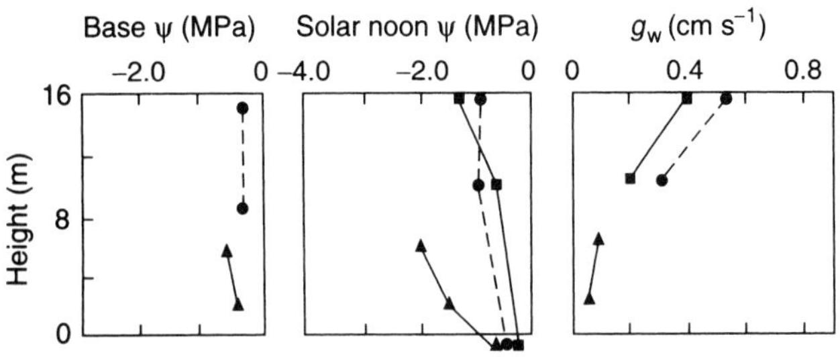

FIGURE 12.28. Vertical gradients in predawn and solar noon leaf water potential (Ψ_w) and leaf conductance to water vapor (g_w) for black walnut (●), white oak (■), and eastern redcedar (▲). Note that eastern redcedar experiences much steeper vertical gradients in Ψ despite having lower leaf conductances than the two angiosperm species. Used with permission of the Society of American Foresters from Ginter-Whitehouse. D. L., Hinckley, T. M., and Pallardy, S. G. (1983). Spatial and temporal aspects of water relations of three tree species with different vascular anatomy. *For. Sci.* **29**, 317–329; permission conveyed through Copyright Clearance Center, Inc.

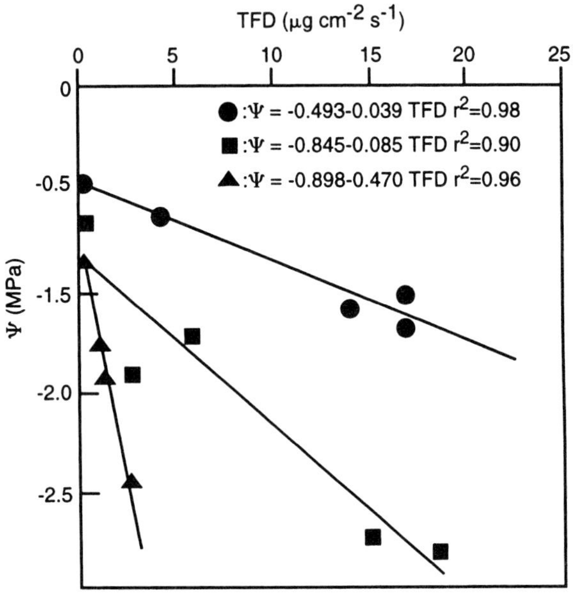

FIGURE 12.29. Variation in leaf water potential (Ψ_w) as a function of transpirational flux density (TFD) on July 10, 1979, for black walnut (●), white oak (■), and eastern red cedar (▲). Note the much steeper slope for eastern red cedar, indicating greater soil-plant flow resistance. Used with permission of the Society of American Foresters from Ginter-Whitehouse. D. L., Hinckley, T. M., and Pallardy, S. G. (1983). Spatial and temporal aspects of water relations of three tree species with different vascular anatomy. *For. Sci.* **29**, 317–329; permission conveyed through Copyright Clearance Center, Inc.

transpiration rate and the drop in Ψ_w across the plant. For example, Tyree et al. (1991) found that tropical *Schefflera morototoni* trees showed much higher LSC values and much smaller internal Ψ_w gradients than did temperate zone sugar maple and northern white cedar trees. Ginter-Whitehouse et al. (1983) compared vertical gradients of Ψ_w (Fig. 12.28) and the relationship between Ψ_l and transpiration rate (Fig. 12.29) among black walnut, white oak, and eastern red cedar trees. Clearly evident was the tendency for the tracheid-bearing eastern red cedar to exhibit larger internal Ψ_w gradients and greater drops in Ψ_l for a given transpiration rate (and hence greater $r_{soil \rightarrow leaf}$, see Eq. 11.8) than was the case in the angiosperms. The low xylem flow capacity of eastern red cedar appeared to be related to the restricted rate of maximal transpiration in this species compared to the others. Evolutionarily, xylem flow capacity must thus be balanced with potential transpiration demands and the possibilities for cavitation.

Although there has been some study of genetic variation of xylem within species, nearly all these studies have been strictly anatomical. Pallardy (1981) reviewed the available literature in this area.

Control of Transpiration

Pallardy (1981) listed several adaptations by which transpiration of plants might be reduced:

- A reduced capacity for growth
- Reductions in leaf size and altered morphology
- Leaf abscission
- High cuticular effectiveness
- Altered stomatal morphology and control

Reduced Capacity for Growth

Grime (1979) asserted that an inherent reduction in capacity for growth was a common evolutionary response to environmental stress. Pallardy (1981) noted that reduced growth rates were characteristic of genetic materials from xeric seed sources of red maple, green ash, balsam fir, loblolly pine, Douglas-fir, eucalyptus, western white pine, Scotch pine, and Caribbean pine.

Reduced Leaf Size and Altered Morphology

As noted earlier in this chapter, there is a tendency even within individual trees (e.g., white oak, Baranski, 1975) for variations in leaf size and shape that result in better energy dissipation, which reduces potential transpiration demands. According to Hinckley et al. (1981), leaves of desert and Mediterranean angiosperm species generally are smaller and more finely dissected than leaves of temperate zone angiosperm trees, and leaves produced during drought are smaller and more highly dissected than those produced when trees receive adequate soil moisture.

Pallardy (1981) noted that within species smaller leaves commonly were associated with more xeric seed sources of temperate zone angiosperm trees. In eastern cottonwood (Fig. 12.30) leaf size showed a clinal trend from large leaves from mesic provenances in the eastern United States to those of more xeric Great Plains provenances (Ying and Bagley, 1976).

FIGURE 12.30. Variation in leaf size of eastern cottonwood leaves from xeric western (left) to mesic eastern (right) U.S. provenances. From Ying and Bagley (1976).

Reductions in leaf size in more xeric seed sources also were found in Caribbean pine (Venator, 1976), Scotch pine (Wright and Bull, 1963; Ruby, 1967), yellow birch (Dancik and Barnes, 1975), and eastern redbud (Donselman, 1976).

Leaf Abscission

There are many reports in the literature of differential leaf abscission among species (Chapter 3). The drought-deciduous characteristic of many desert plants is well known (e.g., Larcher, 1975; Ehleringer, 1985). In temperate, humid regions leaf abscission is not as striking, but may constitute an effective adaptation to drought (Parker and Pallardy, 1985a; Pallardy and Rhoads, 1993; see Chapter 5, Kozlowski and Pallardy, 1997). The leaf abscission response can provide protection against lethal desiccation to vital meristems, but it occurs at the cost of lost photosynthate.

Cuticular Effectiveness

Many plants native to arid regions and those with long summer droughts have heavily cutinized leaves and very low transpiration rates after the stomata close. For example, Oppenheimer (1951) reported that in Israel plants such as carob, laurel, olive, Aleppo pine, and *Arbutus andrachne* have very low transpiration rates when soil moisture is depleted, but almond and fig have high transpiration rates and poor control of transpiration.

Cuticle development usually is more extensive when plants are grown in stressful environments (Chapter 8). Pallardy and Kozlowski (1980) showed that leaves of poplar plants grown in the field had much thicker epicuticular wax deposits than when plants of the same clones were grown in quite moderate conditions in a growth chamber (Fig. 12.31). Additionally, field-grown poplars had some stomata that were completely engulfed by wax, resulting in total occlusion of the stomatal pore. Genetic differences in cuticular development and effectiveness also have been identified (Nagarajah, 1979), suggesting that selection and breeding may improve cuticular effectiveness. In some cases, however, cuticular transpiration does not appear to be closely related to drought tolerance. Pallardy and Rhoads (1993) measured cuticular transpiration of temperate deciduous angiosperm species from dry (white and post oak) and moist (sugar maple and black walnut) habitats. Oak species, although more drought tolerant, exhibited higher cuticular transpiration rates. It should be pointed out that the adaptive value of the cuticle emerges only when other mechanisms that maintain hydration of leaves (e.g., rooting, xylem transport efficiency, stomatal closure) have been ineffective. Hence evolutionary selection for cuticular effectiveness may not be very important if other drought tolerance adaptations are well developed.

Stomatal Control

A large body of work demonstrates variation in the sensitivity of stomata to water stress. The data

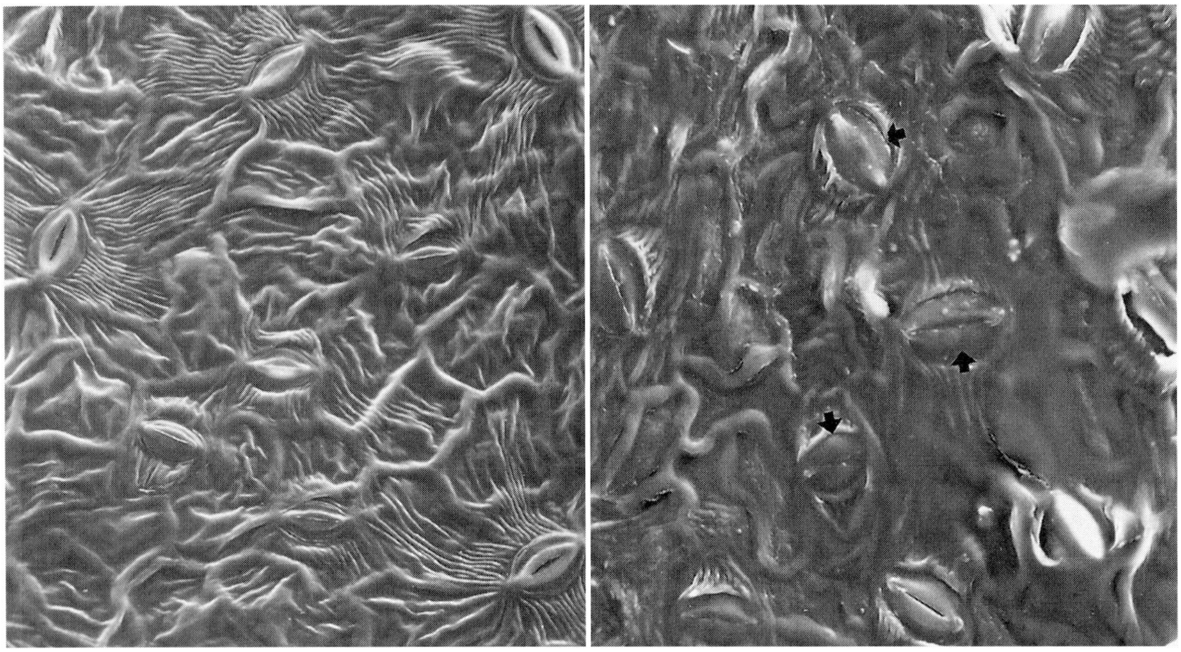

FIGURE 12.31. Scanning electron micrographs (×570) of abaxial surfaces of leaves of a poplar clone grown in a moderate growth chamber environment (left) or in the field (right). Arrows indicate stomata that are completely occluded by waxes. From Pallardy and Kozlowski (1980).

summarized by Lopushinsky (1969) indicated that considerable differences in sensitivity of stomata to water stress exist among conifer species, with stomata of pines generally more sensitive to water stress than those of other conifers. In experiments with two-year-old seedlings Lopushinsky found that stomatal closure occurred at lower water stress (higher Ψ_l) in ponderosa and lodgepole pines and Engelmann spruce than in Douglas-fir and grand fir. Running (1976) also reported differences among conifers with respect to the relationship between leaf water stress and stomatal opening. Hinckley et al. (1978b) showed that during a dry summer the stomata of sugar maple were closed about half the time, those of white and black oak about 20%, and those of northern red oak nearly 30%. This pattern of differential response was supported by Ni and Pallardy (1991), who observed that leaf conductance was reduced at higher Ψ_l in black walnut and sugar maple (between −1.8 and −2.2 MPa) than in white or post oak. Post oak maintained measurable levels of stomatal conductance near −3 MPa.

This general relationship is complicated by the possible influence of root signals in stomatal closure. For example, Parker and Pallardy (1991) found that the high sensitivity of stomata of black walnut to water stress was associated with declining soil Ψ_w rather than leaf Ψ_w (Fig. 12.32). Nearly total stomatal closure was observed over a range of soil drying where leaf Ψ_w was stable. These results suggest that stomata were responding preferentially to some stimulus (likely

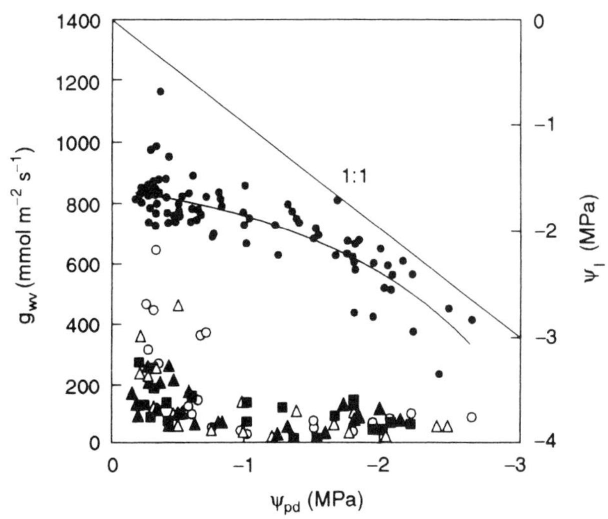

FIGURE 12.32. The response of leaf conductance (g_{wv}) to predawn leaf water potential (Ψ_{pd}) for seedlings of four black walnut families (indicated by ○, ■, △, and ▲ symbols). Leaf conductance values are plotted as a function of Ψ_{pd}. The measured values of leaf water potential (Ψ_l) at a given Ψ_{pd} (●) are plotted directly above the corresponding g_{wv} value. The data are fitted by the following equation: $\Psi_l = 1.56 + 0.22(\Psi_{pd}) + 0.05(\Psi_{pd}^3)$; $R^2 = 0.71$, both slope coefficients are significant ($p < 0.05$). The 1:1 line indicates equivalence of Ψ_{pd} and Ψ_l. From Parker and Pallardy (1991).

ABA) from the roots. These results were supported by field experiments with canopy trees of black walnut (Loewenstein and Pallardy, 1998a) (Fig. 12.6). Thus, stomatal sensitivity to water stress may reflect direct response to leaf water stress or the influence of other, more indirect responses of the plant to water shortage, or (most likely) some combination of the two.

One cannot simply identify stomatal closure at high Ψ_l as a useful adaptation in all species because any supposed beneficial effect on plant water relations comes at the cost of reduced photosynthesis. High rates of photosynthesis can occur only when stomata are wide open, and these conditions also are conducive to high transpiration rates. High water flow rates tend to depress Ψ_l, and photosynthesis and transpiration actually may be negatively correlated with Ψ_l unless soil water is depleted or evaporative demands are extreme (Pallardy and Kozlowski, 1979b; Kubiske and Abrams, 1994). There is evidence that delayed stomatal closure often is associated with species that are native to dry habitats and arid regions (Davies and Kozlowski, 1977; Bunce et al., 1977; Bahari et al., 1985; Abrams, 1990; Ni and Pallardy, 1991; Kubiske and Abrams, 1992). These species also show a greater capacity for dehydration strain tolerance (see later).

Dehydration Strain Tolerance

Dehydration strain tolerance is at the other extreme of drought tolerance, referring to the capacity of protoplasm to sustain partial function or at least avoid irreversible injury as tissue Ψ_w declines. The most extreme examples occur among mosses and lichens, but some flowering plants (often called "resurrection plants") also can be dehydrated to air dryness (Gaff, 1989; Rascio and La Rocca, 2005). Other plants, particularly tropical C_4 grasses and some woody plants from arid regions such as creosote bush, sagebrush, acacias, and shrubs of the Mediterranean maquis and California chaparral have appreciable protoplasmic tolerance of dehydration. However, the vast majority of woody plants show at most moderate tolerance to dehydration.

Bases of Dehydration Strain Tolerance

Two basic responses of plants to reduced environmental Ψ_w can be advanced. In the first type, desiccation avoidance, Figure 12.25, RWC is maintained or excessive reductions are avoided even while Ψ_w falls. This can occur by at least two mechanisms: (1) osmotic adjustment and (2) possession of stiff cell walls (conferring a high modulus of elasticity).

Both effects on water relations of non-growing cells can be deduced from the following equation:

$$\Psi_{w1} = \Psi_{\pi 0}\frac{V_0}{V_1} + \Psi_{p0} + \varepsilon\frac{V_1 - V_0}{V_0} \qquad (12.12)$$

where V_0 and V_1 are the volumes before and after a change in volume, $\Psi_{\pi 0}$ is the initial osmotic potential,

TABLE 12.9. Potential Combinations of Osmotic Potential (Ψ_π) and Tissue Elasticity (ε) and Effects on Turgor Maintenance and Tissue Dehydration (RWC, Relative Water Content)[a]

CASE 1: high Ψ_π, high ε	CASE 2: high Ψ_π, low ε
Turgor loss occurs at very high Ψ_w, RWC RWC declines gradually with Ψ_w	Turgor loss delayed somewhat to lower RWC (and slightly lower Ψ_w) than in CASE 1 RWC declines steeply with Ψ_w
CASE 3: low Ψ_π, high ε	**CASE 4: low Ψ_π, low ε**
Turgor loss at intermediate Ψ_w	Turgor loss delayed to lowest RWC
RWC declines very little as Ψ_w drops	RWC declines steeply as Ψ_w drops

[a]From Kozlowski and Pallardy (2002).

and ε is the elastic modulus of the cells that describes the change in Ψ_p that occurs with a change in volume. For completeness, an initial turgor potential, Ψ_{p0}, is included in the equation. An increase in solute content (other factors being equal) will lower Ψ_π and Ψ_{w1} *without* changing the volume of cell water. Simple dehydration of cells will lower the osmotic potential, but reduce V and RWC at the same time. Thus the accumulation of solutes by cells (i.e., osmotic adjustment) allows maintenance of V and RWC even when the environment is tending to decrease Ψ_w. A high modulus of elasticity reduces volume loss for a given drop in Ψ_w as well, because high ε values will result in large changes in Ψ_p for a given change in cell volume.

The possible patterns of osmotic potential and tissue elasticity in plants are several and each combination has potential ecological significance (Table 12.9) (Kozlowski and Pallardy, 2002). For succulents, high Ψ_s and low ε are associated, which traits confer the capacity to store large amounts of water internally, but without a large solute requirement. These traits are combined with the capacity to isolate the water in the plant body from an environment where Ψ_w is usually quite low. As the soil water potential in arid habitats commonly is much lower than Ψ_π, tissues of succulents would fall far below the point of turgor loss if these plants were unable to isolate themselves from this environment.

The adaptive fitness of other combinations of Ψ_π and ε also may depend on habitat environment. In xeric habitats, where light is often freely available to plants and rapid growth rates are not a necessity, a combination of both low Ψ_π and high ε would appear adaptive, as this suite of traits best protects from tissue dehydration (low RWC) at very low Ψ_w. A combination of high RWC at low Ψ_w may protect metabolic processes from the effects of toxic ion concentrations as tissue water is lost. Such volume maintenance in chloroplasts by osmotic adjustment may preserve photosynthetic capacity (Santakumari and Berkowitz, 1991) and prevent injury from toxic concentrations of ions (Rao et al., 1987).

If a rapid growth rate is adaptive in a particular habitat, elastic tissues that support the maintenance of turgor for growth may be an important drought adaptation growth. Low Ψ_π will similarly support turgor maintenance, but this must be balanced against the metabolic cost of the solutes. Although these patterns are ecologically plausible, the experimental evidence that would support this scheme is still sparse. In crop plants, attempts to improve yields by breeding for lower Ψ_π have produced varying results. For example, whereas Moinuddin and Khanna-Chopra (2004) reported elevated seed yields in droughted chickpea (*Cicer arietinum* L.) in cultivars that showed higher osmotic adjustment, Turner et al. (2007) observed no yield benefit from lines of this species with higher osmotic adjustment. A few other instances of yield improvement attributable to breeding for high levels of osmotic adjustment have been reported for crop plants (e.g., wheat, Morgan, 1983; sorghum, Ludlow et al., 1990; pea, Rodríguez-Maribona et al., 1992), but Serraj and Sinclair (2002) noted that yield increases associated with osmotic adjustment usually were seen only under conditions of severe drought, where yields were so low as to be of little agronomic value. These authors identified the greatest potential benefits of osmotic adjustment as a capacity to sustain root growth under drought so that deeper soil water resources might be accessed.

Osmotic adjustment during drought has been observed in many woody plants (Table 12.10) (Kozlowski and Pallardy, 2002). However, osmotic adjustment is not universally observed under drought. For example, no osmotic adjustment was observed in response to drought in several Chilean evergreen shrubs including *Cryptocarya alba*, *Calliguaya olorifera*, *Lithraea caustica*, and *Quillaja saponaria* (Poole and Miller, 1978). There also was little difference between dry and wet growing seasons in Ψ_s at full turgor and at the turgor loss point in *Fagus sylvatica* and *Quercus petraea* growing in a German hardwood forest (Backes and Leuschner, 2000). Hinckley et al. (1980) also concluded that osmotic adjustment in leaves of several species (including *Cornus mas*, *Crataegus monogyna*, *Olea europaea*, *Sorbus aria*, and *Viburnum lantana*) was of minor importance as an adaptive response to water deficits. Other species that have shown little or no osmotic adjustment to drought include *Alnus sinuata* (Cline and Campbell, 1976), *Acer saccharum* (Bahari et al., 1985), *Cercis canadensis* (Buxton et al., 1985), *Juniperus virginiana* (Bahari et al., 1985), *Liriodendron tulipifera* (Roberts et al., 1980), and *Picea mariana* (Buxton et al., 1985).

The solutes involved in osmotic adjustment appear to be species dependent. Sorbitol, glucose, and fructose accumulated in *Malus domestica* leaves as water stress developed (Wang and Stutte, 1992). In *Populus* hybrids osmotic adjustment was largely attributable to increases in malic acid, K, sucrose, and glucose (Tschaplinski and Tuskan, 1994). Solutes contributing to osmotic adjustment in three *Populus deltoides* clones included sucrose, malic acid, glucose, fructose, myoinositol, and salicin (Gebre et al., 1994). Fructose and glucose increased in *Picea glauca* shoots and *Pinus banksiana* roots under water stress (Koppenaal et al., 1991). Malate and mannitol were involved in osmotic adjustment of *Fraxinus excelsior* leaves (Guicherd et al., 1997).

Elastic Adjustment

As discussed previously, turgor can also be influenced by changes in tissue elasticity. Other things being equal, inelastic tissues lose turgor faster than elastic tissues as water is lost. Hence an increase in tissue elasticity under water stress conditions may be desirable under conditions where turgor maintenance is adaptive (e.g., for growth). On the other hand, decreased elasticity under drought tends to maintain higher levels of tissue relative water content as plant Ψ declines, protecting tissues from low RWC and possibly toxic concentrations of ions. Hence both increases and decreases in elasticity might be advantageous, depending on which response is favored in a particular habitat. Changes in tissue elasticity may be especially important for plants that do not show appreciable osmotic adjustment.

Both increased and decreased elasticity under the influence of drought have been reported. Whereas elasticity decreased in shoots of *Pinus echinata* and leaves of *Betula populifolia* exposed to water stress (Choi, 1992; Morse et al., 1993), it increased in *Celtis occidentalis* and *Quercus muehlenbergii* leaves, roots of *Q. macrocarpa*, and leaves and roots of *Juglans nigra* (Abrams and Knapp, 1986; Parker and Pallardy, 1985b, 1988). Responses to drought of elasticity may diverge even in closely related plants. For example, whereas drought induced increases in ε of leaves of *Eucalyptus nitens*, ε of *E. globulus* leaves decreased (White et al., 1996). The cellular mechanism by which changes in elasticity under water stress arise is not known. Wakabayashi et al. (1997) showed that osmotic stress reduced cell wall stiffening in wheat coleoptiles, a response that was associated with reductions in two

TABLE 12.10. Species of Woody Plants Showing Appreciable Osmotic Adjustment[a]

Species	Source
ANGIOSPERMS	
Alnus glutinosa	Seiler (1985)
Betula populifolia	Morse et al. (1993)
Carya tomentosa	Parker et al. (1982)
Celtis occidentalis	Abrams and Knapp (1986)
Citrus sinensis	Fereres et al. (1979)
Cornus florida	Roberts et al. (1980); Bahari et al. (1985)
Eucalyptus camaldulensis	Lemcoff et al. (1994, 2002)
Eucalyptus behriana	Myers and Neales (1986)
Eucalyptus grandis	Lemcoff et al. (1994)
Eucalyptus microcarpa	Myers and Neales (1986)
Eucalyptus nitens	White et al. (1996)
Eucalyptus polyanthemos	Myers and Neales (1986)
Eucalyptus tereticornis	Lemcoff et al. (1994)
Eucalyptus viminalis	Lemcoff et al. (1994)
Fragaria chiloensis	Zhang and Archbold (1991)
Fragaria virginiana	Zhang and Archbold (1991)
Fraxinus excelsior	Guicherd et al. (1997)
Ilex opaca	Roberts et al. (1980)
Juglans nigra	Parker and Pallardy (1985b)
Malus domestica	Goode and Higgs (1973); Fanjul and Rosher (1984); Lakso et al. (1984); Wang and Stutte (1992); Wang et al. (1995)
Olea europaea	Rieger (1995)
Populus deltoides	Tschapinski et al. (1994)
Populus deltoides x P. nigra	Tschapinski and Blake (1989b)
Populus hybrids	Tschaplinski and Tuskan (1993)
Populus tremuloides	Abrams (1988b)
Populus trichocarpa	Tschaplinski and Tuskan (1994)
ANGIOSPERMS (cont.)	
Quercus alba	Parker et al. (1982); Bahari et al. (1985)
Quercus ellipsoidalis	Abrams (1988)
Quercus ilex	Dreyer et al. (1990)
Quercus macrocarpa	Abrams and Knapp (1986)
Quercus muehlenbergii	Abrams and Knapp (1986)
Quercus petraea	Dreyer et al. (1990)
Quercus pubescens	Dreyer et al. (1990)
Quercus robur	Osonubi and Davies (1978); Dreyer et al. (1990)
Quercus rubra	Parker et al. (1982)
Quercus velutina	Bahari et al. (1985)
Quercus stellata	Parker and Pallardy (1988)
Rosa hybrida	Augé et al. (1986b)
Vitis vinifera	Düring (1985)
Ziziphus mauritania	Clifford et al. (1998)
GYMNOSPERMS	
Picea glauca	Koppenaal et al. (1991)
Picea mariana	Tunstall and Connor (1975); Zwiazek and Blake (1989)
Pinus banksiana	Koppenaal et al. (1991)
Pinus echinata	Pallardy et al. (1982); Choi (1991)
Pinus pinaster	Nguyen and Lamont (1989)
Pinus taeda	Seiler and Johnson (1988); Meier et al. (1992)
Pseudotsuga menziesii	Joly and Zaerr (1987)
Thuja occidentalis	Collier and Boyer (1989)
Tsuga canadensis	Tyree et al. (1978)
Tsuga heterophylla	Kandiko et al. (1980); Buxton et al. (1985)

[a] Modified from Kozlowski and Pallardy (2002).

lignin components, ferulic and diferulic acid. In contrast, Peltier and Marigo (1999) reported a reversible increase in elastic modulus in *Fraxinus excelsior* under water stress, which could be mimicked by application of 20 μmol L^{-1} ABA and counteracted by treatment with buffer solutions with acidic pH. More research is needed to explain the biochemical basis underlying documented changes in tissue elasticity under water stress.

Even in the absence of drought, plants will develop constitutively different levels of Ψ_π and ε that are apparently genetically determined. For example, Bahari et al. (1985) observed a range of Ψ_π between −1.52 and −1.95 MPa and ε between 3.36 and 10.37 MPa in saplings of a variety of temperate deciduous angiosperm species during a summer in which soil moisture was abundant. Similarly, Backes and Leuschner (2000) reported significant differences in Ψ_π at full turgor between *Quercus petraea* and *Fagus sylvatica* trees during a wet growing season.

Eventually relative water content falls to a critical level at which plant survival depends on the degree of dehydration that the protoplasm can endure without undergoing irreversible injury (desiccation tolerance). There seem to be wide differences among species in this respect. Oppenheimer (1932) reported that leaves of almond could be dried to a saturation deficit of 70% before injury occurred, olive to 60%, but fig to only 25%. Different genotypes within a species also may exhibit inherent differences in tolerance of desiccation, as was shown for black spruce and bur oak (Tan and Blake, 1993; Kuhns et al., 1993). Seasonal differences also exist. The leaves of creosote bush produced during moist weather are large and easily injured by water deficit, but the small leaves produced during dry weather can be dried to a saturation deficit of 50% (Runyon, 1936). Pisek and Larcher (1954) found that in several species tolerance to desiccation increases in the winter along with cold tolerance, then decreases in the spring.

At the extreme of cellular water loss are plants in which vegetative parts can remain viable even when in equilibrium with water vapor in the air. Rehydration of these plants results in full recovery over a period of hours or days, a capacity that is responsible for their designation as resurrection plants. Resurrection plants are widespread but not common in plant taxa (Oliver et al., 2000). These authors suggested that vegetative desiccation tolerance is constitutively present in lower plants such as bryophytes and mosses as a legacy of this group's early history of aquatic to dry-land transition. Constitutive desiccation tolerance capacity is metabolically costly, and the consequent slow growth rates of these plants were sufficiently disadvantageous that vegetative desiccation tolerance was lost during tracheophyte evolution. However, genes associated with desiccation tolerance were reintegrated in the seed development process. Thereafter, they remained available for several independent reemergence events of vegetative desiccation tolerance in *Selaginella*, ferns, and the angiosperms. No resurrection plants are known in the gymnosperms.

Primitive desiccation tolerant plants such as the moss *Tortula ruralis* exhibit a suppression of protein synthesis if dried rapidly, suggesting that mechanisms to cope with desiccation are in place before dehydration (Oliver et al., 2000). Although the fine structure of this moss remains relatively intact during drying, rehydration is associated with temporary disruption of membrane systems. The amount of post-dehydration disruption varies with the rate of drying. Post-rehydration events in the moss include down- and up-regulation of synthesis of over 100 proteins, many of the latter of which are already represented in mRNA transcripts when hydration occurs. Oliver et al. (2000) thus suggested that desiccation tolerance in *T. ruralis* was more dependent on preexisting repair systems than on preexisting or induced protection systems.

Plants that survive desiccation only if dried slowly are called modified desiccation-tolerant plants, and this is the only type of tolerance observed in tracheophytes (Oliver et al., 2000; Hoekstra et al., 2001; Rascio and La Rocca, 2005). Desiccation tolerance in this group apparently is induced by dehydration rather than constitutive. Species in this group are separated into those that retain chloroplast structure (homoiochlorophyllous species) and those in which chloroplast disassembly occurs under desiccation (poikilochlorophyllous species). Rascio and La Rocca (2005) identified three types of desiccation-induced cell events that required responses by plants:

- Membrane and protein damage
- Mechanical stress
- Oxidative stress

Accumulation of sugars and a specific class of *l*ate-*e*mbryogenesis-*a*bundant (*lea*) proteins (Baker et al., 1988; Close et al., 1989) may protect membrane and protein configuration because they can first preserve hydration of molecules through preferential exclusion effects and then, under severe dehydration, solvate directly as water replacement molecules (Hoekstra et al., 2001). Sugars have abundant hydroxyl groups that can form hydrogen bonds. *Lea* proteins (some of which have been termed dehydrins) are rich in glycine, lack cysteine and tryptophan, and have specific hydrophobic and hydrophilic regions that can associate with charged molecules. As dehydration reaches extreme levels, high concentrations of sugars may promote the formation of glasses by vitrification, a state in which membrane-breaching crystallization cannot occur and molecular mobility is much reduced.

Mechanical stresses caused by water removal may be accommodated by leaf shrinkage, cell wall folding alteration, possibly with involvement of expansin (Chapter 3) and aquaporin (Chapter 11) proteins. Oxidative stresses arise because of disturbed metabolism in desiccated tissues, with the electron transport systems of the chloroplasts and mitochondria especially prone to promote production of reactive oxygen species (ROS) as the normal substrates for high energy products of these systems disappear. Poikilochlorophyllous species undergo gross chloroplast disorganization, possibly to reduce the sources of ROS. Still, in this group screening anthocyanin pigments accumulate and antioxidant enzymes such as ascorbate peroxidase, glutathione reductase, and superoxide dismutase are up-regulated.

Drought Hardening

It is well known that plants that previously have been subjected to water stress suffer less injury from drought than plants not previously stressed. For example, when potted plants are suddenly transferred from a shaded, humid environment to full sun their leaves often are injured even though the plants are well watered. Rook (1973) found that seedlings of Monterey pine watered daily had higher rates of stomatal and cuticular transpiration than those watered less frequently, but there was no difference in their root–shoot ratios. The seedlings watered daily endured more severe water stress and made less root growth after transplanting than those watered less frequently, probably because the latter had better control of transpiration.

Growers of both herbaceous and woody plants for transplanting to the field commonly "harden" seedlings to increase survival. Often this is done simply by exposing the seedlings to full sun and decreasing the frequency of watering. Forest nursery operators often prune the roots by undercutting and root wrenching their seedlings to produce compact, profusely branched root systems (Chapter 7, Kozlowski and Pallardy, 1997). These treatments also produce temporary water stress.

There likewise is evidence that acclimation responses involving protoplasmic changes are produced that are favorable to survival under water stress. It is well known that drought induces reductions in osmotic potential (i.e., osmotic adjustment), which may lower the Ψ_l associated with stomatal closure (e. g., Parker et al., 1982; Bahari et al., 1985; Guarnaschelli et al., 2003) (Table 12.10). Mooney et al. (1977) indicated that the photosynthetic apparatus of creosote bush growing in a dry habitat is more tolerant of water stress than that of plants in a moist habitat. Matthews and Boyer (1984) demonstrated similar responses in corn.

Membrane properties also appear to respond to water stress preconditioning. For example, Martin et al. (1987) noted that electrolyte leakage from leaves of several species of oak, flowering dogwood, and sugar maple declined noticeably during the growing season, as field-grown trees were subjected to drought (Table 12.11). In contrast, black walnut, which avoids low Ψ_w (Ginter-Whitehouse et al., 1983), showed no reductions in leakage. Gebre and Kuhns (1991) and Kuhns et al. (1993) observed similar reductions in electrolyte leakage with drought in several genotypes of poplar and bur oak, respectively. Kozlowski and Pallardy (2002) reviewed acclimation responses of woody plants to drought and other environmental stresses.

SUMMARY

Transpiration, the loss of water vapor from plants, is a physical process that is under control of both external physical and physiological factors. Solar radiation provides the energy source for transpiration. In general, the rate of transpiration is proportional to the gradient in water vapor concentration between sources of water within the plant and the bulk atmosphere and the total resistance to water vapor diffusion of the plant. Most water loss is from leaves and stomata largely control leaf transpiration. Regulation of stomatal aperture is very complex, as stomata respond to a variety of environmental (e.g., light, humidity, temperature, CO_2 concentration) and endogenous (e.g., root and leaf hormone production and release, age) influences. Other factors that influence transpiration include:

- Leaf area and surface characteristics
- Root–shoot ratio
- Leaf size, shape, and orientation
- Species composition of plant stands
- Silvicultural treatments such as thinning of forest stands

Transpiration varies with changes in controlling factors in a complex fashion, and increases in at least some factors that would be expected to increase transpiration rate (e.g., reduction in boundary layer resistance to water vapor loss) are compensated by other concurrent effects (reduction in leaf temperature and hence vapor concentration gradient).

TABLE 12.11. Species Comparisons of an Index of Electrolyte Leakage (I_d) at Leaf Water Potentials of –3 and –4 MPa for June, July, and August Sample Dates[a,b]

	June I_d		July I_d		August I_d	
Species	–3 MPa	–4 MPa	–3 MPa	–4 MPa	–3 MPa	–4 MPa
White oak	27.7ab	43.6b	17.8ab	28.1b	15.0a	25.1b
Northern red oak	28.6ab	36.3a	9.3a	17.5a	6.7a	15.1a
Black oak	24.0a	33.4a	20.4b	32.2b	8.1a	16.2a
Flowering dogwood	77.0d	88.1c	59.3e	76.2d	9.0a	16.1a
Sugar maple	40.3c	52.1b	45.3d	58.9c	28.7b	41.6c
Black walnut	31.1b	45.1b	32.1c	55.4c	37.2c	51.1c

[a]From Martin et al. (1987).

[b]Within columns, I_d values not followed by the same letter are significantly different ($p < 0.05$).

Relative losses of water vapor by transpiration and evaporation from the soil vary with plant community type and stand density, with communities and stands characterized by sparse cover losing relatively more water by evapotranspiration. Greater evaporative losses in arid areas and understocked forest stands tend to cause lower ratios of dry matter production to total evapotranspiration. Because water often is a limiting resource in plant production, there has been considerable interest in measures of efficiency of water use with regard to photosynthesis and productivity. Water use efficiency generally increases among different photosynthetic types in the following order: $C_3 < C_4 <$ CAM plants, but there also are significant differences among species within each type.

Diurnal and seasonal patterns of water balance of plants depend upon a dynamic interaction of external resources and environment, water absorption and loss patterns of a plant, and redistribution of water within the plant body. Water potential of leaves can vary widely during the day, especially in well-watered plants when solar radiation is variable. As a drought proceeds, plant water potential declines, but the diurnal variation in water potential may be reduced as transpiration is suppressed by stomatal closure. Leaves draw on both internal plant water supplies and soil water as diurnal depression of water potential proceeds, the amount coming from either source being dependent on the amount present and the resistance of the pathway that supplies leaves with water. Although water in stems is released to the transpiration stream, the quantities involved and their rate of release are uncertain; in any case, seasonal water requirements far exceed the amounts stored in nearly all woody plants.

Water stress affects, directly or indirectly, nearly every plant process. Growth is inhibited very early during drought, and water stress has a dramatic impact on physiological processes such as photosynthesis, respiration, carbohydrate metabolism, and protein synthesis. Plants are adapted to drought in a number of ways. Certain herbaceous plants avoid drought by adjustment of their life cycles so that growth and reproduction occur when soil moisture is adequate. Most woody plants are physiologically active during seasons of likely drought and must possess some degree of drought tolerance. In some cases, plants have developed adaptations that allow low water potentials to be avoided or postponed (e.g., deep and extensive root systems, efficient xylem transport, leaf adaptations such as stomatal closure, cuticular development, and abscission). If low water potentials develop, plants may possess dehydration-strain tolerance adaptations that permit maintenance of high relative water content (through osmotic adjustment or high tissue elastic moduli) or have protoplasmic capacity to tolerate removal of water from cells. Although one type of drought tolerance usually dominates in a particular species, these adaptations are not mutually exclusive.

General References

Boyer, J. S. (1995). *Measuring the Water Status of Plants and Soils*. Academic Press, New York.

Flanagan, L. B., Ehleringer, J. R., and Pataki, D. E., eds. (2005). *Stable Isotopes and Biosphere-Atmosphere Interactions: Processes and Biological Controls*. Elsevier Academic, San Diego.

Fry, B. (2006). *Stable Isotope Ecology*. Springer, New York.

Hinckley, T. M., Lassoie, J. P., and Running, S. W. (1978). Temporal and spatial variations in the water status of forest trees. *Forest Science Monograph* **20**, 72 pp.

Kaufmann, M. R. and Landsberg, J. J., eds. (1991). *Advancing toward Closed Models of Forest Ecosystems*. Heron Publishing, Victoria, B. C., Canada.

Kozlowski, T. T., ed. (1968–1983). *Water Deficits and Plant Growth*. Vols. 1–7. Academic Press, New York.

Kozlowski, T. T., Kramer, P. J., and Pallardy, S. G. (1991). *The Physiological Ecology of Woody Plants*. Academic Press, New York.

Kramer, P. J. and Boyer, J. S. (1995). *Water Relations of Plants and Soils*. Academic Press, San Diego.

Kreeb, K. H., Richter, H., and Hinckley, T. M., eds. (1989). *Structural and Functional Responses to Environmental Stresses: Water Shortage*. SPB Academic Publishing, The Hague, The Netherlands.

Lassoie, J. P. and Hinckley, T. M., eds. (1991). *Techniques and Approaches in Forest Tree Ecophysiology*. CRC Press, Boca Raton, Florida.

Levitt, J. (1980). *Responses of Plants to Environmental Stresses*. Volume 2. Water, Radiation, Salt and Other Stresses. Academic Press, New York.

Mooney, H. A., Winner, W. E., and Pell, E. J., eds. (1991). *Response of Plants to Multiple Stresses*. Academic Press, New York.

Nobel, P. S. (2005). *Physicochemical and Environmental Plant Physiology 3rd edition*. Elsevier/Academic Press, San Diego.

Pearcy, R. W., Ehleringer, J., Mooney, H. A., and Rundel, P. W., eds. (1989). *Plant Physiological Ecology: Field Methods and Instrumentation*. Chapman and Hall, New York.

Rundel, P. W., Ehleringer, J. R., and Nagy, K. A., eds. (1989). *Stable Isotopes in Ecological Research*. Springer-Verlag, New York.

Smith, J. A. C. and Griffiths, H., eds. (1993). *Water Deficits: Plant Responses from Cell to Community*. Bios Scientific, Oxford.

CHAPTER

13

Plant Hormones and Other Signaling Molecules

INTRODUCTION

Whereas metabolites such as carbohydrates and N-containing compounds provide the energy and building materials for increase in plant size, the endogenous hormones regulate growth, differentiation, and development of plants at concentrations far below those at which other metabolites affect these processes. Although hormones are synthesized in various parts of woody plants, the primary sources are apical meristems and leaves, as shown by identification of naturally occurring hormones in buds and leaves. Another source of hormones is the root apex, which appears to be a major site of synthesis of cytokinins, gibberellins, and abscisic acid (Little and Savidge, 1987; Cornish and Zeevart, 1985).

The term "hormone" was first used by animal physiologists. They visualized hormones as chemical messengers that were synthesized locally and transported in the blood stream to target tissues where they controlled physiological responses. Auxin, the first plant hormone discovered, was similarly thought to produce a growth response at a distance from its site of synthesis. It is now known, however, that whereas plant hormones are transported in plants and elicit a response at a distance from the source, they also may act in the same tissue or even in the same cell in which they were synthesized (Kozlowski, 1985b). Contemporary notions of plant hormones consider them signal molecules, present in trace quantities, that along with environmental influences mediate plant development (Cleland, 1999; Crozier et al., 2000). There also is recognition that a cell exhibits highly specific responses to hormones depending on its particular developmental history and sensitivity (Osborne and McManus, 2005).

This chapter is an overview of the structure, distribution, and activity of the major groups of plant hormones and some other plant signaling molecules. The regulatory roles of growth hormones on various aspects of vegetative and reproductive growth are discussed in more detail in Chapters 3 and 4 of Kozlowski and Pallardy (1997). The use of synthetic growth regulators to control growth of woody plants is discussed in Chapters 7 and 8 of Kozlowski and Pallardy (1997).

MAJOR CLASSES OF PLANT HORMONES

The five major groups of endogenous plant hormones are the auxins, gibberellins, cytokinins, abscisic acid, and ethylene. These will be discussed separately.

Auxins

The first plant hormones to be identified were the auxins, and for many years they were the only ones known to exist. The discovery of auxin developed from early research on the cause of phototropic bending of plant stems and petioles, which started with the Darwins (1880) and continued until indole-3-acetic acid (IAA) was isolated from fungi in 1934 and from flowering plants in 1946. Auxins usually are defined as substances that in low concentrations (10^{-5} M) stimulate elongation of decapitated oat coleoptiles or segments of pea epicotyl. Although some other naturally occurring indole compounds show a small amount of growth-promoting activity, IAA and IBA (indole-3-butyric acid) seem to be the major naturally occurring auxins. The formula of IAA is shown in Figure 13.1.

The indole-3-acetic acid molecule consists of a benzene ring and an acetic acid side chain linked by a pyrrole ring. Modifications of the benzene ring, the side chain, and a portion of the linking molecule have produced many auxin analogues that have important practical applications in agriculture, horticulture, and forestry (see Chapters 7 and 8, Kozlowski and Pallardy, 1997).

Inputs to the IAA pool in plants include synthesis primarily from the amino acid tryptophan and also from indolylic precursors, hydrolysis of amide and ester conjugates, and translocation from one part of a plant to another. Inactive auxin conjugates that move in the vascular system may be "activated" by enzymatic hydrolysis at unloading zones (Rubery, 1988). Losses of IAA from the pool result from oxidative catabolism, synthesis of conjugates, and use in growth (Reinecke and Bandurski, 1988).

Auxin is readily converted into other compounds, such as indoleacetyl aspartate and glucoside conjugates, and it also is rapidly oxidized by IAA oxidase and peroxidases. Thus, the concentration of free auxin usually is quite low in plant tissue. Auxin also may be bound on proteins in the cytoplasm, often making it difficult to demonstrate a clear correlation between growth and auxin content. Some IAA precursors such as indoleacetaldehyde also exhibit auxin activity. In some plants, other naturally occurring compounds (e.g., indoleacetyl aspartate) show weak auxin activity (Wightman and Lighty, 1982).

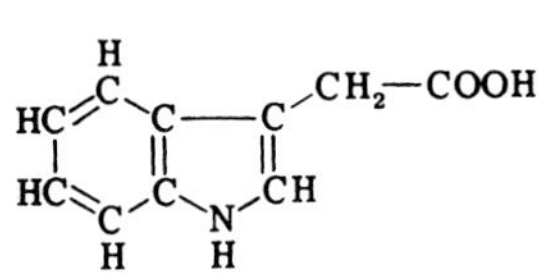

FIGURE 13.1. Structure of indole-3-acetic acid.

A number of synthetic compounds produce effects on plants similar to those produced by naturally occurring auxin. Among them are α-naphthaleneacetic acid (NAA); 2,4,6-trichlorobenzoic acid; and 2,4-dichlorophenoxyacetic acid (2,4-D).

Auxin is involved to different degrees in a wide variety of physiological responses that influence growth of plants (Chapter 3). These include promotion of cell enlargement, cell division, differentiation of vascular tissues, root initiation, apical dominance, phototropism, senescence, abscission of leaves and fruits, fruit set, partitioning of carbohydrates, flowering, and ripening of fruits. Concentrations of auxin higher than those normally found in plant tissues inhibit elongation, and cause abnormal growth such as tumor formation, curling of petioles, and distortion of leaf blades. Still higher concentrations cause death of plants, and this observation led to the use of synthetic auxins as herbicides. In some systems the inhibiting action of high concentrations of auxin is mediated by auxin-induced ethylene production. When synthesis of ethylene is prevented, auxin is no longer inhibitory (Davies, 1988a,b; Cleland, 1999).

Gibberellins

Gibberellins were discovered in Japan before World War II, but did not become known in the West until after that war. They were found by scientists searching for the cause of the *bakanae* (foolish seedling) disease of rice that is characterized by extreme stem elongation and sterility or reduction in yield of rice (Tamura, 1991). The affected plants were infected with the fungus *Gibberella fujikuroi* (known in the imperfect stage as *Fusarium moniliforme*), and an extract from this fungus caused abnormal elongation of rice stems. The active material isolated from the extract was termed *gibberellin*. Research after World War II revealed that gibberellins occur commonly in seed plants and over 70 have been identified. Some examples of gibberellins in various parts of woody plants are given in Table 13.1.

The gibberellins are all diterpenoid acids with the same basic ring structure. Gibberellin A_3, often termed GA_3 for gibberellic acid, is best known. Its formula is shown in Figure 13.2.

The GAs are subdivided into the C_{20} GAs, which contain all 20 carbon atoms and their diterpenoid precursors, and the C_{19} GAs, which have lost one carbon atom (Jones and MacMillan, 1984). Gibberellins commonly exist as conjugates in plants. For example, GA often is linked to a molecule of glucose. The chemical characteristics of gibberellins are discussed in detail by Graebe (1987) and Osborne and McManus (2005), and

TABLE 13.1. Some Gibberellins Found in Woody Plants[a]

Gibberellin	Species
GA_1	*Citrus sinensis*, shoots *Citrus unshiu*, shoots *Corylus avellana*, seeds *Malus domestica*, shoots *Salix pentandra*, shoots *Picea abies*, shoots *Picea sitchensis*, shoots *Pinus radiata*, shoots
GA_3	*Picea abies*, shoots *Picea sitchensis*, shoots *Pinus radiata*, shoots *Pinus attenuata*, pollen
GA_4	*Picea abies*, shoots *Picea sitchensis*, shoots *Pinus radiata*, shoots *Pinus attenuata*, pollen *Pyrus malus*, seeds
GA_7	*Pinus radiata*, shoots *Pyrus malus*, seeds
GA_9	*Picea abies*, shoots *Picea sitchensis*, shoots *Pinus radiata*, shoots *Corylus avellana*, seeds *Pyrus malus*, seeds
GA_{12}	*Pyrus malus*, seeds
GA_{15}	*Picea sitchensis*, shoots *Pinus radiata*, shoots *Pyrus malus*, seeds
GA_{17}	*Citrus unshiu*, shoots *Pyrus communis*, seeds *Pyrus malus*, seeds
GA_{19}	*Citrus unshiu*, shoots *Juglans regia*, shoots *Malus domestica*, shoots *Salix pentandra*, shoots
GA_{20}	*Citrus unshiu*, shoots *Malus domestica*, shoots *Salix pentandra*, shoots *Pyrus malus*, seeds
GA_{25}	*Pyrus communis*, seeds
GA_{29}	*Citrus unshiu*, shoots *Prunus domestica*, fruits
GA_{32}	*Prunus armeniaca*, seeds *Prunus persica*, seeds *Prunus cerasus*, fruits
GA_{45}	*Pyrus communis*, seeds

[a]From "Plant Hormones and Other Growth Substances—Their Background, Structures and Occurrence." (1980). *Encycl. Plant Physiol. N.S.* **9**, 9–112, J. R. Bearder, Table 1.7. and "Gibberellins and the Regulation of Shoot Elongation in Woody Plants." In *Gibberollins* (N. Takahashi, B. O. Phinney, and J. MacMillan, eds.), (1991). pp. 199–210, O. Juntilla, Table 1. © 1980 and 1991 with kind permission of Springer Science and Business Media.

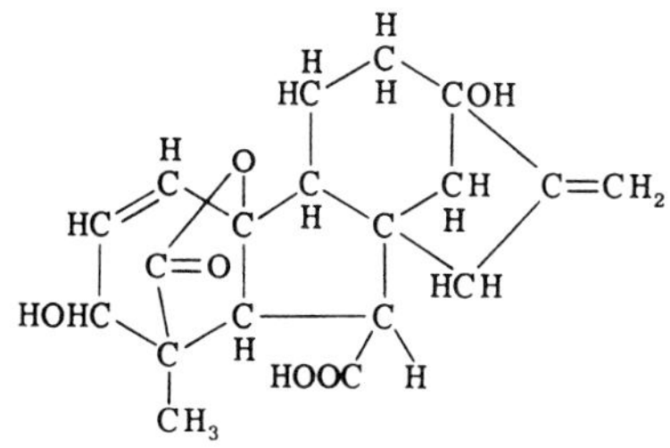

FIGURE 13.2. Structure of gibberellin A_3.

in books edited by Letham et al. (1978), Crozier (1983), Takahashi (1986), Takahashi et al. (1991), Hooykas et al. (1999), and Hedden and Thomas (2006).

In conjunction with other hormones, GAs regulate plant growth. However, a large proportion of the GAs exhibit little or no biological activity, probably because they lack the capacity to fit a receptor molecule. Often many GAs are present in a specific organ or plant but only one may be biologically active (Sponsel, 1988).

Gibberellins influence plant growth at molecular, cellular, organ, and whole-plant levels. Effects at the molecular level include regulation of synthesis of membrane phospholipids, RNA and protein synthesis, cell wall synthesis, cell wall loosening, and hydrolysis of sucrose, proteins, and lipids. At the cellular level, GAs influence division, elongation, and differentiation of cells. At organ and whole-plant levels the GAs influence gravitropism, stem elongation, leaf expansion, initiation of strobili in some gymnosperms, flower initiation, fruit expansion, and seed germination (Graebe, 1987; Brock and Kaufman, 1991; Cleland, 1999; Crozier et al., 2000). An important function of GAs in the hormonal system is the regulation of internode growth. This is evident in dwarfed plants that have short internodes as a result of disturbed GA relations. When treated with GA, the dwarf plants may grow to normal height.

Cytokinins

Skoog and his colleagues at the University of Wisconsin reported in the 1950s that tissue cultures derived from tobacco stem pith enlarged in the presence of auxin but did not divide. Division was induced, however, by adding coconut milk, vascular tissue extracts, autoclaved DNA, or yeast extracts. Subsequently, an active cell division promoter, 6-(furfurylamino) purine ($C_{10}H_9N_5O$), was purified from herring sperm DNA (Miller et al., 1956). The name "kinetin" was given to this compound because it induced cell division (cytokinesis). Later the term "cytokinin" was applied to compounds that promote cell division in the way kinetin does.

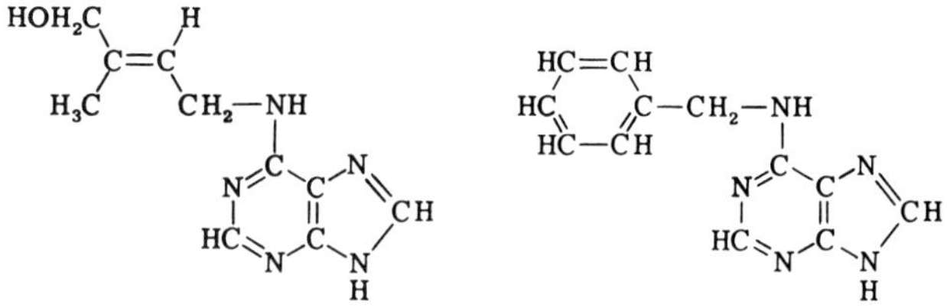

FIGURE 13.3. Structure of zeatin and benzylaminopurine. Used with permission of the American Society of Plant Biologists, from Alvim, R., Hewett, E. W., and Saunders, P. F. (1976). Seasonal variation in the hormone content of willow. I. Changes in abscisic acid content and cytokinin activity in the xylem sap. *Plant Physiol.* **57**, 474–476.; permission conveyed through Copyright Clearance Center, Inc.

Letham (1963) isolated from maize kernels the first naturally occurring cytokinin (zeatin) in higher plants. Since that time a number of cytokinins have been found in various woody plants. They are especially abundant in the milky endosperm tissues of seeds, root tips, xylem sap, and phloem sap, but also are found in leaves and fruits. Roots appear to be major sites of cytokinin synthesis.

Cytokinins are derived from the purine adenine. The structures of zeatin and a synthetic cytokinin, benzylaminopurine (benzyladenine) are given in Figure 13.3.

The amounts of cytokinins in plant tissues and in xylem sap vary with the stage of plant development, season (Fig. 13.4), and environmental conditions. Cytokinin activity in roots is reduced by drought, flooding of soil, low pH, salinity, and high temperature (Letham et al., 1978). Irrigation of droughted coffee trees was followed by increased cytokinin levels in the sap (Browning, 1973).

Both quantitative and qualitative changes in cytokinins occur during leaf development. The amounts of cytokinins were lower in young leaves of ginkgo, willow, and lemon trees than in mature and aging leaves (Van Staden, 1976, 1977; Ilan and Goren, 1979). Cytokinin conjugates also increase during leaf development. For example, cytokinin conjugates of zeatin, ribosylzeatin, and their derivatives were the major cytokinins in mature ginkgo leaves (Van Staden et al., 1983). The amounts of leaf cytokinins could be controlled by (1) partitioning of xylem cytokinins to reproductive tissues and away from the leaf, (2) metabolism in leaves, and (3) changes in amounts of cytokinins in the xylem sap (Singh et al., 1988).

Total cytokinin concentration in the xylem sap of apple trees varied seasonally. It was low from midsummer until late winter. Starting in February several concentration peaks were identified. After trees leafed out, a rapid decline occurred until the original low level was reached in July (Tromp and Ovaa, 1990). The marked decline in cytokinin concentration of the xylem sap was attributed to dilution when the leaves emerged and transpiration increased (Tromp and Ovaa, 1994).

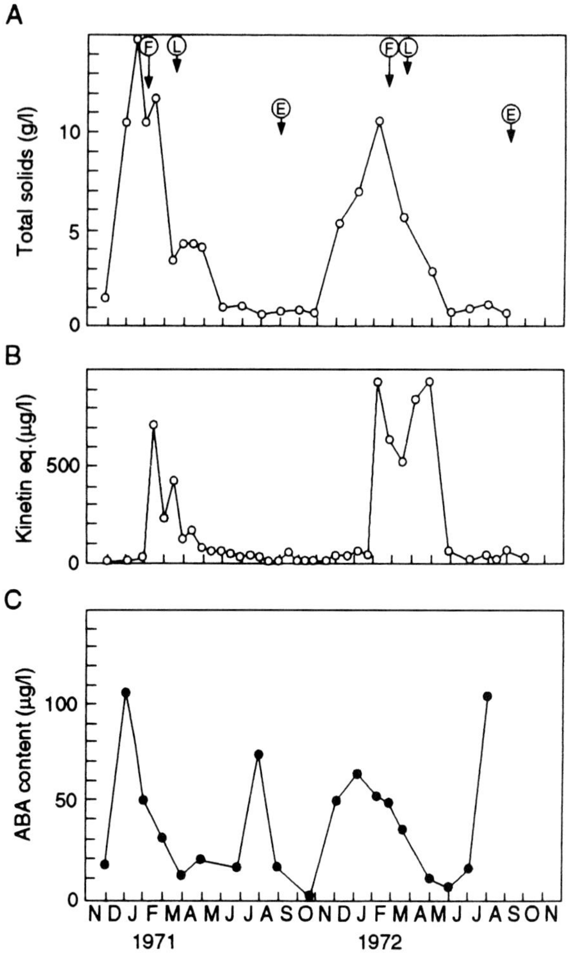

FIGURE 13.4. Seasonal changes in: A, total solids; B, kinetin equivalent; and C, ABA contents in stool shoots of willow; F, flower bud burst; L, leaf bud burst; and E, cessation of extension growth. Used with permission of the American Society of Plant Biologists, from Alvim, R., Hewett, E. W., and Saunders, P. F. (1976). Seasonal variation in the hormone content of willow. I. Changes in abscisic acid content and cytokinin activity in the xylem sap. *Plant Physiol.* **57**, 474–476.; permission conveyed through Copyright Clearance Center, Inc.

Cytokinins have been implicated in several aspects of plant growth and development. Exogenous applications induce cell division in tissue cultures in the presence of auxins, and in crown gall tumors in plants. The occurrence of cytokinins in meristematic tissues (e.g., shoot tips and fruits) suggests that cytokinins naturally control cell division in plants. Cytokinins also regulate cell enlargement; hence they influence leaf expansion. They also are involved in breaking of dormancy of light-sensitive seeds, overcoming apical dominance by releasing inactive lateral buds, inhibiting xylem formation and root growth, delaying senescence, and promoting development of chloroplasts (Letham et al., 1978; Thimann, 1980; McGaw, 1988; Davies and Zhang, 1991; Cleland, 1999).

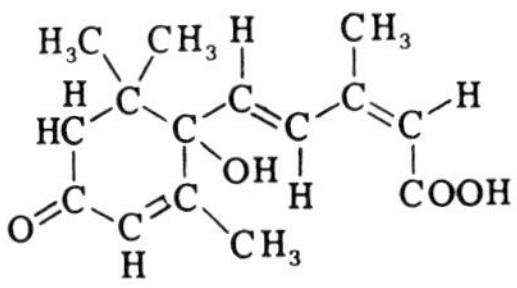

FIGURE 13.5. Structure of abscisic acid.

Abscisic Acid (ABA)

While working on dormancy in trees, Wareing and his colleagues in England isolated a substance from the leaves of sycamore maple that, when placed on stem tips, stopped elongation and caused bud scales to develop (Eagles and Wareing, 1963). The active agent was isolated and called *dormin*. In the United States, Addicott and his colleagues, while investigating leaf abscission, isolated two substances that induced abscission and called them abscisin I and abscisin II (Addicott and Carns, 1983). When purified, abscisin II proved to be identical with dormin and now it is known as abscisic acid or ABA. It is a sesquiterpenoid with mevalonic acid as a precursor. Its formula is shown in Figure 13.5.

Abscisic acid occurs in all seed plants and apparently is synthesized in all plant organs. The amount varies among species and with ontogenetic development. The highest concentrations usually are found in leaves, buds, fruits, and seeds. Abscisic acid commonly exists in varying amounts in both free and conjugated forms. In reproductive organs of citrus the amount of conjugated ABA was four times that of free ABA (Aung et al., 1991). The presence of ABA in both xylem and phloem saps indicates that it is translocated for long distances in both directions.

Like other plant hormones, ABA regulates physiological processes in a variety of plant tissues. Although early research focused on ABA as influencing abscission and dormancy, modern work assigns this hormone a primary role in signaling cell water deficit (Chapter 12) (Osborne and McManus, 2005). For example, ABA sustains root growth under water stress, apparently by suppressing ethylene formation (Spollen et al., 2000). Abscisic acid also is important in seed development (Table 13.2).

Both very rapid and slow responses to ABA have been well documented. An example of a rapid response to ABA is stomatal closure. Endogenous ABA increases rapidly in water-stressed plants, and within minutes export of K^+ ions and loss of guard cell turgor follow, resulting in stomatal closure (Davies et al., 1986). The ABA that originates in the roots can move to leaves in the transpiration stream and induce stomatal closure (Chapter 12). Stomata also can be induced to close by applied ABA (Davies and Kozlowski, 1975a,b). Root drenches of ABA increased the drought tolerance of jack pine seedlings. The seedlings that were pretreated with ABA exhibited rapid stomatal closure, less dehydration, and increased survival when compared with untreated seedlings (Marshall et al., 1991).

TABLE 13.2. **Putative Roles of Endogenous Abscisic Acid in Seeds**[a]

Process	Action	Seed phase
Precocious germination and vivipary	Inhibition	Development
Desiccation tolerance	Promotion	Development
Dry matter accumulation	Inhibition or no effect	Development
Inception of dormancy	Promotion	Development
Maintenance of dormancy	Promotion	Mature
Gene expression		
Reserve proteins	Promotion	Development
Late embryogenesis proteins	Promotion	Development
Dehydration proteins	Promotion	Mature
Reserve mobilizing enzymes	Inhibition	Development Germinated
Cotyledon expansion	Inhibition	Germinated
Cotyledon greening	Inhibition	Mature

[a]From Black (1991).

A relatively slow response (a few hours or longer) to ABA is illustrated in regulation of seed dormancy. Such a slow response is related to cellular differentiation, with ABA usually affecting synthesis of nucleic acid and/or protein synthesis (Ho, 1983). In general, application of ABA to plants counteracts the simultaneous applications of IAA, GA, and cytokinin, but there are exceptions. Often ABA increases plant responses to ethylene.

Ethylene

Ethylene gas ($CH_2{=}CH_2$), the structurally simplest plant growth hormone, is an important component of the hormonal complex that controls growth and development of plants. Ethylene is synthesized primarily as follows:

Methionine → S-Adenosylmethionine (SAM) →
1-Aminocyclopropanecarboxylic acid (ACC) →
Ethylene

Conversion of ACC to ethylene is O_2 dependent. The formation of ACC from SAM by the enzyme ACC synthase, and conversion of ACC to ethylene by the ethylene forming enzyme (EFE), also called ACC oxidase, are crucial steps in regulation of ethylene production. Application of ACC to plant tissues that normally produce little ethylene often dramatically increases ethylene synthesis. This indicates that the rate-limiting step in the pathway is conversion of SAM to ACC. Nonphysiological ethylene production from lipid peroxidation also has been reported. Application of olive oil to developing figs accelerates their development, presumably by the action of ethylene derived from lipid peroxidation (Saad et al., 1969).

Ethylene production is variously influenced by other plant hormones (Abeles et al., 1992). Promotion of ethylene biosynthesis by endogenous auxin is particularly well known. Auxin increases ethylene production in the 100-fold range whereas ABA commonly only doubles it or may even suppress it (e.g., Spollen et al., 2000). Cytokinins also increase ethylene production, usually by about two to four times. Generally the stimulation of ethylene production by combined auxins and cytokinins is greater than by either hormone alone. Gibberellins have only small and variable effects on ethylene production. The effects of hormones on ethylene synthesis are mediated by effects on levels of enzymes in the ethylene production pathway.

Several applied compounds, including ABA, NAA, 2,4-D, and picloram influence ethylene production. However, caution is advised in interpreting results that depend greatly on the concentration of the applied compounds. For example, 10^{-5} M ABA stimulated ethylene production whereas higher concentrations inhibited it (Kondo et al., 1975).

Ethylene is widely distributed in plants. In stems of woody plants the amount differs between the sapwood and heartwood, and it also varies seasonally. In Scotch pine stems ethylene in the sapwood rose to 3 to 7 ppm during the growing season and decreased to 0.1 to 0.3 ppm during the winter (Fig. 13.6). The amount of ethylene in the heartwood was consistently lower than 2 ppm.

Many plant tissues, which normally produce little or no ethylene, synthesize a very large amount (often up to 10 times the normal amount) when exposed to a variety of environmental stresses (Kimmerer and

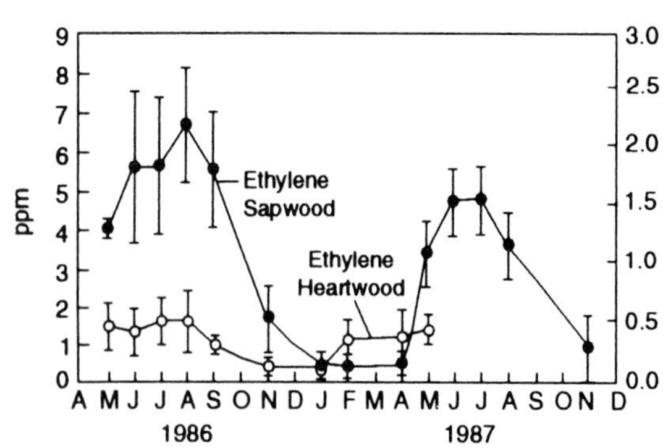

FIGURE 13.6. Seasonal variations in ethylene concentration in the sapwood and heartwood of 70- to 100-year-old Scotch pine trees. From Ingemarsson et al. (1991).

Kozlowski, 1982). Surges of ethylene production have been demonstrated in plants following wounding, chilling, drought, exposure to air pollution, mechanical perturbation, and attack by microorganisms. Some fungi produce large amounts of ethylene as do the fungus-invaded tissues of higher plants. Stress ethylene is produced in intact but physiologically affected cells such as those adjacent to injured tissues (Abeles et al., 1992).

Measurement of stress ethylene often has been used to indicate the onset and degree of plant stress (Chen and Patterson, 1985). However, the use of stress ethylene as a diagnostic tool has some limitations. As mentioned, ethylene is produced by unstressed plants, the amount varying with environmental conditions and age of tissues. When stress ethylene results in cell mortality, ethylene evolution declines. Hence, correlation between environmental stress and ethylene production sometimes is poor.

Increased amounts of ethylene in flooded plants apparently are caused by prevention of ethylene escape from the roots by the surrounding water. When the soil is well aerated, plant-produced ethylene escapes into the soil. Because ethylene is not very soluble in water, the concentration builds up in roots in flooded soil and ethylene moves to the shoots, producing unusually high concentrations there. Ethylene diffuses about as rapidly as CO_2 does and it therefore moves readily through air spaces in plant tissues. Although the concentration of ethylene is increased in shoots of flooded plants, the rate of ethylene production is decreased because of O_2 deficiency in plant tissues. Hence, the ethylene increase in flooded plants is unlike the surge of ethylene production in tissues of plants exposed to other environmental stresses (Yang and Hoffman, 1984).

Influences of ethylene have been demonstrated in many plant responses, including seed germination, abscission, apical dominance, branch angle, bud growth, epinasty, hypertrophy of tissues, latex flow, root growth, stem elongation, flowering, and ripening of fruits. Many enzymes are regulated by ethylene, including those involved in abscission (cellulase and polygalacturonase), aerenchyma formation (cellulase), fruit ripening (cellulase, chlorophyllase, invertase, laccase, malate dehydrogenase, and polygalacturonase), and senescence (ribonuclease) (Abeles, 1985; Cleland, 1999). Ethylene originally was of interest to plant physiologists studying the ripening and storage of fruits because it is produced by ripening fruits (Chapter 6, Kozlowski and Pallardy, 1997). The ethylene concentration must be kept low in cold storage rooms to prevent hastening the ripening of fruits and defoliation of nursery stock.

As far as is known all plant cells continuously produce some ethylene. The capacity of ethylene to function as a growth regulator depends on a change in sensitivity of cells to the ethylene already present (e.g., abscission), or a response caused by a change in the amount of ethylene produced by the tissue (e.g., wound-induced protein synthesis) (Abeles, 1985).

In order to act, ethylene must initially bind to a receptor. Ethylene interacts with several other growth hormones and some ethylene effects involve the action of other hormones. Ethylene inhibits polar auxin transport but IAA stimulates ethylene production. Hence, ethylene may lower the amount of effective IAA, which may decrease the amount of ethylene produced. However, the major interactive effect of IAA and ethylene may be a decrease in tissue sensitivity to ethylene. Other plant hormones also may regulate ethylene action by modifying the sensitivity of tissues to ethylene (Sexton and Roberts, 1982).

OTHER REGULATORY COMPOUNDS

In addition to the major classes of plant hormones, a number of other endogenous compounds influence growth and development of woody plants. These compounds represent important signaling components for plant growth and development, but generally are not included in the traditional grouping of hormones discussed previously.

Brassinosteroids

The brassinosteroids (BRs) are a group of steroidal plant hormones that appear to regulate growth. The first BR, brassinolide, a plant steroid lactone, was isolated from the pollen of rape (*Brassica napus*) in 1970. The structure of brassinolide is shown in Figure 13.7. Since brassinolide was extracted and characterized, more than 60 kinds of BRs have been verified and more than 30 fully characterized. Brassinosteroids have been found in pollen, leaves, flowers, shoots, stems, and insect galls in dicots, monocots, and gymnosperms.

Important processes influenced by BRs include vascular differentiation, reproductive development, and pathogen and abiotic and biotic stress tolerance (Ross et al., 2006). Brassinosteroids, which act at very low concentrations (10^{-9} M), regulate cell division and elongation (Friedrichsen and Chory, 2001). Enhanced growth in woody plants following application of BR compounds has been demonstrated for woody plants (Ono et al., 2000). The mechanisms by which BRs influence growth are several. These growth regulators

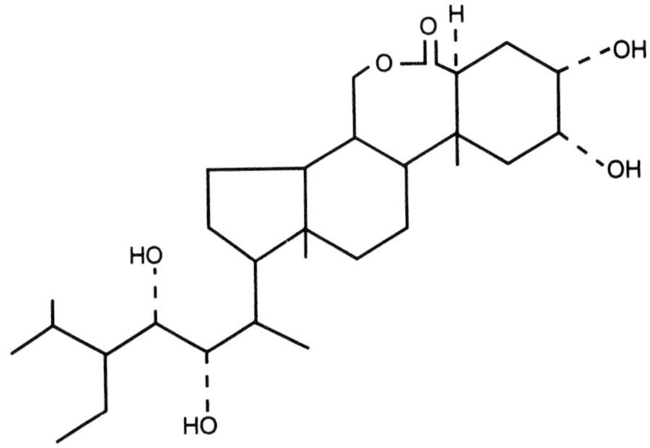

FIGURE 13.7. Structure of brassinolide.

induce gene expression for two types of enzymes that are involved with cell wall modification and extension. As with auxin, BR induces xyloglucan endotransglycosylase that may be involved in cell wall modifications and synthesis in elongating cells. Brassinosteroids also induce endo-1,4-β-glucanases, enzymes that hydrolyze xyloglucan within developing cell walls, and expression of a gene that codes for extracellular invertase. The latter enzyme makes sucrose imported into growing regions available for cell wall synthesis, and possibly increases the sink strength of the region by lowering its sucrose content (Goetz et al., 2000).

There also is good evidence that BRs are involved in differentiation of xylem vascular elements. For example, certain BR-biosynthesis mutants of *Arabidopsis* show reduced vascular cell size, which can be reversed by BR application and also exhibit abnormal bundle development (Choe et al., 1999). Brassinosteroid compounds increase in abundance prior to tracheary element morphogenesis in zinnia cultures. The requirement for BRs appears specific to the final stage of xylem element differentiation, which involves secondary wall thickening and cell death (Yamamoto et al., 2001).

In addition to their effects on growth and differentiation, BRs inhibit abscission of leaves and fruits (Iwahari et al., 1990; Watanabe et al., 1998) and arrest development of adventitious roots (Roddick and Guan, 1991). An important property of BRs is their capacity to increase resistance of plants to various stresses such as cold, fungal infection, herbicide injury, and heat shock (Kim, 1991; Sakurai and Fujioka, 1993; Roth et al., 2000).

The BRs exhibit strong interactions with other endogenous plant hormones and through these interactions regulate plant growth and development (Mandava, 1988; Sakurai and Fujioka, 1993; Friedrichsen and Chory, 2001). Brassinosteroids interact very strongly with auxins, presumably synergistically. When BRs are applied to plants, alone or together with auxins, they stimulate synthesis of ethylene. Abscisic acid also interacts strongly with BRs and prevents effects induced by BRs. By comparison, plant responses to BRs and gibberellins appear to be independent and additive.

Jasmonates

In recent years a wide variety of plant responses have been attributed to the jasmonates, compounds consisting of a cyclopentenone ring variously substituted at several positions. Jasmonates are widely distributed in both angiosperms and gymnosperms, and show highest activity in stem apices, young leaves, root tips, and immature fruits (Sembdner and Parthier, 1993).

Jasmonates regulate leaf senescence (associated with breakdown of chlorophyll, degradation and inhibition of ribulose bisphosphate carboxylase, acceleration of respiration, and inhibition of photosynthesis). They also stimulate the synthesis of certain proteins and play a role in seed dormancy. There is evidence for positive control of jasmonate-induced gene expression at the transcription level as well as negative post-transcriptional control. Some effects of jasmonates resemble those of ABA. As is the case with ABA, formation of jasmonates is induced by wounding, elicitation of phytoalexins, by oligosaccharides, water deficits, or osmotically active compounds (e.g., sorbitol, or mannitol) (Sembdner and Parthier, 1993).

Salicylic Acid

Salycylic acid (SA) is produced in plants by decarboxylation of trans-cinnamic acid to benzoic acid and then addition of a hydroxyl group on the benzene ring structure (Osborne and McManus, 2005). Its structure is shown in Figure 13.8. Salicylic acid plays a key role in flowering of *Arum* lilies by regulating thermogenesis in the flower spadix (Raskin et al., 1987). Purification of the thermogenically active fraction of extracts from floral tissue in voodoo lilly (*Sauromatum guttatum*) indicated that SA was the compound responsible for induction of temperature elevation. Application of 15 μm SA could reproduce the same level of temperature increase as crude extracts. There also is evidence that SA plays a role in leaf senescence (Morris et al., 2000), and can retard senescence in flower petals and induce flowering (Crozier et al., 2000).

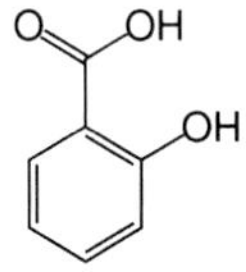

FIGURE 13.8. Structure of salicylic acid.

Salicylic acid has generated interest because of its participation in the disease-resistance mechanisms of plants. Application of SA enhanced the disease resistance of susceptible tobacco plants to tobacco mosaic virus (Crozier et al., 2000). Transgenic tobacco and *Arabidopsis thaliana* plants engineered to block the synthesis of SA were more susceptible to viral, fungal, and bacterial pathogens (Gaffney et al., 1993; Delaney et al., 1994).

Phenolic Compounds

Plants accumulate a wide variety of phenolic compounds (Table 13.3), with phenolic acids and their derivatives the major phenolics in most plants. The

TABLE 13.3. Some Phenolic Compounds Found in Plants[a]

Group	Examples
Simple phenols	Phenol, catechol, hydroxyquinone, phloroglucinol, pyrogallol
Phenolic and benzoic acids	*p*-Hydroxybenzoic, protocatechuic, vanillic, gallic, syringic salicylic, *O*-pyrocatechuic, gentisic
Cinnamic acids	*p*-Coumaric, cinnamic, caffeic, ferulic, sinapic
Acetophenone acids	2-Hydroxyacetophenone, 4-hydroxyacetophenone
Phenylacetic acids	2-Hyroxyphenylacetic acid, 4-hydroxyphenylacetic acid
Coumarins	Umbelliferone, coumarin, bergenin
Flavones	Apigenin, luteolin, tricin
Flavonones	Pinocembrin, naringenin, eridictyol
Isoflavones	Genestein, daidzin, formononetin, coumestrol, biochanin A
Flavonols	Kaempferol, quercetin, myrcetin
Anthocyanins	Pelargonidin, cyanidin, petunidin
Chalcones	Butein, phloretin, methoxychalcone, chrysoeriol
Quinones	Dimethoxybenzoquinone, anthroquinones
Miscellaneous	Biflavonyls, betacyanins, lignin, tannin

[a]Reprinted with permission from Siqueira, J. O., Nair, M. G., Hammerschmidt, R., and Safir, C. R. (1991). Significance of phenolic compounds in plant-soil microbial systems. *Crit Rev Plant Sci*, **10**, 63–121. Copyright CRC Press, Boca Raton, Florida.

majority of the phenolic compounds are synthesized by the shikimate and acetate pathways.

Phenolic compounds, which comprise a large group of endogenous growth inhibitors, can affect plants directly by their involvement in metabolism and function (e.g., in mitosis, nucleic acid and protein metabolism, respiration, photosynthesis, carbohydrate metabolism, membrane function, and hormone physiology), growth and development (e.g., seed germination, root initiation and elongation, leaf expansion, and flowering), and ecological functions (e.g., as allelopathic agents, antimicrobial agents, and protective agents against herbivory) (Siqueira et al., 1991).

The phenolic acids affect plant growth at concentrations many (often 100) times higher than hormones do. They are important for their interactions with hormones, rather than because of their direct action. The two series of phenolic acids, the benzoic series (e.g., *para*-hydroxybenzoic acid and protocatechuic acid) and cinnamic acid series (e.g., cinnamic acid and *para*-coumaric acid) are the most common. Phenolic acids greatly reduce or increase the activity of auxins depending on the specific phenolic acid involved (Thimann, 1977).

Certain naturally occurring aromatic aldehydes (e.g., benzaldehyde and salicylaldehyde) also regulate plant growth. Some flavonones such as naringerin antagonize the action of gibberellins (Phillips, 1962).

Polyamines

The polyamines are universal cell constituents with a wide variety of potentially important functions in plants. Important polyamines include putrescine (H_3N^+-$[CH_2]_4$-NH_3^+), spermidine (H_3N^+-$[CH_2]_3$-NH_2^+-$[CH_2]_4$-NH_3^+), and spermine [H_3N^+-$[CH_2]_3$-NH_2^+-$[CH_2]_4$-NH_2^+-$[CH_2]_3$-NH_3^+). Polyamines often conjugate with other compounds. In most cases they are ionized and associated with such macromolecules as DNA, RNA, phospholipids, and certain proteins. They also are bound to ribosomes (Faust and Wang, 1992). Putrescine is formed directly from ornithine or indirectly through a series of intermediate compounds (Fig. 13.9). Spermidine and spermine are synthesized from putrescine by the addition of aminopropyl groups.

Polyamines have been variously implicated in regulation of cell division, embryogenesis, pollen formation, fruit development, breaking of dormancy, internode elongation, root formation, senescence, and protection against environmental stresses (Galston, 1983; Faust and Wang, 1992; Srivistava, 2001). The postulated role of polyamines is based on their ubiquitous distribution in plant cells, their changes in response to stimuli such as light, growth hormones, and environmental stresses (e.g., drought, low temperature, mineral deficiency); and their effects on plant morphogenesis and growth when applied exogenously. It has been claimed that where polyamine gradients occur in plants, an obligate relation exists between polyamine synthesis and plant growth (Galston, 1983; Smith, 1985).

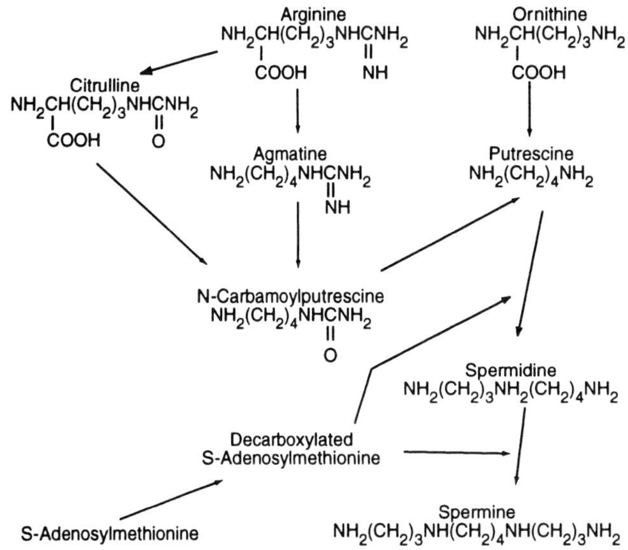

FIGURE 13.9. Biosynthesis of putrescine and the polyamines. From Smith (1985). Reproduced with permission from the *Annual Review of Plant Physiology and Plant Molecular Biology*, Volume 36, © 1985, by Annual Reviews, Inc.

The biological function of polyamines appears to be due to their cationic nature (Altman et al., 1982), and their electrostatic interactions with polyanionic nucleic acids and negatively charged functional groups of membranes, enzymes, or structural proteins in cells (Slocum et al., 1984). Much of the research on polyamines emphasized correlative evidence, and the precise role in regulating plant growth and development has not been firmly established. Evans and Malmberg (1989) concluded that unequivocal evidence is lacking to support the role of polyamines as hormones, second messengers, or other growth regulators. Crozier et al. (2000) suggested that they do not have a hormonal role in plants, but rather participate in certain key metabolic pathways for efficient cell functioning.

Other Compounds

Numerous other compounds have been proposed as signaling molecules with varying amounts of support. In many cases, potential regulatory activity only recently has been reported and much more research will be required before their general distribution in plants and importance to plant growth can be ascertained. These compounds include nitric oxide, methyl jasmonate and methyl salicylate, oligosaccharins, lignans, lipids, and several types of peptides. Refer to the recent text by Osborne and McManus (2005) and the edited volume by Hedden and Thomas (2006) for discussions of these compounds.

MECHANISMS OF HORMONE ACTION

The ways in which endogenous plant hormones influence plant growth are elusive, to a large extent because, unlike animal hormones, plant hormones do not have specific targets. Plant hormones influence many processes and conditions, including enzymatic activity, membrane permeability, relaxation of cell walls, cell division and elongation, and senescence of tissues and organs. Voluminous evidence indicates that growth and development of plants are controlled more by hormonal interactions than by individual hormones, and the relative concentration may be more important than the concentration of any specific hormone.

Each of the major classes of plant hormones has been implicated in regulation of specific processes at the cellular, tissue, and plant levels. In early studies, very distinct functions were assigned to each of the major plant hormones. For example, regulation of cell enlargement was assigned to auxin; of stem growth to gibberellin; of cell division to cytokinin; and of fruit ripening to ethylene. As research progressed it became evident that each of these processes could not be adequately explained as a response to a single hormone. Hence, attention has progressively shifted to regulatory effects of hormone interactions, and four general types of interactions have been identified (Leopold and Nooden, 1984). Hormonal regulation may be achieved by: (1) a balance or ratio between hormones, (2) opposing effects between hormones, (3) alterations of the effective concentration of one hormone by another, and (4) sequential actions of different hormones. Specific examples of each of these types of hormonal control are given by Leopold and Nooden (1984).

In order to have an effect on target cells, a plant hormone signal molecule must bind, if only briefly, to a target site. The resulting hormone-binding site complex alters some biochemical attribute, such as phosphorylation of the receptor (a primary effect), which may then lead to a secondary effect via signal transduction networks:

$$\text{Hormone (H)} + \text{receptor (R)} \rightarrow (\text{H} \cdot \text{R}) \rightarrow \text{biochemical response} \rightarrow \text{signal transduction network}$$

Soluble hormone-binding proteins (receptors) bind chemical signals without altering them chemically (Libbenga and Mennes, 1988). On binding, a receptor molecule undergoes a conformational change and enters an active state. This change activates downstream metabolic changes via a signaling network, leading to a plant response. Although different target cells may have similar methods of perception and induction, their responses to the same signal often differ because of variations among them in their signaling network composition and states. An ethylene receptor has been identified in plants and auxin binding proteins have been isolated, but the latter have not been definitively identified as receptors for auxin-mediated processes (Trewavas, 2000). The framework and function of signal transduction networks of plants is just beginning to receive attention and the reader is referred to Trewavas (2000), Srivistava (2001), and Hedden and Thomas (2006) for further information.

Hormones may initiate biological effects in sensitive tissues through "second messengers" (internal molecules whose influence on cell metabolism depends on external stimuli such as light or hormones). Second messengers may or may not fit the definition of a hormone. For example, exposure of plants to mechanical stress stimulates formation of ethylene, which may become a second messenger. In plants undergoing water deficits, ABA may be a second messenger (Blowers and Trewavas, 1989). Calcium, which occupies a unique position as a second messenger, may modify the functions of each of the five major classes of plant hormones, sometimes increasing a response to hormones and at other times suppressing it. Ca^{2+} influences many physiological processes including auxin-induced elongation, auxin binding, auxin transport, abscission, senescence, ripening, cell division, membrane function, freezing injury, and photosynthesis (Poovaiah and Reddy, 1987).

External signals alter cytosolic levels of Ca^{2+} in plants, leading to cellular responses. Usually the specific triggering signal is not Ca^{2+} but rather a complex between Ca^{2+} and some Ca-binding protein that undergoes a conformational change, leading to modification of its capacity to interact with other proteins and alter their function. The best known widely distributed Ca^{2+}-binding protein is calmodulin. The Ca^{2+}-calmodulin protein complex may act directly on an effector system or indirectly on a regulatory system, generally a protein kinase that stimulates or inhibits the activity of other enzymes. The other enzymes that are regulated by Ca^{2+} and calmodulin include NAD kinase, Ca^{2+}-ATPase, H^+-ATPase, and quinate:NAD oxidase reductase (Poovaiah and Reddy, 1987).

SUMMARY

The major classes of naturally occurring plant growth hormones (including auxins, gibberellins, cytokinins, abscisic acid, and ethylene) regulate plant growth and development at concentrations much lower than those at which nutrients affect plant processes and growth. In addition, a number of other endogenous compounds, including aromatic compounds, N-containing compounds, terpenoids, and aliphatic compounds, also influence growth, often by interacting with hormones.

The mechanisms of action of plant hormones are complex and not fully understood, largely because plant hormones do not have specific targets. Whereas specific and distinct roles in regulation of plant growth and development originally were applied to individual plant hormones, the current view is that plant development is regulated by hormonal interactions. Regulation by hormones may be achieved by (1) a ratio or balance between hormones, (2) opposing effects of hormones, (3) alterations of the effective concentration of one hormone by another, and (4) sequential actions of different hormones.

To affect target cells plant hormone signals must bind to a target site. The hormone-binding site complex induces a primary effect by altering a biochemical process and setting in motion a signal transduction network. On binding, a receptor (protein) molecule enters an active state, thereby activating a set of enzymes, which leads to a plant response. Plant hormones also may act by changing membrane properties. Plant hormones may initiate biological effects in hormone-sensitive tissues through "second messengers" that may or may not be hormones.

In addition to the major classes of hormones plants accumulate other compounds, such as brassinosteroids, jasmonates, salicylic acid, phenolics, and polyamines, that influence plant growth and development. Brassinosteroids primarily affect plant growth and development. Jasmonates produce a variety of effects, at least some of which appear to resemble those of ABA. Salicylic acid plays a role in thermogenesis in *Arum* flowers as well as in senescence and disease resistance in plants. The major phenolic compounds are important for their interactions with hormones rather than for their direct action. The postulated role of polyamines in growth regulation is based on their wide distribution in cells, their changes in response to environmental changes, and effects on growth when applied exogenously. Several other compounds currently are under study as potential signaling molecules in plants.

General References

Abeles, F. B., Morgan, P. W., and Saltveit, M. E., Jr. (1992). *Ethylene in Plant Biology, 2nd ed.* Academic Press, San Diego.

Bosse, W. F. and Morré, D. V., eds. (1989). *Second Messengers in Plant Growth and Development*. Liss, New York.

Chadwick, C. M. and Garrod, D. R., eds. (1986). *Hormones, Receptors, and Cellular Interactions in Plants*. Cambridge Univ. Press, Cambridge.

Cutler, H. G., Yokota, T., and Adam, G., eds. (1991). *Brassinosteroids*. ACS Symposium Series No. 474. American Chemical Society, Washington, D.C.

Davies, P. J., ed. (2004). *Plant Hormones: Biosynthesis, Signal Transduction, Action!* Kluwer, Dordrecht.

Davies, W. J. and Jones, H. G., eds. (1991). *Abscisic Acid*. Information Press, Oxford.

Faust, M. and Wang, S. Y. (1992). Polyamines in horticulturally important plants. *Plant Breeding Rev.* **14**, 333–356.

Flores, H. E., Arteca, R. N., and Shannon, J. C., eds. (1990). *Polyamines and Ethylene: Biochemistry, Physiology, and Interactions*. Amer. Society of Plant Physiologists, Rockville, Maryland.

Fuchs, Y. and Chalutz, F., eds. (1984). *Ethylene. Biochemical, Physiological and Applied Aspects*. Martinus Nijhoff, Dr. W. Junk, The Hague.

Hedden, P. and Thomas, S. G., eds. (2006). *Plant Hormone Signaling*. Blackwell Publishers, Oxford.

Hooley, R. (1994). Gibberellins: Perception, transduction and responses. *Plant Mol Biol* **26**, 1529–1555.

Hooykaas, P. J. J., Hall, M. A., and Libbenga, K. R., eds. (1999). *Biochemistry and Molecular Biology of Plant Hormones*. Elsevier Science, Amsterdam.

Kleczkowski, K. and Schell, J. (1995). Phytohormone conjugates: Nature and function. *Crit Rev Plant Sci* **14**, 283–298.

Kossuth, S. and Ross, S. D., eds. (1987). Hormonal control of plant growth. *Plant Growth Regul.* **6**, 1–215.

Mok, D. W. S. and Mok, M. C., eds. (1994). *Cytokinins, Chemistry, Activity, and Function*. CRC Press, Boca Raton, Florida.

Mattoo, A. K. and Suttle, J. C., eds. (1991). *The Plant Hormone Ethylene*. CRC Press, Boca Raton, Florida.

Sakurai, A. and Fujioka, S. (1993). The current status of physiology and biochemistry of brassinosteroids. *Plant Growth Regul* **13**, 147–159.

Slocum, R. D. and Flores, H. D., eds. (1991). *Biochemistry and Physiology of Polyamines in Plants*. CRC Press, Boca Raton, Florida.

Smith, T. A. (1985). Polyamines. *Annu Rev Plant Physiol* **36**, 117–143.

Srivistava, L. M. (2001). *Plant Growth and Development: Hormones and Environment*. Elsevier–Academic Press, New York.

Stafford, H. A. and Ibrahim, R. K., eds. (1992). *Phenolic Metabolism in Plants*. Plenum, New York.

Takahashi, N., Phinney, B. O., and MacMillan, J., eds. (1991). *Gibberellins*. Springer-Verlag, New York and Berlin.

Bibliography

Abeles, F. B. (1985). Ethylene and plant development: An introduction. In *Ethylene and Plant Development* (J. A. Roberts and G. A. Tucker, eds.), pp. 1–8. Butterworth, London.

Abeles, F. B., Morgan, P. W., and Saltveit, M. E., Jr. (1992). *Ethylene in Plant Biology*. Academic Press, San Diego.

Aber, J. D. and Melillo, J. M. (1980). Litter decomposition: Measuring relative contributions of organic matter and nitrogen to forest soils. *Can. J. Bot.* **58**, 416–421.

Aber, J. D., Nadelhoffer, K. J., Steudler, P., and Mellilo, J. M. (1989). Nitrogen saturation in northern forest ecosystems. *BioScience* **39**, 378–386.

Aber, J. D., Neilson, R. P., McNulty, S., Lenihan, J. M., Bachelet, D., and Drapek, R. J. (2001). Forest processes and global environmental change: Predicting the effects of individual and multiple stressors. *Bioscience* **51**, 735–751.

Abrams, M. D. (1988). Comparative plant and tissue water relations of three successional hardwood species in central Wisconsin. *Tree Physiol.* **4**, 263–273.

Abrams, M. D. (1990). Adaptations and responses to drought in *Quercus* species of North America. *Tree Physiol.* **7**, 227–238.

Abrams, M. D. and Knapp, A. K. (1986). Seasonal water relations of three gallery forest hardwood species in northeast Kansas. *Forest Sci.* **32**, 687–696.

Abrams, M. D. and Mostoller, S. A. (1995). Gas exchange, leaf structure and nitrogen in contrasting successional tree species growing in open and understory sites during a drought. *Tree Physiol.* **15**, 361–370.

Ackerman, E. and Loff, G. (1959). *Technology in American Water Development*. Johns Hopkins Press, Baltimore, Maryland.

Adams, M. A. and Attiwill, P. M. (1986a). Nutrient cycling and nitrogen mineralization in eucalypt forests of south-eastern Australia. II. Indices of nitrogen mineralization. *Plant Soil* **92**, 341–367.

Adams, M. A. and Attiwill, P. M. (1986b). Nutrient cycling and nitrogen mineralization in eucalypt forests of south-eastern Australia. I. Nutrient cycling and nitrogen turnover. *Plant Soil* **92**, 319–339.

Adams, M. S. and Strain, B. R. (1969). Seasonal photosynthetic rates in stems of *Cercidium floridum* Benth. *Photosynthetica* **3**, 55–62.

Adams, W. T. (1983). Application of isozymes in tree breeding. In *Isozymes in Plant Genetics and Breeding* (S. D. Tanksley and T. J. Orton, eds.), Part A, pp. 381–400. Elsevier, Amsterdam and New York.

Addicott, F. T. (1970). Plant hormones in the control of abscission. *Biol. Rev. Cambridge Philos. Soc.* **45**, 485–524.

Addicott, F. T. (1982). *Abscission*. University of California Press, Berkeley, California.

Addicott, F. T. (1991). Abscission: Shedding of parts. In *Physiology of Trees* (A. S. Raghavendra, ed.), pp. 273–300. Wiley, New York.

Addicott, F. T. and Carns, H. R. (1983). History and introduction. In *Abscisic Acid* (F. T. Addicott, ed.), pp. 1–21. Praeger, New York.

Ågren, G. I. and Bosatta, E. (1988). Nitrogen saturation of terrestrial ecosystems. *Environ. Pollut.* **54**, 185–197.

Ahuja, M. R. (2001). Recent advances in molecular genetics of forest trees. *Euphytica* **121**, 173–195.

Ainsworth, E. A. and Long, S. P. (2005). What have we learned from 15 years of free-air CO_2 enrichment (FACE)? A meta-analytic review of the responses of photosynthesis, canopy properties and plant production to rising CO_2. *New Phytol.* **165**, 351–371.

Akinyemiju, O. A. and Dickmann, D. I. (1982). Contrasting effects of simazine on the photosynthetic physiology and leaf morphology of two *Populus* clones. *Physiol. Plant* **35**, 402–406.

Akkermans, A. D., Abdukadir, S., and Trinick, M. J. (1978). Nitrogen-fixing root nodules in Ulmaceae. *Nature (London)* **274**, 190.

Alberte, R. S., McClure, P. R., and Thornber, J. P. (1976). Photosynthesis in trees. Organization of chlorophyll and photosynthetic unit size in isolated gymnosperm chloroplasts. *Plant Physiol.* **58**, 341–344.

Alfieri, F. J. and Evert, R. F. (1968). Seasonal development of the secondary phloem in *Pinus*. *Am. J. Bot.* **55**, 518–528.

Alfieri, F. J. and Evert, R. F. (1973). Structure and seasonal development of the secondary phloem in the Pinaceae. *Bot. Gaz.* **134**, 17–25.

Allard, G., Nelson, C. J., and Pallardy, S. G. (1991). Shade effects on growth of tall fescue. 1. Leaf anatomy and dry matter partitioning. *Crop Sci.* **31**, 163–167.

Allen, E. B. and Allen, M. F. (1986). Water relations of xeric grasses in the field: Interactions of mycorrhizas and competition. *New Phytol.* **104**, 559–571.

Allen, R. M. (1955). Foliage treatments improve survival of longleaf pine plantings. *J. For.* **53**, 724–727.

Allona, I., Quinn, M., Shoop, E., Swope, K., Stcyr, S., Carlis, J., Riedl, J., Retzel, E., Campbell, M. M., Sederoff, R., and Whetten, R. W. (1998). Analysis of xylem formation in pine by cDNA sequencing. *Proc. Nat. Acad. Sci. U. S. A.* **95**, 9693–9698.

Allsopp, A. and Misra, P. (1940). The constitution of the cambium, the new wood, and mature sapwood of the common ash, the common elm and the Scotch pine. *Biochem. J.* **34**, 1078–1084.

Aloni, R. and Zimmermann, M. H. (1983). The control of vessel size and density along the plant axis—A new hypothesis. *Differentiation* **24**, 203–205.

Altman, A., Friedman, R., Amir, D., and Levin, N. (1982). Polyamine effects and metabolism in plants under stress conditions. In *Plant Growth Substances* (P. F. Wareing, ed.), pp. 483–494. Academic Press, London.

Alvim, R., Hewett, E. W., and Saunders, P. F. (1976). Seasonal variation in the hormone content of willow. I. Changes in abscisic acid content and cytokinin activity in the xylem sap. *Plant Physiol.* **57**, 474–476.

Alway, F. J., Kittredge, J., and Methley, W. J. (1933). Components of the forest floor layers under different forest trees on the same soil type. *Soil Sci.* **36**, 387–398.

Amthor, J. S. (1984). The role of maintenance respiration in plant growth. *Plant, Cell Environ.* **7**, 561–569.

Amthor, J. S. (1989). *Respiration and Crop Productivity*. Springer-Verlag, New York and Berlin.

Amthor, J. S. (1994). Plant respiratory responses to the environment and their effects on the carbon balance. In *Plant-Environment Interactions* (R. E. Wilkinson, ed.), pp. 501–554. Marcel Dekker, New York.

Amthor, J. S. (2000). The McCree-de Wit-Penning de Vries-Thornley respiration paradigms—30 years later. *Ann. Bot.* **86**, 1–20.

Amthor, J. S. and Baldocchi, D. D. (2001). Terrestrial higher plant respiration and net primary production. In *Terrestrial Global Productivity* (J. Roy, B. Saugier, and H. A. Mooney, eds.), pp. 33–59. Academic Press, San Diego.

Amthor, J. S., Gill, D. S., and Bormann, F. H. (1990). Autumnal leaf conductance and apparent photosynthesis by saplings and sprouts in a recently disturbed northern hardwood forest. *Oecologia* **84**, 93–95.

Andersen, C. P., Markhart III, A. H., Dixon, R. K., and Sucoff, E. (1988). Root hydraulic conductivity of vesicular-arbuscular mycorrhizal green ash seedlings. *New Phytol.* **109**, 465–472.

Anderson, J. M. and Osmond, C. B. (1987). Shade-sun responses: Compromises between acclimation and photoinhibition. In *Photoinhibition* (D. J. Kyle, C. B. Osmond, and C. J. Arntzen, eds.), pp. 1–38. Elsevier, Amsterdam and New York.

Anderson, J. V., Chao, W. S., and Horvath, D. P. (2001). A current review on the regulation of dormancy in vegetative buds. *Weed Science* **49**, 581–589.

Andrews, M. (1986). The partitioning of nitrate assimilation between root and shoot of higher plants. *Plant, Cell Environ.* **9**, 511–519.

Anekonda, T. S., Criddle, R. S., Libby, W. J., and Hansen, L. D. (1993). Spatial and temporal relationships between growth traits and metabolic heat rates in coast redwood. *Can. J. For. Res.* **23**, 1793–1798.

Anekonda, T. S., Criddle, R. S., Libby, W. J., Breidenbach, R. W., and Hansen, L. D. (1994). Respiration rates predict differences in growth of coast redwood. *Plant, Cell Environ.* **17**, 197–203.

Anfodillo, T., Sigalotti, G. B., Tomasi, M., Semenzato, P., and Valentini, R. (1993). Applications of a thermal imaging technique in the study of the ascent of sap in woody species: Technical note. *Plant, Cell Environ.* **16**, 997–1001.

Angeles, G. (1992). Cambial activity and development of xylem and phloem in flooded trees. *IAWA Bull. N. S.*. **13**, 251–252.

Anonymous (1968). *Forest Fertilization: Theory and Practice*, Symposium on Forest fertilization, Gainesville, Florida, 1967. Tenn. Valley Auth., Muscle Shoals, Alabama.

Appleby, C. A. (1985). Plant hemoglobin properties, function and genetic origin. In *Nitrogen Fixation and CO_2 Metabolism* (P. W. Ludden and J. E. Burris, eds.), pp. 951–953. Elsevier, New York.

Appleby, R. F. and Davies, W. J. (1983). A possible evaporation site in the guard cell wall and the influence of leaf structure on the humidity response by stomata of woody plants. *Oecologia* **56**, 30–40.

Archer, B. L. and Audley, B. G. (1987). New aspects of rubber biosynthesis. *Bot. J. Linnean Soc.* **94**, 181–196.

Armstrong, W. (1968). Oxygen diffusion from the roots of woody species. *Physiol. Plant.* **21**, 539–543.

Armstrong, W., Brandle, R., and Jackson, M. B. (1994). Mechanisms of flood tolerance in plants. *Acta Bot. Neer.* **43**, 307–358.

Arndt, S. K., Wanek, W., Clifford, S. C., and Popp, M. (2000). Contrasting adaptations to drought stress in field-grown *Ziziphus mauritiana* and *Prunus persica*: Water relations, osmotic adjustment and carbon isotope composition. *Aust. J. Plant Physiol.* **27**, 985–996.

Arnebrant, K., Ek, H., Finlay, R. D., and Soderstrom, B. (1993). Nitrogen translocation between *Alnus glutinosa* (L.) Gaertn. seedlings inoculated with *Frankia* sp. and *Pinus contorta* Doug. ex Loud. seedlings connected by a common ectomycorrhizal mycelium. *New Phytol.* **124**, 231–242.

Artus, N. N., Somerville, S. C., and Somerville, C. R. (1986). The biochemistry and cell biology of photorespiration. *Crit. Rev. Plant Sci.* **4**, 121–147.

Ascaso, C., Galvan, J., and Rodriguez–Pascual, C. (1982). The weathering of calcareous rocks by lichens. *Pedobiologia* **24**, 219–229.

Ashton, P. M. S. and Berlyn, G. P. (1994). A comparison of leaf physiology and anatomy of *Quercus* (section of *Erythrobalanus*—Fagaceae) species in different light environments. *Am. J. Bot.* **81**, 589–597.

Ashton, P. S., Givnish, T. J., and Appanah, S. (1988). Staggered flowering in the Dipterocarpaceae: New insights into floral induction and the evolution of mast fruiting in the aseasonal tropics. *Am. Nat.* **132**, 44–66.

Askenasy, E. (1895). *Über das Saftsteigen*. *Bot. Zentralbl.* **62**, 237–238.

Assmann, S. M. and Zeiger, E. (1987). Guard cell bioenergetics. In *Stomatal Function* (E. Zeiger, G. D. Farquhar, and I. R. Cowan, eds.), pp. 163–193. Stanford University Press, Stanford, California.

Assmann, S. M., Snyder, J. A., and Lee, Y. R. J. (2000). ABA-deficient (*aba1*) and ABA-insensitive (*abi1-1*, *abi2-1*) mutants of *Arabidopsis* have a wild-type stomatal response to humidity. *Plant, Cell Environ.* **23**, 387–395.

Atkin, O. K., Bruhn, D., Hurry, V. M., and Tjoelker, M. G. (2005). The hot and the cold: Unraveling the variable response of plant respiration to temperature. *Funct. Plant Biol.* **32**, 87–105.

Atkinson, C. J. (1991). The flux and distribution of xylem sap calcium to adaxial and abaxial epidermal tissue in relation to stomatal behaviour. *J. Exp. Bot.* **42**, 987–993.

Atkinson, C. J., Davies, W. J., and Mansfield, T. A. (1989). Changes in stomatal conductance of intact aging wheat leaves in response to abscisic acid. *J. Exp. Bot.* **40**, 1021–1028.

Atkinson, C. J., Mansfield, T. A., and Davies, W. J. (1990). Does calcium in xylem sap regulate stomatal conductance. *New Phytol.* **116**, 19–28.

Attiwill, P. M. (1995). Nutrient cycling in plants. In *Encyclopedia of Environmental Biology*, (W. A. Nierenberg, ed.), Vol. 2, pp. 624–629. Academic Press, San Diego.

Attiwill, P. M. and Adams, M. A. (1993). Nutrient cycling in forests. *New Phytol.* **124**, 561–582.

Aubertin, G. M. and Patric, J. H. (1974). Water quality after clearcutting a small watershed in West Virginia. *J. Environ. Qual.* **3**, 243–245.

Auclair, D. (1976). *Effets des poussières sur la photosynthese. I. Effets des poussieres de ciment et de chârbon sur la photosynthèse de l'epicéa.* *Ann. Sci. For.* **33**, 247–255.

Auclair, D. (1977). *Effets des poussières sur la photosynthese. II. Influence des polluants particulaires sur la photosynthèse du pin sylvestre et du peuplier.* *Ann. Sci. For.* **34**, 47–57.

Augé, R. M. (2001). Water relations, drought and vesicular-arbuscular mycorrhizal symbiosis. *Mycorrhiza* **11**, 3–42.

Augé, R. M., Schekel, K. A., and Wample, R. L. (1986a). Greater leaf conductance of well-watered VA mycorrhizal rose plants is not related to phosphorus nutrition. *New Phytol.* **103**, 107–116.

Augé, R. M., Schekel, K. A., and Wample, R. L. (1986b). Osmotic adjustment in leaves of VA mycorrhizal and nonmycorrhizal rose plants in response to drought stress. *Plant Physiol.* **82**, 765–770.

Augé, R. M., Green, C. D., Stodola, A. J. W., Saxton, A. M., Olinick, J. B., and Evans, R. M. (2000). Correlations of stomatal conductance with hydraulic and chemical factors in several deciduous tree species in a natural habitat. *New Phytol.* **145**, 483–500.

Aung, L. H., Houck, L. G., and Norman, S. M. (1991). The abscisic acid content of citrus with special reference to lemon. *J. Exp. Bot.* **42**, 1083–1088.

Avery, D. J. (1977). Maximum photosynthetic rate—A case study in apple. *New Phytol.* **78**, 55–63.

Avery, M. E. (1993). Characterization of nitrogenous solutes in tissues and xylem sap of *Leucaena leucocephala*. *Tree Physiol.* **12**, 23–40.

Axelsson, E. and Axelsson, B. (1986). Changes in carbon allocation patterns in spruce and pine trees following irrigation and fertilization. *Tree Physiol.* **2**, 189–204.

Axelos, M., Teyssendier De La Serve, B., and Peaud–Lenoel, C. (1987). Level of messenger RNA encoding small subunit ribulose bisphosphate carboxylase is enhanced by cytokinins in tobacco cell suspension cultures. *Biochimie (Paris)* **69**, 671–676.

Ayers, J. C. and Barden, J. A. (1975). Net photosynthesis and dark respiration of apple leaves as affected by pesticides. *J. Amer. Soc. Hortic. Sci.* **100**, 24–28.

Aylor, D. E., Parlange, J., and Krikorian, A. D. (1973). Stomatal mechanics. *Am. J. Bot.* **60**, 163–171.

Ayres, P. G. (1978). Water relations of diseased plants. In *Water Deficits and Plant Growth* (T. T. Kozlowski, ed.), Vol. 5, pp. 1–60. Academic Press, New York.

Azevedo, J. and Morgan, D. L. (1974). Fog precipitation in coastal California forests. *Ecology* **55**, 1135–1141.

Babu, A. M. and Menon, A. R. S. (1990). Distribution of gum and gum-resin ducts in plant body. Certain familiar features and their significance. *Flora* **184**, 257–261.

Bachelard, E. P. (1969). Studies on the formation of epicormic shoots on eucalypt stem segments. *Aust. J. Biol. Sci.* **22**, 1291–1296.

Bacic, T., van der Eerden, L. J., and Baas, P. (1994). Evidence for recrystallization of epicuticular wax on needles of *Pinus sylvestris*. *Acta Bot. Neerl.* **43**, 271–273.

Backes, K. and Leuschner, C. (2000). Leaf water relations of competitive *Fagus sylvatica* and *Quercus petraea* trees during four years differening in soil drought. *Can. J. For. Res.* **30**, 335–346.

Backhaus, R. A. (1985). Rubber formation in plants—A mini review. *Israel J. Bot.* **34**, 283–293.

Baes, C. F., Jr., Goeller, H. E., Olson, J. S., and Rotty, R. M. (1977). Carbon dioxide and the climate: The uncontrolled experiment. *Am. Sci.* **66**, 310–320.

Bagda, H. (1948). *Morphologische und physiologische Untersuchungen üker Valonia Eichen (*Quercus macrolepsis Ky.*) im Haci-Kadin-Tal bei Ankara. Univ. Ankara Fac. Sci. Commun.* **1**, 89–125.

Bagda, H. (1952). *Untersuchungen über den weiblichen Gametophyten der Valonia Eichen (*Quercus macrolepsis Ky.*). Rev. Fac. Sci. Univ. Istanbul* **17**, 77–94.

Bahari, Z. A., Pallardy, S. G., and Parker, W. C. (1985). Photosynthesis, water relations and drought adaptation in six woody species of oak-hickory forests in central Missouri. *For. Sci.* **31**, 557–569.

Bai, C., Reilly, C. C., and Wood, B. W. (2006). Nickel deficiency disrupts metabolism of ureides, amino acids, and organic acids of young pecan foliage. *Plant Physiol.* **140**, 433–443.

Bailey, I. W. (1920). The cambium and its derivative tissues. II. Size variations of cambium initials in gymnosperms and angiosperms. *Am. J. Bot.* **7**, 255–367.

Baird, L. A. M. and Webster, B. D. (1979). The anatomy and histochemistry of fruit abscission. *Hort. Rev.* **1**, 172–203.

Baker, D. D. and Mullin, B. C. (1992). Actinorhizal symbioses. In *Biological Nitrogen Fixation* (G. Stacey, R. H. Burris, and H. J. Evans, eds.), pp. 259–292. Chapman and Hall, New York.

Baker, E. A. (1974). The influence of environment on leaf wax deposition in *Brassica oleracea* var. *gemmifera*. *New Phytol.* **73**, 955–966.

Baker, E. A. and Bukovac, M. J. (1971). Characterization of the components of plant cuticles in relation to the penetration of 2,4–D. *Ann. Appl. Biol.* **67**, 243–253.

Baker, F. S. (1950). *Principles of Silviculture*. McGraw-Hill, New York.

Baker, J., Steele, C., and Dure, L. (1988). Sequence and characterization of 6 lea proteins and their genes from cotton. *Plant Mol. Biol.* **11**, 277–291.

Baker, M. B. (1986). Effects of ponderosa pine treatments on water yield in Arizona. *Water Resources Res.* **22**, 67–73.

Baker, N. R. and Bowyer, J. R., eds. (1994). *Photoinhibition of Photosynthesis: From Molecular Mechanisms to the Field*. Bios Scientific, Oxford.

Baldocchi, D. D. (2003). Assessing the eddy covariance technique for evaluating carbon dioxide exchange rates of ecosystems: Past, present and future. *Global Change Biol.* **9**, 479–492.

Baldwin, I. T. and Schultz, J. C. (1983). Rapid changes in tree leaf chemistry induced by damage: Evidence for communication between plants. *Science* **221**, 277–279.

Balling, A. and Zimmermann, U. (1990). Comparative measurements of the xylem pressure of *Nicotiana* plants by means of the pressure bomb and pressure probe. *Planta* **182**, 325–338.

Bange, G. G. J. (1953). On the quantitative explanation of stomatal transpiration. *Acta. Bot. Neerl.* **2**, 255–297.

Bannan, M. W. (1955). The vascular cambium and radial growth of *Thuja occidentalis* L. *Can. J. Bot.* **33**, 113–138.

Bannan, M. W. (1962). The vascular cambium and tree-ring development. In *Tree Growth* (T. T. Kozlowski, ed.), pp. 3–21. Ronald, New York.

Bannan, M. W. (1967). Anticlinal divisions and cell length in conifer cambium. *For. Prod. J.* **17**, 63–69.

Baranski, M. J. (1975). An analysis of variation within white oak. *North Carolina Agric. Exp. Station., Technical Bulletin No.* **236**.

Barber, H. N. and Jackson, W. D. (1957). Natural selection in action in *Eucalyptus*. *Nature (London)* **179**, 1267–1269.

Barber, S. A. (1984). *Soil Nutrient Bioavailability*. Wiley Interscience, New York.

Barker, J. E. (1979). Growth and wood properties of *Pinus radiata* in relation to applied ethylene. *N. Z. J. For. Sci.* **9**, 15–19.

Barnard, H. R. and Ryan, M. G. (2003). A test of the hydraulic limitation hypothesis in fast-growing *Eucalyptus saligna*. *Plant, Cell Environ.* **26**, 1235–1245.

Barnes, R. L. (1963a). Nitrogen transport in the xylem of trees. *J. For.* **61**, 50–51.

Barnes, R. L. (1963b). Organic nitrogen compounds in tree xylem sap. *For. Sci.* **9**, 98–102.

Barnes, R. L. (1972). Effects of chronic exposure to ozone on photosynthesis and respiration of pines. *Environ. Pollut.* **3**, 133–138.

Barnett, J. R. (1992). Reactivation of the cambium in *Aesculus hippocastanum* L. A transmission electron microscope study. *Ann. Bot.* **70**, 169–177.

Barros, R. S. and Neill, S. J. (1987). Shoot growth in willow (*Salix viminalis*) in relation to abscisic acid, plant water status and photoperiod. *Physiol. Plant.* **70**, 708–712.

Barrs, H. D. (1968). Determination of water deficits in plant tissues. In *Water Deficits and Plant Growth*, Vol. 1. (T.T. Kozlowski, ed.), pp. 235–368. Academic Press, New York.

Barrs, H. D. and Kramer, P. J. (1969). Water potential increase in diced leaf tissue as a cause of error in vapor phase determinations of water potential. *Plant Physiol.* **44**, 959–964.

Bartley, I. M. and Knee, M. (1982). The chemistry of textural changes in fruit during storage. *Food Chem.* **9**, 47–58.

Basile, B., Reidel, E. J., Weinbaum, S. A., and DeJong, T. M. (2003). Leaf potassium concentration, CO_2 exchange and light interception in almond trees (*Prunus dulcis* (Mill) D.A. Webb). *Sci. Hortic.* **98**, 185–194.

Bassett, J. R. (1964). Tree growth as affected by soil moisture availability. *Proc. Soil Sci. Soc. Am.* **28**, 463–438.

Bassirirad, H., Caldwell, M. M., and Mott, K. A. (1993). Effect of root cooling on photosynthesis of *Artemisia tridentata* seedlings under different light levels. *Bot. Acta* **106**, 223–227.

Bassman, J. H. and Dickmann, D. I. (1982). Effects of defoliation in the developing leaf zone on young *Populus* x *euramericana* plants. I. Photosynthetic physiology, growth, and dry weight partitioning. *For. Sci.* **28**, 599–612.

Bauer, H. and Thoni, W. (1988). Photosynthetic light acclimation in fully developed leaves of the juvenile and adult life phases of *Hedera helix*. *Physiol. Plant.* **73**, 31–37.

Bauerle, W. L., Whitlow, T. H., Setter, T. L., Bauerle, T. L., and Vermeylen, F. M. (2003). Ecophysiology of *Acer rubrum* seedlings from contrasting hydrologic habitats: Growth, gas exchange, tissue water relations, abscisic acid and carbon isotope discrimination. *Tree Physiol.* **23**, 841–850.

Baughn, J. W. and Tanner, C. B. (1976). Leaf water potential: Comparison of pressure chamber and *in situ* hygrometer on five herbaceous species. *Crop Sci.* **16**, 181–184.

Baumgartner, A. (1967). Energetic bases for differential vaporization from forest and agricultural lands. In *Forest Hydrology* (W. E. Sopper and H. W. Lull, eds.), pp. 381–389. Pergamon, Oxford.

Bawa, K. S. (1983). Patterns of flowering in tropical plants. In *Handbook of Experimental Pollination Biology* (C. E. Jones and R. J. Little, eds.), pp. 394–410. Van Nostrand, Reinhold, New York.

Bazzaz, F. A. (1979). The physiological ecology of plant succession. *Annu. Rev. Ecol. Syst.* **10**, 351–371.

Bazzaz, F. A. and Carlson, R. W. (1982). Photosynthetic acclimation to variability in the light environment of early and late successional plants. *Oecologia* **54**, 313–316.

Bazzaz, F. A. and Peterson, D. L. (1984). Photosynthetic and growth responses of silver maple (*Acer saccharinum* L.) seedlings to flooding. *Am. Midl. Nat.* **112**, 261–272.

Bazzaz, F. A., Carlson, R. W., and Harper, J. L. (1979). Contribution to reproductive effort by photosynthesis of flowers and fruits. *Nature (London)* **279**, 554–555.

Beadle, C. L., Neilson, R. E., Talbot, H., and Jarvis, P. G. (1985). Stomatal conductance and photosynthesis in a mature Scots pine forest. I. Diurnal seasonal and spatial variation in shoots. *J. Appl. Ecol.* **22**, 557–571.

Beard, J. S. (1946). The natural vegetation of Trinidad. *Oxford For. Mem.* **20**.

Bearder, J. R. (1980). Plant hormones and other growth substances—Their background, structures and occurrence. *Encycl. Plant Physiol. N.S.* **9**, 9–112.

Becker, W. M. and Deamer, D. W. (1991). *The World of the Cell.* Benjamin Cummings, Redwood City, California.

Becking, J. H. (1972). *Enige gegevens en Kanttekeningen over de betekenis van de zwarte els voorde houttleet in Nederland. Ned. Bosbouw-Tijdschr.* **44**, 128–131.

Beckman, C. H. (2000). Phenolic-storing cells: Keys to programmed cell death and periderm formation in wilt disease resistance and in general defence responses in plants? *Physiol. Molec. Plant Pathol.* **57**, 101–110.

Beckman, C. H., Kuntz, J. E., Riker, A. J., and Berbee, J. G. (1953). Host responses associated with the development of oak wilt. *Phytopathology* **43**, 448–454.

Bedunah, D. and Trlica, M. J. (1979). Sodium chloride effects on carbon dioxide exchange rates and other plant and soil variables of ponderosa pine. *Can. J. For. Res.* **9**, 349–353.

Beevers, H. and Hageman, R. H. (1969). Nitrate reduction in green plants. *Annu. Rev. Plant Physiol.* **20**, 459–522.

Begg, J. E. and Turner, N. C. (1976). Crop water deficits. *Adv. Agron.* **28**, 161–217.

Bentley, B. L. and Carpenter, E. J. (1984). Direct transfer of newly fixed nitrogen from free-living epiphyllous microorganisms to their host plants. *Oecologia* **63**, 52–56.

Benyon, R. G. (1999). Nighttime water use in an irrigated *Eucalyptus grandis* plantation. *Tree Physiol.* **19**, 853–859.

Berg, B., Hannus, K., Popoff, T., and Theander, O. (1982). Changes in organic chemical components of needle litter during decomposition. Long-term decomposition in a Scots pine forest. I. *Can. J. Bot.* **60**, 1310–1319.

Bergh, J. and Linder, S. (1999). Effects of soil warming during spring on photosynthetic recovery in boreal Norway spruce stands. *Global Change Biol.* **5**, 245–253.

Beritognolo, I., Magel, E., Abdel-Latif, A., Charpentier, J. P., Jay-Allemand, C., and Breton, C. (2002). Expression of genes encoding chalcone synthase, flavanone 3-hydroxylase and dihydroflavonol 4-reductase correlates with flavonol accumulation during heartwood formation in *Juglans nigra*. *Tree Physiol.* **22**, 291–300.

Berlyn, G. P. (1962). Developmental patterns in pine polyembryony. *Am. J. Bot.* **49**, 327–333.

Berlyn, G. P. (1967). The structure of germination in *Pinus lambertiana*. *Yale Univ. Sch. For. Bull.* **71**.

Berlyn, R. W. (1964). A method for measuring the strength of the bond between bark and wood: The bark-wood bond meter. *Pulp and Paper Res. Inst. Can. Res. Note 43.*

Berlyn, R. W. (1965). The effect of variations in the strength of the bond between bark and wood in mechanical barking. *Pulp and Paper Res. Inst. Can. Res. Note 54.*

Bernards, M. A. (2002). Demystifying suberin. *Can. J. Bot.* **80**, 227–240.

Berry, J. A. and Björkmann, O. (1980). Photosynthetic response and adaptation to temperature in higher plants. *Annu. Rev. Plant Physiol.* **31**, 491–543.

Bertrand, A., Robitaille, G., Nadeau, P., and Boutin, R. (1994). Effects of soil freezing and drought stress on abscisic acid content of sugar maple sap and leaves. *Tree Physiol.* **14**, 413–425.

Berüter, J. and Droz, P. L. (1991). Studies on locating the signal for fruit abscission in the apple. *Sci. Hortic.* **46**, 201–214.

Beyers, J. L., Riechers, G. H., and Temple, P. J. (1992). Effects of long-term ozone exposure and drought on the photosynthetic capacity of ponderosa pine (*Pinus ponderosa* Laws). *New Phytol.* **122**, 81–90.

Beyschlag, W., Kresse, F., Ryel, R. J., and Pfanz, H. (1994). Stomatal patchiness in conifers—Experiments with *Picea abies* (L.) Karst. and *Abies alba* Mill. *Trees* **8**, 132–138.

Bi, H. Q., Turvey, N. D., and Heinrich, P. (1992). Rooting density and tree size of *Pinus radiata* (D. Don) in response to competition from *Eucalyptus obliqua* (Lherit). *For. Ecol. Manage.* **49**, 31–42.

Biale, J. B. (1950). Postharvest physiology and biochemistry of fruits. *Annu. Rev. Plant Physiol.* **1**, 183–206.

Biale, J. B. (1954). The ripening of fruit. *Sci. Am.* **190**, 40–44.

Biale, J. B. and Young, R. E. (1981). Respiration and ripening in fruits—Retrospect and prospect. In *Recent Advances in the Biochemistry of Fruits and Vegetables* (J. Friend and M. J. C. Rhodes, eds.), pp. 1–39. Academic Press, London.

Bialoglowski, J. (1936). Effect of extent and temperature of roots on transpiration of rooted lemon cuttings. *Proc. Am. Soc. Hortic. Sci.* **34**, 96–102.

Bildusas, I. J., Dixon, R. K., Pfleger, F. L., and Stewart, E. L. (1986). Growth, nutrition and gas exchange of *Bromus inermis* inoculated with *Glomus fasciculatum*. *New Phytol.* **102**, 303–311.

Billings, W. D. and Morris, R. J. (1951). Reflection of visible and infrared radiation from leaves of different ecological groups. *Am. J. Bot.* **38**, 327–331.

Birkeland, P. W. (1984). *Soils and Geomorphology.* Oxford, New York.

Birkhold, K. T., Koch, K. E., and Darnell, R. L. (1992). Carbon and nitrogen economy of developing rabbiteye blueberry fruit. *J. Am. Soc. Hortic. Sci.* **117**, 139–145.

Biswell, H. H. (1935). Effects of environment upon the root habits of certain deciduous forest trees *Bot. Gaz.* **96**, 676–708.

Biswell, H. H., and Schultz, A. M. (1957). Spring flow affected by brush. *Calif. Agric.* **11**, 3–4.

Bittelli, M., Flury, M., Campbell, G. S., and Nichols, E. J. (2001). Reduction of transpiration through foliar application of chitosan. *Agric. For. Meteorol.* **107**, 167–175.

Björkman, O. (1981). Responses to different quantum flux densities. In *Physiological Plant Ecology. I. Responses to the Physical Environment* (O. L. Lange, P. S. Nobel, C. B. Osmond, and H. Ziegler, eds.), pp. 57–107. Springer-Verlag, New York.

Björkman, O., Downton, W. J. S., and Mooney, H. A. (1980). Response and adaptation to water stress in *Nerium oleander*. *Year Book Carnegie Inst. Wash.* **79**, 150–157.

Black, M. (1991). Involvement of ABA in the physiology of developing and mature seeds. In *Abscisic Acid: Physiology and Biochemistry* (W. J. Davies and H. G. Jones, eds.), pp. 99–124. Bios Scientific, Oxford.

Black, V. J. (1984). The effect of air pollutants on apparent respiration. In *Gaseous Air Pollutants and Plant Metabolism* (M. J. Koziol and F. R. Whatley, eds.), pp. 231–248. Butterworth, London.

Blackman, P. G. and Davies, W. J. (1985). Root to shoot communication in maize plants of the effects of soil drying. *J. Exp. Bot.* **36**, 39–48.

Blake, J. and Ferrell, W. K. (1977). The association between soil and xylem water potential, leaf resistance, and abscisic acid content in droughted seedlings of Douglas-fir (*Pseudotsuga menziesii*). *Physiol. Plant* **39**, 106–109.

Blanche, C. A., Lorio, P. L., Jr., Sommers, R. A., Hodges, J. D., and Nebeker, T. E. (1992). Seasonal cambial growth and development of loblolly pine: Xylem formation, inner bark chemistry, resin ducts, and resin flow. *For. Ecol. Manage.* **49**, 151–165.

Blanke, M. M. and Lenz, F. (1989). Fruit photosynthesis. *Plant, Cell Environ.* **12**, 31–46.

Blankenship, R. E. (2002). *Molecular Mechanisms of Photosynthesis.* Blackwell Science Ltd., Oxford.

Blevins, D. G. and Lukaszewski, K. M. (1998). Boron in plant structure and function. *Annu. Rev. Plant Physiol. Plant Mol. Biol.* **49**, 481–500.

Blizzard, W. E. and Boyer, J. S. (1980). Comparative resistance of the soil and the plant to water transport. *Plant Physiol.* **66**, 809–814.

Blowers, D. P. and Trewavas, A. J. (1989). Second messengers: Their existence and relationship to protein kinases. In *Plant Growth and Development* (W. F. Boss and D. J. Morré, eds.), pp. 1–28. Liss, New York.

Bockheim, J. G., Jepsen, E. A., and Heisey, D. M. (1991). Nutrient dynamics in decomposing leaf litter of four tree species on a sandy soil in northwestern Wisconsin. *Can. J. For. Res.* **21**, 803–812.

Bode, H. R. (1959). *Uber den Zusammenhang zwishen Blattenfaltung und Neubildung der Saugwurzeln bei Juglans. Ber. Dtsch. Bot. Ges.* **72**, 93–98.

Bogar, G. D. and Smith, F. H. (1965). Anatomy of seedling roots of *Pseudotsuga menziesii*. *Am. J. Bot.* **52**, 720–729.

Bolazos, L., Lukaszewski, K., Bonilla, I., and Blevins, D. (2004). Why boron? *Plant Physiol. Biochem.* **42**, 907–912.

Bollard, E. G. (1956). Nitrogenous compounds in plant xylem sap. *Nature (London)* **178**, 1189–1190.

Bollard, E. G. (1957). Translocation of organic nitrogen in the xylem. *Aust. J. Biol. Sci.* **10**, 292–301.

Bollard, E. G. (1958). Nitrogenous compounds in tree xylem sap. In *The Physiology of Forest Trees* (K. V. Thimann, ed.), pp. 83–93. Ronald, New York.

Bollard, E. G. (1960). Transport in the xylem. *Annu. Rev. Plant Physiol.* **11**, 141–166.

Bollard, E. G. (1970). The physiology and nutrition of developing fruits. In *The Biochemistry of Fruits and Their Products* (A. C. Hulme, ed.), pp. 387–425. Academic Press, New York.

Bolstad, P. V., Mitchell, K., and Vose, J. M. (1999). Foliar temperature-respiration response functions for broad-leaved tree species in the southern Appalachians. *Tree Physiol.* **19**, 871–878.

Bolstad, P. V., Reich, P., and Lee, T. (2003). Rapid temperature acclimation of leaf respiration rates in *Quercus alba* and *Quercus rubra*. *Tree Physiol.* **23**, 969–976.

Bolstad, P. V., Davis, K. J., Martin, J., Cook, B. D., and Wang, W. (2004). Component and whole-system respiration fluxes in northern deciduous forests. *Tree Physiol.* **24**, 493–504.

Bolton, A. J. and Robson, D. J. (1988). The effect of vessel element structure on element conductivity. *Trees* **2**, 25–31.

Boltz, B. A., Bongarten, B. C., and Teskey, R. O. (1986). Seasonal patterns of net photosynthesis of loblolly pine from diverse origins. *Can. J. For. Res.* **16**, 1063–1068.

Bond, B. J. (2000). Age-related changes in photosynthesis of woody plants. *Trends Plant Sci.* **5**, 349–353.

Bond, W. J. (1989). The tortoise and the hare: Ecology of angiosperm dominance and gymnosperm persistence. *Biol. J. Linn. Soc.* **36**, 227–249.

Bongi, G., Mencuccini, M., and Fontanazza, G. (1987). Photosynthesis of olive leaves: Effect of light flux density, leaf age, temperature, peltate, and H_2O vapor pressure deficit on gas exchange. *J. Am. Soc. Hortic. Sci.* **112**, 143–165.

Bonicel, A. and de Medeivos Raposo, N. V. (1990). Variation of starch and soluble sugars in selected sections of poplar buds during dormancy and post-dormancy. *Plant Physiol. Biochem.* **28**, 577–586.

Bonner, F. T. (1970). Artificial ripening of sweetgum seeds. *Tree Planters Notes* **21**, 23–25.

Bonner, J. (1950). *Plant Biochemistry, 1st ed.* Academic Press, New York.

Bonner, J. and Varner, J. E., eds. (1976). *Plant Biochemistry, 3rd ed.* Academic Press, New York.

Borchert, R. (1969). Unusual shoot growth patterns in a tropical tree, *Oreopanax* (Araliaceae). *Am. J. Bot.* **56**, 1033–1041.

Borchert, R. (1983). Phenology and control of flowering in tropical trees. *Biotropica* **15**, 81–89.

Borger, G. A. (1973). Development and shedding of bark. In *Shedding of Plant Parts* (T. T. Kozlowski, ed.), pp. 205–236. Academic Press, New York.

Borghetti, M., Edwards, W. R. N., Grace, J., Jarvis, P. G., and Raschi, A. (1991). The refilling of embolized xylem in *Pinus sylvestris* L. *Plant, Cell Environ.* **14**, 357–369.

Bormann, F. H. (1957). Moisture transfer between plants through intertwined root systems. *Plant Physiol.* **32**, 48–55.

Bormann, F. H. (1982). The effects of air pollution on the New England landscape. *Ambio* **11**, 338–346.

Bormann, F. H. and Likens, G. E. (1979). *Pattern and Process in a Forested Ecosystem.* Springer-Verlag, New York.

Bormann, F. H., Likens, G. E., and Melillo, J. M. (1977). Nitrogen budget for an aggrading northern hardwood forest ecosystem. *Science* **196**, 981–983.

Boscaglia, A. (1983). The starch content of *Fraxinus ornus* L. during the yearly cycle. Histological observations. *Giornale Botanico Italiano* **116**, 41–49.

Botkin, D. B., Smith, W. H., Carlson, R. W., and Smith, T. L. (1972). Effects of ozone on white pine saplings. Variation in inhibition and recovery of net photosynthesis. *Environ. Pollut.* **3**, 273–289.

Bourdeau, P. (1954). Oak seedling ecology determining segregation of species in Piedmont oak-hickory forests. *Ecol. Monogr.* **24**, 297–320.

Bourdeau, P. F. and Schopmeyer, C. S. (1958). Oleoresin exudation pressure in slash pine: Its measurement, heritability and relation to oleoresin yield. In *The Physiology of Forest Trees* (K. V. Thimann, ed.), pp. 313–319. Ronald, New York.

Bowe, L. M., Coat, G., and dePamphilis, C.W. (2000). Phylogeny of seed plants based on all three genomic compartments: Extant gymnosperms are monophyletic and Gnetales' closest relatives are conifers. *Proc. Natl. Acad. Sci. U.S.A.* **97**, 4092–4097.

Bowen, G. D. (1973). Mineral nutrition of ectomycorrhizae. In *Ectomycorrhizae* (G. C. Marks and T. T. Kozlowski, eds.), pp. 151–205. Academic Press, New York.

Bowen, G. D. (1984). Tree roots and the use of soil nutrients. In *Nutrition of Plantation Forests* (G. D. Bowen and E. K. S. Nambiar, eds.), pp. 147–179. Academic Press, London.

Bowen, G. D. (1985). Roots as a component of tree productivity. In *Attributes of Trees as Crop Plants* (M. G. R. Cannell and J. E. Jackson, eds.), pp. 303–315. Institute of Terrestrial Ecology, Huntingdon, England.

Bowen, G. D. and Theodorou, C. (1967). Studies on phosphorus uptake by mycorrhizas. *Proc. Int. Union Forest Res. Organ.*, 14th, Vol. 5, p. 116, Munich.

Boyd, R., Furbank, R. T., and Read, D. J. (1986). Ecotomycorrhiza and the water relations of trees. In *Physiological and Genetical Aspects of Mycorrhizae: Proc. First European Symp. on Mycorrhizae*, Dijon, 1–5 July 1985, pp. 689–693.

Boyer, J. S. (1969). Measurement of the water status of plants. *Annu. Rev. Plant Physiol.* **20**, 351–364.

Boyer, J. S. (1974). Water transport in plants: Mechanism of apparent changes in resistance during absorption. *Planta* **117**, 187–207.

Boyer, J. S. (1985). Water transport. *Annu. Rev. Plant Physiol.* **36**, 473–516.

Boyer, J. S. (1989). Water potential and plant metabolism: Comments on Dr. P. J. Kramer's article "Changing concepts regarding plant water relations," volume 11, number 7, pp. 565–568, and Dr. J. B. Passioura's response, pp. 569–571. *Plant, Cell Environ.* **12**, 213–216.

Boyer, J. S. (1995). *Measuring the Water Status of Plants and Soils.* Academic Press, San Diego.

Boyer, J. S. and Younis, H. M. (1983). Molecular aspects of photosynthesis at low leaf water potentials. In *Effects of Stress on Photosynthesis* (R. Marcelle et al., eds.) pp. 29–33, Martinus Nijhoff, The Hague.

Boyle, J. R. (1975). Nutrients in relation to intensive culture of forest crops. *Iowa State J. Res.* **49**, 297–303.

Boyle, J. R., Voigt, G. K., and Sawhney, B. L. (1974). Chemical weathering of biotite by organic acids. *Soil Sci.* **117**, 42–45.

Boyle, T. J. B. and Morgenstern, E. K. (1985). Inheritance and linkage relationships of some isozymes of black spruce in New Brunswick. *Can. J. For. Res.* **15**, 992–996.

Bradford, K. J. (1983a). Effects of soil flooding on leaf gas exchange of tomato. *Plant Physiol.* **73**, 475–479.

Bradford, K. J. (1983b). Involvement of plant growth substances in the alteration of leaf gas exchange of flooded tomato plants. *Plant Physiol.* **73**, 480–483.

Brady, C. J. (1987). Fruit ripening. *Annu. Rev. Plant Physiol.* **38**, 155–178.

Braekke, F. H. (1990). Nutrient accumulation and role of atmospheric deposition in coniferous stands. *For. Ecol. Manage.* **30**, 351–359.

Braekke, F. H. and Kozlowski, T. T. (1975). Effect of climatic and edaphic factors on radial stem growth of *Pinus resinosa* and *Betula papyrifera* in northern Wisconsin. *Adv. Frontiers of Pl. Sci.* **30**, 201–221.

Brand, D. G., Weetman, G. F., and Rehsler, P. (1987). Growth analysis of perennial plants: The relative production rate and its yield components. *Ann. Bot.* **59**, 45–53.

Bray, E. A. (1997). Plant responses to water deficit. *Trends Plant Sci.* **2**, 48–54.

Brewer, J. F., Hinesley, L. E., and Snelling, L. K. (1992). Foliage attributes for current-year shoots of Fraser fir. *HortScience* **27**, 920–925.

Briggs, G. E., Kidd, F., and West, C. (1920). A quantitative analysis of plant growth. *Ann. Appl. Biol.* **7**, 202–223.

Briggs, G. M., Jurik, T. W., and Gates, D. M. (1986). Non-stomatal limitation of CO_2 assimilation in three tree species during natural drought conditions. *Physiol. Plant.* **66**, 521–526.

Briggs, L. J. (1949). A new method of measuring limiting negative pressure in liquids. *Science* **109**, 440.

Brinkmann, W. L. F. and Nascimento, J. C. (1973). The effect of slash and burn agriculture on plant nutrients in the tertiary region of Central Amazonia. *Acta Amazonica* **3**, 55–61.

Britt, J. R., Mitchell, R. J., Zutter, B. R., South, D. B., Gjerstad, D. H., and Dickson, J. F. (1991). The influence of herbaceous weed control and seedling diameter on six years of loblolly pine growth—A classical growth analysis approach. *For. Sci.* **37**, 655–668.

Brix, H. (1962). The effect of water stress on the rates of photosynthesis and respiration in tomato plants and loblolly pine seedlings. *Physiol. Plant.* **15**, 10–20.

Brix, H. (1983). Effects of thinning and nitrogen fertilization on growth of Douglas-fir: Relative contribution of foliage quantity and efficiency. *Can. J. For. Res.* **13**, 167–175.

Brock, T. G. and Kaufman, P. G. (1991). Growth regulators: An account of hormones and growth regulation in plants. In *Plant Physiology: A Treatise* (F. C. Steward, ed.), pp. 277–340. Academic Press, San Diego.

Brodribb, T. (1996). Dynamics of changing intercellular CO_2 concentration (C_i) during drought and determination of minimum functional C_i. *Plant Physiol.* **111**, 179–185.

Brodribb, T. and Hill, R. S. (1998). The photosynthetic drought physiology of a diverse group of southern hemisphere conifer species is correlated with minimum seasonal rainfall. *Funct. Ecol.* **12**, 465–471.

Brooks, J. R., Hinckley, T. M., Ford, E. D., and Sprugel, D. G. (1991). Foliage dark respiration in *Abies amabilis* (Dougl.) Forbes: Variation within the canopy. *Tree Physiol.* **9**, 325–338.

Brooks, J. R. and cosigners. (2004). The Cohesion-Tension Theory. *New Phytol.* **163**, 451–452.

Brooks, J. R., Meinzer, F. C., Coulombe, R., and Gregg, J. (2002). Hydraulic redistribution of soil water during summer drought in two contrasting Pacific Northwest coniferous forests. *Tree Physiol.* **22**, 1107–1115.

Brouwer, R. (1953). Water absorption by the roots of *Vicia faba* at various transpiration strengths. *Proc. Kon. Ned. Akad. Wet.* **C56**, 106–115, 129–136.

Brown, C. L. and Sommer, H. E. (1992). Shoot growth and histogenesis of trees possessing diverse patterns of shoot development. *Am. J. Bot.* **79**, 335–346.

Brown, C. L., Sommer, H. E., and Pienaar, L. V. (1995a). The predominant role of the pith in the growth and development of internodes in *Liquidambar styraciflua* (Hamamelidaceae). I. Histological basis of compressive and tensile stresses in developing primary tissues. *Am. J. Bot.* **82**, 769–776.

Brown, C. L., Sommer, H. E., and Pienaar, L. V. (1995b). The predominant role of the pith in the growth and development of internodes in *Liquidambar styraciflua* (Hamamelidaceae). II. Pattern of tissue stress and response of different tissues to specific surgical procedures. *Am. J. Bot.* **82**, 777–781.

Brown, G. W., Gahler, A. R., and Marston, R. B. (1973). Nutrient losses after clear-cut logging and slash burning in the Oregon Coast Range. *Water Resour. Res.* **9**, 1450–1453.

Brown, H. T. and Escombe, F. (1900). Static diffusion of gases and liquids in relation to the assimilation of carbon and translocation in plants. *Phil. Trans. Roy. Soc.* **193**, 223–291.

Brown, J. C. and Jolley, V. D. (1989). Plant metabolic responses to iron-deficiency stress. *BioScience* **39**, 546–551.

Brown, K. M. and Leopold, A. C. (1973). Ethylene and the regulation of growth in pine. *Can. J. For. Res.* **3**, 143–145.

Brown, P. H., Bellaloui, N., Wimmer, M. A., Bassil, E. S., Ruiz, J., Hu, H., Pfeffer, H., Dannel, F., and Romheld, V. (2002). Boron in plant biology. *Plant Biol.* **4**, 205–223.

Browning, G. (1973). Flower bud dormancy in *Coffea arabica* L. II. Relation of cytokinins in xylem sap and flower buds to dormancy-release *J. Hortic. Sci.* **48**, 297–310.

Brownlee, C. (2001). The long and the short of stomatal density signals. *Trends Plant Sci.* **6**, 441–442.

Brownlee, C., Duddridge, J. A., Malibari, A., and Read, D. J. (1983). The structure and function of mycelial systems of ectomycorrhizal roots with special reference to their role in forming inter-plant connections and providing pathways for assimilation and water transport. *Plant Soil* **71**, 433–443.

Bryant, J. P. and Raffa, K. A. (1995). Chemical antiherbivore defense. In *Plant Stress: Physiology and Functional Morphology* (B. L. Gartner, ed.), pp. 365–381. Academic Press, San Diego.

Bryla, D. R., Bouma, T. J., and Eissenstat, D. M. (1997). Root respiration in citrus acclimates to temperature and slows during drought. *Plant, Cell Environ.* **20**, 1411–1420.

Bryla, D. R., Bouma, T. J., Hartmond, U., and Eissenstat, D. M. (2001). Influence of temperature and soil drying on respiration of individual roots in citrus: Integrating greenhouse observations into a predictive model for the field. *Plant, Cell Environ.* **24**, 781–790.

Bucci, S. J., Scholz, F. G., Goldstein, G., Meinzer, F. C., Hinojosa, J. A., Hoffman, W. A., and Franco, A. C. (2004). Processes preventing nocturnal equilibration between leaf and soil water potential in tropical savanna woody species. *Tree Physiol.* **24**, 1119–1127.

Bucci, S. J., Goldstein, G., Meinzer, F. C., Franco, A. C., Campanello, P., and Scholze, F. G. (2005). Mechanisms contributing to seasonal homeostasis of minimum leaf water potential and predawn disequilibrium between soil and plant water potential in Neotropical savanna trees. *Trees* **19**, 296–304.

Bunce, J. A., Miller, L. N., and Chabot, B. F. (1977). Competitive exploitation of soil water by five eastern North American tree species. *Bot. Gaz.* **138**, 168–173.

Bunger, M. T. and Thomson, H. J. (1938). Root development as a factor in the success or failure of windbreak trees in the Southern High Plains. *J. For.* **36**, 790–803.

Burgess, S. S. O., Adams, M. A., Turner, N. C., and Ong, C. K. (1998). The redistribution of soil water by tree root systems. *Oecologia* **115**, 306–311.

Burgess, S. S. O., Adams, M. A., Turner, N. C., White, D. A., and Ong, C. K. (2001). Tree roots: Conduits for deep recharge of soil water. *Oecologia* **126**, 158–165.

Burke, M. K. and Raynal, D. J. (1994). Fine root growth phenology, production, and turnover in a northern hardwood forest ecosystem. *Plant Soil* **162**, 135–146.

Burkhardt, J. and Eiden, R. (1990). The ion concentration of dew condensed on Norway spruce (*Picea abies* (L.) Karst.) and Scots pine (*Pinus sylvestris* L.) needles. *Trees* **4**, 22–26.

Burns, R. M. and Honkala, B. H., eds. (1990a). *Silvics of North America. Vol. 1. Conifers.* U.S.D.A. Forest Service, *Agric. Hdbk.* 654, Washington, D.C.

Burns, R. M. and Honkala, B. H., eds. (1990b). *Silvics of North America. Vol. 2. Hardwoods.* U.S.D.A. Forest Service, *Agric. Hdbk.* 654, Washington, D.C.

Burrows, F. J. and Milthorpe, F. L. (1976). Stomatal conductance in gas exchange control. In *Water deficits and Plant Growth* (T. T. Kozlowski, ed.), Vol. 4. pp. 103–190. Academic Press, New York.

Burtin, P., Jay-Allemand, C., Charpentier, J. P., and Janin, G. (1998). Natural wood colouring process in Juglans sp. (*J. nigra, J. regia* and *hybrid J. nigra x J. regia*) depends on native phenolic compounds accumulated in the transition zone between sapwood and heartwood. *Trees* **12**, 258–264.

Burton, A. J., Pregitzer, K. S., Zogg, G. P., and Zak, D. R. (1998). Drought reduces root respiration in sugar maple forests. *Ecol. Appl.* **8**, 771–778.

Büsgen, M. and Münch, E. (1931). *The Structure and Life of Forest Trees* (transl. by T. Thomson), *3rd ed.* Wiley, New York.

Butler, D. R. and Landsberg, J. J. (1981). Respiration rates of apple trees, estimated by CO_2-efflux measurement. *Plant, Cell Environ.* **4**, 153–159.

Buttery, B. R. and Boatman, S. G. (1976). Water deficits and flow of latex. In *Water Deficits and Plant Growth* (T. T. Kozlowski, ed.), Volume 4, pp. 233–289. Academic Press, New York.

Buxton, G. F., Cyr, D. R., and Dumbroff, E. B. (1985). Physiological responses of three northern conifers to rapid and slow induction of moisture stress. *Can. J. Bot.* **63**, 1171–1176.

Cailloux, M. (1972). Metabolism and the absorption of water by root hairs. *Can. J. Bot.* **50**, 557–573.

Caird, M. A., Richards, J. H., and Donovan, L. A. (2007). Nighttime stomatal conductance and transpiration in C_3 and C_4 plants. *Plant Physiol.* **143**, 4–10.

Cairney, J. W. G. (2000). Evolution of mycorrhiza systems. *Naturwissenschaften* **87**, 467–475.

Caldwell, M. M. (1970). Plant gas exchange at high wind speeds. *Plant Physiol.* **46**, 535–537.

Caldwell, M. M. (1976). Root extension and water absorption. In *Water and Plant Life* (O. L. Lange , L. Callaham, and E. D. Schulze, eds.), pp. 63–85. Springer-Verlag, Berlin and New York.

Caldwell, M. M., Dawson, T. E., and Richards, J. H. (1998). Hydraulic lift—Consequences of water efflux from the roots of plants. *Oecologia* **113**, 151–161.

Calkin, H. W., Gibson, A. C., and Nobel, P. S. (1986). Biophysical model of xylem conductance in tracheids of the fern *Pteris vittata. J. Exp. Bot.* **37**, 1054–1064.

Cameron, S. H. (1941). The influence of soil temperature on the rate of transpiration of young orange trees. *Proc. Am. Soc. Hortic. Sci.* **38**, 75–79.

Cameron, S. H. and Appleman, D. (1933). The distribution of total nitrogen in the orange tree. *Proc. Am. Soc. Hortic. Sci.* **30**, 341–348.

Cameron, S. H. and Compton, O. C. (1945). Nitrogen in bearing orange trees. *Proc. Am. Soc. Hortic. Sci.* **46**, 60–68.

Cameron, S. H. and Schroeder, C. A. (1945). Cambial activity and starch cycle in bearing orange trees. *Proc. Am. Soc. Hortic. Sci.* **46**, 55–59.

Cannell, M. G. R. (1971). Changes in the respiration and growth rates of developing fruits of *Coffea arabica* L. *J. Hortic. Sci.* **46**, 263–272.

Cannell, M. G. R. (1975). Crop physiological aspects of coffee bean yield. *J. Coffee Res.* **5**, 7–20.

Cannell, M. G. R. (1985). Dry matter partitioning in tree crops. In *Attributes of Trees as Crop Plants* (M. G. R. Cannell and J. E. Jackson, eds.), pp. 160–193. Institute of Terrestrial Ecology, Huntingdon, England.

Cannell, M. G. R. (1989). Physiological basis of wood production: A review. *Scand. J. For. Res.* **4**, 459–490.

Cannell, M. G. R. and Dewar, R. C. (1994). Carbon allocation in trees: A review of concepts for modelling. *Adv. Ecol. Res.* **25**, 59–104.

Cannell, M. G. R. and Thornley, J. H. M. (2000). Modelling the components of plant respiration: Some guiding principles. *Ann. Bot.* **85**, 45–54.

Cannell, M.G.R., Thompson, S., and Lines, R. (1976). An analysis of inherent differences in shoot growth within some north temperate conifers. In *Tree Physiology and Yield Improvement* (M. G. R. Cannell and F. T. Last, eds.) pp. 173–205. Academic Press, London.

Cannell, M. G. R., Tabbush, P. M., Deans, J. D., Hollingsworth, M. K., Sheppard, L. J., Philipson, J. J., and Murray, M. B. (1990). Sitka spruce and Douglas-fir seedlings in the nursery and in cold storage: Root growth potential, carbohydrate content, dormancy, frost hardiness, and mitotic index. *Forestry* **63**, 9–27.

Canny, M. J. (1990). What becomes of the transpiration stream? *New Phytol.* **114**, 341–368.

Canny, M. J. (1993a). Transfusion tissue of pine needles as a site of retrieval of solutes from the transpiration stream. *New Phytol.* **123**, 227–232.

Canny, M. J. (1993b). The transpiration stream in the leaf apoplast: Water and solutes. *Phil. Trans. Roy. Soc. Lond.* **B341**, 87–100.

Cao, M. K. and Woodward, F. I. (1998). Net primary and ecosystem production and carbon stocks of terrestrial ecosystems and their responses to climate change. *Global Change Biol.* **4**, 185–198.

Cao, X. F., Linstead, P., Berger, F., Kieber, J., and Dolan, L. (1999). Differential ethylene senistivity of epidermal cells is involved in the establishment of cell pattern in the *Arabidopsis* root. *Physiol. Plant.* **106**, 311–317.

Carlquist, S. (1983). Wood anatomy of Onagraceae: Further species; root anatomy; significance of vestured pits and allied structures in dicotyledons. *Ann. Mo. Bot. Gard.* **69**, 755–769.

Carlquist, S. (1988). *Comparative Wood Anatomy*. Springer-Verlag, New York and Berlin.

Carlson, R. W. (1979). Reduction in the photosynthetic rate of *Acer*, *Quercus*, and *Fraxinus* species caused by sulphur dioxide and ozone. *Environ. Pollut.* **18**, 159–170.

Carlson, R. W. and Bazzaz, F. A. (1977). Growth reduction in American sycamore (*Platanus occidentalis* L.) caused by Pb-Cd interaction. *Environ. Pollut.* **12**, 243–253.

Caron, G. E. and Powell, G. R. (1992). Patterns of cone distribution in crowns of young *Picea mariana*. I. Effects of tree age on seed cones. *Can. J. For. Res.* **22**, 46–55.

Carr, M. K. V. (1977). Changes in the water status of tea clones during dry weather in Kenya. *J. Agric. Sci.* **89**, 297–307.

Cary, A. J., Wennuan, L., and Howell, S. H. (1995). Cytokinin action is coupled to ethylene in its effects on the inhibition of root and hypocotyl elongation in *Arabidopsis thaliana*. *Plant Physiol.* **107**, 1075–1084.

Casperson, G. (1968). Wirkung von Wuchs und Hemmstoffen auf die Kambiumtätigkeit und Reaktionscholzbildung. *Physiol. Plant.* **21**, 1312–1321.

Casson, S. A. and Lindsey, K. (2003). Genes and signalling in root development. *New Phytol.* **158**, 11–38.

Causton, D. and Venus, J. (1981). *The Biometry of Plant Growth*. Edward Arnold, London.

Centritto, M., Lee, H. S. J., and Jarvis, P. G. (1999). Increased growth in elevated [CO_2]: An early, short–term response? *Global Change Biol.* **5**, 623–633.

Ceulemans, R. and Mousseau, M. (1994). Effects of elevated atmospheric CO_2 on woody plants. *New Phytol.* **127**, 425–446.

Ceulemans, R. J. and Saugier, B. (1991). Photosynthesis. In *Physiology of Trees* (A. S. Raghavendra, ed.), pp. 21–50. Wiley, New York.

Ceulemans, R., Gabriels, R., Impens, I., Yoon, P. K., Leong, W., and Ng, A. P. (1984). Comparative study of photosynthesis in several *Hevea brasiliensis* clones and *Hevea* species under tropical field conditions. *Trop. Agric.* **61**, 273–275.

Ceulemans, R., Janssens, I. A., and Jach, M. E. (1999). Effects of CO_2 enrichment on trees and forests: Lessons to be learned in view of future ecosystem studies. *Ann. Bot.* **84**, 577–590.

Chabot, B. F. and Bunce, J. A. (1979). Drought stress effects on leaf carbon balance. In *Topics in Plant Population Biology* (O. T. Solbrig, S. Jain, G. B. Johnson, and P. H. Raven, eds.), pp. 338–355. Columbia University Press, New York.

Chaffey, N., Barnett, J., and Barlow, P. (1999). A cytoskeletal basis for wood formation in angiosperm trees: The involvement of cortical microtubules. *Planta* **208**, 19–30.

Chaffey, N., Barnett, J., and Barlow, P. (2000). A cytoskeletal basis for wood formation in angiosperm trees: The involvement of microfilaments. *Planta* **210**, 890–896.

Chalk, L. (1937). A note on the meaning of the terms earlywood and latewood. *Leeds Philos. Soc. Proc.* **3**, 324–325.

Chalmers, D. J. and Wilson, I. B. (1978). Productivity of peach trees: Tree growth and water stress in relation to fruit growth and assimilate demand. *Ann. Bot.* **42**, 285–294.

Chandler, P. M. and Robertson, M. (1994). Gene expression regulated by abscisic acid and its relation to stress tolerance. *Annu. Rev. Plant Physiol. Plant Mol. Biol.* **45**, 113–141.

Chandler, J. W. and Dale, J. E. (1990). Needle growth in Sitka spruce (*Picea sitchensis*): Effects of nutrient deficiency and needle position within shoots. *Tree Physiol.* **6**, 41–56.

Chandler, R. F., Jr. (1941). The amount and mineral nutrient content of freshly fallen leaf litter in the hardwood forests of central New York. *J. Am. Soc. Agron.* **33**, 859–871.

Chaney, W. R. (1981). Sources of water. In *Water Deficits and Plant Growth* (T. T. Kozlowski, ed.), Vol. 6, pp. 1–47. Academic Press, New York.

Chaney, W. R. and Kozlowski, T. T. (1977). Patterns of water movement in intact and excised stems of *Fraxinus americana* and *Acer saccharum* seedlings. *Ann. Bot.* **41**, 1093–1100.

Chang, J. and Hanover, J. W. (1991). Geographic variation in the monoterpene composition of black spruce. *Can. J. For. Res.* **21**, 1796–1800.

Chang, S. X. (2003). Seedling sweetgum (*Liquidambar styaciflua* L.) half-sib family response to N and P fertilization: Growth, leaf area, net photosynthesis and nutrient uptake. *For. Ecol. Manage.* **173**, 281–291.

Chapin, F. S. III. (1980). The mineral nutrition of wild plants. *Annu. Rev. Ecol. Syst.* **11**, 233–260.

Chapin, F. S. III. (1991). Effects of multiple stresses on nutrient availability and use. In *Response of Plants to Multiple Stresses* (H. A. Mooney, W. E. Winner, and E. J. Pell, eds.), pp. 67–88. Academic Press, San Diego.

Chapin, F. S. III and Kedrowski, R. A. (1983). Seasonal changes in nitrogen and phosphorus fractions and autumn retranslocation in evergreen and deciduous taiga trees. *Ecology* **64**, 376–391.

Chapin, F. S., Vitousek, P. M., and Van Cleve, K. (1986). The nature of nutrient limitation in plant communities. *Am. Nat.* **127**, 48–58.

Chapin, F. S. III, Bloom, A. J., Field, C. B., and Waring, R. H. (1987). Interaction of environmental factors in controlling plant growth. *BioScience* **37**, 49–57.

Chapin, F. S. III, Moilanen, L., and Kielland, K. (1993). Preferential use of organic nitrogen for growth by a non-mycorrhizal arctic sedge. *Nature (London)* **361**, 150–153.

Chapman, A. G. (1935). The effects of black locust on associated species with special reference to forest trees. *Ecol. Monogr.* **5**, 37–60.

Charlesworth, D. (1989). Why do plants produce so many more ovules than seeds. *Nature (London)* **338**, 21–22.

Chattaway, M. M. (1949). The development of tyloses and secretion of gum in heartwood formation. *Austr. J. Sci. Res.*, **Ser. B2**, 227–240.

Chaw, S., Parkinson, C. L., Cheng, Y., Vincent, T. M., and Palmer, J. D. (2000). Seed plant phylogeny inferred from all three plant genomes: Monophyly of extant gymnosperms and origin of Gnetales from conifers. *Proc. Natl. Acad. Sci. U.S.A.* **97**, 4086–4091.

Chazdon, R. L. and Pearcy, R. W. (1986). Photosynthetic responses to light variation in rain forest species. II. Carbon gain and light utilization during sunflecks. *Oecologia* **69**, 524–531.

Chazdon, R. L. and Pearcy, R. W. (1991). The importance of sunflecks for forest understory plants. *BioScience* **41**, 760–766.

Chazen, O. and Neumann, P. M. (1994). Hydraulic signals from the roots and rapid cell-wall hardening in growing maize (*Zea mays* L.) leaves are primary responses to polyethylene glycol-induced water deficits. *Plant Physiol.* **104**, 1385–1392.

Cheliak, W. M. and Pitel, J. A. (1985). Inheritance and linkage of allozymes in *Larix laricina*. *Silvae Genet.* **34**, 142–148.

Cheliak, W. M., Yeh, F. C. H., and Pitel, J. A. (1987). Use of electrophoresis in tree improvement programs. *For. Chron.* **63**, 89–96.

Chen, F., Nonogaki, H., and Bradford, K. J. (2002). A gibberellin-regulated xyloglucan endotransglycosylase gene is expressed in the endosperm cap during tomato seed germination. *J. Exp. Bot.* **53**, 215–223.

Chen, Y.-Z. and Patterson, B. P. (1985). Ethylene and 1-aminocyclopropane-1-carboxylic acid as indicators of chilling sensitivity in various plant species. *Aust. J. Plant Physiol.* **12**, 377–385.

Cheviclet, C. (1987). Effects of wounding and fungus inoculation on terpene producing systems of maritime pine. *J. Exp. Bot.* **38**, 1557–1572.

Childers, N. F. and White, D. G. (1942). Influence of submersion of the roots on transpiration, apparent photosynthesis, and respiration of young apple trees. *Plant Physiol.* **17**, 603–618.

Chilvers, G. A. and Pryor, L. D. (1965). The structure of Eucalypt mycorrhizas. *Aust. J. Bot.* **13**, 245–249.

Ching, T. M. and Ching, K. K. (1962). Physical and physiological changes in maturing Douglas-fir cones and seeds. *For. Sci.* **8**, 21–31.

Ching, T. M. and Ching, K. K. (1972). Content of adenosine phosphates and adenylate energy charge in germinating ponderosa pine seeds. *Plant Physiol.* **50**, 536–540.

Ching, T. M. and Fang, S. C. (1963). Utilization of labeled glucose in developing Douglas-fir seed cones. *Plant Physiol.* **38**, 551–554.

Chiu, S. T. and Ewers, F. W. (1992). Xylem structure and water transport in a twiner, a scrambler, and a shrub of *Lonicera* (Caprifoliaceae). *Trees.* **6**, 216–224.

Chiu, S.-T., Anton, L. H., Ewers, F. W., Hammerschmidt, R., and Pregitzer, K. S. (1992). Effects of fertilization on epicuticular wax morphology of needle leaves of Douglas-fir, *Pseudotsuga menziesii* (Pinaceae). *Am. J. Bot.* **79**, 149–154.

Choe, S., Noguchi, T., Fujioka, S., Takatsuto, S., Tissier, C. P., Gregory, B. D., Ross, A. S., Tanaka, A., Yoshida, S., Tax, F. E., and Feldmann, K. A. (1999). The *Arabidopsis* dw f 7/ste1 mutant is defective in the {delta}7 sterol C-5 desaturation step leading to brassinosteroid biosynthesis. *Plant Cell* **11**, 207–222.

Choi, H. S. (1992). Variation in water potential components among half-sib families of shortleaf pine (*Pinus echinata*) in response to soil drought. *Can. J. For. Res.* **22**, 111–116.

Choudhury, N. K and Behera, R. K. (2001). Photoinhibition of photosynthesis: Role of carotenoids in photoprotection of chloroplast constituents. *Photosynthetica* **39**, 481–488.

Christensen, N. L. (1973). Fire and the nitrogen cycle in California chaparral. *Science* **181**, 66–68.

Christersson, L. (1972). The transpirational rate of unhardened, hardened, and dehardened seedlings of spruce and pine. *Physiol. Plant.* **26**, 258–263.

Chung, H. H. and Barnes, R. L. (1977). Photosynthate allocation in *Pinus taeda* L. I. Substrate requirements for synthesis of shoot biomass. *Can. J. For. Res.* **7**, 106–111.

Chung, H. H. and Kramer, P. J. (1975). Absorption of water and ^{32}P through suberized and unsuberized roots of loblolly pine. *Can. J. For. Res.* **5**, 229–235.

Clark, J. and Gibbs, R. D. (1957). Studies in tree physiology. IV. Further investigations of seasonal changes in moisture content of certain Canadian forest trees. *Can. J. Bot.* **35**, 219–253.

Clark, W. S. (1874). The circulation of sap in plants. *Mass. State Board Agric. Annu. Rep.* **21**, 159–204.

Clark, W. S. (1875). Observations upon the phenomena of plant life. *Mass. State Board Agric. Annu. Rep.* **22**, 204–312.

Clarkson, D. T., Robards, A. W., and Sanderson, J. (1971). The tertiary endodermis in barley roots: Fine structure in relation to radial transport of ions and water. *Planta* **96**, 292–305.

Clausen, J. J. and Kozlowski, T. T. (1967). Food sources for growth of *Pinus resinosa* shoots. *Advan. Front. Plant Sci.* **18**, 23–32.

Clausen, S. and Apel, K. (1991). Seasonal changes in the concentration of the major storage protein and its mRNA in xylem ray cells of poplar trees. *Plant Molec. Biol.* **17**, 669–678.

Clearwater, M. J. and Clark, C. J. (2003). *In vivo* magnetic resonance imaging of xylem vessel contents in woody lianas. *Plant, Cell Environ.* **26**, 1205–1214.

Clearwater, M. J. and Gould, K. S. (1994). Comparative leaf development of juvenile and adult *Pseudopanax crassifolius*. *Can. J. Bot.* 72, 658–670.

Cleland, R. E. (1988). Molecular events of photoinhibitory inactivation in the reaction centre of Photosystem II. *Aust. J. Plant Physiol.* **15**, 135–150.

Cleland, R. E. (1999). Introduction: Nature occurrence, and functioning of plant hormones. In *Biochemistry and Molecular Biology of Plant Hormones* (P. J. J. Hooykaas, M. A. Hall, and K. R. Libbenga, eds.), pp. 3–60. Elsevier Science, Amsterdam.

Cleland, R. E. (2001). Unlocking the mysteries of leaf primordia formation. *Proc. Natl. Acad. Sci. U.S.A.* **98**, 10981–10982.

Clements, F. E. and Martin, E. V. (1934). Effect of soil temperature on transpiration in *Helianthus annuus*. *Plant Physiol.* **9**, 619–630.

Clifford, S. C., Arndt, S. K., Corlett, J. E., Joshi, S., Sankhla, N., Popp, M., and Jones, H. G. (1998). The role of solute accumulation, osmotic adjustment and changes in cell wall elasticity in drought tolerance in *Ziziphus mauritiana* (Lamk.). *J. Exp. Bot.* **49**, 967–977.

Cline, R. G. and Campbell, G. S. (1976). Seasonal and diurnal water relations of selected forest species. *Ecology* **57**, 367–373.

Close, D. C. and Beadle, C. L. (2003). Chilling-dependent photoinhibition, nutrition and growth analysis of *Eucalyptus nitens* seedlings during establishment. *Tree Physiol.* **23**, 217–226.

Close, D. C., Beadle, C. L., Brown, P. H., and Holz, G. K. (2000). Cold-induced photoinhibition affects establishment of *Eucalyptus nitens* (Deane and Maiden) Maiden and *Eucalyptus globulus* Labill. *Trees* **15**, 32–41.

Close, D. D., Groninger, J. W., Mangun, J. C., and Roth, P. L. (2001). Homeowners' opinions on the practice and effects of topping trees. *J. Arbor.* **27**, 160–165.

Close, T. J., Kort, A. A., and Chandler, P. M. (1989). A cDNA-based comparison of dehydration-induced proteins (dehydrins) in barley and corn. *Plant Mol. Biol.* **13**, 95–108.

Clough, B. F. and Sim, R. G. (1989). Changes in gas exchange characteristics and water use efficiency of mangroves in response to salinity and vapour pressure deficit. *Oecologia* **79**, 38–44.

Cochard, H. (1992). Vulnerability of several conifers to air embolism. *Tree Physiol.* **11**, 73–83.

Cochard, H., Cruziat, P., and Tyree, M. T. (1992). Use of positive pressures to establish vulnerability curves. Further support for the air-seeding hypothesis. *Plant Physiol.* **100**, 205–209.

Coe, J. M. and McLaughlin, S. B. (1980). Winter season corticular photosynthesis in *Cornus florida*, *Acer rubrum*, *Quercus alba*, and *Liriodendron tulipifera*. *For. Sci.* **26**, 561–566.

Cohen, Y., Kelliher, F. M., and Black, T. A. (1985). Determination of sap flow in Douglas-fir trees using the heat pulse technique. *Can. J. For. Res.* **15**, 422–428.

Coile, T. S. (1937). Composition of leaf litter of forest trees. *Soil Sci.* **43**, 349–355.

Cole, D. W. (1986). Nutrient cycling in world forests. In *Forest Site and Productivity* (S. P. Gessel, ed.), pp. 103–125. Martinus Nijhoff, Dordrecht.

Cole, D. W. and Rapp, M. (1981). Elemental cycling in forest ecosystems. In *Dynamic Properties of Forest Ecosystems* (D. E. Reichle, ed.), pp. 341–409. Cambridge University Press, Cambridge.

Cole, F. D. and Decker, J. P. (1973). Relation of transpiration to atmospheric vapor pressure. *J. Ariz. Acad Sci.* **8**, 74–75.

Coleman, G. D., Chen, T. H. H., Ernst, S. G., and Fuchigami, L. (1991). Photoperiod control of poplar bark storage protein accumulation. *Plant Physiol.* **96**, 686–692.

Coleman, G. D., Chen, T. H. H., and Fuchigami, L. H. (1992). Complementary DNA cloning of poplar bark storage protein and control of its expression by photoperiod. *Plant Physiol.* **98**, 687–693.

Coleman, G. D., Banados, M. P., and Chen, T. H. H. (1994). Poplar bark storage protein and a related wound-induced gene are differentially induced by nitrogen. *Plant Physiol.* **106**, 211–215.

Coleman, M. D., Bledsoe, C. S., and Smit, B. A. (1990). Root hydraulic conductivity and xylem sap levels of zeatin riboside and abscisic acid in ectomycorrhizal Douglas fir seedlings. *New Phytol.* **115**, 275–284.

Coley, P. D. (1983). Herbivory and defensive characteristics of tree species in a lowland tropical forest. *Ecol. Monogr.* **53**, 209–233.

Collier, D. E. and Boyer, M. G. (1989). The water relations of *Thuja occidentalis* L. from two sites of contrasting moisture availability. *Bot. Gaz.* **150**, 445–448.

Comstock, J. and Ehleringer, J. (1984). Photosynthetic response to slowly decreasing leaf water potentials in *Encelia frutescens*. *Oecologia* **61**, 241–248.

Conn, E. E., ed. (1981). *The Biochemistry of Plants. Vol. 7. Secondary Plant Products*. Academic Press, New York.

Connor, D. J., Begge, N. J., and Turner, N. C. (1977). Water relations of mountain ash (*Eucalyptus regnans* F. Muell.) forests. *Aust. J. Plant Physiol.* **4**, 753–762.

Cook, E. R. and Jacoby, G. C., Jr. (1977). Tree-ring drought relationships in the Hudson Valley, New York. *Science* **198**, 399–401.

Cook, E. R. and Kairiukstis, L., eds. (1990). *Methods of Dendrochronology*. Kluwer Academic Publishers, Dordrecht, Netherlands.

Cooke, J. E. K. and Weih, M. (2005). Nitrogen storage and seasonal nitrogen cycling in *Populus*: Bridging molecular physiology and ecophysiology. *New Phytol.* **167**, 19–30.

Coombe, B. G. (1976). The development of fleshy fruits. *Annu. Rev. Plant Physiol.* **27**, 507–528.

Corak, S. J., Blevins, D. G., and Pallardy, S. G. (1987). Water transfer in an alfalfa-maize association. *Plant Physiol.* **84**, 582–586.

Cornelius, V. R. (1980). Synergistic effects of NaCl and SO_2 on net photosynthesis of trees. *Angew. Bot.* **54**, 329–335.

Corner, E. J. H. (1966). *The Natural History of Palms*. Univ. of California Press, Berkeley.

Cornish, K. and Zeevart, J. A. D. (1985). Abscisic acid accumulation by roots of *Xanthium strumarium* L. and *Lycopersicon esculentum* Mill. in relation to water stress. *Plant Physiol.* **79**, 653–658.

Correia, M. J. and Pereira, J. S. (1994). Abscisic acid in apoplastic sap can account for the restriction in leaf conductance of white lupins during moderate soil drying and after rewatering. *Plant, Cell Environ.* **17**, 845–852.

Correia, M. J., Pereira, J. S., Chaves, M. M., Rodrigues, M. L., and Pacheco, C. A. (1995). ABA xylem concentrations determine maximum daily leaf conductance in field-grown *Vitis vinifera* L. plants. *Plant, Cell Environ.* **18**, 511–521.

Correia, M. J., Rodrigues, M. L., Ferreira, M. I., and Pereira, J. S. (1997). Diurnal change in the relationship between stomatal conductance and abscisic acid in the xylem sap of field-grown peach trees. *J. Exp. Bot.* **48**, 1727–1736.

Cortelli, A. K. and Pochettino, M. L. (1994). Starch grain analysis as a microscopic diagnostic feature in the identification of plant material. *Econ. Bot.* **48**, 171–181.

Cortes, P. M. and Sinclair, T. R. (1985). The role of osmotic potential in spring sap flow of mature sugar maple trees (*Acer saccharum* Marsh). *J. Exp. Bot.* **36**, 12–24.

Coruzzi, G. and Last, R. (2000). Amino acids. In *Biochemistry and Molecular Biology of Plants* (B. Buchanan, B. W. Gruissem, and R. L. Jones, eds.), pp. 358–410. Amer. Soc. Plant Physiol., Rockville, Maryland.

Cosgrove, D. J. (1997). Relaxation in a high-stress environment—The molecular bases of extensible cell walls and cell enlargement. *Plant Cell* **9**, 1031–1041.

Cosgrove, D. J. (1993). How do plant cell walls extend? *Plant Physiol.* **102**, 1–6.

Cosgrove, D. J. (1997). Assembly and enlargement of the primary cell wall in plants. *Annu. Rev. Cell Dev. Biol.* **13**, 171–201.

Cosgrove, D. J. (1998). Cell wall loosening by expansins. *Plant Physiol.* **118**, 333–339.

Cosgrove, D. J. (1999). Enzymes and other agents that enhance cell wall extensibility. *Annu. Rev. Plant Physiol. Plant Molec. Biol.* **50**, 391–417.

Cosgrove, D. J. (2000). Loosening of plant cell walls by expansins. *Nature* **407**, 321–326.

Cosgrove, D. J. (2001). Wall structure and wall loosening. A look backwards and forwards. *Plant Physiol.* **125**, 131–134.

Cosgrove, D. J., Bedinger, P., and Durachko, D. M. (1997). Group I allergens of grass pollen as cell wall-loosening agents. *Proc. Natl. Acad. Sci. U.S.A.* **94**, 6559–6564.

Coté, B. and Dawson, J. O. (1986). Autumnal changes in total nitrogen, salt extractable proteins and amino acids in leaves and adjacent bark of black alder, eastern cottonwood and white basswood. *Physiol. Plant.* **67**, 102–108.

Coté, W. A. (1963). Structural factors affecting the permeability of wood. *J. Polym. Sci.* **Part C 72**, 231–242.

Coté, W. A. (1967). *Wood Ultrastructure*. Univ. of Washington Press, Seattle.

Coutts, M. P. and Grace, J., eds. (1995). *Wind and Trees*. Cambridge Univ. Press, Cambridge.

Coutts, M. P. and Philipson, J. J. (1978a). Tolerance of tree roots to waterlogging. I. Survival of Sitka spruce and lodgepole pine. *New Phytol.* **80**, 63–69.

Coutts, M. P. and Philipson, J. J. (1978b). The tolerance of tree roots to waterlogging. II. Adaptation of Sitka spruce and lodgepole pine to waterlogged soil. *New Phytol.* **80**, 71–77.

Cowan, I. R. (1977). Stomatal behavior and environment. *Adv. Bot. Res.* **4**, 117–227.

Coxson, D. S. and Nadkarni, N. M. (1995). Ecological roles of epiphytes in nutrient cycles of forest ecosystems. In *Forest Canopies* (M. D. Lowman and N. M. Nadkarni, eds.), pp. 495–543. Academic Press, San Diego.

Crafts-Brandner, S. J. and Salvucci, M. E. (2000). Rubisco activase constrains the photosynthetic potential of leaves at high temperature and CO_2 *Proc. Natl. Acad. Sci. U.S.A.* **97**, 13430–13435.

Crane, J. C. and Al-Shalon, I. (1977). Carbohydrate and nitrogen levels in pistachio branches as related to shoot extension and yield. *J. Am. Soc. Hortic. Sci.* **102**, 396–399.

Crane, J. C., Catlin, P. B., and Al-Shalon, I. (1976). Carbohydrate levels in the pistachio as related to alternate bearing. *J. Am. Soc. Hortic. Sci.* **101**, 371–374.

Crane, W. J. B. and Raison, R. J. (1981). Removal of phosphorus in logs when harvesting *Eucalyptus delegatensis* and *Pinus radiata* on short and long rotations. *Aust. For.* **43**, 253–260.

Cranswick, A. M., Rook, D. A., and Zabkiewicz, J. A. (1987). Seasonal changes in carbohydrate concentration and composition of different tissue types of *Pinus radiata* trees. *N. Z. J. For. Sci.* **17**, 229–240.

Crawford, N. M. and Glass, A. D. M. (1998). Molecular and physiological aspects of nitrate uptake in plants. *Trends Plant Sci.* **3**, 389–395.

Crawford, N. M., Kahn, M. L., Leustek, T., and Long, S. R. (2000). Nitrogen and sulfur. In *Biochemistry and Molecular Biology of Plants* (B. Buchanan, B., W. Gruissem, and R. L. Jones, eds.), pp. 786–849. Amer. Soc. Plant Physiol., Rockville, Maryland.

Crawford, R. M. M. and Braendle, R. (1996). Oxygen deprivation stress in a changing environment. *J. Exp. Bot.* **47**, 145–159.

Creasy, L. L. (1985). Biochemical responses of plants to fungal attack. *Rec. Advan. Phytochem.* **19**, 47–79.

Creber, G. T. and Chaloner, W. G. (1984). Influence of environmental factors on the wood structure of living and fossil trees. *Bot. Rev.* **50**, 357–448.

Critchfield, W. B. (1960). Leaf dimorphism in *Populus trichocarpa*. *Am. J. Bot.* **47**, 699–711.

Critchley, C. (1988). The molecular mechanism of photoinhibition—Facts and fiction. *Aust. J. Plant Physiol.* **15**, 27–41.

Croat, T. B. (1969). Seasonal flowering behavior in central Panama. *Ann. Mo. Bot. Gard.* **56**, 295–307.

Crocker, R. L. and Major, J. (1955). Soil development in relation to vegetation and surface age at Glacier Bay, Alaska. *J. Ecol.* **43**, 427–448.

Cromack, K., Sollins, P., Graustein, W. C., Speidel, K., Todd, A. W., Spycher, G., Li, C. Y., and Todd, R. L. (1979). Calcium oxalate accumulation and soil weathering in mats of the hypogenous fungus, *Hysterangium crassum*. *Soil Biol. Bochem.* **11**, 463–468.

Crombie, D. S., Hipkins, M. F., and Milburn, J. A. (1985). Gas penetration of pit membranes in the xylem of *Rhododendron* as the cause of acoustically detectable sap cavitation. *Aust. J. Plant Physiol.* **12**, 445–553.

Cromer, R. N. and Jarvis, P. G. (1990). Growth and biomass partitioning in *Eucalyptus grandis* seedlings in response to nitrogen supply. *Aust. J. Plant Physiol.* **17**, 503–515.

Cronon, C. S. (1980). Controls on leaching from coniferous forest floor microcosms. *Plant Soil* **56**, 301–322.

Crozier, A., ed. (1983). *The Biochemistry and Physiology of Gibberellins*. Vols. 1 and 2. Praeger, New York.

Crozier, A., Kamiya, Y., Bishop, G., and Yakota, T. (2000). Biosynthesis of hormones and elicitor molecules. In *Biochemistry and Molecular Biology of Plants* (B. Buchanan, B. W. Gruissem, and R. L. Jones, eds.), pp. 850–929. Amer. Soc. Plant Physiol., Rockville, Maryland.

Curlin, J. W. (1970). Nutrient cycling as a factor in site productivity and forest fertilization. In *Tree Growth and Forest Soils* (C. T. Younberg and B. B. Davey, eds.), pp. 313–325. Oregon State University Press, Corvallis.

Curtis, P. S., Vogel, C. S., Gough, C. M., Schmid, H. P., Su, H. B., and Bovard, B. D. (2005). Respiratory carbon losses and the carbon-use efficiency of a northern hardwood forest, 1999–2003. *New Phytol.* **167**, 437–455.

Dadswell, H. E. and Hillis, W. E. (1962). Wood. In *Wood Extractives and Their Significance to the Pulp and Paper Industries* (W. E. Hillis, ed.), pp. 3–55. Academic Press, New York.

Dale, J. E. (1992). How do leaves grow? *BioScience* **42**, 423–432.

Dalton, F. N., Raats, P. A. C., and Gardner, W. R. (1975). Simultaneous uptake of water and solutes by plant roots. *Agron. J.* **67**, 334–339.

Daly, J. M. (1976). The carbon balance of diseased plants. Changes in respiration, photosynthesis and translocation. *Encycl. Plant Physiol. N. S.* **4**, 450–479.

Dancik, B. P. and Barnes, B. V. (1975). Leaf variability in yellow birch (*Betula alleghaniensis*) in relation to environment. *Can. J. For. Res.* **5**, 149–159.

Dang, Q. L. and Cheng, S. (2004). Effects of soil temperature on ecophysiological traits in seedlings of four boreal tree species. *For. Ecol. Manage.* **194**, 379–387.

Dansgaard, W. (1964). Stable isotopes in precipitation. *Tellus* **16**, 436–468.

Dathe, W., Sembdrier, G., Yamaguchi, I., and Takahashi, N. (1982). Gibberellins and growth inhibitors in spring bleeding sap, roots and branches of *Juglans regia* L. *Plant Cell Physiol.* **23**, 115–123.

Davies, F. S., and Flore, J. A. (1986). Gas exchange and flooding stress of highbush and rabbiteye blueberries. *J. Am. Soc. Hort. Sci.* **111**, 565–571.

Davies, P. J. (1988a). The plant hormones: Their nature, occurrence and functions. In *Plant Hormones and Their Role in Plant Growth and Development* (P. J. Davies, ed.), pp. 1–11. Kluwer, Dordrecht.

Davies, P. J. (1988b). The plant hormone concept: Transport, concentration, and sensitivity. In *Plant Hormones and Their Role in Plant Growth and Development* (P. J. Davies, ed.), pp. 12–23. Kluwer, Dordrecht.

Davies, W. J. and Kozlowski, T. T. (1974a). Stomatal responses of five woody angiosperm species to light intensity and humidity. *Can. J. Bot.* **52**, 1525–1534.

Davies, W. J. and Kozlowski, T. T. (1974b). Short- and long-term effects of antitranspirants on water relations and photosynthesis of woody plants. *J. Am. Soc. Hortic. Sci.* **99**, 297–304.

Davies, W. J. and Kozlowski, T. T. (1975a). Effect of applied abscisic acid and silicone on water relations and photosynthesis of woody plants. *Can. J. For. Res.* **5**, 90–96.

Davies, W. J. and Kozlowski, T. T. (1975b). Effects of applied abscisic acid and plant water stress on transpiration of woody angiosperms. *For. Sci.* **21**, 191–195.

Davies, W. J. and Kozlowski, T. T. (1975c). Stomatal responses to changes in light intensity as influenced by plant water stress. *For. Sci.* **21**, 129–133.

Davies, W. J. and Kozlowski, T. T. (1977). Variations among woody plants in stomatal conductance and photosynthesis during and after drought. *Plant Soil* **46**, 435–444.

Davies, W. J. and Pereira, J. S. (1992). Plant growth and water use efficiency. In *Crop Photosynthesis: Spatial and Temporal Determinants* (N. R. Baker and H. Thomas, eds.), pp. 213–233. Elsevier, The Hague.

Davies, W. J. and Zhang, J. H. (1991). Root signals and the regulation of growth and development of plants in drying soil. *Annu. Rev. Plant Physiol. Plant Mol. Biol.* **42**, 55–76.

Davies, W. J., Kozlowski, T. T., Chaney, W. R., and Lee, K. J. (1973). Effects of transplanting on physiological responses and growth of shade trees. *Proc. Int. Shade Tree Conf.* **48th, 1972**, 22–30.

Davies, W. J., Kozlowski, T. T., and Lee, K. J. (1974a). Stomatal characteristics of *Pinus resinosa* and *Pinus strobus* in relation to transpiration and antitranspirant efficiency. *Can. J. For. Res.* **4**, 571–574.

Davies, W. J., Kozlowski, T. T., and Pereira, J. (1974b). Effect of wind on transpiration and stomatal aperture of woody plants. In *Mechanisms of Regulation of Plant Growth* (R. L. Bieleski, A. R. Ferguson, and M. M. Creswell, eds.), pp. 433–438. Royal Society of New Zealand, Wellington.

Davies, W. J., Metcalfe, J., Lodge, T. A., and DaCosta, A. R. (1986). Plant growth substances and the regulation of growth under drought. *Aust. J. Plant Physiol.* **13**, 105–125.

Davies, W. J., Mansfield, T. A., and Hetherington, A. M. (1990). Sensing of soil status and the regulation of plant growth and development. *Plant, Cell Environ.* **13**, 709–720.

Davies, W. J., Wilkinson, S., and Loveys, B. (2002). Stomatal control by chemical signalling and the exploitation of this mechanism to increase water use efficiency in agriculture. *New Phytol.* **153**, 449–460.

Davies, W. J., Kudoyarova, G., and Hartung, W. (2005). Long-distance ABA signaling and its relation to other signaling pathways in the detection of soil drying and the mediation of the plant's response to drought. *J. Plant Growth Reg.* **24**, 285–295.

Davis, J. D. and Evert, R. F. (1965). Phloem development in *Populus tremuloides*. *Am. J. Bot.* **52**, 627.

Davis, S. D. and Mooney, H. A. (1986). Tissue water relations of four co-occurring chaparral shrubs. *Oecologia* **70**, 527–535.

Davis, S. D., Ewers, F. W., Sperry, J. S., Portwood, K. A., Crocker, M. C., and Adams, G. C. (2002). Shoot dieback during prolonged drought in *Ceanothus* (Rhamnaceae) chaparral of California: A possible case of hydraulic failure. *Am. J. Bot.* **89**, 820–828.

Davis, T. A. (1961). High root-pressures in palms. *Nature (London)* **192**, 277–278.

Dawson, T. E. (1993). Hydraulic lift and water use by plants—Implications for water balance, performance and plant-plant interactions. *Oecologia* **95**, 565–574.

Dawson, T. E. and Ehleringer, J. R. (1991). Streamside trees that do not use stream water. *Nature (London)* **350**, 335–337.

Dawson, T. E., Mambelli, S., Plamboeck, A. H., Templer, P. H., and Tu, K. P. (2002). Stable isotopes in plant ecology. *Annu. Rev. Ecol. Syst.* **33**, 507–559.

Day, T. A., Heckathorn, S. A., and Delucia, E. H. (1991). Limitations of photosynthesis in *Pinus taeda* L. (loblolly pine) at low soil temperatures. *Plant Physiol.* **96**, 1246–1254.

Dean, B. B. and Kolattukudy, P. E. (1976). Synthesis of suberin during wound-healing in jade leaves, tomato fruit, and bean pods. *Plant Physiol.* **58**, 411–416.

Dean, M. A., Letner, C. A., and Eley, J. H. (1993). Effect of autumn foliar senescence on chlorophyll a : b ratio and respiratory enzymes of *Populus tremuloides*. *Bull. Torrey Bot. Club* **120**, 269–274.

Dean, T. J., Pallardy, S. G., and Cox, G. S. (1982). Photosynthetic responses of black walnut (*Juglans nigra*) to shading. *Can. J. For. Res.* **12**, 725–730.

Decker, W. L. (1983). Probability of drought for humid and subhumid regions. In *Crop Reactions to Water and Temperature Stresses in Humid, Temperate Climates* (C. D. Raper, Jr. and P. J. Kramer, eds.), pp. 1–19. Westview Press, Boulder, Colorado.

DeJong, T. M. (1982). Leaf nitrogen content and CO_2 assimilation capacity in peach. *J. Amer. Soc. Hortic. Sci.* **107**, 955–959.

DeJong, T. M. (1983). CO_2 assimilation characteristics of five *Prunus* tree fruit species. *J. Am. Soc. Hortic. Sci.* **108**, 303–307.

DeJong, T. M. (1986). A whole plant approach to photosynthetic efficiency in fruit trees. In *The Regulation of Photosynthesis in Fruit Trees* (A. N. Lakso and F. Lenz, eds.), pp. 18–22. New York State Agric. Expt. Sta. Cornell Univ., Ithaca, NY.

DeJong, T. M. and Walton, E. F. (1989). Carbohydrate requirements of peach fruit growth and respiration. *Tree Physiol.* **5**, 329–335.

Dekhuijzen, H. M. (1976). Endogenous cytokinins in healthy and diseased plants. *Encycl. Plant Physiol. N. S.* **4**, 526–559.

Delaney, T. P., Uknes, S., Vernooij, B., Friedrich, L., Weymann, K., Negrotto, D., Gaffney, T., Gut-Rella, M., Kessmann, H., Ward, E., and Ryals, J. (1994). A central role of salicylic acid in plant disease resistance. *Science* **266**, 1247–1250.

Dell, B. and Malajczuk, N. (1994). Boron deficiency in eucalypt plantations in China. *Can. J. For. Res.* **24**, 2409–2416.

DeLucia, E. H. (1986). Effect of low root temperature on net photosynthesis, stomatal conductance and carbohydrate concentration in Engelmann spruce (*Picea engelmannii* Parry ex Engelm.) seedlings. *Tree Physiol.* **2**, 143–154.

DeLucia, E. H. and Heckathorn, S. A. (1989). The effect of soil drought on water-use efficiency in a contrasting Great Basin and Sierran montane species. *Plant, Cell Environ.* **12**, 935–940.

DeLucia, E. H. and Smith, W. K. (1987). Air and soil temperature limitations on photosynthesis in Engelmann spruce during summer. *Can. J. For. Res.* **17**, 527–533.

DeLucia, E. H., Sasek, T. W., and Strain, B. R. (1985). Photosynthetic inhibition after long-term exposure to elevated levels of atmospheric carbon dioxide. *Photosynth. Res.* **7**, 175–184.

DeLucia, E. H., Day, T. A., and Öquist, G. (1991). The potential for photoinhibition of *Pinus sylvestris* L. seedlings exposed to high light and low soil temperature. *J. Exp. Bot.* **42**, 611–617.

DeMichele, D. W. and Sharpe, P. J. H. (1973). An analysis of the mechanics of guard cell motion. *J. Theor. Bio.* **41**, 77–96.

Demmig, B. and Björkman, O. (1987). Comparison of the effect of excessive light on chlorophyll fluorescence (77 K) and photon yield of O_2 evolution in leaves of higher plants. *Planta* **171**, 171–174.

Demmig, B., Winter, K., Kruger, A., and Czygan, F. C. (1987). Photoinhibition and zeaxanthin formation in intact leaves. *Plant Physiol.* **84**, 218–224.

Deng, X., Joly, R. J., and Hahn, D. T. (1989). Effects of plant water deficit on the daily carbon balance of leaves of cacao seedlings. *Physiol. Plant* **77**, 407–412.

Dengler, N. G., Mackay, L. B., and Gregory, L. M. (1975). Cell enlargement and tissue differentiation during leaf expansion in beech, *Fagus grandifolia*. *Can. J. Bot.* **53**, 2846–2865.

Dennis, D. T. and Turpin, D. H. (1990). *Plant Physiology and Molecular Biology*. John Wiley and Sons (Longman Scientific and Technical), New York.

Dennis, F. G., Jr. (1979). Factors affecting yield in apple with emphasis on "Delicious." *Hortic. Rev.* **1**, 395–422.

Dennis, F. G., Jr., Archbold, D. D., and Vecino, C. O. (1983). Effects of inhibitors of ethylene synthesis or action, $GA_{4/7}$ and BA on fruit set of apple, sour cherry, and plum. *J. Am. Soc. Hortic. Sci.* **108**, 570–573.

Denny, H. J. and Ridge, I. (1995). Fungal slime and its role in the mycorrhizal amelioration of zinc toxicity to higher plants. *New Phytol.* **130**, 251–257.

Depuit, E. J. and Caldwell, M. M. (1975). Stem and leaf gas exchange of two arid land shrubs. *Am. J. Bot.* **69**, 954–961.

Derr, W. F. and Evert, R. F. (1967). The cambium and seasonal development of the phloem in *Robinia pseudoacacia*. *Am. J. Bot.* **54**, 147–153.

Deshpande, B. P. (1967). Initiation of cambial activity and its relation to primary growth in *Tilia americana* L. Ph.D. Dissertation, University of Wisconsin, Madison.

Desikan, R., Griffiths, R., Hancock, J., and Neill, S. (2002). A new role for an old enzyme: Nitrate reductase-mediated nitric oxide generation is required for abscisic acid-induced stomatal closure in *Arabidopsis thaliana*. *Proc. Natl. Acad. Sci. U.S.A.* **99**, 16314–16318.

DeSilva, D. L., Cox, R. C., Hetherington, A. M., and Mansfield, T. A. (1985). Synergism between calcium ions and abscisic acid in preventing stomatal opening. *New Phytol.* **100**, 473–482.

Desprez-Loustau, M. L., Marcais, B., Nageleisen, L. M., Piou, D., Vannini, A. (2006). Interactive effects of drought and pathogens in forest trees. *Ann. Sci. For.* **63**, 597–612.

Deuel, H. J. (1951). *The Lipids: Their Chemistry and Biochemistry*. Wiley (Interscience), New York.

Devakumar, A. S., Rao, G. G., Rajagopal, R., Rao, P. S., George, M. J., Vijayakumar, K. R., and Sethuraj, M. R. (1988). Studies on soil-plant-atmosphere system in *Hevea*. II. Seasonal effects on water relations and yield. *Indian J. Natural Rubber Res.* **1**, 45–60.

Dewey, D. H., ed. (1977). *Controlled Atmospheres for the Storage and Transport of Perishable Agricultural Commodities*. Hortic. Rep. No. 28. Michigan State University, East Lansing.

Dick, J. McP., Jarvis, P. G., and Leakey, R. R. B. (1990a). Influence of male cones on early season vegetative growth of *Pinus contorta* trees. *Tree Physiol.* **6**, 105–117.

Dick, J. McP., Leakey, R. R. B., and Jarvis, P. G. (1990b). Influence of female cones on the vegetative growth of *Pinus contorta* trees. *Tree Physiol.* **6**, 151–163.

Dickmann, D. I. (1971). Photosynthesis and respiration by developing leaves of cottonwood (*Populus deltoides* Bartr.). *Bot. Gaz.* **132**, 253–259.

Dickmann, D. I. (1985). The ideotype concept applied to forest trees. In *Attributes of Trees as Crop Plants* (M. G. R. Cannell and J. E. Jackson, eds.), pp. 89–101. Institute of Terrestrial Ecology, Huntingdon, England.

Dickmann, D. I. (1991). Role of physiology in forest tree improvement. *Silva Fennica* **25**, 248–256.

Dickmann, D. I. and Kozlowski, T. T. (1968). Mobilization by *Pinus resinosa* cones and shoots of ^{14}C-photosynthate from needles of different ages. *Am. J. Bot.* **55**, 900–906.

Dickmann, D. I. and Kozlowski, T. T. (1969a). Seasonal growth patterns of ovulate strobili of *Pinus resinosa* in Central Wisconsin. *Can. J. Bot.* **47**, 839–848.

Dickmann, D. I. and Kozlowski, T. T. (1969b). Seasonal variations in reserve and structural components of *Pinus resinosa* Ait. cones. *Am. J. Bot.* **56**, 515–521.

Dickmann, D. I. and Kozlowski, T. T. (1970). Photosynthesis by rapidly expanding green strobili of *Pinus resinosa*. *Life Sci.* **9**, 549–552.

Dickmann, D. I. and Kozlowski, T. T. (1971). Cone size and seed yield in red pine. *Am. Midl. Nat.* **85**, 431–436.

Dickmann, D. I. and Stuart, K. W. (1983). *The Culture of Poplars*. Dept. of Forestry, Michigan State University, East Lansing.

Dickmann, D. I., Gjerstad, D. H., and Gordon, J. C. (1975). Developmental patterns of CO_2 exchange, diffusion resistance and protein synthesis in leaves of *Populus x euramericana*. In *Environmental and Biological Control of Photosynthesis* (R. Marcelle, ed.), pp. 171–181. Junk, The Hague.

Dickson, R. E. (1987). Diurnal changes in leaf chemical constituents and ^{14}C partitioning in cottonwood. *Tree Physiol.* **3**, 157–170.

Dickson, R. E. (1991). Assimilate distribution and storage. In *Physiology of Trees* (A. S. Raghavendra, ed.), pp. 51–85. Wiley, New York.

Dickson, R. E., Broyer, T. C., and Johnson, C. M. (1972). Nutrient uptake by tupelo gum and bald cypress from saturated or unsaturated soil. *Plant Soil* **37**, 297–308.

Digby, J. and Wareing, P. F. (1966). The effect of applied growth hormones on cambial division and the differentiation of the cambial derivatives. *Ann. Bot.* **30**, 539–548.

Dighton, J., Poskitt, J. M., and Howard, D. M. (1986). Changes in occurrence of basidiomycetes fruit bodies during forest stand development: With specific reference to mycorrhizal species. *Trans. Br. Mycol. Soc.* **87**, 163–171.

Dina, S. and Klikoff, L. G. (1973). Carbon dioxide exchange by several streamside and scrub oak community species of Red Butte Canyon, Utah. *Am. Midl. Nat.* **89**, 70–80.

Dixon, H. H. (1914). *Transpiration and the Ascent of Sap in Plants*. Macmillan, London.

Dixon, H. H. and Joly, J. (1895). The path of the transpiration current. *Ann. Bot.* **9**, 416–419.

Dixon, R. K., Garrett, H. E., Bixby, J. A., Cox, G. S., and Thompson, J. G. (1981). Growth, ectomycorrizal development and root soluble carbohydrates of black oak seedlings fertilized by two methods. *For. Sci.* **27**, 617–624.

Dixon, R. K., Pallardy, S. G., Garrett, H. E., Cox, G. S., and Sander, I. L. (1983). Comparative water relations of container-grown and bare-root ectomycorrhizal and nonmycorrhizal *Quercus velutina* seedlings. *Can. J. Bot.* **61**, 1559–1565.

Dizengremel, P. (2001). Effects of ozone on the carbon metabolism of forest trees. *Plant Physiol. Biochem.* **39**, 729–742.

Doblin, M. S., Kurek, I., Jacob-Wilk, D., and Delmer, D. P. (2002). Cellulose biosynthesis in plants: From genes to rosettes. *Plant Cell Physiol.* **43**, 1407–1420.

Dodd, I. C. (2003). Hormonal interactions and stomatal responses. *J. Plant Growth Reg.* **22**, 32–46.

Dodd, I. C. (2005). Root-to-shoot signalling: Assessing the roles of "up" in the up and down world of long-distance signalling *in planta*. *Plant Soil* **274**, 251–270.

Doley, D. and Leyton, L. (1968). Effects of growth regulating substances and water potential on the development of secondary xylem in *Fraxinus*. *New Phytol.* **67**, 579–594.

Domanski, R. and Kozlowski, T. T. (1968). Variations in kinetin-like activity in buds of *Betula* and *Populus* during release from dormancy. *Can. J. Bot.* **46**, 397–403.

Donovan, L. A., Grise, D. J., West, J. B., Pappert, R. A., Alder, N. N., and Richards, J. H. (1999). Predawn disequilibrium between plant and soil water potentials in two cold-desert shrubs. *Oecologia* **120**, 209–217.

Donovan, L. A., Linton, M. J., and Richards, J. H. (2001). Predawn plant water potential does not necessarily equilibrate with soil water potential under well-watered conditions. *Oecologia* **129**, 328–335.

Donovan, L. A., Richards, J. H., and Linton, M. J. (2003). Magnitude and mechanisms of disequilibrium between predawn plant and soil water potentials. *Ecology* **84**, 463–470.

Donselman, H. M. (1976). Geographic variation in eastern redbud (*Cercis canadensis* L.). Ph.D. Thesis, Purdue University.

Döring, B. (1935). *Die Temperaturabhängigkeit der Wasseraufnahme und ihre ökologische Bedeutung*. *Z. Bot.* **28**, 305–383.

Dosskey, M. G., Boersma, L., and Linderman, R. G. (1991). Role for the photosynthate demand of ectomycorrhizas in the response of Douglas fir seedlings to drying soil. *New Phytol.* **117**, 327–334.

Dougherty, P. M. and Hinckley, T. M. (1981). The influence of a severe drought on net photosynthesis of white oak (*Quercus alba*). *Can. J. Bot.* **59**, 335–341.

Downton, W. J. S. (1977). Photosynthesis in salt stressed grapevines. *Aust. J. Plant Physiol.* **4**, 183–192.

Downton, W. J. S., Loveys, B. R., and Grant, W. J. R. (1988). Stomatal closure fully accounts for the inhibition of photosynthesis by abscisic acid. *New Phytol.* **108**, 263–266.

Doyle, D. A., Cabral, J. M., Pfuetzner, R. A., Kuo, A. L., Gulbis, J. M., Cohen, S. L., Chait, B. T., and Mackinnon, R. (1998). The structure of the potassium channel—Molecular basis of K^+ conduction and selectivity. *Science* **280**, 69–77.

Doyle, J. (1945). Developmental lines in pollination mechanisms in the Coniferales. *Sci. Proc. Roy. Dublin Soc.* **24**, 43–62.

Drennan, P. M., Smith, M. T., Goldsworthy, D., and Vanstaden, J. (1993). The occurrence of trehalose in the leaves of the desiccation-tolerant angiosperm *Myrothamnus flabellifolius* Welw. *J. Plant Physiol.* **142**, 493–496.

Dreyer, E., Bosquet, F., and Ducrey, M. (1990). Use of pressure-volume curves in water relations analysis in woody shoots: Influence of rehydration and comparison of four European oak species. *Ann. Sci. For.* **47**, 285–297.

Duchesne, L. C. and Larson, D. W. (1989). Cellulose and the evolution of plant life. *BioScience* **39**, 238–241.

Duchesne, L. C., Ouimet, R., Camire, C., and Houle, D. (2001). Seasonal nutrient transfers by foliar resorption, leaching, and litter fall in a northern hardwood forest at Lake Clair Watershed, Quebec, Canada. *Can. J. For. Res.* **31**, 333–344.

Duddridge, J. A., Malibari, A., and Read, D. J. (1980). Structure and function of mycorrhizal rhizomorphs with special reference to their role in water transport. *Nature* (*London*) **287**, 834–836.

Dueker, J. and Arditti, J. (1968). Photosynthetic $^{14}CO_2$ fixation by green *Cymbidium* (Orchidaceae) flowers. *Plant Physiol.* **43**, 130–132.

Duffy, P. D. and Schreiber, J. P. (1990). Nutrient leaching of a loblolly pine forest floor by simulated rainfall. II. Environmental factors. *For. Sci.* **36**, 777–789.

Duffy, P. D., Schreiber, J. D., and McDowell, L. L. (1985). Leaching of nitrogen, phosphorus, and total organic carbon from loblolly pine litter by simulated rainfall. *For. Sci.* **31**, 750–759.

Dumbroff, E. B. and Peirson, D. R. (1971). Probable sites for passive movement of ions across the endodermis. *Can. J. Bot.* **49**, 35–38.

Dunberg, A. (1976). Changes in gibberellin-like substances and indole-3-acetic acid in *Picea abies* during the period of shoot elongation. *Physiol. Plant.* **38**, 186–190.

Duniway, J. M. (1971). Comparison of pressure chamber and thermocouple psychrometer determinations of leaf water status in tomato. *Plant Physiol.* **48**, 106–107.

Durán-Carril, M. V. and Buján, C. R. (1998). Antioxidant systems in the leaf apoplast compartment of *Pinus pinaster* Ait. and *Pinus radiata* D. Don plants exposed to SO_2. *Ann. Appl. Biol.* **133**, 455–466.

Düring, H. (1985). Osmotic adjustment in grapevines. *Acta Hortic.* **171**, 315–322.

Düring, H. (1992). Low air humidity causes non-uniform stomatal closure in heterobaric leaves of *Vitis* species. *Vitis* **31**, 1–7.

Düring, H. (1999). Photoprotection in leaves of grapevines: Responses of the xanthophyll cycle to alterations of light intensity. *Vitis* **38**, 21–24.

Dyer, T. A. (1984). The chloroplast genome: Its nature and role in development. In *Chloroplast Biogenesis* (N. R. Baker and J. Barber, eds.), pp. 23–69. Elsevier, Amsterdam, Netherlands.

Dyson, W. G. and Herbin, G. A. (1970). Variation in leaf wax alkanes in cypress trees grown in Kenya. *Phytochemistry* **9**, 585–589.

Eagles, C. F. and Wareing, P. F. (1963). Dormancy regulators in woody plants. Experimental induction of dormancy in *Betula pubescens*. *Nature* (*London*) **199**, 874–875.

Eagles, C. F. and Wareing, P. F. (1964). The role of growth substances in the regulation of bud dormancy. *Physiol. Plant.* **17**, 697–709.

Eckert, R. T. and Houston, D. B. (1980). Photosynthesis and needle elongation response of *Pinus strobus* clones at low level sulfur dioxide exposures. *Can. J. For. Res.* **10**, 357–361.

Eckstein, K. and Robinson, J. C. (1995a). Physiological responses of banana (*Musa* AAA; Cavendish sub-group) in the tropics. I. Influence of internal plant factors on gas exchange of banana leaves. *J. Hortic. Sci.* **70**, 147–156.

Eckstein, K. and Robinson, J. C. (1995b). Physiological responses of banana (*Musa* AAA; Cavendish sub-group) in the subtropics. II. Influence of climatic conditions on seasonal and diurnal variations in gas exchange of banana leaves. *J. Hortic. Sci.* **70**, 157–167.

Eckstein, K., Robinson, J. C., and Davie, S. J. (1995). Physiological responses of banana (*Musa* AAA; Cavendish sub-group) in the subtropics. III. Gas exchange, growth analysis and source-sink interaction over a complete crop cycle. *J. Hortic. Sci.* **70**, 169–180.

Edmonds, R. (1980). Litter decomposition and nutrient release in Douglas-fir, red alder, western hemlock, and Pacific silver fir in western Washington. *Can. J. For. Res.* **10**, 327–337.

Edwards, D. G. W. (1980). Maturity and quality of tree seeds—A state-of-the-art review. *Seed Sci. Technol.* **8**, 625–657.

Edwards, N. T. (1991). Root and soil respiration responses to ozone in *Pinus taeda* L. seedlings. *New Phytol.* **118**, 315–321.

Edwards, N. T. and Hanson, P. J. (2003). Aboveground autotrophic respiration. In *North American Temperate Deciduous Forest Responses to Changing Precipitaion Regimes* (P. J. Hanson and S. D. Wullschleger, eds.), pp. 48–66. Springer, New York.

Edwards, N. T. and Riggs, J. S. (2003). Automated monitoring of soil respiration: A moving chamber design. *Soil Sci. Soc. Am. J.* **67**, 1266–1271.

Edwards, W. R. N., Jarvis, P. G., Landsberg, J. J., and Talbot, H. (1986). A dynamic model for studying flow of water in single trees. *Tree Physiol.* **1**, 309–324.

Edwards, W. R. N., Jarvis, P. G., Grace, J., and Moncrieff, J. B. (1994). Reversing cavitation in tracheids of *Pinus sylvestris* L. under negative water potentials. *Plant, Cell Environ.* **17**, 389–397.

Egorov, V. P. and Karpushkin, L. T. (1988). Determination of air humidity over evaporating surface inside a leaf by a compensation method. *Photosynthetica* **22**, 394–404.

Ehleringer, J. (1985). Annuals and perennials of warm deserts. In *Physiological Ecology of North American Plant Communities* (B. F. Chabot and H. A. Mooney, eds.), pp. 162–180. Chapman and Hall, New York.

Ehleringer, J. (1993). Gas-exchange implications of isotopic variation in arid-land plants. In *Water Deficits* (J. A. C. Smith and H. Griffiths, eds.), pp. 265–284. BIOS Scientific Publishers, Oxford, U.K.

Ehleringer, J. and Björkman, O. (1978). Pubescence and leaf spectral characteristics in a desert shrub, *Encelia farinosa*. *Oecologia* **36**, 151–162.

Ehleringer, J. R. and Cook, C. S. (1984). Photosynthesis in *Encelia farinosa* Gray in response to decreasing leaf water potential. *Plant Physiol.* **75**, 688–693.

Ehleringer, J. R. and Cooper, T. A. (1988). Correlations between carbon isotope ratio and microhabitat in desert plants. *Oecologia* **76**, 562–566.

Ehleringer, J. R. and Dawson, T. E. (1992). Water uptake by plants—Perspectives from stable isotope composition. *Plant, Cell Environ.* **15**, 1073–1082.

Ehleringer, J. R. and Monson, R. K. (1993). Evolutionary and ecological aspects of photosynthetic pathway variation. *Annu. Rev. Ecol. Syst.* **24**, 411–439.

Ehleringer, J. R. and Mooney, H. A. (1978). Leaf hairs: Effects on physiological activity and adaptive value to a desert shrub. *Oecologia* **37**, 183–200.

Ehleringer, J., Bjorkman, O., and Mooney, H. A. (1976). Leaf pubescence: Effects on absorptance and photosynthesis in a desert shrub. *Science* **192**, 376–377.

Ehleringer, J. R., Mooney, H. A., Rundel, P. R., Evans, R. D., Palma, B., and Delatorre, J. (1992). Lack of nitrogen cycling in the Atacama Desert. *Nature (London)* **359**, 316–318.

Eichenberger, W. and Grob, E. C. (1962). The biochemistry of plant plastids. 1. A study of autumn coloring. *Helv. Chim. Acta* **45**, 974–981.

Eis, S., Garman, E. H., and Ebell, L. F. (1965). Relation between cone production and diameter increment of Douglas-fir (*Pseudotsuga menziesii* (Mirb.) Franco), grand fir (*Abies grandis* (Dougl.) Lindl.), and western white pine (*Pinus monticola* Dougl.). *Can. J. Bot.* **43**, 1553–1559.

Eisenreich, W., Rohdich, F., and Bacher, A. (2001). Deoxyxylulose phosphate pathway to terpenoids. *Trends Plant Sci.* **6**, 78–84.

Eisenreich, W., Bacher, A., Arigoni, D., and Rohdich, F. (2004). Biosynthesis of isoprenoids via the non-mevalonate pathway. *Cell. Mol. Life Sci.* **61**, 1401–1426.

Eklund, L. and Eliasson, L. (1990). Effects of calcium concentration on cell wall synthesis. *J. Exp. Bot.* **41**, 863–867.

Eklund, L. and Little, C. H. A. (1995). Interaction between indole-3-acetic acid and ethylene in the control of tracheid production in detached shoots of *Abies balsamea*. *Tree Physiol.* **15**, 27–34.

Elias, P. (1983). Water relations pattern of understory trees influenced by sunflecks. *Biol. Plant.* **25**, 68–74.

Ellis, R. C. (1971). Rainfall, fog drip and evaporation in a mountainous area of southern Australia. *Aust. For.* **35**, 99–106.

Ellmore, G. S. and Ewers, F. W. (1986). Fluid flow in the outermost xylem increment of a ring-porous tree, *Ulmus americana*. *Am. J. Bot.* **73**, 1771–1774.

Ellsworth, D. S. and Reich, P. B. (1992). Leaf mass per area, nitrogen content and photosynthetic carbon gain in *Acer saccharum* seedlings in contrasting light environments. *Funct. Ecol.* **6**, 423–435.

Ellsworth, D. S. and Reich, P. B. (1993). Canopy structure and vertical patterns of photosynthesis and related leaf traits in a deciduous forest. *Oecologia* **96**, 169–178.

Elmerich, C., Zimmer, W., and Vielle, C. (1992). Associative nitrogen-fixing bacteria. In *Biological Nitrogen Fixation* (G. Stacey, R. H. Burris, and H. J. Evans, eds.), pp. 212–258. Chapman and Hall, New York.

Elston, T., Wang, H., and Oster, G. (1998). Energy transduction in ATP synthase. *Nature* **391**, 510–513.

Epron, D. and Dreyer, E. (1992). Effects of severe dehydration on leaf photosynthesis in *Quercus petraea* (Matt.) Liebl. Photosystem-II efficiency, photochemical and nonphotochemical fluorescence quenching and electrolyte leakage. *Tree Physiol.* **10**, 273–284.

Epron, D. and Dreyer, E. (1993). Long-term effects of drought on photosynthesis of adult oak trees *Quercus petraea* (Matt.) Liebl. and *Quercus robur* L. in a natural stand. *New Phytol.* **125**, 381–389.

Epron, D., Dreyer, E., and Breda, N. (1992). Photosynthesis of oak trees (*Quercus petraea* (Matt.) Liebl.) during drought under field conditions—diurnal course of net co2 assimilation and photochemical efficiency of Photosystem II. *Plant, Cell Environ.* **15**, 809–820.

Epron, D., Godard, D., Cornic, G., and Genty, B. (1995). Limitation of net CO_2 assimilation rate by internal resistances to CO_2 transfer in the leaves of two tree species (*Fagus sylvatica* L. and *Castanea sativa* Mill.). *Plant, Cell Environ.* **18**, 43–51.

Epstein, E. (1972). *Mineral Nutrition of Plants: Principles and Perspectives*. Wiley, New York.

Epstein, E. (1999). Silicon. *Annu. Rev. Plant Physiol. Plant Molec. Biol.* **50**, 641–644.

Epstein, E. and Bloom, A. J. (2005). *Mineral Nutrition of Plants: Principles and Perspectives, 2nd ed.* Sinauer Associates, Inc., Sunderland, Massachusetts.

Erickson, L. C. and Brannaman, B. L. (1960). Abscission of reproductive structures and leaves of orange trees. *Proc. Am. Soc. Hortic. Sci.* **75**, 222–229.

Ericsson, A. and Persson, A. (1980). Seasonal changes in starch reserves and growth of fine roots of 20-year-old Scots pine. In *Structure and Function of Northern Coniferous Forests—An Ecosystem Study* (T. Persson, ed.), pp. 239–250. Ecol. Bull. 32, Swedish Nat. Sci. Res. Council, Stockholm.

Ericsson, T. (1981). Growth and nutrition of three *Salix* clones in low conductivity solutions. *Physiol. Plant.* **52**, 239–244.

Esau, K. (1965). *Plant Anatomy*. Wiley, New York.

Esau, K. and Cheadle, V. I. (1969). Secondary growth in *Bougainvillea*. *Ann. Bot.* **33**, 807–818.

Eschbach, J. M., Roussel, D., Van de Sype, H., Jacob, J.-L., and D'Auzac, J. (1984). Relationships between yield and clonal physiological characteristics of latex from *Hevea brasiliensis*. *Physiol. Veg.* **22**, 295–304.

Eschbach, J. M., Tupy, J., and LaCrotte, R. (1986). Photosynthate allocation and productivity of latex vessels in *Hevea brasiliensis*. *Biol. Plant.* **28**, 321–328.

Eschrich, W., Burchardt, R., and Essiamah, S. (1989). The induction of sun and shade leaves of the European beech (*Fagus sylvatica* L.): Anatomical studies. *Trees* **3**, 1–10.

Esteban, I., Bergmann, F., Gregorius, H.-R., and Huhtinen, O. (1976). Composition and genetics of monoterpenes from cortical oleoresin of Norway spruce and their significance in clone identification. *Silvae Genet.* **25**, 59–66.

Eun, S. O. and Lee, Y. S. (1997). Actin filaments of guard cells are reorganized in response to light and abscisic acid. *Plant Physiol.* **115**, 1491–1498.

Evans, G. C. (1972). *The Quantitative Analysis of Plant Growth*. Univ. of California Press, Berkeley.

Evans, J. R. (1989). Photosynthesis and nitrogen relationships in leaves of C_3 plants. *Oecologia* **78**, 9–19.

Evans, J. R. and Loreto, F. (2000). Acquisition and diffusion of CO_2 in higher plant leaves. *In Photosynthesis: Physiology and Metabolism* (R. C. Leegood, T. D. Sharkey, and S. von Caemmerer, eds.), pp. 321–351. Advances in Photosynthesis, Volume 9, Kluwer Academic Publishers, Dordrecht.

Evans, P. T. and Malmberg, R. L. (1989). Do polyamines have roles in plant development? *Annu. Rev. Plant Physiol. Plant Mol. Biol.* **40**, 235–269.

Evelyn, J. (1670). *Silva*. J. Martyn and J. Allestry, London.

Evert, R. F. (1960). Phloem structure in *Pyrus communis* L. and its seasonal changes. *Univ. Calif., Berkeley, Publ. Bot.* **32**, 127–194.

Evert, R. F. (1963). The cambium and seasonal development of the phloem in *Pyrus malus*. *Am. J. Bot.* **50**, 149–159.

Evert, R. F. and Kozlowski, T. T. (1967). Effect of isolation of bark on cambial activity and development of xylem and phloem in trembling aspen. *Am. J. Bot.* **54**, 1045–1055.

Evert, R. F., Kozlowski, T. T., and Davis, J. D. (1972). Influence of phloem blockage on cambial growth of sugar maple. *Am. J. Bot.* **59**, 632–641.

Ewel, J. J., Berish, C., Brown, B., Price, N., and Raich, J. (1981). Slash and burn impacts on a Costa Rican wet forest site. *Ecology* **62**, 816–829.

Ewel, K. C., Cropper, W. P., Jr., and Gholz, H. L. (1987). Soil CO_2 evolution in Florida slash pine plantations. II. Importance of root respiration. *Can. J. For. Res.* **17**, 330–333.

Ewers, F. W. (1985). Xylem structure and water conduction in conifer trees, dicot trees, and lianas. *IAWA Bull. N.S.* **6**, 309–317.

Ewers, F. W. and Cruziat, P. (1992). Measuring water transport and storage. In *Techniques and Approaches in Forest Tree Ecophysiology* (J. P. Lassoie and T. M. Hinckley, eds.), pp. 91–115. CRC Press, Boca Raton, FL.

Ewers, F. W. and Schmid, R. (1981). Longevity of needle fascicles of *Pinus longaeva* (bristlecone pine) and other North American pines. *Oecologia* **51**, 107–115.

Ewers, F. W. and Zimmermann, M. H. (1984). The hydraulic architecture of balsam fir (*Abies balsamea*). *Physiol. Plant.* **60**, 453–458.

Faber, B. A., Zasoski, R. J., Munns, D. N., and Shackel, K. (1991). A method for measuring hyphal nutrient and water uptake in mycorrhizal plants. *Can. J. Bot.* **69**, 87–94.

Fady, B., Arbez, M., and Marpeau, A. (1992). Geographic variability of terpene composition in *Abies cephalonica* Loudon and *Abies* species around the Aegean: Hypotheses for their possible phylogeny from the Miocene. *Trees* **6**, 162–171.

Faegri, K. and Iversen, J. (1975). *Textbook of Pollen Analysis.* Hafner, New York.

Fahn, A. (1967). *Plant Anatomy*. Pergamon, Oxford.

Fahn, A. (1979). *Secretory Tissues in Plants.* Academic Press, London.

Fahn, A. (1988a). Secretory tissues and factors influencing their development. *Phyton* **28**, 13–26.

Fahn, A. (1988b). Secretory tissues in vascular plants. *New Phytol.* **108**, 229–257.

Fahn, A. (1990). *Plant Anatomy, 4th edition*. Pergamon Press, Oxford and New York.

Fahn, A. and Zamski, E. (1970). The influence of pressure, wind, wounding and growth substances on the rate of resin duct formation in *Pinus halepensis* wood. *Israel J. Bot.* **19**, 429–446.

Fahn, A., Burley, J., Longman, K. A., Mariaux, A., and Tomlinson, P. B. (1981). Possible contributions of wood anatomy to the determination of age of tropical trees. In *Age and Growth of Tropical Trees* (F. H. Bormann and G. Berlyn, eds.), pp. 31–54. Yale Univ. School of Forestry, Bull. 94, New Haven, Connecticut.

Fails, B. S., Lewis, A. J., and Barden, J. A. (1982). Light acclimatization potential of *Ficus benjamina*. *J. Am. Soc. Hortic. Sci.* **107**, 762–766.

Fairbrothers, D. E., Mabry, T. J., Scogin, R. L., and Turner, B. L. (1975). The basis of angiosperm phylogeny: Chemotaxonomy. *Ann. Mo. Bot. Gard.* **62**, 765–800.

Faiz, S. M. A. and Weatherly, P. E. (1982). Root contractions in transpiring plants. *New Phytol.* **92**, 333–343.

Fanjul, L. and Jones, H. G. (1982). Rapid stomatal responses to humidity. *Planta* **154**, 135–138.

Fanjul, L. and Rosher, P. H. (1984). Effect of water stress on internal water relations of apple leaves. *Physiol. Plant.* **62**, 321–328.

Faria, T., Garciaplazaola, J. I., Abadia, A., Cerasoli, S., Pereira, J. S., and Chaves, M. M. (1996). Diurnal changes in photoprotective mechanisms in leaves of cork oak (*Quercus suber*) during summer. *Tree Physiol.* **16**, 115–123.

Farmer, E. E., Almeras, E., and Krishnamurthy, V. (2003). Jasmonates and related oxylipins in plant responses to pathogenesis and herbivory. *Curr. Opin. Plant Biol.* **6**, 372–378.

Farmer, R. E. (1980). Comparative analysis of first year growth in six deciduous trees. *Can. J. For. Res.* **10**, 35–41.

Farnsworth, E. (2004). Hormones and shifting ecology throughout plant development. *Ecology* **85**, 5–15.

Farquhar, G. D. (1978). Feedforward responses of stomata to humidity. *Aust. J. Plant Physiol.* **5**, 787–800.

Farquhar, G. D. and Raschke, K. (1978). On the resistance to transpiration of the sites of evaporation within the leaf. *Plant Physiol.* **61**, 1000–1005.

Farquhar, G. D., O'Leary, H. H., and Berry, J. A. (1982). On the relationships between carbon isotope discrimination and the intercellular carbon dioxide concentration in leaves. *Aust. J. Plant Physiol.* **9**, 121–137.

Farquhar, G. D., Ehleringer, J. R., and Hubick, K. T. (1989). Carbon isotope discrimination and photosynthesis. *Annu. Rev. Plant Physiol. Plant Mol. Biol.* **40**, 503–537.

Farrell, B. D., Dessourd, D. E., and Mitter, C. (1991). Escalation of plant defenses: Do latex and resin canals spur plant diversification? *Am. Nat.* **138**, 881–900.

Faust, M. (1989). *Physiology of Temperate Zone Fruit Trees*. Wiley, New York.

Faust, M. and Wang, S. W. (1992). Polyamines in horticulturally important plants. *Plant Breeding Rev.* **14**, 333–356.

Fayle, D. C. F. (1968). Radial growth in tree roots. *Univ. Toronto, Fac. For. Tech. Rep.* **9**, Toronto, Canada.

Feeny, P. (1976). Plant apparency and chemical defense. *Rec. Advan. Phytochem.* **10**, 1–40.

Feild, T. S., Lee, D. W., and Holbrook, N. M. (2001). Why leaves turn red in autumn. The role of anthocyanins in senescing leaves of red-osier dogwood. *Plant Physiol.* **127**, 566–574.

Fellows, R. J. and Boyer, J. S. (1978). Altered ultrastructure of cells of sunflower leaves having low water potentials. *Protoplasma.* **93**, 381–395.

Fenn, M. E., Poth, M. A., Aber, J. D., Baron, J. S., Bormann, B. T., Johnson, D. W., Lemly, A. D., McNulty, S. G., Ryan, D. E., and Stottlemyer, R. (1998). Nitrogen excess in North American ecosystems: Predisposing factors, ecosystem responses, and management strategies. *Ecol. Appl.* **8**, 706–733.

Fenton, R. H. and Bond, A. R. (1964). The silvics and silviculture of Virginia pine in southern Maryland. *U.S., For. Serv. Res. Pap. NE* NE–27.

Fereres, E., Cruz–Romero, G., Hoffman, G. J., and Rawlins, S. L. (1979). Recovery of orange trees following severe water stress. *J. Appl. Ecol.* **16**, 833–842.

Ferguson, A. R. (1980). Xylem sap from *Actinidia chinensis*: Apparent differences in sap composition arising from the method of collection. *Ann. Bot.* **46**, 791–801.

Field, C. B. (1988). On the role of photosynthetic responses in constraining the habitat distribution of rainforest plants. *Aust. J. Plant Physiol.* **15**, 343–348.

Field, C. B. (1991). Ecological scaling of carbon gain to stress and resource availability. In *Response of Plants to Multiple Stresses* (H. A. Mooney, W. E. Winner, and E. J. Pell, eds.), pp. 35–65. Academic Press, San Diego.

Field, C. B. and Mooney, H. A. (1983). Leaf age and seasonal effects on light, water, and nitrogen use efficiency in a California shrub. *Oecologia* **56**, 348–355.

Field, C. B. and Mooney, H. A. (1986). The photosynthesis-nitrogen relationship in wild plants. In *On the Economy of Plant Form and Function* (T. J. Givnish, ed.), pp. 25–55. Cambridge University Press, New York.

Field, C. B., Behrenfeld, M. J., Randerson, J. T., and Falkowski, P. (1998). Primary production of the biosphere—Integrating terrestrial and oceanic components. *Science* **281**, 237–240.

Fins, L. and Libby, W. J. (1982). Population variation in *Sequoiadendron*. Seed and seedling studies, vegetative propagation, and isozyme variation. *Silvae Genet.* **31**, 102–110.

Finzi, A. C., DeLucia, E. H., Hamilton, J. G., Richter, D. D., and Schlesinger, W. H. (2002). The nitrogen budget of a pine forest under free air CO_2 enrichment. *Oecologia* **132**, 567–578.

Firbas, R. (1931). *Untersuchungen über den Wasserhaushalt der Hochmoorpflanzen. Jahr. Wiss. Bot.* **74**, 457–696.

Fischer, C. and Höll, W. (1991). Food reserves of Scots pine (*Pinus sylvestris* L.). I. Seasonal changes in the carbohydrate and fat reserves of pine needles. *Trees* **5**, 187–195.

Fischer, C. and Höll, W. (1992). Food reserves of Scots pine (*Pinus sylvestris* L.) II. Seasonal changes and radial distribution of carbohydrate and fat reserves in pine wood. *Trees* **6**, 147–155.

Fischer, R. A. (1968). Stomatal opening: Role of potassium uptake by guard cells. *Science* **160**, 784–785.

Fischer, R. A. and Turner, N. C. (1978). Plant productivity in the arid and semiarid zones. *Annu. Rev. Plant Physiol.* **29**, 277–317.

Fiscus, E. and Kramer, P. J. (1975). General model for osmotic and pressure-induced flow in plant roots. *Proc. Nat. Acad. Sci. U.S.A.* **72**, 3114–3118.

Fisher, H. M. and Stone, E. L. (1990a). Air-conducting porosity in slash pine roots from saturated soils. *For. Sci.* **36**, 18–33.

Fisher, H. M. and Stone, E. L. (1990b). Active potassium uptake by slash pine roots from O_2-depleted solutions. *For. Sci.* **36**, 582–598.

Fisher, R. F. (1972). Spodosol development and nutrient distribution under *Hydnaceae* fungal mats. *Soil Sci. Soc. Am. Proc.* **36**, 492–495.

Flexas, J., Bota, J., Loreto, F., Cornic, G., and Sharkey, T. D. (2004). Diffusive and metabolic limitations to photosynthesis under drought and salinity in C_3 plants. *Plant Biol.* **6**, 269–279.

Flore, J. A. and Lakso, A. N. (1989). Environmental and physiological regulation of photosynthesis in fruit crops. *Hort. Rev.* **11**, 111–157.

Fogel, R. (1983). Root turnover and productivity of coniferous forests. *Plant Soil* **71**, 75–86.

Fogel, R. and Cromack, K. (1977). Effect of habitat and substrate quality on Douglas-fir litter decomposition in Western Oregon. *Can. J. Bot.* **55**, 1632–1640.

Fogel, R. and Hunt, G. (1983). Contribution of mycorrhizae and soil fungi to nutrient cycling in a Douglas-fir ecosystem. *Can. J. For. Res.* **13**, 219–231.

Fonseca, S., Monteiro, L., Barreiro, M. G., and Pais, M. S. (2005). Expression of genes encoding cell wall modifying enzymes is induced by cold storage and reflects changes in pear fruit texture. *J. Exp. Bot.* **56**, 2029–2036.

Foote, K. C. and Schaedle, M. (1976). Diurnal and seasonal patterns of photosynthesis and respiration by stems of *Populus tremuloides* Michx. *Plant Physiol.* **58**, 651–655.

Forrest, G. I. (1980). Genotypic variation among native Scots pine populations in Scotland based on monoterpene analysis. *Forestry* **53**, 101–120.

Fownes, J. H. and Harrington, R. A. (1990). Modeling growth and optimal rotations of tropical multipurpose trees using unit leaf rate and leaf area index. *J. Appl. Ecol* **27**, 886–896.

Foyer, C. H. and Noctor, G. (2000). Oxygen processing in photosynthesis: Regulation and signalling. *New Phytol.* **146**, 359–388.

Frak, E., Millard, P., Le Roux, X., Guillaumie, S., and Wendle, R. (2002). Coupling sap flow velocity and amino acid concentrations as an alternative method to N-15 labeling for quantifying nitrogen remobilization by walnut trees. *Plant Physiol.* **130**, 1043–1053.

Francis, R. and Read, D. J. (1984). Direct transfer of carbon between plants connected by vesicular-arbuscular mycorrhizal mycelium. *Nature (London)* **307**, 53–56.

Franco, A. C. and Luttge, U. (2002). Midday depression in savanna trees: Coordinated adjustments in photochemical efficiency, photorespiration, CO_2 assimilation and water use efficiency. *Oecologia* **131**, 356–365.

Franco, A. C., Ball, E., and Luttge, U. (1991). The influence of nitrogen, light and water stress on CO_2 exchange and organic acid accumulation in the tropical C_3-CAM tree, *Clusia minor. J. Exp. Bot.* **42**, 597–603.

Franich, R. A., Wells, L. G., and Barnett, J. R. (1977). Variations with tree age of needle topography and stomatal structure in *Pinus radiata* D. Don. *Ann. Bot.* **41**, 621–626.

Frankie, G. W., Baker, H. G., and Opler, P. (1974). Applications for studies in community ecology. In *Phenology and Seasonality Modeling* (H. Lieth, ed.), pp. 287–296. Springer-Verlag, New York.

Franklin, E. C. (1976). Within-tree variation of monoterpene composition and yield in slash pine clones and families. *For. Sci.* **22**, 185–192.

Franks, P. J. and Farquhar, G. D. (2007). The mechanical diversity of stomata and its significance in gas-exchange control. *Plant Physiol.* **143**, 78–87.

Fredeen, A. L., Griffin, K., and Field, C. B. (1991). Effects of light qualty and quantity and soil nitrogen status on nitrate reductase activity in rainforest species of the genus *Piper*. *Oecologia* **86**, 441–446.

Freedman, B., Morash, R., and Hanson, A. J. (1981). Biomass and nutrient removal by conventional and whole tree clear-cutting of a red spruce-balsam fir stand in Nova Scotia. *Can. J. For. Res.* **11**, 249–257.

Friedland, A. J., Gregory, R. A., Karenlampi, L., and Johnson, A. H. (1984). Winter damage as a factor in red spruce decline. *Can. J. For. Res.* **14**, 963–965.

Friedrich, J. W. and Huffaker, R. C. (1980). Photosynthesis, leaf resistances, and ribulose-1,5-bisphosphate carboxylase degradation in senescing barley leaves. *Plant Physiol.* **65**, 1103–1107.

Friedrichsen, D. and Chory, J. (2001) Steroid signaling in plants from the cell surface to the nucleus. *Bioessays* **23**, 1028–1036.

Friend, D. J. C. (1984). Shade adaptation of photosynthesis in *Cofffea arabica*. *Photosynth. Res.* **5**, 325–334.

Friesner, R. C. (1940). An observation on the effectiveness of root pressure in the ascent of sap. *Butler Univ. Bot. Stud.* **4**, 226–227.

Fritts, H. C. (1976). *Tree Rings and Climate*. Academic Press, New York.

Fryer, J. H. and Ledig, F. T. (1972). Microevolution of the photosynthetic temperature optimum in relation to the elevational complex gradient. *Can. J. Bot.* **50**, 1231–1235.

Fujii, J. A. and Kennedy, R. A. (1985). Seasonal changes in the photosynthetic rate in apple trees. A comparison between fruiting and nonfruiting trees. *Plant Physiol.* **78**, 519–524.

Fujino, M. (1967). Role of adenosinetriphosphate and adenosinetriphosphatase in stomatal movement. *Sci. Bull. Fac. Educ. Nagasaki Univ.* **18**, 1–47.

Fujita, M., Nakagawa, K., Mori, N., and Haroda, H. (1978). The season of tylosis development and changes in parenchyma cell structure in *Robinia pseudoacacia* L. *Bull. Kyoto Univ. Forests* **No. 50**, 183–190.

Fukshansky, L., Remisowsky, A. M., McClendon, J., Ritterbusch, A., Richter, T., and Mohr H. (1993). Absorption spectra of leaves corrected for scattering and distribution error: A radiative transfer and absorption statistics treatment. *Photochem. Photobiol.* **57**, 538–555.

Fuller, G. and Nes, W. D., eds. (1987). *Ecology and Metabolism of Plant Lipids. Am. Chem. Soc. Symp.* **Series No. 325**, Washington, D.C.

Fung, L. E., Wang, S. S., Altman, A., and Hutterman, A. (1998). Effect of NaCl on growth, photosynthesis, ion and water relations of four poplar genotypes. *For. Ecol. Manage.* **107**, 135–146.

Gaastra, P. (1959). Photosynthesis of crop plants as influenced by light, carbon dioxide, temperature, and stomatal diffusion resistance. *Meded. Landb. Wagen.* **59**, 1–68.

Gabriel, W. J. (1968). Dichogamy in *Acer saccharum*. *Bot. Gaz.* **129**, 334–338.

Gaff, D. F. (1989). Responses of desiccation tolerant "resurrection" plants to water stress. In *Structural and Functional Responses to Environmental Stress: Water Shortage* (H. Kreeb, H. Richter, and T. M. Hinckley, eds.), pp. 255–268. SPB Academic Publishing, The Hague, Netherlands.

Gaffney, T., Friedrich, L., Vernooij, B., Negrotto, D., Nye, G., Uknes, S., Ward, E., Kessmann, H., and Ryals, J. (1993). Requirement of salicylic acid for the induction of systemic acquired resistance. *Science* **261**, 754–756.

Gale, J. and Hagan, R. M. (1966). Plant antitranspirants. *Annu. Rev. Plant Physiol.* **17**, 269–282.

Gallardo, M., Turner, N. C., and Ludwig, C. (1994). Water relations, gas exchange and abscisic acid content of *Lupinus cosentinii* leaves in response to drying different proportions of the root system. *J. Exp. Bot.* **45**, 909–918.

Galston, A. W. (1983). Polyamines as modulators of plant development. *BioScience* **33**, 328–388.

Gälweiler, L., Guan, C., Müller, A., Wisman, E., Mendgen, K., Yephremov, A., and Palme, K. (1998). Regulation of polar auxin transport by AtPIN1 in *Arabidopsis* vascular tissue. *Science* **282**, 2226–2230.

Gambles, R. L. and Dengler, R. E. (1982). The anatomy of the leaf of red pine. I. Nonvascular tissues. *Can. J. Bot.* **60**, 2788–2803.

Gamon, J. A. and Pearcy, R. W. (1989). Leaf movement, stress and photosynthesis in *Vitis californica*. *Oecologia* **79**, 475–481.

Gamon, J. A. and Pearcy, R. W. (1990). Photoinhibition in *Vitis californica*. The role of temperature during high-light treatment. *Plant Physiol.* **92**, 487–494.

Garcia-Mata, C. and Lamattina, L. (2002). Nitric oxide and abscisic acid cross talk in guard cells. *Plant Physiol.* **128**, 790–792.

Gardner, V. R., Bradford, F. C., and Hooker, H. C. (1952). *Fundamentals of Fruit Production*. McGraw-Hill, New York.

Garnett, T. P. and Smethurst, P. J. (1999). Ammonium and nitrate uptake by *Eucalyptus nitens*: Effects of pH and temperature. *Plant Soil* **214**, 133–140.

Garnett, T. P., Shabala, S. N., Smethurst, P. J., and Newman, I. A. (2003). Kinetics of ammonium and nitrate uptake by eucalypt roots and associated proton fluxes measured using ion selective microelectrodes. *Funct. Plant Biol.* **30**, 1165–1176.

Garratt, G. A. (1922). Poisonous woods. *J. For.* **20**, 479–487.

Garten, C. T. and Taylor, G. E. (1992). Foliar delta C-13 within a temperate deciduous forest—Spatial, temporal, and species sources of variation. *Oecologia* **90**, 1–7.

Gartner, B. L., Bulloci, S. H., Mooney, H. A., Brown, V. B., and Whitbeck, J. L. (1990). Water transport properties of vine and tree stems in a tropical deciduous forest. *Am. J. Bot.* **77**, 742–749.

Gartner, T. B. and Cardon, Z. G. (2004). Decomposition dynamics in mixed-species leaf litter. *Oikos* **104**, 230–246.

Gary, H. L. and Watkins, R. K. (1985). Snowpack accumulation before and after thinning a dog-hair stand of lodgepole pine. *U.S., For. Serv. Res. Note RM* **RM–450**.

Gates, D. M. (1965). Energy, plants, and ecology. *Ecology* **46**, 1–13.

Gates, D. M. (1980). *Biophysical Ecology*. Springer-Verlag, New York.

Gates, D. M. (1993). *Climate Change and its Biological Consequences*. Sinauer Associates, Sunderland, Massachusetts.

Gäumann, E. (1935). *Der Stoffhaushalt der Büche (Fagus sylvatica L.) im Laufe eines Jahres. Ber. Deut. Bot. Gesell.* **53**, 366–377.

Gebre, G. M. and Kuhns, M. R. (1991). Seasonal and clonal variations in drought tolerance of *Populus deltoides*. *Can. J. For. Res.* **21**, 910–916.

Gebre, G. M., Kuhns, M. R., and Brandle, J. R. (1994). Organic solute accumulation and dehydration tolerance in three water-stressed *Populus deltoides* clones. *Tree Physiol.* **14**, 575–587.

Gedalovich, E. and Fahn, A. (1985a). The development and ultrastructure of gum ducts in *Citrus* plants formed as a result of brown-rot gummosis. *Protoplasma* **127**, 73–81.

Gedalovich, E. and Fahn, A. (1985b). Ethylene and gum duct formation in *Citrus*. *Ann. Bot.* **56**, 571–577.

Gee, G. W. and Federer, C. A. (1972). Stomatal resistance during senescence of hardwood leaves. *Water Resources Res.* **8**, 1456–1460.

Geeske, J., Aplet, G., and Vitousek, P. (1994). Leaf morphology along environmental gradients in Hawaiian *Metrosideros polymorpha*. *Biotropica* **26**, 17–22.

Geiger, D. R. and Servaites, J. C. (1994). Diurnal regulation of photosynthetic carbon metabolism in C_3 plants. *Annu. Rev. Plant Physiol. Plant Mol. Biol.* **45**, 235–256.

Gerendas J., Polacco, J. C., Freyermuth, S. K., and Sattelmacher, B. (1999). Significance of nickel for plant growth and metabolism. *J. Plant Nutr. Soil Sci.* **162**, 241–256.

Gershenzon, J. (1994). Metabolic costs of terpenoid accumulation in higher plants. *J. Chem. Ecol.* **20**, 1281–1328.

Ghanotakis, D. F. and Yocum, C. F. (1990). Photosystem II and the oxygen-evolving complex. *Annu. Rev. Plant Physiol. Plant Mol. Biol.* **41**, 255–276.

Ghouse, A. K. M. and Hashmi, S. (1978). Seasonal cycle of vascular differentiation in *Polyalthia longifolia* (Anonaceae). *Beitr. Biol. Pflanz.* **54**, 375–380.

Ghouse, A. K. M. and Hashmi, S. (1979). Cambium periodicity in *Polyalthia longifolia*. *Phytomorphology* **29**, 64–67.

Ghouse, A. K. M. and Hashmi, S. (1983). Periodicity of cambium and the formation of xylem and phloem in *Mimusops elengi* L., an evergreen member of tropical India. *Flora* **173**, 479–488.

Giaquinta, R. T. (1980). Translocation of sucrose and oligosaccharides. In *The Biochemistry of Plants* (P. K. Stumpf and E. E. Conn, eds.), Vol. 3, pp. 271–320. Academic Press, New York.

Gibbs, J. N. and Burdekin, D. A. (1980). De-icing salts and crown damage to London plane. *Arboric. J.* **6**, 227–237.

Gibbs, R. D. (1935). Studies of wood. II. The water content of certain Canadian trees, and changes in the water-gas system during seasoning and flotation. *Can. J. Res.* **12**, 727–760.

Gibbs, R. D. (1939). Studies in tree physiology. I. General introduction: Water contents of certain Canadian trees. *Can. J. Res. Sect.* **C17**, 460–482.

Gibbs, R. D. (1953). Seasonal changes in water contents of trees. *Proc. Int. Bot. Congr., 7th, 1950.* pp. 230–231.

Gibbs, R. D. (1958). The Mäule reaction, lignins, and the relationships between woody plants. In *The Physiology of Forest Trees* (K. V. Thimann, ed.), pp. 269–312. Ronald, New York.

Gibbs, R. D. (1974). *Chemotaxonomy of Flowering Plants*. McGill Univ. Press, Montreal, Canada.

Gibson, A. C. (1983). Anatomy of photosynthetic old stems of non-succulent dicotyledons from North American deserts. *Bot. Gaz.* **144**, 347–362.

Gibson, J. P. and Hamrick, J. L. (1991). Genetic diversity and structure in *Pinus pungens* (Table Mountain pine) populations. *Can. J. For. Res.* **21**, 635–642.

Gifford, R. M. (1974). A comparison of potential photosynthesis, productivity and yield of plant species with differing photosynthetic metabolism. *Aust. J. Plant Physiol.* **1**, 107–117.

Gifford, R. M. (2003). Plant respiration in productivity models: Conceptualization, representation and issues for global terrestrial carbon-cycle research. *Funct. Plant Biol.* **30**, 171–186.

Gilbert, S. G., Sell, H. M., and Drosdoff, M. (1946). Effect of copper deficiency on nitrogen metabolism and oil synthesis in the tung tree. *Plant Physiol.* **21**, 290–303.

Gillon, J. S. and Yakir, D. (2000). Internal conductance to CO_2 diffusion and $C^{18}OO$ discrimination in C_3 leaves. *Plant Physiol.* **123**, 201–213.

Gimenez, C., Mitchell, V. J., and Lawlor, D. W. (1992). Regulation of photosynthetic rate of two sunflower hybrids under water stress. *Plant Physiol.* **98**, 516–524.

Ginter-Whitehouse, D. L., Hinckley, T. M., and Pallardy, S. G. (1983). Spatial and temporal aspects of water relations of three tree species with different vascular anatomy. *For. Sci.* **29**, 317–329.

Giovannoni, J. (2001). Molecular biology of fruit maturation and ripening. *Annu. Rev. Plant Physiol. Plant Molec. Biol.* **52**, 725–749.

Giraudat, J., Parcy, F., Bertauche, N., Gosti, F., Leung, J., Morris, P. C., Bouvierdurand, M., and Vartanian, N. (1994). Current advances in abscisic acid action and signalling. *Plant Mol. Biol.* **26**, 1557–1577.

Givnish, T. J., Montgomery, R. A., and Goldstein, G. (2004). Adaptive radiation of photosynthetic physiology in the Hawaiian lobeliads: Light regimes, static light responses, and whole-plant compensation points. *Am. J. Bot.* **91**, 228–246.

Glerum, C. (1980). Food sinks and food reserves of trees in temperate climates. *N. Z. J. For. Sci.* **10**, 176–185.

Glerum, C. and Balatinecz, J. J. (1980). Formation and distribution of food reserves during autumn and their subsequent utilization in jack pine. *Can. J. Bot.* **58**, 40–54.

Glock, W. S., Studhalter, R. A., and Agerter, S. R. (1960). Classification and multiplicity of growth layers in the branches of trees at the extreme lower border. *Smithson. Misc. Collect.* **4421**, 1–292.

Gloser, J. (1967). Some problems of the determination of stomatal aperture by the micro-relief method. *Biol. Plant.* **9**, 152–166.

Goddard, R. E. and Hollis, C. A. (1984). The genetic basis of forest tree nutrition. In *Nutrition of Plantation Forests* (G. D. Bowen and E. K. S. Nambiar, eds.), pp. 237–258. Academic Press, London and New York.

Godwin, H. (1935). The effect of handling on the respiration of cherry laurel leaves. *New Phytol.* **34**, 403–406.

Goetz, M., Godt, D. E., and Roitsch, T. (2000). Tissue-specific induction of the mRNA for an extracellular invertase isoenzyme of tomato by brassinosteroids suggests a role for steroid hormones in assimilate partitioning. *Plant Journal* **22**, 515–522.

Golbeck, J. H. (1992). Structure and function of Photosystem I. *Annu. Rev. Plant Physiol. Plant Mol. Biol.* **43**, 293–324.

Gollan, T., Turner, N. C., and Schulze, E.-D. (1985). The responses of stomata and leaf gas exchange to vapour pressure deficits and soil water content. III. In the sclerophyllous woody species *Nerium oleander*. *Oecologia* **65**, 356–362.

Gollan, T., Passioura, J. B., and Munns, R. (1986). Soil water status affects the stomatal conductance of fully turgid wheat and sunflower leaves. *Aust. J. Plant Physiol.* **13**, 459–464.

Gollan, T., Schurr, U., and Schulze, E.-D. (1992). Stomatal response to drying soil in relation to changes in the xylem sap composition of *Helianthus annuus*. 1. The concentration of cations, anions, amino acids in, and pH of, the xylem sap. *Plant, Cell Environ.* **15**, 551–559.

Golombek, S. D. and Lüdders, P. (1993). Effect of short-term salinity on leaf gas exchange of the fig (*Ficus carica* L.). *Plant Soil* **148**, 21–27.

Gonzalez-Meler, M. A., Ribas-Carbo, M., Siedow J. N., and Drake, B. G. (1996). Direct inhibition of plant mitochondrial respiration by elevated CO_2. *Plant Physiol.* **112**, 1349–1355.

Gonzelez-Meler, M. A., Taneva, L., and Trueman, R. J. (2004). Plant respiration and elevated atmospheric CO_2 concentration: Cellular responses and global significance. *Ann. Bot.* **94**, 647–656.

Goode, J. E. and Higgs, K. H. (1973). Water, osmotic and pressure potential relationships in apple leaves. *J. Hort. Sci.* **48**, 203–215.

Goodwin, R. H. and Goddard, D. R. (1940). The oxygen consumption of isolated woody tissues. *Am. J. Bot.* **27**, 234–237.

Goodwin, T. W. (1958). Studies in carotenogenesis. 24. The changes in carotenoid and chlorophyll pigments in the leaves of deciduous trees during autumn necrosis. *Biochem. J.* **68**, 503–511.

Goodwin, T. W. (1980). *The Biochemistry of the Carotenoids. Vol. I. Plants.* Chapman and Hall, London and New York.

Gordon, J. C. (1968). Effect of IAA and sucrose concentration on cell walls of woody stem segments. *Plant Physiol.* **43**, 5–17.

Gordon, J. C. and Larson, P. R. (1970). Redistribution of ^{14}C-labeled reserve food in young red pines during shoot elongation. *For. Sci.* **16**, 14–20.

Gosz, J. R., Likens, G. E., and Bormann, F. H. (1972). Nutrient content of litterfall on the Hubbard Brook Experimental Forest, New Hampshire. *Ecology* **53**, 769–784.

Gosz, J. R., Likens, G. E., and Bormann, F. H. (1973). Nutrient release from decomposing leaf and branch litter in the Hubbard Brook Forest, New Hampshire. *Ecol. Monogr.* **43**, 173–191.

Gould, K. S. (1993). Leaf heteroblasty in *Pseudopanax crassifolius*: Functional significance of leaf morphology and anatomy. *Ann. Bot.* **71**, 61–70.

Goulden, M. L., Munger, W. J., Fan, S. M., Daube, B. C., and Wofsy, S. C. (1996). Exchange of carbon dioxide by a deciduous forest: Response to interannual climate variability. *Science* **271**, 1576–1578.

Govindjee and Govindjee, R. (1975). Introduction to photosynthesis. In *Bioenergetics of Photosynthesis* (Govindjee, ed.), pp. 1–50. Academic Press, New York.

Gower, S. T., Vogt, K. A., and Grier, C. C. (1992). Carbon dynamics of Rocky Mountain Douglas-fir: Influence of water and nutrient availability. *Ecol. Monogr.* **62**, 43–65.

Gower, S. T., Hunter, A., Campbell, J., Vogel, J., Veldhuis, H., Harden, J., Trumbore, S., Norman, J. M., and Kucharik, C. J. (2000). Nutrient dynamics of the southern and northern BOREAS boreal forests. *Ecoscience* **7**, 481–490.

Gowing, D. G., Davies, W. J., and Jones, H. G. (1990). Positive root-sourced signal as an indicator of soil drying in apple, *Malus* x *domestica* Borkh. *J. Exp. Bot.* **41**, 1535–1540.

Gowing, D. G., Jones, H. G., and Davies, W. J. (1993a). Xylem-transported abscisic acid—The relative importance of its mass and its concentration in the control of stomatal aperture. *Plant, Cell Environ.* **16**, 453–459.

Gowing, D. G., Davies, W. J., Trejo, C. L., and Jones, H. G. (1993b). Xylem-transported chemical signals and the regulation of plant growth and physiology. *Phil. Trans. Roy. Soc. Lond.* **B341**, 41–47.

Grace, J. (1977). *Plant Response to Wind.* Academic Press, New York.

Grace, J. (1993). Consequences of xylem cavitation for plant water deficits. In *Water Deficits: Plant Responses from Cell to Community* (J. A. C. Smith and H. Griffiths, eds.), pp. 109–128. Bios Scientific Publishers, Oxford, U.K.

Grace, J., Malcolm, D. C., and Bradbury, I. K. (1975). The effect of wind and humidity on leaf diffusive resistance in Sitka spruce seedlings. *J. Appl. Ecol.* **12**, 931–940.

Graebe, J. (1987). Gibberellin biosynthesis and control. *Annu. Rev. Plant Physiol.* **38**, 419–465.

Grammatikopoulos, G. and Manetas, Y. (1994). Direct absorption of water by hairy leaves of *Phlomis fruticosa* and its contribution to drought avoidance. *Can. J. Bot.* **72**, 1805–1811.

Grant, V. (1981). *Plant Speciation, 2nd ed.* Columbia Univ. Press, New York.

Grantz, D. A. (1990). Plant response to atmospheric humidity. *Plant, Cell Environ.* **13**, 667–680.

Grassi, G., Millard, P., Gioacchini, P., and Tagliavini, M. (2003). Recycling of nitrogen in the xylem of *Prunus avium* trees starts when spring remobilization of internal reserves declines. *Tree Physiol.* **23**, 1061–1068.

Gray, J. C. (1984). Protein synthesis in chloroplasts. In *Chloroplast Biogenesis* (R. J. Ellis, ed.), Cambridge University Press, New York.

Gray, J. T. (1983). Nutrient use by evergreen and deciduous shrubs in Southern California I. Community nutrient cycling and nutrient use efficiency. *J. Ecol.* **71**, 21–41.

Greenidge, K. N. H. (1954). Studies in the physiology of forest trees. I. Physical factors affecting the movement of moisture. *Am. J. Bot.* **41**, 807–811.

Greer, D. H. and Laing, W. A. (1988). Photoinhibition of photosynthesis in intact kiwifruit (*Actinidia deliciosa*) leaves: Recovery and its dependence on temperature. *Planta* **174**, 159–165.

Greer, D. H., Laing, W. A., and Kipnis, T. (1988). Photoinhibition of photosynthesis in intact kiwifruit (*Actinidia deliciosa*) leaves: Effect of temperature. *Planta* **174**, 152–158.

Gregory, G. F. (1971). Correlation of isolability of the oak wilt pathogen with leaf wilt and vascular water flow resistance. *Phytopathology* **64**, 1003–1005.

Gregory, S. C. and Petty, J. A. (1973). Valve action of bordered pits in conifers. *J. Exp. Bot.* **24**, 763–767.

Gregory, R. P. F. (1989a). *Biochemistry of Photosynthesis, 3rd ed.* Wiley, New York.

Gregory, R. P. F. (1989b). *Photosynthesis.* Chapman and Hall, New York.

Grier, C. C., Vogt, K. A., Keyes, M. R., and Edmonds, R. L. (1981). Biomass distribution and above- and below-ground production in young and mature *Abies amabilis* zone ecosystems of the Washington Cascades. *Can. J. For. Res.* **11**, 155–167.

Grierson, W., Soule, J., and Kawada, K. (1982). Beneficial aspects of physiological stress. *Hort. Rev.* **4**, 247–271.

Grieu, P., Aussenac, G., and Larher, F. (1988). *Secheresse edaphique et concentrations en quelques solutes organiques des tissues foliaire et racinaire de trois especes de coniferes: Cedrus atlantica Manotti, Pseudotsuga macrocarpa (Torr.) Mayr, Pseudotsuga menziesii (Mirb.) Franco. Ann. Sci. For.* **45**, 311–322.

Griffin, K. L., Turnbull, M., and Murthy, R. (2002). Canopy position affects the temperature response of leaf respiration in *Populus deltoides*. *New Phytol.* **154**, 609–619.

Griffis, T. J., Black, T. A., Gaumont-Guay, D., Drewitt, G. B., Nesic, Z., Barr, A. G., Morgenstern, K., and Kljun, N. (2004). Seasonal variation and partitioning of ecosystem respiration in a southern boreal aspen forest. *Agric. For. Meteorol.* **125**, 207–223.

Griffiths, H. (1991). Applications of stable isotope technology in physiological ecology. *Funct. Ecol.* **5**, 254–269.

Griffiths, R. P., Castellano, M. A., and Caldwell, B. A. (1991). Hyphal mats formed by two ectomycorrhizal fungi and their association with Douglas-fir seedlings—A case study. *Plant Soil* **134**, 255–259.

Grigal, D. F., Ohmann, L. F., and Brander, R. B. (1976). Seasonal dynamics of tall shrubs in northeastern Minnesota: Biomass and nutrient element changes. *For. Sci.* **22**, 195–208.

Grime, J. P. (1979). *Plant Strategies and Vegetation Processes.* Wiley, New York.

Groth, G. and Strotmann, H. (1999). New results about structure, function and regulation of the chloroplast ATP synthase (CF_0CF_1). *Physiol. Plant.* **106**, 142–148.

Grunwald, C. (1980). Steroids. *Encycl. Plant Physiol.* **8**, 221–256.

Guehl, J. M., Aussenac, G., Bouachrine, J., Zimmermann, R., Pennes, J. M., Ferhi, A., and Grieu, P. (1991). Sensitivity of leaf gas exchange to atmospheric drought, soil drought, and water-use efficiency in some Mediterranean *Abies* species. *Can. J. For. Res.* **21**, 1507–1515.

Gu, L., Baldocchi, D., Verma, S., Black, T. A., Vesala, T., Falge, E. M., and Dowty, P. R. (2002). Advantages of diffuse radiation for terrestrial ecosystem productivity. *J. Geophys. Res.* **107**, D ACL2-1–ACL2-23.

Gu, L. H., Baldocchi, D. D., Wofsy, S. C., Munger, J. W., Michalsky, J. J., Urbanski, S. P., and Boden, T. A. (2003). Response of a deciduous forest to the Mount Pinatubo eruption: Enhanced photosynthesis. *Science* **299**, 2035–2038.

Guak, S., Neilsen, D., Millard, P., Wendler, R., and Neilsen, G. H. (2003). Determining the role of N remobilization for growth of apple (*Malus domestica* Borkh.) trees by measuring xylem-sap N flux. *J. Exp. Bot.* **54**, 2121–2131.

Guarnaschelli, A. B., Lemcoff, J. H., Prystupa, P., and Basci, S. O. (2003). Responses to drought preconditioning in *Eucalyptus globulus* Labill. provenances. *Trees* **17**, 501–509.

Gucci, R., Massai, R., Xiloyannis, C., and Flore, J. A. (1996). The effect of drought and vapour pressure deficit on gas exchange of young kiwifruit (*Actinidia deliciosa* var. *deliciosa*) vines. *Ann. Bot.* **77**, 605–613.

Guicherd, P., Peltier, J. P., Gout, E., Bligny, R., and Marigo, G. (1997). Osmotic adjustment in *Fraxinus excelsior* L.—Malate and mannitol accumulation in leaves under drought conditions. *Trees* **11**, 155–161.

Gulmon, S. L. and Mooney, H. A. (1986). Costs of defense and their effects on plant productivity. In *On the Economy of Plant Form and Function* (T. J. Givnish, ed.), pp. 681–698. Cambridge, New York.

Gunderson, C. A. and Wullschleger, S. D. (1994). Photosynthetic acclimation in trees to rising atmospheric CO_2: A broader perspective. *Photosynth. Res.* **39**, 369–388.

Gunderson, C. A., Norby, R. J., and Wullschleger, S. D. (1993). Foliar gas exchange responses of two deciduous hardwoods during three years of growth in elevated CO_2: No loss of photosynthetic enhancement. *Plant, Cell Environ.* **16**, 797–807.

Haas, A. R. C. (1936). Growth and water losses in citrus affected by soil temperature. *Calif. Citrogr.* **21**, 467–469.

Hacke, U. G., Sperry, J. S., Wheeler, J. K., and Castro, L. (2006). Scaling of angiosperm xylem structure with safety and efficiency. *Tree Physiol.* **26**, 689–701.

Haddad, Y., Clair–Maczulajtys, D., and Bory, D. (1995). Effects of curtain-like pruning on distribution and seasonal patterns of carbohydrate reserves in plane (*Platanus acerifolia* Wild) trees. *Tree Physiol.* **15**, 135–140.

Hagen-Thorn, A., Varnagiryte, I., Nihlgard, B., and Armolaitis, K. (2006). Autumn nutrient resorption and losses in four deciduous forest tree species. *For. Ecol. Manage.* **228**, 33–39.

Hagihara, A. and Hozumi, K. (1991). Respiration. In *Physiology of Trees* (A. S. Raghavendra, ed.), pp. 87–110. Wiley, New York.

Hagstrom, G. R. (1984). Current management practices for correcting iron deficiency with emphasis on soil management. *J. Plant Nutr.* **7**, 23–46.

Hales, S. (1727). *Vegetable Staticks.* W. & J. Innys and T. Woodward, London.

Halevy, A. H. and Wiitwer, S. H. (1965). Foliar uptake and translocation of rubidium in bean plants affected by root-absorbed growth regulators. *Planta* **67**, 375–383.

Hall, D. M., Matus, A. I., Lamberton, J. A., and Barber, H. N. (1965). Intraspecific variation in wax on leaf surfaces. *Aust. J. Biol. Sci.* **18**, 323–332.

Hallam, N. D. (1967). An electron microscope study of the leaf waxes of the genus *Eucalyptus*, L'Heritier. Ph.D. Dissertation, University of Melbourne.

Hallam, N. D. and Chambers, T. C. (1970). The leaf waxes of the genus *Eucalyptus* L'Heritier. *Aust. J. Bot.* **18**, 335–386.

Hallé, F. (1995). Canopy architecture. In *Forest Canopies* (M. D. Lowman and N. M. Nadkarni, eds.), pp. 27–44. Academic Press, San Diego.

Hallé, F., Oldeman, R. A. A., and Tomlinson, P. B. (1978). *Tropical Trees and Forests: An Architectural Analysis*. Springer-Verlag, Berlin, Heidelberg, and New York.

Hallgren, S. W. and Helms, J. A. (1988). Control of height growth components in seedlings of California red and white fir by seed source and water stress. *Can. J. For. Res.* **18**, 521–529.

Hallgren, S. W., Rudinger, M., and Steudle, E. (1994). Root hydraulic properties of spruce measured with the pressure probe. *Plant Soil* **167**, 91–98.

Hampson, M. C. and Sinclair, W. A. (1973). Xylem dysfunction in peach caused by *Cytospora leucostoma*. *Phytopathology* **63**, 676–681.

Han, Y. S. and Kim, Y. M. (1988). Characteristics of photosynthesis and respiration rate in strobili of *Pinus koraiensis* S. et Z. *J. Korean For. Soc.* **77**, 92–99.

Hanover, J. W. (1966). Genetics of terpenes. 1. Gene control of monoterpene levels in *Pinus monticola* Dougl. *Heredity* **21**, 73–81.

Hansen, E. A. and Dickson, R. E. (1979). Water and mineral nutrient transfer between root systems of juvenile *Populus*. *For. Sci.* **25**, 247–252.

Hansen, E. A. and Dawson, J. O. (1982). Effect of *Alnus glutinosa* on hybrid *Populus* height growth in a short-rotation intensively cultured plantation. *For. Sci.* **28**, 49–50.

Hanson, P. J., Sucoff, E. I., and Markhart, A. H. III. (1985). Quantifying apoplastic flux through red pine root systems using trisodium, 3-hydroxy-5,8,10-pyrenetrisulfonate. *Plant Physiol.* **77**, 21–24.

Hanson, P. J., Isebrands, J. G., Dickson, R. E., and Dixon, R. K. (1988). Ontogenetic patterns of CO_2 exchange of *Quercus rubra* L. leaves during three flushes of shoot growth. I. Median flush leaves. *For. Sci.* **34**, 55–68.

Hanson, P. J., Edwards, N. T., Garten, C. T., and Andrews, J. A. (2000). Separating root and soil microbial contributions to soil respiration: A review of methods and observations. *Biogeochemistry* **48**, 115–146.

Hansted, L., Jakobsen, H. B., and Olsen, C. E. (1994). Influence of temperature on the rhythmic emission of volatiles from *Ribes nigrum* flowers. *Plant, Cell Environ.* **17**, 1069–1072.

Harbinson, J. and Woodward, F. I. (1984). Field measurements of the gas exchange of woody plant species in simulated sunflecks. *Ann. Bot.* **53**, 841–851.

Hardie, K. (1985). The effect of removal of extraradical hyphae on water uptake by vesicular-arbuscular mycorrhizal plants. *New Phytol.* **101**, 677–684.

Hardie, K. (1986). The role of extraradical hyphae in water uptake by vesicular-arbuscular mycorrhizal plants. In *Proc. First European Symp. Mycorrhizae: Physiological and Genetical Aspects of Mycorrhizae*, Dijon, France, 1–5 July, 1985, pp. 651–655.

Hardie, K. and Leyton, L. (1981). The influence of vesicular arbuscular mycorrhiza on growth and water relations of red clover. *New Phytol.* **89**, 599–608.

Hardy, R. W. F. and Havelka, U. D. (1976). Photosynthate as a major factor limiting nitrogen fixation by field-grown legumes with emphasis on soybeans. In *Symbiotic Nitrogen Fixation in Plants* (P. S. Nutman, ed.), pp. 421–439. Cambridge Univ. Press, London and New York.

Hargrave, K. R., Kolb, K. J., Ewers, F. W., and Davis, S. D. (1994). Conduit diameter and drought-induced embolism in *Salvia mellifera* Greene (Labiatae). *New Phytol.* **126**, 695–705.

Harley, J. L. (1973). Symbiosis in the ecosystem. *J. Nat. Sci. Council, Sri Lanka* **1**, 31–48.

Harley, J. L. and Smith, S. E. (1983). *Mycorrhizal Symbiosis*. Academic Press, New York.

Harlow, W. M., Harrar, E. S., and White, F. M. (1979). *Textbook of Dendrology, 6th ed.* McGraw-Hill, New York.

Harmer, R. (1992). Relationships between shoot length, bud number and branch production in *Quercus petraea* (Matt.) Liebl. *Forestry* **65**, 61–72.

Harrell, B. A., Jameson, P. E., and Bannister, P. (1990). Growth regulation and phase change in some New Zealand heteroblastic plants. *N. Z. J. Bot.* **28**, 187–193.

Harrell, D. C. and Williams, L. E. (1987). Net CO_2 assimilation rate of grapevine leaves in response to trunk girdling and gibberellic acid application. *Plant Physiol.* **83**, 457–459.

Harrington, C. (1990). *Alnus rubra* Bong. Red alder. In *Silvics of North America. Vol. 2. Hardwoods* (R. M. Burns and B. Honkala, eds.), pp. 116–123. U.S.D.A. For. Serv., Agric. Hdbk. 654, Washington, D.C.

Harrington, J. F. (1972). Seed storage and longevity. In *Seed Biology* (T. T. Kozlowski, ed.), Vol. 3, pp. 145–245. Academic Press, New York.

Harris, J. M. (1954). Heartwood formation in *Pinus radiata* D. Don. *New Phytol.* **53**, 517–524.

Harris, J. R., Bassuk, N. L., Zobel, R. W., and Whitlow, T. H. (1995). Root and shoot growth periodicity of green ash, scarlet oak, turkish hazelnut, and tree lilac. *J. Am. Soc. Hortic. Sci.* **120**, 211–216.

Harris, R. W. (1992). *Arboriculture*. Prentice-Hall, Englewood Cliffs, New Jersey.

Harry, D. E. (1986). Inheritance and linkage of isozyme variation in incense-cedar. *J. Hered.* **77**, 261–266.

Harry, D. E. and Kimmerer, T. W. (1991). Molecular genetics and physiology of alcohol dehydrogenase in woody plants. *For. Ecol. Manage.* **43**, 251–272.

Hart, J. H. (1968). Morphological and chemical differences between sapwood, discolored sapwood and heartwood in black locust and osage orange. *For. Sci.* **24**, 334–338.

Hartman, F. C. and Harpel, M. R. (1994). Structure, function, regulation, and assembly of d-ribulose-1,5-bisphosphate carboxylase oxygenase. *Annu. Rev. Biochem.* **63**, 197–234.

Hartung, W. and Slovik, S. (1991). Physicochemical properties of plant growth regulators and plant tissues determine their distribution and redistribution—Stomatal regulation by abscisic acid in leaves. *New Phytol.* **119**, 361–382.

Harwood, J. L. and Russell, N. J. (1984). *Lipids in Plants and Microbes*. Allen and Unwin, London.

Hassid, W. Z. (1969). Biosynthesis of oligosaccharides and polysaccharides in plants. *Science* **165**, 137–144.

Hatcher, E. S. J. (1959). Auxin relations of the woody shoot. *Ann. Bot.* **23**, 409–423.

Hauck, R. D. (1968). Nitrogen source requirements in different soil-plant systems. In *Forest Fertilization. Theory and Practice*. pp. 47–57. Tenn. Valley Auth., Muscle Shoals, Alabama.

Haupt, A. W. (1953). *Plant Morphology*. McGraw-Hill, New York.

Havaux, M. (1992). Stress tolerance of Photosystem-II *in vivo*—Antagonistic effects of water, heat, and photoinhibition stresses. *Plant Physiol.* **100**, 424–432.

Haynes, R. J. (1986). *Mineral Nitrogen in the Plant-Soil System*. Academic Press, Orlando, Florida.

Hayward, H. E. and Long, E. M. (1942). The anatomy of the seedling and roots of the Valencia orange. *U.S. Dep. Agric., Tech. Bull.* **786**.

Head, G. C. (1964). A study of "exudation" from the root hairs of apple roots by time-lapse cine-photomicrography. *Ann. Bot.* **28**, 495–498.

Head, G. C. (1965). Studies of diurnal changes in cherry root growth and nutational movements of apple root tips by time-lapse cinematography. *Ann. Bot.* **29**, 219–224.

Head, G. C. (1966). Estimating seasonal changes in the quantity of white unsuberized root on fruit trees. *J. Hortic. Sci.* **41**, 197–206.

Head, G. C. (1967). Effects of seasonal changes in shoot growth on the amount of unsuberized root on apple and plum trees. *J. Hortic. Sci.* **42**, 169–180.

Head, G. C. (1968). Seasonal changes in the diameter of secondarily thickened roots of fruit trees in relation to growth of other parts of the tree. *J. Hortic. Sci.* **43**, 275–282.

Head, G. C. (1973). Shedding of roots. In *Shedding of Plant Parts* (T. T. Kozlowski, ed.), pp. 237–293. Academic Press, New York.

Hedden, P. and Thomas, S. G., eds. (2006). *Plant Hormone Signaling*. Blackwell Publishers, Oxford.

Hedrich, R. and Schroeder, J. I. (1989). The physiology of ion channels and electrogenic pumps in higher plants. *Annu. Rev. Plant Physiol. Plant Mol. Biol.* **40**, 539–569.

Heichel, G. H. and Turner, N. C. (1983). CO_2 assimilation of primary and regrowth foliage of red maple (*Acer rubrum* L.) and red oak (*Quercus rubra* L.): Response to defoliation. *Oecologia* **57**, 14–19.

Heikkala, J. J., Papp, J. T. E., Schultz, G. A., and Bewley, J. D. (1984). Induction of heat shock protein messenger RNA in maize mesocotyls by water stress, abscisic acid and wounding. *Plant Physiol.* **76**, 270–274.

Heiner, T. D. and Lavender, D. P. (1972). Early growth and drought avoidance in Douglas-fir seedlings. *For. Res. Lab. Oregon State University Research Paper No. 14.*

Heinicke, D. R. (1966). The effect of natural shade on photosynthesis and light intensity in Red Delicious apple trees. *Proc. Am. Soc. Hortic. Sci.* **88**, 1–8.

Hejnowicz, A. and Tomaszewski, M. (1969). Growth regulators and wood formation in *Pinus silvestris*. *Physiol. Plant.* **22**, 984–992.

Hellkvist, J., Richards, G. P., and Jarvis, P. G. (1974). Vertical gradients of water potential and tissue water relations in Sitka spruce trees measured with the pressure chamber. *J. Appl. Ecol.* **11**, 637–667.

Helms, J. A. (1965). Diurnal and seasonal patterns of assimilation in Douglas-fir [*Pseudotsuga menziesii* (Mirb.) Franco], as influenced by environment. *Ecology* **46**, 698–708.

Hendrickson, L., Ball, M. C., Osmond. C. B., Furbank, R. T., and Chow, W. S. (2003). Assessment of photoprotection mechanisms of grapevines at low temperature. *Funct. Plant Biol.* **30**, 631–642.

Hepting, G. H. (1945). Reserve food storage in shortleaf pine in relation to little-leaf disease. *Phytopathology* **35**, 106–119.

Herold, A. (1980). Regulation of photosynthesis by sink activity—The missing link. *New Phytol.* **86**, 131–144.

Herrero, M. (1992). From pollination to fertilization in fruit trees. *Plant Growth Regul.* **11**, 27–32.

Hetherington, A. M. (2001). Guard cell signaling. *Cell* **107**, 711–714.

Hetherington, A. M. and Quatrano, R. S. (1991). Mechanisms of action of abscisic acid at the cellular level. *New Phytol.* **119**, 9–32.

Hetherington, A. M. and Woodward, F. I. (2003). The role of stomata in sensing and driving environmental change. *Nature* **424**, 901–908.

Hewitt, E. J. and Cutting, C. V., eds. (1979). *Nitrogen Assimilation of Plants*. Academic Press, London.

Hewlett, J. D. and Hibbert, A. R. (1961). Increases in water yield after several types of forest cutting. *Int. Assoc. Sci. Hydrol.* **6**, 5–17.

Hewlett, J. D. and Hibbert, A. R. (1967). Factors affecting the response of small watersheds to precipitation in humid areas. In *Forest Hydrology* (W. E. Sopper and H. W. Lull, eds.), pp. 275–290. Pergamon, Oxford.

Hillis, W. E. (1987). *Heartwood and Tree Exudates*. Springer-Verlag, Berlin, and New York.

Hinckley, T. M. and Ritchie, G. A. (1973). A theoretical model for calculation of xylem sap pressure from climatological data. *Am. Midl. Nat.* **90**, 56–69.

Hinckley, T. M., Lassoie, J. P., and Running, S. W. (1978a). Temporal and spatial variations in the water status of forest trees. *For. Sci. Monograph* **20**, 72 p.

Hinckley, T. M., Aslin, R. G., Aubuchon, R. R., Metcalf, C. L., and Roberts, J. E. (1978b). Leaf conductance and photosynthesis in four species of the oak-hickory forest type. *For. Sci.* **24**, 73–84.

Hinckley, T. M., Dougherty, P. M., Lassoie, J. P., Roberts, J. E., and Teskey, R. O. (1979). A severe drought: Impact on tree growth, phenology, net photosynthetic rate and water relations. *Am. Midl. Nat.* **102**, 307–316.

Hinckley, T. M., Duhme, F., Hinckley, A. R., and Richter, H. (1980). Water relations of drought hardy shrubs: Osmotic potential and stomatal reactivity. *Plant, Cell Environ.* **3**, 131–140.

Hinckley, T. M., Teskey, R. O., Duhme, F., and Richter, H. (1981). Temperate hardwood forests. In *Water Deficits and Plant Growth*, Vol. 6. (T. T. Kozlowski, ed.), pp. 153–208. Academic Press, New York.

Hitchcock, C. (1975). Structure and distribution of plant acyl lipids. In *Recent Advances in the Chemistry of Biochemistry of Plant Lipids* (T. Galliard and I. Mercer, eds.), pp. 1–19. Academic Press, New York.

Hiwasa, K., Rose, J. K. C., Nakano, R., Inaba, A., and Kubo, Y. (2003). Differential expression of seven α-expansin genes during growth and ripening of pear fruit. *Physiol. Plant.* **117**, 564–572.

Ho, L. C. (1988). Metabolism and compartmentation of imported sugars in sink organs in relation to sink strength. *Annu. Rev. Plant Physiol. Plant Mol. Biol.* **39**, 355–378.

Ho, T.-H., D. (1983). Biochemical mode of action of abscisic acid. In *Abscisic Acid* (F. T. Addicott, ed.), pp. 147–169. Praeger, New York.

Hobson, G. E. (1981). Enzymes and texture changes during ripening. In *Recent Advances in the Biochemistry of Fruits and Vegetables* (J. Friend and M. J. C. Rhodes, eds.), pp. 123–132. Academic Press, London.

Hoch, G. and Körner, C. (2003). The carbon charging of pines at the climatic treeline: A global comparison. *Oecologia* **135**, 10–21.

Hoch, G., Richter, A., and Körner, C. (2003). Non-structural carbon compounds in temperate forest trees. *Plant Cell Environ.* **26**, 1067–1081.

Hoch, W. A., Zeldin, E. L., and McCown, B. H. (2001). Physiological significance of anthocyanins during autumnal leaf senescence. *Tree Physiol.* **21**, 1–8.

Hodges, J. D. (1967). Patterns of photosynthesis under natural environmental conditions. *Ecology* **48**, 234–242.

Hoekstra, F. A., Golovina, E. A., and Buitink, J. (2001). Mechanisms of plant desiccation tolerance. *Trends Plant Sci.* **6**, 431–438.

Hofer, R.-M. (1991). Root hairs. In *Plant Roots—The Hidden Half* (Y. Waisel and A. Eshel, eds.), pp. 129–148. Marcel Dekker, New York, Basel, and Hong Kong.

Hoffmann, G. (1966). *Verlauf der Tiefendurchwurzelung und Feinwurzelbildung bei einigen Baumarten. Arch. Forstwes.* **15**, 825–826.

Hoffmann, W. A. and Poorter, H. (2002). Avoiding bias in calculations of relative growth rate. *Ann. Bot.* **90**, 37–42.

Hofmeyer, J. O. J. and Oberholzer, P. C. V. (1948). Genetic aspects associated with the propagation of citrus. *Farming S. Afr.* **23**, 201–208.

Holdheide, W. (1951). *Anatomische mitteleuropischer Gehölzrinden*. In *Handbuch der Mikroskopie in der Technik* (H. Freund, ed.), pp. 195–367. Umsachen-Verlag, Frankfurt.

Höll, W. and Priebe, S. (1985). Storage lipids in the trunk- and rootwood of *Tilia cordata* Mill. from the dormant to the growing period. *Holzforschung* **39**, 7–10.

Holmes, M. G. and Keiller, D. R. (2002). Effects of pubescence and waxes on the reflectance of leaves in the ultraviolet and photosynthetic wavebands: A comparison of a range of species. *Plant, Cell Environ.* **25**, 85–93.

Holmgren, P., Jarvis, P. G., and Jarvis, M. S. (1965). Resistances to CO_2 and water vapor transfer in leaves of different plant species. *Physiol. Plant* **18**, 557–573.

Holmsgaard, E. (1962). Influence of weather on growth and reproduction of beech. *Commun. Inst. For. Fenn.* **55**, 1–5.

Hölttä, T., Vesala, T., Peramäcki, M., Nikinmaa, E. (2006). Refilling of embolised conduits as a consequence of "Münch water" circulation. *Funct. Plant Biol.* **33**, 949–995.

Hom, J. L. and Oechel, W. C. (1983). The photosynthetic capacity, nutrient content and nutrient use efficiency of different needle age classes of black spruce (*Picea mariana*) found in interior Alaska. *Can. J. For. Res.* **13**, 834–839.

Hook, D. H. (1984). Adaptations to flooding with fresh water. In *Flooding and Plant Growth* (T. T. Kozlowski, ed.), pp. 265–294. Academic Press, New York.

Hoover, M. D. (1944). Effect of removal of forest vegetation upon water yields. *Trans. Am. Geophys. Union* **25**, 969–977.

Hooykaas, P. J. J., Hall, M. A., and Libbenga, K. R., eds. (1999). *Biochemistry and Molecular Biology of Plant Hormones*. Elsevier Science, Amsterdam.

Hornbeck, J. W., Martin, C. W., Pierce, R. S., Bormann, F. H., Likens, G. E., and Eaton, J. S. (1986). Clearcutting northern hardwoods: Effects on hydrologic and nutrient ion budgets. *For. Sci.* **32**, 667–686.

Hornbeck, J. W., Smith, C. T., Martin, C. W., Tritton, L. M., and Pierce, R. S. (1990). Effects of intensive harvesting on nutrient capitals of three forest types in New England. *For. Ecol. Manage.* **30**, 55–64.

Horner, J. D., Gosz, J. R., and Cates, R. G. (1988). The role of carbon-based plant secondary metabolites in decomposition in terrestrial ecosystems. *Am. Nat.* **132**, 869–883.

Hose, E., Clarkson, D. T., Steudle, E., Schreiber, L., and Hartung, W. (2001). The exodermis: A variable apoplastic barrier. *J. Exp. Bot.* **52**, 2245–2264.

Howes, F. N. (1949). *Vegetable Gums and Resins*. Chronica Botanica, Waltham, Mass.

Hsiao, T. C. (1973). Plant responses to water stress. *Annu. Rev. Plant Physiol.* **24**, 519–570.

Huang, R., Smith, W. K., and Yost, R. S. (1985). Influence of vesicular-arbuscular mycorrhiza on growth, water relations and leaf orientation in *Leucaena leucocephala* (Lam.) deWit. *New Phytol.* **99**, 229–243.

Huang, X. M., Lakso, A. N., and Eissenstat, D. M. (2005). Interactive effects of soil temperature and moisture on Concord grape root respiration. *J. Exp. Bot.* **56**, 2651–2660.

Hubbard, N. L., Pharr, D. M., and Huber, S. C. (1991). Sucrose phosphate synthase and other sucrose metabolizing enzymes in fruits of various species. *Physiol. Plant.* **82**, 191–196.

Huber, B. (1928). *Weitere quantitative Untersuchungen über das Wasserleitungssystem der Pflanzen. Jahrb. Wiss. Bot.* **67**, 877–959.

Huck, M. G., Klepper, B., and Taylor, H. M. (1970). Diurnal variations in root diameter. *Plant Physiol.* **45**, 529–530.

Hudler, G. W. and Beale, M. A. (1981). Anatomical features of girdling roots. *J. Arbor.* **7**, 29–32.

Hudler, G. W. and Jensen-Tracy, S. (2000). LAC balsam (R) as a treatment to hasten wound closure and minimize discoloration and decay. *J. Arbor.* **28**, 264–269.

Huffaker, R. C., Radin, T., Kleinkopf, G. E., and Cox, E. L. (1970). Effects of mild water stress on enzymes of nitrate assimilation and of the carboxylative phase of photosynthesis in barley. *Crop Sci.* **10**, 471–474.

Hulbert, C., Funkhouser, E. A., Soltes, E. J., and Newton, R. J. (1988). Inhibition of protein synthesis in loblolly pine hypocotyls by mannitol-induced water stress. *Tree Physiol.* **4**, 19–26.

Humble, G. D. and Hsiao, T. C. (1970). Light dependent influx and efflux of potassium of guard cells during stomatal opening and closing. *Plant Physiol.* **46**, 483–487.

Humble, G. D. and Raschke, K. (1971). Stomatal opening quantitatively related to potassium transport. *Plant Physiol.* **48**, 447–453.

Hunt, R. (1978). *Plant Growth Analysis*. Edward Arnold, London.

Hunt, R. (1990). *Basic Growth Analysis: Plant Growth Analysis for Beginners*. Unwin Hyman, London and Boston.

Hunt, R., Causton, D. R., Shipley, B., and Askew, A. P. (2002). A modern tool for classical plant growth analysis. *Ann. Bot.* **90**, 485–488.

Hunt, R. E., Jr., and Nobel, P. S. (1987). Non-steady-state water flow for three desert perennials with different capacitances. *Aust. J. Plant Physiol.* **14**, 363–375.

Hunter, J. R. (1994). Reconsidering the functions of latex. *Trees* **9**, 1–5.

Husch, B., Beers, T. W., and Kershaw, J. A. (2003). *Forest Mensuration, 4th ed.* Wiley, New York.

Hutchison, K. W., Singer, P. .B, McInnis, S., Diaz-Salaz, C., and Greenwood, M. S. (1999). Expansins are conserved in conifers and expressed in hypocotyls in response to exogenous auxin. *Plant Physiol.* **120**, 827–831.

Huxley, J. S. (1932). *The Problems of Relative Growth*. MacVeagh, The Dial Press, New York.

Huxley, P. A. and Van Eck, W. A. (1974). Seasonal changes in growth and development of some woody perennials near Kampala, Uganda. *J. Ecol.* **62**, 579–592.

Hwang, J. U., Suh, S., Yi, H. J., Kim, J., and Lee, Y. (1997). Actin filaments modulate both stomatal opening and inward K^+-channel activities in guard cells of *Vicia faba* L. *Plant Physiol.* **115**, 335–342.

Ikeda, T. and Kiyohara, T. (1995). Water relations, xylem embolism and histological features of *Pinus thunbergii* inoculated with virulent or avirulent pine wood nematode, *Bursaphelenchus xylophilus*. L. *Exp. Bot.* **46**, 441–449.

Ikuma, H. (1972). Electron transport in plant respiration. *Annu. Rev. Plant Physiol.* **23**, 419–436.

Ilan, I. and Goren, R. (1979). Cytokinins and senescence of lemon leaves. *Physiol. Plant.* **45**, 93–95.

Imamura, M. (1943). *Untersuchungen uber den Mechanisms der Turgorschwankung der Spaltoffnungsschliesszellen. Jap. J. Bot.* **12**, 251–346.

Ingemarsson, B. S. M., Lundqvist, E., and Eliasson, L. (1991). Seasonal variation in ethylene concentration in the wood of *Pinus sylvestris* L. *Tree Physiol.* **8**, 273–279.

Inokuchi, R., Kuma, K., Miyata, T., and Okada, M. (2002). Nitrogen-assimilating enzymes in land plants and algae: Phylogenic and physiological perspectives. *Physiol. Plant.* **116**, 1–11.

Iqbal, M. and Ghouse, A. K. M. (1985a) Cell events of radial growth with special reference to cambium of tropical trees. In *Widening Horizons of Plant Sciences* (C. P. Malik, ed.), pp. 218–252. Cosmo, New Delhi.

Iqbal, M. and Ghouse, A. K. M. (1985b). Impact of climatic variation on the structure and activity of vascular cambium in *Prosopis spicigera*. *Flora* **177**, 147–156.

Isebrands, J. G., Ceulemans, R., and Wiard, B. (1988). Genetic variation in photosynthetic traits among *Populus* clones in relation to yield. *Plant Physiol. Biochem.* **26**, 427–437.

Ishibashi, M. and Terashima, I. (1995). Effects of continuous leaf wetness on photosynthesis: Adverse aspects of rainfall. *Plant, Cell Environ.* **18**, 431–438.

Ishida, A., Toma, T., and Marjenah. (1999). Limitation of leaf carbon gain by stomatal and photochemical processes in the top canopy of *Macaranga conifera*, a tropical pioneer tree. *Tree Physiol.* **19**, 467–473.

Ishida, S., Ohtani, K., and Kawarada, T. (1976). Study of tyloses by the scanning electron microscopy. Yearly and seasonal development of tyloses in Harienju *Robinia pseudo-acacia*. *Trans. 8th Mtg. Hokkaido Branch Jpn. Wood Res. Soc.*, 6–9.

Israelsson, M., Siegel, R. S., Young J., Hashimoto, M., Iba, K., and Schroeder, J. I. (2006). Guard cell CO_2 signaling network update and Ca^{+2} sensor priming hypothesis. *Curr. Opinion Plant Biol.* **9**, 654–663.

Ito, M. (1955). On the water amount and distribution in the stems of Akamatsu (*Pinus densiflora* S. and Z.) and Kuri (*Castanea crenata* S. and Z.) grown at Kurakowa district in Gifu Prefecture. *Fac. Arts Sci., Sci. Rep., Gifu Univ.* **3**, 299–307.

Itoh, S. and Barber, S. A. (1983). A numerical solution of whole plant nutrient uptake for soil-root systems with root hairs. *Plant Soil* **70**, 403–413.

Ivanov, A. G., Sane, P. V., Zeinalov, Y., Malmberg, G., Gardestrom, P., Huner, N. P. A., and Öquist, G. (2001). Photosynthetic electron transport adjustments in overwintering Scots pine (*Pinus sylvestris* L.). *Planta* **213**, 575–585.

Ivanov, L. A. (1924). *Uber die Transpiration der Holzgewachse im Winter. Ber. Dtsch. Bot. Ges.* **42**, 44–49.

Iwahari, S., Tominaya, S., and Higuchi, S. (1990). Retardation of abscission of citrus leaf and fruitlet explants by brassinolide. *Plant Growth Regul.* **9**, 119–125.

Jackson, D. S., Gifford, H. H., and Hobbs, I. W. (1973). Daily transpiration rates of radiata pine. *N. Z. J. For. Sci.* **3**, 70–81.

Jackson, G. E., Irvine, J., Grace, J., and Khalil, A. A. M. (1995). Abscisic acid concentrations and fluxes in droughted conifer saplings. *Plant, Cell Environ.* **18**, 13–22.

Jackson, L. W. R. (1967). Effect of shade on leaf structure of deciduous tree species. *Ecology* **48**, 498–499.

Jackson, M. B. and Armstrong, W. (1999). Formation of aerenchyma and the processes of plant ventilation in relation to soil flooding and submergence. *Plant Biol.* **1**, 274–287.

Jackson, M. B., Herman, B., and Goodenough, A. (1982). An examination of the importance of ethanol in causing injury to flooded plants. *Plant, Cell Environ.* **5**, 163–172.

Jackson, R. B., Canadell, J., Ehleringer, J. R., Mooney, H. A., Sala, O. E., and Schulze, E. D. (1996). A global analysis of root distributions for terrestrial biomes. *Oecologia* **108**, 389–411.

Jackson, R. B., Mooney, H. A., and Schulze, E. D. (1997). A global budget for fine root biomass, surface area, and nutrient contents. *Proc. Natl. Acad. Sci. U.S.A.* **94**, 7362–7366.

Jackson, R. B., Sperry, J. S., and Dawson, T. E. (2000). Root water uptake and transport: Using physiological processes in global predictions. *Trends Plant Sci.* **5**, 482–488.

Janssens, I. A., Lankreijer, H., Matteucci, G., Kowalski, A. S., Buchmann, N., Epron, D., Pilegaard, K., Kutsch, W., et al. (2001). Productivity overshadows temperature in determining soil and ecosystem respiration across European forests. *Global Change Biol.* **7**, 269–278.

Jarbeau, J. A., Ewers, F. W., and Davis, S. D. (1995). The mechanism of water-stress-induced embolism in two species of chaparral shrubs. *Plant, Cell Environ.* **18**, 189–196.

Jarvis, P. G. (1975). Water transfer in plants. In *Heat and Mass Transfer in the Plant Environment* (D. A. Devries and N. G. Afgan, eds.), Part 1, pp. 369–394. Scripta Book Co., Washington, D.C.

Jarvis, P. G. (1985). Transpiration and assimilation of tree and agricultural crops: The omega factor. In *Attributes of Trees as Crop Plants* (M. G. R. Cannell and J. E. Jackson, eds.), pp. 460–480. Institute of Terrestrial Ecology, Huntingdon, England.

Jarvis, P. G. (1989). Atmospheric carbon dioxide and forests. *Phil. Trans. Roy. Soc. Lond.* **B324**, 369–392.

Jarvis, P. G. (1993). Water losses of crowns and canopies. In *Water Deficits: Plant Responses from Cell to Community* (J. A. C. Smith and H. Griffiths, eds.), pp. 285–315. Bios Scientific Publishers, Oxford.

Jarvis, P. G. and McNaugton, K. G. (1986). Stomatal control of transpiration: Scaling up from leaf to region. *Adv. Ecol. Res.* **15**, 1–49.

Jarvis, P. G. and Slatyer, R. O. (1970). The role of the mesophyll cell wall in leaf transpiration. *Planta* **90**, 303–322.

Jarvis, P. G., James, G. B., and Landsberg, J. J. (1976). Coniferous forest. In *Vegetation and the Atmosphere*, Vol. 2 (J. L. Monteith, ed.), pp. 171–240. Academic Press, London and New York.

Jeffree, C. E., Johnson, R. P. C., and Jarvis, P. G. (1971). Epicuticular wax in the stomatal antechamber of Sitka spruce and its effects on the diffusion of water vapor and carbon dioxide. *Planta* **98**, 1–10.

Jeffree, C. E., Dale, J. E., and Fry, S. C. (1988). The genesis of intercellular spaces in developing leaves of *Phaseolus vulgaris* L. *Protoplasma* **132**, 90–98.

Jenkins, M. and Pallardy, S. G. (1995). The influence of drought on red oak group species growth and mortality in the Missouri Ozarks. *Can. J. For. Res.* **25**, 1119–1127.

Jensen, K. F. (1983). Growth relationships in silver maple seedlings fumigated with O_3 and SO_2. *Can. J. For. Res.* **13**, 298–302.

Jensen, K. F. and Kozlowski, T. T. (1974). Effects of SO_2 on photosynthesis of quaking aspen and white ash seedlings. *Proc. North Am. For. Biol. Workshop* **3rd**, p. 359.

Jenkins, P. A. (1974). Influence of applied indoleacetic acid and abscisic acid on xylem cell dimensions in *Pinus radiata* D. Don. In *Mechanisms of Regulation of Plant Growth* (R. L. Bieleski, A. R. Ferguson, and M. M. Creswell, eds.), Bull. 12, pp. 737–742. Royal Society of New Zealand, Wellington.

Jenkins, P. A. and Shepherd, K. R. (1974). Seasonal changes in levels of indoleacetic acid and abscisic acid in stem tissues of *Pinus radiata*. *N. Z. J. For. Sci.* **4**, 511–519.

Johansen, L. G. O. P. C. and Junttila, O. (1986). Abscisic acid and cessation of apical growth in *Salix pentandra*. *Physiol. Plant.* **66**, 409–412.

Johnsen, K. H., Seiler, J. R., and Major, J. E. (1996). Growth, shoot phenology and physiology of diverse seed sources of black spruce: II. 23-year-old field trees. *Tree Physiol.* **16**, 375–380.

Johnson, D. W., West, D. C., Todd, D. E., and Mann, L. K. (1982a). Effects of sawlog vs. whole-tree harvesting on the nitrogen, phosphorus, potassium, and calcium budgets of an upland mixed oak forest. *Soil. Sci. Soc. Am. J.* **46**, 1304–1309.

Johnson, D. W., Cole, D. W., Bledsoe, C. S., Cromack, K., Edmonds, R. L., Gessel, S. P., Grier, C. C., Richards, B. N. and Vogt, K. A., eds. (1982b). Nutrient cycling in forests of the Pacific Northwest. In *Analysis of Coniferous Forest Ecosystems in the Western United States* (R. L. Edmonds, ed.), pp. 186–323. Hutchinson Ross, Stroudsburg, Pennsylvania.

Johnson, D. W., Cheng, W., Joslin, J. D., Norby, R. J., Edwards, N. T., and Todd, D. E. (2004). Effects of elevated CO_2 on nutrient cycling in a sweetgum plantation. *Biogeochemistry* **69**, 379–403.

Johnson, E. D. (1926). A comparison of the juvenile and adult leaves of *Eucalyptus globulus*. *New Phytol.* **25**, 202–212.

Johnson-Flanagan, A. M. and Owens, J. N. (1986). Root respiration in white spruce (*Picea glauca*) (Moench) Voss seedlings in relation to morphology and environment. *Plant Physiol.* **81**, 21–25.

Johnson, G. R. and Hauer, R. J. (2000). A practioner's guide to stem girdling roots of trees. *University of Minnesota Extension Bulletin BU–07501.*

Johnson, M. G., Phillips, D. L., Tingey, D. T., and Storm, M. J. (2000). Effects of elevated CO_2, N-fertilization, and season on survival of ponderosa pine fine roots. *Can. J. For. Res.* **30**, 220–228.

Johnson, R. W. and Tyree, M. T. (1992). Effect of stem water content on sap flow from dormant maple and butternut stems—Induction of sap flow in butternut. *Plant Physiol.* **100**, 853–858.

Johnson, R. W., Tyree, M. T., and Dixon, M. A. (1987). A requirement for sucrose in xylem sap flow from dormant maple trees. *Plant Physiol.* **84**, 495–500.

Johnson, T. B., Augé, R. M., Green, C. D., Stodola, A. J. W., Olinick, J. B., and Saxton, A. M. (2001). Correlations of stomatal conductance with hydraulic, chemical and environmental variables in five urban tree species. *Scientia Hortic.* **90**, 305–320.

Joly, R. J. and Zaerr, J. B. (1987). Alteration of cell-wall water content and elasticity in Douglas-fir during periods of water deficit. *Plant Physiol.* **83**, 418–422.

Jones, B. L. and Porter, J. W. (1985). Biosynthesis of carotenes in higher plants. *Crit. Rev. Plant Sci.* **3**, 295–326.

Jones, C. G., Edson, A. W., and Morse, W. J. (1903). The maple sap flow. *Vt., Agric. Exp. Stn., Bull.* **103**.

Jones, H. G. (1983). *Plants and Microclimate.* Cambridge University Press, Cambridge.

Jones, H. G. (1989). Water stress and stem conductivity. In *Environmental Stress in Plants* (J. H. Cherry, ed.), pp. 17–24. Springer-Verlag, New York.

Jones, H. G. (1993). Drought tolerance and water-use efficiency. In *Water Deficits: Plant Responses from Cell to Community* (J. A. C. Smith and H. Griffiths, eds.), pp. 193–203. Bios Scientific Publishers, Oxford.

Jones, K. (1970). Nitrogen fixation in the phyllosphere of Douglas-fir. *Ann. Bot.* **34**, 239–244.

Jones, R. J. and Mansfield, T. A. (1972). Effects of abscisic acid and its esters on stomatal aperture and the transpiration ratio. *Physiol. Plant.* **26**, 321–327.

Jones, R. L. and MacMillan, J. (1984). Gibberellins. In *Advanced Plant Physiology* (M. B. Wilkins, ed.), pp. 21–52. Pitman, Melbourne.

Jordan, C. F. (1985). *Nutrient Cycling in Tropical Forest Ecosystems.* Wiley, New York.

Jordan, C., Golley, F., Hall, J., and Hall, J. (1980). Nutrient scavenging of rainfall by the canopy of an Amazonian rain forest. *Biotropica* **12**, 61–66.

Jose, S., Merritt, S., and Ramsey, C. L. (2003). Growth, nutrition, photosynthesis and transpiration responses of longleaf pine seedlings to light, water and nitrogen. *For. Ecol. Manage.* **180**, 335–344.

Judd, W. S., Campbell, C. S., Kellogg, E. A., and Stevens, P. F. (1999). *Plant Systematics—A Phylogenetic Approach.* Sinauer Assoc., Sunderland, MA.

Juniper, B. E. and Jeffree, C. E. (1983). *Plant Surfaces.* Edward Arnold, London.

Juntilla, O. (1976). Apical growth cessation and shoot tip abscission in *Salix. Physiol. Plant.* **38**, 278–286.

Juntilla, O. (1982). Gibberellin-like activity in shoots of *Salix pentandra* as related to the elongation growth. *Can. J. Bot.* **60**, 1231–1234.

Juntilla, O. (1991). Gibberellins and the regulation of shoot elongation in woody plants. In *Gibberellins* (N. Takahashi, B. O. Phinney, and J. MacMillan, eds.), pp. 199–210. Springer-Verlag, New York and Berlin.

Juntilla, O. and Jensen, E. (1988). Gibberellins and photoperiodic control of shoot elongation in *Salix. Physiol. Plant.* **74**, 371–376.

Jurik, T. W. (1986). Seasonal patterns of leaf photosynthetic capacity in successional northern hardwood species. *Am. J. Bot.* **73**, 131–138.

Kaiser, W. M. (1987). Effects of water deficit on photosynthetic capacity. *Physiol. Plant.* **71**, 142–149.

Kakegawa, K., Ishii, T., and Matsunaga, T. (2005). Effects of boron deficiency in cell suspension cultures of *Populus alba* L. *Plant Cell Rep.* **23**, 573–578.

Kaldenhoff, R. and Fischer, M. (2006). Aquaporins in plants. *Acta Physiol.* **187**, 169–176.

Kandiko, R. A., Timmis, R., and Worrall, J. (1980). Pressure-volume curves of shoots and roots of normal and drought-conditioned western hemlock seedlings. *Can. J. For. Res.* **10**, 10–16.

Kappen, L. and Haeger, S. (1991). Stomatal responses of *Tradescantia albiflora* to changing air humidity in light and in darkness. *J. Exp. Bot.* **42**, 979–986.

Kappes, E. M. and Flore, J. A. (1986). Carbohydrate balance models for "Montmorency" sour cherry leaves, shoots, and fruits during development. *Acta Hortic.* **184**, 123–127.

Kärki, L. and Tigerstedt, P. M. A. (1985). Definition and exploitation of forest tree ideotypes in Finland. In *Attributes of Trees as Crop Plants* (M. G. R. Cannell and J. E. Jackson, eds.), pp. 102–109. Institute of Terrestrial Ecology, Huntingdon, England.

Karl , T., Fall, R., Rosenstiel, T. N., Prazeller, P., Larsen, B., Seufert, G., and Lindinger, W. (2002). On-line analysis of the (CO_2)-C-13 labeling of leaf isoprene suggests multiple subcellular origins of isoprene precursors. *Planta* **215**, 894–905.

Karlovich, D. A., Groninger, J. W., and Close, D. W. (2000). Tree condition associated with topping in southern Illinois communities. *J. Arbor.* **26**, 87–91.

Katainen, H. S., Mäkinen, E., Jokinen, J., Karjalainen, R., and Kellomäki, S. (1987). Effects of SO_2 on the photosynthetic and respiration rates of Scots pine seedlings. *Environ. Pollut.* **46**, 241–251.

Kato, T. (1981). Major nitrogen compounds transported in xylem vessels from roots to top in citrus trees. *Physiol. Plant.* **52**, 275–279.

Katz, C., Oren, R., Schulze, E. D., and Milburn, J. A. (1989). Uptake of water and solutes through twigs of *Picea abies* (L.) Karst. *Trees* **3**, 33–37.

Kaufmann, M. R. (1975). Leaf water stress in Engelmann spruce. Influence of the root and shoot environments. *Plant Physiol.* **56**, 841–844.

Kaufmann, M. R. (1976). Water transport through plants—Current perspectives. In *Transport and Transfer Processes in Plants* (I. Wardlaw and J. Passioura, eds.), pp. 313–317. Academic Press, New York.

Kaufmann, M. R. and Fiscus, E. L. (1985). Water transport through plants—Internal integration of edaphic and atmospheric effects. *Acta Hortic.* **171**, 83–93.

Kaufmann, M. R. and Kelliher, F. M. (1991). Measuring transpiration rates. In *Techniques and Approaches in Forest Tree Ecophysiology*, (J. P. Lassoie and T. M. Hinckley, eds.), pp. 117–140. CRC Press, Boca Raton, Florida.

Kaufmann, M. R. and Troendle, C. A. (1981). The relationship of leaf area and foliage biomass to sapwood conducting area in four subalpine forest tree species. *For. Sci.* **27**, 477–482.

Kauhanen, H. (1986). Stomatal resistance, photosynthesis and water relations in mountain birch in the subarctic. *Tree Physiol.* **2**, 123–130.

Kavanaugh, K. L. (1992). Xylem cavitation: A cause of foliage loss and mortality in newly planted western hemlock (*Tsuga hetero-*

phylla) seedlings. *Proc. 12th North American For. Biol. Workshop.* Abstr. Sault Ste. Marie, Ontario. p. 126.

Ke, B. (2001). *Photosynthesis: Photobiochemistry and Photobiophysics.* Advances in Photosynthesis, Vol. 10. Kluwer Academic Publishers, Dordrecht.

Keck, R. W. and Boyer, J. S. (1974). Chloroplast response to low leaf water potentials. 3. Differing inhibition of electron transport and photophosphorylation. *Plant Physiol.* **53**, 474–479.

Keller, T. (1966). *Uber den Einfluss von transpirationshemmenden Chemikalien (Antitranspiratien) auf Transpiration, CO_2–Aufnahme und Wurzelwachstum van Jungfichten. Forstwiss. Centralbl.* **85**, 65–79.

Keller, T. (1973a). CO_2 exchange of bark of deciduous species in winter. *Photosynthetica* **7**, 320–324.

Keller, T. (1973b). *Über die schädigende Wirkung des Fluors. Schweiz. Zeitsch. Forstw.* **124**, 700–706.

Keller, T. (1977a). On the phytotoxicity of dust-like fluoride compounds. *Staub-Reinhalt. Luft* **33**, 379–381.

Keller, T. (1977b). The effect of long term low SO_2 concentrations upon photosynthesis of conifers. *Proc. Int. Clean Air Congr.* **4th, 1977**, 81–83.

Keller, T. (1980). The simultaneous effect of soil-borne NaF and air pollutant SO_2 on CO_2-uptake and pollutant accumulation. *Oecologia* **44**, 283–285.

Keller, T. (1983). Air pollutant deposition and effects on plants. In *Effects of Accumulation of Air Pollutants in Forest Ecosystems* (B. Ulrich and J. Pankrath, eds.), pp. 285–294. Reidel Publishing Co., Dordrecht.

Keller, T. and Koch, W. (1962). *Der Einfluss der Mineralstoffernährung auf CO_2–Gaswechsel und Blattpigmentgehalt der Pappel. II. Eisen. Mitt. Schweig. Anst. Forstl. Versuchswes.* **38**, 283–318.

Keller, T. and Koch, W. (1964). The effect of iron chelate fertilization of poplar upon CO_2 uptake, leaf size, and contents of leaf pigments and iron. *Plant Soil* **20**, 116–126.

Kelly, G. J., Latzko, E., and Gibbs, M. (1976). Regulatory aspects of photosynthetic carbon metabolism. *Annu. Rev. Plant Physiol.* **27**, 181–205.

Kennedy, R. A. and Johnson, D. (1981). Changes in photosynthetic characteristics during leaf development in apple. *Photosyn. Res.* **2**, 213–223.

Kenrick, J. R. and Bishop, D. G. (1986). Phosphatidylglycerol and sulfoquinovosyldiacyl-glycerol in leaves and fruits of chilling sensitive plants. *Phytochemistry* **25**, 1293–1296.

Kepinski, S. and Leyser, O. (2002). Ubiquitination and auxin signaling: A degrading story. *Plant Cell* **14**, S81–S95.

Kerstiens, G. (1996). Cuticular water permeability and its physiological significance. *J. Exp. Bot.* **47**, 1813–1832.

Kerstiens, G. (2006). Water transport in plant cuticles: An update. *J. Exp. Bot.* **57**, 2493–2499.

Keyes, M. R. and Grier, C. C. (1981). Above- and below-ground net production in 40-year-old Douglas-fir stands on low and high productivity sites. *Can. J. For. Res.* **11**, 599–605.

Khan, M. A., Kalimullah, and Ahmad, Z. (1992). Area occupied by sieve elements in the secondary phloem of some leguminous forest trees of Madhya Pradesh, India. *Flora* **186**, 311–315.

Kidd, P. S., Llugany, M., Poschenrieder, C., Gunsé, B., and Barceló, J. (2001). The role of root exudates in aluminium resistance and silicon-induced amelioration of aluminium toxicity in three varieties of maize (*Zea mays* L.). *J. Exp. Bot.* **52**, 1339–1352.

Kiedaisch, B. M., Blanton, C. R. L., and Haigler, C. H. (2003). Characterization of a novel cellulose synthesis inhibitor. *Planta* **217**, 922–930.

Kielland, K. (1994). Amino acid absorption by arctic plants: Implications for plant nutrition and nitrogen recycling. *Ecology* **75**, 2373–2383.

Kikuzawa, K. (1982). Leaf survival and evolution in Betulaceae. *Ann. Bot.* **50**, 345–353.

Kim, S.-K. (1991). Natural occurrence of brassinosteroids. In *Brassinosteriods* (H. G. Cutler, T. Yokota, and G. Adam, eds.), pp. 26–35. American Chemical Society, Washington, D.C.

Kim, T., Mills, H. A., and Wetzstein, H. Y. (2002). Studies on effects of nitrogen form on growth, development, and nutrient uptake in pecan. *J. Plant Nutr.* **25**, 497–508.

Kimmerer, T. W. and Kozlowski, T. T. (1982). Ethylene, ethane, acetaldehyde, and ethanol production by plants under stress. *Plant Physiol.* **69**, 840–847.

Kimmerer, T. W. and Stringer, M. A. (1988). Alcohol dehydrogenase and ethanol in the stems of trees. *Plant Physiol.* **87**, 693–697.

Kimmins, J. P. (1977). Evaluation of the consequences for future tree productivity of the loss of nutrients in whole-tree harvesting. *For. Ecol. Manage.* **1**, 169–183.

Kinerson, R. S. (1975). Relationships between plant surface area and respiration in loblolly pine. *J. Appl. Ecol.* **12**, 965–971.

Kinerson, R. S., Ralston, C., and Wells, C. (1977). Carbon cycling in a loblolly pine plantation. *Oecologia* **29**, 1–10.

King, J. N. and Dancik, R. P. (1983). Inheritance and linkage of isozymes in white spruce (*Picea glauca*). *Can. J. Gen. Cytol.* **25**, 430–436.

Kira, T. (1975). Primary productivity of forests. In *Photosynthesis and Productivity in Different Environments* (J. P. Cooper, ed.), IBP, Vol. 3, pp. 5–40. Cambridge Univ. Press, Cambridge.

Kirkham, M. B. (2002). The concept of the soil-plant-atmosphere continuum and applications. In *Environmental Mechanics: Water, Mass, and Energy Transfer in the Biosphere* (P. A. C. Raats, D. Smiles, and A. W. Warrick, eds.), pp. 327–335. Geophysical Monograph 129, American Geophysical Union, Washington, D. C.

Kirschbaum, M. U. F. (1988). Recovery of photosynthesis from water stress in *Eucalyptus pauciflora*—A process in two stages. *Plant, Cell Environ.* **11**, 685–694.

Kirschbaum, M. U. F. and Pearcy, R. W. (1988). Gas exchange analysis of the relative importance of stomatal and biochemical factors in the photosynthetic induction in *Alocasia macrorrhiza*. *Plant Physiol.* **86**, 782–785.

Kitajima, K. (1994). Relative importance of photosynthetic traits and allocation patterns as correlates of seedling shade tolerance of 13 tropical trees. *Oecologia* **98**, 419–428.

Kite, G. A. (1981). Annual variations in soluble sugars, starch, and total food reserves in *Eucalyptus obliqua* roots. *For. Sci.* **27**, 449–454.

Kjellbom, P., Larsson, C., Johansson, I., Karlsson, M., and Johanson, U. (1999). Aquaporins and water homeostasis in plants. *Trends Plant Sci.* **4**, 308–314.

Klebs, G. (1913). *Über das Verhaltniss der Aussenwelt zur Entwicklung der Pflanzen. Sitzungber. Heidelb. Akad. Wiss., Abt. B.* **5**, 1–47.

Klebs, G. (1914). *Über das Treiben der einheimischen Baume, speziell der Buche. Sitzungber. Heidelb. Akad. Wiss., Abh. Math.–Naturwiss. Kl.* **3**, reviewed in *Plant World* **18**, 19 (1915).

Klepper, B. (1968). Diurnal pattern of water potential in woody plants. *Plant Physiol.* **43**, 1931–1934.

Klepper, B. and Kaufmann, M. R. (1966). Removal of salt spray from xylem sap by leaves and stems of guttating plants. *Plant Physiol.* **41**, 1743–1747.

Kleinman, P. J. A., Pimentel, D., and Bryant, R. B. (1995). The ecological sustainability of slash-and-burn agriculture. *Agric. Ecosys. Environ.* **52**, 235–249.

Knee, M. (1972). Anthocyanin, carotenoid, and chlorophyll changes in the peel of Cox's Orange Pippin apples during ripening on and off the tree. *J. Exp. Bot.* **23**, 184–196.

Knoerr, K. R. (1965). Partitioning of the radiant heat load by forest stands. *Proc. Soc. Am. For. (1964)*, pp. 105–109.

Knoerr, K. R. (1967). Contrasts in energy balances between individual leaves and vegetated surfaces. In *Forest Hydrology* (W. E. Sopper and H. W. Lull, eds.), pp. 391–401. Pergamon, Oxford.

Knudsen, J. T., Tollsten, L., and Burgström, L. G. (1993). Floral scents—A checklist of volatile compounds isolated by head-space techniques. *Phytochemistry* **33**, 253–280.

Koch, G. W., Sillett, S. C., Jennings, G. M., and Davis, S. D. (2004). The limits to tree height. *Nature* **428**, 851–854.

Koch, P. (1985). Utilization of hardwoods growing on southern pine sites. Vol. II. Processing. *U.S., Dep. Agric., Agric. Hdbk.* 605. Washington, D.C.

Koch, W. and Keller, T. (1961). *Der Einfluss von Alterung und Abschneiden auf den CO_2-Gaswechsel von Pappelblättern. Ber. Deut. Bot. Ges.* **74**, 64–74.

Kochian, L. V. (1995). Cellular mechanisms of aluminum toxicity and resistance in plants. *Annu. Rev. Plant Physiol. Plant Mol. Biol.* **46**, 237–260.

Kochian, L. V., Pineros, M. A., and Hoekenga, O. A. (2005). The physiology, genetics and molecular biology of plant aluminum resistance and toxicity. *Plant Soil* **274**, 175–195.

Koelling, M. R., Blum, B. M., and Gibbs, C. B. (1968). A summary and evaluation of research on the use of plastic tubing in maple sap production. *U.S. For. Serv., Res. Pap. NE* **NE–116**.

Koide, R. (1985). The nature and location of variable hydraulic resistance to *Helianthus annuus* L. (Sunflower). *J. Exp. Bot.* **36**, 1430–1440.

Koike, T. (1990). Autumn coloring, photosynthetic performance and leaf development of deciduous broad-leaved trees in relation to forest succession. *Tree Physiol.* **7**, 21–32.

Kolattukudy, P. E. (1980). Biopolyester membranes of plants: Cutin and suberin. *Science* **208**, 990–1000.

Kolattukudy, P. E. (1981). Structure, biosynthesis, and biodegradation of cutin and suberin. *Annu. Rev. Plant Physiol.* **32**, 539–567.

Kolattukudy, P. E. (1984). Biochemistry and function of cutin and suberin. *Can. J. Bot.* **62**, 2918–2933.

Kolattukudy, P. E., Crawford, M. S., Woloshuk, C. P., Ettinger, W. F., and Saliday, C. L. (1987). The role of cutin, the plant cuticular hydroxy fatty acid polymer, in the fungal interaction with plants. In *Ecology and Metabolism of Plant Lipids* (G. Fuller and W. D. Nes, eds.), pp. 152–175. Am. Chem. Soc., Washington, D.C.

Kolesnikov, V. A. (1966). *Fruit Biology*. Mir Publishers, Moscow.

Konar, R. N. and Oberoi, Y. P. (1969). Recent work on the reproductive structures of living conifers and taxads—A review. *Bot. Rev.* **35**, 89–116.

Kondo, K., Watanabe, A., and Imaseki, H. (1975). Relationships in actions of indoleacetic acid, benzyladenine and abscisic acid in ethylene production. *Plant, Cell Physiol.* **16**, 1001–1007.

Koppenaal, R. S., Tschaplinski, T. J., and Colombo, S. J. (1991). Carbohydrate accumulation and turgor maintenance in seedling shoots and roots of two boreal conifers subjected to water stress. *Can. J. Bot.* **69**, 2522–2528.

Korcak, R. F. (1987). Iron deficiency chlorosis. *Hort. Rev.* **9**, 133–186.

Koriba, K. (1958). On the periodicity of tree growth in the tropics, with reference to the mode of branching, the leaf fall, and the formation of the resting bud. *Gardens Bull.* **17**, 11–81.

Körner, C. (2003). Carbon limitation in trees. *J. Ecol.* **91**, 4–17.

Korol, R. L., Running, S. W., Milner, K. S., and Hunt, E. R., Jr. (1991). Testing a mechanistic carbon balance model against observed tree growth. *Can. J. For. Res.* **21**, 1098–1105.

Kossuth, S. V. and Koch, P. (1989). Paraquat and CEPA stimulation of oleoresin production in lodgepole pine central stump-root system. *Wood Fiber Sci.* **21**, 263–273.

Kozlowski, T. T. (1943). Transpiration rates of some forest tree species during the dormant season. *Plant Physiol.* **16**, 252–260.

Kozlowski, T. T. (1949). Light and water in relation to growth and competition of Piedmont forest tree species. *Ecol. Monogr.* **19**, 207–231.

Kozlowski, T. T. (1957). Effect of continuous high light intensity on photosynthesis of forest tree seedlings. *For. Sci.* **3**, 220–224.

Kozlowski, T. T. (1962). Photosynthesis, climate, and growth of trees. In *Tree Growth* (T. T. Kozlowski, ed.), pp. 149–170. Ronald Press, New York.

Kozlowski, T. T. (1964). Shoot growth in woody plants. *Bot. Rev.* **30**, 335–392.

Kozlowski, T. T. (1969). Tree physiology and forest pests. *J. For.* **69**, 118–122.

Kozlowski, T. T. (1971a). *Growth and Development of Trees. Vol. I. Seed Germination, Ontogeny, and Shoot Growth*. Academic Press, New York.

Kozlowski, T. T. (1971b). *Growth and Development of Trees. Vol II. Cambial Growth, Root Growth, and Reproductive Growth*. Academic Press, New York.

Kozlowski, T. T. (1972). Physiology of water stress. *U.S. For. Serv., Gen. Tech. Rep.* INT **INT–1**, 229–244.

Kozlowski, T. T. (1973). Extent and significance of shedding of plant parts. In *Shedding of Plant Parts* (T. T. Kozlowski, ed.), pp. 1–44. Academic Press, New York.

Kozlowski, T. T. (1976a). Water supply and leaf shedding. In *Water Deficits and Plant Growth* (T. T. Kozlowski, ed.), Vol. 4, pp. 191–231. Academic Press, New York.

Kozlowski, T. T. (1976b). Water relations and tree improvement. In *Tree Physiology and Yield Improvement* (M. G. R. Cannell and F. T. Last, eds.), pp. 307–327. Academic Press, San Diego.

Kozlowski, T. T. (1979). *Tree Growth and Environmental Stresses*. University of Washington Press, Seattle.

Kozlowski, T. T. (1982a). Water supply and tree growth. Part I. Water deficits. *For. Abstr.* **43**, 57–95.

Kozlowski, T. T. (1982b). Water supply and tree growth. Part II. Flooding. *For. Abstr.* **43**, 145–161.

Kozlowski, T. T. (1984a). Plant responses to flooding of soil. *BioScience* **34**, 162–167.

Kozlowski, T. T. (1984b). Responses of woody plants to flooding. In *Flooding and Plant Growth* (T. T. Kozlowski, ed.), pp. 129–163. Academic Press, New York.

Kozlowski, T. T. (1985a). Soil aeration, flooding, and tree growth. *J. Arboric.* **11**, 85–96.

Kozlowski, T. T. (1985b). Tree growth in response to environmental stresses. *J. Arboric.* **11**, 97–111.

Kozlowski, T. T. (1986a). The impact of environmental pollution on shade trees. *J. Arboric.* **12**, 29–37.

Kozlowski, T. T. (1986b). Soil aeration and growth of forest trees (review article). *Scand. J. For. Res.* **1**, 113–123.

Kozlowski, T. T. (1992). Carbohydrate sources and sinks in woody plants. *Bot. Rev.* **58**, 107–222.

Kozlowski, T. T. (1995). The physiological ecology of forest stands. In (W. J. Nierenberg, ed.) *Encyclopedia of Environmental Biology*, Vol. 3, pp. 81–91. Academic Press, San Diego.

Kozlowski, T. T. and Ahlgren, C. E., eds. (1974). *Fire and Ecosystems*. Academic Press, New York.

Kozlowski, T. T. and Clausen, J. J. (1965). Changes in moisture contents and dry weights of buds and leaves of forest trees. *Bot. Gaz.* **126**, 20–26.

Kozlowski, T. T. and Clausen, J. J. (1966a). Anatomical responses of pine needles to herbicides. *Nature (London)* **209**, 486–487.

Kozlowski, T. T. and Clausen, J. J. (1966b). Shoot growth characteristics of heterophyllous woody plants. *Can. J. Bot.* **44**, 827–843.

Kozlowski, T. T. and Constantinidou, H. A. (1986a). Responses of woody plants to environmental pollution. Part I. Sources, types of pollutants, and plant responses. *For. Abstr.* **47**, 5–51.

Kozlowski, T. T. and Constantinidou, H. A. (1986b). Responses of woody plants to environmental pollution. Part II. Factors affecting responses to pollution. *For. Abstr.* **47**, 105–132.

Kozlowski, T. T. and Gentile, A. C. (1958). Respiration of white pine buds in relation to oxygen availability and moisture content. *For. Sci.* **4**, 147–152.

Kozlowski, T. T. and Greathouse, T. E. (1970). Shoot growth characteristics of tropical pines. *Unasylva* **24**, 1–10.

Kozlowski, T. T. and Keller, T. (1966). Food relations of woody plants. *Bot. Rev.* **32**, 293–382.

Kozlowski, T. T. and Pallardy, S. G. (1979). Stomatal responses of *Fraxinus pennsylvanica* seedlings during and after flooding. *Physiol. Plant.* **46**, 155–158.

Kozlowski, T. T. and Pallardy, S. G. (1984). Effect of flooding on water, carbohydrate and mineral relations. In *Flooding and Plant Growth* (T. T. Kozlowski, ed.), pp. 165–194. Academic Press, New York.

Kozlowski, T. T. and Pallardy, S. G. (1997). *Growth Control of Woody Plants*. Academic Press, San Diego.

Kozlowski, T. T. and Pallardy, S. G. (2002). Acclimation and adaptive responses of woody plants to environmental stresses. *Bot. Rev.* **68**, 270–334.

Kozlowski, T. T. and Peterson, T. A. (1962). Seasonal growth of dominant, intermediate, and suppressed red pine trees. *Bot. Gaz.* **124**, 146–154.

Kozlowski, T. T. and Scholtes, W. H. (1948). Growth of roots and root hairs of pine and hardwood seedlings in the Piedmont. *J. For.* **46**, 750–754.

Kozlowski, T. T. and Winget, C. H. (1963). Patterns of water movement in forest trees. *Bot. Gaz.* **124**, 301–311.

Kozlowski, T. T. and Winget, C. H. (1964). Diurnal and seasonal variation in radii of tree stems. *Ecology* **45**, 149–155.

Kozlowski, T. T., Winget, C. H., and Torrie, J. H. (1962). Daily radial growth of oak in relation to maximum and minimum temperature. *Bot. Gaz.* **124**, 9–17.

Kozlowski, T. T., Hughes, J. F., and Leyton, L. (1966). Patterns of water movement in dormant gymnosperm seedlings. *Biorheology* **3**, 77–85.

Kozlowski, T. T., Hughes, J. F., and Leyton, L. (1967). Dye movement in gymnosperms in relation to tracheid alignment. *Forestry* **40**, 209–227.

Kozlowski, T. T., Torrie, J. H., and Marshall, P. E. (1973). Predictability of shoot length from bud size in *Pinus resinosa* Ait. *Can. J. For. Res.* **3**, 34–38.

Kozlowski, T. T., Davies, W. J., and Carlson, S. D. (1974). Transpiration rates of *Fraxinus americana* and *Acer saccharum* leaves. *Can. J. For. Res.* **4**, 259–267.

Kozlowski, T. T., Kramer P. J., and Pallardy, S. G. (1991). *The Physiological Ecology of Woody Plants*. Academic Press, San Diego.

Kramer, P. J. (1937). The relation between rate of transpiration and rate of absorption of water in plants. *Am. J. Bot.* **24**, 10–15.

Kramer, P. J. (1940). Sap pressure and exudation. *Am. J. Bot.* **27**, 929–931.

Kramer, P. J. (1969). *Plant and Soil Water Relationships: A Modern Synthesis*. McGraw-Hill, New York.

Kramer, P. J. (1980). Drought, stress and the origins of adaptations. In *Adaptation of Plants to Water and High Temperature Stresses* (N. C. Turner and P. J. Kramer, eds.), pp. 7–20. Wiley-Interscience, New York.

Kramer, P. J. (1981). Carbon dioxide concentration, photosynthesis, and dry matter production. *BioScience* **31**, 29–33.

Kramer, P. J. (1983). *Water Relations of Plants*. Academic Press, New York.

Kramer, P. J. (1988). Opinion: Changing concepts regarding plant water relations. *Plant, Cell Environ.* **11**, 565–569.

Kramer, P. J. and Boyer, J. S. (1995). *Water Relations of Plants and Soils*. Academic Press, San Diego.

Kramer, P. J. and Bullock, H. C. (1966). Seasonal variations in the proportions of suberized and unsuberized roots of trees in relation to the absorption of water. *Am. J. Bot.* **53**, 200–204.

Kramer, P. J. and Clark, W. S. (1947). A comparison of photosynthesis in individual pine needles and entire seedlings at various light intensities. *Plant Physiol.* **22**, 51–57.

Kramer, P. J. and Decker, J. P. (1944). Relation between light intensity and rate of photosynthesis of loblolly pine and certain hardwoods. *Plant Physiol.* **19**, 350–358.

Kramer, P. J. and Kozlowski, T. T. (1960). *Physiology of Trees*. McGraw-Hill, New York.

Kramer, P. J. and Kozlowski, T. T. (1979). *Physiology of Woody Plants*. Academic Press, New York.

Kramer, P. J. and Wilbur, K. M. (1949). Absorption of radioactive Phosphorus by mycorrhizal roots of pine. *Science* **110**, 8–9.

Kramer, P. J., Riley, W. S., and Bannister, T. T. (1952). Gas exchange of cypress (*Taxodium distichum*) knees. *Ecology* **33**, 117–121.

Krause, G. H., Grube, E., Virgo, A., and Winter, K. (2003). Sudden exposure to solar UV-B radiation reduces net CO_2 uptake and Photosystem I efficiency in shade-acclimated tropical tree seedlings. *Plant Physiol.* **131**, 745–752.

Kreutzer, K. (1972). The effect of Mn deficiency on colour, pigments, and gas exchange of Norway spruce needles. *Forstwiss. Cbl.* **91**, 80–98.

Kriebel, H. (1963). Selection for drought resistance in sugar maple. *Proc. World Consult. For. Genet. For. Tree Breed., Food Agric. Organ./FORGEN* **63 Ser. 3/9 No. 2**, 1–5.

Kriedemann, P. E., Kliewer, W. M., and Harris, J. M. (1970). Leaf age and photosynthesis in *Vitis vinifera*. *Vitis* **9**, 97–104.

Kriedemann, P. E., Loveys, B. R., Fuller, G. L., and Leopold, A. C. (1972). Abscisic acid and stomatal regulation. *Plant Physiol.* **79**, 842–847.

Krömer, S. (1995). Respiration during photosynthesis. *Annu. Rev. Plant Physiol. Plant Molec. Biol.* **46**, 45–70.

Kronzucker, H. J., Siddiqi, M. Y., and Glass, A. D. M. (1997). Conifer root discrimination against soil nitrate and the ecology of forest succession. *Nature (London)* **385**, 59–61.

Krotkov, G. (1941). The respiratory metabolism of McIntosh apples during ontogeny as determined at 22°C. *Plant Physiol.* **16**, 799–812.

Kubiske, M. E. and Abrams, M. D. (1992). Photosynthesis, water relations, and leaf morphology of xeric versus mesic *Quercus rubra* ecotypes in central Pennsylvania in relation to moisture stress. *Can. J. For. Res.* **22**, 1402–1407.

Kubiske, M. E. and Abrams, M. D. (1993). Stomatal and nonstomatal limitations of photosynthesis in 19 temperate tree species on contrasting sites during wet and dry years. *Plant, Cell Environ.* **16**, 1123–1129.

Kubiske, M. E. and Abrams, M. D. (1994). Ecophysiological analysis of woody species in contrasting temperate communities during wet and dry years. *Oecologia* **98**, 303–312.

Kubler, H. (1990). Natural loosening of the wood/bark bond: A review and synthesis. *For. Prod. J.* **40**, 25–31.

Kubler, H. (1991). Function of spiral grain in trees. *Trees* **5**, 125–135.

Kühlbrandt, W., Wang, D. N., and Fujiyoshi, Y. (1994). Atomic model of plant light-harvesting complex by electron crystallography. *Nature (London)* **367**, 614–621.

Kuhns, M. R. and Gjerstad, D. H. (1988). Photosynthate allocation in loblolly pine (*Pinus taeda*) seedlings as affected by moisture stress. *Can. J. For. Res.* **18**, 285–291.

Kuhns, M. R. and Gjerstad, D. H. (1991). Distribution of ^{14}C-labeled photosynthate in loblolly pine (*Pinus taeda*) seedlings as affected by season and time after exposure. *Tree Physiol.* **8**, 259–271.

Kuhns, M. R., Stroup, W. W., and Gebre, G. M. (1993). Dehydration tolerance of 5 bur oak (*Quercus macrocarpa*) seed sources from Texas, Nebraska, Minnesota, and New York. *Can. J. For. Res.* **23**, 387–393.

Kull, O. (2002). Acclimation of photosynthesis in canopies: Models and limitations. *Oecologia* **133**, 267–279.

Kulman, H. M. (1971). Effects of insect defoliation on growth and mortality of trees. *Annu. Rev. Entomol.* **16**, 289–324.

Kummerow, J., Alexander, J. V., Neel, J. W., and Fishbeck, K. (1978). Symbiotic nitrogen fixation in *Ceanothus* roots. *Am. J. Bot.* **65**, 63–69.

Küppers, M. and Schneider, H. (1993). Leaf gas exchange of beech (*Fagus sylvatica* L.) seedlings in lightflecks: Effects of fleck length and leaf temperature in leaves grown in deep and partial shade. *Trees* **7**, 160–168.

Küppers, M., Wheeler, A. M., Küppers, B. I. L., Kirschbaum, M. U. F., and Farquhar, G. D. (1986). Carbon fixation in eucalypts in the field. Analysis of diurnal variations in photosynthetic capacity. *Oecologia* **70**, 273–282.

Kurdali, F. (2000). Seasonal nitrogen changes in *Alnus orientalis* and *Populus nigra* and N_2 fixation by exotic alder species in Syria. *Commun. Soil Sci. Plant Anal.* **31**, 2509–2522.

Kursar, T. A. and Coley, P. D. (1992a). Delayed development of the photosynthetic apparatus in tropical rain forest species. *Funct. Ecol.* **6**, 411–422.

Kursar, T. A. and Coley, P. D. (1992b). Delayed greening in tropical leaves: An anti-herbivore defense? *Biotropica* **24**, 256–262.

Kwak, J. M., Nguyen, V., and Schroeder, J. I. (2006). The role of reactive oxygen species in hormonal responses. *Plant Physiol.* **141**, 323–329.

Kwon, K. W. and Pallardy, S. G. (1989). Temporal changes in tissue water relations of seedlings of *Quercus acutissima*, *Q. alba*, and *Q. stellata* subjected to chronic water stress. *Can. J. For. Res.* **19**, 622–626.

Kyle, D. J. (1987). The biochemical basis for photoinhibition of Photosystem II. In *Photoinhibition* (D. J. Kyle, C. B. Osmond, and C. J. Arntzen, eds.), pp. 197–226. Elsevier, Amsterdam, Netherlands.

Lachaud, S. (1983). Xylogenèse chez les Dicotylèdones arborescentes. IV. Influence des bourgeous, de l' acide β-indolylacètique et de l'acide gibberellique sur la rèactivation cambial et la xylogénèse dans les jeunes tiges de Hêtre. *Can. J. Bot.* **61**, 1768–1774.

Lachaud, S. (1989). Participation of auxin and abscisic acid in the regulation of seasonal variations in cambial activity and xylogenesis. *Trees* **3**, 125–137.

Ladefoged, K. (1952). The periodicity of wood formation. *Dan. Biol. Skr.* **7**, 1–98.

Ladipo, D. O., Grace, J., Sanford, A. P., and Leakey, R. R. B. (1984). Clonal variation in photosynthetic and respiration rates and diffusion resistances in the tropical hardwood *Triplochiton scleroxylon* K. Schum. *Photosynthetica* **18**, 20–27.

Lake, J. A., Quick, W. P., Beerling, D. J., and Woodward, F. I. (2001). Signals from new to mature leaves. *Nature (London)* **411**, 154.

Lakso, A. N. (1979). Seasonal changes in stomatal response to leaf water potential in apple. *J. Am. Soc. Hortic. Sci.* **104**, 58–60.

Lakso, A. N., Geyer, A. S., and Carpenter, S. G. (1984). Seasonal osmotic relations in apple leaves of different ages. *J. Am. Soc. Hortic. Sci.* **109**, 541–547.

Lambers, H. (1985). Respiration in intact plants and tissues: Its regulation and dependence on environmental factors, metabolism, and invaded organisms. *Encycl. Plant Physiol. N. S.* **18**, 418–473.

Lamoreaux, R. J. and Chaney, W. R. (1977). Growth and water movement in silver maple seedlings affected by cadmium. *J. Environ. Qual.* **6**, 201–205.

Lamoreaux, R. J. and Chaney, W. R. (1978a). The effect of cadmium on net photosynthesis, transpiraiton, and dark respiration of excised silver maple leaves. *Physiol. Plant* **43**, 231–236.

Lamoreaux, R. J., and Chaney, W. R. (1978b). Photosynthesis and transpiration of excised silver maple leaves exposed to cadmium and sulphur dioxide. *Environ. Pollut.* **17**, 259–268.

Lancaster, J. E. (1992). Regulation of skin color in apples. *Crit. Rev. Plant Sci.* **10**, 487–502.

Landsberg, J. J. (1986). *Physiological Ecology of Forest Production*. Academic Press, New York.

Landsberg, J. J. (1995). Forest canopies. In *Encyclopedia of Environmental Biology* (W. A. Nierenberg, ed.), Vol. 3, pp. 81–94. Academic Press, San Diego.

Lang, G. A., Early, J. D., Arroyave, N. J., Darnell, R. L., Martin, G. C., and Stutte, G. W. (1985). Dormancy: Toward a reduced universal terminology. *HortScience* **20**, 809–812.

Lang, G. A., Early, J. D., Martin, G. C., and Darnell, R. L. (1987). Endo-, para- and ecodormancy: Physiological terminology and classifications for dormancy research. *HortScience* **22**, 371–377.

Lange, O., Lösch, R., Schulze, E. D., and Kappen, L. (1971). Responses of stomata to changes in humidity. *Planta* **100**, 76–86.

Lange, O., Schulze, E. -D., Evenari, M., Kappen, L., and Buschbom, U. (1974). The temperature-related photosynthetic capacity of plants under desert conditions. I. Seasonal changes in the photosynthetic response to temperature. *Oecologia* **17**, 97–110.

Langenheim, J. H., Osmond, C. B., Brooks, A., and Ferrar, P. J. (1984). Photosynthetic responses to light in seedlings of selected Amazonian and Australian rainforest tree species. *Oecologia* **63**, 215–224.

Langheinrich, R. and Tischner, R. (1991). Vegetative storage proteins in poplar. Induction and characterization of a 32- and a 36-kilodalton polypeptide. *Plant Physiol.* **97**, 1017–1025.

Langer, K., Levchenko, V., Fromm, J., Geiger, D., Steinmeyer, R., Lautner, S., Ache, P., and Hedrich, R. (2004). The poplar K^+ channel KPT1 is associated with K^+ uptake during stomatal opening and bud development. *Plant J.* **37**, 828–838.

Lanner, R. M. (1964). Temperature and the diurnal rhythm of height growth in pines. *J. For.* **62**, 493–495.

Lanner, R. M. (1966). The phenology and growth habits of pines in Hawaii. *U.S. For. Serv., Res. Pap. PSW* **PNW–29**.

Lanner, R. M. (1976). Patterns of shoot development in *Pinus* and their relationship to growth potential. In *Tree Physiology and Yield Improvement* (M. G. R. Cannell and F. T. Last, eds.), pp. 223–243. Academic Press, New York.

Larcher, W. (1969). The effect of environmental and physiological variables on the carbon dioxide gas exchange of trees. *Photosynthetica* **3**, 167–198.

Larcher, W. (1975). *Physiological Plant Ecology*. Springer-Verlag, Berlin and New York.

Larcher, W. (1983). *Physiological Plant Ecology*. Springer-Verlag, Berlin.

Larcher, W. and Bauer, H. (1981). Ecological significance of resistance to low temperature. *Encycl. Plant Physiol. N.S.* **12A**, 404–437.

Larcher, W. and Nagele, M. (1992). Changes in photosynthetic activity of buds and stem tissues of *Fagus sylvatica* during winter. *Trees* **6**, 91–95.

Larson, K. D., Schaffer, B., and Davies, F. S. (1989). Flooding, carbon assimilation and growth of mango trees. *Am. Soc. Hortic. Sci. Ann. Mtg. Tulsa, Okla. Prog. and Abstr.* p. 126.

Larson, P. R. (1960). A physiological consideration of the springwood-summerwood transition in red pine. *For. Sci.* **6**, 110–122.

Larson, P. R. (1962). Auxin gradients and the regulation of cambial activity. In *Tree Growth* (T. T. Kozlowski, ed.), pp. 97–117. Ronald, New York.

Larson, P. R. (1969). Wood formation and the concept of wood quality. *Yale Sch. For. Bull.* **74**.

Larson, P. R. (1994). *The Vascular Cambium: Development and Structure.* Springer-Verlag. New York and Berlin.

Larson, P. R. and Isebrands, J. G. (1978). Functional significance of the nodal constriction zone in *Populus deltoides*. *Can. J. Bot.* **56**, 801–804.

Lasheen, A. M. and Chaplin, C. E. (1971). Biochemical comparisons of seasonal variations in three peach cultivars differing in cold hardiness. *J. Am. Soc. Hortic. Sci.* **96**, 154–159.

Lassoie, J. P. and Hinckley, T. M., eds. (1991). *Techniques and Approaches in Forest Tree Ecophysiology*. CRC Press, Boca Raton, Florida.

Lassoie, J. P., Dougherty, P. M., Reich, P. B., Hinckley, T. M., Metcalf, C. M., and Dina, S. J. (1983). Ecophysiological investigations of understory eastern red pcedar in central Missouri. *Ecology* **64**, 1355–1366.

Lassoie, J. P., Hinckley, T. M., and Grier, C. C. (1985). Coniferous forests of the Pacific Northwest. In *Physiological Ecology of North American Plant Communities* (B. F. Chabot and H. A. Mooney, eds.), pp. 127–161. Chapman and Hall, New York.

Last, F. T., Mason, P. A., Ingleby, K., and Fleming, L. V. (1984). Succession of fruit bodies of sheathing mycorrhizal fungi associated with *Betula pendula*. *For. Ecol. Manage.* **9**, 229–234.

Lavigne, M. B. (1987). Differences in stem respiration responses to temperature between balsam fir trees in unthinned and thinned stands. *Tree Physiol.* **3**, 225–233.

Lavigne, M. B. (1988). Stem growth and respiration of young balsam fir trees in thinned and unthinned stands. *Can. J. For. Res.* **18**, 483–489.

Law, B. E., Ryan, M. G., and Anthoni, P. M. (1999). Seasonal and annual respiration of a ponderosa pine ecosystem. *Global Change Biol.* **5**, 169–182.

Lawlor, D. W. (1987). *Photosynthesis*. Wiley, New York.

Lawlor, D. W. (2001). *Photosynthesis, 3rd ed.* BIOS Sci. Publ., Oxford.

Lawlor, D. W. and Cornic, G. (2002). Photosynthetic carbon assimilation and associated metabolism in relation to water deficits in higher plants. *Plant, Cell Environ.* **25**, 275–294.

Lawrence, W. T. and Oechel, W. C. (1983). Effects of soil temperature on the carbon exchange of taiga seedlings. I. Root respiration. *Can. J. For. Res.* **13**, 840–849.

Layne, D. R. and Flore, J. A. (1992). Photosynthetic compensation to partial leaf area reduction in sour cherry. *J. Am. Soc. Hortic. Sci.* **117**, 279–286.

Layne, D. R. and Flore, J. A. (1993). Physiological responses of *Prunus cerasus* to whole-plant source manipulation—Leaf gas exchange, chlorophyll fluorescence, water relations and carbohydrate concentrations. *Physiol. Plant.* **88**, 44–51.

Lea, P. J. and Miflin, B. J. (2003). Glutamate synthase and the synthesis of glutamate in plants. *Plant Physiol. Biochem.* **41**, 555–564.

Leach, R. W. A. and Wareing, P. F. (1967). Distribution of auxin in horizontal woody stems in relation to gravimorphism. *Nature (London)* **214**, 1025–1027.

Ledig, F. T. (1976). Physiological genetics, photosynthesis and growth models. In *Tree Physiology and Yield Improvement* (M. G. R. Cannell and F. T. Last, eds.), pp. 21–54. Academic Press, New York.

Ledig, F. T. and Kurbobo, D. R. (1983). Adaptation of sugar maple populations along altitudinal gradients: Photosynthesis, respiration and specific leaf weight. *Am. J. Bot.* **70**, 256–265.

Ledig, F. T. and Perry, T. O. (1965). Physiological genetics of the shoot-root ratio. In *Proc. Soc. Amer. For. Annu. Meetings*, pp. 39–43. Minneapolis, MN.

Lee, J. E., Oliveira, R. S., Dawson, T. E., and Fung, I. (2005). Root functioning modifies seasonal climate. *Proc. Natl. Acad. Sci. U.S.A.* **102**, 17576–17581.

Lee, K. J. and Kozlowski, T. T. (1974). Effect of silicone antitranspirants on woody plants. *Plant Soil* **40**, 493–506.

Lee, S. H. and Chung, G. C. (2005). Sensitivity of root system to low temperature appears to be associated with the root hydraulic properties through aquaporin activity. *Sci. Hortic.* **105**, 1–11.

Lee, T. D. (1988). Patterns of fruit and seed production. In *Plant Reproductive Ecology* (J. L. Doust and L. L. Doust, eds.), pp. 179–202. Oxford, New York.

Lee, T. D., Reich, P. B., and Bolstad, P. V. (2005). Acclimation of leaf respiration to temperature is rapid and related to specific leaf area, soluble sugars and leaf nitrogen across three temperate deciduous tree species. *Funct. Ecol.* **19**, 640–647.

Lee, Y. and Kende, H. (2001). Expression of beta-expansins is correlated with internodal elongation in deepwater rice. *Plant Physiol.* **127**, 645–654.

Lee, Y. and Kende, H. (2002). Expression of alpha-expansin and expansin-like genes in deepwater rice. *Plant Physiol.* **130**, 1396–1405.

Leech, R. M. (1984). Chloroplast development in angiosperms. In *Chloroplast Biogenesis* (N. R. Baker and J. Barber, eds.), pp. 1–21. Elsevier, Amsterdam, Netherlands.

Leike, H. (1967). Wirkung von Gibberellinsäure und Kinetin auf ruhende Knospen verschiedener Gehölze. *Flora A* **158**, 351–362.

Lemcoff, J. H., Guarnaschelli, A. B., Garau, A. M., Bascialli, M. E., and Ghersa, C. M. (1994). Osmotic adjustment and its use as a selection criterion in eucalyptus seedlings. *Can. J. For. Res.* **24**, 2404–2408.

Lemcoff, J. H., Guarnaschelli, A. B., Garau, A. M., and Prystupa, P. (2002). Elastic and osmotic adjustments in rooted cuttings of several clones of *Eucalyptus camaldulensis* Dehnh. from southeastern Australia after drought. *Flora* **197**, 134–142.

Leong, S. K., Leong, W., and Yoon, P. K. (1982). Harvesting of shoots for rubber extraction in *Hevea*. *J. Rubber Res. Inst. Malaysia* **30**, 117–122.

Leopold, A. C. and Nooden, L. D. (1984). Hormonal regulatory systems in plants. *Encycl. Plant Physiol. N. S.* **10**, 4–22.

Leroux, D., Stock, W. D., Bond, W. J., and Maphanga, D. (1996). Dry mass allocation, water use efficiency and $\delta^{13}C$ in clones of *Eucalyptus grandis, E. grandis* × *E. camaldulensis* and *E. grandis* × *nitens* grown under two irrigation regimes. *Tree Physiol.* **16**, 497–502.

LeRoux, X., Walcroft, A. S., Daudet, F. A., Sinoquet, H., Chaves, M. M., Rodrigues, A., and Osorio, L. (2001). Photosynthetic light acclimation in peach leaves: Importance of changes in mass:area ratio, nitrogen concentration, and leaf nitrogen partitioning. *Tree Physiol.* **21**, 377–386.

Letham, D. S. (1963). Zeatin, a factor inducing cell division from *Zea mays*. *Life Sci.* **8**, 569–573.

Letham, D. S., Goodwin, P. B., and Higgins, T. J. V. (1978). *Phytohormones and Related Compounds—A Comprehensive Treatise.* Elsevier, Amsterdam, Oxford, and New York.

Levi, M. P. and Cowling, E. B. (1968). Role of nitrogen in wood deterioration. V. Change in decay susceptibility of oak sapwood with season of cutting. *Phytopathology* **58**, 246–249.

Levitt, J. (1980a). *Responses of Plants to Environmental Stresses. Vol. 1. Chilling, Freezing, and High Temperature Stresses*. Academic Press, New York.

Levitt, J. (1980b). *Responses of Plants to Environmental Stresses. Vol. 2. Water, Radiation, Salt and Other Stresses*. Academic Press, New York.

Lev–Yadun, S. and Aloni, R. (1995). Differentiation of the ray system in woody plants. *Bot. Rev.* **61**, 45–84.

Lewandowska, M. and Öquist, G. (1980). Structural and functional relationships in developing *Pinus silvestris* chloroplasts. *Physiol. Plant.* **48**, 39–46.

Lewandowska, M., Hart, J. W., and Jarvis, P. G. (1976). Photosynthetic electron transport in plants of Sitka spruce subjected to differing light environments during growth. *Physiol. Plant.* **37**, 269–274.

Lewis, T. E., ed. (1995). *Tree Rings as Indicators of Ecosystem Health*. CRC Press, Boca Raton, Florida.

Lewis, W. M. (1981). Precipitation chemistry and nutrient loading by precipitation in a tropical watershed. *Water Resources Res.* **17**, 169–181.

Leyser, O. (2002). Molecular genetics of auxin signaling. *Annu. Rev. Plant Biol.* **53**, 377–398.

Leyton, L. (1975). *Fluid Behaviour in Biological Systems*. Clarendon Press, Oxford.

Leyton, L. and Armitage, I. P. (1968). Cuticle structure and water relations of the needles of *Pinus radiata* (D. Don). *New Phytol.* **67**, 31–38.

Li, B., McKeand, S. E., and Allen, H. L. (1991a). Genetic variation in nitrogen use efficiency of loblolly pine seedlings. *For. Sci.* **37**, 613–626.

Li, B., Allen, H. L., and McKeand, S. E. (1991b). Nitrogen and family effects on biomass allocation of loblolly pine seedlings. *For. Sci.* **37**, 271–283.

Li, S., Pezeshki, S. R., Goodwin, S., and Shields, F. D. (2004). Physiological responses of black willow (*Salix nigra*) cuttings to a range of soil moisture regimes. *Photosynthetica* **42**, 585–590.

Lian, H. L., Yu, X., Ye, Q., Ding, X. S., Kitagawa, Y., Kwak, S. S., Su, W. A., and Tang, Z. C. (2004). The role of aquaporin RWC3 in drought avoidance in rice. *Plant Cell Physiol.* **45**, 481–489.

Liang, J., Zhang, J., and Wong, M. H. (1997). Can stomatal closure caused by xylem ABA explain the inhibition of leaf photosynthesis under soil drying? *Photosyn. Res.* **51**, 149–159.

Libbenga, K. K. and Mennes, A. M. (1988). Hormone binding and its role in hormone action. In *Plant Hormones and Their Role in Plant Growth and Development* (P. J. Davies, ed.), pp. 194–221. Kluwer, Dordrecht.

Lichtenthaler, H. K., Buschmann, C., Doll, M., Fietz, H.-J., Bach, T., Kozel, U., Meier, D., and Rahmsdorf, U. (1981). Photosynthetic activity, chloroplast ultrastructure, and leaf characteristics of high-light and low-light plants and of sun and shade leaves. *Photosynth. Res.* **2**, 115–141.

Liepman, A. H., Wilkerson, C. G., and Keegstra, K. (2005). Expression of cellulose synthase-like (*Csl*) genes in insect cells reveals that *CslA* family members encode mannan synthases. *Proc. Natl. Acad. Sci. U. S. A.* **102**, 2221–2226

Likens, G. E., Bormann, F. H., Johnson, N. M., Fisher, D. W., and Pierce, R. S. (1970). Effects of forest cutting and herbicide treatment on nutrient budgets in the Hubbard Brook watershed ecosystem. *Ecol. Monogr.* **40**, 23–47.

Lin, C. and Powell, L. E. (1981). The effect of bud scales in dormancy of apple buds. *HortScience* **16(Abstr.)**, 441.

Lin, T. Y., Sucoff, E., and Brenner, M. (1986). Abscisic acid content and components of water status in leaves of *Populus deltoides*. *Can. J. Bot.* **64**, 2295–2298.

Lindberg, S. E., Harriss, R. C., and Turner, R. R. (1982). Atmospheric deposition of metals to forest vegetation. *Science* **215**, 1609–1611.

Lindberg, S. E., Lovett, G. M., Richter, P. D., and Johnson, D. W. (1986). Atmospheric deposition and canopy interactions of major ions in a forest. *Science* **231**, 141–145.

Linder, S. and Axelsson, B. (1982). Change in carbon uptake and allocation patterns as a result of irrigation and fertilization in a young *Pinus sylvestris* stand. In *Carbon Uptake and Allocation: Key to Management of Subalpine Forest Ecosystems* (R. H. Waring, ed.), pp. 38–44. IUFRO Workshop, For. Res. Lab., Oregon State University, Corvallis.

Linder, S. and Rook, D. A. (1984). Effects of mineral nutrition on carbon dioxide exchange and partitioning of carbon in trees. In *Nutrition of Plantation Crops* (G. D. Bowen and E. K. S. Nambiar, eds.), pp. 211–236. Academic Press, London.

Linder, S. and Troeng, E. (1980). Photosynthesis and transpiration of 20-year old Scots pine. In *Structure and Function of Northern Coniferous Forests—An Ecosystem Study* (T. Persson, ed.), pp. 165–181. Ecol. Bull. 32, Swedish Nat. Sci. Res. Council, Stockholm.

Linder, S. and Troeng, E. (1981a). The seasonal course of respiration and photosynthesis in strobili of Scots pine. *For. Sci.* **27**, 267–276.

Linder, S. and Troeng, E. (1981b). The seasonal variation in stem and coarse root respiration of a 20-year-old Scots pine (*Pinus sylvestris* L.). *Mitt. Forstl. Bundesversuch., Wien.* **142**, 125–140.

Linder, S., McDonald, J., and Lohammar, T. (1981). Effect of nitrogen status and irradiance during cultivation on photosynthesis and respiration in birch seedlings. *Energy Forestry Project, Uppsala. Tech. Rept.* **No. 12**.

Little, C. H. A. (1970). Apical dominance in long shoots of white pine (*Pinus strobus*). *Can. J. Bot.* **48**, 239–253.

Little, C. H. A. (1975). Inhibition of cambial activity in Abies balsamea by internal water stress: Role of abscisic acid. *Can. J. Bot.* **53**, 3041–3050.

Little, C. H. A. (1981). Effect of cambial dormancy state on the transport of [1–^{14}C]indol-3-ylacetic acid in *Abies balsamea* shoots. *Can. J. Bot.* **59**, 342–348.

Little, C. H. A. and Bonga, J. M. (1974). Rest in the cambium of *Abies balsamea*. *Can. J. Bot.* **52**, 1723–1730.

Little, C. H. A. and Eidt, D. C. (1968). Effects of abscisic acid on budbreak and transpiration in woody species. *Nature (London)* **220**, 498–499.

Little, C. H. A. and Loach, K. (1975). Effect of gibberellic acid on growth and photosynthesis in *Abies balsamea*. *Can. J. Bot.* **53**, 1805–1810.

Little, C. H. A. and Savidge, R. A. (1987). The role of plant growth regulators in forest tree cambial growth. *Plant Growth Regul.* **6**, 137–169.

Little, C. H. A. and Wareing, P. F. (1981). Control of cambial activity and dormancy in *Picea sitchensis* by indole-3-acetic acid and abscisic acid. *Can. J. Bot.* **59**, 1480–1493.

Liu, L., McDonald, A. J. S., Stadenberg, I., and Davies, W. J. (2001a). Abscisic acid in leaves and roots of willow: Significance for stomatal conductance. *Tree Physiol.* **21**, 759–764.

Liu, L., McDonald, A. J. S., Stadenberg, I., and Davies, W. J. (2001b). Stomatal and leaf growth responses to partial drying of root tips in willow. *Tree Physiol.* **21**, 765–770.

Lloyd, F. E. (1914). Morphological instability, especially in *Pinus radiata*. *Bot. Gaz.* **57**, 314–319.

Lloyd, J., Kriedemann, P. E., and Syvertsen, J. P. (1987). Gas exchange, water relations, and ion concentrations of leaves of salt stressed "Valencia" orange, *Citrus sinensis* (L.) Osbeck. *Aust. J. Plant Physiol.* **14**, 387–396.

Lloyd, J., Kriedemann, P. E., and Aspinall, D. (1990). Contrasts between *Citrus* species in response to salinisation: An analysis of photosynthesis and water relations for different rootstock—scion combinations. *Physiol. Plant.* **78**, 236–246.

Loach, K. (1967). Shade tolerance in tree seedlings. I. Leaf photosynthesis and respiration in plants raised under artificial shade. *New Phytol.* **66**, 607–621.

Loescher, W. H., Roper, T. R., and Keller, J. (1986). Carbohydrate partitioning in sweet cherry. *Proc. Wash. State Hortic. Assn. 1985.* **81**, 240–248.

Loescher, W. H., McCamant, T., and Keller, J. D. (1990). Carbohydrate reserves, translocation, and storage in woody plant roots. *HortScience* **25**, 274–281.

Lockhart, J. A. (1965). An analysis of irreversible plant cell elongation. *J. Theor. Biol.* **8**, 264–275.

Loewenstein, N. J. and Pallardy, S. G. (1998a). Drought tolerance, xylem sap abscisic acid and stomatal conductance during soil drying: A comparison of young plants of four temperate deciduous angiosperms. *Tree Physiol.* **18**, 421–430.

Loewenstein, N. J. and Pallardy, S. G. (1998b). Drought tolerance, xylem sap abscisic acid and stomatal conductance during soil drying: A comparison of canopy trees of three temperate deciduous angiosperms. *Tree Physiol.* **18**, 431–440.

Loewenstein, N. J. and Pallardy, S. G. (2002). Influence of a drying cycle on post-drought xylem sap abscisic acid and stomatal responses in young temperate deciduous angiosperms. *New Phytol.* **156**, 351–361.

Logan, K. T. and Krotkov, G. (1968). Adaptations of the photosynthetic mechanism of sugar maple (*Acer saccharum*) seedlings grown in various light intensities. *Physiol. Plant.* **22**, 104–116.

Long, S. P. and Baker, N. R. (1986). Saline terrestrial environments. In *Photosynthesis in Contrasting Environments* (N. R. Baker and S. P. Long, eds.), pp. 63–102. Elsevier, New York.

Long, S. P., Humphries, S., and Falkowski, P. G. (1994). Photoinhibition of photosynthesis in nature. *Annu. Rev. Plant Physiol. Plant Mol. Biol.* **45**, 633–662.

Long, S. P., Ainsworth, E. A., Rogers, A., and Ort, D. R. (2004). Rising atmospheric carbon dioxide: Plants face the future. *Annu. Rev. Plant Biol.* **55**, 591–628.

Long, T. J. and Jones, R. H. (1996). Seedling growth strategies and seed size effects in fourteen oak species native to different soil moisture habitats. *Trees* **11**, 1–8.

Longman, K. A. and Jenik, J. (1974). *Tropical Forest and Its Environment.* Longman, New York.

Longman, K. A. and Jenik, J. (1987). *Tropical Forest and Its Environment, 2nd ed.* Longman, New York.

Lopez, M., Bousser, A. S., Sissoeff, I., Gaspar, M., Lachaise, B., Hoarau, J., and Mahe, A. (2003). Diurnal regulation of water transport and aquaporin gene expression in maize roots: Contribution of PIP2 proteins. *Plant Cell Physiol.* **44,** 1384–1395.

Lopushinsky, W. (1964). Effect of water movement on ion movement into the xylem of tomato roots. *Plant Physiol.* **39**, 494–501.

Lopushinsky, W. (1969). Stomatal closure in conifer seedlings in response to moisture stress. *Bot. Gaz.* **130**, 258–263.

Lopushinsky, W. (1980). Occurrence of root pressure exudation in Pacific Northwest conifer seedlings. *For. Sci.* **26**, 275–279.

Lopushinsky, W. (1986). Seasonal and diurnal trends of heat pulse velocity in Douglas-fir and ponderosa pine. *Can. J. For. Res.* **16**, 814–821.

Lopushinsky, W. and Beebe, T. (1976). Relationship of shoot-root ratio to survival and growth of outplanted Douglas-fir and ponderosa pine seedlings. *U.S. For. Serv., Res. Note PNW* **PNW–274.**

Lopushinsky, W. and Klock, G. O. (1974). Transpiration of conifer seedlings in relation to soil water potential. *For. Sci.* **20**, 181–186.

Lorenc-Plucinska, G. (1982). Effect of sulphur dioxide on CO_2 exchange in SO_2-tolerant and SO_2-susceptible Scots pine seedlings. *Photosynthetica* **16**, 140–144.

Lorenc-Plucinska, G. (1988). Effect of nitrogen dioxide on CO_2 exchange in Scots pine seedlings. *Photosynthetica* **22**, 108–111.

Lorenc-Plucinska, G. (1989). Some effects of exposure to sulphur dioxide on the metabolism of Scots pine in the winter. I. Effects on photosynthesis and respiration. *Arboretum Kornickie* **31**, 229–236.

Loreto, F. and Sharkey, T. D. (1990). A gas-exchange study of photosynthesis and isoprene emission in *Quercus rubra L. Planta* **182**, 523–531.

Loreto, F., Mannozzi, M., Maris, C., Nascetti, P., Ferranti, F., and Pasqualini, S. (2001). Ozone quenching properties of isoprene and its antioxidant role in leaves. *Plant Physiol.* **126**, 993–1000.

Lorio, P. L., Jr. (1986). Growth-differentiation balance: A basis for understanding southern pine beetle–tree interactions. *For. Ecol. Manage.* **14**, 259–273.

Lorio, P. L. and Hodges, J. D. (1968). Oleoresin exudation pressure and relative water content of inner bark as indicators of moisture stress in loblolly pines. *For. Sci.* **14**, 392–398.

Lösch, R. and Schenk, B. (1978). Humidity response of stomata and the potassium content of guard cells. *J. Exp. Bot.* **29**, 781–787.

Lotocki, A. and Zelawski, W. (1973). Effect of ammonium and nitrate source of nitrogen on productivity of photosynthesis in Scots pine (*Pinus silvestris* L.) seedlings. *Acta Soc. Bot. Pol.* **42**, 599–605.

Loustalot, A. J. (1945). Influence of soil moisture conditions on apparent photosynthesis and transpiration of pecan leaves. *J. Agric. Res.* **71**, 519–532.

Loustalot, A. J., Burrows, F. W., Gilbert, S. G., and Nason, A. (1945). Effect of copper and zinc deficiencies on the photosynthetic activity of the foliage of young tung trees. *Plant Physiol.* **20**, 283–288.

Lovett, G. M., Reiners, W. A., and Olson, R. K. (1982). Cloud droplet deposition in subalpine fir forests: Hydrologic and chemical inputs. *Science* **218**, 1303–1304.

Lovett, G. M., Lindberg, S. E., Richter, D. D., and Johnson, D. W. (1985). Effects of acidic deposition on cation leaching from three deciduous forest canopies. *Can. J. For. Res.* **15**, 1055–1060.

Loveys, B. R., Atkinson, L. J., Sherlock, D. J., Roberts, R. L., Fitter, A. H., and Atkin, O. K. (2003). Thermal acclimation of leaf and root respiration: An investigation comparing inherently fast- and slow-growing plant species. *Global Change Biol.* **9**, 895–910.

Lowman, M. D. (1986). Light interception and its relation to structural differences in three Australian rainforest canopies. *Austr. J. Ecol.* **11**, 163–170.

Lowman, M. D. (1992). Leaf growth dynamics and herbivory in five species of Australian rain-forest canopy trees. *J. Ecol.* **80**, 433–447.

Luckwill, L. C. (1953). Studies of fruit development in relation to plant hormones. I. Hormone production by the developing apple seed in relation to fruit drop. *J. Hortic. Sci.* **28**, 14–24.

Ludlow, M. M. (1989). Strategies of response to water stress. In *Structural and Functional Responses to Environmental Stress: Water Shortage* (K. H. Kreeb, H. Richter, and T. M. Hinckley, eds.), pp. 269–281. SPB Academic Publishing, The Hague, Netherlands.

Ludlow, M. M. and Björkmann, O. (1984). Paraheliotropic leaf movement in siratro as a protective mechanism against drought-induced damage to primary photosynthetic reactions: Damage caused by excessive light and heat. *Planta* **161**, 508–518.

Ludlow, M. M., Santamaria, J. M., and Fukai, S. (1990). Contribution of osmotic adjustment to grain yield under water-limited conditions. II. Water stress after anthesis. *Aust. J. Agric. Res.* **41**, 67–78.

Lulai, E. C. and Corsini, D. L. (1998). Differential deposition of suberin phenolic and aliphatic domains and their roles in resistance to infection during potato tuber (*Solanum tuberosum* L.) wound-healing. *Physiol. Molec. Plant Pathol.* **53**, 209–222.

Lundmark, T., Hällgren, J. E., and Degermark, C. (1990). Effects of summer frost on the gas exchange of field-grown *Pinus sylvestris* L. seedlings. *Scand. J. For. Res.* **3**, 441–448.

Luo, Y., Su, B., Currie, W. S., Dukes, J. S., Finzi, A., Hartwig, U., Hungate, B., McMurtrie, R. E. et al. (2004). Progressive nitrogen limitation of ecosystem responses to rising atmospheric carbon dioxide. *Bioscience* **54**, 731–739.

Lutz, J. F. (1978). Wood veneer: Log selection, cutting, and drying. *U.S.D.A. For. Serv. Tech. Bull. 1577*. U.S. Govt. Printing Office, Washington, D.C.

Luu, D. T. and Maurel, C. (2005). Aquaporins in a challenging environment: Molecular gears for adjusting plant water status. *Plant Cell Environ.* **28**, 85–96.

Luukkanen, O. and Kozlowski, T. T. (1972). Gas exchange in six *Populus* clones. *Silvae Genet.* **21**, 220–229.

Luxford, R. F. (1930). Distribution and amount of moisture in virgin redwood trees *J. For.* **28**, 770–772.

Luxmoore, R. J., Gizzard, T., and Strand, R. H. (1981). Nutrient translocation in the outer canopy and understory of an eastern deciduous forest. *For. Sci.* **27**, 505–518.

Lyford, W. H. (1975). Rhizography of the non-woody roots of trees in the forest floor. In *The Development and Function of Roots* (J. G. Torrey and D. T. Clarkson, eds.), pp. 179–196. Academic Press, London.

Lyford, W. H. and Wilson, B. F. (1964). Development of the root system of *Acer rubrum* L. *Harv. For. Pap.* **10**, 1–10.

Lynch, J. M. and Bragg, E. (1985). Microorganisms and soil aggregate stability. *Adv. Soil. Sci.* **2**, 133–171.

Lyons, A. (1956). The seed production capacity and efficiency of red pine cones (*Pinus resinosa* Ait.). *Can. J. Bot.* **34**, 27–36.

Lyons, J. M. (1973). Chilling injury in plants. *Annu. Rev. Plant Physiol.* **24**, 445–466.

Lyr, H. and Hoffmann, G. (1967). Growth rates and growth periodicity of tree roots. *Int. Rev. For. Res.* **2**, 181–206.

Ma, J. F. (2005). Plant root responses to three abundant soil minerals: Silicon, aluminum and iron. *Crit. Rev. Plant Sci.* **24**, 267–281.

Maathuis, J. M., Ichida, A. M., Sanders, D., and Schroeder, J. I. (1997). Roles of higher plant K^+ channels. *Plant Physiol.* **114**, 1141–1149.

McAinsh, M. R., Gray, J. E., Hetherington, A. M., Leckie, C. P., and Ng, C. (2000). Ca^{+2} signalling in stomatal guard cells. *Biochem. Soc. Trans.* **28**, 476–481.

MacCallum, A. G. (1905). On the distribution of potassium in animal and vegetable cells. *J. Gen. Physiol.* **52**, 95–128.

McCarty, R. E., Evron, Y., and Johnson, E. A. (2000). The chloroplast ATP synthase: A rotary enzyme? *Annu. Rev. Plant Physiol. Mol. Biol.* **51**, 83–109.

McClendon, J. H. (1992). Photographic survey of the occurrence of bundle-sheath extensions in deciduous dicots. *Plant Physiol.* **99**, 1677–1679.

MacDonald, R. C. and Kimmerer, T. W. (1991). Ethanol in the stems of trees. *Physiol. Plant.* **82**, 582–588.

McDougall, D. N. and Blanchette, R. A. (1996). Polyethylene plastic wrap for tree wounds: A promoter of wound closure on fresh wounds. *J Arbor.* **22**, 206–210.

MacDougal, D. T. (1938). *Tree Growth*. Chronica Botanica, Waltham, Massachusetts.

MacDowall, F. D. H. (1963). Mid-day closure of stomata in aging tobacco leaves. *Can. J. Bot.* **41**, 1289–1300.

MacFall, J. S., Johnson, G. A., and Kramer, P. J. (1990). Observation of a water-depletion region surrounding loblolly pine roots by magnetic resonance imaging. *Proc. Natl. Acad. Sci. USA* **87**, 1203–1207.

MacFall, J. S., Slack, S. A., and Iyer, J. (1991). Effects of *Hebeloma arenosa* and phosphorus fertility on growth of red pine (*Pinus resinosa*) seedlings. *Can. J. Bot.* **69**, 372–379.

MacFall, J. S., Slack, S. A., and Wehrli, S. (1992). Phosphorus distribution in red pine roots and the ectomycorrhizal fungus *Hebeloma arenosa*. *Plant Physiol.* **100**, 713–717.

McGaw, B. A. (1988). Cytokinin biosynthesis and metabolism. In *Plant Hormones and Their Role in Plant Growth and Development* (P. J. Davies, ed.), pp. 76–93. Kluwer, Dordrecht.

McGregor, W. H. D. and Kramer, P. J. (1963). Seasonal trends in rates of photosynthesis and respiration of loblolly pine. *Am. J. Bot.* **50**, 760–765.

McGregor, W. H. D., Allen, R. M., and Kramer, P. J. (1961). The effect of photoperiod on growth, photosynthesis, and respiration of loblolly pine seedlings from two geographic sources. *For. Sci.* **7**, 342–348.

McKell, C. M., Blaisdell, J. P., and Goodin, J. R., eds. (1972). *Useful Wildland Shrubs—Their Biology and Utilization*. U.S.D.A. For. Serv. Intermountain For. Range Exp. Stn., Gen. Tech. Rep., **INT–1**, 1–194.

McLaughlin, S. B. and Barnes, R. L. (1975). Effects of fluoride on photosynthesis and respiration of some south-east American forest trees. *Environ. Pollut.* **8**, 91–96.

McLaughlin, S. B. and Shriner, D. S. (1980). Allocation of resources to defense and repair. In *Plant Disease* (J. G. Horsfall and E. B. Cowling, eds.), pp. 407–431. Academic Press, New York.

McLaughlin, S. B. and Wimmer, R. (1999). Calcium physiology and terrestrial ecosystem processes. *New Phytol.* **142**, 373–417.

McLaughlin, S. B., McConathy, R. K., and Dinger, B. E. (1978). Seasonal changes in respiratory metabolism of yellow-poplar (*Liriodendron tulipifera*) branches. *J. Appl. Ecol.* **15**, 327–334.

McLaughlin, S. B., McConathy, R. K., Barnes, R. L., and Edwards, N. T. (1980). Seasonal changes in energy allocation by white oak (*Quercus alba*). *Can. J. For. Res.* **10**, 379–388.

McLaughlin, S. B., McConathy, R. K., Duvick, D., and Mann, L. K. (1982). Effects of chronic air pollution stress on photosynthesis, carbon, allocation and growth of white pine trees. *For. Sci.* **28**, 60–70.

McLeod, K. W., Donovan, L. A., Stumpff, N. J., and Sherrod, K. C. (1986). Biomass, photosynthesis and water use efficiency of woody swamp species subjected to flooding and elevated water temperature. *Tree Physiol.* **2**, 341–346.

McMillen, G. G. and McClendon, J. H. (1983). Dependence of photosynthetic rates on leaf density thickness in deciduous woody plants grown in sun and shade. *Plant Physiol.* **72**, 674–678.

McMurtrie, R. E. (1985). Forest productivity in relation to carbon partitioning and nutrient cycling: A mathematical model. In *Attributes of Trees as Crop Plants* (M. G. R. Cannell and J. E. Jackson, eds.), pp. 194–207. Institute of Terrestrial Ecology, Huntingdon, England.

McNaughton, K. G. and Jarvis, P. G. (1983). Predicting effects of vegetation changes on transpiration and evaporation. In *Water Deficits and Plant Growth* (T. T. Kozlowski, ed.), Vol. 7, pp. 1–47. Academic Press, New York.

McNulty, S. G., Aber, J. D., and Newman, S. D. (1996). Nitrogen saturation in a high elevation New England spruce-fir stand. *For. Ecol. Manage.* **84**, 109–121.

McPherson, H. G., Snelgar, W. P., Manson, P. J., and Snowball, A. M. (1997). Bud respiration and dormancy of kiwifruit (*Actinidia deliciosa*). *Ann. Bot.* **80**, 411–418.

McQueen-Mason, S. J. and Cosgrove, D. J. (1995) Expansin mode of action on cell walls—Analysis of wall hydrolysis, stress relaxation, and binding. *Plant Physiol.* **107**, 87–100.

McQueen-Mason, S. J., Durachko, D. M., and Cosgrove, D. J. (1992). Two endogenous proteins that induce cell wall extension in plants. *Plant Cell* **4**, 1425–1433.

McRae, J. and Thor, E. (1982). Cortical monoterpene variation in 12 loblolly pine provenances planted in Tennessee. *For. Sci.* **28**, 732–736.

McWilliam, J. R. (1958). The role of the micropyle in the pollination of *Pinus*. *Bot. Gaz.* **120**, 109–117.

Madgwick, H. A. I. (1970). The nutrient content of old-field *Pinus virginiana* stands. *Tree Growth For. Soils, Proc. North Am. For. Soils Conf., 3rd, 1968*, pp. 275–282.

Madgwick, H. A. I. and Ovington, J. D. (1959). The chemical composition of precipitation in adjacent forest and open plots. *Forestry* **32**, 14–22.

Magel, E. A. (2000). Biochemistry and physiology of heartwood formation. In *Cell and Molecular Biology of Wood Formation* (R. A. Savidge, J. R. Burnett, and R. Napier, eds.), pp. 363–376. BIOS Sci. Publ. Ltd., Oxford, U.K.

Magel, E. and Huebner, B. (1997). Distribution of phenylalanine ammonia lyase and chalcone synthase within trunks of *Robinia pseudoacacia* L. *Bot. Acta.* **110**, 314–322.

Magel, E., Jay-Allemand, C., and Ziegler, H. (1994). Formation of heartwood substances in the stemwood of *Robinia pseudoacacia* L. II.. Distribution of nonstructural carbohydrates and wood extractives across the trunk. *Trees* **8**, 165–171.

Magistad, O. C. and Breazeale, J. F. (1929). Plant and soil relations at and below the wilting percentage. *U. Ariz., College of Agric., Agric. Expt. Sta., Tech. Bull.* **25**.

Magness, J. R. and Regeimbal, L. O. (1938). The nitrogen requirement of the apple. *Proc. Am. Soc. Hortic. Sci.* **36**, 51–55.

Mahall, B. E. and Callaway, R. M. (1991). Root communication among desert shrubs. *Proc. Nat. Acad. Sci. U. S. A.* **88**, 874–876.

Maheshwari, P. (1950). *An Introduction to the Embryology of Angiosperms*. McGraw-Hill, New York.

Maheshwari, P. and Rangaswamy, N. S. (1965). Embryology in relation to physiology and genetics. In *Advances in Botanical Research* (R. D. Preston, ed.), pp. 219–321. Academic Press, New York.

Maheshwari, P. and Sachar, R. P. (1963). Polyembryony. In *Recent Advances in the Embryology of Angiosperms* (P. Maheshwari, ed.), pp. 265–296, Catholic Press, Ranchi, India.

Maillette, I. (1982a). Structural dynamics of silver birch. I. Fates of buds. *J. Appl. Ecol.* **19**, 203–218.

Maillette, I. (1982b). Structural dynamics of silver birch. II. A matrix model of the bud population. *J. Appl. Ecol.* **19**, 219–238.

Malaguti, D., Millard, P., Wendler, R., Hepburn, A., and Tagliavini, M. (2001). Translocation of amino acids in the xylem of apple (*Malus domestica* Borkh.) trees in spring as a consequence of both N remobilization and root uptake. *J. Exp. Bot.* **52**, 1664–1671.

Mamaev, V. V. (1984). Respiration of tree roots in the *Pinetum* and *Betuletum oxalidoso—myrtillosum*. *Lesovedenie* **6**, 53–60.

Mandava, N. B. (1988). Plant growth-promoting brassinosteroids. *Annu. Rev. Plant Physiol. Plant Mol. Biol.* **39**, 23–52.

Mann, L. K., Johnson, D. W., West, D. C., Cole, D. W., Hornbeck, J. W., Martin, C. W., Rieberk, H., Smith, C. T. et al. (1988). Effects of whole-tree and stem only clearcutting on postharvest hydrologic losses, nutrient capital, and regrowth. *For. Sci.* **34**, 412–428.

Mansfield, T. A. and Atkinson, C. J. (1990). Stomatal behaviour in water stressed plants. In *Stress Responses in Plants: Adaptation and Acclimation Mechanisms* (R. G. Alscher and J. R. Cumming, eds.), pp. 241–264. Wiley, New York.

Mansfield, T. A. and Davies, W. J. (1985). Mechanisms for leaf control of gas exchange. *BioScience* **46**, 158–161.

Manter, D. K. and Kerrigan, J. (2004). A/C_i curve analysis across a range of woody plant species: Influence of regression analysis parameters and mesophyll conductance. *J. Exp. Bot.* **55**, 2581–2588.

Marcus, A., ed. (1981). *The Biochemistry of Plants. Vol 6. Proteins and Nucleic Acids*. Academic Press, New York.

Marek, M., Masarovicova, E., Kratochvilova, I., Elias, P., and Janous, D. (1989). Stand microclimate and physiological activity of tree leaves in an oak–hornbeam forest. II. Leaf photosynthetic activity. *Trees* **4**, 234–240.

Margolis, H., Oren, R., Whitehead, D., and Kaufmann, M. R. (1995). Leaf area dynamics of conifer forests. In *Ecophysiology of Coniferous Forests* (W. K. Smith and T. M. Hinckley, eds.), pp. 181–223. Academic Press, San Diego.

Marini, R. P. and Marini, M. C. (1983). Seasonal changes in specific leaf weight, net photosynthesis, and chlorophyll content of peach leaves as affected by light penetration and canopy position. *J. Am. Soc. Hortic. Sci.* **108**, 600–605.

Marion, G. M. (1979). Biomass and nutrient removal in long-rotation stands. In *Impact of Intensive Harvesting on Forest Nutrient Cycling* (A. L. Leaf, ed.), pp. 98–110. College of Environmental Science and Forestry, State Univ. of New York. Syracuse, New York.

Marjanović, Ž., Uehlin, N., Kaldenhoff, R., Zwiazek, J. J., Weiß, M., Hampp, R., and Nehls, U. (2005). Aquaporins in poplar: What a difference a symbiont makes! *Planta* **222,** 258–268.

Markhart III, A. H. and Smit, B. (1990). Measurement of root hydraulic conductance. *HortScience* **25**, 282–287.

Marks, G. C. and Kozlowski, T. T. (1973). *Ectomycorrhizae*. Academic Press, New York.

Marpeau, W., Walter, J., Launay, J., Charon, J., Baradat, P., and Gleizes, M. (1989). Effects of wounds on the terpene content of twigs of maritime pine. II. Changes in the volatile terpene hydrocarbon composition. *Trees* **4**, 220–226.

Marquard, R. D. and Hanover, J. W. (1984). Relationship between gibberellin $GA_{4/7}$ concentration, time of treatment, and crown position on flowering of *Picea glauca*. *Can. J. For. Res.* **14**, 547–553.

Marschner, H. (1995). *The Mineral Nutrition of Higher Plants*. Academic Press, London.

Marshall, J. G., Scarratt, J. B., and Dumbroff, E. B. (1991). Induction of drought resistance by abscisic acid and paclobutrazol in jack pine. *Tree Physiol.* **8**, 415–421.

Martin, C. E., Loeschen, V. S., and Borchert, R. (1994). Photosynthesis and leaf longevity in trees in a tropical deciduous forest in Costa Rica. *Photosynthetica* **30**, 341–351.

Martin, C. W., and Harr, R. D. (1989). Logging of mature Douglas-fir in western Oregon has little effect on nutrient output budgets. *Can. J. For. Res.* **19**, 35–43.

Martin-Cabrejas, M. A., Waldron, K. W., Selvendran, R. R., Parker, M. L., and Moates, G. K. (1994). Ripening-related changes in the cell walls of Spanish pear (*Pyrus communis*). *Physiol. Plant.* **91**, 671–679.

Martin, E. V. (1943). Studies of evaporation and transpiration under controlled conditions. *Carnegie Inst. Washington Publ.* **550**.

Martin, G. C. (1991). Bud dormancy in deciduous fruit trees. In *Plant Physiology: A Treatise Vol. X* (F. C. Steward, ed.), pp. 183–225. Academic Press, San Diego.

Martin, J. G., Kloeppel, B. D., Schaefer, T. L., Kimbler, D. L., and McNulty, S. G. (1998). Aboveground biomass and nitrogen allocation of ten deciduous southern Appalachian tree species. *Can. J. For. Res.* **28**, 1648–1659.

Martin, J. P. and Juniper, B. E. (1970). *The Cuticles of Plants*. Arnold, London.

Martin, U., Pallardy, S. G., and Bahari, Z. A. (1987). Dehydration tolerance of leaf tissues of six woody angiosperm species. *Physiol. Plant.* **69**, 182–186.

Martinez–Ferri, E., Balaguer, L., Valladares, F., Chico, J. M., and Manrique, E. (2000). Energy dissipation in drought-avoiding and drought-tolerant tree species at midday during the Mediterranean summer. *Tree Physiol.* **20**, 131–138.

Marvin, J. W. and Green, M. T. (1951). Temperature-induced sap flow in excised stem of *Acer*. *Plant Physiol.* **26**, 565–580.

Marx, D. H. (1969). The influence of ectotrophic mycorrhizal fungi on the resistance of pine roots to pathogenic infection. II. Production, identification, and biological activity of antibiotics produced by *Leucopaxillus cerealus* var. *piceina*. *Phytopathology* **59**, 411–417.

Massiere, F. and Badetdenisot, M. A. (1998). The mechanism of glutamine-dependent aminotransferases. *Cell. Molec. Life Sci.* **54**, 205–222.

Matthews, M. A. and Boyer, J. S. (1984). Acclimation of photosynthesis to low leaf water potentials. *Plant Physiol.* **74**, 161–166.

Matile, P. (2000). Biochemistry of Indian summer: Physiology of autumnal leaf coloration. *Exper. Geront.* **35**, 145–158.

Mattsson, J., Sung, Z. R., and Berleth, T. (1999). Responses of plant vascular systems to auxin transport inhibitors. *Development* **126**, 2979–2991.

Matyssek, R., Reich, P., Oren, R., and Winner, W. E. (1995). Response mechanisms of conifers to air pollutants. In *Ecophysiology of Coniferous Forests* (W. K. Smith and T. M. Hinckley, eds.), pp. 255–308. Academic Press, San Diego.

Maxwell, D. P., Laudenbach, D. E., and Huner, N. P. A. (1995). Redox regulation of light-harvesting complex II and cab mRNA abundance in *Dunaliella salina*. *Plant Physiol.* **109**, 787–795.

Mazliak, P. and Kader, J. C. (1980). Phospholipid—Exchange systems. In *The Biochemistry of Plants* (P. K. Stumpf and E. E. Conn, eds.), pp. 283–300. Academic Press, New York.

Mazliak, P., Robert, D., and Decotte-Justin, A. M. (1977). *Metabolisme des lipides dans les noyaux isoles d'hypotyles de germination de tournesol*. *Plant Sci. Lett.* **9**, 211–223.

Mbah, B. N., McWilliams, E. L., and McCree, K. J. (1983). Carbon balance of *Peperomia obtusifolia* during acclimation to low PPFD. *J. Am. Soc. Hortic. Sci.* **108**, 769–773.

Mebrahtu, T. and Hanover, J. W. (1991). Family variation in gas exchange, growth and leaf traits of black locust half-sib families. *Tree Physiol.* **8**, 185–193.

Medhurst, J. L., Battaglia, M., and Beadle, C. L. (2002). Measured and predicted changes in tree and stand water use following high-intensity thinning of an 8-year-old *Eucalyptus nitens* plantation. *Tree Physiol.* **22**, 775–784.

Meier, C. E., Newton, R. J., Puryear, J. D., and Sen, S. (1992). Physiological responses of loblolly pine (*Pinus taeda* L.) seedlings to drought stress—Osmotic adjustment and tissue elasticity. *J. Plant Physiol.* **140**, 754–760.

Meijer, H. J. G. and Munnik, T. (2003). Phospholipid-based signaling in plants. *Annu. Rev. Plant Biol.* **54**, 265–306.

Meinzer, F. G., Goldstein, G., and Jaimes, M. (1984). The effect of atmospheric humidity on stomatal control of gas exchange in two tropical coniferous species. *Can. J. Bot.* **62**, 591–595.

Meinzer, F. C., Saliendra, N. Z., and Crisosto, C. H. (1992). Carbon isotope discrimination and gas exchange in *Coffea arabica* during adjustment to different soil moisture regimes. *Aust. J. Plant Physiol.* **19**, 171–184.

Meinzer, F. C., Goldstein, G., Holbrook, N. M., Jackson, P., and Cavelier, J. (1993). Stomatal and environmental control of transpiration in a lowland tropical forest tree *Plant, Cell Environ.* **16**, 429–436.

Meinzer, F. C., Hinckley, T. M., and Ceulemans, R. (1997). Apparent responses of stomata to transpiration and humidity in a hybrid poplar canopy. *Plant, Cell Environ.* **20**, 1301–1308.

Meinzer, F. C., Brooks, J. R., Bucci, S., Goldstein, G., Scholz, F. G., and Warren, J. M. (2004). Converging patterns of uptake and hydraulic redistribution of soil water in contrasting woody vegetation types. *Tree Physiol.* **24**, 919–928.

Meir, P., Grace, J., and Miranda, A. C. (2001). Leaf respiration in two tropical rainforests: Constraints on physiology by phosphorus, nitrogen and temperature. *Funct. Ecol.* **15**, 378–387.

Melching, J. B., and Sinclair, W. A. (1975). Hydraulic conductivity of stem internodes related to resistamce of American elms to *Ceratocyctis ulmi*. *Phytopathology* **65**, 645–647.

Melillo, J. M., Aber, J., and Muratore, J. F. (1982). Nitrogen and lignin control of hardwood leaf litter decomposition dynamics. *Ecology* **63**, 621–626.

Melin, E. and Nilsson, H. (1950a). Transfer of radioactive phosphorus to pine seedlings by means of mycorrhizal fungi. *Physiol. Plant* **3**, 88–92.

Melin, E. and Nilsson, H. (1950b). Transport of labelled nitrogen from an ammonium source to pine seedlings through mycorrhizal mycelium. *Sven. Bot. Tidskr.* **46**, 281–285.

Melin, E., Nilsson, H., and Hacskaylo, E. (1958). Translocation of cations to seedlings of *Pinus virginiana* through mycorrhizal mycelium. *Bot. Gaz.* **119**, 243–245.

Mendes, A. J. T. (1941). Cytological observations in *Coffea*. VI. Embryo and endosperm development in *Coffea arabica* L. *Am. J. Bot.* **28**, 784–789.

Mengel, K. and Kirkby, E. A. (1978). Boron. In *Principles of Plant Nutrition*, pp. 559–572. International Potash Inst., Worblaufen–Bern, Switzerland.

Mengel, K. and Kirkby, E. A. (1982). *Principles of Plant Nutrition, 3rd ed.* International Potash Institute, Bern, Switzerland.

Mengel, K., Breininger, M. T., and Lutz, H. J. (1990). Effect of simulated acidic fog on carbohydrate leaching, CO_2 assimilation and development of damage symptoms in young spruce trees (*Picea abies* (L.) Karst.). *Env. Exp. Bot.* **30**, 165–173.

Mercier, S. and Langlois, C. G. (1992). Indices of maturity and storage of white spruce seeds as a function of time of harvesting in Quebec. *Can. J. For. Res.* **22**, 1516–1523.

Merrill, W. and Cowling, E. B. (1966). Role of nitrogen in wood deterioration. Amounts and distribution of nitrogen in tree stems. *Can. J. Bot.* **44**, 1555–1580.

Metcalf, C. R. (1967). Distribution of latex in the plant kingdom. *Econ. Bot.* **21**, 115–125.

Metzger, J. D. (1988). Hormones and reproductive development. In *Plant Hormones and Their Role in Plant Growth and Development* (P. J. Davies, ed.), pp. 431–462. Kluwer, Dordrecht.

Meyer, B. S., Anderson, D. B., Bohning, R. H., and Fratianne, D. G. (1973). *Introduction to Plant Physiology*. Van Nostrand-Reinhold, Princeton, New Jersey.

Michaelis, P. (1934). *Ökologische Studien an der alpinen Baumgrenze. IV. Zur Kenntnis des winterlichen Wasserhaushaltes*. *Jahr. Wiss. Bot.* **80**, 169–247.

Michelana, V. A. and Boyer, J. S. (1982). Complete turgor maintenance at low water potentials in the elongating region of maize leaves. *Plant Physiol.* **69**, 1145–1149.

Mielke, E. A. and Dennis, F. G., Jr. (1978). Hormonal control of flower bud dormancy in sour cherry (*Prunus cerasus* L.). II. Effects of leaves, defoliation and temperature on levels of abscisic acid in flower primordia. *J. Am. Soc. Hort. Sci.* **103**, 446–449.

Mielke, M. S., de Almeida, A. A. F., Gomes, F. P., Aguilar, M. A. G., and Mangabeira, P. O. (2003). Leaf gas exchange, chlorophyll fluorescence and growth responses of *Genipa americana* seedlings to soil flooding. *Environ. Exp. Bot.* **50**, 221–231.

Miflin, B. J., ed. (1981). *The Biochemistry of Plants. Vol. 5. Amino Acids and Derivatives*. Academic Press, New York.

Milburn, J. A. and Davis, T. A. (1973). Role of pressure in xylem transport of coconut and other palms. *Physiol. Plant.* **29**, 415–420.

Milburn, J. A. and Kallarackal, J. (1991). Sap exudation. In *Physiology of Trees* (A. S. Raghavendra, ed.), pp. 385–402. Wiley, New York.

Milburn, J. A. and O'Malley, P. E. R. (1984). Freeze-induced sap absorption in *Acer pseudoplatanus*: A possible mechanism. *Can. J. Bot.* **62**, 2101–2106.

Milla, R., Castro-Diez, P., Maestro-Martinez, M., and Montserrat-Marti, G. (2005). Relationships between phenology and the remobilization of nitrogen, phosphorus and potassium in branches of eight Mediterranean evergreens. *New Phytol.* **168**, 167–178.

Millard, P. (1996). Ecophysiology of the internal cycling of nitrogen for tree growth. *Z. Pflanzenernahrung Bodenkunde* **159**, 1–10.

Millard, P. and Proe, M. F. (1991). Leaf demography and the seasonal internal cycling of nitrogen in sycamore (*Acer pseudoplatanus* L.) seedlings in relation to nitrogen supply. *New Phytol.* **117**, 587–596.

Millard, P. and Thomson, C. M. (1989). The effect of the autumn senescence of leaves on the internal cycling of nitrogen for the spring growth of apple trees. *J. Exp. Bot.* **40**, 1285–1290.

Millard, P., Wendler, R., Hepburn, A., and Smith, A. (1998). Variations in the amino acid composition of xylem sap of *Betula pendula* Roth. trees due to remobilization of stored N in the spring. *Plant Cell Environ.* **21**, 715–722.

Millard, P., Hester, A., Wendler, R., and Baillie, G. (2001). Interspecific defoliation responses of trees depend on sites of winter nitrogen storage. *Funct. Ecol.* **15**, 535–543.

Millenaar, F. F. and Lambers, H. (2003). The alternative oxidase: *In vivo* regulation and function. *Plant Biol.* **5**, 2–15.

Miller, C. O., Skoog, F., Okomura, F. S., Saltza, M. H. von, and Strong, F. M. (1956). Isolation, structure, and synthesis of kinetin, a substance promoting cell division. *J. Am. Chem. Soc.* **78**, 1345–1350.

Miller, D. R., Vavrina, C. A., and Christensen, T. W. (1980). Measurement of sap flow and transpiration in ring-porous oaks using a heat pulse velocity technique. *For. Sci.* **26**, 485–494.

Miller, E. C. (1938). *Plant Physiology*. McGraw-Hill, New York.

Miller, P. R., Parmeter, J. R., Flick, B. H., and Martinez, C. W. (1969). Ozone damage response of ponderosa pine seedlings. *J. Air Pollut. Contr. Assoc.* **19**, 435–438.

Miller, R. and Rüsch, J. (1960). *Zur Frage der Kohlensäureversorgung des Waldes. Forstwiss. Centralb.* **79**, 42–62.

Millington, W. F. and Chaney, W. R. (1973). Shedding of shoots and branches. In *Shedding of Plant Parts* (T. T. Kozlowski, ed.), pp. 149–204. Academic Press, New York.

Milne, R. (1989). Diurnal water storage in the stems of *Picea sitchensis* (Bong.) Carr. *Plant, Cell Environ.* **12**, 63–72.

Minoletti, M. L. and Boerner, R. E. (1994). Drought and site fertility effects on foliar nitrogen and phosphorus dynamics and nutrient resorption by the forest understory shrub *Viburnum acerifolium* L. *Am. Midl. Nat.* **131**, 109–119.

Mirov, N. T. (1954). Chemical composition of gum turpentines of pines of the United States and Canada. *J. For. Prod. Res. Soc.* **4**, 1–7.

Mirov, N. T. (1967). *The Genus Pinus*. Ronald, New York.

Mitchell, A. K. and Hinckley, T. M. (1993). Effects of foliar nitrogen concentration on photosynthesis and water use efficiency in Douglas-fir. *Tree Physiol.* **12**, 403–410.

Mitchell, H. L. and Chandler, R. F., Jr. (1939). The nitrogen nutrition and growth of certain deciduous trees of northeastern United States. *Black Rock For. Bull.* **11**.

Mitchell, K. A., Bolstad, P. V., and Vose, J. M. (1999). Interspecific and environmentally induced variation in foliar dark respiration among eighteen southeastern deciduous tree species. *Tree Physiol.* **19**, 861–870.

Mitchell, M. J., Foster, N. W., Shepard, J. P., and Morrison, I. K. (1992). Nutrient cycling in Huntingdon Forest and Turkey Lakes deciduous stands: Nitrogen and sulfur. *Can. J. For. Res.* **22**, 457–464.

Moinuddin and Khanna-Chopra, R. (2004). Osmotic adjustment in chickpea in relation to seed yield and yield parameters. *Crop Sci.* **44**, 449–455.

Molina, R., Massicote, H., and Trappe, J. M. (1992). Specificity phenomena in mycorrhizal symbioses: Community—Ecological consequences and practical implications. In *Mycorrhizal Functioning* (M. F. Allen, ed.), pp. 357–423. Chapman Hall, New York.

Molisch, H. (1902). *Über localen Blutungsdruck und seine Ursachen. Bot. Zg.* **60**, 45–63.

Möller, C. M. (1946). *Untersuchungen über Laubmenge, Stoffverlust und Stoffproduktion des Waldes. Fostl. Forsoegsvaes. Dan.* **17**, 1–287.

Møller, I. M. (2001). Plant mitochondria and oxidative stress: Electron transport, NADPH turnover, and metabolism of reactive oxygen species. *Annu. Rev. Plant Physiol. Plant Molec. Biol.* **52**, 561–591.

Molz, F. J. and Peterson, C. M. (1976). Water transport from roots to soil. *Agron. J.* **68**, 901–904.

Monk, C. D. (1966). An ecological significance of evergreenness. *Ecology* **47**, 504–505.

Monk, C. D. (1971). Leaf decomposition and loss of ^{45}Ca from deciduous and evergreen trees. *Am. Midl. Nat.* **86**, 370–384.

Monson, R. K. (1989). On the evolutionary pathways resulting in C_4 photosynthesis and Crassulacean acid metabolism. *Adv. Ecol. Res.* **19**, 57–110.

Monson, R. K. (1999). The origins of C_4 genes and evolutionary pattern in the C_4 metabolic phenotype. In *C_4 Plant Biology* (R. F. Sage and R. K. Monson, eds.), pp. 377–410. Academic Press, San Diego.

Monson, R. K. (2003). Gene duplication, neofunctionalization, and the evolution of C_4 photosynthesis. *Intl. J. Plant. Sci.* **164**, S43–S54.

Monson, R. K. and Fall, R. (1989). Isoprene emission from aspen leaves. The influence of environment and relation to photosynthesis and photorespiration. *Plant Physiol.* **90**, 267–274.

Monteith, J. L. (1995). A reinterpretation of stomatal responses to humidity. *Plant, Cell Environ.* **18**, 357–364.

Mooney, H. A. (1969). Dark respiration of related evergreen and deciduous Mediterranean plants during induced drought. *Bull. Torrey Bot. Club* **96**, 550–555.

Mooney, H. A. and Shropshire, F. (1967). Population variability in temperature related photosynthetic acclimation. *Oecol. Plant.* **2**, 1–13.

Mooney, H. A., Björkman, O., and Collatz, G. J. (1977). Photosynthetic acclimation to temperature and water stress in the desert shrub *Larrea divaricata*. *Year Book—Carnegie Inst. Washington* **76**, 328–335.

Mooney, H. A., Ferrar, P. J., and Slatyer, R. D. (1978). Photosynthetic capacity and carbon allocation patterns in diverse growth forms of *Eucalyptus*. *Oecologia* **36**, 103–111.

Moore, D. J., Nowak, R. S., and Tausch, R. J. (1999). Gas exchange and carbon isotope discrimination of *Juniperus osteosperma* and *Juniperus occidentalis* across environmental gradients in the Great Basin of western North America. *Tree Physiol.* **19**, 421–423.

Moore, P. D. (1974). Misunderstandings over C_4 carbon fixation. *Ann. Bot.* **252**, 439–441.

Moore, T. S., Jr., ed. (1993). *Lipid Metabolism in Plants*. CRC Press, Boca Raton, Florida.

Morey, P. R. (1973). *How Trees Grow.* Arnold, London.

Morgan, J. M. (1983). Osmoregulation as a selection criterion for drought tolerance in wheat. *Aust. J. Agric. Res.* **34**, 607–614.

Morgan, J. M. (1984). Osmoregulation and water stress in higher plants. *Annu. Rev. Plant Physiol.* **35**, 299–319.

Morgan, P. W. (1990). Effects of abiotic stresses on plant hormone systems. In *Stress Responses in Plants: Adaptation and Acclimation Mechanisms* (R. G. Alscher and J. R. Cumming, eds.), pp. 113–146. Wiley, New York.

Moriguchi, T. M., Abe, K., Sanada, T., and Yamaki, S. (1992). Levels and role of sucrose synthase, sucrose-phosphate synthase, and acid invertase in sucrose accumulation in fruit of Asian pear. *J. Am. Soc. Hortic. Sci.* **117**, 274–278.

Morikawa, Y., Hattori, S., and Kiyono, Y. (1986). Transpiration of a 31-year-old *Chamaecyparis obtusa* Endl. stand before and after thinning. *Tree Physiol.* **2**, 105–114.

Mork, E. (1928). *Die Qualität des Fichtenholzes unter besonderer Rücksichtnahme auf Schleif—und Papierholz. Pap. Fabrik* **26**, 741–747.

Morris, D. A. and Johnson, C. F. (1990). The role of efflux carriers in the reversible loss of polar auxin transport in the pea (*Pisum sativum* L.) stem. *Planta* **181**, 117–124.

Morris, D. M., Gordon, A. G., and Gordon, A. (2003). Patterns of canopy interception and throughfall along a topographic sequence for black spruce dominated forest ecosystems in northwestern Ontario. *Can. J. For. Res.* **33**, 1046–1060.

Morrison, J. C. and Polito, V. S. (1985). Gum duct development in almond fruit, *Prunus dulcis* (Mill.) D. A. Webb. *Bot. Gaz.* **146**, 15–25.

Morse, S. R., Wayne, P., Miao, S. L., and Bazzaz, F. A. (1993). Elevated CO_2 and drought alter tissue water relations of birch (*Betula populifolia* Marsh) seedlings. *Oecologia* **95**, 599–602.

Morselli, M. F., Marvin, J. W., and Laing, F. M. (1978). Image-analyzing computer in plant science: More and larger vascular rays in sugar maples of high sap and sugar yield. *Can. J. Bot.* **56**, 983–986.

Mott, K. A. and Parkhurst, D. F. (1991). Stomatal responses to humidity in air and helox. *Plant, Cell Environ.* **14**, 509–515.

Mudd, J. B. (1980). Phospholipid biosynthesis. In *The Biochemistry of Plants* (P. K. Stumpf, ed.), pp. 250–282. Academic Press, New York.

Mulder, E. G. (1975). Physiology and ecology of free-living nitrogen-fixing bacteria. In *Nitrogen Fixation by Free-Living Micro-Organisms* (W. D. P. Stewart, ed.), pp. 3–28. Cambridge Univ. Press, London and New York.

Müller, J., Aeschbacher, R. A., Wingler, A., Boller, T., and Wiemken, A. (2001). Trehalose and trehalase in *Arabidopsis. Plant Physiol.* **125**, 1086–1093.

Mullins, M. G., Plummer, J. A., and Snowball, A. M. (1989). Flower initiation: New approaches to the study of flowering in perennial fruit plants. In *Manipulation of Fruiting* (C. J. Wright, ed.), pp. 65–77. Butterworth, London.

Munns, R. and King, R. W. (1988). Abscisic acid is not the only stomatal inhibitor in the transpiration stream of wheat plants. *Plant Physiol.* **88**, 703–708.

Murata, K., Mitsuoka, K., Hirai, T., Walz, T., Agre, P. Heymann, J. B., Engel, A., and Fujiyoshi, Y. (2000). Structural determinants of water permeation through aquaporin-1. *Nature (London)* **407**, 599–605.

Murneek, A. E. (1930). Quantitative distribution and seasonal fluctuation of nitrogen in apple trees. *Proc. Am. Soc. Hortic. Sci.* **27**, 228–231.

Murneek, A. E. (1933). Carbohydrate storage in apple trees. *Proc. Am. Soc. Hortic. Sci.* **30**, 319–321.

Murneek, A. E. (1942). Quantitative distribution of nitrogen and carbohydrates in apple trees. *Missouri Agr. Expt. Sta. Res. Bull.* **348**.

Myers, B. A. and Neales, T. F. (1986). Osmotic adjustment induced by drought in seedlings of three *Eucalyptus* species. *Aust. J. Plant Physiol.* **13**, 597–604.

Nadelhoffer, K. J. and Raich, J. W. (1992). Fine root production estimates and belowground carbon allocation in forest ecosystems. *Ecology* **73**, 1139–1147.

Nagarajah, S. (1979). Differences in cuticular resistance in relation to transpiration in tea (*Camellia sinensis*). *Plant Physiol.* **46**, 89–92.

Nagarajah, S. and Ratnasooriya, G. B. (1977). Studies with antitranspirants on tea (*Camellia sinensis* L.). *Plant Soil* **48**, 185–197.

Nambiar, E. K. S. and Fife, D. N. (1991). Nutrient retranslocation in temperate conifers. *Tree Physiol.* **9**, 185–207.

Nairn, C. J. and Haselkorn, T. (2005). Three loblolly pine *CesA* genes expressed in developing xylem are orthologous to secondary cell wall *CesA* genes of angiosperms. *New Phytol.* **166**, 907–915.

Näsholm, T., Ekblad, A., Nordin, A., Giesler, R., Högberg, M., and Högberg, P. (1998). Boreal forest plants take up organic nitrogen. *Nature (London)* **392**, 914–916.

Natr, L. (1975). Influence of mineral nutrition on photosynthesis and the use of assimilates. In *Photosynthesis and Productivity in Different Environments* (J. P. Cooper, ed.), pp. 537–555. Cambridge University Press, Cambridge.

Neale, D. B. and Adams, W. T. (1981). Inheritance of isozyme variants in seed tissues of balsam fir (*Abies balsamea*). *Can. J. Bot.* **59**, 1285–1291.

Neale, D. B. and Williams, C. G. (1991). Restriction fragment length polymorphism mapping in conifers and applications to forest genetics and tree improvement. *Can. J. For. Res.* **21**, 545–554.

Neave, I. A., Dawson, J. O., and DeLucia, E. H. (1989). Autumnal photosynthesis is extended in nitrogen-fixing European black alder compared with white basswood: Possible adaptive significance. *Can. J. For. Res.* **19**, 12–17.

Nebel, B. and Matile, P. (1992). Longevity and senescence of needles in *Pinus cembra* L. *Trees* **6**, 156–161.

Neely, D. E. (1970). Healing of wounds on trees. *J. Am. Soc. Hortic. Sci.* **95**, 536–540.

Negisi, K. (1977). Respiration in forest trees. In *Primary Productivity of Japanese Forests. Vol. 16. Productivity of Terrestrial Communities* (T. Shidei and T. Kira, eds.), pp. 86–93. University of Tokyo Press, Tokyo.

Negisi, K. and Satoo, T. (1954). The effect of drying of soil on apparent photosynthesis, transpiration, carbohydrate reserves and growth of seedlings of akamatu (*Pinus densiflora* Sieb. et Zucc.). *J. Jpn. For. Soc.* **36**, 66–71.

Neher, H. V. (1993). Effects of pressures inside Monterey pine trees. *Trees* **8**, 9–17.

Nelson, E. A. and Dickson, R. E. (1981). Accumulation of food reserves in cottonwood stems during dormancy induction. *Can. J. For. Res.* **11**, 145–154.

Nelson, L. E., Switzer, G. L., and Smith, W. H. (1970). Dry matter and nutrient accumulation in young loblolly pine (*Pinus taeda* L.). *Tree Growth For. Soils, Proc. North Am. For. Soils Conf., 3rd, 1968.* pp. 261–273.

Nelson, N. D. (1984). Woody plants are not inherently low in photosynthetic capacity. *Photosynthetica* **18**, 600–605.

Nelson, N. D. and Ehlers, P. (1984). Comparative carbon dioxide exchange for two *Populus* clones grown in growth room, greenhouse, and field environments. *Can. J. For. Res.* **14**, 924–932.

Nelson, N. D. and Isebrands, J. G. (1983). Late season photosynthesis and photosynthate distribution in an intensively-cultured *Populus nigra* x *laurifolia* clone. *Photosynthetica* **17**, 537–549.

Nelson, N. D., Dickmann, D. I., and Gottschalk, K. W. (1982). Autumnal photosynthesis in short-rotation intensively cultured *Populus* clones. *Photosynthetica* **16**, 321–333.

Nevell, T. P. and Zeronian, S. H. (1985). Cellulose chemistry fundamentals. In *Cellulose Chemistry and Its Application* (T. P. Newell and S. H. Zeronian, eds.), pp. 15–29. Halsted Press, New York.

Newbanks, D., Bosch, A, and Zimmermann, M. H. (1983). Evidence for xylem dysfunction by embolism in Dutch elm disease. *Phytopathology* **73**, 1060–1063.

Newell, E. A., Mulkey, S. S., and Wright, S. J. (2002). Seasonal patterns of carbohydrate storage in four tropical tree species. *Oecologia* **131**, 333–342.

Newman, E. I. (1976). Water movement through root systems. *Phil. Trans. R. Soc. Lond.* **B273**, 463–478.

Newman, S. E. and Davies, F. T., Jr. (1988). High root-zone temperatures, mycorrhizal fungi, water relations, and root hydraulic conductivity of container-grown woody plants. *J. Amer. Soc. Hortic. Sci.* **113**, 138–146.

Newstrom, L. E., Frankie, G. W., Baker, H. G., and Colwell, R. K. (1993). Diversity of flowering patterns at La Selva. In *La Selva: Ecology and Natural History of a Lowland Tropical Rainforest* (L. A. McDade, K. S. Bawa, G. S. Hortshorn, and H. A. Hespenheide, eds.). Univ. of Chicago Press, Illinois.

Newstrom, L. E., Frankie, G. W., and Baker, H. G. (1994). A new classification for plant phenology based on flowering patterns in lowland tropical rain forest trees at La Selva, Costa Rica. *Biotropica* **26**, 141–159.

Newton, A. C., Allnutt, T. R., Gillies, A. C. M., Lowe, A. J., and Ennos, R. A. (1999). Molecular phylogeography, intraspecific variation and the conservation of tree species. *Trends Ecol. Evol.* **14**, 140–145.

Nguyen, A. and Lamont, A. (1989). Variation in growth and osmotic regulation of roots of water-stressed maritime pine (*Pinus pinaster* Ait.) provenances. *Tree Physiol.* **5**, 123–133.

Nguyen, P. V., Dickmann, D. I., Pregitzer, K. S., and Hendrick, R. (1990). Late-season changes in allocation of starch and sugar to shoots, coarse roots, and fine roots in two hybrid poplar clones. *Tree Physiol.* **7**, 95–105.

Ni, B. R. and Pallardy, S. G. (1991). Response of gas exchange to water stress in seedlings of woody angiosperms. *Tree Physiol.* **8**, 1–10.

Ni, B. R. and Pallardy, S. G. (1992). Stomatal and nonstomatal limitations to net photosynthesis in seedlings of woody angiosperms. *Plant Physiol.* **99**, 1502–1508.

Niimi, Y. and Torikata, H. (1979). Changes in photosynthesis and respiration during berry development in relation to the ripening of Delaware grapes. *J. Jpn. Soc. Hortic. Sci.* **47**, 448–453.

Niinemets, U., Kull, O., and Tenhunen, J. D. (1998). An analysis of light effects on foliar morphology, physiology, and light interception in temperate deciduous woody species of contrasting shade tolerance. *Tree Physiol.* **18**, 681–696.

Niinemets, U., Kull, O., and Tenhunen, J. D. (2004). Within-canopy variation in the rate of development of photosynthetic capacity is proportional to integrated quantum flux density in temperate deciduous trees. *Plant Cell Environ.* **27**, 293–313.

Nissan, S. J. and Foley, M. E. (1986). No latex starch utilization in *Euphorbia esula*. *Plant Physiol.* **81**, 696–698.

Nitsch, J. P. (1953). The physiology of fruit growth. *Annu. Rev. Plant Physiol.* **4**, 199–236.

Niyogi, K. K. (1999). Photoprotection revisited: Genetic and molecular approaches. *Annu. Rev. Plant Physiol. Plant Molec. Biol.* **50**, 333–359.

Nobel, P. S. (1976). Photosynthetic rates of sun versus shade leaves of *Hyptis emoryi* Torr. *Plant Physiol.* **58**, 218–223.

Nobel, P. S. (1985). Desert succulents. In *Physiological Ecology of North American Plant Communities* (B. F. Chabot and H. A. Mooney, eds.), pp. 181–197. Chapman and Hall, New York.

Nobel, P. S. (1991). *Physicochemical and Environmental Plant Physiology*. Academic Press, New York.

Nobel, P. S. (2005). *Physicochemical and Biophysical Plant Physiology, 3rd ed.* Elsevier Academic Press, Amsterdam.

Nobel, P. S. and Cui, M. Y. (1992). Hydraulic conductances of the soil, the root soil air gap, and the root—Changes for desert succulents in drying soil. *J. Exp. Bot.* **43**, 319–326.

Nobel, P. S. and Lee, C. H. (1991). Variations in root water potentials. Influence of environmental factors for two succulent species. *Ann. Bot.* **67**, 549–554.

Nobel, P. S. and Sanderson, J. (1984). Rectifier-like activities of two desert succulents. *J. Exp. Bot.* **154**, 727–737.

Nobel, P. S., Schulte, P. J., and North, G. B. (1990). Water influx characteristics and hydraulic conductivity for roots of *Agave deserti* Engelm. *J. Exp. Bot.* **41**, 409–415.

Nobles, D. R., Romanovicz, D. K., and Brown, R. M. (2001). Cellulose in cyanobacteria. Origin of vascular plant cellulose synthase? *Plant Physiol.* **127**, 529–542.

Nobuchi, T. and Harada, H. (1983). Physiological features of the "white zone" of sugi (*Cryptomeria japonica* D. Don). Cytological feature and moisture content. *Mokuzai Gakkaishi* **29**, 824–832.

Nobuchi, T., Takahara, S., and Harada, H. (1979). Studies on the survival rate of ray parenchyma cells with aging process in coniferous secondary xylem. *Bull. Kyoto Univ. For.* **51**, 239–246.

Noel, A. R. A. (1968). Callus formation and differentiation at an exposed cambial surface. *Ann. Bot.* **32**, 347–359.

Noland, T. L. and Kozlowski, T. T. (1979). Effect of SO_2 on stomatal aperture and sulfur uptake of woody angiosperm seedlings. *Can. J. For. Res.* **9**, 57–62.

Nonami, H. and Boyer, J. S. (1989). Turgor and growth at low water potentials. *Plant Physiol.* **89**, 798–804.

Nonami, H. and Boyer, J. S. (1990a). Primary events regulating stem growth at low water potentials. *Plant Physiol.* **93**, 1601–1609.

Nonami, H. and Boyer, J. S. (1990b). Wall extensibility and cell hydraulic conductivity decrease in enlarging stem tissues at low water potentials *Plant Physiol* **93**, 1610–1619.

Nonami, H., Schulze, E. D., and Ziegler, H. (1990). Mechanisms of stomatal movement in response to air humidity, irradiance and xylem water potential. *Planta* **183**, 57–64.

Nooden, L. D. and Leopold, A. C., eds. (1988). *Senescence and Aging in Plants*. Academic Press, San Diego.

Noormets, A., Sober, A., Pell, E. J., Dickson, R. E., Podila, G. K, Sober, J., Isebrands, J. G., and Karnosky, D. F. (2001). Stomatal and non-stomatal limitation to photosynthesis in two trembling aspen (*Populus tremuloides* Michx.) clones exposed to elevated CO_2 and/or O_3. *Plant, Cell Environ.* **24**, 327–336.

Norby, R. J. and O'Neill, E. G. (1989). Growth dynamics and water use of seedling of *Quercus alba* L. in CO_2-enriched atmospheres. *New Phytol.* **111**, 491–500.

Norby, R. J. and O'Neill, E. G. (1991). Leaf area compensation and nutrient interactions in CO_2-enriched seedlings of yellow-poplar (*Liriodendron tulipifera* L.). *New Phytol.* **117**, 515–528.

Norby, R. J., O'Neill, E. G., and Luxmoore, R. J. (1986). Effects of atmospheric CO_2 enrichment on the growth and mineral nutrition of *Quercus alba* seedlings in nutrient-poor soil. *Plant Physiol.* **82**, 83–89.

Norby, R. J., O'Neill, E. G., Hood, W. G., and Luxmoore, R. J. (1987). Carbon allocation, root exudation, and mycorrhizal colonization of *Pinus echinata* seedlings grown under CO_2 enrichment. *Tree Physiol.* **3**, 203–210.

Norby, R. J., Gunderson, C. A., Wullschleger, S. D., O'Neill, E. G., and McCracken, M. K. (1992). Productivity and compensatory responses of yellow-poplar trees in elevated CO_2. *Nature* (*London*) **357**, 322–324.

Norby, R. J., Wullschleger, S. D., Gunderson, C. A., Johnson, D. W., and Ceulemans, R. (1999). Tree responses to rising CO_2 in field experiments: Implications for the future forest. *Plant, Cell Environ.* **22**, 683–714.

Norby, R. J., Long, T. M., Hartz-Rubin, J. S., and O'Neill, E. G. (2000). Nitrogen resorption in senescing tree leaves in a warmer, CO_2-enriched atmosphere. *Plant Soil* **224**, 15–29.

Norby, R. J., Ogle, K., Curtis, P. S., Badeck, F. W., Huth, A., Hurtt, G. C., Kohyama, T., and Penuelas, J. (2001). Aboveground growth and competition in forest gap models: An analysis for studies of climatic change. *Climatic Change* **51**, 415–447.

Nordin, A., Högberg, P., and Nasholm, T. (2001). Soil nitrogen form and plant nitrogen uptake along a boreal forest productivity gradient. *Oecologia* **129**, 125–132.

Norisada, M. and Kojima, K. (2005). Nitrogen form preference of six dipterocarp species. *For. Ecol. Manage.* **216**, 175–186.

North, G. B. and Nobel, P. S. (1991). Changes in hydraulic conductivity and anatomy caused by drying and rewetting roots of *Agave deserti* (Agavaceae). *Am. J. Bot.* **78**, 906–915.

North, G. B. and Nobel, P. S. (1997). Root-soil contact for the desert succulent agave deserti in wet and drying soil. *New Phytol.* **135**, 21–29.

Norris, R. F. and Bukovac, M. J. (1968). Structure of the pear leaf cuticle with special reference to cuticular penetration. *Am. J. Bot.* **55**, 975–983.

Noskowiak, A. S. (1963). Spiral grain in trees. A review. *For. Prod. J.* **13**, 266–275.

Nowak, R. S., Ellsworth, D. S., and Smith, S. D. (2004). Functional responses of plants to elevated atmospheric CO_2—Do photosynthetic and productivity data from FACE experiments support early predictions? *New Phytol.* **162**, 253–280.

Nsimba-Lubacki, M., and Peumans, W. J. (1986). Seasonal fluctuations of lectins in barks of elderberry (*Sambucus nigra*) and black locust (*Robinia pseudoacacia*). *Plant Physiol.* **80**, 747–751.

Nye, P. H. and Greenland, D. J. (1960). The soil under shifting cultivation. *Tech. Commun. 51, Commonw. Bur. of Soils, Commonwealth Agric. Bur., Farnham Royal, England.*

Nye, P. H. and Tinker, P. B. (1977). *Solute Movement in the Soil-Root System.* Blackwell Scientific, Oxford.

O'Kennedy, B. F. and Titus, J. S. (1979). Isolation and mobilization of storage proteins from apple shoot bark. *Physiol. Plant.* **45**, 419–424.

O'Malley, P. E. R. and Milburn, J. A. (1983). Freeze-induced fluctuations in xylem sap pressure of *Acer pseudoplatanus*. *Can. J. Bot.* **61**, 3100–3106.

O'Neill, S. D. and Leopold, A. C. (1982). An assessment of phase transitions in soybean membranes. *Plant Physiol.* **70**, 1405–1409.

Oaks, A. and Hirel, B. (1985). Nitrogen metabolism in roots. *Annu. Rev. Plant Physiol.* **36**, 345–365.

Obaton, M. (1960). *Les lianes ligneuses, a structure anormale des forets denses d'Afrique occidentale. Ann. Sci. Nat., Bot. Biol. Veg.* [12] **1**, 1–220.

Oberbauer, S. F. and Strain, B. R. (1986). Effects of canopy position and irradiance on the leaf physiology and morphology of *Pentaclethra macroloba* (Mimosaceae). *Am. J. Bot.* **73**, 409–416.

Oberbauer, S. F., Clark, D. B., Clark, D. A., and Quesada, M. (1988). Crown light environments of saplings of two species of rain forest emergent trees. *Oecologia* **75**, 207–212.

Odum, E. P. (1985). Trends expected in stressed ecosystems. *BioScience* **35**, 419–423.

Oechel, W. C. and Lawrence, W. T. (1985). Taiga. In *Physiological Ecology of North American Plant Communities* (B. F. Chabot and H. A. Mooney, eds.), pp. 66–94. Chapman and Hall, New York.

Ogawa, K., Hagihara, A., and Hazumi, K. (1988). Photosynthesis and respiration in cones of hinoki. *J. Jap. For. Soc.* **70**, 220–226.

Ögren, E. (1988). Photoinhibition of photosynthesis in willow leaves under field conditions. *Planta* **175**, 229–236.

Oh, M. H., Romanow, W. G., Smith, R. C., Zamski, E., Sasse, J., and Clouse, S. D. (1998). Soybean BRU1 encodes a functional xyloglucan endotransglycosylase that is highly expressed in inner epicotyl tissues during brassinosteroid-promoted elongation. *Plant Cell Physiol.* **39**, 124–130.

Ohya, T. and Suzuki, H. (1991). The effects of benzyladenine on the accumulation of messenger RNAs that encode the large and small subunits of ribulose-1, 5-bisphosphate carboxylase-oxygenase and light-harvesting chlorophyll a-b protein in excised cucumber cotyledons. *Plant Cell Physiol.* **32**, 577–580.

Oland, K. (1963). Changes in the content of dry matter and major nutrient elements of apple foliage during senescence and abscission. *Physiol. Plant.* **16**, 682–694.

Oleksyn, J. (1984). Effects of SO_2, HF and NO_2 on net photosynthetic and dark respiration rates of Scots pine needles of various ages. *Photosynthetica* **18**, 259–262.

Oleksyn, J. and Bialobok, S. (1986). Net photosynthesis, dark respiration and susceptibility to air pollution of 20 European provenances of Scots pine *Pinus sylvestris* L. *Environ. Pollut.* (*Series A*) **40**, 287–302.

Oliveira, C. M. and Priestley, C. A. (1988). Carbohydrate reserves in deciduous fruit trees. *Hort. Rev.* **10**, 403–430.

Oliver, M. J., Tuba, Z., and Mishler, B. D. (2000). The evolution of vegetative desiccation tolerance in land plants. *Plant Ecol.* **151**, 85–100.

Oliver, W. W. (1974). Seed maturity in white fir and red fir. *U.S. Forest Service, Res. Pap. PSW* **PSW-99**, 1–12.

Olofinboba, M. O., Kozlowski, T. T., and Marshall, P. E. (1974). Effects of antitranspirants on distribution and utilization of photosynthate in *Pinus resinosa* seedlings. *Plant Soil* **40**, 619–635.

Olsen, J. E., Junttila, O., Nilsen, J., Eriksson, M. E., Martinussen, I., Olsson, O., Sandberg, G., and Moritz, T. (1997). Ectopic expression of oat phytochrome a in hybrid aspen changes critical daylength for growth and prevents cold acclimatization. *Plant J.* **12**, 1339–1350.

Ono, E. O., Nakamura, T., Machado, S. R., and Rodrigues, J. D. (2000). Application of brassinosteroid to *Tabebuia alba* (Bignoniaceae) plants. *Revista Brasileira de Fisiologia Vegetal* **12**, 187–194.

Ono, S., Kudo, K., and Daito, H. (1978). Studies on the photosynthesis and productivity of satsumas (*Citrus unshiu*). I. Effects of environmental factors on photosynthetic activity. *Bull. Shikoku Agric. Expt. Sta.* **31**, 147–157.

Opik, H. (1980). *The Respiration of Higher Plants.* Arnold, London.

Opler, P. A., Frankie, G. W., and Baker, H. G. (1980). Comparative phenological studies of treelet and shrub species in tropical wet and dry forests in the lowlands of Costa Rica. *J. Ecol.* **68**, 167–188.

Oppenheimer, H. R. (1932). *Zur Kenntnis der hochsomerlichen Wasserbilanz mediterranean Gehölze. Ber. Dtsch. Bot. Ges.* **50**, 185–243.

Oppenheimer, H. R. (1951). Summer drought and water balance of plants growing in the near east. *J. Ecol.* **39**, 357–362

Öquist, G. (1983). Effects of low temperature on photosynthesis. *Plant, Cell Environ.* **6**, 281–300.

Öquist, G. and Huner, N. P. A. (2003). Photosynthesis of overwintering evergreen plants. *Annu. Rev. Plant Biol.* **54**, 329–355.

Orchard, J. E., Collin, H. A., and Hardwick, K. (1981). Biochemical and physiological aspects of leaf development in cocoa (*Theo-*

broma cacao L.). V. Changes in auxins and cytokinins. *Cafe Cacao Thé* **25**, 25–28.

Orlov, A. J. (1960). *Rost i vozrastnye izmenenija sosuschih kornej eli Picea excelsa Link. Bot. Ztg.* **45**, 888–896.

Orr-Ewing, A. L. (1957). Possible occurrence of viable unfertilized seeds in Douglas-fir. *For. Sci.* **3**, 243–248.

Ort, D. R. (2001). When there is too much light. *Plant Physiol.* **125**, 29–32.

Osborne, D. J. and McManus, M. T. (2005). *Hormones, Signals and Target Cells in Plant Development.* Cambridge University Press, New York.

Osmond, C. B. (1987). Photosynthesis and carbon economy of plants. *New Phytol.* **106**, **Suppl.**, 161–175.

Osmond, C. B. and Chow, W. S. (1988). Ecology of photosynthesis in the sun and shade. *Aust. J. Plant Physiol.* **15**, 1–9.

Osmond, C. B., Smith, S. D., Gui-Ying, B., and Sharkey, T. D. (1987). Stem photosynthesis in a desert ephemeral, *Eriogonum inflatum*. Characterization of the leaf and stem CO_2 fixation and H_2O vapor exchange under controlled conditions. *Oecologia* **72**, 542–549.

Osonubi, O. and Davies, W. J. (1978). Solute accumulation in leaves and roots of woody plants subjected to water stress. *Oecologia* **32**, 323–332.

Othieno, C. O. (1978a). Supplementary irrigation of young clonal tea in Kenya. 1. Survival, growth, and yield. *Exp. Agric.* **14**, 229–230.

Othieno, C. O. (1978b). Supplementary irrigation of young clonal tea in Kenya. 2. Internal water status. *Exp. Agric.* **14**, 309–316.

Ottosen, C. O. (1990). Growth versus net photosynthesis in clones of *Ficus benjamina. HortScience* **25**, 956–957.

Outlaw, W. H., Jr. (1987). An introduction to carbon metabolism in guard cells. In *Stomatal Function* (E. Zeiger, G. D. Farquhar, and I. R. Cowan, eds.), pp. 151–123. Stanford University Press, Stanford, CA.

Ovington, J. D. (1956). The form, weights, and productivity of tree species grown in close stands. *New Phytol.* **55**, 289–388.

Owens, J. N. and Blake, M. D. (1985). Forest tree seed production. *Info. Rept. PI-X-53. Petawawa National For. Inst., Canadian Forestry Service, Agriculture Canada.*

Owens, J. N. and Hardev, V. (1990). Sex expression in gymnosperms. *Crit. Rev. Plant Sci.* **9**, 281–294.

Owens, J. N. and Molder, M. (1984a). The reproductive cycles of western red cedar and yellow cedar. *B.C. Min. Forest Info. Serv. Branch, Victoria, B.C. Canada.* 28 p.

Owens, J. N. and Molder, M. (1984b). The reproductive cycle of lodgepole pine. *B.C. Min. Forest Info. Service Branch, Victoria, B.C. Canada.* 29 p.

Owens, J. N. and Molder, M. (1984c). The reproductive cycle of interior spruce. *B.C. Min. Forest Info. Service Branch, Victoria B.C. Canada.* 30 p.

Owens, J. N. and Smith, F. H. (1964). The initiation and early development of the seed cone of Douglas-fir. *Can. J. Bot.* **42**, 1031–1047.

Owens, J. N. and Smith, F. H. (1965). Development of the seed cone of Douglas-fir following dormancy. *Can. J. Bot.* **43**, 317–332.

Padgett, M. and Morrison, J. C. (1990). Changes in grape berry exudates during fruit development and their effect on mycelial growth of *Botrytis cinerea. J. Am. Soc. Hortic. Sci.* **115**, 269–273.

Paembonan, S. A., Hagihara, A., and Hozumi, K. (1992). Long-term respiration in relation to growth and maintenance processes of the aboveground parts of a hinoki forest tree. *Tree Physiol.* **10**, 101–110.

Paiva, E. and Robitaille, H. A. (1978). Breaking bud rest on detached apple shoots: Effects of wounding and ethylene. *J. Am. Soc. Hort. Sci.* **103**, 101–104.

Pallardy, S. G. (1981). Closely related woody plants. In *Water Deficits and Plant Growth* (T. T. Kozlowski, ed.), Vol. 6, pp. 511–548. Academic Press, New York.

Pallardy, S. G. (1989). Hydraulic architecture and conductivity: An overview. In *Structural and Functional Responses to Environmental Stresses: Water Shortage* (K. H. Kreeb, H. Richter and T. M. Hinckley, eds.), pp. 3–19. SPB Academic Publishing, The Hague.

Pallardy, S. G. and Kozlowski, T. T. (1979a). Stomatal response of *Populus* clones to light intensity and vapor pressure deficit. *Plant Physiol.* **64**, 112–114.

Pallardy, S. G. and Kozlowski, T. T. (1979b). Relationship of leaf diffusion resistance of *Populus* clones to leaf water potential and environment. *Oecologia* **40**, 371–380.

Pallardy, S. G. and Kozlowski, T. T. (1980). Cuticle development in the stomatal region of *Populus* clones. *New Phytol.* **85**, 363–368.

Pallardy, S. G. and Kozlowski, T. T. (1981). Water relations of *Populus* clones. *Ecology* **62**, 159–169.

Pallardy, S. G. and Loewenstein, N. J. (2004). The role of xylem sap abscisic acid in leaf abscission of droughted seedlings of isohydric and anisohydric temperate deciduous angiosperms. In *Proceedings of the 14th Central Hardwood Conference* (D. Yaussy, D. M. Hix, P. C. Goebel, and R. P. Long, eds.), pp. 365–372. U.S. Dept. Agric., Forest Serv., GTR NE-316, Wooster, OH.

Pallardy, S. G. and Rhoads, J. L. (1993). Morphological adaptations to drought in seedlings of deciduous angiosperms. *Can. J. For. Res.* **23**, 1766–1774.

Pallardy, S. G., Parker, W. C., Dixon, R. K., and Garrett, H. E. (1982). Tissue water relations of roots and shoots of droughted ectomycorrhizal shortleaf pine seedlings. *Proc. Seventh North American Forest Biology Workshop* (B. A. Thielges, ed.), pp. 368–373. University of Kentucky, Lexington.

Pallardy, S. G., Pereira, J. S., and Parker, W. C. (1991). Measuring the state of water in tree systems. In *Techniques and Approaches in Forest Tree Ecophysiology* (J. P. Lassoie and T. M. Hinckley, eds.), pp. 28–76. CRC Press, Boca Raton.

Palmer, J. W. (1986). The regulation of photosynthesis in fruit trees. *Symp. Proc. Publ. N.Y. State Agric. Expt. Sta. Geneva, New York* (A. N. Lakso and F. Lenz, eds.), pp. 30–33.

Pandey, U. and Singh, J. S. (1982). Leaf-litter decomposition in an oak-conifer forest in Himalaya: The effects of climate and chemical composition. *Forestry* **55**, 47–60.

Panshin, A. J., DeZeeuw, C., and Brown, H. P. (1966). *Textbook of Wood Technology.* McGraw-Hill, New York.

Pardossi, A., Pritchard, J., and Tomos, A. D. (1994). Leaf illumination and root cooling inhibit bean leaf expansion by decreasing turgor pressure. *J. Exp. Bot.* **45**, 415–422.

Parker, G. G. (1983). Throughfall and stemflow in the forest nutrient cycle. *Adv. Ecol. Res.* **13**, 57–133.

Parker, G. G. (1995). Structure and microclimate of forest canopies. In *Forest Canopies* (M. D. Lowman and N. M. Nadkarni, eds.), pp. 73–106. Academic Press, San Diego.

Parker, J. (1949). Effects of variations in the root-leaf ratio on transpiration rate. *Plant Physiol.* **24**, 739–743.

Parker, J. (1950). The effects of flooding on the transpiration and survival of southern forest tree species. *Plant Physiol.* **25**, 453–460.

Parker, J. (1952). Desiccation in conifer leaves: Anatomical changes and determination of the lethal level. *Bot. Gaz.* **114**, 189–198.

Parker, J. (1956). Variations in copper, boron, and manganese in leaves of *Pinus ponderosa. For. Sci.* **2**, 190–198.

Parker, J. (1964). Autumn exudation from black birch. *Sci. Tree Top* **12**, 97–248.

Parker, J. and Patton, R. L. (1975). Effects of drought and defoliation on some metabolites in roots of black oak seedlings. *Can. J. For. Res.* **5**, 457–463.

Parker, W. C. and Pallardy, S. G. (1985a). Drought-induced leaf abscission and whole-plant drought tolerance of seedlings of seven black walnut families. *Can. J. For. Res.* **15**, 818–821.

Parker, W. C. and Pallardy, S. G. (1985b). Genotypic variation in tissue water relations of leaves and roots of black walnut (*Juglans nigra*) seedlings. *Physiol. Plant.* **64**, 105–110.

Parker, W. C. and Pallardy, S. G. (1988). Leaf and root osmotic adjustment in *Quercus alba*, *Q. macrocarpa* and *Q. stellata* seedlings. *Can. J. For. Res.* **18**, 1–5.

Parker, W. C. and Pallardy, S. G. (1991). Gas exchange during a soil drying cycle in seedlings of four black walnut (*Juglans nigra* L.) families. *Tree Physiol.* **9**, 339–348.

Parker, W. C., Pallardy, S. G., Hinckley, T. M., and Teskey, R. O. (1982). Seasonal changes in tissue water relations of three woody species of the *Quercus-Carya* forest type. *Ecology* **63**, 1259–1267.

Parkhurst, D. F. (1994). Diffusion of CO_2 and other gases inside leaves. *New Phytol.* **126**, 449–479.

Parsons, L. R. and Kramer, P. J. (1974). Diurnal cycling in root resistance to water movement. *Physiol. Plant* **30**, 19–23.

Passioura, J. B. (1988a). Response to Dr. P. J. Kramer's article "Changing concepts regarding plant water relations," Vol. 11, Num. 7, pp. 565–568. *Plant, Cell Environ.* **11**, 569–572.

Passioura, J. B. (1988b). Water transport in and to roots. *Annu. Rev. Plant Physiol. Plant Mol. Biol.* **39**, 245–265.

Passioura, J. B. (1991). An impasse in plant water relations? *Bot. Acta* **104**, 405–411.

Passioura, J. B. (1994). The physical chemistry of the primary cell wall: Implications for the control of expansion rate. *J. Exp. Bot.* **45**, 1675–1682.

Passioura, J. B. and Fry, S. C. (1992). Turgor and cell expansion—Beyond the Lockhart equation. *Aust. J. Plant Physiol.* **19**, 565–576.

Passioura, J. and Boyer, J. S. (2003). Tissue stresses and resistance to water flow conspire to uncouple the water potential of the epidermis from that of the xylem in elongating plant stems. *Funct. Plant Biol.* **30**, 325–334.

Pate, J. S. (1980). Transport and partitioning of nitrogenous solutes. *Annu. Rev. Plant Physiol.* **31**, 313–340.

Pate, J. S. (1994). The mycorrhizal association: Just one of many nutrients acquiring specializations in natural ecosystems. *Plant Soil* **159**, 1–10.

Pathre, U., Sinha, A. K., Shirke, P. A., and Sane, P. V. (1998). Factors determining the midday depression of photosynthesis in trees under monsoon climate. *Trees* **12**, 472–481.

Patiño, S., Tyree, M. T., and Herre, E. A. (1995). Comparison of hydraulic architecture of woody plants of differing phylogeny and growth form with special reference to freestanding and hemi-epiphytic *Ficus* species from Panama. *New Phytol.* **129**, 125–134.

Patric, J. H., Douglass, J. E., and Hewlett, J. D. (1965). Soil water absorption by mountain and Piedmont forests. *Soil Sci. Soc. Am. Proc.* **29**, 303–308.

Patton, R. F. and Johnson, D. W. (1970). Mode of penetration of needles of eastern white pine by *Cronartium ribicola*. *Phytopathology* **60**, 977–982.

Paul, M., Pellny, T., and Goddijn, O. (2001). Enhancing photosynthesis with sugar signals. *Trends Plant Sci.* **6**, 197–200.

Paul, M. J. and Foyer, C. H. (2001). Sink regulation of photosynthesis. *J. Exp. Bot.* **52**, 1383–1400.

Pavel, E. W. and DeJong, T. M. (1993). Estimating the photosynthetic contribution of developing peach (*Prunus persica*) fruits to their growth and maintenance carbohydrate requirements. *Physiol. Plant.* **88**, 331–338.

Pavel, E. W. and DeJong, T. M. (1995). Seasonal patterns of nonstructural carbohydrates of apple (*Malus pumila* Mill.) fruits. Relationship with relative growth rates and contribution to solute potential. *J. Hortic. Sci.* **70**, 127–134.

Pavel, E. W. and Fereres, E. (1998). Low soil temperatures induce water deficits in olive (*Olea europaea*) trees. *Physiol. Plant.* **104**, 525–532.

Pearcy, R. W. (1977). Acclimation of photosynthetic and respiratory carbon dioxide exchange to growth temperature in *Atriplex lentiformis* (Torr.) Wats. *Plant Physiol.* **59**, 795–799.

Pearcy, R. W. (1983). The light enviroment and growth of C_3 and C_4 species in the understory of a Hawaiian forest. *Oecologia* **58**, 26–32.

Pearcy, R. W. (1988). Photosynthetic utilization of lightflecks by understory plants. *Aust. J. Plant Physiol.* **15**, 223–238.

Pearcy, R. W. (1990). Sunflecks and photosynthesis in plant canopies. *Annu. Rev. Plant Physiol. Plant Mol. Biol.* **41**, 421–453.

Pearcy, R. W. and Calkin, H. W. (1983). Carbon dioxide exchange of C_3 and C_4 tree species in the understory of a Hawaiian rainforest. *Oecologia* **58**, 26–31.

Pearcy, R. W. and Troughton, J. H. (1975). C_4 photosynthesis in tree form *Euphorbia* species from Hawaiian rainforest sites. *Plant Physiol.* **55**, 1054–1056.

Pearcy, R. W., Harrison, A. T., Mooney, H. A., and Björkman, O. (1974). Seasonal changes in net photosynthesis of *Atriplex hymenelytra* shrubs growing in Death Valley, Calif. *Oecologia* **17**, 111–121.

Pearcy, R. W., Osteryoung, K., and Calkin, H. W. (1985). Photosynthetic responses to dynamic light environments by Hawaiian trees: Time course of CO_2 uptake and carbon gain during sunflecks. *Plant Physiol.* **79**, 896–902.

Pearcy, R. W., Ehleringer, J., Mooney, H. A., and Rundel, P. W., eds. (1989). *Plant Physiological Ecology: Field Methods and Instrumentation.* Chapman and Hall, New York.

Pearcy, R. W., Chazdon, R. L., Gross, L. J., and Mott, K. A. (1994). Photosynthetic utilization of sunflecks: A temporally patchy resource on a time scale of seconds to minutes. In *Exploitation of Environmental Heterogeneity by Plants* (M. M. Caldwell and R. W. Pearcy, eds.), pp. 175–208. Academic Press, San Diego.

Pelkonen, P. and Luukkanen, O. (1974). Gas exchange in three populations of Norway spruce. *Silvae Genet.* **23**, 160–164.

Peltier, J. P. and Marigo, G. (1999). Drought adaptation in *Fraxinus excelsior* L.: Physiological basis of the elastic adjustment. *J. Plant Physiol.* **154**, 529–535.

Penfold, A. R. and Willis, J. L. (1961). *The Eucalypts.* Wiley (Interscience), New York.

Penning, de Vries, F. W. T. (1975a). The cost of maintenance processes in plant cells. *Ann. Bot.* **39**, 77–92.

Penning, de Vries, F. W. T. (1975b). The use of assimilates in higher plants. In *Photosynthesis and Productivity in Different Environments* (J. P. Cooper, ed.), pp. 459–480. Cambridge University Press, New York.

Perchorowicz, J. T., Raynes, D. A., and Jensen, R. G. (1981). Light limitation of photosynthesis and regulation of ribulose bisphosphate carboxylase in wheat seedlings. *Proc. Natl. Acad. Sci.* **78**, 2895–2989.

Pereira, J. S. and Kozlowski, T. T. (1976). Leaf anatomy and water relations of *Eucalyptus camaldulensis* and *E. globulus* seedlings. *Can. J. Bot.* **54**, 2868–2880.

Pereira, J. S. and Kozlowski, T. T. (1977a). Variations among woody angiosperms in response to flooding. *Physiol. Plant.* **41**, 184–192.

Pereira, J. S. and Kozlowski, T. T. (1977b). Influence of light intensity, temperature, and leaf area on stomatal aperture and water potential of woody plants. *Can. J. For. Res.* **7**, 145–153.

Pereira, J. S., Tenhunen, J. D., Lange, O. L., Beyschlag, W., Meyer, A., and David, M. M. (1986). Seasonal and diurnal patterns in leaf gas exchange of *Eucalyptus globulus* trees growing in Portugal. *Can. J. For. Res.* **16**, 177–184.

Perry, D. A., Molina, R., and Amaranthus, M. P. (1987). Mycorrhizae, mycorrhizospheres, and reforestation: Current knowledge and research needs. *Can. J. For. Res.* **17**, 929–940.

Perry, T. O. (1971). Winter season photosynthesis and respiration by twigs and seedlings of deciduous and evergreen trees. *For. Sci.* **17**, 41–43.

Perttunen, J., Sievänen, R., and Nikinmaa, E. (1998). LIGNUM: A model combining the structure and the functioning of trees. *Ecol. Model.* **108**, 189–198.

Perumalla, C. J., Peterson, C. A., and Enstone, D. E. (1990). A survey of angiosperm species to detect hypodermal Casparian bands. I. Roots with a uniseriate hypodermis and epidermis. *Bot. J. Linnean Soc.* **103**, 93–112.

Peterson, A. G., Ball, J. T., Luo, Y. Q., Field, C. B., Reich, P. B., Curtis, P. S., Griffin, K. L., Gunderson, C. A. et al. (1999). The photosynthesis leaf nitrogen relationship at ambient and elevated atmospheric carbon dioxide: A meta-analysis. *Global Change Biol.* **5**, 331–346.

Peterson, B. J. and Fry, B. (1987). Stable isotopes in ecosystem studies. *Annu. Rev. Ecol. System.* **18**, 293–320.

Peterson, C. A. and Perumalla, C. J. (1990). A survey of angiosperm species to detect hypodermal Casparian bands. II. Roots with a multiseriate hypodermis or epidermis. *Bot. J. Linnean Soc.* **103**, 113–126.

Peterson, C. A., Enstone, D. E., and Taylor, J. H. (1999). Pine root structure and its potential significance for root function. *Plant Soil* **217**, 205–213.

Pezeshki, S. R. (1993). Differences in patterns of photosynthetic responses to hypoxia in flood-tolerant and flood-sensitive tree species. *Photosynthetica* **28**, 423–430.

Pezeshki, S. R. and Chambers, J. L. (1985a). Stomatal and photosynthetic response of sweet gum (*Liquidambar styraciflua*) to flooding. *Can. J. For. Res.* **15**, 371–375.

Pezeshki, S. R. and Chambers, J. L. (1985b). Responses of cherrybark oak (*Quercus falcata* var. *pagodaefolia*) seedlings to short-term flooding. *For. Sci.* **31**, 760–771.

Pezeshki, S. R. and Chambers, J. L. (1986). Effect of soil salinity on stomatal conductance and photosynthesis of green ash (*Fraxinus pennsylvanica*). *Can. J. For. Res.* **16**, 569–573.

Pezeshki, S. R., DeLaune, R. D., and Patrick, W. H. (1987). Effect of salinity on leaf ion content and photosynthesis of *Taxodium distichum* L. *Am. Midl. Nat.* **119**, 185–192.

Pfanz, H., Aschan, G., Langenfeld-Heyser, R., Wittmann, C., and Loose, M. (2002). Ecology and ecophysiology of tree stems: Corticular and wood photosynthesis. *Naturwissenschaften* **89**, 147–162.

Pharis, R. P., Jenkins, P. A., Aoki, H., and Sassa, T. (1981). Hormonal physiology of wood growth in *Pinus radiata* D. Don: Effects of gibberellin A_4 and the influence of abscisic acid upon (^{3}H) gibberellin A_4 metabolism. *Aust. J. Plant Physiol.* **8**, 559–570.

Phelps, J. E., McGinnes, E. A., Smolinski, M., Saniewski, M., and Pieniazek, J. (1977). A note on the formation of compression wood induced by morphactin IT 3456 in *Thuja* shoots. *Wood and Fiber* **8**, 223–227.

Phelps, J. E., McGinnes, E. A., Jr., Saniewski, M., Pieniazek, J., and Smolinski, M. (1980). Some anatomical observations on the effect of morphactin IT 3456 and Ethrel on wood formation in *Salix fragilis* L. *IAWA Bull. N. S.* **1–2**, 76–82.

Philip, J. R. (1966). Plant water relations: Some physical aspects. *Annu. Rev. Plant Physiol.* **17**, 245–268.

Philipson, J. J. and Coutts, M. P. (1978). The tolerance of tree roots to waterlogging. III. Oxygen transport in lodgepole pine and Sitka spruce roots of primary structure. *New Phytol.* **80**, 341–349.

Philipson, J. J. and Coutts, M. A. (1980a). Effects of growth hormone application on the secondary growth of roots and stems in *Picea sitchensis* (Bong.) Carr. *Ann. Bot.* **46**, 747–755.

Philipson, J. J. and Coutts, M. P. (1980b). Tolerance of tree roots to waterlogging. *New Phytol.* **85**, 489–494.

Phillips, I. D. J. (1962). Some interactions of gibberellic aid with naringerin (5,7,4′-trihydroxy-flavanone) in the control of dormancy and growth in plants. *J. Exp. Bot.* **13**, 213–226.

Phillips, I. D. J. and Van Loon, D. H. (1984). Biomass removal and nutrient drain as affected by total tree harvest in southern pine and hardwood stands. *J. For.* **82**, 547–550.

Phillips, I. D. J. and Wareing, P. F. (1958). Studies in the dormancy of sycamore. I. Seasonal changes in growth substance content of the shoot. *J. Exp. Bot.* **9**, 350–364.

Phillips, I. D. J., Miners, J., and Roddick, J. G. (1980). Effects of light and photoperiodic conditions on abscisic acid in leaves and roots of *Acer pseudoplatanus*. *Planta* **149**, 118–122.

Phillipson, R., Putnam, R. J., Steel, J., and Woodall, S. R. J. (1975). Litter input, litter decomposition and the evolution of carbon dioxide in a beech woodland. Wytham Woods, Oxford. *Oecologia* **20**, 203–217.

Phung, H. T. and Knipling, E. B. (1976). Photosynthesis and transpiration of citrus seedlings under flooded conditions. *HortScience* **11**, 131–133.

Piatkowski, D., Schneider, K., Salamini, F., and Bartels, D. (1990). Characterization of five abscisic-acid responsive cDNA clones isolated from the desiccation tolerant plant *Craterostigma plantagineum* and their relationship to other water-stress genes. *Plant Physiol.* **94**, 1682–1688.

Pickard, W. F. (1981). The ascent of sap in plants. *Prog. Biophys. Mol. Biol.* **37**, 181–230.

Picon, C., Guehl, J. M., and Ferhi, A. (1996). Leaf gas exchange and carbon isotope composition responses to drought in a drought-avoiding (*Pinus pinaster*) and a drought-tolerant (*Quercus petraea*) species under present and elevated atmospheric CO_2 concentrations. *Plant, Cell Environ.* **19**, 182–190.

Picon-Cochard, C., Coll, L., and Balandier, P. (2006). The role of below-ground competition during early stages of secondary succession: The case of 3-year-old Scots pine (*Pinus sylvestris* L.) seedlings in an abandoned grassland. *Oecologia* **148**, 373–383.

Pien, P., Wyrzykowska, J., McQueen-Mason, S., Smart, C., and Fleming, A. (2001). Local expression of expansin induces the entire process of leaf development and modifies leaf shape. *Proc. Natl. Acad. Sci. U.S.A.* **98**, 11812–11817.

Pieniazek, J., Smolinski, M., and Saniewski, M. (1970). Induced structural changes in anatomy of apple shoots after treatment with morphactin IT 3456 and other growth regulators (NAA, GA, BA). *Acta Agrobot.* **23**, 387–396.

Pierce, M. and Raschke, K. (1980). Correlation between loss of turgor and accumulation of abscisic acid in detached leaves. *Planta* **148**, 174–182.

Pieterse, C. M. and Van Loon, L. (2004). NPR1: The spider in the web of induced resistance signaling pathways. *Curr. Opin. Plant Biol.* **7**, 456–464.

Piironen, V., Lindsay, D. G., Miettinen, T. A., Toivo, J., and Lampi, A. M. (2000). Plant sterols: Biosynthesis, biological function and their importance to human nutrition. *J. Sci. Food Agric.* **80**, 939–966.

Pillsbury, A. F., Pelishek, R. E., Osborn, J. F., and Szuszkiewicz, T. E. (1961). Chaparral to grass conversion doubles watershed runoff. *Calif. Agric.* **15**, 12–13.

Pirson, A. (1958). *Mineralstoffe and Photosynthese. Encycl. Plant Physiol.* **4**, 355–381.

Pisek, A. and Larcher, W. (1954). *Zusammenhang zwischen Austrocknungsresistenz und Frosthärte bei Immergrünen. Protoplasma* **44**, 30–46.

Pisek, A. and Winkler, E. (1958). *Assimilations—vermogen und Respiration der Fichte (Picea excelsa) in verschiedener Hohenlage und der Zirbe (Pinus cembra L.) an der alpinen Waldgrenze. Planta* **51**, 518–543.

Piskornik, Z. (1969). Effect of industrial air pollutants on photosynthesis in hardwoods. *Bull. Inst. Paper Chemistry* **43**, 8509.

Pittermann. J., Sperry, J. S., Hacke, U. G., Wheeler, J. K., and Sikkema, E. H. (2005). Torus-margo pits help conifers compete with angiosperms. *Science* **310**, 1924.

Podila, G. P., Dickinson, M. B., and Kolattukudy, P. E. (1988). Transcriptional activation of a cutinase gene in isolated fungal nuclei by plant cutin monomers. *Science* **242**, 922–925.

Polhamous, L. G. (1962). *Rubber: Botany, Production, and Utilization.* Interscience, New York.

Pollard, D. F. W. (1970). Leaf area development of different shoot types in a young aspen stand and its effect on production. *Can. J. Bot.* **48**, 1801–1804.

Pollard, D. F. W. and Logan, K. T. (1974). The role of tree growth in the differentiation of provenances of black spruce *Picea mariana* (Mill.) B.S.P. *Can. J. For. Res.* **4**, 308–311.

Pollard, D. F. W. and Logan, K. T. (1976). Inherent variation in "fall growth" in relation to numbers of needles produced by provenances of *Picea mariana*. In *Tree Physiology and Yield Improvement* (M. G. R. Cannell and F. T. Last, eds.), pp. 245–251. Academic Press, New York.

Pollard, J. K. and Sproston, T. (1954). Nitrogenous constituents of sap exuded from the sapwood of *Acer saccharum*. *Plant Physiol.* **29**, 360–364.

Pollock, B. M. (1953). The respiration of *Acer* buds in relation to the inception and termination of the winter rest. *Physiol. Plant.* **6**, 47–64.

Polster, H. (1955). *Vergleichende Untersuchungen über die Kohlendioxidassimilation und Atmung der Douglasie, Fichte und Weymouthskiefer. Arch. Forstwes.* **4**, 689–714.

Ponnamperuma, F. N. (1972). The chemistry of submerged soils. *Adv. Agron.* **24**, 29–96.

Poole, D. K. and Miller, P. C. (1978). Water related characteristics of some evergreen sclerophyll shrubs in central Chile. *Oecol. Plant.* **13,** 289–299.

Poorter, H. and Garnier, E. (1996). Plant growth analysis: An evaluation of experimental design and computational methods. *J. Exp. Bot.* **47**, 1343–1351.

Poorter, H., Van der Werf, A., Atkin, O. K., and Lambers, H. (1991). Respiratory energy requirements of roots vary with the potential growth rate of plant species. *Physiol. Plant.* **83**, 469–475.

Poovaiah, B. W. and Reddy, A. S. N. (1987). Calcium messenger system in plants. *Crit. Rev. Plant Sci.* **6**, 47–103.

Popham, R. A. (1952). *Developmental Plant Anatomy.* Long's College Book Co., Columbus, Ohio.

Popp, M., Kramer, D., Lee, H., Diaz, M., Ziegler, H., and Luttge, U. (1987). Crassulacean acid metabolism in tropical dicotyledonous trees of the genus *Clusia*. *Trees* **1**, 238–247.

Porandowski, J., Rakowski, K., and Wodgicki, T. J. (1982). Apical control of xylem formation in the pine stem. II. Responses of differentiating tracheids. *Acta Soc. Bot. Pol.* **51**, 203–214.

Porpiglia, P. J. and Barden, J. A. (1980). Seasonal trends in net photosynthetic potential, dark respiration, and specific leaf weight of apple leaves as affected by canopy position. *J. Am. Soc. Hortic. Sci.* **105**, 920–923.

Portis, A. R. (2003). Rubisco activase—Rubisco's catalytic chaperone. *Photosyn. Res.* **75**, 11–27.

Portis, A. R., Salvucci, M. E., and Ogren, W. L. (1986). Activation of ribulose bisphosphate carboxylase/oxygenase at physiological CO_2 and ribulose bisphosphate concentrations by Rubisco activase. *Plant Physiol.* **82**, 967–971.

Pospisilova, J. and Santrucek, J. (1994). Stomatal patchiness. *Biol. Plant.* **36**, 481–510.

Possingham, J. V. (1970). Aspects of the physiology of grape vines. In *Physiology of Tree Crops* (L. C. Luckwill and C. V. Cutting, eds.), pp. 335–349. Academic Press, London.

Postgate, J. R. (1998). *Nitrogen Fixation, 3rd ed.* Cambridge University Press, Cambridge and New York.

Postlethwaite, S. N. and Rogers, B. (1958). Tracing the path of transpiration stream in trees by the use of radioactive isotopes. *Am. J. Bot.* **45**, 753–757.

Powell, G. R. (1991). Preformed and neoformed extension of shoots and sylleptic branching in relation to shoot length in *Tsuga canadensis*. *Trees* **5**, 107–116.

Powell, G. R. (1992). Patterns of leaf size and morpholgy in relation to shoot length in *Tsuga canadensis*. *Trees* **7**, 59–66.

Powell, G. R., Tosh, K. J., and MacDonald, J. E. (1982). Indeterminate shoot elongation and heterophylly in *Acer saccharum*. *Can. J. For. Res.* **12**, 166–170.

Powell, L. E. (1972). Naturally occurring plant growth regulators and their physiological roles in fruit trees. *Acta Hortic.* **34**, 33–39.

Powell, L. E. (1982). Shoot growth in woody plants and possible participation of abscisic acid. In *Plant Growth Substances* (P. F. Wareing, ed.), pp. 363–372. Academic Press, London.

Powell, L. E. (1987). Hormonal aspects of bud and seed dormancy in temperate-zone woody plants. *HortScience* **22**, 845–850.

Powles, S. B. (1984). Inhibition of photosynthesis induced by visible light. *Annu. Rev. Plant Physiol.* **35**, 15–44.

Prentice, I., Farquhar, G., Fasham, M., Goulden, M., Heinmann, M. et al. (2001). The carbon cycle and atmospheric carbon dioxide. In *Climate Change 2001: The Scientific Basis. Contributions of Working Group I to the Third Assessment Report of the Intergovernmental Panel on Climate Change* (J. T. Houghton, Y. Ding, D. J. Griggs, M. Noguer, P. J. van der Linden et al., eds.), pp. 183–238. Cambridge University Press, Cambridge, U.K.

Priestley, C. A. (1962). The location of carbohydrate resources within the apple tree. *Proc. 16th Int. Hortic. Congr.* pp. 319–327.

Priestley, C. A. (1977). The annual turnover of resources in young olive trees. *J. Hortic. Sci.* **52**, 105–112.

Primack, R. B. (1985). Longevity of individual flowers. *Annu. Rev. Ecol. Syst.* **16**, 15–32.

Pritchard, J. (1994). The control of cell expansion in roots. *New Phytol.* **127**, 3–26.

Proebsting, E. L. (1934). A fertilizer trial with Bartlett pears. *Proc. Amer. Soc. Hortic. Sci.* **30**, 55–57.

Proebsting, E. L. (1958). A quantitative evaluation of the effect of fruiting on growth of Elberta peach trees. *Proc. Am. Soc. Hortic. Sci.* **71**, 103–109.

Putuhena, W. M. and Cordery, I. (2000). Some hydrological effects of changing forest cover from eucalypts to *Pinus radiata*. *Agric. For Meteorol.* **100**, 59–72.

Pyankov, P. I., Gunin, P. D., Tsoog, S., and Black, C. C. (2000). C_4 plants in the vegetation of Mongolia: Their natural occurrence and geographical distribution in relation to climate. *Oecologia* **123**, 15–31.

Queen, W. H. (1967). Radial movement of water and ^{32}P through the suberized and unsuberized roots of grape. Ph.D. Dissertation, Duke University, Durham, North Carolina.

Quinn, P. J. and Harwood, J. S., eds. (1990). *Plant Lipid Biochemistry: Structure and Utilization*. Portland Press, London.

Raber, O. (1937). Water utilization by trees, with special reference to the economic forest species of the north temperate zone. *U.S., Dep. Agric. Misc. Publ.* **257**.

Radford, P. J. (1967). Growth analysis formulae: Their use and abuse. *Crop Sci.* **7**, 171–175.

Raghavendra, A. S. (1991). Latex exudation from rubber tree, *Hevea brasiliensis*. In *Physiology of Trees* (A. S. Raghavendra, ed.), pp. 403–417. Wiley, New York.

Rai, A. N., Soderback, E., and Bergman, B. (2000). Cyanobacterium-plant symbioses. *New Phytol.* **147**, 449–481.

Raja Harun, R. M. and Hardwick, K. (1987). The effects of prolonged exposure to different light intensities on the photosynthesis of cocoa leaves. *Proc. Int. Cocoa Res. Conf., 10th. Cocoa Producers Alliance, Lagos, Nigeria*, pp. 205–209.

Ralston, C. W. and Prince, A. B. (1965). Accumulation of dry matter and nutrients by pine and hardwood forests in the lower Piedmont of North Carolina. In *Forest–Soil Relationships of North America* (C. T. Youngberg, ed.), pp. 77–94. Oregon State Univ. Press, Corvallis.

Ramos, J. and Grace, J. (1990). The effects of shade on the gas exchange of seedlings of four tropical trees from Mexico. *Funct. Ecol.* **4**, 667–677.

Rao, G. G., Devakumar, A. S., Rajagopal, R., Annama, R., Vijayakumar, K. R., and Sethuraj, M. R. (1988). Clonal variation in leaf epicuticular waxes and reflectance: Possible role in drought tolerance in *Hevea*. *Indian J. Natural Rubber Res.* **1**, 84–87.

Rao, I. M., Sharp, R. E., and Boyer, J. S. (1987). Leaf magnesium alters photosynthetic response to low water potentials in sunflower. *Plant Physiol.* **84**, 1214–1219.

Rappaport, H. F. and Rallo, L. (1991). Postanthesis flower and fruit abscission in "Manzanillo" olive. *J. Am. Soc. Hortic. Sci.* **116**, 720–723.

Rapport, D. L., Regier, H. A., and Hutchinson, T. C. (1985). Ecosystem behavior under stress. *Am. Nat.* **125**, 617–640.

Raschke, K. (1976). How stomata resolve the problem of opposing priorities. *Phil. Trans. Roy. Soc. Lond.* **B273**, 551–560.

Raschke, K. and Humble, G. D. (1973). No uptake of anions required by opening stomata of *Vicia faba*: Guard cells release hydrogen ions. *Planta* **115**, 47–57.

Rascio, N. and La Rocca, N. (2005). Resurrection plants: The puzzle of surviving extreme vegetative desiccation. *Crit. Rev. Plant Sci.* **24**, 209–225.

Raskin, I., Ehmann, A., Melander, W. R., and Meeuse, B. J. D. (1987). Salicylic acid: A natural inducer of heat production in *Arum* lillies. *Science* **237**, 1601–1602.

Rasmussen, R. A. (1970). Isoprene: Identified as a forest-type emission to the atmosphere. *Environ. Sci. Tech.* **4**, 667–671.

Rasmussen, R. A. and Went, F. W. (1965). Volatile organic matter of plant origin in the atmosphere. *Proc. Natl. Acad. Sci. U.S.A.* **53**, 215–220.

Rasmusson, A. G., Soole, K. L., and Elthon, T. E. (2004). Alternative NAD(P)H dehydrogenases of plant mitochondria. *Annu. Rev. Plant Biol.* **55**, 23–39.

Rau, W. (1983). Photoregulation of carotenoid biosynthesis. In *Biosynthesis of Isoprenoid Compounds* (J. W. Porter and S. L. Spurgeon, eds.), pp. 123–157. Wiley-Interscience, New York.

Rauscher, H. M., Isebrands, J. G., Host, G. E., Dickson, R. E., Dickmann, D. I., Crow, T. R., and Michael, D. A. (1990). ECOPHYS: An ecophysiological growth process model for juvenile poplar. *Tree Physiol.* **7**, 255–281.

Raval, M. K., Biswal, B., and Biswal, U. C. (2005). The mystery of oxygen evolution: Analysis of structure and function of Photosystem II, the water-plastoquinone oxido-reductase. *Photosynth. Res.* **85**, 267–293.

Raven, P., Evert, R. F., and Eichorn, S. (1992). *Plant Biology, 5th ed.* Worth Publishers, New York.

Read, D. J., Francis, R., and Finlay, R. D. (1985). Mycorrhizal mycelia and nutrient cycling in plant communities. In *Ecological Interactions in Soil* (A. H. Fitter, ed.), pp. 193–217. Blackwell Scientific, Oxford.

Rebbeck, J. and Loats, K. V. (1997). Ozone effects on seedling sugar maple (*Acer saccharum*) and yellow-poplar (*Liriodendron tulipifera*)—Gas exchange. *Can. J. For. Res.* **27**, 1595–1605.

Reckmann, U., Scheibe, R., and Raschke, K. (1990). Rubisco activity in guard cells compared with the solute requirement for stomatal opening. *Plant Physiol.* **92**, 246–253.

Redgwell, R., Melton, L. D., and Brasch, D. J. (1992). Cell wall dissolution in ripening kiwi fruit (*Actinidia deliciosa*). *Plant Physiol.* **98**, 71–81.

Rediske, J. H. and Nicholson, D. C. (1965). Maturation of noble fir seed—A biochemical study. *Weyerhaeuser For. Pap.* **2**, 1–15.

Regehr, D. L., Bazzaz, F. A., and Boggess, W. R. (1975). Photosynthesis, transpiration and leaf conductance of *Populus deltoides* in relation to flooding and drought. *Photosynthetica* **9**, 52–61.

Reich, P. B. (1983). Effects of low concentrations of ozone on net photosynthesis, dark respiration, and chlorophyll contents in aging hybrid poplar leaves. *Plant Physiol.* **73**, 291–296.

Reich, P. B. (1984a). Loss of stomatal function in aging hybrid poplar leaves. *Ann. Bot.* **53**, 691–698.

Reich, P. B. (1984b). Relationships between leaf age, irradiance, leaf conductance, CO_2 exchange and water use efficiency in hybrid poplar. *Photosynthetica* **18**, 445–453.

Reich, P. B. (1987). Quantifying plant response to ozone: A unifying theory. *Tree Physiol.* **3**, 63–91.

Reich, P. B. (1995). Phenology of tropical forests: Patterns, causes, and consequences. *Can. J. Bot.* **73**, 164–174.

Reich, P. B. and Amundson, R. G. (1985). Ambient levels of O_3 reduce net photosynthesis in tree and crop species. *Science* **230**, 566–570.

Reich, P. B. and Schoettle, A. W. (1988). Role of phosphorus and nitrogen in photosynthetic and whole plant carbon gain and nutrient use efficiency in eastern white pine. *Oecologia* **77**, 25–33.

Reich, P. B., Walters, M. B., and Tabone, T. J. (1989). Response of *Ulmus americana* seedlings to varying nitrogen and water status. 2. Water and nitrogen use efficiency in photosynthesis. *Tree Physiol.* **5**, 173–184.

Reich, P. B., Ellsworth, D. S., Kloeppel, B. D., Fownes, J. H., and Gower, S. T. (1990). Vertical variation in canopy structure and CO_2 exchange of oak-maple forests: Influence of ozone, nitrogen, and other factors on simulated canopy carbon gain. *Tree Physiol.* **7**, 329–345.

Reich, P. B., Uhl, C., Walters, M. B., and Ellsworth, D. S. (1991a). Leaf lifespan as a determinant of leaf structure and function among 23 Amazonian tree species. *Oecologia* **86**, 16–24.

Reich, P. B., Walters, M. B., and Ellsworth, D. S. (1991b). Leaf age and season influence the relationships between leaf nitrogen, leaf mass per area and photosynthesis in maple and oak trees. *Plant, Cell Environ.* **14**, 251–259.

Reich, P. B., Koike, T., Gower, S. T., and Schoettle, A. W. (1995). Causes and consequences of variation in conifer leaf life span. In *Ecophysiology of Coniferous Forests* (W. K. Smith and T. M. Hinckley, eds.), pp. 225–254. Academic Press, San Diego.

Reich, P. B., Walters, M. B., and Ellsworth, D. S. (1997). From tropics to tundra: Global convergence in plant functioning. *Proc. Natl. Acad. Sci. U.S.A.* **94**, 13730–13734.

Reich, P. B., Walters, M. B., Ellsworth, D. S., Vose, J. M., Volin, J. C., Gresham, C., and Bowman, W. D. (1998). Relationships of leaf dark respiration to leaf nitrogen, specific leaf area and leaf lifespan—A test across biomes and functional groups. *Oecologia* **114**, 471–482.

Reich, P. B., Ellsworth, D. S., Walters, M. B., Vose, J. M., Gresham, C., Volin, J. C., and Bowman, W. D. (1999). Generality of leaf trait relationships: A test across six biomes. *Ecology* **80**, 1955–1969.

Reid, C. P. P. (1984). Mycorrhizae: A root-soil interface in plant nutrition. In *Microbial Plant Interactions* (R. L. Todd and J. E. Giddens, eds.), pp. 29–50. Am. Soc. Agron. Special Publ. 47, Madison, Wisconsin.

Reid, C. P. P., Kidd, F. A., and Ekwebelam, J. A. (1983). Nitrogen nutrition, photosynthesis and carbon allocation in ectomycorrhizal pines. *Plant Soil* **71**, 415–432.

Reid, J. B. and Huck, M. G. (1990). Diurnal variation of crop hydraulic resistance: A new analysis. *Agron. J.* **82**, 827–834.

Reid, J. B., Sorensen, I., and Petrie, R. A. (1993). Root demography in kiwifruit (*Actinidia deliciosa*). *Plant, Cell Environ.* **16**, 949–957.

Reinecke, D. M. and Bandurski, R. S. (1988). Auxin biosynthesis and metabolism. In *Plant Hormones and Their Role in Plant Growth and Development* (P. J. Davies, ed.), pp. 24–42. Kluwer, Dordrecht.

Reinert, R. A., Heagle, A. S., and Heck, W. W. (1975). Plant responses to pollutant combinations. In *Responses of Plants to Air Pollution* (J. B. Mudd and T. T. Kozlowski, eds.), pp. 159–177. Academic Press, New York.

Remphrey, W. R. (1989). Shoot ontogeny in *Fraxinus pennsylvanica* (green ash). I. Seasonal cycle of terminal meristem activity. *Can. J. Bot.* 6–7, 1624–1632.

Remphrey, W. R. and Davidson, C. G. (1994a). Shoot and leaf growth in *Fraxinus pennsylvanica* and its relation to crown location and pruning. *Can. J. For. Res.* **24**, 1997–2005.

Remphrey, W. R. and Davidson, C. G. (1994b). Shoot preformation in clones of *Fraxinus pennsylvanica* in relation to site and year of bud formation. *Trees* **8**, 126–131.

Renner, O. (1912). *Versuche zur Mechanik der Wasserversorgung. 2. Über Wurzeltätigkeit. Ber. Deutsch. Bot. Ges.* **30**, 576–580; 642–648.

Retzlaff, W. A., Williams, L. E., and DeJong, T. M. (1991). The effect of different atmospheric ozone partial pressures on photosynthesis and growth of nine fruit and nut species. *Tree Physiol.* **8**, 93–105.

Reuther, W. and Burrows, F. W. (1942). The effect of manganese sulfate on the photosynthetic activity of frenched tung foliage. *Proc. Am. Soc. Hortic. Sci.* **40**, 73–76.

Rey, M., Diaz-Sala, C., and Rodriguez, R. (1994). Comparison of endogenous polyamine content in hazel leaves and buds between the annual dormancy and flowering phases of growth. *Physiol. Plant.* **91**, 45–50.

Reynolds, J. F., Chen, J. L., Harley, P. C., Hilbert, D. W., Dougherty, R. L., and Tenhunen, J. D. (1992). Modeling the effects of elevated CO_2 on plants—Extrapolating leaf response to a canopy. *Agric. For. Meteorol.* **61**, 69–94.

Rhoades, D. F. (1983). Responses of alder and willow to attack by tent caterpillars and webworms: Evidence for pheromonal sensitivity of willows. *Am. Chem. Soc. Symp. Ser.* **208**, 55–68.

Rhoades, D. F. (1985). Offensive defensive interactions between herbivores and plants: Their relevance in herbivore population dynamics and ecological theory. *Am. Nat.* **125**, 205–238.

Rhodes, M. J. C. (1980). The maturation and ripening of fruits. In *Senescence in Plants* (K. V. Thimann, ed.), pp. 157–205. CRC Press, Boca Raton, Florida.

Ribas-Carbo, M., Taylor, N. L., Giles, L., Busquets, S., Finnegan, P. M., Day, D. A., Lambers, H., Medrano, H., Berry, J. A., and Flexas J. (2005). Effects of water stress on respiration in soybean leaves. *Plant Physiol.* **139**, 466–473.

Ricard, B., Couée, I., Raymond, P., Saglio, P. H., Saint-Ges, V., and Pradet, A. (1994). Plant metabolism under hypoxia and anoxia. *Plant Physiol. Biochem.* **32**, 1–10.

Rice, E. L. (1974). *Allelopathy*. Academic Press, New York.

Rice, E. L. (1984). *Allelopathy, 2nd ed.* Academic Press, Orlando, Florida.

Richards, L. A. and Weaver, L. R. (1944). Moisture retention by some irrigated soils as related to soil-moisture tension. *J. Agric. Res.* **69**, 215–235.

Richards, P. W. (1966). *Tropical Rain Forest*. Cambridge Univ. Press, London and New York.

Richmond, K. E. and Sussman M. (2003). Got silicon? The non-essential beneficial plant nutrient. *Curr. Opin. Plant Biol.* **6**, 268–272.

Richter, D. D., Ralston, C. W., and Harms, W. R. (1982). Prescribed fire: Effects on water quality and forest nutrient cycling. *Science* **215**, 661–663.

Richter, H. (1973a). Frictional potential losses and total water potential in plants: A reevaluation. *J. Exp. Bot.* **23**, 983–994.

Richter, H. (1973b). *Wie entstehen Saugspannungsgradienten in Bäumen? Ber. Deutsch. Bot. Ges.* **85**, 341–351.

Richter, H. (1974). *Erhote Saugspannungswerte und morphologische Veranderungen durch transverde Einschnitte in einem Taxus-stamm* [Elevated suction tensions and morphological alterations caused by transverse cuts in a *Taxus* stem]. *Flora* **163**, 291–309.

Richter, H. (1978). Water relations of single drying leaves: Evaluation with a dewpoint hygrometer. *J. Exp. Bot.* **29**, 277–280.

Richter, H. F., Duhme, F., Glatzel, G., Hinckley, T. M., and Karlic, H. (1981). Some limitations and applications of the pressure volume curve technique in ecophysiological research. In *Plants and Their Atmospheric Environment*. (J. Grace, E. D. Ford, and P. G. Jarvis, eds.), pp. 263–272. Blackwell Scientific Publishers, Oxford.

Rider, N. E. (1957). Water losses from various land surfaces. *Q.J.R. Meteorol. Soc.* **83**, 181–193.

Riederer, M. and Müller, C., eds. (2006). *Biology of the Cuticle*. Annual Plant Reviews, Vol. 23, Blackwell Publishers, Oxford.

Rieger, M. (1995). Offsetting effects of reduced root hydraulic conductivity and osmotic adjustment following drought. *Tree Physiol.* **15**, 379–385.

Rimbawanto, A., Coolbear, P., Dourado, A. M., and Firth, A. (1988a). Seed maturation precedes cone ripening in New Zealand *Pinus radiata*. *N. Z. J. For. Sci.* **18**, 139–148.

Rimbawanto, A., Coolbear, P., and Firth, A. (1988b). Artificial ripening of prematurely harvested cones of New Zealand *Pinus radiata* and its effect on seed quality. *N. Z. J. For. Sci.* **18**, 149–160.

Ringoet, A. (1952). *Recherches sur la transpiration et le bilan d'eau de quelques plantes tropicales. Publ. Inst. Natl. Etude Agron. Congo Belge, Ser. Sci.* **56**.

Rinne, P., Saarelainen, A., and Juntilla, O. (1994a). Growth cessation and bud dormancy in relation to ABA level in seedlings and coppice shoots of *Betula pubescens* as affected by a short photoperiod, water-stress and chilling. *Planta* **90**, 451–458.

Rinne, P., Tuominen, H., and Juntilla, O. (1994b). Seasonal changes in bud dormancy in relation to bud morphology, water and starch content, and abscisic acid concentration in adult trees of *Betula pubescens*. *Tree Physiol.* **14**, 549–561.

Ritchie, G. A. and Hinckley, T. M. (1975). The pressure chamber as an instrument for ecological research. *Adv. Ecol. Res.* **9**, 165–254.

Robards, A. W., Davidson, E., and Kidwai, P. (1969). Short-term effects of some chemicals on cambial activity. *J. Exp. Bot.* **20**, 912–920.

Robbie, F. A. and Atkinson, C. J. (1994). Wood and tree age as factors influencing the ability of apple flowers to set fruit. *J. Hortic. Sci.* **69**, 609–623.

Roberds, J. H., Namkoong, G., and Davey, C. B. (1976). Family variation in growth response of loblolly pine to fertilizing with urea. *For. Sci.* **22**, 291–299.

Roberntz, P. (2001). Atmospheric carbon dioxide concentration, nitrogen availability, temperature and the photosynthetic capacity of current-year Norway spruce shoots. *Tree Physiol.* **21**, 931–940.

Roberts, B. R., Kramer, P. J., and Karl, C. M., Jr. (1963). Long-term effects of gibberellin on the growth of loblolly pine seedlings. *For. Sci.* **9**, 202–205.

Roberts, B. R., Townsend, A. M., and Dochinger, L. S. (1971). Photosynthetic response to SO_2 fumigation in red maple. *Plant Physiol.* **47**, **Suppl.**, 30.

Roberts, D. R. and Peters, W. J. (1977). Chemically induced lightwood formation in southern pines. *For. Prod. J.* **27**, 28–30.

Roberts, J. (1976a). A study of root distribution and growth in a *Pinus sylvestris* L. plantation in East Anglia. *Plant Soil* **44**, 607–621.

Roberts, J. (1976b). An examination of the quantity of water stored in mature *Pinus sylvestris* L. trees. *J. Exp. Bot.* **27**, 473–479.

Roberts, L. W. and Miller, A. R. (1983). Is ethylene involved in xylem differentiation? *Vistas Plant Sci.* **6**, 1–24.

Roberts, S. K. (2006). Plasma membrane anion channels in higher plants and their putative functions in roots. *New Phytol.* **169**, 647–666.

Roberts, S. W., Strain, B. R., and Knoerr, K. R. (1980). Seasonal patterns of leaf water relations in four co-occurring forest tree species: Parameters from pressure-volume curves. *Oecologia* **46**, 330–337.

Robertson, J. M., Pharis, R. P., Huang, Y. Y., Reid, D. M., and Yeung, E. C. (1985). Drought-induced changes in abscisic acid levels in the root apex of sunflower. *Plant Physiol.* **79**, 1086–1089.

Robichaux, R. H. and Pearcy, R. W. (1984). Evolution of C_3 and C_4 plants along an environmental moisture gradient: Patterns of photosynthetic differentiation in Hawaiian *Scaevola* and *Euphorbia* species. *Am. J. Bot.* **71**, 121–129.

Robitaille, H. A. and Leopold, A. C. (1974). Ethylene and the regulation of apple stem growth under stress. *Physiol. Plant.* **32**, 301–304.

Roddick, J. G. and Guan, M. (1991). Brassinosteroids and root development. In *Brassinosteroids* (H. G. Cutler, T. Yokota, and G. Adam, eds.), pp. 231–245. Am. Chem. Soc., Washington, D.C.

Roden, J. S. and Pearcy, R. W. (1993a). The effect of leaf flutter on the flux of CO_2 in poplar leaves. *Funct. Ecol.* **7**, 669–675.

Roden, J. S. and Pearcy, R. W. (1993b). Effect of leaf flutter on the light environment of poplars. *Oecologia* **93**, 201–207.

Roden, J., Van Volkenberg, E., and Hinckley, T. M. (1990). Cellular basis for limitation of poplar leaf growth by water deficit. *Tree Physiol.* **6**, 211–220.

Rodríguez-Maribona, B., Tenorio, J. L., Conde, J. R., and Ayerbe, L. (1992). Correlation between yield and osmotic adjustment of peas (*Pisum sativum* L.) under drought stress. *Field Crops Res.* **29**, 15–22.

Roelfsema, M. R. G. and Hedrich, R. (2005). In the light of stomatal opening: New insights into "the Watergate." *New Phytol.* **167**, 665–691.

Roelfsema, M. R. G., Hanstein, S., Felle, H. H., and Hedrich, R. (2002). CO_2 provides an intermediate link in the red light response of guard cells. *Plant J.* **32**, 65–75.

Rogers, W. S. and Booth, G. A. (1959–1960). The roots of fruit trees. *Sci. Hortic.* **14**, 27–34.

Rogers, W. S. and Head, G. C. (1966). The roots of fruit trees. *J. Hortic. Sci.* **91**, 198–205.

Rogers, W. S. and Head, G. C. (1969). Factors affecting the distribution of roots in perennial woody plants. In *Root Growth* (W. J. Whittington, ed.), pp. 280–295. Butterworth, London.

Rohde, A., Howe, G. T., Olsen, J. E., Moritz, T., Von Montagu, M., Junttila, O., and Boerjan, W. (2000). Molecular aspects of bud dormancy in trees. In *Molecular Biology of Woody Plants*, Vol. 1. (S. M. Jain and S. C. Minocha, eds.). pp. 89–134. Kluwer Academic Pub., Dordrecht, Netherlands.

Rohde, A. and Boerjan, W. (2001). Insights into bud development and dormancy in poplar. In *Trends in European Forest Tree Physiology Research* (S. Huttunen, H. Heikkilä, J. Bucher, B. Sundberg, P. Jarvis, and R. Matyssek, R., eds.), pp. 33–52. Kluwer Academic Publishers, Dordrecht and Boston.

Rohmer, M., Seemann, M., Horbach, S., Bringermeyer, S., and Sahm, H. (1996). Glyceraldehyde 3-phosphate and pyruvate as precursors of isoprenic units in an alternative non-mevalonate pathway for terpenoid biosynthesis. *J. Am. Chem. Soc.* **118**, 2564–2566.

Romberger, J. A. (1963). Meristems, growth, and development in woody plants. *U.S., Dept. Agric., Tech. Bull.* 1293.

Romberger, J. A., Hejnowicz, Z., and Hill, J. F. (1993). *Plant Structure: Function and Development*. Springer-Verlag, Berlin.

Ronco, F. (1973). Food reserves of Engelmann spruce planting stock. *For. Sci.* **19**, 213–219.

Rook, D. A. (1971). Effect of undercutting and wrenching on growth of *Pinus radiata* D. Don seedlings. *J. Appl. Ecol.* **8**, 477–490.

Rook, D. A. (1973). Conditioning radiata pine seedlings to transplanting by restricted watering. *N. Z. J. For. Sci.* **3**, 54–59.

Rook, D. A. (1985). Physiological constraints on yield. In *Crop Physiology of Forest Trees* (P. M. A. Tigerstedt, P. Puttonen, and V. Koski, eds.), pp. 1–19. Helsinki Univ. Press, Helsinki, Finland.

Rook, D. A. and Corson, M. J. (1978). Temperature and irradiance and the total daily photosynthetic production of the crown of a *Pinus radiata* tree. *Oecologia* **36**, 371–382.

Rose, J. K. C., Cosgrove, D. J., Albersheim, P., Darvill, A. G., and Bennett, A. B. (2000). Detection of expansin proteins and activity during tomato fruit ontogeny. *Plant Physiol.* **123**, 1583–1592.

Rosen, C. J. and Carlson, R. M. (1984). Influence of root zone oxygen stress on potassium and ammonium absorption by myrobalan plum rootstock. *Plant Soil* **80**, 345–353.

Ross, J. J., Symons, G. M., Abas, L., Reid, J. B., and Luschnig, C. (2006). Hormone distribution and transport. In *Plant Hormone Signaling* (P. Hedden and S. G. Thomas, eds.), pp. 257–291. Blackwell Publishers, Oxford.

Ross, S. D., Pharis, R. P., and Binder, W. D. (1983). Growth regulators and conifers: Their physiology and potential uses in forestry. In *Plant Growth Regulating Chemicals*, Vol. 2, (L. G. Nickell, ed.), pp. 35–78. CRC Press, Boca Raton, Florida.

Roth, U., Friebe, A., and Schnabl, H. (2000). Resistance induction in plants by a brassinosteroid-containing extract of *Lychnis viscaria* L. *Z. Naturforsch. Sec. C, Biosciences* **55**, 552–559.

Rougier, M. (1981). Secretory activity of the root cap. *Encycl. Plant Physiol. N.S.* **13B**, 542–574.

Rovira, A. D. (1969). Plant root exudates. *Bot. Rev.* **35**, 35–57.

Rovira, A. D., Foster, R. C., and Martin, J. K. (1979). Note on terminology: Origin, nature and nomenclature of the organic materials in the rhizosphere. In *Soil-Root Interface* (J. L. Harley and R. S. Russell, eds.), pp. 1–4. Academic Press, New York.

Rowe, J. W., Conner, A. H., Diehl, M. A., and Wroblewska, H. (1976). Effects of treating northern and western conifers with paraquat. *Light-wood Coord. Counc. Ann. Meet., 1976*, pp. 66–76.

Roy Chowdhury, C. (1961). The morphology and embryology of *Cedrus deodara* Loud. *Phytomorphology* **11**, 283–304.

Rubery, P. H. (1988). Auxin transport. In *Plant Hormones and Their Role in Plant Growth and Development* (P. J. Davies, ed.), pp. 341–362. Kluwer, Dordrecht.

Ruby, J. L. (1967). The correspondence between genetic morphological, and climatic variation patterns in Scotch pine. 1. Variations in parental characteristics. *Silvae Genet.* **16**, 50–56.

Rudinsky, J. A. and Vité, J. P. (1959). Certain ecological and phylogenetic aspects of the pattern of water conduction in conifers. *For. Sci.* **5**, 259–266.

Rudolph, T. D. (1964). Lammas growth and prolepsis in jack pine in the Lake States. *For. Sci. Monogr.* **6**. 70p.

Ruinen, J. (1965). The phyllosphere. III. Nitrogen fixation in the phyllosphere. *Plant Soil.* **22**, 375–394.

Ruiz, L. P., Atkinson, C. J., and Mansfield, T. A. (1993). Calcium in the xylem and its influence on the behaviour of stomata. *Phil. Trans. Roy. Soc. Lond.* **B341**, 67–74.

Rundel, P. W. (1982). Water uptake by organs other than roots. *Encycl. Plant Physiol., N. S.* **12B**, 111–134.

Runion, G. B., Mitchell, R. J., Green, T. H., Prior, S. A., Rogers, H. H., and Gjerstad, D. H. (1999). Longleaf pine photosynthetic response to soil resource availability and elevated atmospheric carbon dioxide. *J. Env. Qual.* **28**, 880–887.

Runjin, L. (1989). Effects of vesicular-arbuscular mycorrhizas and phosphorus on water status and growth of apple. *J. Plant Nutr.* **12**, 997–1017.

Running, S. W. (1976). Environmental control of leaf water conductance in conifers. *Can. J. For. Res.* **6**, 104–112.

Running, S. W. (1979). Environmental and physiological control of water flux through *Pinus contorta*. Ph.D. Dissertation, Colorado State University, Fort Collins.

Runyon, E. H. (1936). Ratio of water content to dry weight in leaves of the creosote bush. *Bot. Gaz.* **97**, 518–553.

Russell, E. W. (1973). *Soil Conditions and Plant Growth, 10th ed.* Longman, London.

Ruter, J. M. and Ingram, D. L. (1992). High root-zone temperatures influence RuBisCo activity and pigment accumulation in leaves of "Rotundifolia" holly. *J. Am. Soc. Hortic. Sci.* **117**, 154–157.

Rutter, A. J. (1957). Studies in the growth of young plants of *Pinus sylvestris* L. I. The annual cycle of assimilation and growth. *Ann. Bot.* **21**, 399–426.

Rutter, A. J. (1968). Water consumption by forests. In *Water Deficits and Plant Growth* (T. T. Kozlowski, ed.), Vol. 1, pp. 23–84. Academic Press, New York.

Ryan, M. G. (1990). Growth and maintenance respiration in stems of *Pinus contorta* and *Picea engelmannii*. *Can. J. For. Res.* **20**, 48–57.

Ryan, M. G. (1991). Effects of climate change on plant respiration. *Ecol. Appl.* **1**, 157–167.

Ryan, M. G., Lavigne, M. B., and Gower, S. T. (1997). Annual carbon cost of autotrophic respiration in boreal forest ecosystems in relation to species and climate. *J. Geophys. Res.-Atmos.* **102**, 28871–28883.

Rygiewicz, P. T. and Andersen, C. P. (1994). Mycorrhizae alter quality and quantity of carbon allocated below ground. *Nature (London)* **369**, 58–60.

Saad, F. A., Crane, J. C., and Maxie, E. C. (1969). Timing of olive oil application and its probable role in hastening maturation of fig fruits. *J. Am. Soc. Hortic. Sci.* **94**, 335–337.

Sachs, T. (1981). The control of patterned differentiation of vascular tissues. *Adv. Bot. Res.* **9**: 151–262.

Sack, L. and Holbrook, N. M. (2006). Leaf hydraulics. *Annu. Rev. Plant Biol.* **57**, 361–381.

Safir, G. R. and Nelsen, C. E. (1985). VA mycorrhizas: Plant and fungal water relations. *Proc Sixth N. Amer. Conf. Mycorrhizae: June 25–29, 1984, Bend, Oregon.* pp. 161–164.

Safir, G. R., Boyer, J. S., and Gerdemann, J. W. (1972). Nutrient status and mycorrhizal enhancement of water transport in soybean. *Plant Physiol.* **49**, 799–803.

Sage, R. F. (1994). Acclimation of photosynthesis to increasing atmospheric CO_2: The gas exchange perspective. *Photosynth. Res.* **39**, 351–368.

Sage, R. F. (2004). The evolution of C_4 photosynthesis. *New Phytol.* **161**, 341–370.

Sahlen, K. and Gjelsvik, S. (1993). Determination of *Pinus sylvestris* seed maturity using leachate conductivity measurements. *Can. J. For. Res.* **2**, 864–870.

Sakai, S. (1992). Regulatory functions of soluble auxin-binding proteins. *Int. Rev. Cytol.* **135**, 239–265.

Sakurai, A. and Fujioka, S. (1993). The current status of physiology and biochemistry of brassinosteroids. *Plant Growth Regul.* **13**, 147–159.

Salleo, S., Lo Gullo, M. A., and Siracusano, L. (1984). Distribution of vessel ends in stems of some diffuse-porous and ring-porous trees the nodal regions of safety zones of the water conducting system. *Ann. Bot.* **54**, 543–552.

Salleo, S., Lo Gullo, M. A., Trifilo, P., and Nardini, A. (2004). New evidence for a role of vessel-associated cells of phloem in the rapid xylem refilling of cavitated stems of *Laurus nobilis* L. *Plant, Cell Environ.* **27**, 1065–1076.

Salvucci, M. E. and Crafts-Brandner, S. J. (2004). Inhibition of photosynthesis by heat stress: The activation state of Rubisco as a limiting factor in photosynthesis. *Physiol. Plant* **120**, 179–186.

Salvucci, M. E., Portis, A. R., and Ogren, W. L. (1986). Light and CO_2 response of ribulose-1,5-bisphosphate carboxylase/oxygenase activation in *Arabidopsis* leaves. *Plant Physiol.* **80**, 655–659.

Sampson, A. W. and Samisch, R. (1935). Growth and seasonal changes in composition of oak leaves. *Plant Physiol.* **10**, 739–751.

Sams, C. E. and Flore, J. A. (1982). The influence of age, position, and environmental variables on net photosynthetic rate of sour cherry leaves. *J. Am. Soc. Hortic. Sci.* **107**, 339–344.

Samsuddin, Z. and Impens, I. (1978). Comparative net photosynthesis of four *Hevea brasiliensis* clonal seedlings. *Exp. Agric.* **14**, 337–340.

Samsuddin, Z. and Impens, I. (1979). Relationship between leaf age and some carbon dioxide exchange characteristics of four *Hevea* Muell. Arg. clones. *Photosynthetica* **13**, 208–210.

Sanadze, G. A. (2004). Biogenic isoprene (a review). *Russ. J. Plant Physiol.* **51**, 729–741.

Sanchez, P. A. (1973). A review of soils research in tropical Latin America. *N.C., Agric. Exp. Stn. Tech. Bull.* **219**.

Sanchez, P. A., Villachica, J. H., and Bandy, D. E. (1983). Soil fertility dynamics after clearing a tropical rain forest in Peru. *Soil Sci. Soc. Am. J.* **47**, 1171–1178.

Sanders, D. and Bethke, P. (2000). Membrane transport. In *Biochemistry and Molecular Biology of Plants* (B. Buchanan, B. W. Gruissem, and R. L. Jones, eds.), pp. 110–158. Amer. Soc. Plant Physiol., Rockville, Maryland.

Sands, R. and Theodorou, C. (1978). Water uptake by mycorrhizal roots of radiata pine seedlings. *Aust. J. Plant Physiol.* **5**, 301–309.

Sands, R., Fiscus, E. L., and Reid, C. P. P. (1982). Hydraulic properties of pine and bean roots with varing degrees of suberization, vascular differentiation, and mycorrhizal infection. *Aust. J. Plant Physiol.* **9**, 559–569.

Saniewski, M., Ueda, J., Miyamoto, K., Okubo, H., and Puchalski, J. (2004). Interaction of methyl jasmonate and ethephon in gum formation in tulip bulbs. *J. Fac. Agric. Kyushu Univ.* **49**, 207–215.

Santakumari, M. and Berkowitz, G. A. (1991). Chloroplast volume—Cell water potential relationships and acclimation of photosynthesis to leaf water deficits. *Photosyn. Res.* **28**, 9–20.

Santamour, F. S., Jr. and McArdle, A. J. (1982). Checklist of cultivated maples. I. *Acer rubrum* L. *J. Arboric.* **8**, 110–112.

Santantonio, D. and Santantonio, E. (1987). Effects of thinning on production and mortality of fine roots in a *Pinus radiata* plantation on a fertile site in New Zealand. *Can. J. For. Res.* **17**, 919–928.

Sanz, A., Monerri, G., Gonzalez-Ferrer, J., and Guardiola, J. L. (1987). Changes in carbohydrates and mineral elements in citrus leaves during flowering and fruit set. *Physiol. Plant.* **69**, 93–98.

Saranpää, P. (1990). Heartwood formation in stems of *Pinus sylvestris* L. Lipids and carbohydrates of sapwood and heartwood and ultrastructure of ray parenchyma cells. *Publ. No. 14. Dept. of Botany, Univ. of Helsinki.*

Saranpää, P. and Höll, W. (1989). Soluble carbohydrates of *Pinus sylvestris* L. sapwood and heartwood. *Trees* **3**, 138–143.

Saranpää, P. and Nyberg, H. (1987). Lipids and sterols of *Pinus sylvestris* L. sapwood and heartwood. *Trees* **1**, 82–87.

Sarvas, R. (1955a). Investigations into the flowering and seed quality of forest trees. *Commun. Inst. Forst. Fenn.* **45**.

Sarvas, R. (1955b). *Ein Beitrag zur Fernverbreitung des Blütenstaubes einiger Waldbäume. 2. Forstgenet. Forstpflanz.* **4**, 137–142.

Sarvas, R. (1962). Investigations on the flowering and seed crop of *Pinus silvestris*. *Commun. Inst. For. Fenn.* **53**, 1–198.

Sasaki, S. and Kozlowski, T. T. (1967). Effects of herbicides on carbon dioxide uptake of pine seedlings. *Can. J. Bot.* **45**, 961–971.

Saur, E., Nambiar, E. K. S., and Fife, D. N. (2000). Foliar nutrient retranslocation in *Eucalyptus globulus*. *Tree Physiol.* **20**, 1105–1112.

Saure, M. C. (1985). Dormancy release in deciduous fruit trees. *Hort. Rev.* **7**, 239–300.

Saurer, M., Aellen, K., and Siegwolf, R. (1997). Correlating $\delta^{13}C$ and delta $\delta^{18}O$ in cellulose of trees. *Plant, Cell Environ.* **20**, 1543–1550.

Sauter, A., Davies, W. J., and Hartung, W. (2001). The long-distance abscisic acid signal in the droughted plant: The fate of the hormone on its way from root to shoot. *J. Exp. Bot.* **52**, 1991–1997.

Sauter, J. J. (1980). Seasonal variation of sucrose content in the xylem sap of *Salix*. *Z. Pflanzenphysiol.* **98**, 377–391.

Sauter, J. J. and Neumann, U. (1994). The accumulation of storage materials in ray cells of poplar wood (*Populus* × *canadensis* <*robusta*>): Effect of ringing and defoliation. *J. Plant Physiol.* **143**, 21–26.

Sauter, J. J. and van Cleve, B. (1990). Biochemical, immunochemical, and ultrastructural studies of protein storage in poplar (*Populus* × *canadensis* "Robusta") wood. *Planta* **173**, 31–34.

Sauter, J. J. and van Cleve, B. (1992). Seasonal variation of amino acids in the xylem sap of "*Populus* x *canadensis*" and its relation to protein body mobilization. *Trees* **7**, 26–32.

Sauter, J. J., Iten, W., and Zimmermann, M. H. (1973). Studies on the release of sugar into the vessels of sugar maple (*Acer saccharum*). *Can. J. Bot.* **51**, 1–8.

Sauter, J. J., van Cleve, B., and Apel, K. (1988). Protein bodies in ray cells of *Populus* × *canadensis* Moench "Robusta." *Planta* **183**, 92–100.

Savidge, R. A. (1988). Auxin and ethylene regulation of diameter growth in trees. *Tree Physiol.* **4**, 401–414.

Savidge, R. A. and Wareing, P. F. (1981). Plant-growth regulators and the differentiation of vascular elements. In *Plant-Growth Regulators and the Differentiation of Vascular Elements* (J. R. Barnett, ed.), pp. 192–235. Castle House, Tunbridge Wells.

Savidge, R. A. and Wareing, P. F. (1984). Seasonal cambial activity and xylem development in *Pinus contorta* in relation to endogenous indol-3-yl-acetic acid and (S)-abscisic acid levels. *Can. J. For. Res.* **14**, 676–682.

Savill, P. S. (1983). Silviculture in windy climates. *For. Abstr.* **44**, 473–488.

Scandalios, J. G. (1974). Isozymes in development and differentiation. *Annu. Rev. Plant Physiol.* **25**, 225–258.

Schaberg, P. G., Wilkinson, R. C., Share, J. B., Donnelly, J. R., and Cali, P. F. (1995). Winter photosynthesis of red spruce from three Vermont seed sources. *Tree Physiol.* **15**, 345–350.

Schaedle, M. (1975). Tree photosynthesis. *Annu. Rev. Plant Physiol.* **26**, 101–115.

Scheidegger, Y., Saurer, M., Bahn, M., and Siegwolf, R. (2000). Linking stable oxygen and carbon isotopes with stomatal conductance and photosynthetic capacity: A conceptual model. *Oecologia* **125**, 350–357.

Scherbatskoy, T. and Klein, R. M. (1983). Response of spruce and birch foliage to leaching by acidic mists. *J. Environ. Qual.* **12**, 189–195.

Schiefelbein, J. W. (2000). Constructing a plant cell. The genetic control of root hair development. *Plant Physiol.* **124**, 1525–1531.

Schlesinger, W. H. and Chabot, B. F. (1977). The use of water and minerals by evergreen and deciduous shrubs in Okefenokee Swamp. *Bot. Gaz.* **138**, 490–497.

Schlesinger, W. H. and Hasey, M. M. (1981). Decomposition of chaparral shrub foliage: Losses of organic and inorganic constituents from deciduous and evergreen leaves. *Ecology* **62**, 762–774.

Schluepmann, H., Pellny, T., van Dijken, A., Smeekens, S., and Paul, M. (2003). Trehalose 6-phosphate is indispensable for carbohydrate utilization and growth in *Arabidopsis thaliana*. *Proc. Natl. Acad. Sci. USA.* **100**, 6849–6854.

Schmid, C. and Ziegler, H. (1992). Permeation characteristics of isolated cuticles of *Citrus aurantium* L. for monoterpenes. *Trees* **6**, 172–177.

Scholander, P. F. (1958). The rise of sap in lianas. In *The Physiology of Forest Trees* (K. V. Thimann, ed.), pp. 3–17. Ronald Press, New York.

Scholander, P. F., Dam, L. van, and Scholander, S. I. (1955). Gas exchange in the roots of the mangrove. *Am. J. Bot.* **42**, 92–98.

Scholander, P. F., Ruud, B., and Leivestad, H. (1957). The rise of sap in a tropical liana. *Plant Physiol.* **41**, 529–532.

Scholander, P. F., Hammel, H. T., Hemmingsen, E. A., and Bradstreet, E. D. (1964). Hydrostatic pressure and osmotic potential in leaves of mangroves and some other plants. *Proc. Nat. Acad. Sci. USA* **52**, 1919–1925.

Schönherr, J. (1976). Water permeability of isolated cuticular membranes: The effect of cuticular waxes on diffusion of water. *Planta* **131**, 159–164.

Schopmeyer, C. S., ed. (1974). *Seeds of Woody Plants in the United States*. Agric. Handb. No. 450. U.S. For. Serv., Washington, D.C.

Schreiber, J. D., Duffy, P. D., and McClurkin, D. C. (1974). Dissolved nutrient losses to storm runoff from five southern pine watersheds. *J. Environ. Qual.* **5**, 201–205.

Schreiber, J. D., Duffy, P. D., and McDowell, L. L. (1990). Nutrient leaching of a loblolly pine forest floor by simulated rainfall. I. Intensity effects. *For. Sci.* **36**, 765–776.

Schroeder, J. I., Hedrich, R., and Fernandez, J. M. (1984). Potassium-selective single channels in guard cell protoplasts of *Vicia faba*. *Nature (London)* **312**, 361–362.

Schroeder, J. I., Raschke, K., and Neher, E. (1987). Voltage dependence of K^+ channels in guard-cell protoplasts. *Proc. Natl. Acad. Sci. USA* **84**, 4108–4112.

Schroeder, J. I., Allen, G. J., Hugouvieux, V., Kwak, J. M., and Waner, D. (2001). Guard cell signal transduction. *Annu. Rev. Plant Physiol. Plant Mol. Biol.* **52**, 627–658.

Schuck, H. J. (1972). *Die Zusammensetzung der Nadelwachse von Pinus silvestris in Abhängigkeit von Herkunft und Nadelalter sowie ihre Bedeutung für das Anfälligkeit gegenüber Lophodermium pinastri (Schrad.). Chev. Flora* **161**, 604–622.

Schulman, E. (1958). Bristlecone pine, oldest known living thing. *Natl. Geogr. Mag.* **113**, 355–372.

Schultz, H. R. and Matthews, M. A. (1993). Growth, osmotic adjustment, and cell-wall mechanics of expanding grape leaves during water deficits. *Crop Sci.* **33**, 287–294.

Schulze, E. D. (1986). Carbon dioxide and water vapor exchange in response to drought in the atmosphere and in the soil. *Annu. Rev. Plant Physiol.* **37**, 247–274.

Schulze, E. D. (1993). Soil water deficits and atmospheric humidity as environmental signals. In *Water Deficits: Plant Responses from Cell to Community* (J. A. C. Smith and H. Griffiths, eds.), pp. 129–145. Bios Scientific Publishers, Oxford.

Schulze, E.-D. and Mooney, H. A. (1993). Ecosystem function of biodiversity: A summary. In *Biodiversity and Ecosystem Function* (E.-D. Schulze and H. A. Mooney, eds.), pp. 497–510. Springer-Verlag, Berlin and New York.

Schulze, E. D., Lange, O. L., Buschbom, V., Kappen, L., and Evenari, M. (1972). Stomatal response to changes in humidity in plants growing in the desert. *Planta* **108**, 259–270.

Schulze, E.-D., Fuchs, M., and Fuchs, M. I. (1977). Spatial distribution of photosynthetic capacity and performance in a mountain spruce forest of northern Germany. I. Biomass distribution and daily CO_2 uptake in different crown layers. *Oecologia* **29**, 43–61.

Schulze, E. D., Steudle, E., Gollan, T., and Schurr, U. (1988). Response to Dr. P. J. Kramer's article, "Changing concepts regarding plant water relations," Vol. 11, No. 7, pp. 565–568. *Plant Cell Environ.* **11**, 573–576.

Schurr, U., Gollan, T., and Schulze, E. D. (1992). Stomatal response to drying soil in relation to changes in the xylem sap composition of *Helianthus annuus*. 2. Stomatal sensitivity to abscisic acid imported from the xylem sap. *Plant Cell Environ.* **15**, 561–567.

Schwartz, O. J. (1983). Paraquat-induced lightwood formation in pine. In *Plant Growth Regulating Chemicals* (L. G. Nickell, ed.), pp. 79–97. Vol. II. CRC Press, Boca Raton, Florida.

Schwarz, P. A., Fahey, T. J., and Dawson, T. E. (1997). Seasonal air and soil temperature effects on photosynthesis in red spruce (*Picea rubens*) saplings. *Tree Physiol.* **17**, 187–194.

Schweingruber, F. H. (1988). *Tree Rings: Basics and Applications of Dendrochronology*. Kluwer Academic Publishers, Dordrecht, Netherlands.

Schwintzer, C. R. (1983). Primary productivity and nitrogen, carbon, and biomass distribution in a dense *Myrica gale* stand. *Can. J. Bot.* **61**, 2943–2948.

Scott, F. M. (1950). Internal suberization of tissues. *Bot. Gaz.* **110**, 492–495.

Scott, F. M., Schroeder, M. R., and Turrell, F. M. (1948). Development, cell shape, suberization of internal surface, and abscission in the leaf of the Valencia orange, *Citrus sinensis*. *Bot. Gaz.* **109**, 381–411.

Sedgley, M. and Griffin, A. R. (1989). *Sexual Reproduction of Tree Crops*. Academic Press, London and New York.

Seeley, S. D. and Powell, L. E. (1981). Seasonal changes of free and hydrolyzable abscisic acid in vegetative apple buds. *J. Am. Soc. Hort. Sci.* **106**, 405–409.

Seemann, J. R., Sharkey, T. D., Wang, J., and Osmond, C. B. (1987). Environmental effects on photosynthesis, nitrogen-use efficiency, and metabolite pools in leaves of sun and shade plants. *Plant Physiol.* **84**, 796–802.

Seiler, J. R. and Johnson, J. D. (1985). Photosynthesis and transpiration of loblolly pine seedlings as influenced by moisture-stress conditioning. *For. Sci.* **31**, 742–749.

Seiler, J. R. and Johnson, J. D. (1988). Physiological and morphological responses of three half-sib families of loblolly pine to water-stress conditioning. *For. Sci.* **34**, 487–495.

Sembdner, G. and Parthier, B. (1993). The biochemistry and the physiological and molecular actions of jasmonates. *Annu. Rev. Plant Physiol. Plant Mol. Biol.* **44**, 569–589.

Sena Gomes, A. R. and Kozlowski, T. T. (1980). Growth responses and adaptations of *Fraxinus pennsylvanica* seedlings to flooding. *Plant Physiol.* **66**, 267–271.

Sena Gomes, A. R. and Kozlowski, T. T. (1986). The effects of flooding on water relations and growth of *Theobroma cacao* var. *catongo* seedlings. *J. Hort. Sci.* **61**, 265–276.

Sena Gomes, A. R. and Kozlowski, T. T. (1987). Effects of temperature on growth and water relations of cacao (*Theobroma cacao* var. *comum*) seedlings. *Plant Soil* **103**, 3–11.

Sena Gomes, A. R., Kozlowski, T. T., and Reich, P. B. (1987). Some physiological responses of *Theobroma cacao* var. *catongo* seedlings to air humidity. *New Phytol.* **107**, 591–602.

Senevirathna, A. M. W. K., Stirling, C. M., and Rodrigo, V. H. L. (2003). Growth, photosynthetic performance and shade adaptation of rubber (*Hevea brasiliensis*) grown in natural shade. *Tree Physiol.* **23**, 705–712.

Serraj, R. and Sinclair, T. R. (2002). Osmolyte accumulation: Can it really help increase crop yield under drought conditions? *J. Exp. Bot.* **25**, 333–341.

Serraj, R., Sinclair, T. R., and Purcell, L. C. (1999). Symbiotic N_2 fixation response to drought. *J. Exp. Bot.* **50**, 143–155.

Sexton, R. and Roberts, J. A. (1982). Cell biology of abscission. *Annu. Rev. Plant Physiol.* **33**, 133–162.

Shah, J. J. and Babu, A. M. (1986). Vascular occlusions in stems of *Ailanthus excelsa* Roxb. *Ann. Bot.* **57**, 603–611.

Shainsky, L. J., Newton, M., and Radosevich, S. R. (1992). Effects of intra-specific and inter-specific competition on root and shoot biomass of young Douglas-fir and red alder. *Can. J. For. Res.* **22**, 101–110.

Shaltout, A. D. and Unrath, C. R. (1983). Rest completion prediction model for "Starkrimson Delicious" apples. *J. Am. Soc. Hort. Sci.* **108**, 957–961.

Shaner, D. L. and Boyer, J. S. (1976). Nitrate reductase activity in maize (*Zea mays* L.) leaves. I. Regulation by nitrate flux. *Plant Physiol.* **58**, 499–504.

Sharkey, T. D. (1985). Photosynthesis in intact leaves of C_3 plants: Physics, physiology and rate limitations. *Bot. Rev.* **51**, 53–105.

Sharkey, T. D. (1996). Isoprene synthesis by plants and animals. *Endeavour* **20**, 74–78.

Sharkey, T. D. and Ogawa, T. (1987). Stomatal response to light. In *Stomatal Function* (E. Zeiger, G. D. Farquhar, and I. R. Cowan, eds.), pp. 195–208. Stanford University Press, Stanford, California.

Sharkey, T. D. and Seemann, J. R. (1989). Mild water stress effects on carbon-reduction-cycle intermediates, ribulose bisphosphate carboxylase activity, and spatial homogeneity of photosynthesis in intact leaves. *Plant Physiol.* **89**, 1060–1065.

Sharkey, T. D. and Singsaas, E. L. (1995). Why plants emit isoprene. *Nature (London)* **374**, 769.

Sharkey, T. D., Seemann, J. R., and Pearcy, R. W. (1986). Contribution of metabolites of photosynthesis to post-illumination CO_2 assimilation in response to lightflecks. *Plant Physiol.* **82**, 1063–1068.

Sharkey, T. D., Chen, X. Y., and Yeh, S. (2001). Isoprene increases thermotolerance of fosmidomycin-fed leaves. *Plant Physiol.* **125**, 2001–2006.

Sharma, M. L. (1976). Contribution of dew in the hydraulic balance of a semi-arid grassland. *Agr. Meteorol.* **17**, 321–331.

Sharp, R. E. and Davies, W. J. (1979). Solute regulation and growth by roots and shoots of water-stressed maize plants. *Planta* **147**, 43–49.

Sharples, A. and Gunnery, H. (1933). Callus formation in *Hibiscus rosa-sinensis* L. and *Hevea brasiliensis* Mull. Arg. *Ann. Bot.* **47**, 827–840.

Shear, C. B. and Faust, M. (1980). Nutritional ranges in deciduous tree fruits and nuts. *Hort. Rev.* **2**, 142–163.

Sheen, J. (1994). Feedback control of gene expression. *Photosynth. Res.* **39**, 427–438.

Shepherd, T. and Griffiths, D. W. (2006). The effects of stress on plant cuticular waxes. *New Phytol.* **171**, 469–499.

Sheriff, D. W. (1977). The effect of humidity on water uptake by, and viscous flow resistance of, excised leaves of a number of species: Physiological and anatomical observations. *J. Exp. Bot.* **28**, 1399–1407.

Sheriff, D. W. (1983). Control by indole-3-acetic acid of wood production in *Pinus radiata* D. Don segments in culture. *Aust. J. Plant Physiol.* **10**, 131–135.

Sheriff, D. W. (1995). Gas exchange of field-grown *Pinus radiata*—Relationships with foliar nutrition and water potential, and with climatic variables. *Aust. J. Plant Physiol.* **22**, 1015–1026.

Shieh, M. W. and Cosgrove, D. J. (1998). Expansins. *J. Plant Res.* **111**, 149–157.

Shigo, A. L. and Wilson, C. L. (1977). Wound dressings on red maple and American elm: Effectiveness after five years. *J. Arboric.* **3**, 81–87.

Shigo, A. L., Gregory, G. F., Campana, R. J., Dudzik, K. R., and Zimel, D. M. (1986). Patterns of starch reserves in healthy and diseased American elms. *Can. J. For. Res.* **16**, 204–210.

Shimshi, D. (1963). Effect of soil moisture and phenylmercuric acetate upon stomatal aperture, transpiration, and photosynthesis. *Plant Physiol.* **38**, 713–721.

Shinohara, N., Demura, T., and Fukuda, H. (2000). Isolation of a vascular cell wall-specific monoclonal antibody recognizing a cell polarity by using a phage display subtraction method. *Proc. Nat. Acad. Sci. U. S. A* **97**, 2585–2590.

Shive, J. B. and Brown, K. W. (1978). Quaking and gas exchange in leaves of cottonwood. *Plant Physiol.* **61**, 331.

Shoulders, E. and Ralston, C. W. (1975). Temperature, root aeration and light influence slash pine nutrient uptake rates. *For. Sci.* **21**, 401–410.

Siau, J. F. (1972). Non-isothermal moisture movement in wood. *Wood Sci.* **13**, 11–13.

Siedow, J. N. and Day, D. A. (2000). Respiration and photorespiration. In *Biochemistry and Molecular Biology of Plants* (B. Buchanan, B. W. Gruissem, and R. L. Jones, eds.), pp. 676–728. Amer. Soc. Plant Physiol., Rockville, Maryland.

Sibley, J. L., Eakes, D. J., Gilliam, C. H., Keever, G. J., and Dozier, W. A. (1995). Growth and fall color of red maple selections in the southeastern United States. *J. Environ. Hortic.* **13**, 51–53.

Siefritz, F., Tyree, M. T., Lovisolo, C., Schubert, A., and Kaldenhoff, R. (2002). PIP1 plasma membrane aquaporins in tobacco: From cellular effects to function in plants. *Plant Cell* **14**, 869–876.

Siler, D. J. and Cornish, K. (1993). A protein from *Ficus elastica* rubber particles is related to proteins from *Hevea brasiliensis* and *Parthenium argentatum*. *Phytochemistry* **32**, 1097–1102.

Simard, S. W., Jones, M. D., Durall, D. M., Perry, D. A., Myrold, D. D., and Molina, R. (1997). Reciprocal transfer of carbon isotopes between ectomycorrhizal *Betula papyrifera* and *Pseudotsuga menziesii*. *New Phytol.* **137**, 529–542.

Sinclair, T. R. and Ludlow, M. M. (1985). Who taught plants thermodynamics? The unfulfilled potential of plant water potential. *Aust. J. Plant Physiol.* **12**, 213–217.

Sinclair, T. R. and Purcell, L. C. (2005). Is a physiological perspective relevant in a 'genocentric' age? *J. Exp. Bot.* **56**, 2777–2782.

Singh, H. P., Batish, D. R., and Kohli, R. K. (1999). Autotoxicity: Concept, organisms, and ecological significance. *Crit. Rev. Plant Sci.* **18**, 757–772.

Singh, M. (2000). Turnover of D1 protein encoded by psbA gene in higher plants and cyanobacteria sustains photosynthetic efficiency to mainain plant productivity under photoinhibitory irradiance. *Photosynthetica* **38**, 161–169.

Singh, S., Letham, D. S., Jameson, P. E., Zhang, R., Parker, C. W., Badenoch-Jones, J., and Nooden, L. D. (1988). Cytokinin biochemistry in relation to leaf senescence. IV. Cytokinin metabolism in soybean explants. *Plant Physiol.* **88**, 788–794.

Sionit, N., Strain, B. R., Hellmers, H., Riechers, G. H., and Jaeger, C. H. (1985). Long-term atmospheric CO_2 enrichment affects the growth and development of *Liquidambar styraciflua* and *Pinus taeda* seedlings. *Can. J. For. Res.* **15**, 468–471.

Siqueira, J. O., Nair, M. G., Hammerschmidt, R., and Safir, C. R. (1991). Significance of phenolic compounds in plant-soil microbial systems. *Crit. Rev. Plant Sci.* **10**, 63–121.

Siwecki, R. and Kozlowski, T. T. (1973). Leaf anatomy and water relations of excised leaves of six *Populus* clones. *Arboretum Kornickie* **8**, 83–105.

Skeffington, R. A. and Wilson, E. J. (1988). Excess nitrogen deposition: Issues for consideration. *Environ. Pollut.* **54**, 159–184.

Skene, D. S. (1972). The kinetics of tracheid development in *Tsuga canadensis* Carr. and its relation to tree vigor. *Ann. Bot.* **36**, 179–187.

Skinner, R. H. and Radin, J. W. (1994). The effect of phosphorus nutrition on water flow through the apoplastic bypass in cotton roots. *J. Exp. Bot.* **45**, 423–428.

Skrzypek, E., Miyamoto, K., Saniewski, M., and Ueda, J. (2005). Identification of jasmonic acid and its methyl ester as gum-inducing factors in tulips. *J. Plant Res.* **118**, 27–30.

Slatyer, R. O. (1956). Absorption of water from atmospheres of different humidity and its transport through plants. *Aust. J. Biol. Sci.* **9**, 552–558.

Slatyer, R. O. (1957). The significance of the permanent wilting percentage in studies of plant and soil water relations. *Bot. Rev.* **23**, 585–636.

Slatyer, R. O. (1967). *Plant-Water Relationships*. Academic Press, London.

Slatyer, R. O. (1977). Altitudinal variation in the photosynthetic characteristics of snow gum, *Eucalyptus pauciflora* Sieb. ex Spreng. IV. Temperature response of four populations grown at different temperatures. *Aust. J. Plant Physiol.* **4**, 583–594.

Slatyer, R. O. and Bierhuizen, J. F. (1964). Transpiration from cotton leaves under a range of environmental conditions in relation to internal and external diffusive resistances. *Aust. J. Biol. Sci.* **17**, 115–130.

Slatyer, R. O. and Ferrar, P. J. (1977). Altitudinal variation in the photosynthetic characteristics of snow gum, *Eucalyptus pauciflora* Sieb. ex Spreng. V. Rate of acclimation to an altered growth environment. *Aust. J. Plant Physiol.* **4**, 595–609.

Slavik, B. (1974). *Methods of Studying Plant Water Relations*. Ecological Studies Vol. 9, Springer, Berlin and New York.

Slocum, R. D., Kaur-Sawhney, R., and Galston, A. W. (1984). The physiology and biochemistry of polyamines in plants. *Arch. Biochem. Biophys.* **235**, 283–303.

Small, E. (1972). Photosynthetic rates in relation to nitrogen recycling as an adaptation to nutrient deficiency in peat bog plants. *Can. J. Bot.* **50**, 2227–2233.

Smart, C. M. (1994). Gene expression during leaf senescence. *New Phytol.* **126**, 419–448.

Smirnoff, N., Todd, P., and Stewart, G. R. (1984). The occurrence of nitrate reduction in the leaves of woody plants. *Ann. Bot.* **54**, 363–374.

Smith, A. M. (1994). Xylem transport and the negative pressure sustainable by water. *Ann. Bot.* **74**, 647–651.

Smith, C. J. S., Watson, C. F., Ray, J., Bird, C. R., Morris, P. C., Schuch, W., and Grierson, D. (1988). Antisense RNA inhibition of polygalacturonase gene expression in transgenic tomatoes. *Nature (London)* **334**, 724–726.

Smith, D. J. and Schwabe, W. W. (1980). Cytokinin activity in oak (*Quercus robur*) with particular reference to transplanting. *Physiol. Plant.* **48**, 27–32.

Smith, H. C. (1966). Epicormic branching on eight species of Appalachian hardwoods. *U. S. For. Serv., Res. Note NE* **NE–53**.

Smith, J. A. C. and Griffiths, H., eds. (1993). *Water Deficits.* BIOS Scientific Publishers, Oxford, U.K.

Smith, M. W. and Ager, P. L. (1988). Effects of soil flooding on leaf gas exchange of seedling pecan trees. *HortScience* **23**, 370–372.

Smith, P. F. (1976). Collapse of "Murcott" tangerine trees. *J. Am. Soc. Hortic. Sci.* **101**, 23–25.

Smith, R. C. and Fry, S. C. (1991). Endotransglycosylation of xyloglucans in plant cell suspension cultures. *Biochem. J.* **279**, 529–535.

Smith, S. E. and Read, D. J. (1997). *Mycorrhizal Symbiosis.* Academic Press, San Diego.

Smith, T. A. (1985). Polyamines. *Annu. Rev. Plant Physiol.* **36**, 117–143.

Smith, W. H. (1974). Air pollution—Effects on the structure and function of temperate forest ecosystems. *Environ. Pollut.* **6**, 111–129.

Smith, W. H. (1981). *Air Pollution and Forests.* Springer-Verlag, New York.

Smith, W. H. (1990). *Air Pollution and Forests, 2nd ed.* Springer-Verlag, New York.

Smith-Ramirez, C. and Arnesto, J. J. (1994). Flowering and fruiting patterns in the temperate rainforest of Chiloé, Chile-Ecologies and climatic restraints. *J. Ecol.* **82**, 353–365.

Smock, R. M. (1979). Controlled atmosphere storage of fruits. *Hort. Rev.* **1**, 301–336.

Snaith, P. J. and Mansfield, T. A. (1982). Stomatal sensitivity to abscisic acid: Can it be defined? *Plant Cell Environ.* **5**, 309–311.

Snaydon, R. W. (1991). The productivity of C_3 and C_4 plants—A reassessment. *Funct. Ecol.* **5**, 321–330.

Sobeih, W. Y., Dodd, I. C., Bacon, M. A., Grierson, D., and Davies, W. J. (2004). Long-distance signals regulating stomatal conductance and leaf growth in tomato (*Lycopersicon esculentum*) plants subjected to partial root-zone drying. *J. Exp. Bot.* **55**, 2353–2363.

Sobrado, M. A. (1993). Trade-off between water transport efficiency and leaf life-span in a tropical dry forest. *Oecologia* **96**, 19–23.

Sobrado, M. A., Grace, J., and Jarvis, P. G. (1992). The limits to xylem embolism recovery in *Pinus sylvestris* L. *J. Exp. Bot.* **43**, 831–836.

Soe, K. (1959). Anatomical studies of bark regeneration following scarring. *J. Arnold Arbor. Harv. Univ.* **40**, 260–267.

Solbrig, O. T. (1991). *Biodiversity: Scientific Issues and Collaborative Research Proposals.* UNESCO, Paris.

Solbrig, O. T. (1993). Plant traits and adaptive strategies. In *Biodiversity and Ecosystem Function* (E.-D. Schulze and H. A. Mooney, eds.), pp. 97–116. Springer-Verlag, Berlin and New York.

Solbrig, O. T. and Cantino, P. D. (1975). Reproductive adaptations in *Prosopis* (Leguminosae, Mimosoideae). *J. Arnold Arboretum* **56**, 185–210.

Solbrig, O. T., Goldstein, G., Medina, E., Sormiento, G., and Silva, J. (1992). Responses of savannas to stress and disturbance: A research approach. In *Environmental Rehabilitation* (M. K. Wali, ed.), pp. 63–73. SPB Academic Publishers, The Hague.

Sollins, P. and McCorison, F. M. (1981). Nitrogen and carbon solution chemistry of an old-growth coniferous forest watershed before and after cutting. *Water Resourses Res.* **17**, 1409–1418.

Solomis, T. (1977). Cyanide-resistant respiration in higher plants. *Annu. Rev. Plant Physiol.* **28**, 279–297.

Sparks, D. (1991). Geographical origin of pecan cultivars influences time required for fruit development and nut size. *J. Am. Soc. Hortic. Sci.* **116**, 627–631.

Spermulli, L. (2000). Protein synthesis, assembly, and degradation. In *Biochemistry and Molecular Biology of Plants* (B. Buchanan, B. W. Gruissem, and R. L. Jones, eds.), pp. 412–454. Amer. Soc. Plant Physiol., Rockville, Maryland.

Sperry, J. S. (1986). Relationship of xylem pressure potential, stomatal closure, and shoot morphology in the palm *Rhapis excelsa.* *Plant Physiol.* **80**, 110–116.

Sperry, J. S. (2003). Evolution of water transport and xylem structure. *Intl. J. Plant Sci.* **164**, S115–S127.

Sperry, J. S. and Hacke, U. G. (2002). Desert shrub water relations with respect to soil characteristics and plant functional type. *Funct. Ecol.* **16**, 367–378.

Sperry, J. S. and Sullivan, J. E. M. (1992). Xylem embolism in response to freeze-thaw cycles and water stress in ring-porous, diffuse-porous, and conifer species. *Plant Physiol.* **100**, 605–613.

Sperry, J. S. and Tyree, M. T. (1988). Mechanism of water stress-induced xylem embolism. *Plant Physiol.* **88**, 581–587.

Sperry, J. S., Holbrook, N. M., Zimmermann, M. H., and Tyree, M. T. (1987). Spring filling of xylem vessels in wild grapevine. *Plant Physiol.* **83**, 414–417.

Sperry, J. S., Donnelly, J. R., and Tyree, M. T. (1988a). Seasonal occurrence of xylem embolism in sugar maple (*Acer saccharum*). *Am. J. Bot.* **75**, 1212–1218.

Sperry, J. S., Tyree, M. T., and Donnelly, J. R. (1988b). Vulnerability of xylem to embolism in a mangrove vs. an inland species of Rhizophoraceae. *Physiol. Plant.* **74**, 276–283.

Sperry, J. S., Perry, A. H., and Sullivan, J. E. M. (1991). Pit membrane degradation and air–embolism formation in ageing xylem vessels of *Populus tremuloides* Michx. *J. Exp. Bot.* **42**, 1399–1406.

Spollen, W. G., LeNoble, M. E., Samuels, T. D., Bernstein, N., and Sharp, R. E. (2000). Abscisic acid accumulation maintains maize primary root elongation at low water potentials by restricting ethylene production. *Plant Physiol.* **122**, 967–976.

Sponsel, V. M. (1988). Gibberellin biosynthesis and metabolism. In *Plant Hormones and Their Role in Plant Growth and Development* (P. J. Davies, ed.), pp. 43–75. Kluwer, Dordrecht.

Sprackling, J. A. and Read, R. A. (1979). Tree root systems in eastern Nebraska. *Nebraska Con. Bull. 37. Cons. and Survey Div.*, Inst. *Agric. and Nat. Resources, U. of Nebraska, Lincoln.*

Sprent, J. I. and Sprent, P. (1990). *Nitrogen Fixing Organisms: Pure and Applied Aspects.* Chapman and Hall, London, London.

Sprugel, D. G. (1990). Components of woody tissue respiration in young *Abies amabilis* (Dougl.) Forbes trees. *Trees* **4**, 88–95.

Sprugel, D. G. and Benecke, U. (1991). Measuring woody-tissue respiration and photosynthesis. In *Techniques and Approaches in Forest Tree Ecophysiology* (J. P. Lassoie and T. M. Hinckley, eds.), pp. 329–355. CRC Press, Boca Raton, Florida.

Sprugel, D. G., Ryan, M. G., Brooks, J. R., Vogt, K. A., and Martin, T. A. (1995). Respiration from the organ level to the stand. In *Resource Physiology of Conifers: Acquisition Allocation, and Utilization* (W. K. Smith and T. M. Hinckley, eds.), pp. 255–299. Academic Press, San Diego.

Squillace, A. E. (1971). Inheritance of monoterpene composition in cortical oleoresin of slash pine. *For. Sci.* **17**, 381–387.

Squillace, A. E. and Swindel, B. F. (1986). Linkage among genes controlling monoterpene constituent levels in loblolly pine. *For. Sci.* **32**, 97–112.

Srivastava, H. S., Jolliffe, R. A., and Runeckles, V. C. (1975). The effects of environmental conditions on the inhibition of leaf gas exchange by NO_2. *Can. J. Bot.* **53**, 475–482.

Srivastava, L. M. (1963). Secondary phloem in the Pinaceae. *Univ. Calif. Berkeley, Publ. Bot.* **36**, 1–142.

Srivistava, L. M. (2001). *Plant Growth and Development: Hormones and the Environment*. Academic Press, San Diego.

Stacey, G., Burris, R. H., and Evans, H. J., eds. (1992). *Biological Nitrogen Fixation*. Routledge, Chapman and Hall, New York.

Stadler, J. and Gebauer, G. (1992). Nitrate reduction and nitrate content in ash trees (*Fraxinus excelsior* L.): Distribution between compartments, site comparison and seasonal variation. *Trees* **6**, 236–240.

Stadler, R., Buttner, M., Ache, P., Hedrich, R., Ivashikina, N., Melzer, M., Shearson, S. M., Smith, S. M., and Sauer, N. (2003). Diurnal and light-regulated expression of AtSTP1 in guard cells of *Arabidopsis*. *Plant Physiol.* **133**, 528–537.

Ståfelt, M. G. (1932). *Der stomatare Regulator in der pflanzlichen Transpiration*. *Planta* **17**, 22–85.

Stålfelt, M. G. (1963). On the distribution of the precipitation in a spruce stand. In *The Water Relations of Plants* (A. J. Rutter and F. Whitehead, eds.), pp. 116–126. Blackwell, Oxford.

Stammitti, L., Garrec, J.-P., and Derridj, S. (1995). Permeability of isolated cuticles of *Prunus laurocerasus* to soluble carbohydrate. *Plant Physiol. Biochem.* **33**, 319–326.

Stansell, J. R., Klepper, B., Browning, V. D., and Taylor, H. M. (1973). Plant water status in relation to clouds. *Agron. J.* **65**, 677–678.

Starr, G., Neuman, D. S., and Oberbauer, S. F. (2004). Ecophysiological analysis of two arctic sedges under reduced root temperatures. *Physiol. Plant.* **120**, 458–464.

Stassen, P. J. C. (1984). *Seisoenveranderinge in die koolhidraatingshoud von jong appelbome*. *S. Afr. J. Plant Soil* **1**, 92–95.

Stead, A. D. (1992). Pollination-induced flower senescence: A review. *Plant Growth Reg.* **11**, 13–20.

Stebbins, G. L. (1989). Introduction. In *Isozymes in Plant Biology* (D. E. Soltis and P. S. Soltis, eds.), pp. 1–4. Dioscorides Press, Portland, Oregon.

Steinbeck, K. (1966). Site, height, and mineral nutrient content relations of Scotch pine provenances. *Silvae Genet.* **15**, 42–50.

Steinhoff, R. J., Joyce, D. G., and Fins, L. (1983). Isozyme variation in *Pinus monticola*. *Can. J. For. Res.* **13**, 1122–1132.

Stephenson, A. G. (1981). Flower and fruit abortion: Proximate causes and ultimate functions. *Ann. Rev. Ecol. Syst.* **12**, 253–279.

Steudle, E. (1994). Water transport across roots. *Plant Soil* **167**, 79–90.

Steudle, E. (2000). Water uptake by roots: Effects of water deficit. *J. Exp. Bot.* **51**, 1531–1542.

Steudle, E. (2001). The cohesion-tension mechanism and the acquisition of water by plant roots. *Annu. Rev. Plant Physiol. Plant Mol. Biol.* **52**, 847–875.

Steudle, E. and Jeschke, W. D. (1983). Water transport in barley roots. *Planta* **158**, 237–248.

Steudle, E. and Peterson, C. A. (1998). How does water get through roots *J. Exp. Bot.* **49**, 775–788.

Steudle, E., Oren, R., and Schulze, E. D. (1987). Water transport in maize roots. Measurement of hydraulic conductivity, solute permeability, and of reflection coefficients of excised roots using the root pressure probe. *Plant Physiol.* **84**, 1220–1232.

Stevens, C. L. and Eggert, R. L. (1945). Observations on the causes of the flow of sap in red maple. *Plant Physiol.* **20**, 636–648.

Stewart, C. M. (1967). Moisture content of living trees. *Nature (London)* **214**, 138.

Stewart, J. D., Elabidine, A. Z., and Bernier, P. Y. (1995). Stomatal and mesophyll limitations of photosynthesis in black spruce seedlings during multiple cycles of drought. *Tree Physiol.* **15**, 57–64.

Stone, E. C. and Fowells, H. A. (1955). Survival value of dew as determined under laboratory conditions with *Pinus ponderosa*. *For. Sci.* **1**, 183–188.

Stone, E. C., Went, F. W., and Young, C. L. (1950). Water absorption from the atmosphere by plants growing in dry soil. *Science* **111**, 546–548.

Stone, E. L. and Kalisz, P. J. (1991). On the maximum extent of tree roots. *For. Ecol. Manage.* **46**, 59–102.

Stone, J. E. and Stone, E. L. (1975a). Water conduction in lateral roots of red pine. *For. Sci.* **21**, 53–60.

Stone, J. E. and Stone, E. L. (1975b). The communal root system of red pine: Water conduction through root grafts. *For. Sci.* **22**, 255–262.

Stösser, R. (1979). Investigations on gum duct formation in cherries using plastic embedding medium. *Sci. Hortic.* **11**, 247–252.

Strain, B. R., Higginbotham, K. O., and Mulroy, J. C. (1976). Temperature preconditioning and photosynthetic capacity of *Pinus taeda* L. *Photosynthetica* **10**, 47–52.

Strand, M., Lundmark, T., Söderbergh, I., and Mellander, P.-E. (2002). Impacts of seasonal air and soil temperatures on photosynthesis in Scots pine trees. *Tree Physiol.* **22**, 839–847.

Strong, W. L. and La Roi, G. H. (1983). Root-system morphology of common boreal forest trees in Alberta, Canada. *Can. J. For. Res.* **13**, 1164–1173.

Struckmeyer, B. E., Beckman, C. E., Kuntz, J. E., and Riker, A. J. (1954). Plugging of vessels by tyloses and gums in wilting oaks. *Phytopathology* **44**, 148–153.

Strugger, S. (1949). *Praktikum der Zell–und Gewebephysiologie der Pflanzen*. Springer-Verlag, Berlin and New York.

Studholme, W. P. and Philipson, W. R. (1966). A comparison of the cambium in two woods with included phloem: *Heimerliriodendron brunonianum* and *Avicennia resinifera*. *N. Z. J. Bot.* **4**, 355–365.

Stumpf, P. K. and Conn, E. E., eds. (1980–1987). *The Biochemistry of Plants*. Vols. 4–9. Academic Press, New York.

Subbarao, G. V., Ito, O., Berry, W. L., and Wheeler, R. M. (2003). Sodium—A functional plant nutrient. *Crit. Rev. Plant Sci.* **22**, 391–416.

Sucoff, E., Thornton, F. C., and Joslin, J. D. (1990). Sensitivity of tree seedlings to aluminum. I. Honeylocust. *J. Environ. Qual.* **19**, 163–171.

Sullivan, P. F. and Welker, J. M. (2007). Variation in leaf physiology of *Salix arctica* within and across ecosystems in the High Arctic: Test of a dual isotope (Δ^{13}C and Δ^{18}O) conceptual model. *Oecologia* **151**, 372–386.

Sundberg, B., Uggla, C., and Tuominen, H. (2000). Cambial growth and auxin gradients. In *Cell and Molecular Biology of Wood Formation* (R. A. Savidge, J. R. Burnett, and R. Napier, eds.), pp. 169–188. BIOS Sci. Publ. Ltd., Oxford, U.K.

Sutton, R. F. (1980). Root system morphogenesis. *N. Z. J. For. Sci.* **10**, 264–292.

Suzuki, A. and Knaff, D. B. (2005). Glutamate synthase: Structural, mechanistic and regulatory properties, and role in the amino acid metabolism. *Photosynth. Res.* **83**, 191–217.

Swank, W. T. and Douglass, J. E. (1974). Steamflow greatly reduced by converting deciduous hardwood stands to pine. *Science* **185**, 857–859.

Swanson, R., Edlund, A. F., and Preuss, D. (2004). Species specificity in pollen-pistil interactions. *Annu. Rev. Genet.* **38**, 793–818.

Sweatt, M. R. and Davies Jr., F. T. (1984). Mycorrhizae, water relations, growth, and nutrient uptake of geranium grown under moderately high phosphorus regimes. *J. Am. Soc. Hortic. Sci.* **109**, 210–213.

Sweet, G. B. and Thulin, I. J. (1969). The abortion of conelets in *Pinus radiata*. *N. Z. J. For.* **14**, 59–67.

Swietlik, D. and Faust, M. (1984). Foliar nutrition of fruit crops. *Hort. Rev.* **6**, 287–355.

Swietlik, D., Korcak, R. F. and Faust, M. (1983). Stomatal and non-stomatal inhibition of photosynthesis in water-stressed apple seedlings. *J. Am. Soc. Hortic. Sci.* **108**, 54–58.

Swift, M. J., Russell-Smith, A. and Perfect, T. J. (1981). Decomposition and mineral-nutrient dynamics of plant litter in a regenerating bush-fallow in sub-humid tropical Nigeria. *J. Ecol.* **69**, 981–995.

Switzer, G. L., Nelson, L. E. and Smith, W. H. (1968). The mineral cycle in forest stands. In *Forest Fertilization: Theory and Practice*, pp. 1–9. Tenn. Valley Auth., Muscle Shoals, Alabama.

Switzer, G. L., Nelson, L. E. and Baker, J. B. (1976). Accumulation and distribution of dry matter and nutrients in *Aigeiros* poplar plantations. In *Proceedings: Symposium on Eastern Cottonwood and Related Species*, pp. 359–369. Div. of Continuing Education, Louisiana State Univ., Baton Rouge.

Syvertsen, J. P. (1984). Light acclimation in citrus leaves. II. CO_2 assimilation and light, water, and nitrogen use efficiency. *J. Am. Soc. Hortic. Sci.* **109**, 812–817.

Taber, R. A. and Taber, W. A. (1984). Evidence for ectomycorrhizal fungus-mediated transfer between *Pinus* and *Tradescantia*. *For. Sci.* **30**, 892–896.

Takahashi, N. ed. (1986). *Chemistry of Plant Hormones*. CRC Press, Boca Raton, Florida.

Takahashi, N., Phinney, B. O., and MacMillan, J., eds. (1991). *Gibberellins*. Springer-Verlag, New York and Berlin.

Talbert, C. M. and Holch, A. E. (1957). A study of the lobing of sun and shade leaves. *Ecology* **38**, 655–658.

Talbott, L. D. and Zeiger, E. (1996). Central roles for potassium and sucrose in guard-cell osmoregulation. *Plant Physiol.* **111**, 1051–1057.

Talboys, P. W. (1978). Disfunction of the water system. In *Plant Disease* (J. G. Horsfall and E. C. Tanford, eds.), pp. 141–162. Academic Press, New York.

Tamm, C. O. (1964). Determination of nutrient requirement of forest stands. *Int. Rev. For. Res.* **1**, 115–170.

Tamura, S. (1991). Historical aspects of gibberellins. In *Giberellins* (N. Takahashi, B. O. Phinney, and J. MacMillan, eds.), pp. 1–8. Springer-Verlag, Berlin and New York.

Tan, W. X. and Blake, T. J. (1993). Drought tolerance, abscisic acid and electrolyte leakage in fast- and slow-growing black spruce (*Picea mariana*) progenies. *Physiol. Plant.* **89**, 817–823.

Tan, W. and Hogan, G. (1995). Limitations to net photosynthesis as affected by nitrogen status in jack pine (*Pinus banksiana* Lamb.) seedlings. *J. Exp. Bot.* **46**, 407–413.

Taneda, H. and Tateno, M. (2005). Hydraulic conductivity, photosynthesis and leaf water balance in six evergreen woody species from fall to winter. *Tree Physiol.* **25**, 299–306.

Tang, Z. C. and Kozlowski, T. T. (1982). Some physiological and growth responses of *Betula papyrifera* seedlings to flooding. *Physiol. Plant.* **55**, 415–420.

Tansey, M. R. (1977). Microbial facilitation of plant mineral nutrition. In *Microorganisms and Minerals* (E. D. Weinberg, ed.), pp. 343–385. Dekker, New York.

Tardieu, F. and Davies, W. J. (1992). Stomatal response to abscisic acid is a function of current plant water status. *Plant Physiol.* **98**, 540–545.

Tardieu, F. and Davies, W. J. (1993). Integration of hydraulic and chemical signalling in the control of stomatal conductance and water status of droughted plants. *Plant, Cell Environ.* **16**, 341–349.

Tardieu, F. and Simonneau, T. (1998). Variability among species of stomatal control under fluctuating soil water status and evaporative demand—Modelling isohydric and anisohydric behaviours. *J. Exp. Bot.* **49**, 419–432.

Tardieu, F., Katerji, N., Bethenod, O., Zhang, J., and Davies, W. J. (1991). Maize stomatal conductance in the field—Its relationship with soil and plant water potentials, mechanical constraints and ABA concentration in the xylem sap. *Plant, Cell Environ.* **14**, 121–126.

Tardieu, F., Zhang, J., and Gowing, D. J. G. (1993). Stomatal control by both [ABA] in the xylem sap and leaf water status—A test of a model for droughted or ABA-fed field-grown maize. *Plant, Cell Environ.* **16**, 413–420.

Tartachnyk, I. I. and Blanke, M. M. (2004). Effect of delayed fruit harvest on photosynthesis, transpiration and nutrient remobilization of apple leaves. *New Phytol.* **164**, 441–450.

Taylor, B. K. and May, L. H. (1967). The nitrogen nutrition of the peach tree. II. Storage and mobilization of nitrogen in young trees. *Aust. J. Biol. Sci.* **20**, 389–411.

Taylor, F. H. (1956). Variation in sugar content of maple sap. *Vermont Agr. Expt. Sta. Bull.* **587**.

Taylor, H. M. and Willatt, S. T. (1983). Shrinkage of soybean roots. *Agron. J.* **75**, 818–820.

Telewski, F. W. and Jaffe, M. J. (1986). Thigmomorphogenesis: The role of ethylene in the response of *Pinus taeda* and *Abies fraseri* to mechanical perturbation. *Physiol. Plant.* **66**, 227–233.

Templer, P. H. and Dawson, T. E. (2004). Nitrogen uptake by four tree species of the Catskill Mountains, New York: Implications for forest N dynamics. *Plant Soil* **262**, 251–261.

Teng, Y. and Timmer, V. R. (1990a). Phosphorus-induced micronutrient disorders in hybrid poplar. I. Preliminary diagnosis. *Plant Soil* **126**, 19–29.

Teng, Y. and Timmer, V. R. (1990b). Phosphorus-induced micronutrient disorders in hybrid poplar. III. Prevention and correction in nursery culture. *Plant Soil* **126**, 41–51.

Teng, Y. and Timmer, V. R. (1993). Growth and nutrition of hybrid poplar in response to phosphorus, zinc, and gibberellic acid treatment. *For. Sci.* **39**, 252–259.

Tenhunen, J. D., Lange, O. L., and Braun, M. (1981). Midday stomatal closure in Mediterranean type sclerophylls under simulated habitat conditions in an environmental chamber. II. Effect of the complex of leaf temperature and air humidity on gas exchange of *Arbutus unedo* and *Quercus ilex*. *Oecologia* **50**, 5–11.

Tenhunen, J. D., Lange, O. L., and Jahner, D. (1982). The control by atmospheric factors and water stress of midday stomatal closure in *Arbutus unedo* growing in a natural macchia. *Oecologia* **55**, 165–169.

Tepper, H. B. (1963). Leader growth of young pitch and shortleaf pines. *For. Sci.* **9**, 344–353.

Tepper, H. B. and Hollis, C. A. (1967). Mitotic reactivation of the terminal bud and cambium of white ash. *Science* **156**, 1635–1636.

Terashima, I. and Hikosaka, K. (1995). Comparative ecophysiology of leaf and canopy photosynthesis. *Plant, Cell Environ.* **18**, 1111–1128.

Terry, M. E., McGraw, D., and Jones, R. L. (1982). Effect of IAA on growth and soluble cell wall polysaccharides contrifuged from pine hypocotyl sections. *Plant Physiol.* **69**, 323–326.

Teskey, R. O., Grier, C. C., and Hinckley, T. M. (1984a). Changes in photosynthesis and water relations with age and season in *Abies amabilis*. *Can. J. For. Res.* **14**, 77–84.

Teskey, R. O., Hinckley, T. M., and Grier, C. C. (1984b). Temperature induced changes in the water relations of *Abies amabilis* (Dougl.) Forbes. *Plant Physiol.* **74**, 77 80.

Teskey, R. O., Fites, J. A., Samuelson, L. J., and Bongarten, B. C. (1986). Stomatal and non-stomatal limitations to net photosynthesis in *Pinus taeda* L. under different environmental conditions. *Tree Physiol.* **2**, 131–142.

Tevini, M. and Lichtenthaler, H. R., eds. (1977). *Lipids and Lipid Polymers in Higher Plants*. Springer-Verlag, Berlin and New York.

Tezara, W., Mitchell, V. J., Driscoll, S. D., and Lawlor, D. W. (1999). Water stress inhibits plant photosynthesis by decreasing coupling factor and ATP. *Nature (London)* **401**, 914–917.

Thijsse, G. and Baas, P. (1990). "Natural" and NH_3-induced variation in epicuticular needle wax morphology of *Pseudotsuga menziesii* (Mirb.) Franco. *Trees* **4**, 111–114.

Thimann, K. V. (1977). *Hormone Action in the Whole Life of Plants*. Univ. of Massachusetts Press, Amherst.

Thimann, K. V. (1980). The senescence of leaves. In *Senescence in Plants* (K. V. Thimann, ed.), pp. 85–115. CRC Press, Boca Raton, Florida.

Thomas, D. S., Eamus, D., and Shanahan, S. (2000). Influence of season, drought and xylem ABA on stomatal responses to leaf-to-air vapour pressure difference of trees of the Australian wet-dry tropics. *Aust. J. Bot.* **48**, 143–151.

Thomas, P., Paul, P., Nagaraja, N., and Dalal, V. B. (1983). Physiochemical and respiratory changes in Dwarf Cavendish variety of bananas during growth and maturation. *J. Food Sci. Tech., India* **20**, 51–56.

Thomas, W. (1927). Nitrogenous metablism of *Pyrus malus*. III. The partition of nitrogen in the leaves, one- and two-year branch growth and non-bearing spurs throughout a year's cycle. *Plant Physiol.* **2**, 109–137.

Thomas, W. A. and Grigal, D. F. (1976). Phosphorus conservation by evergreenness of mountain laurel. *Oikos* **27**, 19–26.

Thompson, M. M. (1979). Growth and development of the pistillate flower and nut in "Barcelona" filbert. *J. Am. Soc. Hortic. Sci.* **104**, 427–432.

Thomson, C. J. and Greenway, H. (1991). Metabolic evidence for stelar anoxia in maize roots exposed to low O_2 concentrations. *Plant Physiol.* **96**, 1294–1301.

Thomson, W. W., Mudd, J. B., and Gibbs, M., eds. (1983). *Biosynthesis and Function of Plant Lipids*. Am. Soc. Plant Physiologists, Rockville, Maryland.

Thorn, A. J. and Robertson, E. D. (1987). Zinc deficiency in *Pinus radiata* at Cape Karikari, New Zealand. *N. Z. J. For. Sci.* **17**, 129–132.

Thornley, J. H. M. and Cannell, M. G. R. (2000). Modelling the components of plant respiration: Representation and realism. *Ann. Bot.* **85**, 55–67.

Tibbals, E. C., Carr, E. K., Gates, D. M., and Kreith, F. (1964). Radiation and convection in conifers. *Am. J. Bot.* **51**, 529–538.

Tibbitts, T. W. and Kozlowski, T. T., eds. (1979). *Controlled Environment Guidelines for Plant Research*. Academic Press, New York.

Tieszen, L. L. (1978). Photosynthesis in the principal Barrow, Alaska species. A summary of field and laboratory responses. In *Vegetation and Production Ecology of an Alaskan Arctic Tundra* (L. L. Tieszen, ed.), pp. 241–268. Springer-Verlag, Berlin and New York.

Tilman, D. (1988). *Plant Strategies and the Dynamics and Structure of Plant Communities*. Princeton Univ. Press, Princeton, New Jersey.

Timmons, D. R., Verry, E. S., Burwell, R. E., and Holt, R. F. (1977). Nutrient transport in surface runoff and interflow from an aspen forest. *J. Environ. Qual.* **6**, 188–192.

Tingey, D. T. and Ratsch, H. C. (1978). Factors influencing isoprene emissions from live oak. *Plant Physiol.* **61(4) Suppl.**, 86.

Tingey, D. T., Phillips, D. L., and Johnson, M. G. (2000). Elevated CO_2 and conifer roots: Effects on growth, life span and turnover. *New Phytol.* **147**, 87–103.

Tinker, P. B. and Läuchli, A., eds. (1984–1988). *Advances in Plant Nutrition*. Vols. 1–3, Prager, New York.

Tinker, P. B. and Nye, P. H. (2000). *Solute Movement in the Rhizosphere*. Oxford University Press, New York.

Tinoco-Ojanguren, C. and Pearcy, R. W. (1993). Stomatal dynamics and its importance to carbon gain in 2 rainforest *Piper* species. 2. Stomatal versus biochemical limitations during photosynthetic induction. *Oecologia* **94**, 395–402.

Tinoco-Ojanguren, C. and Vazquez-Yanes, C. (1983). *Especies CAM in la selva humeda tropical de Los Tuxtlas, Veracruz. Boletin de la Sociedad Botanica de Mexico* **45**, 150–153.

Tisdale, R. A. and Nebeker, T. E. (1992). Resin flow as a function of height along the bole of loblolly pine. *Can. J. Bot.* **70**, 2509–2511.

Tissue, D. T., Lewis, J. D., Wullschleger, S. D., Amthor, J. S., and Griffin, K. L., Anderson R. (2002). Leaf respiration at different canopy positions in sweetgum (*Liquidambar styraciflua*) grown in ambient and elevated concentrations of carbon dioxide in the field. *Tree Physiol.* **22**, 1157–1166.

Titus, J. S. and Kang, S. M. (1982). Nitrogen metabolism, translocation and recycling in apple trees. *Hort. Rev.* **4**, 204–246.

Tiwari, S. B., Hagen, G., and Guilfoyle, T. (2003). The roles of auxin response factor domains in auxin-responsive transcription. *Plant Cell* **15**, 533–543.

Tjepkema, J. D. (1985). Utilization of photosynthate for nitrogen fixation in seedlings of *Myrica gale* and *Alnus rubra*. In *Nitrogen Fixation and CO_2 Metabolism* (P. W. Ludden and V. E. Burns, eds.), pp. 183–192. Elsevier, New York.

Tobiessen, P., Rundel, P. W., and Steckar, R. E. (1971). Water potential gradient in a tall *Sequoiadendron*. *Plant Physiol.* **48**, 303–304.

Tobón, C., Sevink, J., and Verstraten, J. M. (2004). Solute fluxes in throughfall and stemflow in four forest ecosystems in northwest Amazonia. *Biogeochemistry* **70**, 1–25.

Todd, G. W. and Garber, M. J. (1958). Some effects of air pollutants on the growth and productivity of plants. *Bot. Gaz.* **120**, 75–80.

Toivonen, A., Rikala, R., Repo, T., and Smolander, H. (1991). Autumn colouration of first year *Pinus sylvestris* seedlings during frost hardening. *Scand. J. For. Res.* **6**, 31–39.

Tolley, L. C. and Strain, B. R. (1984). Effects of atmospheric CO_2 enrichment and water stress on growth of *Liquidambar styraciflua* L. and *Pinus taeda* L. seedlings. *Can. J. Bot.* **62**, 2135–2139.

Tomlinson, G. H. (2003). Acidic deposition, nutrient leaching and forest growth. *Biogeochemistry* **65**, 51–81.

Tomlinson, P. B. and Gill, A. M. (1973). Growth habits of tropical trees: Some guiding principles. In *Tropical Forest Ecosystems in Africa and South America: A Comparative Review* (B. J. Eggers, E. S. Ayensu, and W. D. Duckworth, eds.), pp. 129–143. Smithson. Inst. Press, Washington, D.C.

Tomos, D. and Pritchard, J. (1994). Biophysical and biochemical control of cell expansion in roots and leaves. *J. Exp. Bot.* **45**, 1721–1731.

Topa, M. A. and McLeod, K. W. (1986). Aerenchyma and lenticel formation in pine seedlings: A possible avoidance mechanism to anaerobic growth conditions. *Physiol. Plant.* **68**, 540–550.

Tornroth-Horsefield, S., Wang, Y., Hedfalk, K., Johanson, U., Karlsson, M., Tajkhorshid, E., Neutze, R., and Kjellbom, P. (2006). Structural mechanism of plant aquaporin gating. *Nature (London)* **439**, 688–694.

Torrey, J. G. (1976). Root hormones and plant growth. *Annu. Rev. Plant Physiol.* **27**, 435–459.

Torrey, J. G., Fosket, D. E., and Hepler, P. K. (1971). Xylem formation: A paradigm of cytodifferentiation in higher plants. *Am. Sci.* **59**, 338–352.

Toumadje, A., Crane, J. C., and Kader, A. A. (1980). Respiration and ethylene production of the developing "Kerman" pistachio fruit. *HortScience* **15**, 725–727.

Tournaire-Roux, C., Sutka, M., Javot, H., Gout, E., Gerbeau, P., Luu, D. T., Bligny, R., and Maurel, C. (2003). Cytosolic pH regulates root water transport during anoxic stress through gating of aquaporins. *Nature* **425**, 393–397.

Townsend, A. M. (1977). Characteristics of red maple progenies from different geographic areas. *J. Am. Soc. Hortic. Sci.* **102**, 461–468.

Trainotti, L., Bonghi, C., Ziliotto, F., Zanin, D., Rasori, A., Casadoro, G., Ramina, A., and Tonutti, P. (2006). The use of microarray μPEACH1.0 to investigate transcriptome changes during transition from pre-climacteric to climacteric phase in peach fruit. *Plant Sci.* **170**, 606–613.

Trethewey, R. N. and Smith, A. M. (2000). Starch metabolism in leaves. In *Photosynthesis: Physiology and Metabolism* (R. C. Leegood, T. D. Sharkey, and S. von Caemmerer, eds.), pp. 205–231. Advances in Photosynthesis, Volume 9, Kluwer Academic Publishers, Dordrecht.

Trewavas, A. (2000). Signal perception and transduction. In *Biochemistry and Molecular Biology of Plants* (B. Buchanan, B. W. Gruissem, and R. L. Jones, eds.), pp. 930–985. Amer. Soc. Plant Physiol., Rockville, Maryland.

Trimble, G. R., Jr. and Seegrist, D. W. (1973). Epicormic branching on hardwood trees bordering forest openings. *U.S. For. Serv., Res. Pap. NE* **NE–261**.

Tripepi, R. P. and Mitchell, C. A. (1984). Stem hypoxia and root respiration of flooded maple and birch seedlings. *Physiol. Plant.* **60**, 567–571.

Trockenbrodt, M. (1995). Calcium oxalate crystals in the bark of *Quercus robur, Ulmus glabra, Populus tremula,* and *Betula pendula*. *Ann. Bot.* **75**, 281–284.

Troendle, C. A. (1988). Effect of partial cutting and thinning on the water balance of the subalpine forest. *U.S. For. Serv., Gen. Tech. Rep. INT* **INT–243**, pp. 108–116.

Tromp, J. (1970). Storage and mobilization of nitrogenous compounds in apple trees with special reference to arginine. In *Physiology of Tree Crops* (L. C. Luckwill and C. V. Cutting, eds.), pp. 143–159. Academic Press, London.

Tromp, J. (1983). Nutrient reserves of roots of fruit trees, in particular carbohydrates and nitrogen. *Plant Soil* **71**, 401–413.

Tromp, J. and Ovaa, J. C. (1990). Seasonal changes in the cytokinin composition of xylem sap of apple. *J. Plant Physiol.* **136**, 606–610.

Tromp, J. and Ovaa, J. C. (1994). Spring cytokinin composition of xylem sap of apple at two root temperatures. *Sci. Hortic.* **57**, 1–6.

Trumbore, S. (2006). Carbon respired by terrestrial ecosystems—Recent progress and challenges. *Global Change Biol.* **12**, 141–153.

Tschaplinski, T. J. and Blake, T. J. (1989). Water stress tolerance and late-season organic solute accumulation in hybrid poplar. *Can. J. Bot.* **67**, 1681–1688.

Tschaplinski, T. J. and Tuskan, G. A. (1994). Water-stress tolerance of black and eastern cottonwood clones and four hybrid progeny. 2. Metabolites and inorganic ions that constitute osmotic adjustment. *Can. J. Forest Res.* **24**, 681–687.

Tucker, G. A. and Grierson, D. (1987). Fruit ripening. In *The Biochemistry of Plants*. Vol. 12 (D. D. Davies, ed.), pp. 265–318. Academic Press, New York.

Tucker, G. F. and Emmingham, W. H. (1977). Morphological changes in leaves of residual western hemlock after clear and shelterwood cutting. *For. Sci.* **23**, 195–203.

Tukey, H. B. (1935). Growth of the embryo, seed, and pericarp of the sour cherry (*Prunus cerasus*) in relation to season of fruit ripening. *Proc. Am. Soc. Hortic. Sci.* **31**, 125–144.

Tukey, H. B., Jr. (1970a). The leaching of substances from plants. *Annu. Rev. Plant Physiol.* **21**, 305–324.

Tukey, H. B., Jr. (1970b). Leaching of metabolites from foliage and its implication in the tropical rain forest. In *A Tropical Rain Forest* (H. T. Odum, ed.), pp. H155–H160. U.S. Atomic Energy Commission, Washington, D.C.

Tukey, H. B., Jr. (1971). Leaching of substances from plants. In *Ecology of Leaf Surface Microorganisms* (T. F. Preece and L. H. Dickinson, eds.), pp. 67–80. Academic Press, London.

Tukey, H. B., Jr. (1980). Some effects of rain and mist on plants with implications for acid precipitation. In *Effects of Acid Precipitation on Terrestrial Ecosystems* (T. C. Hutchinson and M. Havas, eds.), pp. 141–150. Plenum Press, New York.

Tukey, H. B. and Young, J. O. (1942). Gross morphology and histology of developing fruit of the apple. *Bot. Gaz.* **104**, 3–25.

Tukey, H. B., Jr., Mecklenberg, R. A., and Morgan, J. V. (1965). A mechanism for leaching of metabolites from foliage. In *Isotopes and Radiation in Soil-Plant Nutrition Studies*, pp. 371–385. IAEA, Vienna.

Tukey, L. D. (1989). Growth factors and plant regulators in the manipulation of plant development and cropping in tree fruits. In *Manipulation of Fruiting* (C. J. Wright, ed.), pp. 343–361. Butterworth, London.

Tunstall, B. R. and Connor, D. J. (1975). Internal water balance of brigalow (Acacia harpophylla F. Muell.) under natural conditions. *Aust. J. Plant Physiol.* **2**, 489–499.

Tuomi, J., Niemala, P., and Mannila, R. (1982). Resource allocation in dwarf shoots of birch (*Betula pendula*): Reproduction and leaf growth. *New Phytol.* **91**, 483–487.

Tuominen, H., Puech, L., Fink, S., and Sundberg, B. (1997). A radial concentration gradient of indole-3-acetic acid is related to secondary xylem development in hybrid aspen. *Plant Physiol.* **115**, 577–585.

Turnbull, C. R. A., Beadle, C. L. West, P. W., and Cramer, R. N. (1994). Copper deficiency, a probable cause of stem deformity in fertilized *Eucalyptus nitens*. *Can. J. For. Res.* **24**, 1434–1439.

Turnbull, M. H., Whitehead, D., Tissue, D. T., Schuster, W. S. F., Brown K. J., and Griffin, K. L. (2003). Scaling foliar respiration in two contrasting forest canopies. *Funct. Ecol.* **17**, 101–114.

Turner, N. C. (1981). Techniques and experimental approaches for the measurement of plant water status. *Plant Soil* **58**, 339–366.

Turner, N. C. (1986). Adaptation to water deficits: A changing perspective. *Aust. J. Plant Physiol.* **13**, 175–190.

Turner, N. C. (1988). Measurement of plant water status by the pressure chamber technique. *Irrig. Sci.* **9**, 289–308.

Turner, N. C. and DeRoo, H. C. (1974). Hydration of eastern hemlock as influenced by waxing and weather. *For. Sci.* **20**, 19–24.

Turner, N. C. and Heichel, G. H. (1977). Stomatal development and seasonal changes in diffusive resistance of primary and regrowth foliage of red oak and red maple. *New Phytol.* **78**, 71–81.

Turner, N. C. and Jarvis, P. G. (1975). Photosynthesis in Sitka spruce (*Picea sitchensis* (Bong.) Carr.). IV. Response to soil temperature. *J. Appl. Ecol.* **12**, 561–576.

Turner, N. C., Schulze, E.-D., and Gollan, T. (1984). The responses of stomata and leaf gas exchange to vapour pressure deficits and soil water content. I. Species comparisons at high soil water contents. *Oecologia* **63**, 338–342.

Turner, N. C., Abbo, S., Berger, J. D., Chaturvedi, S. K., French, R. J., Ludwig, C., Mannur, D. M., Singh, S. J., and Yadava, H. S. (2007). Osmotic adjustment in chickpea (*Cicer arietinum* L.) results in no yield benefit under terminal drought. *J. Exp. Bot.* **58**, 187–194.

Turvey, N. D. (1984). Copper deficiency in *Pinus radiata* planted in a podsol in Victoria, Australia. *Plant Soil* **77**, 73–86.

Turvey, N. D., Carlyle, C., and Downes, G. M. (1992). Effects of micronutrients on the growth form of two families of *Pinus radiata* (D. Don) seedlings. *Plant and Soil* **139**, 59–65.

Tuskan, G. A., DiFazio, S., Jansson, S., Bohlmann, J., Grigoriev, I., Hellsten, U., Putnam, N., Ralph, S., Rombauts, S., Salamov, A., Schein, J., Sterck, L., Aerts, A., Bhalerao, R. R., Bhalerao, R. P., Blaudez, D., Boerjan, W., Brun, A., Brunner, A., Busov, V., Campbell, M., Carlson, J., Chalot, M., Chapman, J., Chen, G. L., and et al. (2006). The genome of black cottonwood, *Populus trichocarpa* (Torr. & Gray). *Science* **313**, 1596–1604.

Tyree, M. T. (1983). Maple sap uptake, exudation, and pressure changes correlated with freezing exotherms and thawing endotherms. *Plant Physiol.* **73**, 277–285.

Tyree, M. T. (1988). A dynamic model for water flow in a single tree: Evidence that models must account for hydraulic architecture. *Tree Physiol.* **4**, 195–217.

Tyree, M. T. and Ewers, F. W. (1991). The hydraulic architecture of trees and other woody plants. *New Phytol.* **119**, 345–360.

Tyree, M. T. and Jarvis, P. G. (1982). Water in tissues and cells. *Encycl. Plant Physiol. N. S.* **12B**, 35–77.

Tyree, M. T. and Sperry, J. S. (1988). Do woody plants operate near the point of catastrophic xylem dysfunction caused by dynamic water stress? Answers from a model. *Plant Physiol.* **88**, 574–580.

Tyree, M. T. and Sperry, J. S. (1989). Vulnerability of xylem to cavitation and embolism. *Annu. Rev. Plant Physiol. Plant Mol. Biol.* **40**, 19–38.

Tyree, M. T. and Yang, S. D. (1990). Water-storage capacity of *Thuja*, *Tsuga* and *Acer* stems measured by dehydration isotherms—The contribution of capillary water and cavitation. *Planta* **182**, 420–426.

Tyree, M. T. and Yang, S. D. (1992). Hydraulic conductivity recovery versus water pressure in xylem of *Acer saccharum*. *Plant Physiol.* **100**, 669–676.

Tyree, M. T. and Zimmermann, M. H. (2002). *Xylem Structure and the Ascent of Sap*. 2nd ed. Springer, Berlin and New York.

Tyree, M. T., Cheung, Y. N. S., MacGregor, M. E., and Talbot, A. J. B. (1978). The characteristics of seasonal and ontogenetic changes in the tissue-water relations of *Acer, Tsuga* and *Picea*. *Can. J. Bot.* **56,** 635–647.

Tyree, M. T., Graham, M. E. D., Cooper, K. E., and Bazos, L. J. (1983). The hydraulic architecture of *Thuja occidentalis*. *Can. J. Bot.* **61**, 2105.

Tyree, M. T., Snyderman, D. A., Wilmot, T. R., and Machado, J. L. (1991). Water relations and hydraulic architecture of a tropical tree (*Schefflera morototoni*)—Data, models, and a comparison with two temperate species (*Acer saccharum* and *Thuja occidentalis*). *Plant Physiol.* **96**, 1105–1113.

Tyree, M. T., Davis, S. D., and Cochard, H. (1994). Biophysical perspectives of xylem evolution. Is there a tradeoff of hydraulic efficiency for vulnerability to dysfunction? *IAWA J.* **15**, 335–360.

Uehlein, N. and Kaldenhoff, R. (2006). Aquaporins and biological rhythm. *Biol. Rhythm Res.* **37,** 315–322.

Uehlein, N., Lovisolo, C., Siefritz, F., and Kaldenhoff, R. (2003). The tobacco aquaporin NtAQP1 is a membrane CO_2 pore with physiological functions. *Nature* (*London*) **425**, 734–737.

Uggla, C., Mellerowicz, E. J., and Sundberg, B. (1998). Indole-3-acetic acid controls cambial growth in Scots pine by positional signaling. *Plant Physiol.* **117**, 113–121.

Uggla, C., Moritz, T., Sandberg, G., and Sundberg, B. (1996). Auxin as a positional signal in pattern formation in plants. *Proc. Natl. Acad. Sci. U.S.A.* **93**, 9282–9286.

Uhl, C., Clarck, K., Dezzeo, N., and Maquirino, P. (1988). Vegetation dynamics in Amazonian treefall gaps. *Ecology* **69**, 751–763.

Updegraff, K., Pastor, J., Bridgham, S. D., and Johnston, C. A. (1995). Environmental and substrate controls over carbon and nitrogen mineralization in northern wetlands. *Ecol. Appl.* **5**, 151–163.

Uren, N. C. and Reisenauer, H. M. (1986). The role of root exudates in nutrient acquisition. *Adv. Plant Nutr.* **3**, 79–114.

Uritani, I. (1976). Protein metabolism. *Encycl. Plant Physiol. N. S.* **4**, 509–525.

Uriu, K., Davenport, D., and Hagan, R. M. (1975). Preharvest antitranspirant spray on cherries. I. Effect on fruit size. II. Postharvest fruit benefits. *Calif. Agric.* **29**, 7–11.

Ursprung, A. (1915). *Über die Kohäsion des Wassers in Farnanulus. Ber. Dtsch. Bot. Ges.* **33**, 153–162.

Utkhede, R. S. (2006). Soil sickness, replant problem or replant disease and its integrated control. *Allelopath. J.* **18**, 23–38.

Valentini, R., Matteucci, G., Dolman, A. J., Schulze, E. D., Rebmann, C., Moors, E. J., Granier, A., Gross, P. et al. (2000). Respiration as the main determinant of carbon balance in European forests. *Nature* (*London*) **404**, 861–865.

Valladares, F. and Pearcy, R. W. (1997). Interactions between water stress, sun-shade acclimation, heat tolerance and photoinhibition in the sclerophyll *Heteromeles arbutifolia*. *Plant, Cell Environ.* **20**, 25–36.

Van Buijtenen, J. P., Bilan, M. V., and Zimmerman, R. H. (1976). Morphophysiological characteristics related to drought resistance in *Pinus taeda* L. In *Tree Physiology and Yield Improvement* (M. G. Cannell and F. T. Last, eds.), pp. 349–359. Academic Press, London.

Van Cleve, B. and Appel, K. (1993). Induction by nitrogen and low temperature of storage-protein synthesis in poplar trees exposed to long days. *Planta* **189**, 157–160.

Van Cleve, K., Oliver, L., Schlentnee, R., Viereck, L. A., and Dyrness, C. T. (1983). Production and nutrient cycling in taiga forest ecosystems. *Can. J. For. Res.* **13**, 747–766.

Van Cleve, K., Oechel, W. C., and Hom, J. L. (1990). Response of black spruce (*Picea mariana*) ecosystems to soil temperature modification in interior Alaska. *Can. J. For. Res.* **20**, 1530–1535.

van den Driessche, R. (1979). Respiration rate of cold-stored nursery stock. *Can. J. For. Res.* **9**, 15–18.

van den Driessche, R. (1984). Nutrient storage, retranslocation and relationship of stress to nutrition. In *Nutrition of Plantation Forests* (G. D. Bowen and E. K. S. Nambiar, eds.), pp. 181–209. Academic Press, London.

Van Die, J. and Tammes, P. M. L. (1975). Phloem exudation from monocotyledonous axes. In *Transport in Plants. I. Phloem Transport* (M. H. Zimmermann and J. A. Milburn, eds.), pp. 196–222. Springer-Verlag, Berlin, Heidelberg and New York.

van Hove, C. (1976). Bacterial leaf symbiosis and nitrogen fixation. In *Symbiotic Nitrogen Fixation in Plants* (P. S. Nutman, ed.), pp. 551–560. Cambridge University Press, London.

Van Hove, L. W. A., van Kooten, O., van Wijk, K. J., Vredenberg, W. J., Adema, E. H., and Pieters, G. A. (1991). Physological effects of long term exposure to low concentrations of SO_2 and NH_3 on poplar leaves. *Physiol. Plant.* **82**, 32–40.

Van Oosten, J. J., Wilkins, D., and Besford, R. T. (1994). Regulation of the expression of photosynthetic nuclear genes by CO_2 is mimicked by regulation by carbohydrates: A mechanism for the acclimation of photosynthesis to high CO_2? *Plant, Cell Environ.* **17**, 913–923.

Van Rees, K. C. J. and Comerford, N. B. (1990). The role of woody roots of slash pine seedlings in water and potassium absorption. *Can. J. For. Res.* **20**, 1183–1191.

Van Staden, J. (1976). Seasonal changes in the cytokinin content of *Ginkgo biloba* leaves. *Physiol. Plant.* **38**, 1–5.

Van Staden, J. (1977). Seasonal changes in the cytokinin content of the leaves of *Salix babylonica*. *Physiol. Plant.* **40**, 296–299.

Van Staden, J., Hutton, M. J., and Drewes, S. E. (1983). Cytokinins in the leaves of *Ginkgo biloba*. I. The complex in mature leaves. *Plant Physiol.* **73**, 223–227.

Varnell, L. M. (1998). The relationship between inundation history and baldcypress stem form in a Virginia floodplain swamp. *Wetlands* **18**, 176–183.

Vaughan, M. A. (1988). Is ribulose bisphosphate carboxylase present in guard cell chloroplasts? *Physiol. Plant* **74**, 409–413.

Vegis, A. (1964). Dormancy in higher plants. *Annu. Rev. Plant Physiol.* **15**, 185–224.

Venator, C. R. (1976). Natural selection for drought resistance in *Picea caribaea* Morelet. *Turrialba* **26**, 381–387.

Verhoeven, A. S., Adams, W. W., Demmig-Adams, B., Croce, R., and Bassi, R. (1999). Xanthophyll cycle pigment localization and dynamics during exposure to low temperatures and light stress in *Vinca major. Plant Physiol.* **120**, 727–737.

Véry, A. A. and Sentenac, H. (2003). Molecular mechanisms and regulation of K^+ transport in higher plants. *Annu. Rev. Plant Biol.* **54**, 575–603.

Vessey, J. K., Pawlowski, K., and Bergman, B. (2004). Root-based N_2-fixing symbioses: Legumes, actinorhizal plants, *Parasponia* sp. and cycads. *Plant Soil* **266**, 205–230.

Vierling, E. (1990). Heat shock protein function and expression in plants. In *Stress Responses in Plants: Adaptation and Acclimation Mechanisms* (R. G. Alscher and J. R. Cumming, eds.), pp. 357–375. Wiley, New York.

Viets, F. G., Jr. (1972). Water deficits and nutrient availability. In *Water Deficits and Plant Growth* (T. T. Kozlowski, ed.), pp. 217–239. Academic Press, New York.

Viro, P. J. (1974). Effects of fire on soil. In *Fire and Ecosystems* (T. T. Kozlowski and C. E. Ahlgren, eds.), pp. 39–45. Academic Press, New York.

Virtanen, A. I. (1957). Investigations on nitrogen fixation by the alder. II. Associated culture of spruce and inoculated alder without combined nitrogen. *Physiol. Plant.* **10**, 164–169.

Vischer, W. (1923). *Über die Konstanz anatomischer und physiologischer Eigenschaften von* Hevea brasiliensis *Müller Arg. (Euphorbiaceae). Verh. Naturf. Ges. Basel* **35**, 174–185.

Vissenberg, K., Fry, S. C., and Verbelen, J. P. (2001). Root hair initiation is coupled to a highly localized increase of xyloglucan endotransglycosylase action in *Arabidopsis* roots. *Plant Physiol.* **127**, 1125–1135.

Vitousek, P. M. and Melillo, J. M. (1979). Nitrate losses from disturbed forests: Patterns and mechanisms. *For. Sci.* **25**, 605–619.

Vogelmann, H. W. (1982). Fog precipitation in the cloud forests of eastern Mexico. *BioScience* **23**, 96–100.

Vogelmann, T. C. (1993). Plant tissue optics. *Annu. Rev. Plant Physiol. Plant Molec. Biol.* **44**, 231–251.

Vogelmann, T. C. and Martin, G. (1993). The functional significance of palisade tissue—Penetration of directional versus diffuse light. *Plant, Cell Environ.* **16**, 65–72.

Vogelmann, T. C., Dickson, R. E., and Larson, P. R. (1985). Comparative distribution and metabolism of xylem-borne amino compounds and sucrose in shoots of *Populus deltoides. Plant Physiol.* **77**, 418–428.

Vogelmann, T. C., Knapp, A. K., McClean, T. M., and Smith, W. K. (1988). Measurement of light within thin plant tissues with fiber optic microprobes. *Physiol. Plant.* **72**, 623–630.

Vogelmann, T. C., Nishio, J. N., and Smith, W. K. (1996). Leaves and light capture—Light propagation and gradients of carbon fixation within leaves. *Trends Plant Sci.* **1**, 65–70.

Vogt, K. A., Grier, C. C., and Vogt, D. J. (1986). Production, turnover and nutrient dynamics of above- and below-ground detritus of world forests. *Adv. Ecol. Res.* **15**, 303–377.

von Fircks, Y., Ericsson, T., and Sennerby-Forsse, L. (2001). Seasonal variation of macronutrients in leaves, stems and roots of *Salix dasyclados* Wimm. grown at two nutrient levels. *Biomass Bioenerg.* **21**, 321–334.

von Caemmerer, S. and Farquhar, G. D. (1981). Some relationships between the biochemistry of photosynthesis and the gas exchange of leaves. *Planta* **153**, 376–387.

von Höhnel F. (1881). Über den Wasserverbrauch der Holzegewächse mit Beziehung auf die meteorologischen Faktoren. *Wollny's Forsch. Geb. Agr.-Phys.* **4**, 435–445.

von Höhnel, F. (1884). Über das Wasserbedürfniss der Walder. *Zentralbl. der Pflanzenzelle* **10**, 384–409.

Von Rudloff, E. (1972). Seasonal variation in the composition of the volatile oil of the leaves, buds, and twigs of the white spruce (*Picea glauca*). *Can. J. Bot.* **50**, 1595–1603.

Von Rudloff, E. (1975). Seasonal variation of the terpenes of the leaves, buds, and twigs of blue spruce (*Picea pungens*). *Can. J. Bot.* **53**, 2978–2982.

Voronkov, V. V. (1956). The dying off of the feeder root system in the tea plant. *Dokl. Vses. Akad. Seljsk. Nauk* **21**, 22–24; *Hortic. Abstr.* **27**, 2977 (1957).

Vose, J. M. (1988). Patterns of leaf area distribution within crowns of nitrogen- and phosphorus-fertilized loblolly pine trees. *For. Sci.* **34**, 564–573.

Vose, J. M. and Allen, H. L. (1988). Leaf area, stemwood growth and nutrition relationships in loblolly pine. *For. Sci.* **34**, 547–563.

Wadleigh, C. H., Gauch, H. G., and Magistad, O. C. (1946). Growth and rubber accumulation in guayule as conditioned by soil salinity and irrigation regime. *U.S., Dept. Agric., Tech. Bull.* **925**, 1–34.

Waggoner, P. E. and Bravdo, B. (1967). Stomata and the hydrologic cycle. *Proc. Natl. Acad. Sci. U.S.A.* **57**, 1096–1102.

Waggoner, P. E. and Reifsnyder, W. E. (1968). Simulation of the temperature, humidity, and evaporation profiles in a leaf canopy. *J Appl. Meterol.* **7**, 400–409.

Waggoner, P. E. and Turner, N. C. (1971). Transpiration and its control by stomata in a pine forest. *Bull. Conn. Agric. Expt. Sta. No.* 726.

Waisel, Y., Noah, I., and Fahn, A. (1966). Cambial activity in *Eucalyptus camaldulensis* Dehn. II. The production of phloem and xylem elements. *New Phytol.* **65**, 319–324.

Waisel, Y., Borger, G. H., and Kozlowski, T. T. (1969). Effects of phenylmercuric acetate on stomatal movement and transpiration of excised *Betula papyrifera* leaves. *Plant Physiol.* **44**, 685–690.

Wakabayashi, K., Hoson, T., and Kamisaka, S. (1997). Abscisic acid suppresses the increases in cell wall-bound ferulic and diferulic acid levels in dark-grown wheat (*Triticum aestivum* L.) coleoptiles. *Plant Cell Physiol.* **38**, 811–817.

Wakeley, P. C. and Marrero, J. (1958). Five-year intercept as site index in southern pine plantations. *J. For.* **56**, 332–336.

Walker, R. B. (1991). Measuring mineral nutrient utilization. In *Techniques and Approaches in Forest Tree Ecophysiology* (J. P. Lassoie and T. M. Hinckley, eds.), pp. 183–206. CRC Press, Boca Raton, Florida.

Walker, R. F., West, D. C., McLaughlin, S. B., and Amundsen, C. C. (1989). Growth, xylem pressure potential, and nutrient absorption of loblolly pine on a reclaimed surface mine as affected by an induced *Pisolithus tinctorius* infection. *For. Sci.* **35**, 569–581.

Walker, R. R., Törökfalvy, E., Scott, N. E., and Kriedemann, P. E. (1981). An analysis of photosynthetic response to salt treatment in *Vitis vinifera. Aust. J. Plant Physiol.* **8**, 359–374.

Walker, R. R., Törökfalvy, E., and Downton, W. J. S. (1982). Photosynthetic responses of the citrus varieties Rangpur lime and Etrog citron to salt treatment. *Aust. J. Plant Physiol.* **9**, 783–790.

Walser, R. H., Walker, D. R., and Seeley, S. D. (1981). Effect of temperature, fall defoliation, and gibberellic acid on the rest period of peach leaf buds. *J. Am. Soc. Hort. Sci.* **106**, 91–94.

Walter, J., Charon, J., Marpeau, A., and Launay, J. (1989). Effect of wounding on the terpene content of twigs of maritime pine (*Pinus pinaster* Ait.). I. Changes in the concentration of diterpene resin acids and ultrastructural modifications of the resin duct epithelial cells following injuries. *Trees* **4**, 210–219.

Walters, M. B. and Reich, P. B. (1989). Response of *Ulmus americana* seedlings to varying nitrogen and water status. I. Photosynthesis and growth. *Tree Physiol.* **5**, 159–172.

Wan, X. C., Landhausser, S. M., Zwiazek, J. J., and Lieffers, V. J. (1999). Root water flow and growth of aspen (*Populus tremuloides*) at low root temperatures. *Tree Physiol.* **19**, 879–884.

Wan, X. C., Landhausser, S. M., Zwiazek, J. J., and Lieffers, V. J. (2004). Stomatal conductance and xylem sap properties of aspen (*Populus tremuloides*) in response to low soil temperature. *Physiol. Plant.* **122**, 79–85.

Wang, C. Y. (1982). Physiological and biochemical responses of plants to chilling stress. *HortScience* **17**, 173–186.

Wang, H., Inukai, Y., and Yamauchi, A. (2006). Root development and nutrient uptake. *Crit. Rev. Plant Sci.* **25**, 279–301.

Wang, S. Y., Faust, M., and Steffens, G. L. (1985). Metabolic changes in cherry flower buds associated with breaking of dormancy in early and late blooming cultivars. *Physiol. Plant.* **65**, 89–94.

Wang, X. M. (2004). Lipid signaling. *Curr. Opin. Plant Biol.* **7**, 329–336.

Wang, Z. and Stutte, G. W. (1992). The role of carbohydrates in active osmotic adjustment in apple under water stress. *J. Am. Soc. Hortic. Sci.* **117**, 816–823.

Wang, Z., Quebedeaux, B., and Stutte, G. W. (1995). Osmotic adjustment—Effect of water stress on carbohydrates in leaves, stems and roots of apple. *Aust. J. Plant Phys.* **22**, 747–754.

Ward, D. A. and Bunce, J. A. (1986). Responses of net photosynthesis and conductance to independent changes in the humidity environments of the upper and lower surfaces of leaves of sunflower and soybean. *J. Exp. Bot.* **37**, 1842–1853.

Ward, W. W. (1966). Epicormic branching of black and white oaks. *For. Sci.* **12**, 290–296.

Wardrop, A. B. (1962). Cell wall organization in higher plants. I. The primary wall. *Bot. Rev.* **28**, 241–285.

Wareing, P. F., Hanney, C. E. A., and Digby, J. (1964). The role of endogenous hormones in cambial activity and xylem differentiation. In *The Formation of Wood in Forest Trees* (M. H. Zimmermann, ed.), pp. 323–344. Academic Press, New York.

Wargo, P. M. (1976). Variation of starch content among and within roots of red and white oak trees. *For. Sci.* **22**, 468–471.

Waring, R. H. (1983). Estimating forest growth and efficiency in relation to canopy leaf area. *Adv. Ecol. Res.* **13**, 327–354.

Waring, R. H. (1985). Imbalanced forest ecosystems: Assessments and consequences. *For. Ecol. Manage.* **12**, 93–112.

Waring, R. H. (1987). Characteristics of trees predisposed to die. *BioScience* **37**, 569–574.

Waring, R. H. and Franklin, J. F. (1979). Evergreen coniferous forests of the Pacific Northwest. *Science* **204**, 1380–1386.

Waring, R. H. and Running, S. W. (1978). Sapwood water storage: Its contribution to transpiration and effect upon water conductance through the stems of old-growth Douglas fir. *Plant, Cell Environ.* **1**, 131–140.

Waring, R. H. and Running, S. W. (1998). *Forest Ecosystems, 2nd ed.* Academic Press, San Diego.

Waring, R. H. and Schlesinger, W. H. (1985). *Forest Ecosystems: Concepts and Management*. Academic Press, Orlando, Florida.

Waring, R. H., Whitehead, D., and Jarvis, P. G. (1979). The contribution of stored water to transpiration in Scots pine. *Plant, Cell Environ.* **2**, 309–317.

Waring, R. H., Newman, K., and Bell, J. (1981). Efficiency of tree crowns and stemwood production at different canopy leaf densities. *Forestry* **54**, 129–137.

Warren, C. R., McGrath, J. F, and Adams, M. A. (2001). Water availability and carbon isotope discrimination in conifers. *Oecologia* **127**, 476–486.

Warren, J. M., Meinzer, F. C., Brooks, J. R., and Domec, J. C. (2005). Vertical stratification of soil water storage and release dynamics in Pacific Northwest coniferous forests. *Agric. For. Meteorol.* **130**, 39–58.

Warren, J. M., Meinzer, F. C., Brooks, J. R., Domec, J. C., and Coulombe, R. (2007). Hydraulic redistribution of soil water in two old-growth coniferous forests: Quantifying patterns and controls. *New Phytol.* **173**, 753–765.

Wartinger, A., Heilmeier, H., Hartung, W., and Schulze, E.-D. (1990). Daily and seasonal courses of leaf conductance and abscisic acid in the xylem sap of almond trees [*Prunus dulcis* (Miller) D. A. Webb] under desert conditions. *New Phytol.* **116**, 581–587.

Watada, A. E., Herner, R. C., Kader, A. A., Romani, R. J., and Staby, G. L. (1984). Terminology for the description of developmental stages of horticultural crops. *HortScience* **19**, 20–21.

Watanabe, T., Noguchi, T., Kuriyama, H., Kadota, M., Takatsuto, S., and Kamuro, Y. (1998). Effects of brassinosteroid compound (TS303) on fruit-setting, fruit-growth taking roots and cold-resistance. *Acta Horticulturae* 267–271.

Watling, J. R., Robinson, S. A., Woodrow, I. E., and Osmond, C. B. (1997). Responses of rainforest understorey plants to excess light during sunflecks. *Aust. J. Plant Physiol.* **24**, 17–25.

Watmough, S. A., Aherne, J., Alewell, C., Arp, P., Bailey, S., Clair, T., Dillon, P., Duchesne, L. et al. (2005). Sulphate, nitrogen and base cation budgets at 21 forested catchments in Canada, the United States and Europe. *Environ. Monit. Assess.* **109**, 1–36.

Watson, R. T. and The Core Writing Team, eds. (2001). *Climate Change 2001: Synthesis Report*. Cambridge University Press, Cambridge and New York.

Watts, J. E. and DeVilliers, O. T. (1980). Seasonal changes in the sorbitol, sugar, and starch content of Packhams's Triumph pear trees. *South. Afr. J. Sci.* **76**, 276–277.

Weatherly, P. E. (1950). Studies in the water relations of the cotton plant. 1. The field measurements of water deficits in leaves. *New Phytol.* **49**, 81–97.

Weatherley, P. E. (1976). Introduction: Water movement through plants. *Phil. Trans. Roy Soc. Lond.* **B273**, 435–444.

Weaver, J. E. and Mogensen, A. (1919). Relative transpiration of conifers and broadleaved trees in autumn and winter. *Bot. Gaz.* **68**, 393–424.

Weaver, R. F. (2005). *Molecular Biology, 3rd ed.* McGraw-Hill, Boston.

Webb, W., Szarek, S., Lauenroth, W., Kinerson, R., and Smith, M. (1978). Primary productivity and water use in native forest, grassland and desert ecosystems. *Ecology* **59**, 1239–1247.

Webster, D. H. and Brown, G. L. (1980). Trunk growth of apple tree as affected by crop load. *Can. J. Plant Sci.* **60**, 1383–1391.

Wedding, R. T., Reihl, L. A., and Rhoads, W. A. (1952). Effect of petroleum oil spray on photosynthesis and respiration in citrus leaves. *Plant Physiol.* **27**, 269–278.

Weete, J. D., Leek, G. L., Peterson, C. M., Currie, H. E., and Branch, W. D. (1978). Lipid and surface wax synthesis in water stressed cotton leaves. *Plant Physiol.* **62**, 675–677.

Weetman, G. F. and Algar, D. (1983). Low-site class black spruce and jack pine nutrient removals after full-tree and tree length logging. *Can. J. For. Res.* **13**, 1030–1036.

Weichmann, J. (1986). The effect of controlled-atmosphere storage on the sensory and nutritional quality of fruits and vegetables. *Hort. Rev.* **8**, 101–127.

Weinstein, L. H. (1977). Fluoride and plant life. *J. Occupational Medicine* **19**, 49–78.

Wei, C. F., Tyree, M. T., and Steudle, E. (1999a). Direct measurement of xylem pressure in leaves of intact maize plants. A test of the cohesion-tension theory taking hydraulic architecture into consideration. *Plant Physiol.* **121**, 1191–1205.

Wei, C., Steudle, E., and Tyree, M. T. (1999b). Water ascent in plants: Do ongoing controversies have a sound basis? *Trends Plant, Sci.* **4**, 371–375.

Wei, C., Steudle, E., Tyree, M. T., and Linthlac, P. M. (2001). The essentials of direct xylem pressure measurement. *Plant, Cell Environ.* **24**, 549–555.

Weis, K. G., Webster, B. D., Goren, R., and Martin, G. C. (1991). Inflorescence abscission in olive: Anatomy and histochemistry in response to ethylene and ethephon. *Bot. Gaz.* **152**, 51–58.

Wells, C. G. (1968). Techniques and standards for foliar diagnosis of N deficiency in loblolly pine. In *Forest Fertilization: Theory and Practice*, Tenn. Valley Auth., Muscle Shoals, Alabama. pp. 72–76.

Wells, C. G., Jorgensen, J. R., and Burnette, C. E. (1975). Biomass and mineral elements in a thinned loblolly pine plantation at age 16. *U.S. For. Serv. Res. Pap. SE* **SE–126.**

Wellwood, R. W. (1955). Sapwood-heartwood relationships in second growth Douglas-fir. *For. Prod. J.* **5**, 108–111.

Wenger, K. F. (1953). The sprouting of sweetgum in relation to season of cutting and carbohydrate content. *Plant Physiol.* **28**, 35–49.

Wenham, M. W. and Cusick, F. (1975). The growth of secondary wood fibers. *New Phytol.* **74**, 247–271.

Wenkert, W. (1983). Water transport and balance within the plant: An overview. In *Limitations to Efficient Water Use in Crop Production* (H. M. Taylor, W. R. Jordan, and T. R. Sinclair, eds.), pp. 137–172. Am. Soc. Agron., Madison, Wisconsin.

Went, F. W. (1975). Water vapor absorption in *Prosopis*. In *Physiological Adaptation to the Environment* (F. J. Vernberg, ed.), pp. 65–75. Intext Educational Publications, New York.

Went, F. W. and Babu, V. R. (1978). The effect of dew on plant water balance in *Citrullus vulgaris* and *Cucumis melo*. *Physiol. Plant.* **44**, 307–311.

Went, F. W. and Stark, N. (1968). The biological and mechanical role of soil fungi. *Proc. Natl. Acad. Sci. U.S.A.* **60**, 497–504.

Westwood, M. N. (1978). *Temperate-Zone Pomology*. Freeman, San Francisco and California.

Wetzel, S. and Greenwood, J. S. (1991). A survey of seasonal bark proteins in light temperate hardwoods. *Trees* **5**, 153–157.

Wetzel, S., Demmers, C., and Greenwood, J. S. (1989a). Seasonally fluctuating bark proteins are a potential form of nitrogen storage in 3 temperate hardwoods. *Planta* **178**, 275–281.

Wetzel, S., Demmers, C., and Greenwood, J. S. (1989b). Spherical organelles analogous to seed protein bodies fluctuate seasonally in parenchymatous cells of hardwoods. *Can. J. Bot.* **67**, 3439–3445.

Weyers, J. and Meidner, H. (1990). *Methods in Stomatal Research*. Longman Scientific and Technical, Essex, U.K.

Wheeler, J. K., Sperry, J. S., Hacke, U. G., and Hoang, N. (2005). Intervessel pitting and cavitation in woody Rosaceae and other vesselled plants: A basis for a safety versus efficiency trade-off in xylem transport. *Plant, Cell Environ.* **28**, 800–812.

Wheldale, M. (1916). *The Anthocyanin Pigments of Plants*. Cambridge University Press, Cambridge.

White, C. S. and Gosz, J. R. (1987). Factors controlling nitrogen mineralization and nitrification in forest ecosystems in New Mexico. *Biol. Fertil. Soils* **5**, 195–202.

White, D. A., Beadle, C. L., and Worledge, D. (1996). Leaf water relations of *Eucalyptus globulus* spp. *globulus* and *E. nitens*—Seasonal, drought and species effects. *Tree Physiol.* **16**, 469–476.

White, D. L., Haines, B. L., and Boring, L. R. (1988). Litter decomposition in southern Appalachian black locust and pine-hardwood stands: Litter quality and nitrogen dynamics. *Can. J. For. Res.* **18**, 54–63.

White, E. E. and Nilsson, J.-E. (1984). Foliar terpene heritability in *Pinus contorta*. *Silvae Genet.* **33**, 16–22.

White, J. W. C. (1989). Stable hydrogen isotope ratios in plants: A review of current theory and some potential applications. In *Stable Isotopes in Ecological Research* (P. W. Rundel, J. R. Ehleringer, and K. A. Nagy, eds.), pp. 142–162. Springer-Verlag, New York.

White, J. W. C., Cook, E. R., Lawrence, J. R., and Broecker, W. S. (1985). The D/H ratios of sap in trees: Implications for water sources and tree ring D/H ratios. *Geochim. Cosmochim. Acta* **49**, 237–246.

Whitehead, D. R. (1984). Wind pollination: Some ecological and evolutionary perspectives. In *Pollination Biology* (L. Real, ed.), pp. 97–108. Academic Press, Orlando, Florida.

Whitehead, D. R. and Hinckley, T. M. (1991). Models of water flux through forest stands. Critical leaf and stand parameters. *Tree Physiol.* **9**, 35–57.

Whitehead, D. R., and Jarvis, P. G. (1981). Coniferous forest plantations. In *Water Deficits and Plant Growth* (T. T. Kozlowski, ed.), Vol. 6, pp. 49–152. Academic Press, New York.

Whitfield, C. J. (1932). Ecological aspects of transpiration. II. Pikes Peak and Santa Barbara regions: Edaphic and climatic aspects. *Bot. Gaz.* **94**, 183–196.

Whitford, L. A. (1956). A theory on the formation of cypress knees. *J. Elisha Mitchell Sci. Soc.* **72**, 80–83.

Whitmore, F. W. and Zahner, R. (1966). Development of the xylem ring in stems of young red pine trees. *For. Sci.* **12**, 198–210.

Whitmore, T. C. (1966). The social status of *Agathis* in a rain forest in Melanesia. *J. Ecol.* **54**, 285–301.

Whitmore, T. C. (1984). *Tropical Rain Forests of the Far East*. Oxford Univ. Press (Clarendon), London and New York.

Whitney, S. E. C., Gidley, M. J., and McQueen-Mason, S.J. (2000). Probing expansin action using cellulose/hemicellulose composites. *Plant J.* **22**, 327–334.

Whittaker, R. H. and Woodwell, G. M. (1971). Measurement of net primary production of forests. In *Productivity of Forest Ecosystems* (P. Duvigneaud, ed.), pp. 159–175. UNESCO, Paris.

Wieckowski, S. (1958). Studies on the autumnal breakdown of chlorophyll. *Acta Biol. Cracov., Ser. Bot.* **1**, 131–135.

Wieler, A. (1893). *Das Bluten der Pflanzen. Beitr. Biol. Pflanz.* **6**, 1–211.

Wiggans, C. C. (1936). The effect of orchard plants on subsoil moisture. *Proc. Am. Soc. Hortic. Sci.* **33**, 103–107.

Wiggans, C. C. (1937). Some further observations on the depletion of subsoil moisture by apple trees. *Proc. Am. Soc. Hortic. Sci.* **34**, 160–163.

Wightman, F. and Lighty, D. G. (1982). Identification of phenylacetic acid as a natural auxin in the shoots of higher plants. *Physiol. Plant.* **55**, 17–24.

Wilkinson, S. and Davies, W. J. (2002). ABA-based chemical signalling: The co-ordination of responses to stress in plants. *Plant, Cell Environ.* **25**, 195–210.

Wilcox, H. (1954). Primary organization of active and dormant roots of noble fir, *Abies procera*. *Am. J. Bot.* **41**, 812–821.

Wilcox, H. (1962). Cambial growth characteristics. In *Tree Growth* (T. T. Kozlowski, ed.), pp. 57–88. Ronald, New York.

Wild, A. (1989). *Russell's Soil Conditions and Plant Growth*. Wiley, New York.

Wildman, H. G. and Parkinson, D. (1981). Seasonal changes in water-soluble carbohydrates of *Populus tremuloides* leaves. *Can. J. Bot.* **59**, 862–869.

Williams, J. E., Davis, S. D., and Portwood, K. (1997). Xylem embolism in seedlings and resprouts of *Adenostoma fasciculatum* after fire. *Aust. J. Bot.* **45**, 291–300.

Williamson, M. J. (1966). Premature abscissions and white oak acorn crops. *For. Sci.* **12**, 19–21.

Wilm, H. G. and Dunford, E. G. (1948). Effect of timber cutting on water available for stream flow from a lodgepole pine forest. *U.S. Dep. Agric. Tech. Bull.* 968.

Wilson, B. F. (1963). Increase in cell surface area enlargement of cambial derivatives in *Abies concolor*. *Am. J. Bot.* **50**, 95–102.

Wilson, B. F. (1964). Structure and growth of woody roots of *Acer rubrum* L. *Harvard For. Pap.* No. 11.

Wilson, B. F. (1967). Root growth around barriers. *Bot. Gaz.* **128**, 79–82.

Wilson, B. F. (1975). Distribution of secondary thickening in tree root systems. In *The Development and Function of Roots* (J. G. Torrey and D. T. Clarkson, eds.), pp. 197–219. Academic Press, New York.

Wilson, B. F., Wodzicki, T., and Zahner, R. (1966). Differentiation of cambial derivatives: Proposed terminology. *For. Sci.* **12**, 438–440.

Wilson, C. C. (1948). Fog and atmospheric carbon dioxide as related to apparent photosynthetic rate of some broadleaf evergreens. *Ecology* **29**, 507–508.

Wilson, E. O. (1989). Threats to biodiversity. *Sci. Amer.* **261**, 108–116.

Wilson, K. B., Baldocchi, D. D., and Hanson, P. J. (2000a). Quantifying stomatal and non-stomatal limitations to carbon assimilation resulting from leaf aging and drought in mature deciduous tree species. *Tree Physiol.* **20**, 787–797.

Wilson, K. B., Baldocchi, D. D., and Hanson, P. J. (2000b). Seasonal and spatial variability of photosynthetic parameters and their relationship to leaf nitrogen in a deciduous forest. *Tree Physiol.* **20**, 565–578.

Winget, C. H. and Kozlowski, T. T. (1965). Seasonal basal area growth as an expression of competition in northern hardwoods. *Ecology* **46**, 786–793.

Winter, K. (1981). C_4 plants of high biomass in arid regions of Asia—Occurrence of C_4 photosynthesis in Chenopodiaceae and Polygonaceae from the Middle East. *Oecologia* **48**, 100–106.

Wise, L. E. and Jahn, E. C. (1952). *Wood Chemistry*, Vol. 1, Van Nostrand Reinhold, Princeton, New Jersey.

Wisniewski, M., Bassett, C., and Arora, R. (2004). Distribution and partial characterization of seasonally expressed proteins in different aged shoots and roots of "Loring" peach (*Prunus persica*). *Tree Physiol.* **24**, 339–345.

Wodzicki, T. J. and Brown, C. L. (1973). Organization and breakdown of the protoplast during maturation of pine tracheids. *Am. J. Bot.* **60**, 631–640.

Wodzicki, T. J. and Wodzicki, A. B. (1980). Seasonal abscisic acid accumulation in the stem cambial region of *Pinus silvestris*, and its contribution to the hypothesis of a latewood control system in conifers. *Physiol. Plant.* **48**, 443–447.

Wodzicki, T. J. and Wodzicki, A. B. (1981). Modulation of the oscillatory system involved in polar transport of auxin by other phytohormones. *Physiol. Plant.* **53**, 176–180.

Wodzicki, T. J. and Zajaczkowski, S. (1974). Effect of auxin on xylem tracheid differentiation in decapitated stems of *Pinus silvestris* L. and its interaction with some vitamins and growth regulators. *Acta Soc. Bot. Polon.* **43**, 129–148.

Wolf, F. T. (1956). Changes in chlorophylls A and B in autumn leaves. *Am. J. Bot.* **43**, 714–718.

Wolfertz, M., Sharkey, T. D., Boland, W., Kuhnemann, F., Yeh, S., and Weise, S. E. (2003). Biochemical regulation of isoprene emission. *Plant, Cell Environ.* **26**, 1357–1364.

Wolfertz, M., Sharkey, T. D., Boland, W., and Kuhnemann, F. (2004). Rapid regulation of the methylerythritol 4-phosphate pathway during isoprene synthesis. *Plant Physiol.* **135**, 1939–1945.

Wolff, I. A. (1966). Seed lipids. *Science* **154**, 1140–1149.

Wolter, K. E. (1977). Ethylene-potential alternative to bipyridilium herbicides for inducing lightwood in pine. *Proc. Lightwood Res. Coord. Counc., 1977*, 90–99.

Wolter, K. E. and Zinkel, D. F. (1984). Observations on the physiological mechanisms and chemical constituents of induced oleoresin synthesis in *Pinus resinosa*. *Can. J. For. Res.* **14**, 452–458.

Wong, S. C. and Dunin, F. X. (1987). Photosynthesis and transpiration of trees in a eucalypt forest stand: CO_2, light and humidity responses. *Aust. J. Plant Physiol.* **14**, 619–632.

Wood, B. W. (1983). Changes in indoleacetic acid, abscisic acid, gibberellins, and cytokinins during bud break of pecan. *J. Am. Soc. Hort. Sci.* **108**, 333–338.

Wood, B. W. (1986). Cold injury susceptibility of pecan as influenced by cultivar, carbohydrates, and crop load. *HortScience* **21**, 285–286.

Wood, B. W., Reilly, C. C., and Nyczepir, A. P. (2004). Mouse-ear of pecan: A nickel deficiency. *Hortscience* **39**, 1238–1242.

Woodman, J. N. (1971). Variation of net photosynthesis within the crown of a large forest-grown conifer. *Photosynthetica* **5**, 50–54.

Woodmansee, R. G. and Wallach, L. S. (1981). Effects of fire regimes on biogeochemical cycles. *U.S. For. Serv., Gen. Tech. Rep. WO* **WO–26**, 379–400.

Woods, D. B. and Turner, N. C. (1971). Stomatal responses to changing light by four tree species of varying shade tolerance. *New Phytol.* **70**, 77–84.

Woodward, F. I. (1987). *Climate and Plant Distribution*. Cambridge University Press, Cambridge.

Woodward, F. I. (1995). Ecophysiological controls of conifer distributions. In *Ecophysiology of Conifer Forests* (W. K. Smith and T. M. Hinckley, eds.), pp. 79–94. Academic Press, San Diego.

Woodward, F. I. and Kelly, C. K. (1995). The influence of CO_2 concentration on stomatal density. *New Phytol.* **131**, 311–327.

Woolley, J. T. (1961). Mechanisms by which wind influences transpiration. *Plant Physiol.* **36**, 112–114.

Worrell, E., Price, L., Martin, N., Hendriks, C., and Meida, L. O. (2001). Carbon dioxide emissions from the global cement industry. *Annu. Rev. Energy Environ.* **26**, 303–329.

Wright, H. A. and Bailey, A. W. (1982). *Fire Ecology*. Wiley, New York.

Wright, J. W. (1952). Pollen dispersion of some forest trees. *U.S. For. Serv., Northeast For. Exp. Stn.* **Pap. 46**.

Wright, J. W. (1953). Pollen dispersion studies: Some practical applications. *J. For.* **51**, 114–118.

Wright, J. W. and Bull, W. T. (1963). Geographic variation in Scotch pine: Results of a 3-year Michigan study. *Silvae Genet.* **12**, 1–25.

Wright, R. D. and Aung, L. H. (1975). Carbohydrates in two *Rhododendron* cultivars. *J. Am. Soc. Hortic. Sci.* **100**, 527–529.

Wright, I. J., Reich, P. B., Westoby, M., Ackerly, D. D., Baruch, Z., Bongers, F., Cavender-Bares, J., Chapin, T. et al. (2004). The worldwide leaf economics spectrum. *Nature* (*London*) **428**, 821–827.

Wu, L., Joshi, C. P., and Chiang, V. L. (2000). A xylem-specific cellulose synthase gene from aspen (*Populus tremuloides*) is responsive to mechanical stress. *Plant J.* **22**, 495–502.

Wullschleger, S. D. (1993). Biochemical limitations to carbon assimilation in C_3 plants—A retrospective analysis of the A/C_i curves from 109 species. *J. Exp. Bot.* **44**, 907–920.

Wullschleger, S. D. and Norby, R. J. (1992). Respiratory cost of leaf growth and maintenance in white oak saplings exposed to atmospheric CO_2 enrichment. *Can. J. For. Res.* **22**, 1717–1721.

Würth, M. K. R., Pelaez-Riedl, S., Wright, S. J., and Korner, C. (2005). Non-structural carbohydrate pools in a tropical forest. *Oecologia* **143**, 11–24.

Xie, X., Wang, Y., Williamson, L., Holroyd, G. H., Tagliava, C., Murchie, E., Theobald, J., Knight, M. R., Davies, W. J., Leyser, H. M. O., and Hetherington, A. M. (2006). The identification of genes involved in the stomatal response to reduced atmospheric relative humidity. *Curr. Biol.* **16**, 882–887.

Xu, L. K. and Baldocchi, D. D. (2003). Seasonal trends in photosynthetic parameters and stomatal conductance of blue oak (*Quercus douglasii*) under prolonged summer drought and high temperature. *Tree Physiol.* **23**, 865–877.

Yadava, D. L. and Doud, S. L. (1980). The short life and replant problems of deciduous fruit trees. *Hortic. Rev.* **2**, 1–116.

Yamaguchi, K., Shimaji, K., and Itoh, T. (1983). Simultaneous inhibition and induction of compression wood formation by morphactin in artificially inclined stems of Japanese larch (*Larix leptolepis* Gordon). *Wood Sci. Technol.* **17**, 81–89.

Yamamoto, F. and Kozlowski, T. T. (1987a). Effect of ethrel on stem anatomy of *Ulmus americana* seedlings. *IAWA Bull. N. S.* **8**, 3–9.

Yamamoto, F. and Kozlowski, T. T. (1987b). Effect of ethrel on growth and stem anatomy of *Pinus halepensis* seedlings. *IAWA Bull. N. S.* **8**, 11–19.

Yamamoto, F. and Kozlowski, T. T. (1987c). Effects of flooding, tilting of stems, and ethrel application on growth, stem anatomy and ethylene production of *Pinus densiflora* seedlings. *J. Exp. Bot.* **38**, 293–310.

Yamamoto, F. and Kozlowski, T. T. (1987d). Effects of flooding, tilting of stems, and ethrel application on growth, stem anatomy, and ethylene production of *Acer platanoides* seedlings. *Scand. J. For. Res.* **2**, 141–156.

Yamamoto, F., Angeles, G., and Kozlowski, T. T. (1987). Effect of ethrel on stem anatomy of *Ulmus americana* seedlings. *IAWA Bull. N. S.* **8**, 3–9.

Yamamoto, K. (1982). Yearly and seasonal process of maturation of ray parenchyma cells in *Pinus* species. *Res. Bull. Extl. For. Hokkaido Univ.* **39**, 245–296.

Yamamoto, R., Fujioka, S., Demura, T., Takatsuto, S., Yoshida, S., and Fukuda, H. (2001). Brassinosteroid levels increase drastically prior to morphogenesis of tracheary elements. *Plant Physiol.* **125**, 556–563.

Yandow, T. S. and Klein, R. M. (1986). Nitrate reductase of primary roots of red spruce seedlings: Effects of acidity and metal ions. *Plant Physiol.* **81**, 723–725.

Yang, K. C. (1993). Survival rate and nuclear irregularity index of sapwood ray parenchyma cells in four tree species. *Can. J. For. Res.* **23**, 673–679.

Yang, K. C., Hazenberg, G., Bradfield, G. E., and Maze, J. R. (1985). Vertical variation of sapwood thickness in *Pinus banksiana* and *Larix laricina*. *Can. J. For. Res.* **15**, 822–828.

Yang, S. and Tyree, M. T. (1992). A theoretical model of hydraulic conductivity recovery from embolism with comparison to experimental data on *Acer saccharum*. *Plant, Cell Environ.* **15**, 633–643.

Yang, S. D. and Tyree, M. T. (1994). Hydraulic architecture of *Acer saccharum* and *A. rubrum*: Comparison of branches to whole trees and the contribution of leaves to hydraulic resistance. *J. Exp. Bot.* **45**, 179–186.

Yang, S. F. and Hoffman, N. E. (1984). Ethylene biosynthesis and regulation in higher plants. *Annu. Rev. Plant Physiol.* **35**, 155–189.

Yani, A., Pauly, G., Faye, M., Salin, F., and Gleizes, M. (1993). The effect of a long-term water stress on the metabolism and emission of terpenes of the foliage of *Cupressus sempervirens*. *Plant, Cell Environ.* **16**, 975–981.

Yazawa, K. (1960). The seasonal water content of sapwood and heartwood of broad-leaved trees, especially beech (*Fagus crenata* Blume). *J. Jap. Wood Res. Soc.* **6**, 170–175.

Yazawa, K. and Ishida, S. (1965). On the existence of the intermediate wood in some broad-leaved trees grown in Hokkaido, Japan. *J. Fac. Agriculture, Hokkaido Univ.* **54**, 137–150.

Yazawa, K., Ishida, S., and Miyajima, H. (1965). On the wet heartwood of some broad-leaved trees grown in Japan. *J. Jap. Wood Res. Soc.* **11**, 71–75.

Yazdani, R., Rudin, D., Alden, T., Lindgren, D., Harborn, B., and Ljung, K. (1982). Inheritance pattern of five monoterpenes in Scots pine (*Pinus sylvestris* L.). *Hereditas* **97**, 261–272.

Yelenosky, G. (1964). Tolerance of trees to deficiencies of soil aeration. *Proc. Int. Shade Tree Conf.* **40**, 127–147.

Yelenosky, G. and Guy, C. L. (1977). Carbohydrate accumulation in leaves and stems of "Valencia" orange at progressively colder temperatures. *Bot. Gaz.* **138**, 13–17.

Ying, C. C. and Bagley, W. T. (1976). Genetic variation of eastern cottonwood in an eastern Nebraska provenance study. *Silvae Genet.* **25**, 67–73.

Ylimartimo, A., Pääkkonen, E., Holopainen, T., and Rita, H. (1994). Unbalanced nutrient status and epicuticular wax of Scots pine needles. *Can. J. For. Res.* **24**, 522–532.

Yoon, J., Abe-Suzuki M., Eko, P., Tamai, H., Hanamitsu, S., and Nakane, K. (2006). Negative effects of hydroxyl radical-generating mists (simulated dew water) on the photosynthesis and growth of Japanese apricot seedlings (*Prunus mume*). *Ecol. Res.* **26**, 117–125.

Young, A. and Stephen, I. (1965). Rock weathering and soil formation of high altitude plateaux of Malawi. *J. Soil Sci.* **16**, 322–333.

Young, E. (1989). Cytokinin and soluble carbohydrate concentration in xylem sap of apple during dormancy and budbreak. *J. Am. Soc. Hort. Sci.* **114**, 297–300.

Zaerr, J. B. (1971). Moisture stress and stem diameter in young Douglas-fir. *For. Sci.* **17**, 466–469.

Zaerr, J. B. (1983). Short-term flooding and net photosynthesis in seedlings of three conifers. *For. Sci.* **29**, 71–78.

Zahner, R. (1968). Water deficits and growth of trees. In *Water Deficits and Plant Growth* (T. T. Kozlowski, ed.), Vol. 2, pp. 191–254. Academic Press, New York.

Zajaczkowska, J. (1973). Gas exchange and organic substance production of Scots pine (*Pinus silvestris* L.) seedlings grown in soil cultures with ammonium or nitrate form of nitrogen. *Acta Soc. Bot. Pol.* **43**, 103–116.

Zajaczkowski, S. (1973). Auxin stimulation of cambial activity in *Pinus silvestris* L. I. The differential cambial response. *Physiol. Plant.* **29**, 231–287.

Zak, B. (1964). Role of mycorrhizae in root disease. *Ann. Rev. Phytopathol.* **2**, 377–392.

Zak, D. R. and Pregitzer, K. S. (1990). Spatial and temporal variability of nitrogen cycling in northern lower Michigan. *For. Sci.* **36**, 367–380.

Zakrzewski, J. (1983). Hormonal control of cambial activity and vessel differentiation in *Quercus robur*. *Physiol. Plant.* **57**, 537–542.

Zarska-Maciezewska, B. and Lewak, S. (1983). The role of proteolytic enzymes in the release from dormancy of apple seeds. *Z. Pflanzenphysiol.* **110**, 409–417.

Zavitkovski, J. and Ferrell, W. K. (1970). Effect of drought upon rates of photosynthesis, respiration, and transpiration of seedlings of two ecotypes of Douglas-fir. II. Two-year old seedlings. *Photosynthetica* **4**, 58–67.

Zech, W. (1990). Mineral deficiencies in forest plantations of North-Luzon, Philippines. *Trop. Ecol.* **31**, 22–31.

Zelawski, W. and Goral, I. (1966). Seasonal changes in the photosynthetic rate of Scots pine seedlings grown from seed of various provenances. *Acta Soc. Bot. Pol.* **35**, 587–598.

Zeller, O. (1955). *Entwicklungsverlauf auf der Infloreszenknospen einiger Kernund Steinobstsorten. Angew. Bot.* **29**, 69–89.

Zeppel, M. J. B., Yunusa, I. A. M., and Eamus, D. (2006). Daily, seasonal and annual patterns of transpiration from a stand of remnant vegetation dominated by a coniferous *Callitris* species and a broad-leaved *Eucalyptus* species. *Physiol. Plant.* **127**, 413–422.

Zha, T. S., Wang, K. Y., Ryyppo, A., and Kellomaki, S. (2002). Needle dark respiration in relation to within-crown position in Scots pine trees grown in long-term elevation of CO_2 concentration and temperature. *New Phytol.* **156**, 33–41.

Zhang, J. and Davies, W. J. (1989). Abscisic acid produced in dehydrating roots may enable the plant to measure the water status of the soil. *Plant, Cell Environ.* **12**, 73–81.

Zhang, J. and Davies, W. J. (1990a). Changes in the concentration of ABA in xylem sap as a function of changing soil water status can account for changes in leaf conductance and growth. *Plant, Cell Environ.* **13**, 277–286.

Zhang, J. and Davies, W. J. (1990b). Does ABA in the xylem control the rate of leaf growth in soil-dried maize and sunflower plants? *J. Exp. Bot.* **41**, 1125–1132.

Zhang, J. and Davies, W. J. (1991). Antitranspirant activity in xylem sap of maize plants. *J. Exp. Bot.* **42**, 317–321.

Zhang, J., Schurr, U., and Davies, W. J. (1987). Control of stomatal behaviour by abscisic acid which apparently originates in the roots. *J. Exp. Bot.* **38**, 1174–1181.

Zhang, S. and Gao, R. (1999). Diurnal changes of gas exchange, chlorophyll fluorescence, and stomatal aperture of hybrid poplar clones subjected to midday light stress. *Photosynthetica* **37**, 559–571.

Zhu, B. L. and Coleman, G. D. (2001). Phytochrome-mediated photoperiod perception, shoot growth, glutamine, calcium, and protein phosphorylation influence the activity of the poplar bark storage protein gene promoter (*bspA*). *Plant Physiol.* **126**, 342–351.

Ziegler, H. (1964). Storage, mobilization, and distribution of reserve material in trees. In *The Formation of Wood in Forest Trees* (M. H. Zimmermann, ed.), pp. 303–320. Academic Press, New York.

Zimmermann, M. H. (1957). Translocation of organic substances in trees. I. The nature of sugars in the sieve tube exudate of trees. *Plant Physiol.* **32**, 288–291.

Zimmermann, M. H. (1961). Movement of organic substances in trees. *Science* **133**, 73–79.

Zimmermann, M. H. (1983). *Xylem Structure and the Ascent of Sap.* Springer-Verlag, Berlin.

Zimmermann, M. H. and Brown, C. L. (1971). *Trees: Structure and Function.* Springer-Verlag, Berlin and New York.

Zimmermann, M. H. and Jeje, A. A. (1981). Vessel-length distribution in stems of some American woody plants. *Can. J. Bot.* **59**, 1882–1892.

Zimmermann, M. H. and Milburn, J. A., eds. (1975). *Transport in Plants. I. Phloem Transport.* Springer-Verlag, Berlin and New York.

Zimmermann, M. H. and Potter, D. (1982). Vessel-length distribution in branches, stems and roots of *Acer rubrum* L. *IAWA Bull. N.S.* **3**, 103–106.

Zimmermann, U., Haase, A., Langbein, D., and Meinzer, F. (1993). Mechanisms of long-distance water transport in plants: A reexamination of some paradigms in the light of new evidence. *Phil. Trans. Roy. Soc. Lond.* **B344**, 19–31.

Zimmermann, U., Schneider, H., Wegner, L. H., and Haase, A. (2004). Water ascent in tall trees: Does evolution of land plants rely on a highly metastable state? *New Phytol.* **162**, 575–615.

Zinke, P. A., Stangenberger, A. G., Post, W. M., Emanuel, W. R., and Olson, J. S. (1984). Worldwide organic soil carbon and nitrogen data. *Oak Ridge National Laboratory, Environmental Sciences Div. Publ. 2212, U.S. Dept. of Energy.*

Ziska, L. H., Seemann, L. H., and Dejong, T. M. (1990). Salinity induced limitations on photosynthesis in *Prunus salicina*, a deciduous tree species. *Plant Physiol.* **93**, 864–870.

Zotz, G., Tyree, M. T., and Cochard, H. (1994). Hydraulic architecture, water relations and vulnerability to cavitation of *Clusia uvitana* Pittier: A C_3-CAM tropical hemiepiphyte. *New Phytol.* **127**, 287–295.

Zwiazek, J. J. and Blake, T. J. (1989). Effects of preconditioning on subsequent water relations, stomatal sensitivity, and photosynthesis in osmotically stressed black spruce. *Can. J. Bot.* **67**, 2240–2244.

Zwieniecki, M. A. and Holbrook, N. M. (1998). Diurnal variation in xylem hydraulic conductivity in white ash (*Fraxinus americana* L.), red maple (*Acer rubrum* L.) and red spruce (*Picea rubens* Sarg). *Plant, Cell Environ.* **21**, 1173–1180.

Index

D

Q

R

S

Y

Z